Welding Theory and Application
TC 9-237
A US Military Welding Textbook

edited by
Brian Greul

Welding is the process of fusing two materials into one. It is most often used with metal, but there are processes for other materials such as plastic. This is a comprehensive training manual used by the US Military to teach welding theory and application. It covers a variety of techniques and has been often been considered the go-to welding book.

Should you have suggestions or feedback on ways to improve this book please send email to Books@OcotilloPress.com

Edited 2021 Ocotillo Press
ISBN 978-1-954285-50-7

Ocotillo Press
Houston, TX 77017
Books@OcotilloPress.com

Disclaimer: The user of this book is responsible for following safe and lawful practices at all times. The publisher assumes no responsibility for the use of the content of this book. The publisher has made an effort to ensure that the text is complete and properly typeset, however omissions, errors, and other issues may exist that the publisher is unaware of.

OPERATOR'S CIRCULAR
WELDING THEORY AND APPLICATION

REPORTING ERRORS AND RECOMMENDING IMPROVEMENTS

MAY 1993

HEADQUARTERS, DEPARTMENT OF THE ARMY

DISTRIBUTION RESTRICTION: Approved for public release; distribution is unlimited.

TC 9-237
HEADQUARTERS
DEPARTMENT OF THE ARMY
Washington, DC, 7 May 1993

Training Circular
No. 9-237

Operator's Circular
Welding Theory and Application
REPORTING ERRORS AND RECOMMENDING IMPROVEMENTS

You can help improve this circular. If you find any mistakes or if you know of a way to improve the procedures, please let us know. Mail your letter or DA Form 2028 (Recommended Changes to Publications and Blank Forms), located in the back of this manual, direct to: Commander, US Army Ordnance Center and School, ATTN: ATSL-CD-CS, Aberdeen Proving Ground, MD 21005-5201. A reply will be furnished to you.

Table of Contents

Table of Contents (cont)

LIST OF ILLUSTRATIONS

LIST OF ILLUSTRATIONS (cont)

LIST OF ILLUSTRATIONS

LIST OF ILLUSTRATIONS

LIST OF ILLUSTRATIONS

LIST OF TABLES

LIST OF TABLES (cont)

WARNINGS

Cyanide and cyanide fumes are dangerous poisons. The cyaniding method of case hardening requires expert supervision and adequate ventilation.

Oil or grease in the presence of oxygen will ignite violently, especially in an enclosed pressurized area.

Do not substitute oxygen for compressed air in pneumatic tools. Do not use oxygen to blow out pipe lines, test radiators, purge tanks or containers, or to "dust" clothing.

Welding machine Model 301, AC/DC, Heliarc with inert gas attachment, NSN 3431-00-235-4728, may cause electrical shock if not properly grounded. If one is being used, contact Castolin Institute, 4462 York St., Denver, Colorado 80216 ATTN: Mr. Lent.

The vapors from some chlorinated solvents (e.g. carbon tetrachloride, trichloroethylene, and perchloroethylene) break down under the ultra-violet radiation of an electric arc and form a toxic gas. Avoid welding where such vapors are present. These solvents vaporize easily and prolonged inhalation of the vapor can be hazardous. These organic vapors should be removed from the work area before welding is begun.

Do not assume that a container that has held combustibles is clean and safe until proven so by proper tests. Do not weld in places where dust or other combustible particles are suspended in air or where explosive vapors are present. Removal of flammable material from vessels/containers may be done either by steaming out or boiling.

The automotive exhaust method of cleaning should be conducted only in well ventilated areas to ensure levels of toxic gases are kept below hazardous levels.

Welding polyurethane foam-filled parts can produce toxic gases. Welding should not be attempted on parts filled with polyurethane foam. If repair of such parts by welding is necessary, the foam must be removed from the heat affected area, including the residue, prior to welding.

Do not stand facing cylinder valve outlets of oxygen, acetylene, or other compressed gases when opening them.

If it is necessary to blow out the acetylene hose, do it in a well ventilated place, free of sparks, flame, or other sources of ignition.

WARNINGS (cont)

Purge both acetylene and oxygen lines (hoses) prior to igniting torch. Failure to do this can cause serious injury to personnel and damage to the equipment.

Regulators with gas leakage between the regulator seat and the nozzle should be repaired immediately to avoid damaged to other parts of the regulator or injury to personnel. With acetylene regulators, this leakage is particularly dangerous. Acetylene at high pressure in the hose is an explosion hazard.

Defects in oxyacetylene welding torches which are sources of gas leaks must be corrected immediately, as they may result in flashbacks, or backfires, with resultant injury to the operator and/or damage to the welding apparatus.

Damaged inlet connection threads may cause fires by ignition of the leaking gas, resulting in injury to the welding operator and/or damaged to the equipment.

Dry cleaning solvent and mineral spirits paint thinner are highly flammable. Do not clean parts near an open flame or in a smoking area. Dry cleaning solvent and mineral spirits paint thinner evaporate quickly and have a defatting effect on the skin. When used without protective gloves, these chemicals may cause irritation or cracking of the skin. Cleaning operations should be performed only in well ventilated areas.

The acid solutions used to remove aluminum welding and brazing fluxes after welding or brazing are toxic and highly corrosive. Goggles, rubber gloves, and rubber aprons should be worn when handling the acids and solutions. Do not inhale fumes. When spilled on the body or clothing, wash immediately with large quantities of cold water.

Never pour water into acid when preparing solutions; instead, pour acid into water. Always mix acid and water slowly. These operations should only be performed in well ventilated areas.

Precleaning and postcleaning acids used in magnesium welding and brazing are highly toxic and corrosive. Goggles, rubber gloves, and rubber aprons should be worn when handling the acids and solutions. Do not inhale fumes and mists. When spilled on the body or clothing, wash immediately with large quantities of cold water, and seek medical attention. Do not pour water into acid when preparing solution; instead, pour acid into water. Always mix acid and water slowly. Cleaning operations should be performed only in well ventilated areas.

If the electrode becomes frozen to the base metal during the process of starting the arc, all work to free the electrode while the current is on must be done with the eyes shielded.

The nitric acid used to preclean titanium for inert gas shielded arc welding is highly toxic and corrosive. Goggles, rubber gloves, and rubber aprons should be worn when handling the acid and the acid solution. Do not inhale gases and mists. When spilled on the body or clothing, wash immediately with large quantities of cold water, and seek medical help. Do not pour water into acid when preparing the solution; instead, pour acid into water. Always mix acid and water slowly. Perform cleaning operations only in well ventilated areas.

The caustic chemicals (including sodium hydride) used to preclean titanium for inert gas shielded arc welding are highly toxic and corrosive. Goggles, rubber gloves, and rubber aprons should be worn when handling these chemicals. Do not inhale gases or mists. When caustics are spilled on the body or clothing, wash immediately with large quantities of cold water, and seek medical help. Special care should be taken at all times to prevent any water from coming in contact with the molten bath or any other large amount of sodium hydride, as this will cause the evolution of highly explosive hydrogen gas.

When using weld backup tape, the weld must be allowed to cool for several minutes before attempting to remove the tape from the workpiece.

Safety precautions must be exercised in underwater cutting and welding. Electrode holder and cable must be insulated, current must be shut off when changing electrodes, and the diver should avoid contact between the electrode and grounded work.

In thermit welding, the mold must be thoroughly dried before the charge in the crucible is ignited. When the charge has been ignited, the operator should stand a safe distance away and should wear goggles. Painful burns may occur from splashing metal, upsetting of the crucible, breaking of the mold, or allowing the molten metal to come in contact with moisture in the mold.

Before welding on equipment painted with CARC paint, remove the paint from an area larger than that which will be heated during welding.

Do not operate welding machines in an enclosed area unless the exhaust gases are piped to the outside. Inhalation of exhaust fumes will result in serious illness or death.

When filling the fuel tank, always provide a metal-to-metal contact between the container and the fuel tank. This will prevent a spark from being generated as fuel flows over the metallic surfaces.

Do not fill the fuel tanks while the engine is running. Fuel spilled on a hot engine may explode and cause injury to personnel.

Do not attempt any maintenance on the welding machine while it is in operation. The voltage generated by it can cause injury or death.

Ensure that all welding machines are properly grounded. Failure to properly ground welding machines could result in electrical shock.

Always use ear plugs. Diesel engines exceed a permissible decibel level. Failure to observe this warning could result in a permanent hearing injury.

Always wear arc proof glasses or a welder's helmet when welding to prevent serious eye burns or possible blindness.

Use only approved cleaning solvents to avoid the possibility of fire or poisoning.

Inert gas, metal-arc welding processes produce intense ultra-violet radiation which can be harmful to the eyes and skin. Therefore, certain precautions must be observed to protect the operator from injury.

Skin must be completely covered. Leather gloves are recommended for hand protection. Heavy, dark colored clothing should be worn to prevent the radiation from penetrating to the skin or reflecting onto the neck under the helmet. Lightweight leather clothing is recommended because of its durability and resistance to deterioration from radiation. Cotton clothing will deteriorate rapidly when subjected to ultra-violet radiation.

Adequate ventilation should be provided to remove fumes which are produced by welding processes. American standard Z-49.1 on welding safety covers such ventilation procedures. Highly toxic gases are formed when the vapors from halogenated solvents are subjected to ultra-violet radiation. Therefore, it is recommended that degreasers and other sources of these vapors should be located so that the vapors cannot reach the welding operation.

Under no circumstances should acetylene cylinders be positioned or stored in other than an upright position. Storage of the cylinder in a horizontal or reclining position could create a hazardous condition.

Stand to the side of gas and oxygen cylinders when turning on the pressure release valves. The cylinders contain extreme pressure. Injury could occur if a defective flowmeter or pressure regulator valve ruptures when subjected to these pressures.

Ensure that all gages are removed from gas and oxygen cylinders before transporting. Failure to observe this warning could create a hazardous condition.

Wear head and eye protection, rubber gloves, boots, and aprons when handling steam, hot water, and caustic solutions. When handling dry caustic soda or soda ash, wear approved respiratory protective equipment, long sleeves, and gloves. Wear fire resistant hand pads or gloves to handle hot drums.

Brazing filler metals containing cadmium may form poisonous fumes on heating. Do not breathe fumes. Use only with adequate ventilation, such as fume collectors, exhaust ventilators, or air-supplied respirators. See American National Standards Institute Standard Z49.1-1973. If chest pain, cough, or fever develops after use, call physician immediately.

Acetylene, stored in a free state under pressure greater than 15 psi (103.4 kPa), can break down from heat or shock, and possible explode. Under pressure of 29.4 psi (203 kPa), acetylene becomes self-explosive, and a slight shock can cause it to explode spontaneously.

Acetylene which may accumulate in a storage room or in a confined space is a fire and explosion hazard. All acetylene cylinders should be checked, using a soap solution, for leakage at the valves and safety fuse plugs.

Do not stand facing cylinder valve outlets of oxygen, acetylene, or other compressed gases when opening them.

Always have suitable fire extinguishing equipment at hand when doing and welding.

CHAPTER 1
INTRODUCTION

Section I. GENERAL

1-1. SCOPE

This training circular is published for use by personnel concerned with welding and other metal joining operations in the manufacture and maintenance of materiel.

1-2. DESCRIPTION

a. This circular contains information as outlined below:

(1) Introduction
(2) Safety precautions in welding operations
(3) Print reading and welding symbols
(4) Joint design and preparation of metals
(5) Welding and cutting equipment
(6) Welding techniques
(7) Metals identification
(8) Electrodes and filler metals
(9) Maintenance welding operations for military equipment
(10) Arc welding and cutting processes
(11) Oxygen fuel gas welding processes
(12) Special applications
(13) Destructive and nondestructive testing

b. Appendix A contains a list of current references, including supply and technical manuals and other available publications relating to welding and cutting operations.

c. Appendix B contains procedure guides for welding.

d. Appendix C contains a troubleshooting chart.

e. Appendix D contains tables listing materials used for brazing. welding. soldering, arc cutting, and metallizing.

f. Appendix E contains miscellaneous data as to temperature ranges, melting points, and other information not contained in the narrative portion of this manual.

Section II. THEORY

1-3. GENERAL

Welding is any metal joining process wherein coalescence is produced by heating the metal to suitable temperatures with or without the application of pressure and with or without the use of filler metals. Basic welding processes are described and illustrated in this manual. Brazing and soldering, procedures similar to welding, are also covered.

1-4. METALS

a. Metals are divided into two classes, ferrous and nonferrous. Ferrous metals are those in the iron class and are magnetic in nature. These metals consist of iron, steel,and alloys related to them. Nonferrous metals are those that contain either no ferrous metals or very small amounts. These are generally divided into the aluminum, copper, magnesium, lead, and similar groups.

b. Information contained in this circular covers theory and application of welding for all types of metals including recently developed alloys.

CHAPTER 2
SAFETY PRECAUTIONS IN WELDING OPERATIONS

Section I. GENERAL SAFETY PRECAUTIONS

2-1. GENERAL

a. To prevent injury to personnel, extreme caution should be exercised when using any types of welding equipment. Injury can result from fire, explosions, electric shock, or harmful agents. Both the general and specific safety precautions listed below must be strictly observed by workers who weld or cut metals.

b. Do not permit unauthorized persons to use welding or cutting equipment.

c. Do not weld in a building with wooden floors, unless the floors are protected from hot metal by means of fire resistant fabric, sand, or other fireproof material. Be sure that hot sparks or hot metal will not fall on the operator or on any welding equipment components.

d. Remove all flammable material, such as cotton, oil, gasoline, etc., from the vicinity of welding.

e. Before welding or cutting, warm those in close proximity who are not protected to wear proper clothing or goggles.

f. Remove any assembled parts from the component being welded that may become warped or otherwise damaged by the welding process.

g. Do not leave hot rejected electrode stubs, steel scrap, or tools on the floor or around the welding equipment. Accidents and/or fires may occur.

h. Keep a suitable fire extinguisher nearby at all times. Ensure the fire extinguisher is in operable condition.

i. Mark all hot metal after welding operations are completed. Soapstone is commonly used for this purpose.

2-2. PERSONAL PROTECTIVE EQUIPMENT

a. General. The electric arc is a very powerful source of light, including visible, ultraviolet, and infrared. Protective clothing and equipment must be worn during all welding operations. During all oxyacetylene welding and cutting processes, operators must use safety goggles to protect the eyes from heat, glare, and flying fragments of hot metals. During all electric welding processes, operators must use safety goggles and a hand shield or helmet equipped with a suitable filter glass to protect against the intense ultraviolet and infrared rays. When others are in the vicinity of the electric welding processes, the area must be screened so the arc cannot be seen either directly or by reflection from glass or metal.

b. Helmets and Shields.

(1) Welding arcs are intensely brilliant lights. They contain a proportion of ultraviolet light which may cause eye damage. For this reason, the arc should never be viewed with the naked eye within a distance of 50.0 ft (15.2 m). The brilliance and exact spectrum, and therefore the danger of the light, depends on the welding process, the metals in the arc, the arc atmosphere, the length of the arc, and the welding current. Operators, fitters, and those working nearby need protection against arc radiation. The intensity of the light from the arc increases with increasing current and arc voltage. Arc radiation, like all light radiation, decreases with the square of the distance. Those processes that produce smoke surrounding the arc have a less bright arc since the smoke acts as a filter. The spectrum of the welding arc is similar to that of the sun. Exposure of the skin and eyes to the arc is the same as exposure to the sun.

(2) Being closest, the welder needs a helmet to protect his eyes and face from harmful light and particles of hot metal. The welding helmet (fig. 2-1) is generally constructed of a pressed fiber insulating material. It has an adjustable headband that makes it usable by persons with different head sizes. To minimize reflection and glare produced by the intense light, the helmet is dull black in color. It fits over the head and can be swung upward when not welding. The chief advantage of the helmet is that it leaves both hands free, making it possible to hold the work and weld at the same time.

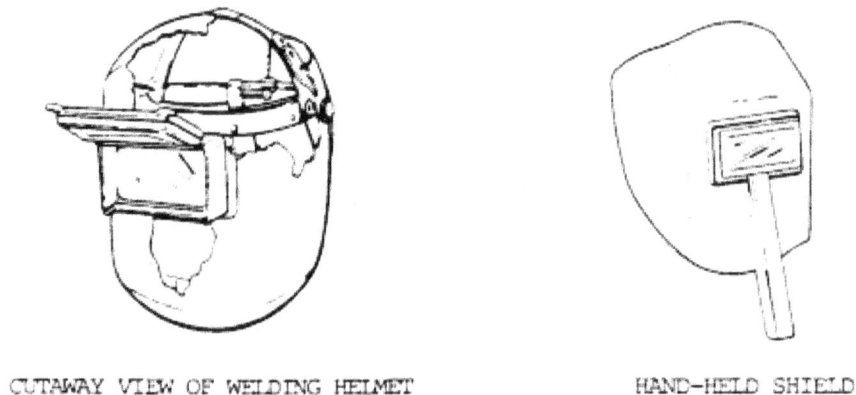

CUTAWAY VIEW OF WELDING HELMET   HAND-HELD SHIELD

Figure 2-1. Welding helmet and hand-held shield.

(3) The hand-held shield (fig. 2-1) provides the same protection as the helmet, but is held in position by the handle. This type of shield is frequently used by an observer or a person who welds for a short period of time.

(4) The protective welding helmet has lens holders used to insert the cover glass and the filter glass or plate. Standard size for the filter plate is 2 x 4-1/4 in. (50 x 108 mm). In some helmets lens holders open or flip upwards. Lenses are designed to prevent flash burns and eye damage by absorption of the infrared and ultraviolet rays produced by the arc. The filter glasses or plates come in various optical densities to filter out various light intensities, depending on the welding process, type of base metal, and the welding current. The color of the lens, usually green, blue, or brown, is an added protection against the intensity of white light or glare. Colored lenses make it possible to clearly see the metal and weld. Table 2-1 lists the proper filter shades to be used. A magnifier lens placed behind the filter glass is sometimes used to provide clear vision.

Table 2-1. Lens Shades for Welding and Cutting

Welding or Cutting Operation	Electrode Size Metal Thickness or Welding Current	Filter Shade Number
Torch soldering	-	2
Torch brazing	-	3 or 4
Oxygen cutting		
Light	Under 1 in., 25 mm	3 or 4
Medium	1 to 6 in., 25 to 150 mm	4 or 5
Heavy	Over 6 in., 150 mm	5 or 6
Gas welding		
Light	Under 1/8 in., 3 mm	4 or 5
Medium	1/8 to 1/2 in., 3 to 12 mm	5 or 6
Heavy	Over 1/2 in., 12 mm	6 or 8
Shielded metal-arc	Under 5/32 in., 4 mm	10
welding (stick)	5/32 to 1/4 in., 4 to 6.4 mm	12
electrodes	Over 1/4 in., 6.4 mm	14
Gas metal-arc welding (MIG)		
Non-ferrous base metal	All	11
Ferrous base metal	All	12
Gas tungsten arc welding (TIG)	All	12
Atomic hydrogen welding	All	12
Carbon arc welding	All	12
Plasma arc welding	All	12
Carbon arc air gouging		
Light	-	12
Heavy	-	14
Plasma arc cutting		
Light	Under 300 Amp	9
Medium	300 to 400 Amp	12
Heavy	Over 400 Amp	14

A cover plate should be placed outside the filter glass to protect it from weld spatter. The filter glass must be tempered so that is will not break if hit by flying weld spatter. Filter glasses must be marked showing the manufacturer, the shade number, and the letter "H" indicating it has been treated for impact resistance.

NOTE

Colored glass must be manufactured in accordance with specifications detailed in the "National Safety Code for the Protection of Hands and Eyes of Industrial Workers", issued by the National Bureau of Standards, Washington DC, and OSHA Standards, Subpart Q, "Welding, Cutting, and Brazing", paragraph 1910.252, and American National Standards Institute Standard (ANSI) Z87.1-1968, "American National Standard Practice for Occupational and Educational Eye and Face Protection".

(5) Gas metal-arc (MIG) welding requires darker filter lenses than shielded metal-arc (stick) welding. The intensity of the ultraviolet radiation emitted during gas metal-arc welding ranges from 5 to 30 times brighter than welding with covered electrodes.

(6) Do not weld with cracked or defective shields because penetrating rays from the arc may cause serious burns. Be sure that the colored glass plates are the proper shade for arc welding. Protect the colored glass plate from molten metal spatter by using a cover glass. Replace the cover glass when damaged or spotted by molten metal spatter.

(7) Face shields (fig. 2-2) must also be worn where required to protect eyes. Welders must wear safety glasses and chippers and grinders often use face shields in addition to safety glasses.

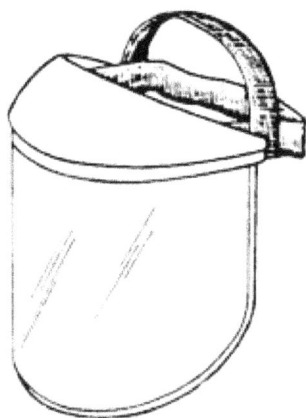

CLEAR FACE SHIELD   HELMET WITH RESPIRATOR

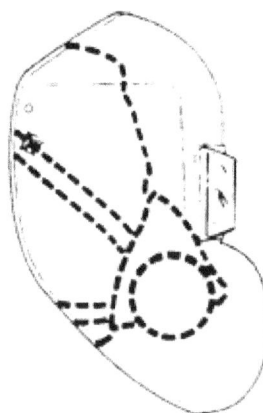

Figure 2-2. Welding helmets and shields.

(8) In some welding operations, the use of mask-type respirators is required. Helmets with the "bubble" front design can be adapted for use with respirators.

c. Safety Goggles. During all electric welding processes, operators must wear safety goggles (fig. 2-3) to protect their eyes from weld spatter which occasionally gets inside the helmet. These clear goggles also protect the eyes from slag particles when chipping and hot sparks when grinding. Contact lenses should not be worn when welding or working around welders. Tinted safety glasses with side shields are recommended, especially when welders are chipping or grinding. Those working around welders should also wear tinted safety glasses with side shields.

TYPE GC-2 CHIPPER'S GOGGLES   TYPE GC CHIPPER'S GOGGLES

Figure 2-3. Safety goggles.

d. Protective Clothing.

(1) Personnel exposed to the hazards created by welding, cutting, or brazing operations shall be protected by personal protective equipment in accordance with OSHA standards, Subpart I, Personal Protective Equipment, paragraph 1910.132. The appropriate protective clothing (fig. 2-4) required for any welding operation will vary with the size, nature, and location of the work to be performed. Welders should wear work or shop clothes without openings or gaps to prevent arc rays from contacting the skin. Those working close to arc welding should also wear protective clothing. Clothing should always be kept dry, including gloves.

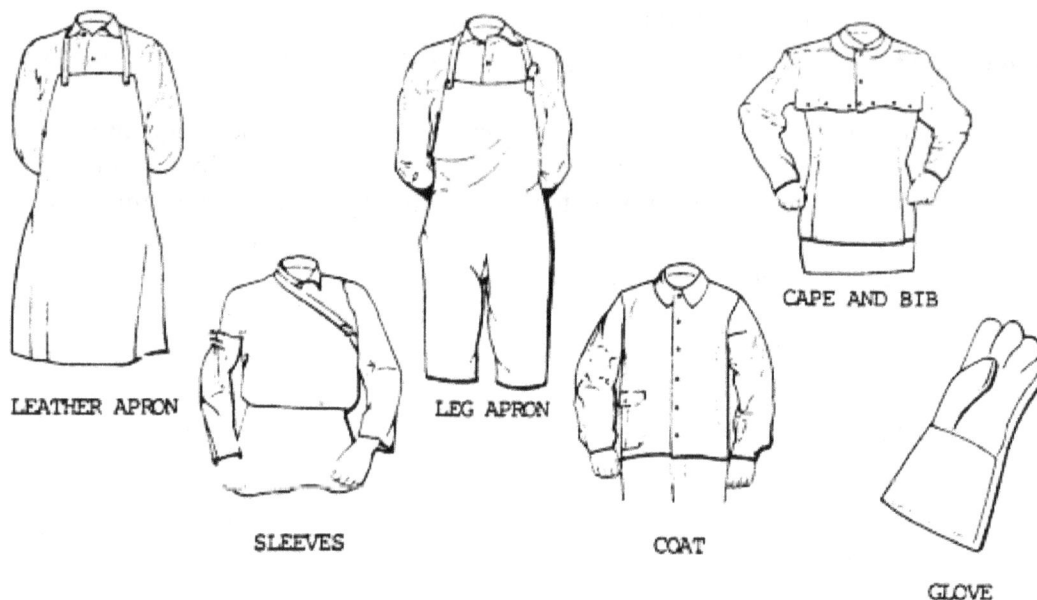

Figure 2-4. Protective clothing.

(2) Woolen clothing should be worn instead of cotton since wool is not easily burned or damaged by weld spatter and helps to protect the welder from changes in temperature. Cotton clothing, if used, should be chemically treated to reduce its combustibility. All other clothing, such as jumpers or overalls, should be reasonably free from oil or grease.

(3) Flameproof aprons or jackets made of leather, fire resistant material, or other suitable material should be worn for protection against spatter of molten metal, radiated heat, and sparks. Capes or shoulder covers made of leather or other suitable materials should be worn during overhead welding or cutting operations. Leather skull caps may be worn under helmets to prevent head burns.

(4) Sparks may lodge in rolled-up sleeves, pockets of clothing, or cuffs of overalls and trousers. Therefore, sleeves and collars should be kept buttoned and pockets should be eliminated from the front of overalls and aprons. Trousers and overalls should not be turned up on the outside. For heavy work, fire-resistant leggings, high boots, or other equivalent means should be used. In production work, a sheet metal screen in front of the worker's legs can provide further protection against sparks and molten metal in cutting operations.

(5) Flameproof gauntlet gloves, preferably of leather, should be worn to protect the hands and arms from rays of the arc, molten metal spatter, sparks, and hot metal. Leather gloves should be of sufficient thickness so that they will not shrivel from the heat, burn through, or wear out quickly. Leather gloves should not be used to pick up hot items, since this causes the leather to become stiff and crack. Do not allow oil or grease to cane in contact with the gloves as this will reduce their flame resistance and cause them to be readily ignited or charred.

e. <u>Protective Equipment</u>.

(1) Where there is exposure to sharp or heavy falling objects or a hazard of bumping in confined spaces, hard hats or head protectors must be used.

(2) For welding and cutting overhead or in confined spaces, steel-toed boots and ear protection must also be used.

(3) When welding in any area, the operation should be adequately screened to protect nearby workers or passers-by from the glare of welding. The screens should be arranged so that no serious restriction of ventilation exists. The screens should be mounted so that they are about 2.0 ft above the floor unless the work is performed at such a low level that the screen must be extended closer to the floor to protect adjacent workers. The height of the screen is normally 6.0 ft (1.8 m) but may be higher depending upon the situation. Screen and surrounding areas must be painted with special paints which absorb ultraviolet radiation yet do not create high contrast between the bright and dark areas. Light pastel colors of a zinc or titanium dioxide base paint are recommended. Black paint should not be used.

2-3. FIRE HAZARDS

a. Fire prevention and protection is the responsibility of welders, cutters, and supervisors. Approximately six percent of the fires in industrial plants are caused by cutting and welding which has been done primarily with portable equipment or in areas not specifically designated for such work. The elaboration of basic precautions to be taken for fire prevention during welding or cutting is found in the Standard for Fire Prevention in Use of Cutting and Welding Processes, National Fire Protection Association Standard 51B, 1962. Some of the basic precautions for fire prevention in welding or cutting work are given below.

b. During the welding and cutting operations, sparks and molten spatter are formal which sometimes fly considerable distances. Sparks have also fallen through cracks, pipe holes, or other small openings in floors and partitions, starting fires in other areas which temporarily may go unnoticed. For these reasons, welding or cutting should not be done near flammable materials unless every precaution is taken to prevent ignition.

c. Hot pieces of base metal may come in contact with combustible materials and start fires. Fires and explosions have also been caused when heat is transmitted through walls of containers to flammable atmospheres or to combustibles within containers. Anything that is combustible or flammable is susceptible to ignition by cutting and welding.

d. When welding or cutting parts of vehicles, the oil pan, gasoline tank, and other parts of the vehicle are considered fire hazards and must be removed or effectively shielded from sparks, slag, and molten metal.

e. Whenever possible, flammable materials attached to or near equipment requiring welding, brazing, or cutting will be removed. If removal is not practical, a suitable shield of heat resistant material should be used to protect the flammable material. Fire extinguishing equipment, for any type of fire that may be encountered, must be present.

2-4. HEALTH PROTECTION AND VENTILATION

a. General.

(1) All welding and thermal cutting operations carried on in confined spaces must be adequately ventilated to prevent the accumulation of toxic materials, combustible gases, or possible oxygen deficiency. Monitoring instruments should be used to detect harmful atmospheres. Where it is impossible to provide adequate ventilation, air-supplied respirators or hose masks approved for this purpose must be used. In these situations, lookouts must be used on the outside of the confined space to ensure the safety of those working within. Requirements in this section have been established for arc and gas welding and cutting. These requirements will govern the amount of contamination to which welders may be exposed:

(a) Dimensions of the area in which the welding process takes place (with special regard to height of ceiling).

(b) Number of welders in the room.

(c) Possible development of hazardous fumes, gases, or dust according to the metals involved.

(d) Location of welder's breathing zone with respect to rising plume of fumes.

(2) In specific cases, there are other factors involved in which respirator protective devices (ventilation) should be provided to meet the equivalent requirements of this section. They include:

(a) Atmospheric conditions.

(b) Generated heat.

(c) Presence of volatile solvents.

(3) In all cases, the required health protection, ventilation standards, and standard operating procedures for new as well as old welding operations should be coordinated and cleaned through the safety inspector and the industrial hygienist having responsibility for the safety and health aspects of the work area.

b. <u>Screened Areas</u>. When welding must be performed in a space entirely screened on all sides, the screens shall be arranged so that no serious restriction of ventilation exists. It is desirable to have the screens mounted so that they are about 2.0 ft (0.6 m) above the floor, unless the work is performed at such a low level that the screen must be extended closer to the floor to protect workers from the glare of welding. See paragraph 2-2 e (3).

c. <u>Concentration of Toxic Substances</u>. Local exhaust or general ventilating systems shall be provided and arranged to keep the amount of toxic frees, gas, or dusts below the acceptable concentrations as set by the American National Standard Institute Standard 7.37; the latest Threshold Limit Values (TLV) of the American Conference of Governmental Industrial Hygienists; or the exposure limits as established by Public Law 91-596, Occupational Safety and Health Act of 1970. Compliance shall be determined by sampling of the atmosphere. Samples collected shall reflect the exposure of the persons involved. When a helmet is worn, the samples shall be collected under the helmet.

NOTE

Where welding operations are incidental to general operations, it is considered good practice to apply local exhaust ventilation to prevent contamination of the general work area.

d. <u>Respiratory Protective Equipment</u>. Individual respiratory protective equipment will be well retained. Only respiratory protective equipment approved by the US Bureau of Mines, National Institute of Occupational Safety and Health, or other government-approved testing agency shall be utilized. Guidance for selection, care, and maintenance of respiratory protective equipment is given in Practices for Respiratory Protection, American National Standard Institute Standard 788.2 and TB MED 223. Respiratory protective equipment will not be transferred from one individual to another without being disinfected.

e. <u>Precautionary Labels</u>. A number of potentially hazardous materials are used in flux coatings, coverings, and filler metals. These materials, when used in welding and cutting operations, will become hazardous to the welder as they are released into the atmosphere. These include, but are not limited to, the following materials: fluorine compounds, zinc, lead, beryllium, cadmium, and mercury. See paragraph 2-4 i through 2-4 n. The suppliers of welding materials shall determine the hazard, if any, associated with the use of their materials in welding, cutting, etc.

(1) All filler metals and fusible granular materials shall carry the following notice, as a minimum, on tags, boxes, or other containers:

CAUTION

Welding may produce fumes and gases hazardous to health. Avoid breathing these fumes and gases. Use adequate ventilation. See American National Standards Institute Standard Z49.1-1973, Safety in Welding and Cutting published by the American Welding Society.

(2) Brazing (welding) filler metals containing cadmium in significant amounts shall carry the following notice on tags, boxes, or other containers:

WARNING
CONTAINS CADMIUM - POISONOUS FUMES MAY BE FORMED ON HEATING

Do not breathe fumes. Use only with adequate ventilation, such as fume collectors, exhaust ventilators, or air-supplied respirators. See American National Standards Institute Standard Z49.1-1973. If chest pain, cough, or fever develops after use, call physician immediately.

(3) Brazing and gas welding fluxes containing fluorine compounds shall have a cautionary wording. One such wording recommended by the American Welding Society for brazing and gas welding fluxes reads as follows:

CAUTION
CONTAINS FLUORIDES

This flux, when heated, gives off fumes that may irritate eyes, nose, and throat.
Avoid fumes--use only in well-ventilated spaces.
Avoid contact of flux with eyes or skin.
Do not take internally.

f. Ventilation for General Welding and Cutting.

(1) General. Mechanical ventilation shall be provided when welding or cutting is done on metals not covered in subparagraphs i through p of this section, and under the following conditions:

(a) In a space of less than 10,000 cu ft (284 cu m) per welder.

(b) In a roan having a ceiling height of less than 16 ft (5 m).

(c) In confined spaces or where the welding space contains partitions, balconies, or other structural barriers to the extent that they significantly obstruct cross ventilation.

(2) Minimum rate. Ventilation shall be at the minimum rate of 200 cu ft per minute (57 cu m) per welder, except where local exhaust heeds, as in paragraph 2-4 g below, or airline respirators approved by the US Bureau of Mines, National Institute of Occupational Safety and Health, or other government-approved testing agency, are used. When welding with rods larger than 3/16 in. (0.48 cm) in diameter, the ventilation shall be higher as shown in the following:

Rod diameter (inches)	Required ventilation (cfm)
1/4 (0.64 cm)	3500
3/8 (0.95 cm)	4500

Natural ventilation is considered sufficient for welding or cutting operations where the conditions listed above are not present. Figure 2-5 is an illustration of a welding booth equipped with mechanical ventilation sufficient for one welder.

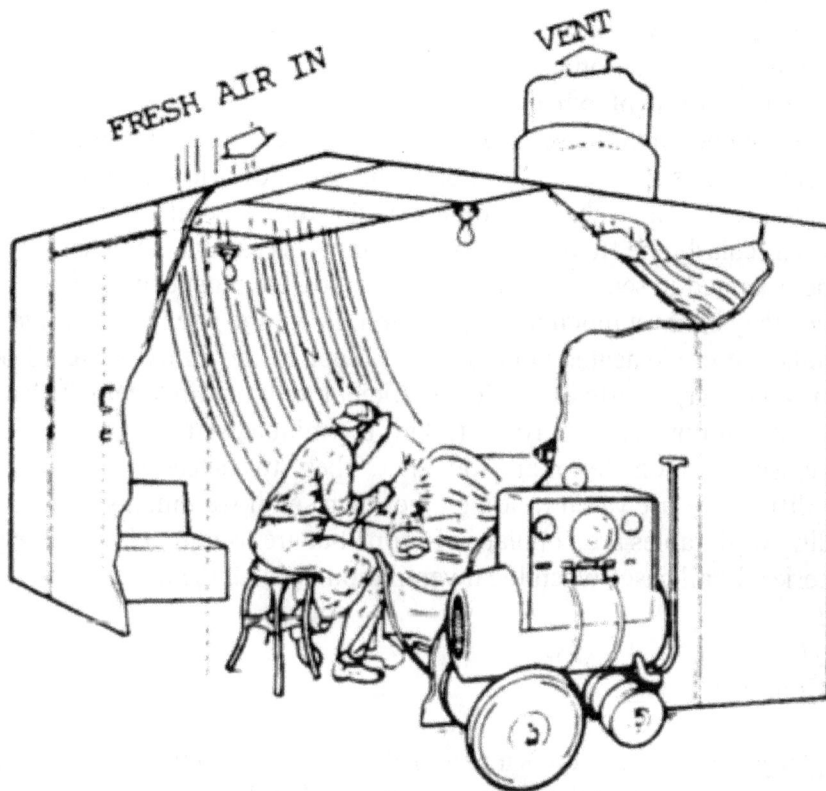

Figure 2-5. Welding booth with mechanical ventilation.

g. Local Exhaust Ventilation. Mechanical local exhaust ventilation may be obtained by either of the following means:

(1) Hoods. Freely movable hoods or ducts are intended to be placed by the welder as near as practicable to the work being welded. These will provide a rate of airflow sufficient to maintain a velocity the direction of the hood of 100 in linear feet per minute in the zone of welding. The ventilation rates required to accomplish this control velocity using a 3-in. wide flanged suction opening are listed in table 2-2.

Table 2-2. Required Exhaust Ventilation

Welding zone	Minimum air flow, cu ft per min	Duct diameter, in.
4 to 6 in. from arc or torch	150	3
6 to 8 in. from arc or torch	275	3-1/2
8 to 10 in. from arc or torch	425	4-1/2
10 to 12 in. from arc or torch	600	6-1/2

(2) Fixed enclosure. A fixed enclosure with a top and two or more sides which surrounds the welding or cutting operations will have a rate of airflow sufficient to maintain a velocity away from the welder of not less than 100 linear ft per minute. Downdraft ventilation tables require 150 cu ft per minute per square foot of surface area. This rate of exhausted air shall be uniform across the face of the grille. A low volume, high-density fume exhaust device attached to the welding gun collects the fumes as close as possible to the point of origin or at the arc. This method of fume exhaust has become quite popular for the semiautomatic processes, particularly the flux-cored arc welding process. Smoke exhaust systems incorporated in semiautomatic guns provide the most economical exhaust system since they exhaust much less air they eliminate the need for massive air makeup units to provide heated or cooled air to replace the air exhausted. Local ventilation should have a rate of air flow sufficient to maintain a velocity away from the welder of not less than 100 ft (30 m) per minute. Air velocity is measurable using a velometer or air flow inter. These two systems can be extremely difficult to use when welding other than small weldments. The down draft welding work tables are popular in Europe but are used to a limited degree North America. In all cases when local ventilation is used, the exhaust air should be filtered.

h. Ventilation in Confined Spaces.

(1) Air replacement. Ventilation is a perquisite to work in confined spaces. All welding and cutting operations in confined spaces shall be adequately ventilated to prevent the accumulation of toxic materials -or possible oxygen deficiency. This applies not only to the welder but also to helpers and other personnel in the immediate vicinity.

(2) Airline respirators. In circumstances where it is impossible to provide adequate ventilation in a confined area, airline respirators or hose masks, approved by the US Bureau of Mines, National Institute of Occupational Safety and Health, or other government-approved testing agency, will be used for this purpose. The air should meet the standards established by Public Law 91-596, Occupational Safety and Health Act of 1970.

(3) Self-contained units. In areas immediately hazardous to life, hose masks with blowers or self-contained breathing equipment shall be used. The breathing

equipment shall be approved by the US Bureau of Mines or National Institute of Occupational Safety and Health, or other government-approved testing agency.

(4) Outside helper. Where welding operations are carried on in confined spaces and where welders and helpers are provided with hose masks, hose masks with blowers, or self-contained breathing equipment, a worker shall be stationed on the outside of such confined spaces to ensure the safety of those working within.

(5) Oxygen for ventilation. Oxygen must never be used for ventilation.

i. Fluorine Compounds.

(1) General. In confined spaces, welding or cutting involving fluxes, coverings, or other materials which fluorine compounds shall be done in accordance with paragraph 2-4 h, ventilation in confined spaces. A fluorine compound is one that contains fluorine as an element in chemical combination, not as a free gas.

(2) Maximum allowable concentration. The need for local exhaust ventilation or airline respirators for welding or cutting in other than confined spaces will depend upon the individual circumstances. However, experience has shown that such protection is desirable for fixed-location production welding and for all production welding on stainless steels. When air samples taken at the welding location indicate that the fluorides liberated are below the maximum allowable concentration, such protection is not necessary.

j. Zinc.

(1) Confined spaces. In confined spaces, welding or cutting involving zinc-bearing filler metals or metals coated with zinc-bearing materials shall be done in accordance with paragraph 2-4 h, ventilation in confined spaces.

(2) Indoors. Indoors, welding or cutting involving zinc-bearing metals or filler metals coated with zinc-bearing materials shall be done in accordance with paragraph 2-4 g.

k. Lead.

(1) Confined spaces. In confined spaces, welding involving lead-base metals (erroneously called lead-burning) shall be done in accordance with paragraph 2-4 h.

(2) Indoors. Indoors, welding involving lead-base metals shall be done in accordance with paragraph 2-4 g, local exhaust ventilation.

(3) Local ventilation. In confined spaces or indoors, welding or cutting involving metals containing lead or metals coated with lead-bearing materials, including

paint, shall be done using local exhaust ventilation or airline respirators. Outdoors, such operations shall be done using respirator protective equipment approved by the US Bureau of Mines, National Institute of Occupational Safety and Health, or other government-approved testing agency. In all cases, workers in the immediate vicinity of the cutting or welding operation shall be protected as necessary by local exhaust ventilation or airline respirators.

l. Beryllium. Welding or cutting indoors, outdoors, or in confined spaces involving beryllium-bearing material or filler metals will be done using local exhaust ventilation and airline respirators. This must be performed without exception unless atmospheric tests under the most adverse conditions have established that the workers' exposure is within the acceptable concentrations of the latest Threshold Limit Values (TLV) of the American Conference of Governmental Industrial Hygienists, or the exposure limits established by Public Law 91-596, Occupational Safety and Health Act of 1970. In all cases, workers in the immediate vicinity of the welding or cutting operations shall be protected as necessary by local exhaust ventilation or airline respirators.

m. Cadmium.

(1) General. Welding or cutting indoors or in confined spaces involving cadmium-bearing or cadmium-coated base metals will be done using local exhaust ventilation or airline respirators. Outdoors, such operations shall be done using respiratory protective equipment such as fume respirators, approved by the US Bureau of Mines, National Institute of Occupational Safety and Health, or other government-approved testing agency, for such purposes.

(2) Confined space. Welding (brazing) involving cadmium-bearing filler metals shall be done using ventilation as prescribed in paragraphs 2-4 g, local exhaust ventilation, and 2-4 h, ventilation in confined spaces, if the work is to be done in a confined space.

NOTE

Cadmium-free rods are available and can be used in most instances with satisfactory results.

n. Mercury. Welding or cutting indoors or in a confined space involving metals coated with mercury-bearing materials, including paint, shall be done using local exhaust ventilation or airline respirators. Outdoors, such operations will be done using respiratory protective equipment approved by the National Institute of Occupational Safety and Health, US Bureau of Mines, or other government-approved testing agency.

o. Cleaning Compounds.

(1) Manufacturer's instructions. In the use of cleaning materials, because of their toxicity of flammability, appropriate precautions listed in the manufacturer's instructions will be followed.

(2) Degreasing. Degreasing or other cleaning operations involving chlorinated hydrocarbons will be located so that no vapors from these operations will reach or be drawn into the area surrounding any welding operation. In addition, trichloroethylene and perchloroethylene should be kept out of atmospheres penetrated by the ultraviolet radiation of gas-shielded welding operations.

p. Cutting of Stainless Steels. Oxygen cutting, using either a chemical flux or iron powder, or gas-shielded arc cutting of stainless steel will be done using mechanical ventilation adequate to remove the fumes generated.

q. First-Aid Equipment. First-aid equipment will be available at all times. On every shift of welding operations, there will be personnel present who are trained to render first-aid. All injuries will be reported as soon as possible for medical attention. First-aid will be rendered until medical attention can be provided.

2-5. WELDING IN CONFINED SPACES

a. A confined space is intended to mean a relatively small or restricted space such as a tank, boiler, pressure vessel, or small compartment of a ship or tank.

b. When welding or cutting is being performed in any confined space, the gas cylinders and welding machines shall be left on the outside. Before operations are started, heavy portable equipment mounted on wheels shall be securely blocked to prevent accidental movement.

c. Where a welder must enter a confined space through a manhole or other all opening, means will be provided for quickly removing him in case of emergency. When safety belts and life lines are used for this purpose, they will be attached to the welder's body so that he cannot be jammed in a small exit opening. An attendant with a preplanned rescue procedure will be stationed outside to observe the welder at all times and be capable of putting rescue operations into effect.

d. When arc welding is suspended for any substantial period of time, such as during lunch or overnight, all electrodes will be removed from the holders with the holders carefully located so that accidental contact cannot occur. The welding machines will be disconnected from the power source.

e. In order to eliminate the possibility of gas escaping through leaks or improperly closed valves when gas welding or cutting, the gas and oxygen supply valves will be closed, the regulators released, the gas and oxygen lines bled, and the valves on the torch shut off when the equipment will not be used for a substantial period of time. Where practical, the torch and hose will also be removed from the confined space.

f. After welding operations are completed, the welder will mark the hot metal or provide some other means of warning other workers.

Section II. SAFETY PRECAUTIONS IN OXYFUEL WELDING

2-6. GENERAL

a. In addition to the information listed in section I of this chapter, the following safety precautions must be observed.

b. Do not experiment with torches or regulators in any way. Do not use oxygen regulators with acetylene cylinders. Do not use any lubricants on regulators or tanks.

c. Always use the proper tip or nozzle, and always operate it at the proper pressure for the particular work involved. This information should be taken from work sheets or tables supplied with the equipment.

d. When not in use, make sure the torch is not burning. Also, release the regulators, bleed the hoses, and tightly close the valves. Do not hang the torch with its hose on the regulator or cylinder valves.

e. Do not light a torch with a match or hot metal, or in a confined space. The explosive mixture of acetylene and oxygen might cause personal injury or property damage when ignited. Use friction lighters or stationary pilot flames.

f. When working in confined spaces, provide adequate ventilation for the dissipation of explosive gases that may be generated. For ventilation standards, refer to paragraph 2-4, Health Protection and Ventilation.

g. Keep a clear space between the cylinder and the work so the cylinder valves can be reached easily and quickly.

h. Use cylinders in the order received. Store full and empty cylinders separately and mark the empty ones with "MT".

i. Compressed gas cylinders owned by commercial companies will not be painted regulation Army olive drab.

j. Never use cylinders for rollers, supports, or any purpose other than that for which they are intended.

k. Always wear protective clothing suitable for welding or flame cutting.

l. Keep work area clean and free from hazardous materials. When flame cutting, sparks can travel 30 to 40 ft (9 to 12 m). Do not allow flare cut sparks to hit hoses, regulators, or cylinders.

m. Use oxygen and acetylene or other fuel gases with the appropriate torches and only for the purpose intended.

n. Treat regulators with respect. Do not turn valve handle using force.

o. Always use the following sequence and technique for lighting a torch:

> (1) Open acetylene cylinder valve.

> (2) Open acetylene torch valve 1/4 turn.

> (3) Screw in acetylene regulator adjusting valve handle to working pressure.

> (4) Turn off the acetylene torch valve (this will purge the acetylene line).

> (5) Slowly open oxygen cylinder valve all the way.

> (6) Open oxygen torch valve 1/4 turn.

> (7) Screw in oxygen regulator screw to working pressure.

> (8) Turn off oxygen torch valve (this will purge the oxygen line).

> (9) Open acetylene torch valve 1/4 turn and light with lighter.

NOTE

Use only friction type lighter or specially provided lighting device.

> (10) Open oxygen torch valve 1/4 turn.

> (11) Adjust to neutral flame.

p. Always use the following sequence and technique for shutting off a torch:

> (1) Close acetylene torch valve first, then the oxygen valve.

> (2) Close acetylene cylinder valve, then oxygen cylinder valve.

> (3) Open torch acetylene and oxygen valves to release pressure in the regulator and hose.

> (4) Back off regulator adjusting valve handle until no spring tension is left.

> (5) Close torch valves.

q. Use mechanical exhaust at the point of welding when welding or cutting lead, cadmium, chromium, manganese, brass, bronze, zinc, or galvanized steel.

r. Do not weld or flame cut on containers that have held combustibles without taking special precautions.

s. Do not weld or flame cut into sealed container or compartment without providing vents and taking special precautions.

t. Do not weld or cut in a confined space without taking special precautions.

2-7. ACETYLENE CYLINDERS

CAUTION

If acetylene cylinders have been stored or transported horizontally (on their sides), stand cylinders vertically (upright) for 45 minutes prior to (before) use.

a. Always refer to acetylene by its full name and not by the word "gas" alone. Acetylene is very different from city or furnace gas. Acetylene is a compound of carbon and hydrogen, produced by the reaction of water and calcium carbide.

b. Acetylene cylinders must be handled with care to avoid damage to the valves or the safety fuse plug. The cylinders must be stored upright in a well ventilated, well protected, dry location at least 20 ft from highly combustible materials such as oil, paint, or excelsior. Valve protection caps must always be in place, hand tight, except when cylinders are in use. Do not store the cylinders near radiators, furnaces, or in any are with above normal temperatures. In tropical climates, care must be taken not to store acetylene in areas where the temperature is in excess of 137°F (58°C). Heat will increase the pressure, which may cause the safety fuse plug in the cylinder to blow out. Storage areas should be located away from elevators, gangways, or other places where there is danger of cylinders being knocked over or damaged by falling objects.

c. A suitable truck, chain, or strap must be used to prevent cylinders from falling or being knocked over while in use. Cylinders should be kept at a safe distance from the welding operation so there will be little possibility of sparks, hot slag, or flames reaching them. They should be kept away from radiators, piping systems, layout tables, etc., which may be used for grounding electrical circuits. Nonsparking tools should be used when changing fittings on cylinders of flammable gases.

d. Never use acetylene without reducing the pressure with a suitable pressure reducing regulator. Never use acetylene at pressures in excess of 15 psi.

e. Before attaching the pressure regulators, open each acetylene cylinder valve for an instant to blow dirt out of the nozzles. Wipe off the connection seat with a clean cloth. Do not stand in front of valves when opening them.

f. Outlet valves which have become clogged with ice should be thawed with warm water. Do not use scalding water or an open flame.

g. Be sure the regulator tension screw is released before opening the cylinder valve. Always open the valve slowly to avoid strain on the regulator gage which records the cylinder pressure. Do not open the valve more than one and one-half turns. Usually, one-half turn is sufficient. Always use the special T-wrench provided for the acetylene cylinder valve. Leave this wrench on the stem of the valve tile the cylinder is in use so the acetylene can be quickly turned off in an emergency.

h. Acetylene is a highly combustible fuel gas and great care should be taken to keep sparks, flames, and heat away from the cylinders. Never open an acetylene cylinder valve near other welding or cutting work.

i. Never test for an acetylene leak with an open flame. Test all joints with soapy water. Should a leak occur around the valve stem of the cylinder, close the valve and tighten the packing nut. Cylinders leaking around the safety fuse plug should be taken outdoors, away from all fires and sparks, and the valve opened slightly to permit the contents to escape.

j. If an acetylene cylinder should catch fire, it can usually be extinguished with a wet blanket. A burlap bag wet with calcium chloride solution is effective for such an emergency. If these fail, spray a stream of water on the cylinder to keep it cool.

k. Never interchange acetylene regulators, hose, or other apparatus with similar equipment intended for oxygen.

l. Always turn the acetylene cylinder so the valve outlet will point away from the oxygen cylinder.

m. When returning empty cylinders, see that the valves are closed to prevent escape of residual acetylene or acetone solvent. Screw on protecting caps.

n. Make sure that all gas apparatus shows UL or FM approval, is installed properly, and is in good working condition.

o. Handle all compressed gas with extreme care. Keep cylinder caps on when not in use.

p. Make sure that all compressed gas cylinders are secured to the wall or other structural supports. Keep acetylene cylinders in the vertical condition.

q. Store compressed gas cylinders in a safe place with good ventilation. Acetylene cylinders and oxygen cylinders should be kept apart.

r. Never use acetylene at a pressure in excess of 15 psi (103.4 kPa). Higher pressure can cause an explosion.

s. Acetylene is nontoxic; however, it is an anesthetic and if present in great enough concentrations, is an asphyxiant and can produce suffocation.

2-8. OXYGEN CYLINDERS

a. Always refer to oxygen by its full name and not by the word "air" alone.

b. Oxygen should never be used for "air" in any way.

WARNING

Oil or grease in the presence of oxygen will ignite violently, especially in an enclosed pressurized area.

c. Oxygen cylinders shall not be stored near highly combustible material, especially oil and grease; near reserve stocks of carbide and acetylene or other fuel gas cylinders, or any other substance likely to cause or accelerate fire; or in an acetylene generator compartment.

d. Oxygen cylinders stored in outside generator houses shall be separated from the generator or carbide storage rooms by a noncombustible partition having a fire resistance rating of at least 1 hour. The partition shall be without openings and shall be gastight.

e. Oxygen cylinders in storage shall be separated from fuel gas cylinders or combustible materials (especially oil or grease) by a minimum distance of 20.0 ft (6.1 m) or by a noncombustible barrier at least 5.0 ft (1.5 m) high and having a fire-resistance rating of at least one-half hour.

f. Where a liquid oxygen system is to be used to supply gaseous oxygen for welding or cutting and a bulk storage system is used, it shall comply with the provisions of the Standard for Bulk Oxygen Systems at Consumer Sites, NFPA No. 566-1965, National Fire Protection Association.

g. When oxygen cylinders are in use or being roved, care must be taken to avoid dropping, knocking over, or striking the cylinders with heavy objects. Do not handle oxygen cylinders roughly.

h. All oxygen cylinders with leaky valves or safety fuse plugs and discs should be set aside and marked for the attention of the supplier. Do not tamper with or attempt to repair oxygen cylinder valves. Do not use a hammer or wrench to open the valves.

i. Before attaching the pressure regulators, open each oxygen cylinder valve for an instant to blow out dirt and foreign matter from the nozzle. Wipe off the connection seat with a clean cloth. Do not stand in front of the valve when opening it.

WARNING

Do not substitute oxygen for compressed air in pneumatic tools. Do not use oxygen to blow out pipe lines, test radiators, purge tanks or containers, or to "dust" clothing or work.

j. Open the oxygen cylinder valve slowly to prevent damage to regulator high pressure gage mechanism. Be sure that the regulator tension screw is released the before opening the valve. When not in use, the cylinder valve should be closed and the protecting caps screwed on to prevent damage to the valve.

k. When the oxygen cylinder is in use, open the valve to the full limit to prevent leakage around the valve stem.

l. Always use regulators on oxygen cylinders to reduce the cylinder pressure to a low working pressure. High cylinder pressure will burst the hose.

m. Never interchange oxygen regulators, hoses, or other apparatus with similar equipment intended for other gases.

2-9. MAPP GAS CYLINDERS

a. MAPP gas is a mixture of stabilized methylacetylene and propadiene.

b. Store liquid MAPP gas around 70°F (21°C) and under 94 psig pressure.

c. Repair any leaks immediately. MAPP gas vaporizes when the valve is opened and is difficult to detect visually. However, MAPP gas has an obnoxious odor detectable at 100 parts per million, a concentration 1/340th of its lower explosive limit in air. If repaired when detected, leaks pose little or no danger. However, if leaks are ignored, at very high concentrations (5000 parts per million and above) MAPP gas has an anesthetic effect.

d. Proper clothing must be worn to prevent injury to personnel. Once released into the open air, liquid MAPP gas boils at -36 to -4°F (-54 to -20°C). This causes frost-like burns when the gas contacts the skin.

e. MAPP gas toxicity is rated very slight, but high concentrations (5000 part per million) may have an anesthetic affect.

f. MAPP gas has some advantages in safety which should be considered when choosing a process fuel gas, including the following:

(1) MAPP gas cylinders will not detonate when dented, dropped, or incinerated.

(2) MAPP gas can be used safely at the full cylinder pressure of 94 psig.

(3) Liquefied fuel is insensitive to shock.

(4) Explosive limits of MAPP gas are low compared to acetylene.

(5) Leaks can be detected easily by the strong smell of MAPP gas.

(6) MAPP cylinders are easy to handle due to their light weight.

2-10. FUEL GAS CYLINDERS

a. Although the most familiar fuel gas used for cutting and welding is acetylene, propane, natural gas, and propylene are also used. Store these fuel gas cylinders in a specified, well-ventilated area or outdoors, and in a vertical condition.

b. Any cylinders must have their caps on, and cylinders, either filled or empty, should have the valve closed.

c. Care must be taken to protect the valve from damage or deterioration. The major hazard of compressed gas is the possibility of sudden release of the gas by removal or breaking off of the valve. Escaping gas which is under high pressure will cause the cylinder to act as a rocket, smashing into people and property. Escaping fuel gas can also be a fire or explosion hazard.

d. In a fire situation there are special precautions that should be taken for acetylene cylinders. All acetylene cylinders are equipped with one or more safety relief devices filled with a low melting point metal. This fusible metal melts at about the killing point of water (212°F or 100°C). If fire occurs on or near an acetylene cylinder the fuse plug will melt. The escaping acetylene may be ignited and will burn with a roaring sound. Immediately evacuate all people from the area. It is difficult to put out such a fire. The best action is to put water on the cylinder to keep it cool and to keep all other acetylene cylinders in the area cool. Attempt to remove the burning cylinder from close proximity to other acetylene cylinders, from flammable or hazardous materials, or from combustible buildings. It is best to allow the gas to burn rather than to allow acetylene to escape, mix with air, and possibly explode.

e. If the fire on a cylinder is a small flame around the hose connection, the valve stem, or the fuse plug, try to put it out as quickly as possible. A wet glove, wet heavy cloth, or mud slapped on the flame will frequently extinguish it. Thoroughly wetting the gloves and clothing will help protect the person approaching the cylinder. Avoid getting in line with the fuse plug which might melt at any time.

f. Oxygen cylinders should be stored separately from fuel gas cylinders and separately from combustible materials. Store cylinders in cool, well-ventilated areas. The temperature of the cylinder should never be allowed to exceed 130°F (54°C).

g. When cylinders are empty they should be marked empty and the valves must be closed to prohibit contamination from entering.

h. When the gas cylinders are in use a regulator is attached and the cylinder should be secured to prevent falling by means of chains or clamps.

i. Cylinders for portable apparatuses should be securely mounted in specially designed cylinder trucks.

j. Cylinders should be handled with respect. They should not be dropped or struck. They should never be used as rollers. Hammers or wrenches should not be used to open cylinder valves that are fitted with hand wheels. They should never be moved by electromagnetic cranes. They should never be in an electric circuit so that the welding current could pass through them. An arc strike on a cylinder will damage the cylinder causing possible fracture, requiring the cylinder to be condemned and discarded from service.

2-11. HOSES

a. Do not allow hoses to come in contact with oil or grease. These will penetrate and deteriorate the rubber and constitute a hazard with oxygen.

b. Always protect hoses from being walked on or run over. Avoid kinks and tangles. Do not leave hoses where anyone can trip over them. This could result in personal injury, damaged connections, or cylinders being knocked over. Do not work with hoses over the shoulder, around the legs, or tied to the waist.

c. Protect hoses from hot slag, flying sparks, and open flames.

d. Never force hose connections that do not fit. Do not use white lead, oil, grease, or other pipe fitting compounds for connections on hose, torch, or other equipment. Never crimp hose to shut off gases.

e. Examine all hoses periodically for leaks by immersing them in water while under pressure. Do not use matches to check for leaks in acetylene hose. Repair leaks by cutting hose and inserting a brass splice. Do not use tape for mending. Replace hoses if necessary.

f. Make sure that hoses are securely attached to torches and regulators before using.

g. Do not use new or stored hose lengths without first blowing them out with compressed air to eliminate talc or accumulated foreign matter which might otherwise enter and clog the torch parts.

h. Only approved gas hoses for flame cutting or welding should be used with oxyfuel gas equipment. Single lines, double vulcanized, or double multiple stranded lines are available.

i. The size of hose should be matched to the connectors, regulators, and torches.

j. In the United States, the color green is used for oxygen, red for acetylene or fuel gas, and black for inert gas or compressed air. The international standard calls for blue for oxygen and orange for fuel gas.

k. Connections on hoses are right-handed for inert gases and oxygen, and left-handed for fuel gases.

l. The nuts on fuel gas hoses are identified by a groove machined in the center of the nuts.

m. Hoses should be periodically inspected for burns, worn places, or leaks at the connections. They must be kept in good repair and should be no longer than necessary.

Section III. SAFETY IN ARC WELDING AND CUTTING

2-12. ELECTRIC CIRCUITS

a. A shock hazard is associated with all electrical equipment, including extension lights, electric hand tools, and all types of electrically powered machinery. Ordinary household voltage (115 V) is higher than the output voltage of a conventional arc welding machine.

b. Although the ac and dc open circuit voltages are low compared to voltages used for lighting circuits and motor driven shop tools, these voltages can cause severe shock, particularly in hot weather when the welder is sweating. Consequently, the precautions listed below should always be observed.

(1) Check the welding equipment to make certain that electrode connections and insulation on holders and cables are in good condition.

(2) Keep hands and body insulated from both the work and the metal electrode holder. Avoid standing on wet floors or coming in contact with grounded surfaces.

(3) Perform all welding operations within the rated capacity of the welding cables. Excessive heating will impair the insulation and damage the cable leads.

WARNING

Welding machine, Model 301, AC/DC, Heliarc with inert gas attachment, NSN 3431-00-235-4728, may cause electrical shock if not properly grounded. If one is being used, contact Castolin Institute, 4462 York St. Denver, Colorado 80216.

c. Inspect the cables periodically for looseness at the joints, defects due to wear, or other damage. Defective or loose cables are a fire hazard. Defective electrode holders should be replaced and connections to the holder should be tightened.

d. Welding generators should be located or shielded so that dust, water, or other foreign matter will not enter the electrical windings or the bearings.

e. Disconnect switches should be used with all power sources so that they can be disconnected from the main lines for maintenance.

2-13. WELDING MACHINES

a. When electric generators powered by internal combustion engines are used inside buildings or in confined areas, the engine exhaust must be conducted to the outside atmosphere.

b. Check the welding equipment to make sure the electrode connections and the insulation on holders and cables are in good condition. All checking should be done with the machine off or unplugged. All serious trouble should be investigated by a trained electrician.

c. Motor-generator welding machines feature complete separation of the primary power and the welding circuit since the generator is mechanically connected to the electric rotor. A rotor-generator type arc welding machine must have a power ground on the machine. Metal frames and cases of motor generators must be grounded since the high voltage from the main line does come into the case. Stray current may cause a severe shock to the operator if he should contact the machine and a good ground.

d. In transformer and rectifier type welding machines, the metal frame and cases must be grounded to the earth. The work terminal of the welding machine should not be grounded to the earth.

e. Phases of a three-phase power line must be accurately identified when paralleling transformer welding machines to ensure that the machines are on the same phase and in phase with one another. To check, connect the work leads together and measure the voltage between the electrode holders of the two machines. This voltage should be practically zero. If it is double the normal open circuit voltage, it means that either the primary or secondary connections are reversed. If the voltage is approximately 1-1/2 times the normal open circuit voltage it means that the machines are connected to different phases of the three phase power line. Corrections must be made before welding begins.

f. When large weldments, like ships, buildings, or structural parts are involved, it is normal to have the work terminal of many welding machines connected to it. It is important that the machines be connected to the proper phase and have the same polarity. Check by measuring the voltage between the electrode holders of the different machines as mentioned above. The situation can also occur with respect to direct current power sources when they are connected to a common weldment. If one machine is connected for straight polarity and one for reverse polarity, the voltage between the electrode holders will be double the normal open circuit voltage. Precautions should be taken to see that all machines are of the same polarity when connected to a common weldment.

g. Do not operate the polarity switch while the machine is operating under welding current load. Consequent arcing at the switch will damage the contact surfaces and the flash may burn the person operating the switch.

h. Do not operate the rotary switch for current settings while the machine is operating under welding current load. Severe burning of the switch contact surfaces will result. Operate the rotary switch while the machine is idling.

i. Disconnect the welding machines from the power supply when they are left unattended.

j. The welding electrode holders must be connected to machines with flexible cables for welding application. Use only insulated electrode holders and cables. There can be no splices in the electrode cable within 10 feet (3 meters) of the electrode holder. Splices, if used in work or electrode leads, must be insulated. Wear dry protective covering on hands and body.

k. Partially used electrodes should be removed from the holders when not in use. A place will be provided to hang up or lay down the holder where it will not come in contact with persons or conducting objects.

l. The work clamp must be securely attached to the work before the start of the welding operation.

m. Locate welding machines where they have adequate ventilation and ventilation ports are not obstructed.

2-14. PROTECTIVE SCREENS

a. When welding is done near other personnel, screens should be used to protect their eyes from the arc or reflected glare. See paragraph 2-2 e for screen design and method of use.

b. In addition to using portable screens to protect other personnel, screens should be used, when necessary, to prevent drafts of air from interfering with the stability of the arc.

c. Arc welding operations give off an intense light. Snap-on light-proof screens should be used to cover the windows of the welding truck to avoid detection when welding at night.

2-15. PLASMA ARC CUTTING AND WELDING

a. Plasma arc welding is a process in which coalescence is produced by heating with a constricted arc between an electrode and the work piece (transfer arc) or the electrode and the constricting nozzle (nontransfer arc). Shielding is obtained from the hot ionized gas issuing from the orifice which may be supplemented by an auxiliary source of shielding gas. Shielding gas may be an inert gas or a mixture of gases; pressure may or may not be used, and filler metal may or may not be supplied. Plasma welding is similar in many ways to the tungsten arc process. Therefore, the safety considerations for plasma arc welding are the same as for gas tungsten arc welding.

b. Adequate ventilation is required during the plasma arc welding process due to the brightness of the plasma arc, which causes air to break down into ozone.

c. The bright arc rays also cause fumes from the hydrochlorinated cleaning materials or decreasing agents to break down and form phosgene gas. Cleaning operations using these materials should be shielded from the arc rays of the plasma arc.

d. When welding with transferred arc current up to 5A, safety glasses with side shields or other types of eye protection with a No. 6 filter lens are recommended. Although face protection is not normally required for this current range, its use depends on personal preference. When welding with transferred arc currents between 5 and 15A, a full plastic face shield is recommended in addition to eye protection with a No. 6 filter lens. At current levels over 15A, a standard welder's helmet with proper shade of filter plate for the current being used is required.

e. When a pilot arc is operated continuously, normal precautions should be used for protection against arc flash and heat burns. Suitable clothing must be worn to protect exposed skin from arc radiation.

f. Welding power should be turned off before electrodes are adjusted or replaced.

g. Adequate eye protection should be used when observation of a high frequency discharge is required to center the electrode.

h. Accessory equipment, such as wire feeders, arc voltage heads, and oscillators should be properly grounded. If not grounded, insulation breakdown might cause these units to become electrically "hot" with respect to ground.

i. Adequate ventilation should be used, particularly when welding metals with high copper, lead, zinc, or beryllium contents.

2-16. AIR CARBON ARC CUTTING AND WELDING

a. Air carbon arc cutting is an arc cutting process in which metals to be cut are melted by the heat of a carbon arc and the molten metal is removed by a blast of air. The process is widely used for back gouging, preparing joints, and removing defective metal.

b. A high velocity air jet traveling parallel to the carbon electrode strikes the molten metal puddle just behind the arc and blows the molten metal out of the immediate area. Figure 2-6 shows the operation of the process.

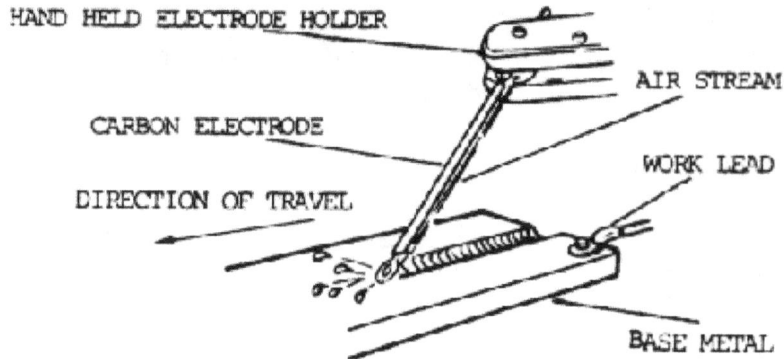

Figure 2-6. Process diagram for air carbon arc cutting.

c. The air carbon arc cutting process is used to cut metal and to gouge out defective metal, to remove old or inferior welds, for root gouging of full penetration welds, and to prepare grooves for welding. Air carbon arc cutting is used when slightly ragged edges are not objectionable. The area of the cut is small, and since the metal is melted and removed quickly, the surrounding area does not reach high temperatures. This reduces the tendency towards distortion and cracking. The air carbon arc can be used for cutting or gouging most of the common metals.

d. The process is not recommended for weld preparation for stainless steel, titanium, zirconium, and other similar metals without subsequent cleaning. This cleaning, usually by grinding, must remove all of the surface carbonized material adjacent to the cut. The process can be used to cut these materials for scrap for remelting.

e. The circuit diagram for air carbon arc cutting or gouging is shown by figure 2-7. Normally, conventional welding machines with constant current are used. Constant voltage can be used with this process.

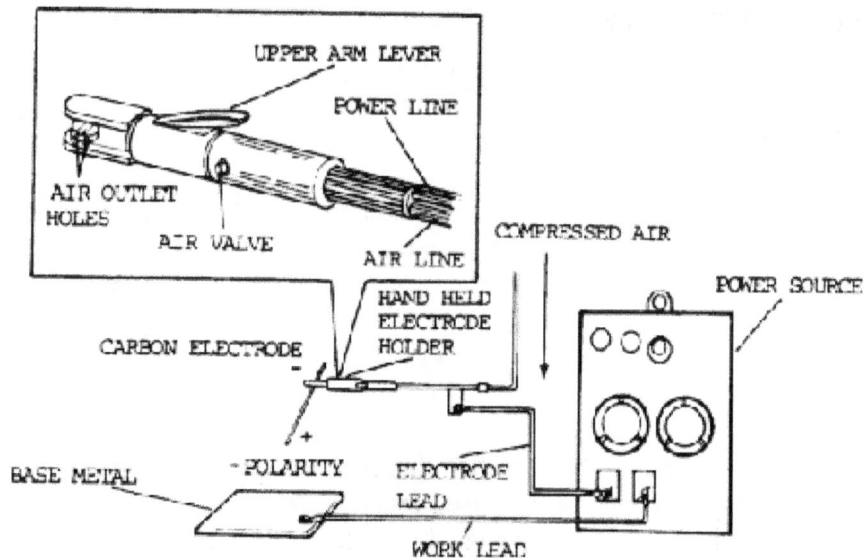

Figure 2-7. Circuit block diagram AAC.

f. When using a constant voltage (CV) power source precautions must be taken to operate it within its rated output of current and duty cycle.

g. Alternating current power sources having conventional drooping characteristics can also be used for special applications. AC type carbon electrodes must be used.

h. Special heavy duty high current machines have been made specifically for the air carbon arc process. This is because of extremely high currents used for the large size carbon electrodes.

i. The air pressure must range from 80 to 100 psi (550 to 690 kPa). The volume of compressed air required ranges from as low as 5.0 cu ft/min. (2.5 liter/rein.) up to 50 cu ft/min. (24 liter/min.) for the largest-size carbon electrodes.

j. The air blast of air carbon arc welding will cause the molten metal to travel a very long distance. Metal deflection plates should be placed in front of the gouging operation, and all combustible materials should be moved away from the work area. At high-current levels, the mass of molten metal removed is quite large and will become a fire hazard if not properly contained.

k. A high noise level is associated with air carbon arc welding. At high currents with high air pressure a very loud noise occurs. Ear protection, ear muffs or ear plugs must be worn by the arc cutter.

Section IV. SAFETY PRECAUTIONS FOR GAS SHIELDED ARC WELDING

2-17. POTENTIAL HAZARDS

When any of the welding processes are used, the shielded from the air in order to obtain a high molten puddle of metal should be quality weld deposit. In shielded metal arc welding, shielding from the air is accomplished by gases produced by the disintegration of the coating in the arc. With gas shielded arc welding, shielding from the air is accomplished by surrounding the arc area with a localized gaseous atmosphere throughout the welding operation at the molten puddle area.

Gas shielded arc welding processes have certain dangers associated with them. These hazards, which are either peculiar to or increased by gas shielded arc welding, include arc gases, radiant energy, radioactivity from throated tungsten electrodes, and metal fumes.

2-18. PROTECTIVE MEASURES

a. Gases.

   (1) Ozone. Ozone concentration increases with the type of electrodes used, amperage, extension of arc tine, and increased argon flow. If welding is carried out in confined spaces and poorly ventilated areas, the ozone concentration may increase to harmful levels. The exposure level to ozone is reduced through good welding practices and properly designed ventilation systems, such as those described in paragraph 2-4.

   (2) Nitrogen Oxides. Natural ventilation may be sufficient to reduce the hazard of exposure to nitrogen oxides during welding operations, provided all three ventilation criteria given in paragraph 2-4 are satisfied. Nitrogen oxide concentrations will be very high when performing gas tungsten-arc cutting of stainless steel using a 90 percent nitrogen-10 percent argon mixture. Also, high concentrations have been found during experimental use of nitrogen as a shield gas. Good industrial hygiene practices dictate that mechanical ventilation, as defined in paragraph 2-4, be used during welding or cutting of metals.

   (3) Carbon Dioxide and Carbon Monoxide. Carbon dioxide is disassociated by the heat of the arc to form carbon monoxide. The hazard from inhalation of these gases will be minimal if ventilation requirements found in paragraph 2-4 are satisfied.

WARNING

The vapors from some chlorinated solvents (e.g., carbon tetrachloride, trichloroethylene, and perchloroethylene) break down under the ultra-violet radiation of an electric arc and forma toxic gas. Avoid welding where such vapors are present. Furthermore, these solvents vaporize easily and prolonged inhalation of the vapor can be hazardous. These organic vapors should be removed from the work area before welding is begun. Ventilation, as prescribed in paragraph 2-4, shall be provided for control of fumes and vapors in the work area.

(4) <u>Vapors of Chlorinated Solvents</u>. Ultraviolet radiation from the welding or cutting arc can decompose the vapors of chlorinated hydrocarbons, such as perchloroethylene, carbon tetrachloride, and trichloroethylene, to form highly toxic substances. Eye, nose, and throat irritation can result when the welder is exposed to these substances. Sources of the vapors can be wiping rags, vapor degreasers, or open containers of the solvent. Since this decomposition can occur even at a considerable distance from the arc, the source of the chlorinated solvents should be located so that no solvent vapor will reach the welding or cutting area.

b. <u>Radiant Energy</u>. Electric arcs, as well as gas flames, produce ultraviolet and infrared rays which have a harmful effect on the eyes and skin upon continued or repeated exposure. The usual effect of ultraviolet is to "sunburn" the surface of the eye, which is painful and disabling but generally temporary. Ultraviolet radiation may also produce the same effects on the skin as a severe sunburn. The production of ultraviolet radiation doubles when gas-shielded arc welding is performed. Infrared radiation has the effect of heating the tissue with which it comes in contact. Therefore, if the heat is not sufficient to cause an ordinary thermal burn, the exposure is minimal. Leather and Wool clothing is preferable to cotton clothing during gas-shielded arc welding. Cotton clothing disintegrates in one day to two weeks, presumably because of the high ultraviolet radiation from arc welding and cutting.

c. <u>Radioactivity from Thoriated Tungsten Electrodes</u>. Gas tungsten-arc welding using these electrodes may be employed with no significant hazard to the welder or other room occupants. Generally, special ventilation or protective equipment other than that specified in paragraph 2-4 is not needed for protection from exposure hazards associated with welding with thoriated tungsten electrodes.

d. <u>Metal Fumes</u>. The physiological response from exposure to metal fumes varies depending upon the metal being welded. Ventilation and personal protective equipment requirements as prescribed in paragraph 2-4 shall be employed to prevent hazardous exposure.

Section V. SAFETY PRECAUTIONS FOR WELDING AND CUTTING CONTAINERS THAT HAVE HELD COMBUSTIBLES

2-19. EXPLOSION HAZARDS

a. Severe explosions and fires can result from heating, welding, and cutting containers which are not free of combustible solids, liquids, vapors, dusts, and gases. Containers of this kind can be made safe by following one of the methods described in paragraphs 2-22 through 2-26. Cleaning the container is necessary in all cases before welding or cutting.

WARNING

Do not assume that a container that has held combustibles is clean and safe until proven so by proper tests. Do not weld in places where dust or other combustible particles are suspended in air or where explosive vapors are present. Removal of flammable material from vessels and/or containers may be done either by steaming out or boiling.

b. Flammable and explosive substances may be present in a container because it previously held one of the following substances:

(1) Gasoline, light oil, or other volatile liquid that releases potentially hazardous vapors at atmospheric pressure.

(2) An acid that reacts with metals to produce hydrogen.

(3) A nonvolatile oil or a solid that will not release hazardous vapors at ordinary temperatures, but will release such vapors when exposed to heat.

(4) A combustible solid; i. e., finely divided particles which may be present in the form of an explosive dust cloud.

c. Any container of hollow body such as a can, tank, hollow compartment in a welding, or a hollow area on a casting, should be given special attention prior to welding. Even though it may contain only air, heat from welding the metal can raise the temperature of the enclosed air or gas to a dangerously high pressure, causing the container to explode. Hollow areas can also contain oxygen-enriched air or fuel gases, which can be hazardous when heated exposed to an arc or flame. Cleaning the container is necessary in all cases before cutting or welding.

2-20. USING THE EXPLOSIMETER

a. The explosimeter is an instrument which can quickly measure an atmosphere for concentrations of flammable gases and vapors.

b. It is important to keep in mind that the explosimeter measures only flammable gases and vapors. For example, an atmosphere that is indicated non-hazardous from the standpoint of fire and explosion may be toxic if inhaled by workmen for some time.

c. Model 2A Explosimeter is a general purpose combustible gas indicator. It will not test for mixtures of hydrogen, acetylene, or other combustibles in which the oxygen content exceeds that of normal air (oxygen-enriched atmospheres). Model 3 Explosimeter is similar except that it is equipped with heavy duty flashback arresters which are capable of confining within the combustion chambers explosions of mixtures of hydrogen or acetylene and oxygen in excess of its normal content in air. Model 4 is designed for testing oxygen-acetylene mixtures and is calibrated for acetylene.

d. Testing Atmospheres Contaminated with Leaded Gasoline. When an atmosphere contaminated with lead gasoline is tested with a Model 2A Explosimeter, the lead produces a solid product of combustion which, upon repeated exposure, may develop a coating upon the detector filament resulting in a loss of sensitivity. To reduce this possibility, an inhibitor-filter should be inserted in place of the normal cotton filter in the instrument. This device chemically reacts with the tetraethyl lead vapors to produce a more volatile lead compound. One inhibitor-filter will provide protection for an instrument of eight hours of continuous testing.

CAUTION

Silanes, silicones, silicates, and other compounds containing silicon in the test atmosphere may seriously impair the response of the instrument. Some of these materials rapidly "poison" the detector filament so that it will not function properly. When such materials are even suspected to be in the atmosphere being tested, the instrument must be checked frequently (at least after 5 tests). Part no. 454380 calibration test kit is available to conduct this test. If the instrument reads low on the test gas, immediately replace the filament and the inlet filter.

e. Operation Instructions. The MSA Explosimeter is set in its proper operating condition by the adjustment of a single control. This control is a rheostat regulating the current to the Explosimeter measuring circuit. The rheostat knob is held in the "OFF" position by a locking bar. This bar must be lifted before the knob can be turned from "OFF" position.

To test for combustible gases or vapors in an atmosphere, operate the Model 2A Explosimeter as follows:

(1) Lift the left end of the rheostat knob "ON-OFF" bar and turn the rheostat knob one quarter turn clockwise. This operation closes the battery circuit. Because of unequal heating or circuit elements, there will be an initial deflection of the meter pointer. The meter pointer may move rapidly upscale and then return to point below "ZERO", or drop directly helm "ZERO".

(2) Flush fresh air through the instrument. The circuit of the instrument must be balanced with air free of combustible gases or vapors surrounding the detector filament. Five squeezes of the aspirator bulb are sufficient to flush the combustion chamber. If a sampling line is used, an additional two squeezes will be required for each 10 ft (3m) of line.

(3) Adjust rheostat knob until meter pointer rests at "ZERO". Clockwise rotation of the rheostat knob causes the meter pointer to move up scale. A clockwise rotation sufficient to move the meter pointer considerably above "ZERO" should be avoided as this subjects the detector filament to an excessive current and may shorten its life.

(4) Place end of sampling line at, or transport the Model 2A Explosimeter to, the point where the sample is to be taken.

(5) Readjust meter pointer to "ZERO" if necessary by turning rheostat knob.

(6) Aspirate sample through instrument until highest reading is obtained. Approximately five squeezes of the bulb are sufficient to give maximum deflection. If a sampling line is used, add two squeezes for each 10 ft (3 m) of line. This reading indicates the concentration of combustible gases or vapors in the sample.

The graduations on the scale of the indicating inter are in percent of the lower explosive limit. Thus, a deflection of the meter pointer between zero and 100 percent shins how closely the atmosphere being tested approaches the minimum concentration required for the explosion. When a test is made with the instrument and the inter pointer is deflected to the extreme right side of the scale and remains there, the atmosphere under test is explosive.

If the meter pointer moves rapidly across the scale, and on continued aspiration quickly returns to a position within the scale range or below "ZERO", it is an indication that the concentration of flammable gases or vapors may be above the upper explosive limit. To verify this, immediately aspirate fresh air through the sampling line or directly into the instrument. Then, if the meter pointer moves first to the right and then to the left of the scale, it is an indication that the concentration of flammable gas or vapor in the sample is above the upper explosive limit.

When it is necessary to estimate or compare concentrations of combustible gases above the lower explosive limit a dilution tube may be employed. See paragraph 2-20 f (1).

The meter scale is red above 60 to indicate that gas concentrations within that range are very nearly explosive. Such gas-air mixtures are considered unsafe.

(7) To turn instrument off: Rotate rheostat knob counterclockwise until arrow on knob points to "OFF". The locking bar will drop into position in its slot indicating that the rheostat is in the "OFF" position.

NOTE

When possible, the bridge circuit balance should be checked before each test. If this is not practical, the balance adjustment should be made at 3-minute internals during the first ten minutes of testing and every 10 minutes thereafter.

f. Special Sampling Applications

(1) Dilution tube. For those tests in which concentrations of combustible gases in excess of liner explosive limit concentrations (100 percent on instrument inter) are to be compared, such as in testing bar holes in the ground adjacent to a leak in a buried gas pipe, or in following the purging of a closed vessel that has contained f flammable gases or vapors, a special air-dilution tube must be used. Such dilution tubes are available in 10:1 and 20:1 ratios of air to sample, enabling rich concentrations of gas to be compared.

In all tests made with the dilution tube attached to the instrument, it is necessary that the instrument be operated in fresh air and the gaseous sample delivered to the instrument through the sampling line in order to permit a comparison of a series of samples beyond the normal range of the instrument to determine which sample contains the highest concentration of combustible gases. The tube also makes it possible to follow the progress of purging operation when an atmosphere of combustibles is being replaced with inert gases.

(2) Pressure testing bar holes. In sane instances when bar holes are drilled to locate pipe line leaks, a group of holes all containing pure gas may be found. This condition usually occurs near a large leak. It is expected that the gas pressure will be greatest in the bar hole nearest the leak. The instrument may be used to locate the position of the leak by utilizing this bar hole pressure. Observe the time required for this pressure to force gas through the instrument sampling line. A probe tube equipped with a plug for sealing off the bar hole into which it is inserted is required. To remove the flow regulating orifice from the instrument, aspirate fresh air through the Explosimeter and unscrew the aspirator bulb coupling. Adjust the rheostat until the meter pointer rests on "ZERO".

The probe tube is now inserted in the bar hole and sealed off with the plug. Observe the time at which this is done. Pressure developed in the bar hole will force gas through the sampling line to the instrument, indicated by an upward deflection of the meter pointer as the gas reaches the detector chamber.

Determine the time required for the gas to pass through the probe line. The bar hole showing the shortest time will have the greatest pressure.

When the upward deflection of the meter pointer starts, turn off the instrument, replace the aspirator bulb and flush out the probe line for the next test.

2-21. PREPARING THE CONTAINER FOR CLEANING

CAUTION

Do not use chlorinated hydrocarbons, such as trichloroethylene or carbon tetrachloride, when cleaning. These materials may be decomposed by heat or radiation from welding or cutting to form phosgene. Aluminum and aluminum alloys should not be cleaned with caustic soda or cleaners having a pH above 10, as they may react chemically. Other nonferrous metals and alloys should be tested for reactivity prior to cleaning.

NOTE

No container should be considered clean or safe until proven so by tests. Cleaning the container is necessary in all cases before welding or cutting.

a. Disconnect or remove from the vicinity of the container all sources of ignition before starting cleaning.

b. Personnel cleaning the container must be protected against harmful exposure. Cleaning should be done by personnel familiar with the characteristics of the contents.

c. If practical, move the container into the open. When indoors, make sure the room is well ventilated so that flammable vapors may be carried away.

d. Empty and drain the container thoroughly, including all internal piping, traps, and standpipes. Removal of scale and sediment may be facilitated by scraping, hammering with a nonferrous mallet, or using a nonferrous chain as a scrubber. Do not use any tools which may spark and cause flammable vapors to ignite. Dispose of the residue before starting to weld or cut.

e. Identify the material for which the container was used and determine its flammability and toxicity characteristics. If the substance previously held by the container is not known, assure that the substance is flammable, toxic, and insoluble in water.

f. Cleaning a container that has held combustibles is necessary in all cases before any welding or cutting is done. This cleaning may be supplemental by filling the container with water or an inert gas both before and during such work.

g. Treat each compartment in a container in the same manner, regardless of which compartment is be welded or cut.

2-22. METHODS OF PRECLEANING CONTAINERS WHICH HAVE HELD FLAMMABLE LIQUIDS

a. General. It is very important for the safety of personnel to completely clean all tanks and containers which have held volatile or flammable liquids. Safety precautions cannot be over-emphasized because of the dangers involved when these items are not thoroughly purged prior to the application of heat, especially open flame.

b. Accepted Methods of Cleaning. Various methods of cleaning containers which have held flammable liquids are listed in this section. However, the automotive exhaust and steam cleaning methods are considered by military personnel to be the safest and easiest methods of purging these containers.

2-23. AUTOMOTIVE EXHAUST METHOD OF CLEANING

WARNING

Head and eye protection, rubber gloves, boots, and aprons must be worn when handling steam, hot water, and caustic solutions. When handling dry caustic soda or soda ash, wear approved respiratory protective equipment, long sleeves, and gloves. Fire resistant hand pads or gloves must be worn when handling hot drums.

The automotive exhaust method of cleaning should be conducted only in well-ventilated areas to ensure levels of toxic exhaust gases are kept below hazardous levels.

CAUTION

Aluminum and aluminum alloys should not be cleaned with caustic soda or cleaners having a pH above 10, as they may react chemically. Other nonferrous metals and alloys should be investigated for reactivity prior to cleaning.

a. Completely drain the container of all fluid.

b. Fill the container at least 25 percent full with a solution of hot soda or detergent (1 lb per gal of water (0.12 kg per 1)) and rinse it sufficiently to ensure that the inside surface is thoroughly finished.

c. Drain the solution and rinse the container again with clean water.

d. Open all inlets and outlets of the container.

e. Using a flexible tube or hose, direct a stream of exhaust gases into the container. Make sure there are sufficient openings to allow the gases to flow through the container.

f. Allow the gases to circulate through the container for 30 minutes.

g. Disconnect the tube from the container and use compressed air (minimum of 50 psi (345 kPa)) to blow out all gases.

h. Close the container openings. After 15 minutes, reopen the container and test with a combustible gas indicator. If the vapor concentration is in excess of 14 percent of the lower limit of flammability, repeat cleaning procedure.

2-24. STEAM METHOD OF CLEANING

WARNING

Head and eye protection, rubber gloves, boots, and aprons must be worn when handling steam, hot water, and caustic solutions. When handling dry caustic soda or soda ash, wear approved respiratory protective equipment, long sleeves, and gloves. Fire resistant hand pads or gloves must be worn when handling hot drums.

The automotive exhaust method of cleaning should be conducted only in well-ventilated areas to ensure levels of toxic exhaust gases are kept below hazardous levels.

CAUTION

Aluminum and aluminum alloys should not be cleaned with caustic soda or cleaners having a pH above 10, as they may react chemically. Other nonferrous metals and alloys should be investigated for reactivity prior to cleaning.

a. Completely drain the container of all fluid.

b. Fill the container at least 25 percent full with a solution of hot soda, detergent, or soda ash (1 lb per gal of water (0.12 kg per 1)) and agitate it sufficiently to ensure that the inside surfaces are thoroughly flushed.

NOTE

Do not use soda ash solution on aluminum.

c. Drain the solution thoroughly.

d. Close all openings in the container except the drain and filling connection or vent. Use damp wood flour or similar material for sealing cracks or other damaged sections.

e. Use steam under low pressure and a hose of at least 3/4-in. (19.05 mm) diameter. Control the steam pressure by a valve ahead of the hose. If a metal nozzle is used at the outlet end, it should be made of nonsparking material and should be electrically connected to the container. The container, in turn, should be grounded to prevent an accumulation of static electricity.

f. The procedure for the steam method of cleaning is as follows:

(1) Blow steam into the container, preferably through the drain, for a period of time to be governed by the condition or nature of the flammable substance previously held by the container. When a container has only one opening, position it so the condensate will drain from the same opening the steam inserted into.

(When steam or hot water is used to clean a container, wear suitable clothing, such as boots, hood, etc., to protect against burns.)

(2) Continue steaming until the container is free from odor and the metal parts are hot enough to permit steam vapors to flow freely out of the container vent or similar opening. Do not set a definite time limit for steaming containers since rain, extreme cold or other weather conditions may condense the steam as fast as it is introduced. It may take several hours to heat the container to such a temperature that steam will flow freely from the outlet of the container.

(3) Thoroughly flush the inside of the container with hot, preferably boiling, water.

(4) Drain the container.

(5) Inspect the inside of the container to see if it is clean. To do this, use a mirror to reflect light into the container. If inspection shows that it is not clean, repeat steps (1) through (4) above and inspect again. (Use a nonmetal electric lantern or flashlight which is suitable for inspection of locations where flammable vapors are present.)

(6) Close the container openings. In 15 minutes, reopen the container and test with a combustible gas indicator. If the vapor concentration is in excess of 14 percent of the lower limit of flammability, repeat the cleaning procedure.

2-25. WATER METHOD OF CLEANING

a. Water-soluble substances can be removed by repeatedly filling and draining the container with water. Water-soluble acids, acetone, and alcohol can be removed in this manner. Diluted acid frequently reacts with metal to produce hydrogen; care must be taken to ensure that all traces of the acid are removed.

b. When the original container substance is not readily water-soluble, it must be treated by the steam method or hot chemical solution method.

2-26. HOT CHEMICAL SOLUTION METHOD OF CLEANING

WARNING

Wear head and eye protection, rubber gloves, boots, and aprons when handling steam, hot water, and caustic solutions. When handling dry caustic soda or soda ash, wear approved respiratory protective equipment, long sleeves, and gloves. Wear fire resistant hand pads or gloves to handle hot drums.

CAUTION

Aluminum and aluminum alloys should not be cleaned with caustic soda or cleaners having a pH above 10, as they may react chemically. Other nonferrous metals and alloys should be investigated for reactivity prior to cleaning.

a. The chemicals generally used in this method are trisodium phosphate (strong washing powder) or a commercial caustic cleaning compound dissolved in water to a concentration of 2 to 4 oz (57 to 113 g) of chemical per gallon of water.

b. The procedure for the hot chemical solution method of cleaning is as follows:

(1) Close all container openings except the drain and filling connection or vent. Use damp wood flour or similar material for sealing cracks or other damaged sections.

(2) Fill the container to overflowing with water, preferably letting the water in through the drains. If there is no drain, flush the container by inserting the hose through the filling connection or vent. Lead the hose to the bottom of the container to get agitation from the bottom upward. This causes any remaining liquid, scum, or sludge to be carried upward and out of the container.

(3) Drain the container thoroughly.

(4) Completely dissolve the amount of chemical required in a small amount of hot or boiling water and pour this solution into the container. Then fill the container with water.

(5) Make a steam connection to the container either through the drain connection or by a pipe entering through the filling connection or vent. Lead the steam to the bottom of the container. Admit steam into the chemical solution and maintain the solution at a temperature of 170 to 180°F (77 to 82°C). At intervals during the steaming, add enough water to permit overflying of any volatile liquid, scum, or sludge that may have collected at the top. Continue steaming to the point where no appreciable amount of volatile liquid, scum, or sludge appears at the top of the container.

(6) Drain the container.

(7) Inspect the inside of the container as described in paragraph 2-24 f (5). If it is not clean, repeat steps (4) thru (6) above and inspect again.

(8) Close the container openings. In 15 minutes, test the gas concentration in the container as described in paragraph 2-24 f (6).

c. If steaming facilities for heating the chemical solution are not available, a less effective method is the use of a cold water solution with the amount of cleaning compound increased to about 6 oz (170 g) per gal of water. It will help if the solution is agitated by rolling the container or by blowing air through the solution by means of an air line inserted near the bottom of the container.

d. Another method used to clean the container is to fill it 25 percent full with cleaning solution and clean thoroughly, then introduce low pressure steam into the container, allowing it to vent through openings. Continue to flow steam through the container for several hours.

2-27. MARKING OF SAFE CONTAINERS

After cleaning and testing to ensure that a container is safe for welding and cutting, stencil or tag it. The stencil or tag must include a phrase, such as "safe for welding and cutting," the signature of the person so certifying, and the date.

2-28. FILLING TREATMENT

It is desirable to fill the container with water during welding or cutting as a supplement to any of the cleaning methods (see fig. 2-8). Where this added precaution is taken, place the container so that it can be kept find to within a few inches of the point where the work is to be done. Make sure the space above the water level is vented so the heated air can escape from the container.

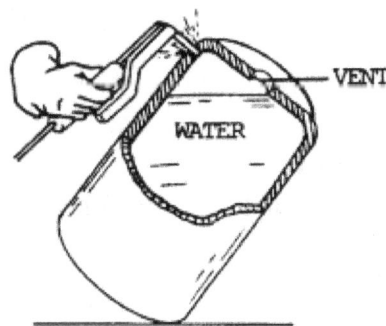

Figure 2-8.   Safe way to weld container that held combustibles.

2-29. PREPARING THE CLEAN CONTAINER FOR WELDING OR CUTTING-- INERT GAS TREATMENT

a. General. Inert gas may be used as a supplement to any of the cleaning methods and as an alternative to the water filling treatment. If sufficient inert gas is mixed with flammable gases and vapors, the mixture will come non-flammable. A continuous flow of steam may also be used. The steam will reduce the air concentration and make the air flammable gas mixture too lean to burn. Permissible inert gases include carbon dioxide and nitrogen.

b. <u>Carbon Dioxide and Nitrogen</u>.

(1) When carbon dioxide is used, a minimum concentration of 50 percent is required, except when the flammable vapor is principally hydrogen, carbon monoxide, or acetylene. In these cases, a minimum concentration of 80 percent carbon dioxide is required. Carbon dioxide is heavier than air, and during welding or cutting operations will tend to remain in containers having a top opening.

(2) When nitrogen is used, the concentrations should be at least 10 percent greater than those specified for carbon dioxide.

(3) Do not use carbon monoxide.

c. <u>Procedure</u>. The procedure for inert gas, carbon dioxide, or nitrogen treatment is as follows:

(1) Close all openings in the container except the filling connection and vent. Use damp wood flour or similar material for sealing cracks or other damaged sections.

(2) Position the container so that the spot to be welded or cut is on top. Then fill it with as much water as possible.

(3) Calculate the volume of the space above the water level and add enough inert gas to meet the minimum concentration for nonflammability. This will usually require a greater volume of gas than the calculated minimum, since the inert gas may tend to flow out of the vent after displacing only part of the previously contained gases or vapors.

(4) Introduce the inert gas, carbon dioxide, or nitrogen from the cylinder through the container drain at about 5 psi (34.5 kPa). If the drain connection cannot be used, introduce the inert gas through the filling opening or vent. Extend the hose to the bottom of the container or to the water level so that the flammable gases are forced out of the container.

(5) If using solid carbon dioxide, crush and distribute it evenly over the greatest possible area to obtain a rapid formation of gas.

d. <u>Precautions When Using Carbon Dioxide</u>. Avoid bodily contact with solid carbon dioxide, which may produce "burns". Avoid breathing large amounts of carbon dioxide since it may act as a respiratory stimulant, and, in sufficient quantities, can act as an asphyxiant.

e. <u>Inert Gas Concentration</u>. Determine whether enough inert gas is present using a combustible gas indicator instrument. The inert gas concentration must be maintained during the entire welding or cutting operation. Take steps to maintain a high inert gas concentration during the entire welding or cutting operation by one of the following methods:

(1) If gas is supplied from cylinders, continue to pass the gas into the container.

(2) If carbon dioxide is used in solid form, add small amounts of crushed solid carbon dioxide at intervals to generate more carbon dioxide gas.

Section VI. SAFETY PRECAUTIONS FOR WELDING AND CUTTING POLYURETHANE FOAM FILLED ASSEMBLIES

2-30. HAZARDS OF WELDING POLYURETHANE FOAM FILLED ASSEMBLIES

WARNING

Welding polyurethane foam-filled parts can produce toxic gases. Welding should not be attempted on parts filled with polyurethane foam. If repair by welding is necessary, the foam must be removed from the heat-affected area, including the residue, prior to welding.

a. General. Welding polyurethane foam filled parts is a hazardous procedure. The hazard to the worker is due to the toxic gases generated by the thermal breakdown of the polyurethane foam. The gases that evolve from the burning foam depend on the amount of oxygen available. Combustion products of polyurethane foam in a clean, hot fire with adequate oxygen available are carbon dioxide, water vapor, and varying amounts of nitrogen oxides, carbon monoxide, and traces of hydrogen cyanide. Thermal decomposition of polyurethanes associated with restricted amounts of oxygen as in the case of many welding operations results in different gases being produced. There are increased amounts of carbon monoxide, various aldehydes, isocyanates and cyanides, and small amounts of phosgene, all of which have varying degrees of toxicity.

b. Safety Precautions.

(1) It is strongly recommended that welding on polyurethane foam filled parts not be performed. If repair is necessary, the foam must be removed from the heat affected zone. In addition, all residue must be cleaned from the metal prior to welding.

(2) Several assemblies of the M113 and M113A1 family of vehicles should not be welded prior to removal of polyurethane foam and thorough cleaning.

CHAPTER 3
PRINT READING AND WELDING SYMBOLS

Section I. PRINT READING

3-1. GENERAL

a. <u>Drawings</u>. Drawing or sketching is a universal language used to convey all necessary information to the individual who will fabricate or assemble an object. Prints are also used to illustrate how various equipment is operated, maintained, repaired, or lubricated. The original drawings for prints are made either by directly drawing or tracing a drawing on a translucent tracing paper or cloth using waterproof (India) ink or a special pencil. The original drawing is referred to as a tracing or master copy.

b. <u>Reproduction Methods</u>. Various methods of reproduction have been developed which will produce prints of different colors from the master copy.

(1) One of the first processes devised to reproduce a tracing produced white lines on a blue background, hence the term "blueprints".

(2) A patented paper identified as "BW" paper produces prints with black lines on a white background.

(3) The ammonia process, or "Ozalids" produces prints with either black, blue, or maroon lines on a white background.

(4) Vandyke paper produces a white line on a dark brown background.

(5) Other reproduction methods are the mimeograph machine, ditto machine, and photostatic process.

3-2. PARTS OF A DRAWING

a. <u>Title Block</u>. The title block contains the drawing number and all the information required to identify the part or assembly represented. Approved military prints will include the name and address of the Government Agency or organization preparing the drawing, the scale, the drafting record, authentication, and the date.

b. <u>Revision Block</u>. Each drawing has a revision block which is usually located in the upper right corner. All changes to the drawing are noted in this block. Changes are dated and identified by a number or letter. If a revision block is not used, a revised drawing may be shown by the addition of a letter to the original number.

c. <u>Drawing Number</u>. All drawings are identified by a drawing number. If a print has more than one sheet and each sheet has the same number, this information is included in the number block, indicating the sheet number and the number of sheets in the series.

3-2. PARTS OF A DRAWING (cent)

d. Reference Numbers and Dash Numbers. Reference numbers that appear in the title block refer to other print numbers. When more than one detail is shown on a drawing, dashes and numbers are frequently used. If two parts are to be shown in one detail drawing, both prints will have the same drawing nunber plus a dash and an individual number such as 7873102-1 and 7873102-2.

e. Scale. The scale of the print is indicated in one of the spaces within the title block. It indicates the size of the drawing as compared with the actual size of the part. Never measure a drawing--use dimensions. The print may have been reduced in size from the original drawing.

f. Bill of Material. A special block or box on the drawing may contain a list of necessary stock to make an assembly. It also indicates the type of stock, size, and specific amount required.

3-3. CONSTRUCTION LINES

a. Full Lines (A, fig. 3-1). Full lines represent the visible edges or outlines of an object.

b. Hidden Lines (A, fig. 3-1). Hidden lines are made of short dashes which represent hidden edges of an object.

c. Center Lines (B, fig. 3-1). Center lines are made with alternating short and long dashes. A line through the center of an object is called a center line.

d. Cutting Plane Lines (B, fig. 3-1). Cutting plane lines are dashed lines, generally of the same width as the full lines extending through the area being cut. Short solid wing lines at each end of the cutting line project at 90 degrees to that line and end in arrowheads which point in the direction of viewing. Capital letters or numcrals are placed just beyond the points of the arrows to designate the section.

e. Dimension Lines (A, fig. 3-1). Dimension lines are fine full lines ending in arrowheads. They are used to indicate the measured distance between two points.

f. Extension Lines (A, fig. 3-1). Extension lines are fine lines from the outside edges or intermediate points of a drawn object. They indicate the limits of dimension lines.

g. Break Lines (C, fig. 3-1). Break lines are used to show a break in a drawing and are used when it is desired to increase the scale of a drawing of uniform cross section while showing the true size by dimension lines. There are two kinds of break lines: short break and long break. Short break lines are usually heavy, wavy, semiparallel lines cutting off the object outline across a uniform section. Long break lines are long dash parallel lines with each long dash in the line connected to the next by a "2" or sharp wave line.

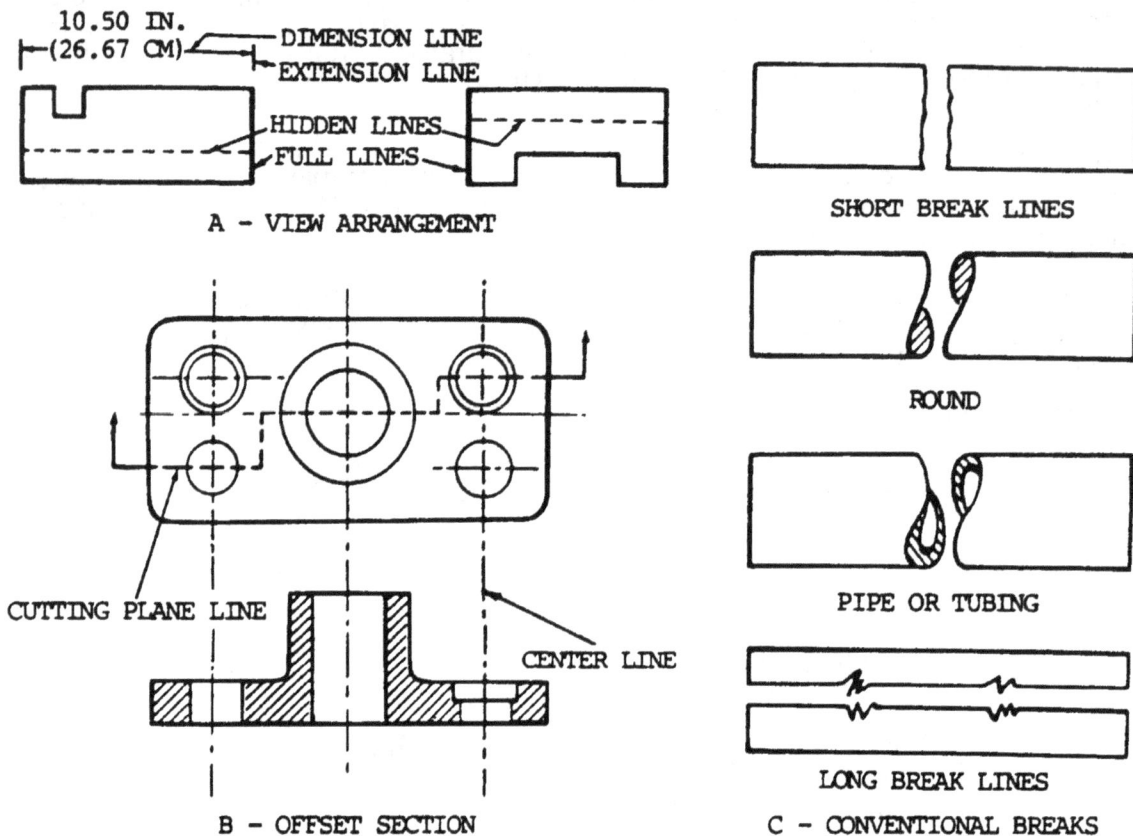

Figure 3-1. Construction lines.

Section II. WELD AND WELDING SYMBOLS

3-4. GENERAL

Welding cannot take its proper place as an engineering tool unless means are provided for conveying the information from the designer to the workmen. Welding symbols provide the means of placing complete welding information on drawings. The scheme for symbolic representation of welds on engineering drawings used in this manual is consistent with the "third angle" method of projection. This is the method predominantly used in the United States.

3-4. GENERAL (cont)

The joint is the basis of reference for welding symbols. The reference line of the welding symbol (fig. 3-2) is used to designate the type of weld to be made, its location, dimensions, extent, contour, and other supplementary information. Any welded joint indicated by a symbol will always have an arrow side and an other side. Accordingly, the terms arrow side, other side, and both sides are used herein to locate the weld with respect to the joint.

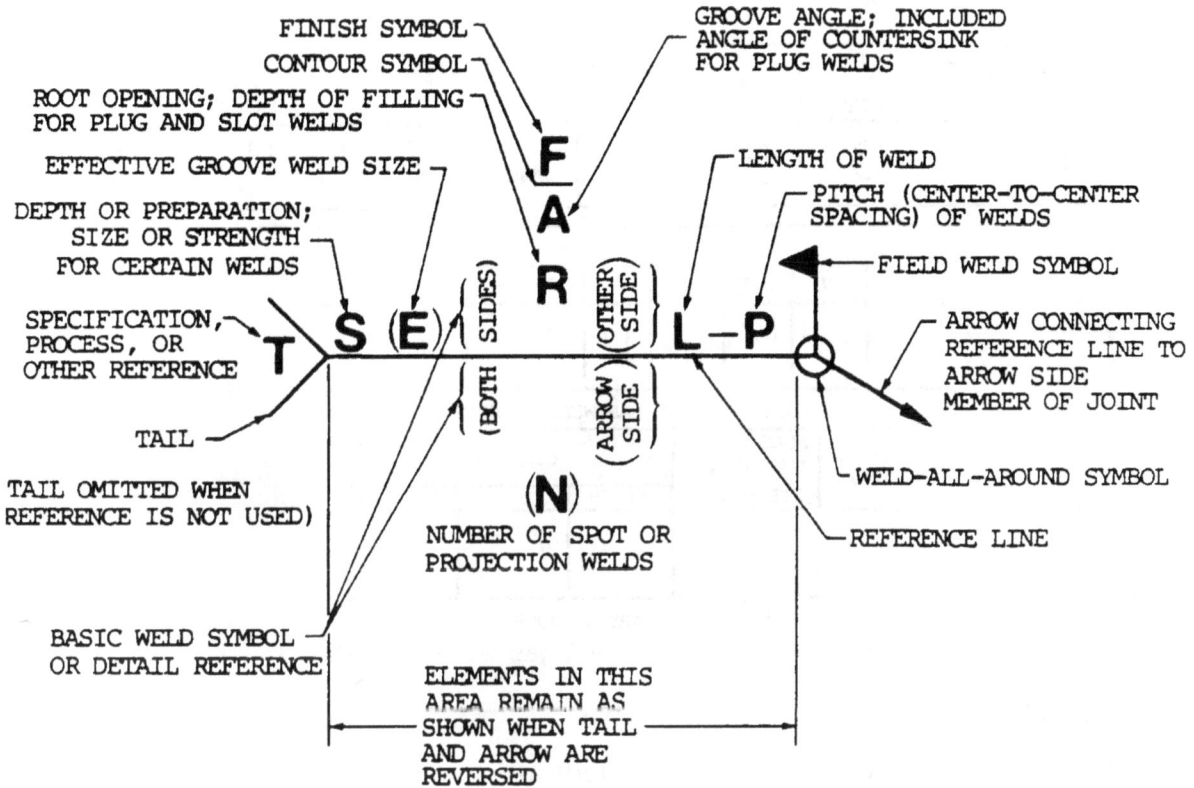

Figure 3-2. Standard locations of elements of a welding symbol.

The tail of the symbol is used for designating the welding and cutting processes as well as the welding specifications, procedures, or the supplementary information to be used in making the weld. If a welder knows the size and type of weld, he has only part of the information necessary for making the weld. The process, identification of filler metal that is to be used, whether or not peening or root chipping is required, and other pertinent data must be related to the welder. The notation to be placed in the tail of the symbol indicating these data is to be establish by each user. If notations are not used, the tail of the symbol may be omitted.

3-5. ELMENTS OF A WELDING SYMBOL

A distinction is made between the terms "weld symbol" and "welding symbol". The weld symbol (fig. 3-3) indicates the desired type of weld. The welding symbol (fig. 3-2) is a method of representing the weld symbol on drawings. The assembled "welding symbol" consists of the following eight elements, or any of these elements as necessary: reference line, arrow, basic weld symbols, dimensions and other data, supplementary symbols, finish symbols, tail, and specification, process, or other reference. The locations of welding symbol elements with respect to each other are shown in figure 3-2.

3-3. Basic and supplementary arc and gas weld symbols.

3-6. BASIC WELD SYMBOLS

a. <u>General</u>. Weld symbols are used to indicate the welding processes used in metal joining operations, whether the weld is localized or "all around", whether it is a shop or field weld, and the contour of welds. These basic weld symbols are summrized below and illustrate in figure 3-3.

b. <u>Arc and Gas Weld Symbols</u>. See figure 3-3.

c. <u>Resistance Weld Symbols</u>. See figure 3-3.

d. <u>Brazing, Forge, Thermit, Induction, and Flow Weld Symbols</u>.

(1) These welds are indicated by using a process or specification reference in the tail of the welding symbol as shown in figure 3-4.

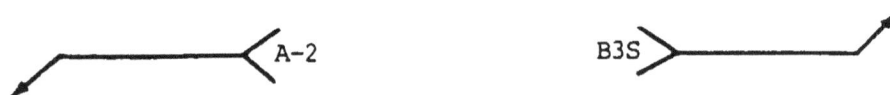

(2) When the use of a definite process is required (fig. 3-5) , the process may be indicated by one or more of the letter designations shown in tables 3-1 and 3-2.

3-6. BASIC WELD SYMBOLS (cont)

GMAW - AU   SMAW   GTAW

Figure 3-5. Definite process reference.

(3) When no specification, process, or other reference is used with a welding symbol, the tail may be omitted (fig. 3-6).

Figure 3-6. No process or specification reference.

Table 3-1. Designation of Welding Processes by Letters*

Welding Process	Letter Designation
Brazing	
Torch brazing	TB
Twin carbon-arc brazing	TCAB
Furnace brazing	FB
Induction brazing	IB
Resistance brazing	RB
Dip brazing	DB
Block brazing	BB
Flow brazing	FLB
Flow welding	FLOW
Resistance welding	
Flash welding	FW
Upset welding	UW
Percussion welding	PEW
Induction welding	IW
Arc welding	
Bare metal-arc welding	BMAW
Stud welding	SW
Gas shielded stud welding	GSSW
Submerged arc welding	SAW
Gas tungsten-arc welding	GTAW
Gas metal-arc welding	GMAW
Atomic hydrogen welding	AHW
Shielded metal-arc welding	SMAW
Twin carbon-arc welding	TCAW
Carbon-arc welding	CAW
Gas carbon-arc welding	GCAW
Shielded carbon-arc welding	SCAW
Flux cored-arc welding	FCAW

Table 3-1. Designation of Welding Processes by Letters* (cont)

Welding Process	Letter Designation
Thermit welding	
Nonpressure thermit welding	NTW
Pressure thermit welding	PTW
Gas welding	
Pressure gas welding	PGW
Oxyhydrogen welding	OHW
Oxyacetylene welding	OAW
Air-acetylene welding	AAW
Forge welding	
Roll welding	RW
Die welding	DW
Hammer welding	HW

*The following suffixes may be used to indicate the method of applying the above processes:

Automatic welding	AU
Machine welding	ME
Manual welding	MA
Semi-automatic welding	SA

NOTE

Letter designations have not been assigned to arc spot, resistance spot, arc seam, resistance seam, and projection welding since the weld symbols used are adequate.

Table 3-2. Designation of Cutting Processes by Letters*

Cutting Process	Letter Designation
Arc cutting	AC
Air-carbon-arc cutting	AAC
Carbon-arc cutting	CAC
Metal-arc cutting	MAC
Oxygen cutting	OC
Chemical flux cutting	FOC
Metal powder cutting	POC
Arc-oxygen cutting	AOC

*The following suffixes may be used to indicate the methods of applying the above processes:

Automatic cutting	AU
Machine cutting	ME
Manual cutting	MA
Semi-automatic cutting	SA

3-6. BASIC WELD SYMBOLS (cont)

e. <u>Other Common Weld Symbols</u>. Figures 3-7 and 3-8 illustrate the weld-all-around and field weld symbol, and resistance spot and resistance seam welds.

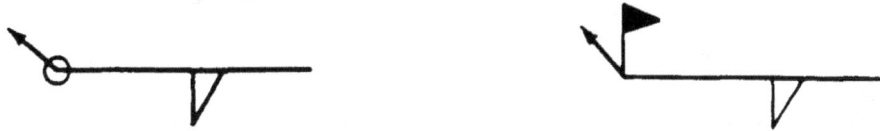

Figure 3-7. Weld-all-around and field weld symbols.

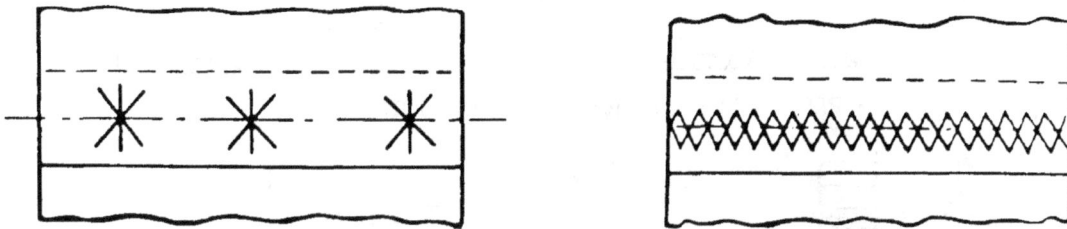

Figure 3-8. Resistance spot and resistance seam welds.

f. <u>Supplermntary Symbols</u>.These symbols are used in many welding processes in congestion with welding symbols and are used as shown in figure 3-3, p 3-5.

3-7. LOCATION SIGNIFICANCE OF ARROW

a. <u>Fillet, Groove, Flange, Flash, and Upset welding symbols</u>For these symbols, the arrow connects the welding symbol reference line to one side of the joint and this side shall be considered the arrow side of the joint (fig. 3-9). The side opposite the arrow side is considered the other side of the joint (fig. 3-10).

DESIRED WELD   SECTION OR END VIEW

Figure 3-9. Arrow side fillet welding symbol.

DESIRED WELD   SECTION OR END VIEW

Figure 3-10. Other side fillet welding symbol.

b. <u>Plug, Slot, Arc Spot, Arc Seam, Resistance Spot, Resistance Seam, and Projection Welding Symbols.</u> For these symbols, the arrow connects the welding symbol reference line to the outer surface of one member of the joint at the center line of the desired weld. The member to which the arrow points is considered the arrow side member. The other member of the joint shall be considered the other side member (fig. 3-11).

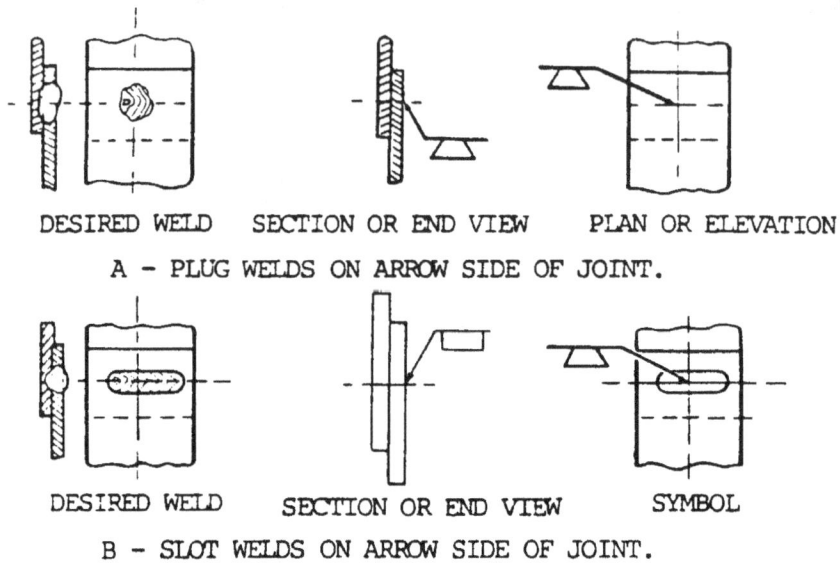

DESIRED WELD   SECTION OR END VIEW   PLAN OR ELEVATION

A - PLUG WELDS ON ARROW SIDE OF JOINT.

DESIRED WELD   SECTION OR END VIEW   SYMBOL

B - SLOT WELDS ON ARROW SIDE OF JOINT.

Figure 3-11. Plug and slot welding symbols indicating location and dimensions of the weld.

c. <u>Near Side.</u> When a joint is depicted by a single line on the drawing and the arrow of a welding symbol is directed to this line, the arrow side of the joint is considered as the near side of the joint, in accordance with the usual conventions of drafting (fig. 3-12 and 3-13).

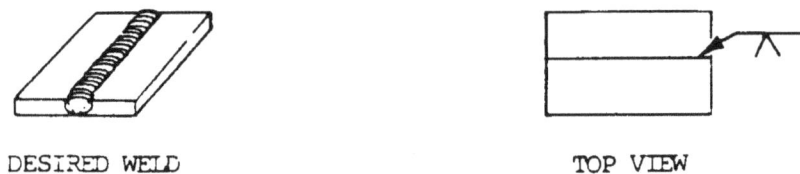

DESIRED WELD   TOP VIEW

Figure 3-12. Arrow side V-groove welding symbol.

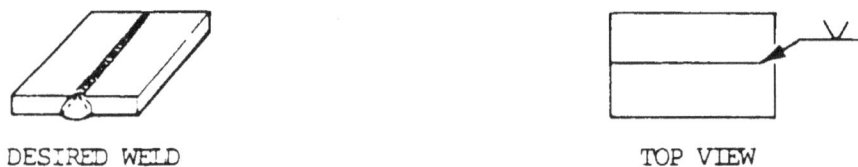

DESIRED WELD   TOP VIEW

Figure 3-13. Other side V-groove welding symbol.

d. <u>Near Member.</u> When a joint is depictd as an area parallel to the plane of projection in a drawing and the arrow of a welding symbol is directed to that area, the arrow side member of the joint is considered as the near member of the joint, in accordance with the usual conventions of drafting (fig. 3-11).

3-8. LOCATION OF THE WELD WITH RESPECT TO JOINT

a. <u>Arrow Side</u>. Welds on the arrow side of the joint are shown by placing the weld symbol on the side of the reference line toward the reader (fig. 3-14).

Figure 3-14. Welds on the arrow side of joint.

b. <u>Other Side</u>. Welds on the other side of the joint are shown by placing the weld symbol on the side of the reference line away from the reader (fig. 3-15).

Figure 3-15. Welds on the other side of joint.

c. <u>Both Sides</u>. Welds on both sides of the joint are shown by placing weld symbols on both sides of the reference line, toward and away from the reader (fig. 3-16).

Figure 3-16. Welds on both sides of joint.

d. <u>No Side Significance</u>. Resistance spot, resistance seam, flash, and upset weld symbols have no arrow side or other side significance in themselves, although supplementary symbols used in conjunction with these symbols may have such significance. For example, the flush contour symbol (fig. 3-3) is used in conjunction with the spot and seam symbols (fig. 3-17) to show that the exposed surface of one member of the joint is to be flush. Resistance spot, resistance seam, flash, and upset weld symbols shall be centered on the reference line (fig. 3-17).

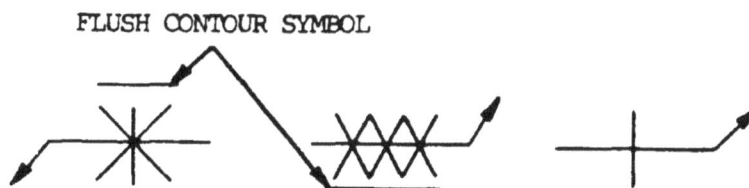

FLUSH CONTOUR SYMBOL

Figure 3-17. Spot, seam, and flash or upset weld symbols.

3-9. REFERENCES AND GENERAL NOTES

a. <u>Symbols With References</u>. When a specification, process, or other reference is used with a welding symbol, the reference is placed in the tail fig. 3-4, p 3-5).

b. <u>Symbols Without References</u>. Symbols may be used without specification, process, or other references when:

(1) A note similar to the following appears on the drawing: "Unless otherwise designated, all welds are to be made in accordance with specification no...."

(2) The welding procedure to be used is described elsewhere, such as in shop instructions and process sheets.

c. <u>General Notes</u>. General notes similar to the follwing may be placed on a drawing to provide detailed information pertaining to the predominan welds. This information need not be repeated on the symbols:

(1) "Unless otherwise indicated, all fillet welds are 5/16 in. (0.80 cm) size."

(2) "Unless otherwise indicated, root openings for all groove welds are 3/16 in. (0.48 cm)."

d. <u>Process Indication</u>. When use of a definite process is required, the process may be indicated by the letter designations listed in tables 3-1 and 3-2 (fig. 3-5, p 3-6).

e. <u>Symbol Without a Tail</u>. When no specification, process, or other reference is used with a welding symbol, the tail may be omitted (fig. 3-6, p 3–6).

3-10. WELD-ALL-AROUND AND FIELD WELD SYMBOLS

a. Welds extending completely around a joint are indicated by mans of the weld-all-around symbol (fig. 3-7, p 3-8). Welds that are completely around a joint which includes more than one type of weld, indicated by a combination weld symbol, are also depicted by the weld–all-around symbol. Welds completely around a joint in which the metal intersections at the points of welding are in more than one plane are also indicated by the weld-all-around symbol.

b. Field welds are welds not made in a shop or at the place of initial construction and are indicated by means of the field weld symbol (fig. 3–7, p 3–8).

3-11.   EXTENT OF WELDING DENOTED BY SYMBOLS

   a. <u>Abrupt Changes</u>. Symbols apply between abrupt changes in the direction of the welding or to the extent of hatching of dimension lines, except when the weld-all-around symbol (fig. 3-3, p 3-5) is used.

   b. <u>Hidden Joints</u>. Welding on hidden joints may be covered when the welding is the same as that of the visible joint. The drawing indicates the presence of hidden members. If the welding on the hidden joint is different from that of the visible joint, specific information for the welding of both must be given.

3-12.   LOCATION OF WELD SYMBOLS

   a. Weld symbols, except resistance spot and resistance seam, must be shown only on the welding symbol reference line and not on the lines of the drawing.

   b. Resistance spot and resistance seam weld symbols may be placed directly at the locations of the desired welds (fig. 3-8, p 3-8).

3-13.   USE OF INCH, DEGREE, AND POUND MARKS

NOTE

   Inch marks are used for indicating the diameter of arc spot, resistance spot, and circular projection welds, and the width of arc seam and resistance seam welds when such welds are specified by decimal dimensions.

In general, inch, degree, and pound marks may or may not be used on welding symbols, as desired.

3-14.   CONSTRUCTION OF SYMBOLS

   Fillet, bevel and J-groove, flare bevel groove, and corner flange symbols shall be shown with the perpendicular leg always to the left (fig. 3-18) .

Figure 3-18. Construction of symbols, perpendicular leg always to the left.

   b. In a bevel or J-groove weld symbol, the arrow shall point with a definite break toward the member which is to be chamfered (fig. 3-19) . In cases where the member to be chamfered is obvious, the break in the arrow may be omitted.

Figure 3-19. Construction of symbols, arrow break toward chamfered member.

c. Information on welding symbols shall be placed to read from left to right along the reference line in accordance with the usual conventions of drafting (fig. 3-20) .

NOTE
ALL DIMENSIONS SHOWN ARE IN INCHES.

Figure 3-20. Construction of symbols, symbols placed to read left to right.

d. For joints having more than one weld, a symbol shall be shown for each weld (fig. 3-21).

SINGLE BEVEL GROOVE
AND BACK OR BACKING
WELD SYMBOLS

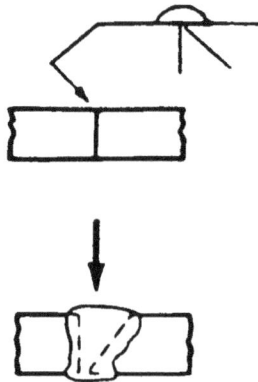

BACK OR BACKING
SINGLE J-GROOVE,
AND FILLET WELD SYMBOLS

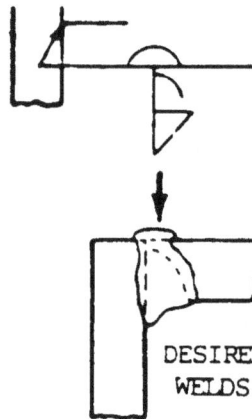

SINGLE BEVEL GROOVE
AND DOUBLE FILLET
WELD SYMBOL

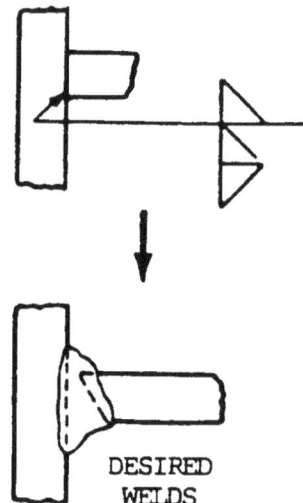

DESIRED WELDS

DESIRED WELDS

DESIRED WELDS

Figure 3-21. Combinations of weld symbols.

e. The letters CP in the tail of the arrow indicate a complete penetration weld regardless of the type of weld or joint preparation (fig. 3-22).

CP

Figure 3-22. Complete penetration indication.

3-14. CONSTRUCTION OF SYMBOLS (cont)

f. When the basic weld symbols are inadequate to indicate desired weld, the weld shall be shown by a cross section, detail, or other data with a reference on the welding symbol according to location specifications given in para 3-7 (fig. 3-23).

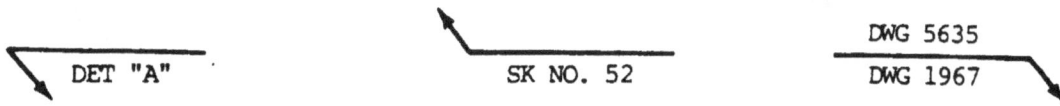

Figure 3-23. Construction of symbols, special types of welds.

g. Two or more reference lines may be used to indicate a sequence of operations. The first operation must be shown on the reference line nearest the arrow. Subsequent operations must be shown sequentially on other reference lines (fig. 3-24). Additional reference lines may also be used to show data supplementary to welding symbol information included on the reference line nearest the arrow. Test information may be shown on a second or third line away from the arrow (fig. 3-25). When required, the weld-all-around symbol must be placed at the junction of the arrow line and reference line for each operation to which it applies (fig. 3-26). The field weld symbol may also be used in this manner.

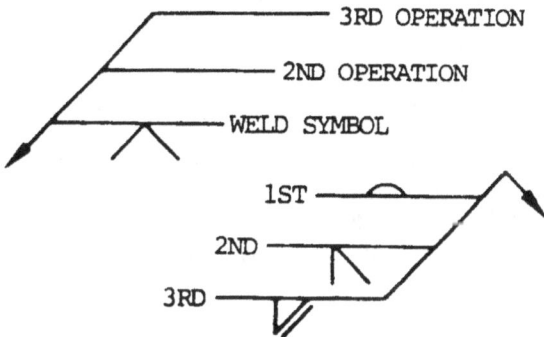

Figure 3-24. Multiple reference lines.

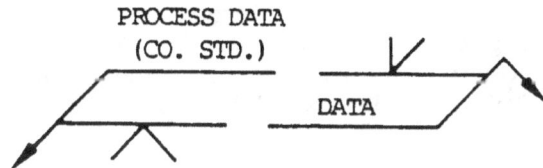

Figure 3-25. Supplementary data.

Figure 3-26. Supplementary symbols.

3-15. FILLET WELDS

Dimensions of fillet welds must be shown on the same side of the reference line as the weld symbol (A, fig. 3-27).

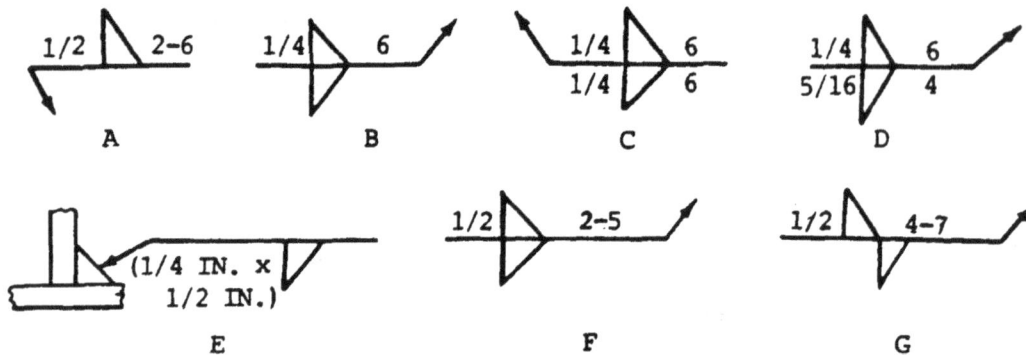

NOTE
ALL DIMENSIONS SHOWN ARE IN INCHES.

Figure 3-27. Dimensions of fillet welds.

b. When fillet welds are indicated on both sides of a joint and no general note governing the dimensions of the welds appears on the drawing, the dimensions are indicated as follows:

(1) When both welds have the same dimensions, one or both may be dimensioned (B or C, fig. 3-27).

(2) When the welds differ in dimensions, both must be dimensioned (D, fig. 3-27).

When fillet welds are indicated on both sides of a joint and a general note governing the dimensions of the welds appears on the drawing, neither weld need be dimensioned. However, if the dimensions of one or both welds differ from the dimensions given in the general note, both welds must be dimensioned (C or D, fig. 3-27).

3-16. SIZE OF FILLET WELDS

The size of a fillet weld must be shown to the left of the weld symbol (A, fig. 3-27).

b. The size of a fillet weld with unequal legs must be shown in parentheses to the left of the weld symbol. Weld orientation is not shown by the symbol and must be shown on the drawing when necessary (E, fig. 3-27).

c. Unless otherwise indicated, the deposited fillet weld size must not be less than the size shown on the drawing.

d. When penetration for a given root opening is specified, the inspection method for determining penetration depth must be included in the applicable specification.

3-17. LENGTH OF FILLET WELDS

a. The length of a fillet weld, when indicated on the welding symbol, must be shown to the right of the weld symbol (A through D, fig. 3-27).

3-17.   LENGTH OF FILLET WELDS (cont)

b. When fillet welding extends for the full distance between abrupt changes in the direction of the welding, no length dimension need be shown on the welding symbol.

c. Specific lengths of fillet welding may be indicated by symbols in conjunction with dimension lines (fig. 3-28).

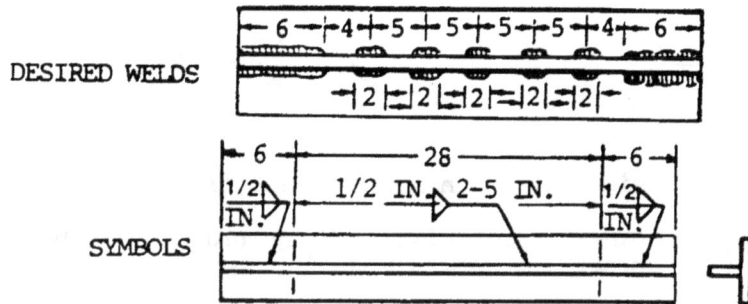

Figure 3-28.   Combined intermittent and continuous welds.

3-18.   EXTENT OF FILLET WELDING

a.   Use one type of hatching(with or without definite lines) to show the extent of fillet welding graphically.

b.   Fillet welding extendingbeyond abrupt changes in the direction of the welding must be indicated by additional arrows pointing to each section of the joint to be welded (fig. 3-29) except when the weld-all-around symbol is used.

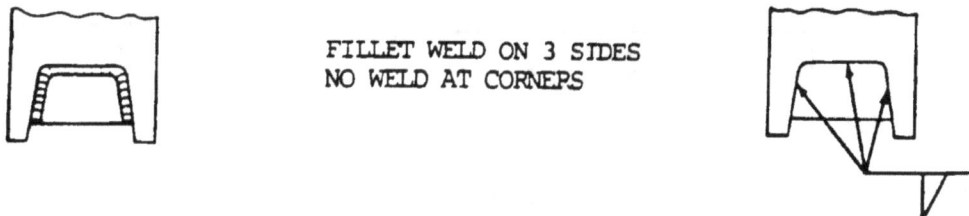

FILLET WELD ON 3 SIDES
NO WELD AT CORNERS

Figure 3-29.   Extent of fillet welds.

3-19. DIMENSIONING OF INTERMITTENT FILLET WELDING

a. The pitch (center-to-center spacing) of intermittent fillet welding shall be shown as the distance between centers of increments on one side of the joint.

b. The pitch of intermittent fillet welding shall be shown to the right of the length dimension (A, fig 3-27, p 3-15).

c. Dimensions of chain intermittent fillet welding must be shown on both sides of the reference line. Chain intermittent fillet welds shall be opposite each other (fig. 3-30).

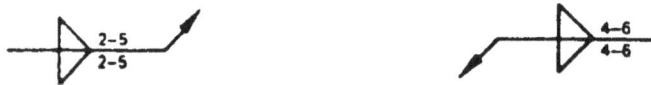

Figure 3-30. Dimensions of chain intermittent fillet welds.

d. Dimensions of staggered intermittent fillet welding must be shown on both sides of the reference line as shown in figure 3-31.

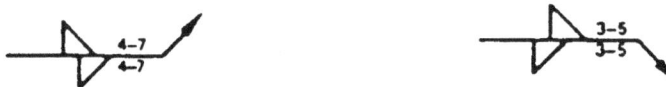

3-31. Dimensions of staggered intermittent fillet welds.

Unless otherwise specified, staggered intermittent fillet welds on both sides shall be symmetrically spaced as in figure 3-32.

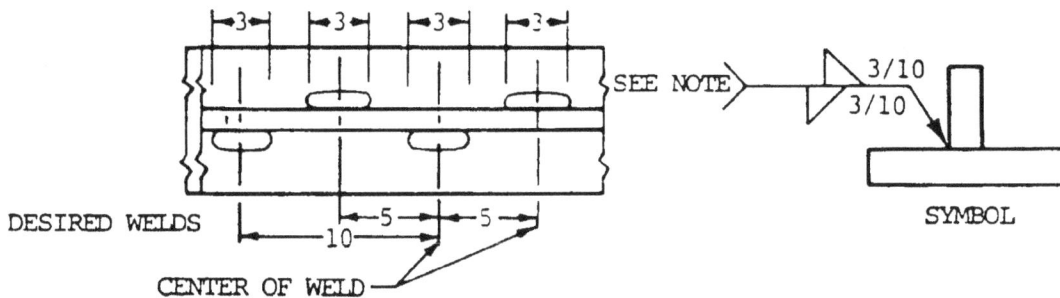

LENGTH AND PITCH OF INCREMENTS OF STAGGERED INTERMITTENT WELDING

NOTE
IF REQUIRED BY ACTUAL LENGTH OF THE JOINT, THE LENGTH OF THE INCREMENT OF THE WELDS AT THE END OF THE JOINT SHOULD BE INCREASED TO TERMINATE THE WELD AT THE END OF THE JOINT.

Figure 3-32. Application of dimensions to intermittent fillet weld symbols.

3-20. TERMINATION OF INTERMITTENT FILLET WELDING

a. When intermittent fillet welding is used by itself, the symbol indicates that increments are located at the ends of the dimensioned length.

b. When intermittent fillet welding is used between continuous fillet welding, the symbol indicates that spaces equal to the pitch minus the length of one increment shall be left at the ends of the dimensioned length.

c. Separate symbols must be used for intermittent and continuous fillet welding when the two are combined along one side of the joint (fig. 3-28, p 3-16).

3-21. SURFACE CONTOUR OF FILLET WELDS

a. Fillet welds that are to be welded approximately flat, convex, or concave faced without recourse to any method of finishing must be shown by adding the flush, convex, or concave contour symbol to the weld symbol, in accordance with the location specifications given in paragraph 3-7 (A, fig. 3-33).

b. Fillet welds that are to be made flat faced by mechanical means must be shown by adding both the flush contour symbol and the user's standard finish symbol to the weld symbol, in accordance with location specifications given in paragraph 3-7 (B, fig. 3-33).

c. Fillet welds that are to be mechanically finished to a convex contour shall be shown by adding both the convex contour symbol and the user's standard finish symbol to the weld symbol, in accordance with location specifications given in paragraph 3-7 (C, fig. 3-33).

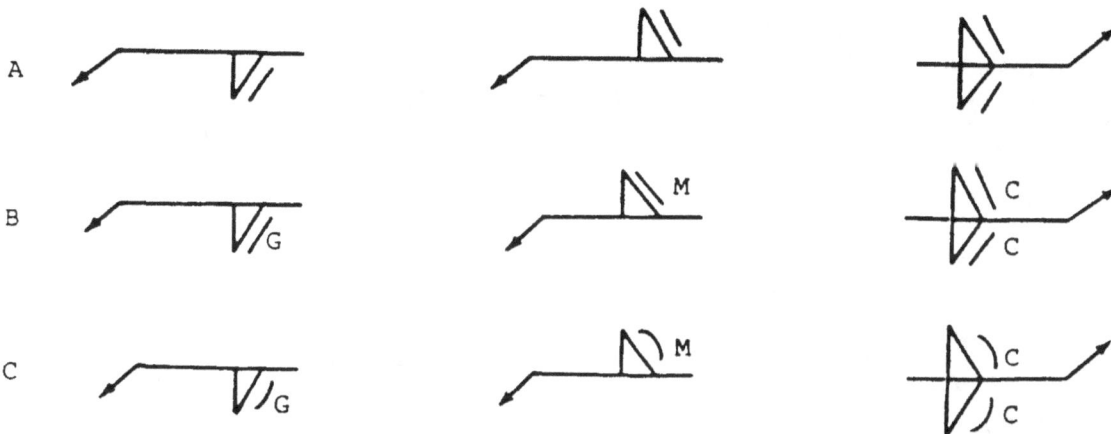

Figure 3-33. Surface contour of fillet welds.

d. Fillet welds that are to be mechanically finished to a concave contour must be shown by adding both the concave contour symbol and the user's standard finish symbol to the weld symbol in accordance with location specification given in paragraph 3-7.

e. In cases where the angle between fusion faces is such that the identification of the type of weld and the proper weld symbol is in question, the detail of the desired joint and weld configuration must be shown on the drawing.

NOTE

Finish symbols used here indicate the method of finishing ("c" = chipping, "G" = grinding, "H" = hammering, "M" = machining), not the degree of finish.

3-22. PLUG AND SLOT WELDING SYMBOLS

a. <u>General</u>. Neither the plug weld symbol nor the slot weld symbol may be used to designate fillet welds in holes.

b. <u>Arrow Side and Other Side Indication of Plug and Slot Welds</u>. Holes or slots in the arrow side member of a joint for plug or slot welding must be indicated by placing the weld symbol on the side of the reference line toward the reader (A, fig. 3-11, p 3-9). Holes or slots in the other side member of a joint shall be indicated by placing the weld symbol on the side of the reference line away from the reader (B, fig. 3-11, p 3-9).

c. <u>Plug Weld Dimensions</u>. Dimensions of plug welds must be shown on the same side of the reference line as the weld symbol. The size of a weld must be shown to the left of the weld symbol. Included angle of countersink of plug welds must be the user's standard unless otherwise indicated. Included angle of countersink, when not the user's standard, must be shown either above or below the weld symbol (A and C, fig. 3-34). The pitch (center-to-center spacing) of plug welds shall be shown to the right of the weld symbol.

d. <u>Depth of Filling of Plug and Slot Welds</u>. Depth of filling of plug and slot welds shall be completed unless otherwise indicated. When the depth of filling is less than complete, the depth of filling shall be shown in inches inside the weld symbol (B, fig. 3-34).

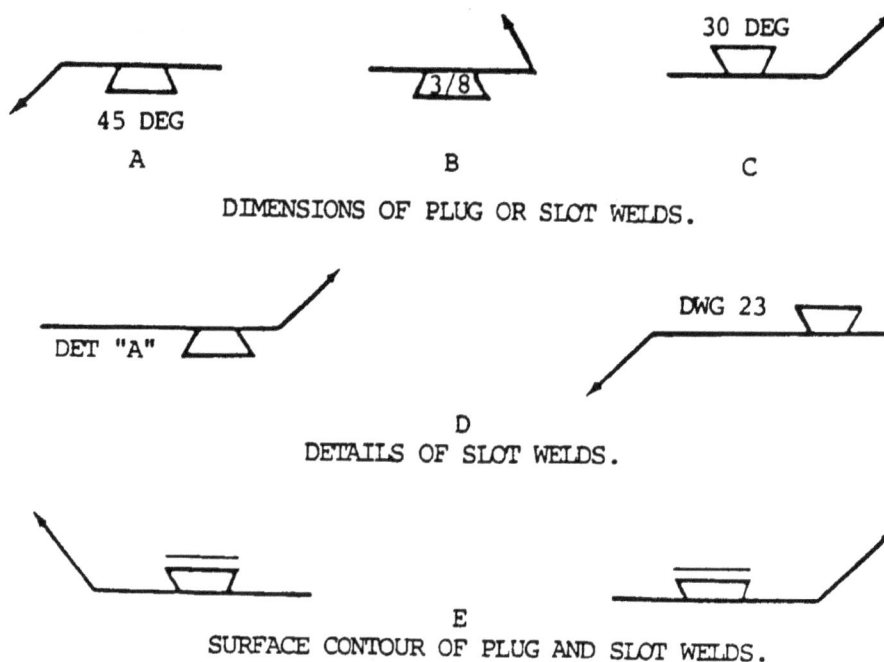

Figure 3-34. Plug and slot welding symbols indicating location and dimensions of the weld.

3-22. PLUG AND SLOT WELDING SYMBOLS (cont)

e. <u>Surface Contour of Plug Welds and Slot Welds.</u>Plug welds that are to be welded approximately flush without recourse to any method of finishing must be shown by adding the finish contour symbol to the weld symbol (fig. 3-35). Plug welds that are to be welded flush by mechanical means must be shown by adding both the flush contour symbol and the user's standard finish symbol to the weld symbol (fig. 3-36).

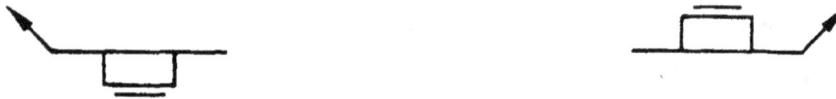

Figure 3-35. Surface contour of plug welds and slot welds.

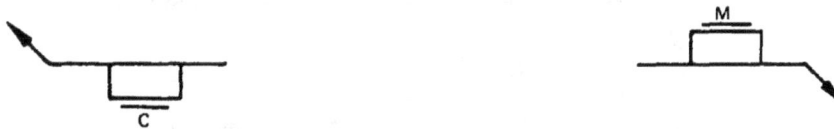

Figure 3-36. Surface contour of plug welds and slot welds with user's standard finish symbol.

f. <u>Slot Weld Dimensions.</u> Dimensions of slot welds must be shown on the same side of the reference line as the the symbol (fig. 3-37).

Figure 3-37. Slot weld dimensions.

g. <u>Details of Slot Welds.</u> Length, width, spacing, included angle of counter-sink, orientation and location of slot welds cannot be shown on the welding symbols. This data must be shown on the drawing or by a detail with a reference to it on the welding symbol, in accordance with location specifications given in paragraph 3-7 (D, fig. 3-3, p 3-19).

3-23. ARC SPOT AND ARC SEAM WELDs

a. <u>General.</u> The spot weld symbol, in accordance with its location in relation to the reference line, may or may not have arrow side or other side significance. Dimensions must be shown on the same side of the reference line as the symbol or on either side when the symbol is located astride the reference line and has no arrow side or other side significance. The process reference is indicated in the tail of the welding symbol. Then projection welding is to be used, the spot weld symbol shall be used with the projection welding process reference in the tail of the welding symbol. The spot weld symbol must be centered above or below the, reference line.

3-20

b. Size of Arc Spot and Arc Seam Welds.

(1) These welds may be dimensioned by either size or strength.

(2) The size of arc spot welds must be designated as the diameter of the weld. Arc seam weld size shall be designated as the width of the weld. Dimensions will be expressed in fractions or in decimals in hundredths of an inch and shall be shown, with or without inch marks, to the left of the weld symbol (A, fig. 3-38).

(3) The strength of arc spot welds must be designated as the minimum acceptable shear strength in pounds or newtons per spot In arc seam welds, strength is designated in pounds per linear inch. Strength is shown to the left of the weld symbol (B, fig. 3-38).

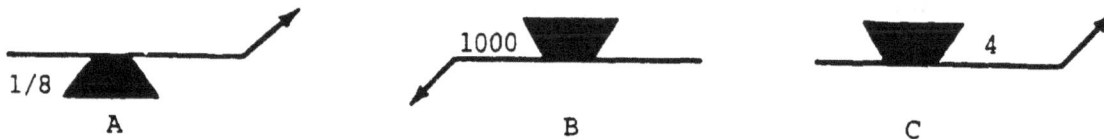

Figure 3-38. Dimensions of arc spot and arc seam welds.

c. Spacing of Arc Spot and Arc Seam Welds.

(1) The pitch (center-to-center spacing) of arc spot welds and, when indicated, the length of arc seam welds, must be shown to the right of the weld symbol (C, fig. 3-38).

(2) When spot welding or arc seam welding extends for the full distance between abrupt changes in the direction of welding, no length dimension need be shown on the welding symbol.

d. Extent and Number of Arc Spot Welds and Arc Seam Welds.

(1) When arc spot welding extends less than the distance between abrupt changes in the direction of welding or less than the full length of the joint, the extent must be dimensioned (fig. 3-39).

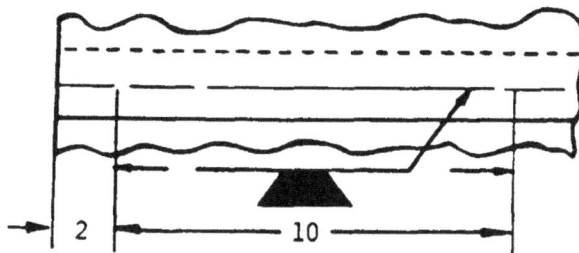

Figure 3-39. Extent of arc spot welding.

3-23. ARC SPOT AND ARC SEAM WELDS (cont)

(2) When a definite number of arc spot welds is desired in a certain joint, the number must be shown in parentheses either above or below the weld symbol (fig. 3-40) .

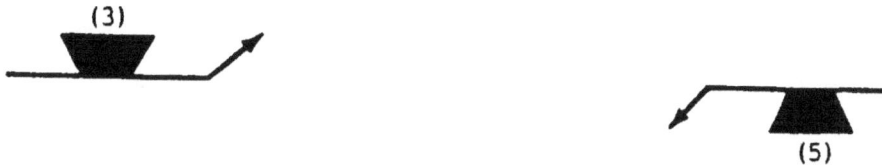

Figure 3-40. Number of arc spot welds in a joint.

(3) A group of spot welds may be located on a drawing by intersecting center lines. The arrows point to at least one of the centerline passing through each weld location.

e. Flush Arc Spot and Arc Seam Welded Joint. When the exposed surface of one member of an arc spot or arc seam welded joint is to be flush, that surfae must be indicated by adding the flush contour symbol (fig. 3-41) in the same manner as that for fillet welds (para 3-21).

Figure 3-41. Surface contour of arc spot and arc seam welds.

f. Details of Arc Seam Welds. Spacing, extent, orientation, and location of arc seam welds cannot be shown on the welding symbols. This data must be shown on the drawing.

3-24. GROOVE WELDS

a. General.

(1) Dimensions of groove welds must be shown on the same side of the reference line as the weld symbol (fig. 3-42).

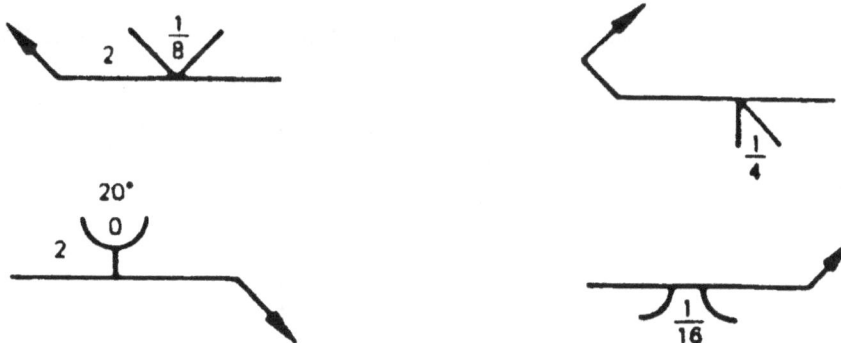

Figure 3-42. Groove weld dimensions.

(2) When no general note governing the dimensions of double groove welds appears, dimensions shall be shown as follows:

(a) When both welds have the same dimensions, one or both may be dimensioned (fig. 3-43).

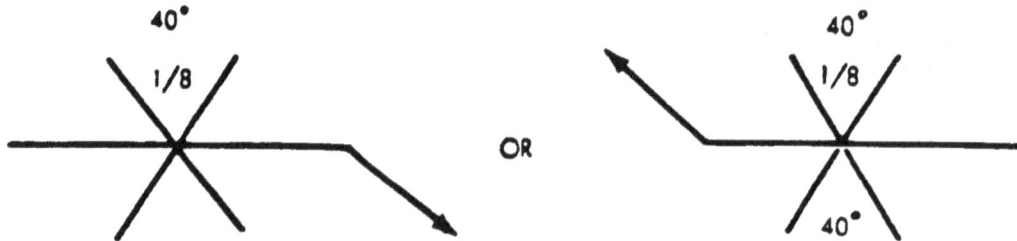

Figure 3-43. Groove weld dimensions having no general note.

(b) When the welds differ in dimensions, both shall be dimensioned (fig. 3-44).

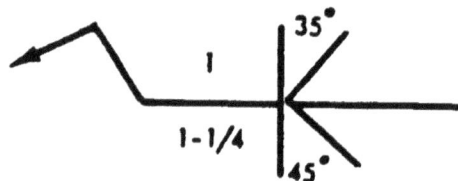

Figure 3-44. Groove welds with differing dimensions.

(3) When a general note governing the dimensions of groove welds appears, the dimensions of double groove welds shall be indicated as follows:

(a) If the dimensions of both welds are as indicated in the note, neither symbol need be dimensioned.

(b) When the dimensions of one or both welds differ from the dimensions given in the general note, both welds shall be dimensioned (fig. 3-44).

b. Size of Groove Welds.

(1) The size of groove welds shall be shown to the left of the weld symbol (fig. 3-44).

(2) Specifications for groove welds with no specified root penetration are shown as follows:

(a) The size of single groove and symmetrical double groove welds which extend completely through the member or members being joined need not be shown on the welding symbol (A and B, fig. 3-45).

Figure 3-45. Groove weld dimensions for welds extending through the members joined.

3-24.   GROOVE WELDS (cont)

(b) The size of groove welds which extend only partly through the member members being joined must be shown on the welding symbol (A and B, fig. 3-46).

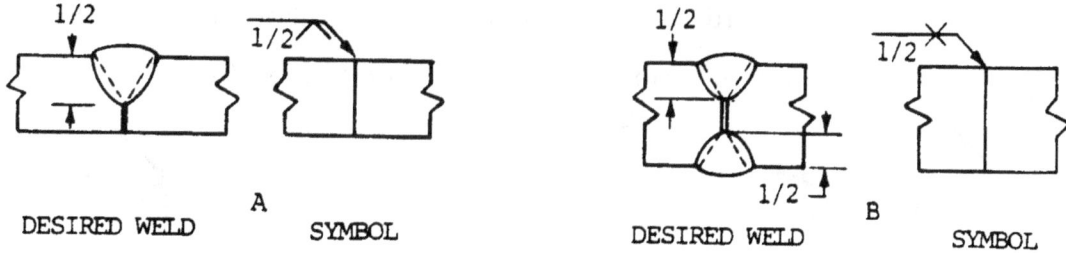

Figure 3-46.   Groove weld dimensions for welds extending partly through the members joined.

(3) The size of groove welds with specified root penetration, except square groove welds, must be indicated by showing the depth of chamfering and the root penetration separated by a plus mark and placed to the left of the weld symbol. The depth of chamfering and the root penetration must read in that order from left to right along the reference line (A and B, fig. 3-47). The size of square groove welds must be indicated by showing only the root penetration.

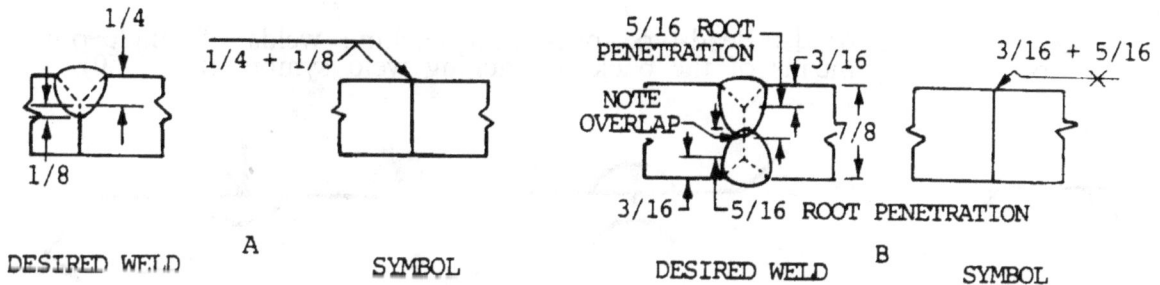

Figure 3-47.   Dimensions of groove welds with specified root penetration.

(4) The size of flare groove welds is considered to extend only to the tangent points as indicated by dimension lines (fig. 3-48).

FLARE BEVEL GROOVE

FLARE V-GROOVE

Figure 3-48.   Flare groove welds.

c. Groove Dimensions

(1) Root opening, groove angle, groove radii, and root faces of the U and J groove welds are the user's standard unless otherwise indicated.

(2) When the user's standard is not used, the weld symbols are as follows:

(a) Root opening is shown inside the weld symbol (fig. 3-49) .

Figure 3-49. Root opening.

(b) Groove angle of groove welds is shown outside the weld symbol (fig. 3-42, p 3-22).

(c) Groove radii and root faces of U and J groove welds are shown by a cross section, detail, or other data, with a reference to it on the welding symbol, in accordance with location specifications given in paragraph 3-7 (fig. 3-22, p 3-13).

d. Back and Backing Welds. Bead-type back and backing welds of single~oove welds shall be shown by means of the back or backing weld symbol (fig. 3-50).

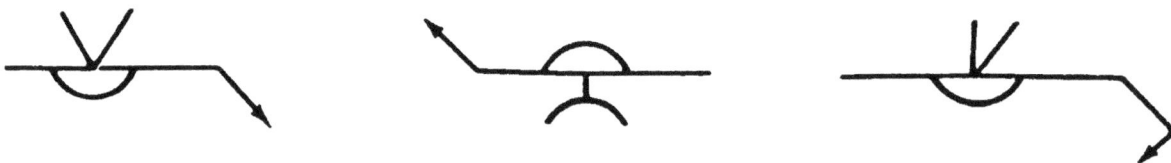

Figure 3-50. Back or backing weld symbol.

e. Surface Contour of Groove Welds. The contour symbols for groove welds (F, fig. 3-51) are indicated in the same manner as that for fillet welds (para 3-21).

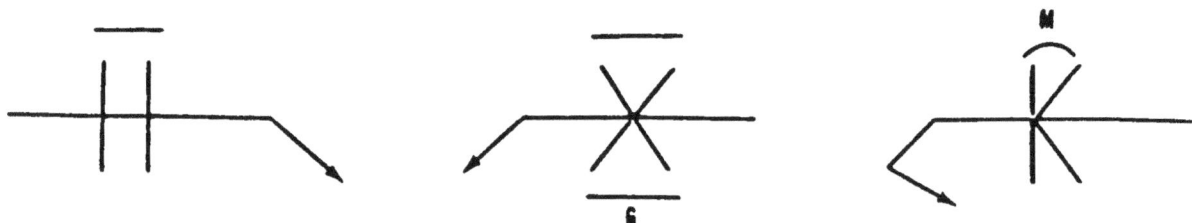

Figure 3-51. Surface contour of groove welds.

3-24. GROOVE WELDS (cont)

(1) Groove welds that are to be welded approximately flush without recourse to any method of finishing shall be shown by adding the flush contour symbol to the weld symbol, in accordance with the location specifications given in paragraph 3-7 (fig. 3-52).

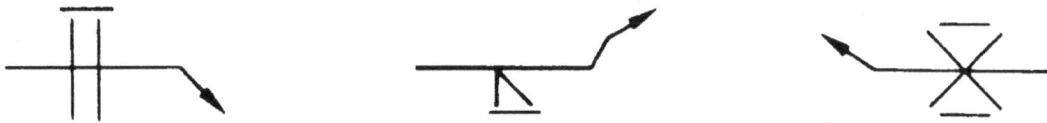

Figure 3-52. Contours obtained by welding.

(2) Groove welds that are to be made flush by mechanical means shall be shown by adding the the flush contour symbol and the user's standard finish symbol to the weld symbol, in accordance with the location specifications given in paragraph 3-7 (fig. 3-53).

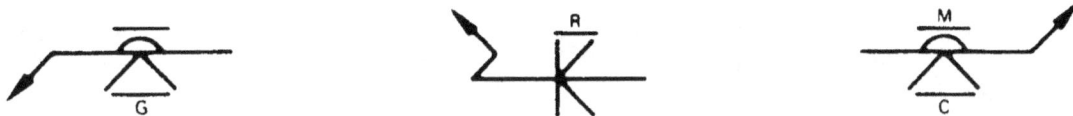

Figure 3-53. Flush contour by machining.

(3) Groove welds that are to be mechanically finished to a convex contour shall be shown by adding both the convex contour symbol and the user's standard finish symbol to the weld symbol, in accordance with the location specifications given in para 3-7 (fig. 3-54).

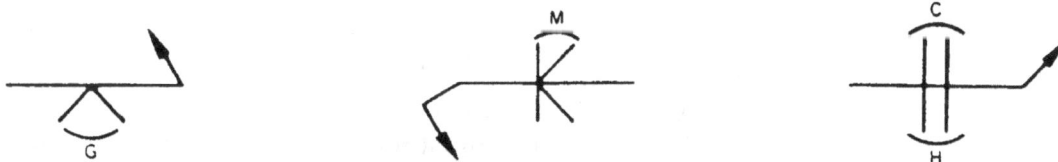

Figure 3-54. Convex contour by machining.

3-25. BACK OR BACKING WELDS

a. General.

(1) The back or backing weld symbol (fig. 3-50, p 3-25) must be used to indicate bead-type back or backing welds of single-groove welds.

(2) Back or backing welds of single-groove welds must be shown by placing a back or backing weld symbol on the side of the reference line opposite the groove weld symbol (fig. 3-50, p 3-25).

(3) Dimensions of back or backing welds should not be shown on the welding symbol . If it is desired to specify these dimensions, they must be shown on the drawing .

b. <u>Surface Contour of Back or Backing Welds</u>. The contour symbols (fig. 3-55) for back or backing welds are indicated in the same manner as that for fillet welds (para 3-21).

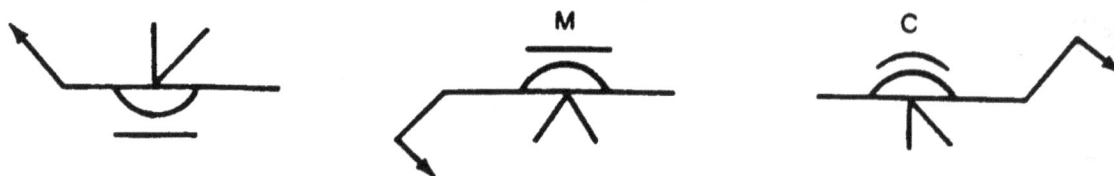

Figure 3-55. Surface contour of back or backing welds.

3-26. MELT-THRU WELDS

a. <u>General</u>.

(1) The melt-thru symbol shall be used where at least 100 percent joint penetration of the weld through the material is required in welds made from one side only (fig. 3-56).

(2) Melt-thru welds shall be shown by placing the melt-thru weld symbol on the side of the reference line opposite the groove weld, flange, tee, or corner weld symbol (fig. 3-56).

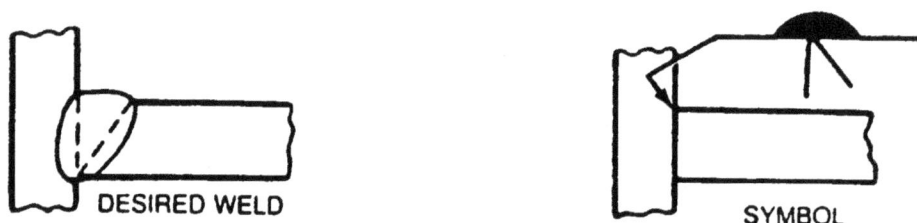

DESIRED WELD

SYMBOL

Figure 3-56. Melt-thru weld symbol.

(3) Dimensions of melt-thru welds should rot be shown on the welding symbol. If it is desired to specify these dimensions, they must be shown on the drawing.

b. <u>Surface Contour of Melt-thru Welds</u>. The contour symnbols for melt-thru welds are indicated in the same manner as that for fillet welds (fig. 3-57).

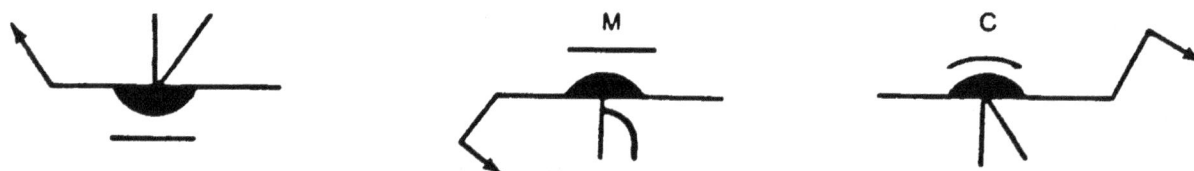

Figure 3-57. Surface contour of melt-thru welds.

3-27. SURFACING WELDS

a. General.

(1) The surfacing weld symbol shall be used to indicate surfaces built up by welding (fig. 3-58), whether built up by single- or multiple-pass surfacing welds.

(2) The surfacing weld symbol does not indicate the welding of a joint and thus has no arrow or other side significance. This symbol shall be drawn on the side of the reference line toward the reader and the arrow shall point clearly to the surface on which the weld is to be deposited.

b. Size of Built-up Surfaces. The size (height) of a surface built up by welding shall be indicated by showing the minimum height of the weld deposit to the left of the weld symbol. The dimensions shall always be on the same side of the reference line as the weld symbol (fig. 3-58). When no specific height of weld deposit is desired, no size dimension need be shown on the welding symbol.

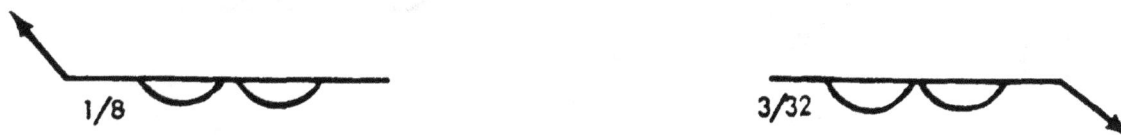

Figure 3-58. Size of surfaces built up by welding.

c. Extent, Location, and Orientation of Surfaces Built up by Welding. When the entire area of a plane or curved surface is to be built up by welding, no dimension, other than size, need be shown on the welding symbol. If only a portion of the area of a plane or curved surface is to be built up by welding, the extent, location, and orientation of the area to be built up shall be indicated on the drawing.

3-28. FLANGE WELDS

a. General.

(1) The following welding symbols are used for light gage metal joints involving the flaring or flanging of the edges to be joined (fig. 3-59). These symbols have no arrow or other side significance.

(2) Edge flange welds shall be shown by the edge flange weld symbol (A, fig. 3-59).

(3) Corner flange welds shall be shown by the corner flange weld symbol (B, fig. 3-59). In cases where the corner flange joint is not detailed, a break in the arrow is required to show which member is flanged (fig. 3-59).

b. Dimensions of Flange Welds.

(1) Dimensions of flange welds are shown on the same side of the reference line as the weld symbol.

(2) The radius and the height above the point of tangency must be indicated by showing the radius and height, separated by a plus mark, and placed to the left of the weld symbol. The radius and height must read in that order from left to right along the reference line (C, fig. 3-59).

(3) The size (thickness) of flange welds must be shown by a dimension placed outward of the flange dimensions (C, fig. 3-59).

(4) Root opening of flange welds are not shown on the welding symbol. If specification of this dimension is desired, it must be shown on the drawing.

c. <u>Multiple-Joint Flange Welds</u>. For flange welds in which one or more pieces are inserted between the two outer pieces, the same symbol shall be used as for the two outer pieces, regardless of the number of pieces inserted.

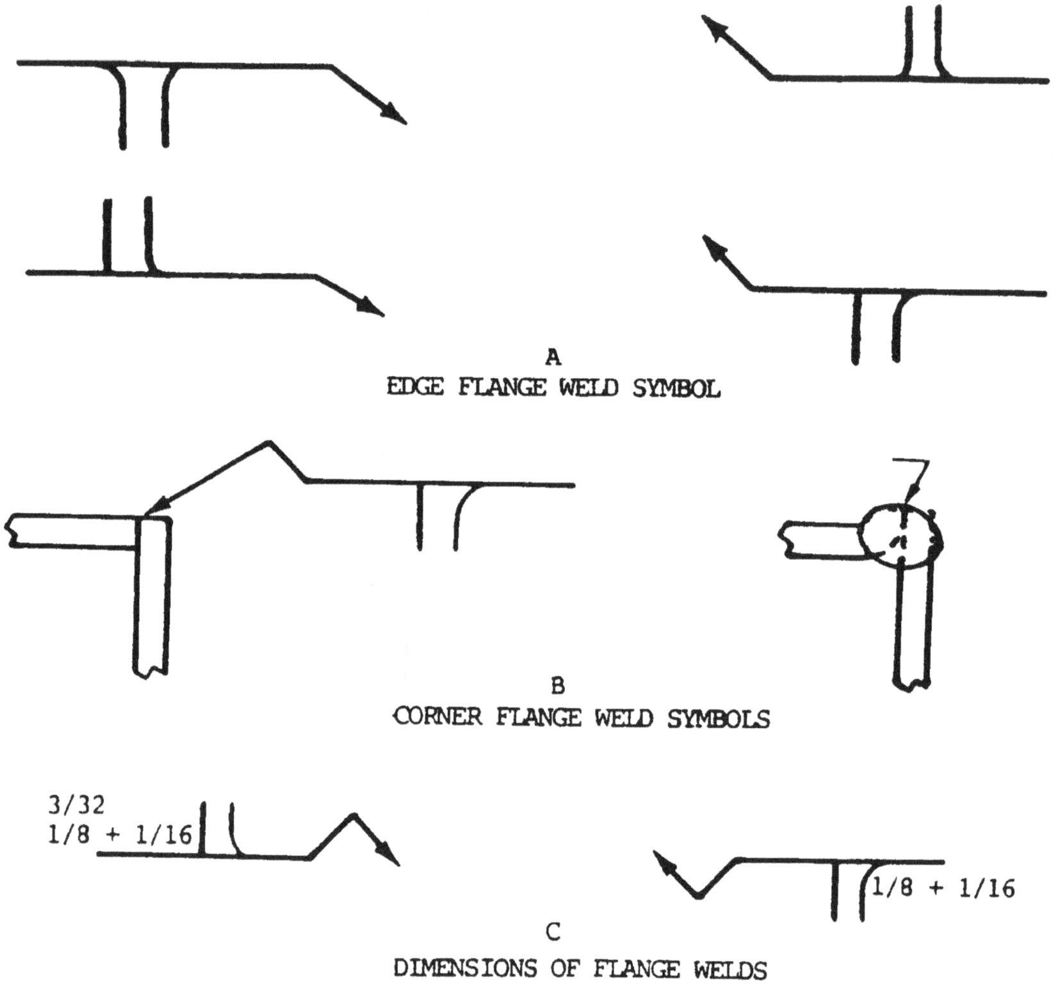

Figure 3-59. Flange weld symbols.

3-29. RESISTANCE SPOT WELDS

a. General. Resistance spot weld symbols (fig. 3-3, p 3-5) have no arrow or other side significance in themselves, although supplementary symbols used in conjunction with them may have such significance. Resistance spot weld symbols shall be centered on the reference line. Dimensions may be shown on either side of the reference line.

b. Size of Resistance Spot Welds. Resistance spot welds are dimensioned by either size or strength as follows:

(1) The size of resistance spot welds is designated as the diameter of the weld expressed in fractions or in decimals in hundredths of an inch and must be shown, with or without inch marks, to the left of the weld symbol (fig. 3-60).

Figure 3-60. Size of resistance spot welds.

(2) The strength of resistance spot welds is designated as the minimum acceptable shear strength in pounds per spot and must be shown to the left of the weld symbol (fig. 3-61).

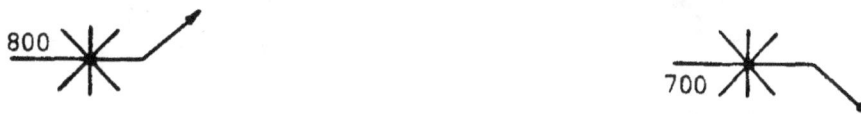

Figure 3-61. Strength of resistance spot welds.

c. Spacing of Resistance Spot Welds.

(1) The pitch of resistance spot welds shall be shown to the right of the weld symbol (fig. 3-62).

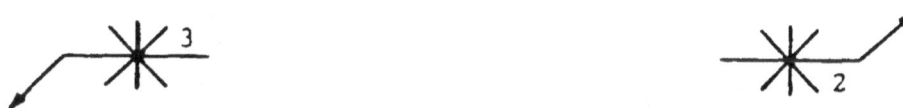

Figure 3-62. Spacing of resistance spot welds.

(2) When the symbols are shown directly on the drawing, the spacing is shown by using dimension lines.

(3) When resistance spot welding extends less than the distance between abrupt changes in the direction of the welding or less than the full length of the joint, the extent must be dimensioned (fig. 3-63).

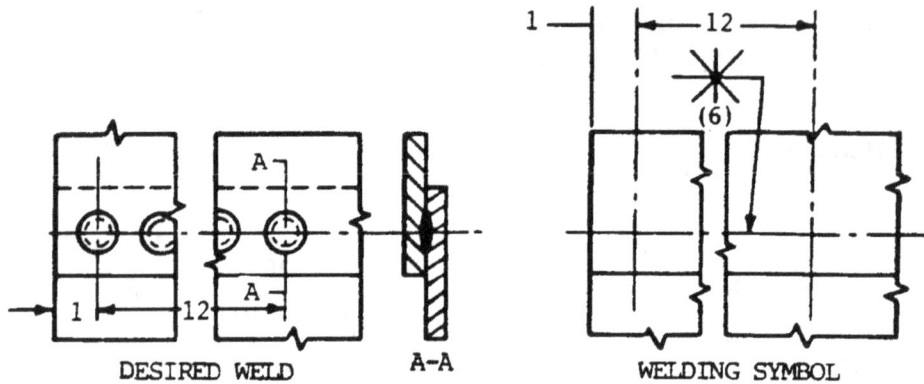

Figure 3-63. Extent of resistance spot weld.

d. Number of Resistance Spot Welds. When a definite number of welds is desired in a certain joint, the number must be shown in parentheses either above or below the weld symbol (fig. 3-64).

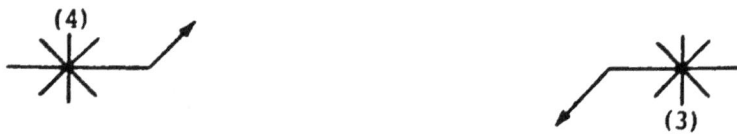

Figure 3-64. Number of resistance spot welds.

e. Flush Resistance Spot Welding Joints. When the exposed surface of one member of a resistance spot welded joint is to be flush, that surface shall be indicated by adding the flush contour symbol (fig. 3-3, p 3-5) to the weld symbol, (fig. 3-65) in accordance with location specifications given in paragraph 3-7.

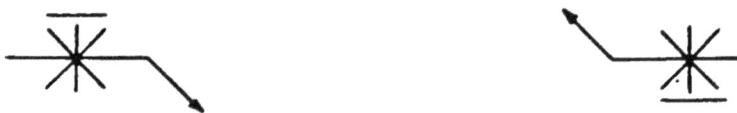

Figure 3-65. Contour of resistance spot welds.

3-30. RESISTANCE SEAM WELDS

a. General.

(1) Resistance seam weld symbols have no arrow or other side significance in themselves, although supplementary symbols used in injunction with them may have such significance. Resistance seam weld symbols must be centered on the reference line.

(2) Dimensions of resistance seam welds may be shown on either side of the reference line.

3-30.   RESISTANCE SEAM WELDS (cont)

b. Size of Resistance Seam Welds. Resistance seam welds must be dimensioned by either size or strength as follows:

(1) The size of resistance seam welds must be designated as the width of the weld expressed in fractions or in decimals in hundredths of an inch and shall be shown, with or without inch marks, to the left of the weld symbol (fig. 3-66).

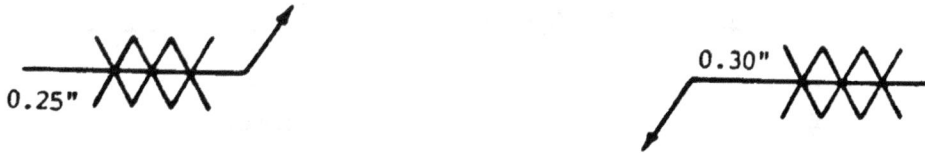

Figure 3-66.   Size of resistance seam welds.

(2) The strength of resistance seam welds must be designated as the minimum acceptable shear strength in pounds per linear inch and must be shown to the left of the weld symbol (fig. 3-67).

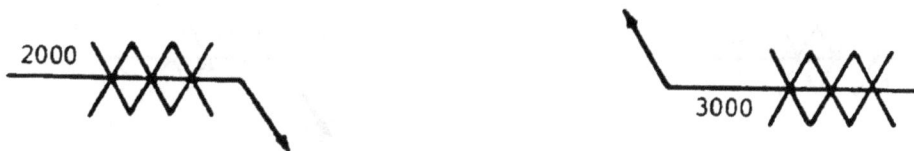

Figure 3-67.   Strength of resistance seam welds.

c. Length of Resistance Seam Welds.

(1) The length of a resistance seam weld, when indicated on the welding symbol, must be shown to the right of the welding symbol (fig. 3-68).

Figure 3-68.   Length of resistance seam welds.

(2) When resistance seam welding extends for the full distance between abrupt changes in the direction of the welding, no length dimension need be shown on the welding symbol.

(3) When resistance seam welding extends less than the distance between abrupt changes in the direction of the welding or less than the full length of the joint, the extent must be dimensioned (fig. 3-69).

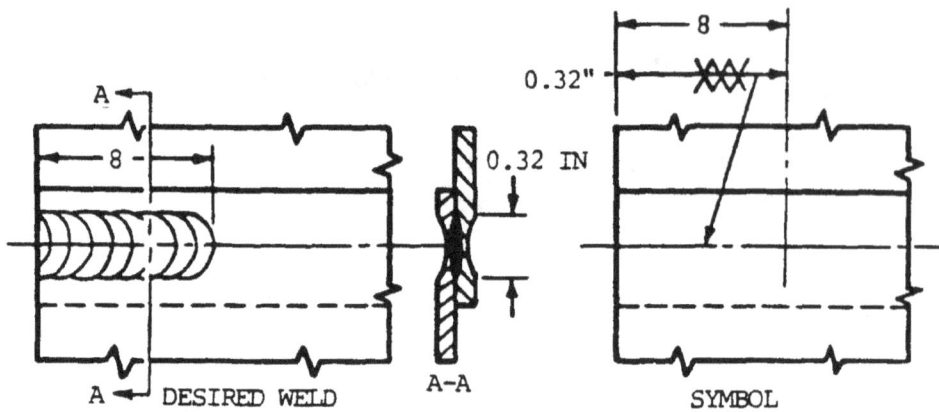

Figure 3-69. Extent of resistance seam welds.

d. Pitch of Resistance Seam Welds. The pitch of intermittent resistance seam welding shall be designated as the distance between centers of the weld increments and must be shown to the right of the length dimension (fig. 3-70).

Figure 3-70. Dimensioning of intermittent resistance seam welds.

e. Termination of Intermittent Resistance Seam Welding. When intermittent resistance seam welding is used by itself, the symbol indicates that increments are located at the ends of the dimensioned length. When used between continuous resistance seam welding, the symbol indicates that spaces equal to the pitch minus the length of one increment are left at the ends of the dimensional length. Separate symbols must be used for intermittent and continuous resistance seam welding when the two are combined.

f. Flush Projection Welded Joints. When the exposed surface of one member of a projection welded joint is to be made flush, that surface shall be indicated by adding the flush contour symbol (fig. 3-3, p 3-5) to the weld symbol, observing the usual location significance (fig. 3-79).

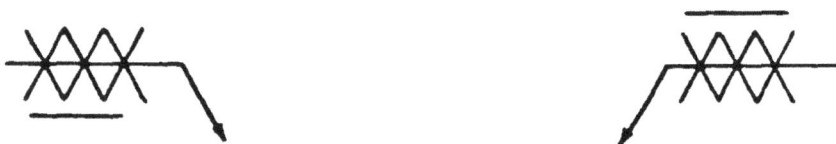

Figure 3-71. Contour of resistance seam welds.

3-31. PROJECTION WELDS

a. General.

(1) When using projection welding, the spot weld symbol must be used with the projection welding process reference in the tail of the welding symbol. The spot weld symbol must be centered on the reference line.

(2) Embossments on the arrow side member of a joint for projection welding shall be indicated by placing the weld symbol on the side of the reference line toward the reader (fig. 3-72).

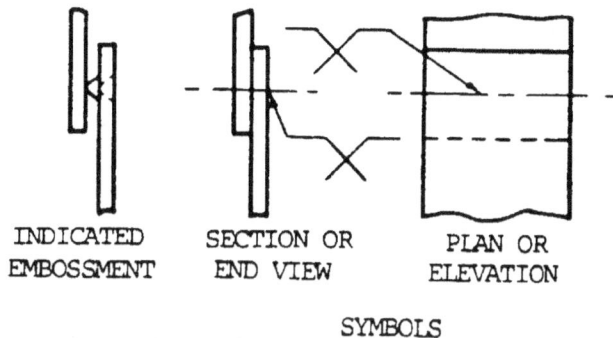

Figure 3-72. Embossment on arrow-side member of joint for projection welding.

(3) Embossment on the other side member of joint for projection welding shall be indicated by placing the weld symbol on the side of the reference line away from the reader (fig. 3-73).

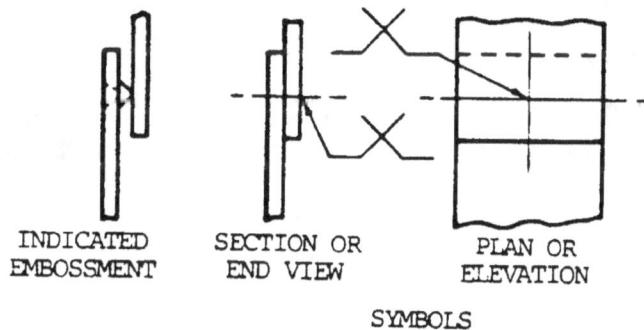

Figure 3-73. Embossment on other-side member of joint for projection welding.

(4) Proportions of projections must be shown by a detail or other suitable means.

(5) Dimensions of projection welds must be shown on the same side of the reference line as the weld symbol.

b. Size of Projection Welds.

(1) Projection welds must be dimensioned by strength. Circular projection welds may be dimensioned by size.

3-34

(2) The size of circular projection welds shall be designated as the diameter of the weld expressed in fractions or in decimals in hundredths of an inch and shall be shown, with or without inch marks, to the left of the weld symbol (fig. 3-74).

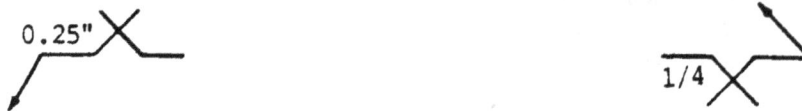

Figure 3-74. Diameter of projection welds.

(3) The strength of projection welds shall be designated as the minimum acceptable shear strength in pounds per weld and shall be shown to the left of the weld symbol (fig. 3-75).

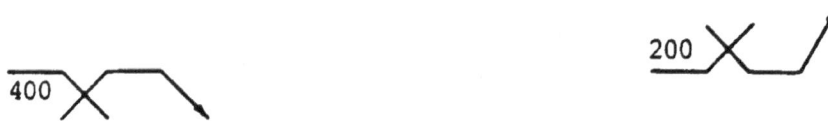

Figure 3-75. Strength of projection welds.

c. <u>Spacing of Projection Welds</u>. The pitch of projection welds shall be shown to the right of the weld symbol (fig. 3-76).

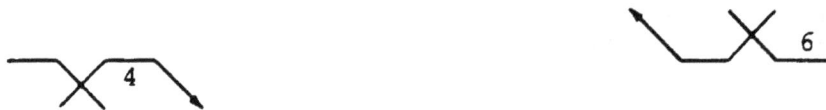

Figure 3-76. Spacing of projection welds.

d. Number of Projection Welds. When a definite number of projection welds is desired in a certain joint, the number shall be shown in parentheses (F, fig. 3-77).

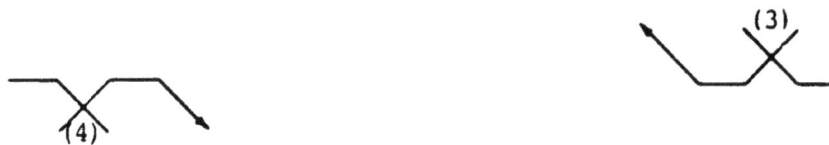

Figure 3-77. Number of projection welds.

3-31. PROJECTION WELDS (cont)

e. <u>Extent of Projection Welding</u>. When the projection welding extends less than the distance between abrupt changes in the direction of the welding or less than the full length of the joint, the extent shall be dimensioned (fig. 3-78).

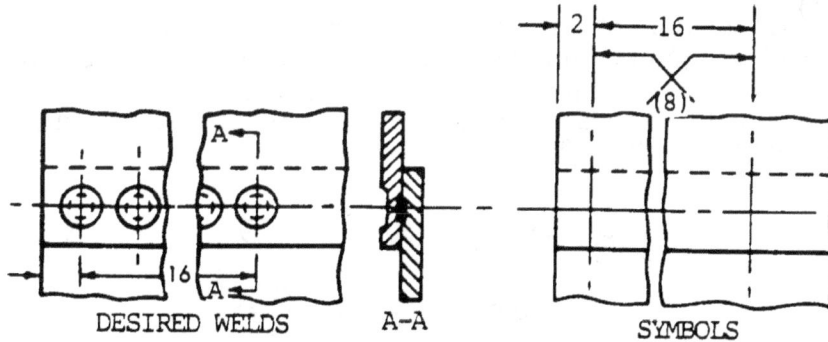

Figure 3-78. Extent of projection welds.

f. <u>Flush Resistance Seam Welded Joints</u>. When the exposed surface of one member of a resistance seam welded joint is to be flush, that surface shall be indicated by adding the flush contour symbol (fig. 3-3, p 3-5) to the weld symbol, observing the usual location significance (fig. 3-71).

Figure 3-79. Contour of projection welds.

3-32. FLASH OR UPSET WELDS

a. <u>General</u>. Flash or upset weld symbols have no arrow side or other side significance in themselves, although supplementary symbols used in conjunction with then may have such significance. The weld symnbols for flash or upset welding must be centered on the reference line. Dimensions need not be shown on the welding symbol .

b. <u>Surface Contour of Flash or Upset Welds</u>. The contour symbols (fig. 3-3, p 3-5) for flash or upset welds (fig. 3-80) are indicated in the same manner as that for fillet welds (paragraph 3-21).

Figure 3-80. Surface contour of flash or upset welds.

CHAPTER 4
JOINT DESIGN AND PREPARATION OF METALS

4-1. JOINT TYPES

Welds are made at the junction of the various pieces that make up the weldment.
The junctions of parts, or joints, are defined as the location where two or more
members are to be joined. Parts being joined to produce the weldment may be in the
form of rolled plate, sheet, shapes, pipes, castings, forgings, or billets. The
five basic types of welding joints are listed below.

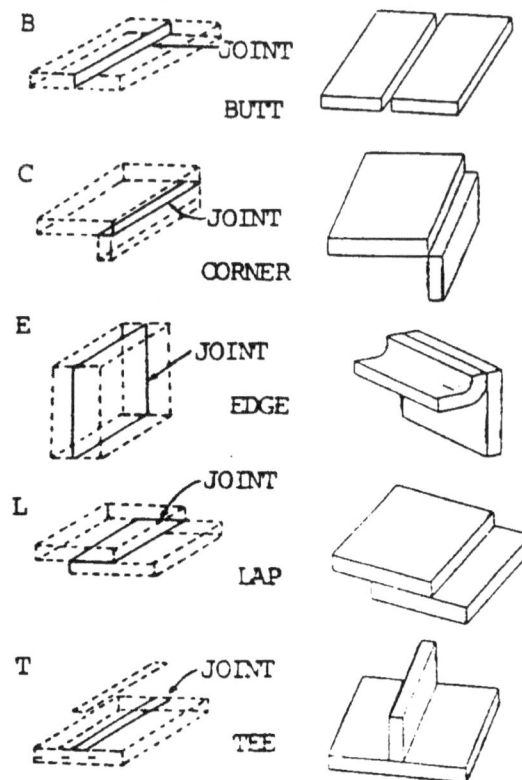

Figure 4-1. The five basic types of joints.

a. B, <u>Butt Joint</u>. A joint between two members lying approximately in the same
plane.

b. C, <u>Corner Joint</u>. A joint between two members located approximately at right
angles to each other in the form of an angle.

c. E, <u>Edge Joint</u>. A joint between the edges of two or more parallel or mainly
parallel members.

d. L, <u>Lap Joint</u>. A joint between two overlapping members.

e. T, <u>Tee Joint</u>. A joint between two members located approximately at right
angles to each other in the form of a T.

4-2. WELD JOINTS

In order to produce weldments it is necessary to combine the joint types with weld types to produce weld joints for joining the separate members. Each weld type cannot always be combined with each joint type to make a weld joint. Table 4-1 shows the welds applicable to the basic joints.

Table 4-1. WELDS APPLICABLE TO THE BASIC JOINT COMBINATIONS

Weld Type	Symbol	Basic Joint Types				
		B Butt	C Corner	E Edge	L Lap	T Tee
Fillet		Special	Yes	Special	Yes	Yes
Plug or slot		-	-	-	Yes	Yes
Spot or projection		-	-	-	Yes	Special
Seam		-	Special	-	Yes	Special
Square groove		Yes	Yes	Yes	-	Yes
Vee groove		Yes	Yes	Yes	-	Yes
Bevel groove		Yes	Yes	Yes	Yes	Yes
U groove		Yes	Yes	Yes	-	-
J groove		Yes	Yes	Yes	Yes	Yes
Flare V groove		Yes	Yes	-	-	-
Flare bevel groove		Yes	Yes	-	Yes	Yes
Backing weld		Combin.	Combin.	-	-	Combin.
Surfacing		-	-	-	-	-
Flange edge		-	-	Yes	-	-
Flange corner		-	Yes	-	-	-

4-3. WELD JOINT DESIGN AND PREPARATION

a. Purpose Weld joints are designed to transfer the stresses between the members of the joint and throughout the weldment. Forces and loads are introduced at different points and are transmitted to different areas throughout the weldment. The type of loading and service of the weldment have a great bearing on the joint design required.

b. <u>Categories</u>. All weld joints can be classified into two basic categories: full penetration joints and partial penetration joints.

(1) A full penetration joint has weld metal throughout the entire cross section of the weld joint.

(2) A partial penetration joint has an unfused area and the weld does not completely penetrate the joint. The rating of the joint is based on the percentage of weld metal depth to the total joint i.e., a 50 percent partial penetration joint would have weld metal halfway through the joint.

NOTE
When joints are sub jetted to dynamic loading, reversing loads, and impact leads, the weld joint must be very efficient. This is more important if the weldment is sub jetted to cold-temperature service. Such services require full-penetration welds. Designs that increase stresses by the use of partial-penetration joints are not acceptable for this type of service.

c. <u>Strength</u> The strength of weld joints depends not only on the size of the weld, but also on the strength of the weld metal.

(1) Mild and low alloy steels are generally stronger than the materials being joined.

(2) When welding high-alloy or heat-treated materials, special precautions must be taken to ensure the welding heat does not cancel the heat treatment of the base metal, causing it to revert to its lower strength adjacent to the weld.

d. <u>Design.</u> The weld joint must be designed so that its cross-sectional area is the minimum possible. The cross-sectional area is a measurement of the amount or weight of weld metal that must be used to make the joint Joints may be prepared by shearing, thermal cutting, or machining.

(1) Carbon and lw alloy joint design and prepaation. These weld joints are prepared either by flame cutting or mechanically by machining or grinding, depending on the joint details. Before welding, the joint surfaces must be cleared of all foreign materials such as paint, dirt, scale, or must Suitable solvents or light grinding can be used for cleaning. The joint surface should not be nicked or gouged since nicks and gouges may interfere with the welding operation. Specific information on welding carbon and low alloy metals may be found in chapter 7, paragraph 7-10.

CAUTION
Aluminum and aluminum alloys should not be cleaned with caustic soda or strong cleaner with a pH above 10. The aluminum or aluminum alloy will react chemically with these types of cleaners. Other nonferrous metals and alloys should be investigated prior to using these cleaners to determine their reactivity.

4-3. WELD JOINT DESIGN AND PREPARATION (cent)

(2) _Aluminum and aluminum alloy joint design and preparation._ Weld joint designs often unintentionally require welds that cannot be made. Check your design to avoid these and similar errors. Before welding, the joint surfaces must be cleared of all foreign materials such as paint, dirt, scale, or oxide; solvent cleaning, light grinding, or etching can be used. The joint surfaces should not be nicked or gouged since nicks and gouges may interfere with welding operations. Specific information regarding welding aluminum and aluminum alloy metals may be found in chapter 7, paragraph 7-17.

(3) _Stainless steel alloy joint design and preparation._ These weld joints are prepared either by plasma arc cutting or by machining or grinding, depending on the alloy. Before welding, the joint surfaces must be cleaned of all foreign material, such as paint, dirt, scale, or oxides. Cleaning may be done with suitable solvents (e.g., acetone or alcohol) or light grinding. Care should be taken to avoid nicking or gouging the joint surface since such flaws can interfere with the welding operation. Specific information regarding welding stainless steel alloy metals may he found in chapter 7, paragraph 7-14.

4-4. WELD ACCESSIBILITY

The weld joint must be accessible to the welder using the process that is employed. Weld joints are often designed for welds that cannot be made. Figure 4-2 illustrates several types of inaccessible welds.

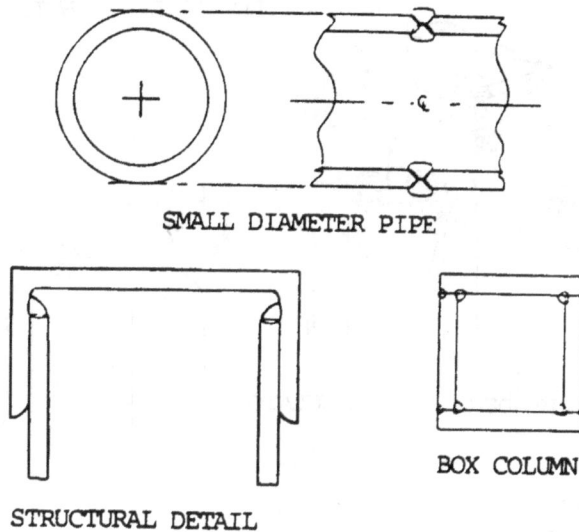

SMALL DIAMETER PIPE

STRUCTURAL DETAIL

BOX COLUMN

Figure 4-2. Inaccessible welds.

CHAPTER 5
WELDING AND CUTTING EQUIPMENT

Section I. OXYACETYLENE WELDING EQUIPMENT

5-1. GENERAL

The equipment used for oxyacetylene welding consists of a source of oxygen and a source of acetylene from a portable or stationary outfit, along with a cutting attachment or a separate cutting torch. Other equipment requirements include suitable goggles for eye protection, gloves to protect the hands, a method to light the torch, and wrenches to operate the various connections on the cylinders, regulators, and torches.

5-2. STATIONARY WELDING EQUIPMENT

Stationay welding equipment is installed where welding operations are conducted in a fixed location. Oxygen and acetylene are provided in the welding area as outlined below.

   a. Oxygen. Oxygen is obtained from a number of Cylinders manifolded and equipped with a master regulator. The regulator and manifold control the pressure and the flow together (fig. 5-1). The oxygen is supplied to the welding stations through a pipe line equipped with station outlets (fig. 5-2, p 5-2).

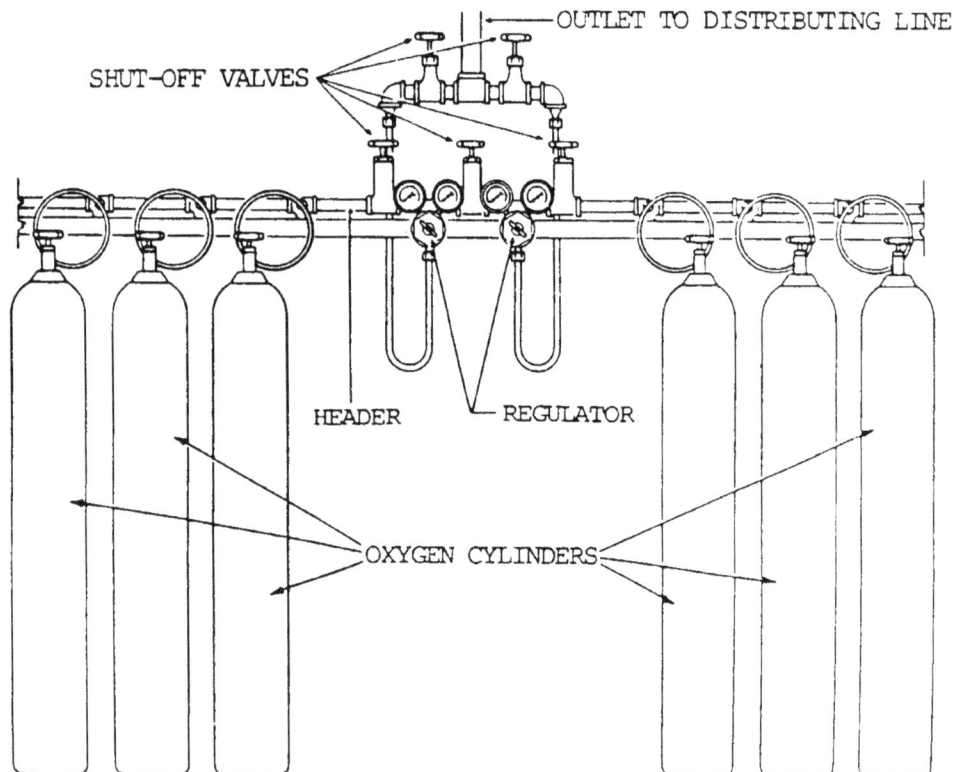

Figure 5-1. Stationary oxygen cylinder manifold and other equipment.

5-2. STATIONARY WELDING EQUIPMENT (cont)

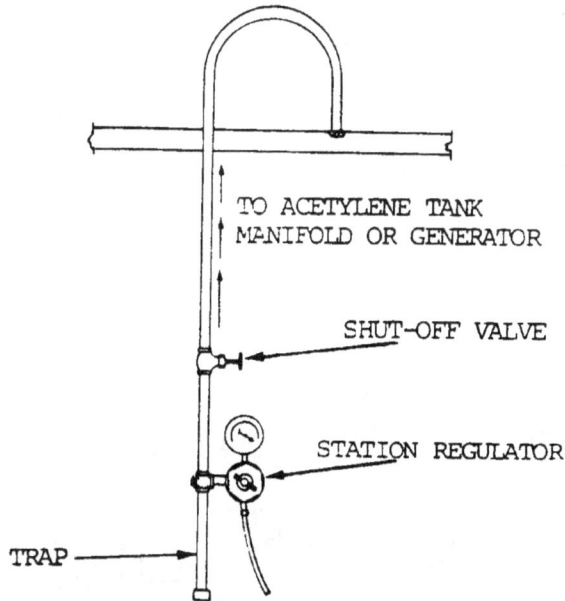

Figure 5-2. Station outlet for oxygen or acetylene.

b. Acetylene. Acetylene is obtained either from acetylene cylinders set up as shown in figure 5-3, or an acetylene generator (fig. 5-4). The acetylene is supplied to the welding stations through a pipe line equipped with station outlets as shown in figure 5-2.

A--LINE VALVE

B--RELEASE VALVE

C--FILLER PLUG

D--HEADER PIPE

E--REGULATOR

F--FLASH ARRESTOR CHAMBER

G--ESCAPE PIPE

H--CYLINDER CONNECTION PIPE

J--CHECK VALVE AND DRAIN PLUG

K--ACETYLENE CYLINDERS

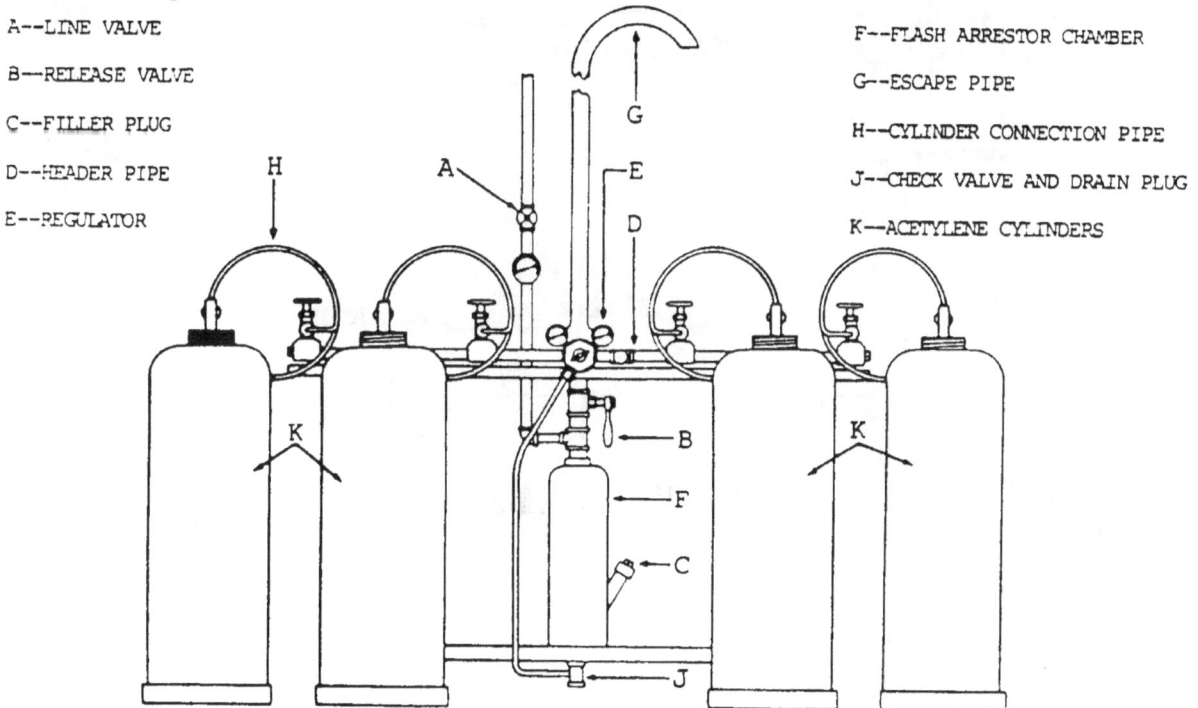

Figure 5-3. Stationary acetylene cylinder manifold and other equipment.

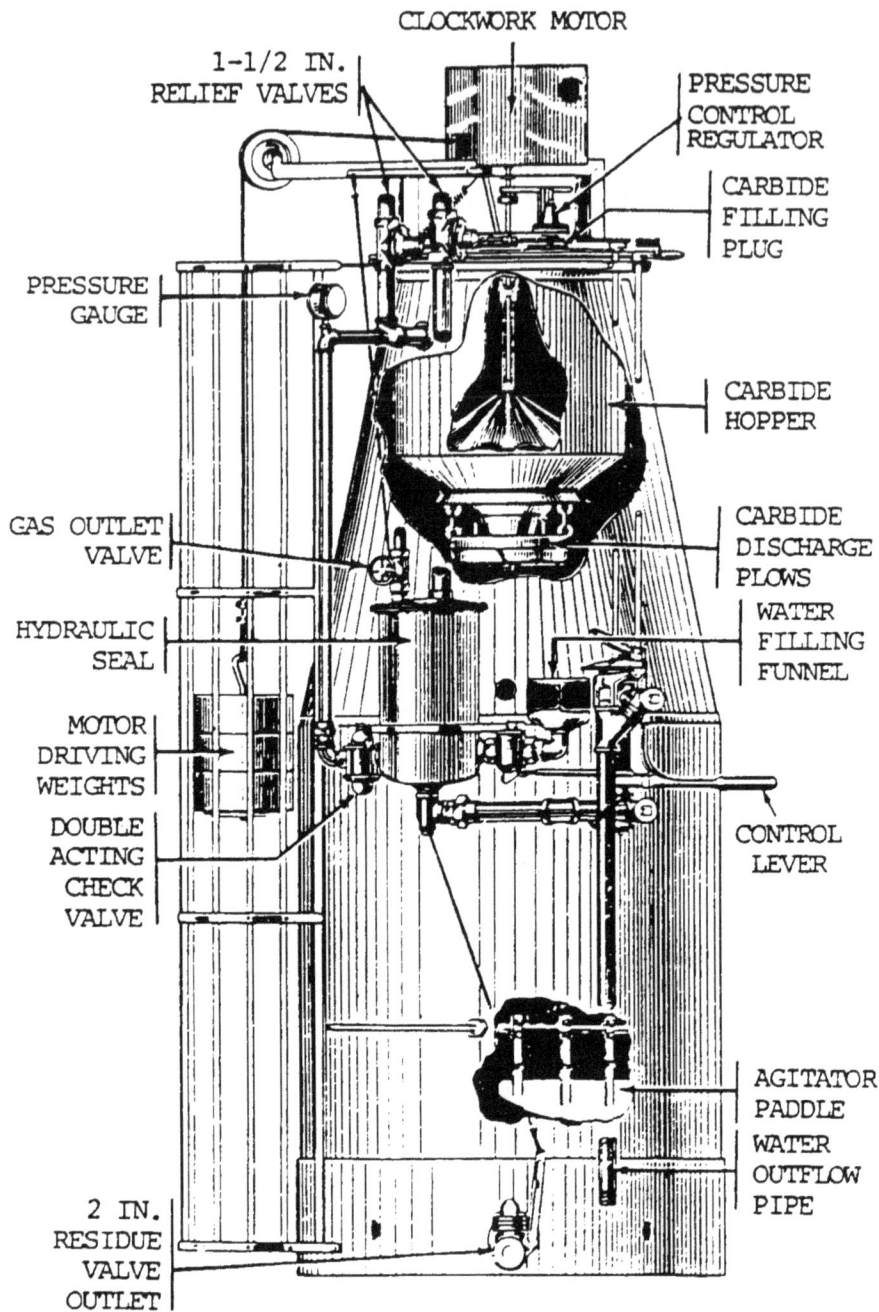

Figure 5-4. Acetylene generator and operating equipment.

5-3. PORTABLE WELDING EQUIPMENT

The portable oxyacetylene welding outfit consists of an oxygen cylinder and an acetylene cylinder with attached valves, regulators, gauges, and hoses (fig. 5-5). This equipment may be temporarily secured on the floor or mounted on an all welded steel truck. The trucks are equipped with a platform to support two large size cylinders. The cylinders are secured by chains attached to the truck frame. A metal toolbox, welded to the frame, provides storage space for torch tips, gloves, fluxes, goggles, and necessary wrenches.

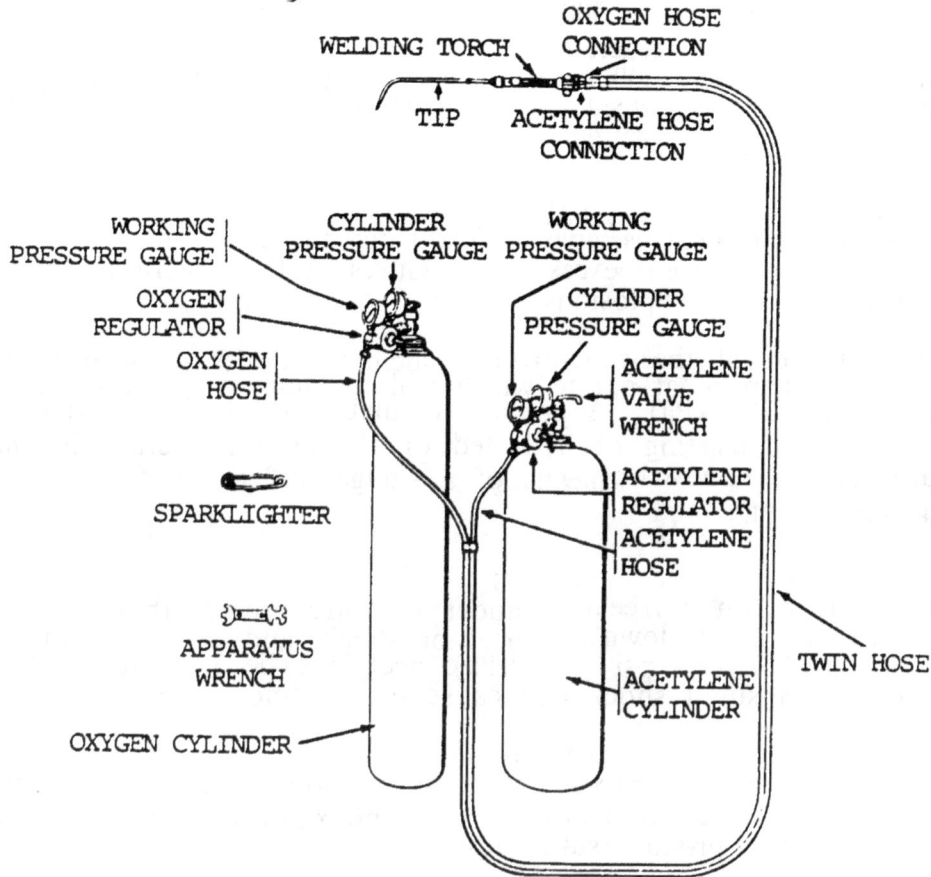

Figure 5-5. Portable oxyacetylene welding and cutting equipment.

5-4. ACETYLENE GENERATOR

NOTE
Acetylene generator equipment is not a standard item of issue and is included in this manual for information only.

a. Acetylene is a fuel gas composed of carbon and hydrogen. C_2H_2), generated by the action of calcium carbide, a gray stonelike substance, and water in a generating unit. Acetylene is colorless, but has a distinctive odor that can be easily detected.

b. Mixtures of acetylene and air, containing from 2 to 80 percent acetylene by volume, will explode when ignited. However, with suitable welding equipment and proper precautions, acetylene can be safely burned with oxygen for heating, welding, and cutting putposes.

Acetylene, when burned with oxygen, produces an oxyacetylene flame with inner; cone tip temperatures of approximately 6300°F (3482 °C), for an oxidizing flame; 5850 °F (3232 °C) for a neutral flame; and 5700°F (3149 °C) for a carburizing flame.

d. The generator shown in figure 5–4 is a commonly used commercial type. A single rated 300-lb generator uses 300 lb of calcium carbide and 300 gal. of water. This amount of material will generate 4.5 cu ft of acetylene per pound; the output for this load is approximately 300 cu ft per hour for 4.5 hours. A double rated generator uses 300 lb of finer sized calcium carbide fed through a special hopper and will deliver 600 cu ft of acetylene per hour for 2.5 hours.

CAUTION
Since considerable heat is given off during the reaction, precautions must be taken to prevent excessive pressures in the generator which might cause fires or explosions.

e. In the operation of the generator the calcium carbide is added to the water through a hopper mechanism at a rate which will maintain a working pressure of approximately 15 psi (103.4 kPa). A pressure regulator is a built-in part of this equipment. A sludge, consisting of hydrated or slaked lime, settles in the bottom of the generator and is removed-by means of a sludge outlet.

5-5. ACETYLENE CYLINDERS

WARNING
Acetylene, stored in a free state under pressure greater than 15 psi (103.4 kPa), can break down from heat or shock, and possibly explode. Under pressure of 29.4 psi (203 kPa), acetylene becomes self-explosive, and a slight shock can cause it to explode spontaneously.

CAUTION
Although acetylene is nontoxic, it is an anesthetic, and if present in a sufficiently high concentration, is an asphyxiant in that it replaces oxygen and can produce suffocation.

a. Acetylene is a colorless, flammable gas composed of carbon and hydrogen, manufactured by the reaction of water and calcium carbide. It is slightly lighter than air. Acetylene burns in the air with an intensely hot, yellow, luminous, smoky flare.

b. Although acetylene is stable under low pressure, if compressed to 15 psi (103.4 kPa), it becomes unstable. Heat or shock can cause acetylene under pressure to explode. Avoid exposing filled cylinders to heat, furnaces, radiators, open fires, or sparks (from a torch). Avoid striking the cylinder against other objects and creating sparks. To avoid shock when transporting cylinders, do not drag, roll, or slide them on their sides. Acetylene can be compressed into cylinders when dissolved in acetone at pressures up to 250 psi (1724 kPa) .

c. For welding purposes, acetylene is contained in three common cylinders with capacities of 1, 60, 100, and 300 cu ft. Acetylene must not be drawn off in volumes greater than 1/7 of the cylinder's rated capacity.

5-5. ACETYLENE CYLINDERS (cont)

d. In order to decrease the size of the open spaces in the cylinder, acetylene cylinders (fig. 5-6) are filled with porous materials such as balsa wood, charcoal, corn pith, or portland cement. Acetone, a colorless, flammable liquid, is added to the cylinder until about 40 percent of the porous material is saturated. The porous material acts as a large sponge which absorbs the acetone, which then absorbs the acetylene. In this process, the volume of acetone increases as it absorbs the acetylene, while acetylene, being a gas, decreases in volume.

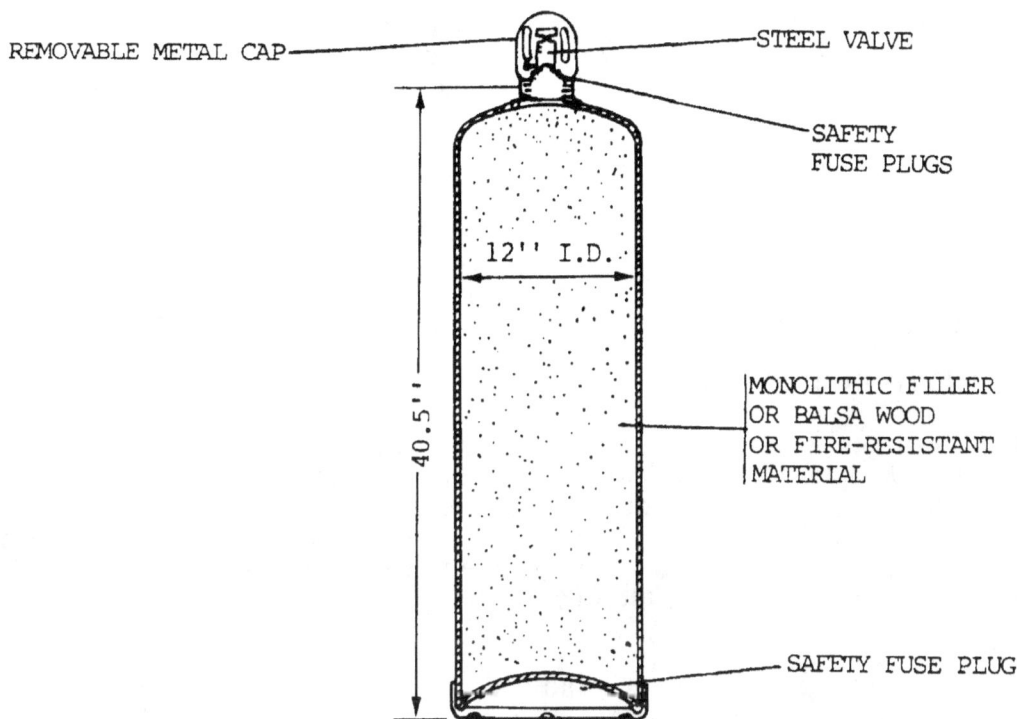

Figure 5-6. Acetylene cylinder construction.

CAUTION
Do not fill acetylene cylinders at a rate greater than 1/7 of their rated capacity, or about 275 cu ft per hour. To prevent drawing off of acetone and consequent impairment of weld quality and damage to the welding equipment, do not draw acetylene from a cylinder at continuous rates in volumes greater than 1/7 of the rated capacity of the cylinder, or 32.1 cu ft per hour. When more than 32.1 cu ft per hour are required, the cylinder manifold system must be used.

e. Acetylene cylinders are equipped with safety plugs (fig. 5-6) which have a small hole through the center. This hole is filled with a metal alloy which melts at approximately 212°F (100 °C), or releases at 500 psi (3448 kPa). When a cylinder is overheated, the plug will melt and permit the acetylene to escape before dangerous pressures can be developed. The plug hole is too small to premit a flame to burn back into the cylinder if escaping acetylene is ignited.

f. The brass acetylene cylinder valves have squared stainless steel valve stems. These stems can be fitted with a cylinder wrench and opened or closed when the cylinder is in use. The outlet of the valve is threaded for connection to an acetylene pressure regulator by means of a union nut.The regulator inlet connection gland fits against the face of the threaded cylinder connection, and the union nut draws the two surfaces together. Whenever the threads on the valve connections are damaged to a degree that will prevent proper assembly to the regulator, the cylinder should be marked and set aside for return to the manufacturer.

WARNING

Acetylene which mᵃy ᵃccumulᵃte in ᵃ storᵃge room or in ᵃ confined space is a fire arid explosion hazard. All acetylene cylinders should be checked, using a soap solution, for leakage at the valves and safety fuse plugs.

g. A protective metal cap (fig. 5-6) screws onto the valve to prevent damage during shipment or storage.

h. Acetylene, when used with oxygen, produces the highest flame temperature of any of the fuel gases. It also has the most concentrated flame, but produces less gross heat of combustion than the liquid petroleum gases and the synthetic gases.

5-6. OXYGEN AND ITS PRODUCTION

a. General. Oxygen is a colorless, tasteless, odorless gas that is slightly heavier than air. It is nonflammable but will support combustion with other elements. In its free state, oxygen is one of the most common elements. The atmosphere is made up of approximately 21 parts of oxygen and 78 parts of nitrogen, the remainder being rare gases. Rusting of ferrous metals, discoloration of copper, and the corrosion of aluminum are all due to the action of atmospheric oxygen, known as oxidation.

b. Production of Oxygen. Oxygen is obtained commercially either by the liquid air process or by the electrolytic process.

(1) In the liquid air process, air is compressed and cooled to a point where the gases become liquid. As the temperature of the liquid air rises, nitrogen in a gaseous form is given off first, since its boiling point is lower than that of liquid oxygen. These gases, having been separated, are then further purified and compressed into cylinders for use. The liquid air process is by far the most widely used to produce oxygen.

(2) In the electrolytic process, water is broken down into hydrogen and oxygen by the passage of an electric current. The oxygen collects at the positive terminal and the hydrogen at the negative terminal. Each gas is collected and compressed into cylinders for use.

5-7. OXYGEN CYLINDER

CAUTION

Always refer to oxygen as oxygen, never as air. Combustibles should be kept away from oxygen, including the cylinder, valves, regulators, and other hose apparatus. Oxygen cylinders and apparatus should not be handled with oily hands or oily gloves. Pure oxygen will support and accelerate combustion of almost any material, and is especially dangerous in the presence of oil and grease. Oil and grease in the presence of oxygen may spontaneously ignite and burn violently or explode. Oxygen should never be used in any air tools or for any of the purposes for which compressed air is normally used.

A typical oxygen cylinder is shown in figure 5-7. It is made of steel and has a capacity of 220 cu ft at a pressure of 2000 psi (13,790 kPa) and a temperature of 70 °F (21 °C). Attached equipment provided by the oxygen supplier consists of an outlet valve, a removable metal cap for the protection of the valve, and a low melting point safety fuse plug and disk. The cylinder is fabricated from a single plate of high grade steel so that it will have no seams and is heat treated to achieve maximum strength. Because of their high pressure, oxygen cylinders undergo extensive testing prior to their release for work, and must be periodically tested thereafter.

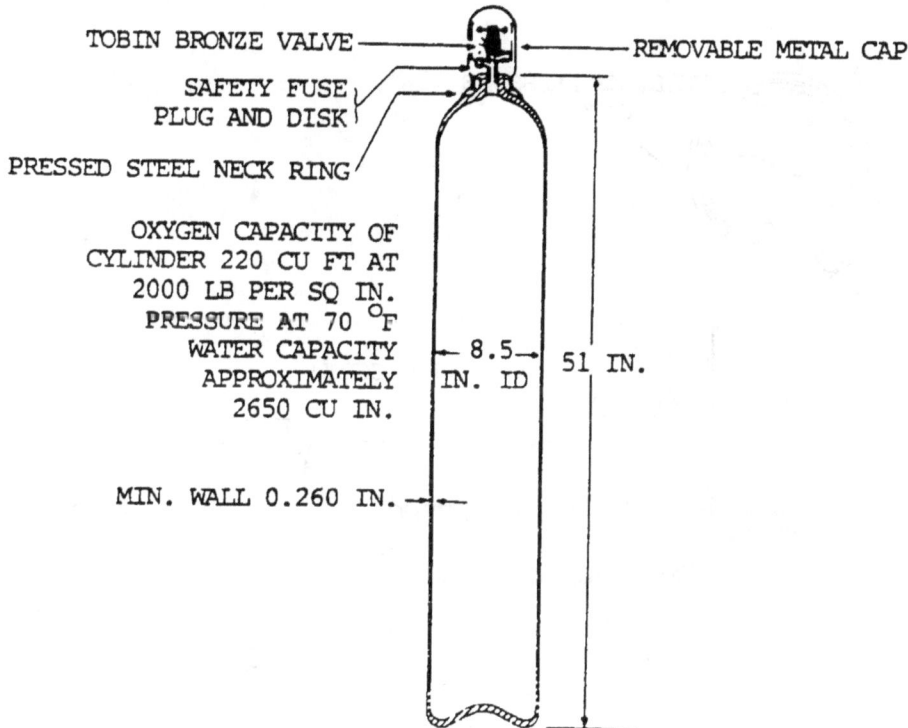

Figure 5-7. Oxygen cylinder construction.

5-8. OXYGEN AND ACETYLENE REGULATORS

a. General. The gases compressed in oxygen and acetylene cylinders are held at pressures too high for oxyacetylene welding. Regulators reduce pressure and control the flow of gases from the cylinders. The pressure in an oxygen cylinder can be as high as 2200 psi (15,169 kPa), which must be reduced to a working pressure of 1 to 25 psi (6.90 to 172.38 kPa). The pressure of acetylene in an acetylene cylinder can be as high as 250 psi (1724 kPa) and must be reduced to a working pressure of from 1 to 12 psi (6.90 to 82.74 kPa). A gas pressure regulator will automatically deliver a constant volume of gas to the torch at the adjusted working pressure.

NOTE
The regulators for oxygen, acetylene, and liquid petroleum fuel gases are of different construction. They must be used only for the gas for which they were designed.

Most regulators in use are either the single stage or the two stage type. Check valves must be installed between the torch hoses and the regulator to prevent flashback through the regulator.

b. Single Stage Oxygen Regulator. The single stage oxygen regulator reduces the cylinder pressure of a gas to a working pressure in one step. The single stage oxygen regulator mechanism (fig. 5-8) has a nozzle through which the high pressure gas passes, a valve seat to close off the nozzle, and balancing springs. Some types have a relief valve and an inlet filter to exclude dust and dirt. Pressure gauges are provided to show the pressure in the cylinder or pipe line and the working pressure.

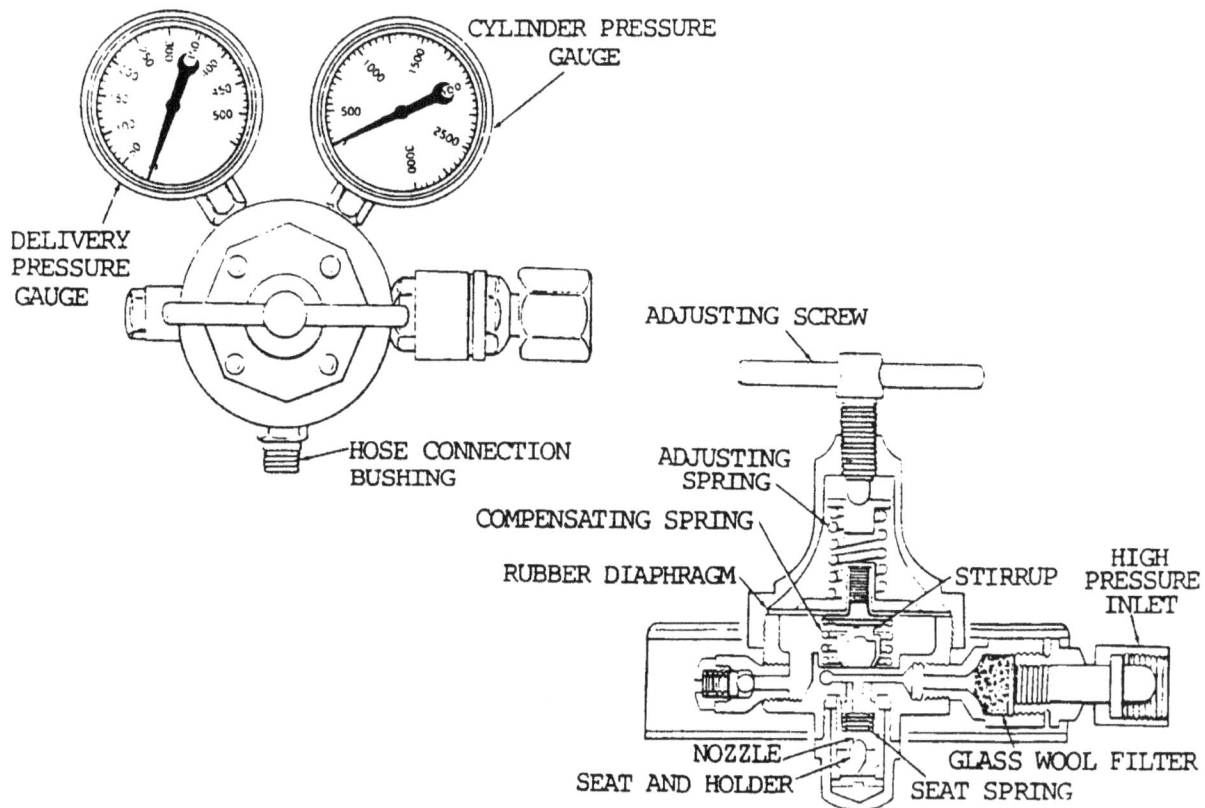

Figure 5-8. Single stage oxygen regulator.

5-8. OXYGEN AND ACETYLENE REGULATORS (cont)

NOTE

In operation, the working pressure falls as the cylinder pressure falls, which occurs gradually as gas is withdrawn. For this reason, the working pressure must be adjusted at intervals during welding operations when using a single stage oxygen regulator.

The oxygen regulator controls and reduces the oxygen pressure from any standard commercial oxygen cylinder containing pressures up to 3000 psi. The high pressure gauge, which is on the inlet side of the regulator, is graduated from 0 to 3000 psi. The low or working pressure gauge, which is on the outlet side of the regulator, is graduated from O to 500 psi.

c. Operation of Single Stage Oxygen Regulator.

(1) The regulator consists of a flexible diaphragm, which controls a needle valve between the high pressure zone and the working zone, a compression spring, and an adjusting screw, which compensates for the pressure of the gas against the diaphragm. The needle valve is on the side of the diaphragm exposed to high gas pressure while the compression spring and adjusting screw are on the opposite side in a zone vented to the atomsphere.

(2) The oxygen enters the regulator through the high pressure inlet connection and passes through a glass wool filter, which removes dust and dirt. The seat, which closes off the nozzle, is not raised until the adjusting screw is turned in. Pressure is applied to the adjusting spring by turning the adjusting screw, which bears down on the rubber diaphragm. The diaphragm presses downward on the stirrup and overcomes the pressure on the compensating spring. When the stirrup is forced downward, the passage through the nozzle is open. Oxygen is then allowed to flow into the low pressure chamber of the regulator. The oxygen then passes through the regulator outlet and the hose to the torch. A certain set pressure must be maintained in the low pressure chamber of the regulator so that oxygen will continue to be forced through the orifices of the torch, even if the torch needle valve is open. This pressure is indicated on the working pressure gage of the regulator, and depends on the position of the regulator adjusting screw. Pressure is increased by turning the adjusting screw to the right and decreased by turning this screw to the left.

(3) Regulators used at stations to which gases are piped from an oxygen manifold, acetylene manifold, or acetylene generator have only one low pressure gage because the pipe line pressures are usually set at 15 psi (103.4 kPa) for acetylene and approximately 200 psi (1379 kPa) for oxygen. The two stage oxygen regulator (fig. 5–9) is similar in operation to the one stage regulator, but reduces pressure in two steps. On the high pressure side, the pressure is reduced from cylinder pressure to intermediate pressure. On the low pressure side the pressure is reduced from intermediate pressure to work pressure. Because of the two stage pressure control, the working pressure is held constant and pressure adjustment during welding operations is not required.

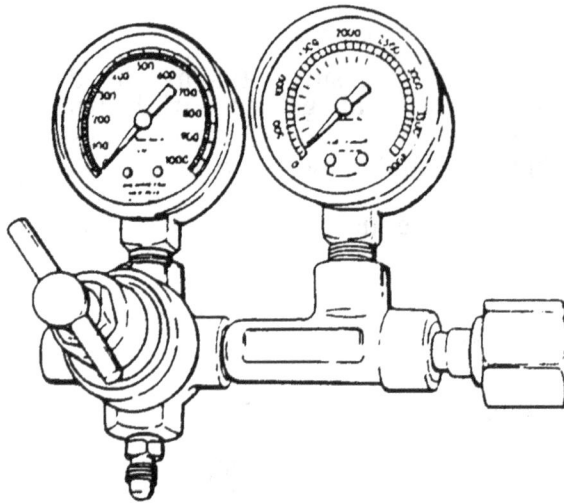

Figure 5-9. Two stage oxygen regulator.

e. <u>Acetylene Regulator</u>.

CAUTION
Acetylene should never be used at pressures exceeding 15 psi (103.4 kPa).

This regulator controls the acetylene pressure from any standard commercial cylinder containing pressures up to 500 psi (3447.5 kPa). The acetylene regulator design is generally the same as that of the oxygen regulator, but will not withstand such high pressures. The high pressure gage, on the inlet side of the regulator, is graduated from O to 500 psi (3447.5 kPa). The low pressure gage, on the outlet side of the regulator, is graduated from O to 30 psi (207 kPa). Acetylene should not be used at pressures exceeding 15 psi (103.4 kPa).

5-9. OXYACETYLENE WELDING TORCH

a. <u>General</u>. The oxyacetylene welding torch is used to mix oxygen and acetylene in definite proportions. It also controls the volume of these gases burning at the welding tip, which produces the required type of flame. The torch consists of a handle or body which contains the hose connections for the oxygen and the fuel gas. The torch also has two needle valves, one for adjusting the flow of oxygen and one for acetylene, and a mixing head. In addition, there are two tubes, one for oxygen, the other for acetylene; inlet nipples for the attachment of hoses; a tip; and a handle. The tubes and handle are of seamless hard brass, copper-nickel alloy, stainless steel. For a description and the different sized tips, see paragraph 5-10.

b. <u>Types of Torches</u>. There are two general types of welding torches; the low pressure or injector type, and the equal pressure type.

5-9. OXYACETYLENE WELDING TORCH (cont)

(1) In the low pressure or injector type (fig. 5-10), the acetylene pressure is less than 1 psi (6.895 kPa). A jet of high pressure oxygen is used to produce a suction effect to draw in the required amount of acetylene. Any change in oxygen flow will produce relative change in acetylene flow so that the proportion of the two gases remains constant. This is accomplishd by designing the mixer in the torch to operate on the injector principle.The welding tips may or may not have separate injectors designed integrally with each tip.

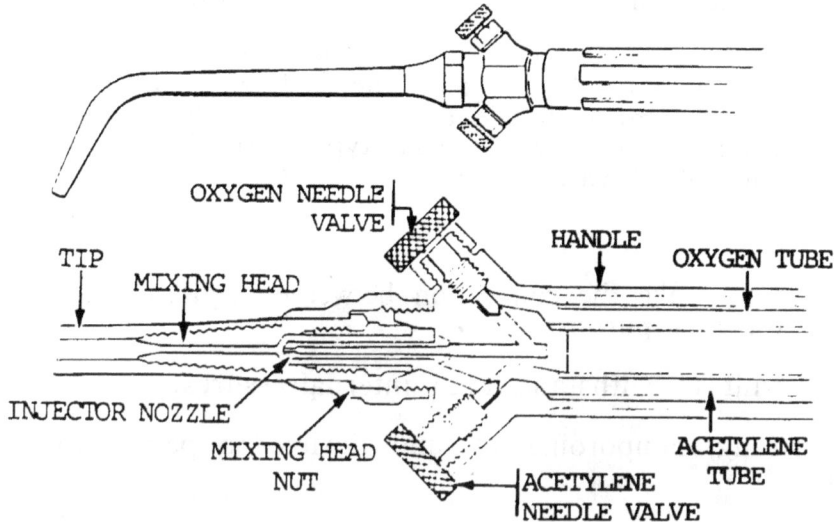

Figure 5-10. Mixing head for injector type welding torch.

(2) The equal pressure torch (fig.5-11) is designed to operate with equal pressures for the oxygen and acetylene. The pressure ranges from 1 to 15 psi (6.895 to 103.4 kPa). This torch has certain advantages over the \lopressure type. It can be more readily adjusted,and since equal pressures are used for each gas, the torch is less susceptible to flashbacks.

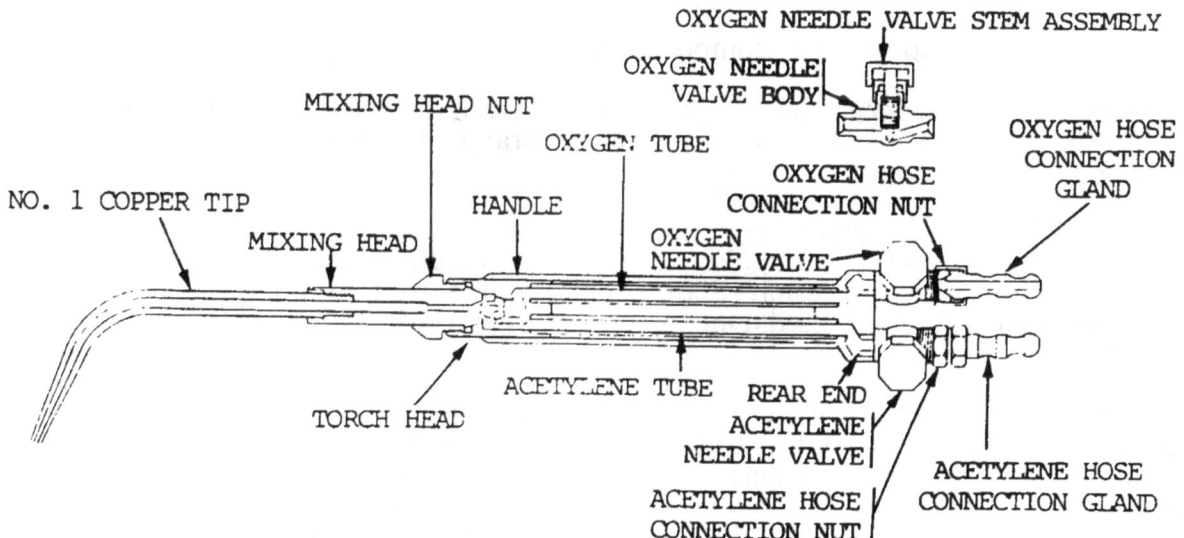

Figure 5-11. Equal pressure type general purpose welding torch.

5-10. WELDING TIPS AND MIXERS

a. The welding tips (fig. 5-10 and 5-11) are made of hard drawn electrolytic copper or 95 percent copper and 5 percent tellurium.They are made in various styles and types, some having a one-piece tip either with a single orifice or a number of orifices. The diameters of the tip orifices differ in order to control the quantity of heat and the type of flame.These tip sizes are designated by numbers which are arranged according to the individual manufacturer's system. Generally, the smaller the number, the smaller the tip orifice.

b. Mixers (fig. 5-10 and 5-11) are frequently provided in tip tier assemblies which assure the correct flow of mixed gases for each size tip. In this tip mixer assembly, the mixer is assembled with the tip for which it has been drilled and then screwed onto the torch head.The universal type mixer is a separate unit which can be used with tips of various sizes.

5-11. HOSE

a. The hoses used to make the connection between the regulators and the torch are made especially for this purpose.

(1) Hoses are built to withstand high internal pressures.

(2) They are strong, nonporous, light, and flexible to permit easy manipulation of the torch.

(3) The rubber used in the manufacture of hose is chemically treated to remove free sulfur to avoid possible spontaneous combustion.

(4) The hose is not impaired by prolonged exposure to light.

CAUTION
Hose should never be used for one gas if it was previously used for another .

b. Hose identification and composition.

(1) In North America, the oxygen hose is green and the acetylene hose is red. In Europe, blue is used for oxygen and orange for acetylene. Black is sometimes also used for oxygen.

(2) The hose is a rubber tube with braided or wrapped cotton or rayon reinforcements and a rubber covering. For heavy duty welding and cutting operations, requiring 1/4- to 1/2-in. internal diameter hose,three to five plies of braided or wrapped reinforcements are used. One ply is used in the 1/8- to 3/16-in. hose for light torches.

c. Hoses are provided with connections at each end so that they may be connected to their respective regulator outlet and torch inlet connections. To prevent a dangerous interchange of acetylene and oxygen hoses, all threaded fittings used for the acetylene hook up are left hand, and all threaded fittings for the oxygen hook up are right hand. Notches are also placed on acetylene fittings to prevent a mixup .

TC 9-237

5-11. HOSE (cont)

d. Welding and cutting hoses are obtainable as a single hose for each gas or with the hoses bonded together along their length under a common outer rubber jacket. The latter type prevents the hose from kinking or becoming tangled during the welding operation.

5-12. SETTING UP THE EQUIPMENT

WARNING
Always have suitable fire extinguishing equipment at hand when doing any welding.

When setting up welding and cutting equipment, it is important that all operations be performed systematically in order to avoid mistakes and possible trouble. The setting up procedures given in a through d below will assure safety to the operator and the apparatus.

a. Cylinders.

WARNING
Do not stand facing cylinder valve outlets of oxygen, acetulene, or other compressed gases when opening them.

(1) Place the oxygen and the acetylene cylinders on a level floor (if they are not mounted on a truck)and tie them firmly to a work bench, post, wall, or other secure anchorage to prevent their being knocked or pulled over.

(2) Remove the valve protecting caps.

(3) "Crack" both cylinder valves by opening first the acetylene and then the oxygen valve slightly for an instant to blow out any dirt or foreign matter that may have accumulated during shipment or storage.

(4) Close the valves and wipe the connection seats with a clean cloth.

b. Pressure Regulators.

(1) Check the regulator fittings for dirt and obstructions. Also check threads of cylinders and regulators for imperfections.

(2) Connect the acetylene regulator to the acetylene regulator and the oxygen regulator to the oxygen cylinder. Use either a regulator wrench or a close fitting wrench and tighten the connecting nuts sufficiently to prevent leakage.

(3) Check hose for burns, nicks, and bad fittings.

(4) Connect the red hose to the acetylene regulator and the green hose to the oxygen regulator. Screw the connecting nuts tightly to insure leakproof seating. Note that the acetylene hose connection has left hand threads.

WARNING
If it is necessary to blow out the acetylene hose, do it in a well ventilated place which is free of sparks, flame, or other sources of ignition.

(5) Release the regulator screws to avoid damage to the regulators and gages. Open the cylinder valves slowly. Read the high pressure gages to check the cylinder gas pressure. Blow out the oxygen hose by turning the regulator screw in and then release the regulator screw. Flashback suppressors must be attached to the torch whenever possible.

c. Torch. Connect the red acetylene hose to the torch needle valve which is stamped "AC or flashback suppressor".Connect the green oxygen hose to the torch needle valve which is stamped "OX or flashback suppressor".Test all hose connections for leaks at the regulators and torch valves by turning both regulators' screws in with the torch needle valves closed. Use a soap and water solution to test for leaks at all connections. Tighten or replace connections where leaks are found. Release the regulator screws after testing and drain both hose lines by opening the torch needle valves. Slip the tip nut over the tip, and press the tip into the mixing head. Tighten by hand and adjust the tip to the proper angle. Secure this adjustment by tightening with the tip nut wrench.

WARNING
Purge both acetylene and oxygen lines (hoses) prior to igniting torch. Failure to do this can cause serious injury to personnel and damage to the equipment.

d. Adjustment of Working Pressure. Adjust the acetylene working pressure by opening the acetylene needle valve on the torch and turning the regulator screw to the right. Then adjust the acetylene regulator to the required pressure for the tip size to be used (tables 5-1 and 5-2). Close the needle valve. Adjust the oxygen working pressure in the same manner.

Table 5-1. Low Pressure or Injector Type Torch

Tip Size No.	Oxygen psi	Acetylene psi

NOTE
Tips are provided by a number of manufacturers, and sizes may vary slightly.

Tip Size No.	Oxygen psi	Acetylene psi
0	9	1
1	9	1
2	10	1
3	10	1
4	11	1
5	12	1
6	14	1
7	16	1
8	19	1
10	21	1
12	25	1
15	30	1

5-12.   SETTING UP THE WELDING EQUIPMENT (cont)

Table 5-2.   Balanced Pressure Type Torch

Tip Size No.	Oxygen psi	Acetylene psi

NOTE

Tips are provided by a number of manufacturers, and sizes may vary slightly.

Tip Size No.	Oxygen psi	Acetylene psi
1	2	2
2	2	2
3	3	3
3	3	3
5	3.5	3.5
6	3.5	3.5
7	5	5
8	7	7
9	9	9
10	12	12

5-13.   SHUTTING DOWN WELDING APPARATUS

a.   Shut off the gases. Close the acetylene valve first, then the oxygen valve on the torch. Then close the acetylene and oxygen cylinder valves.

b.   Drain the regulators and hoses by the following procedures:

(1) Open the torch acetylene valve until the gas stops flowing and the gauges read zero, then close the valve.

(2) Open the torch oxygen valve to drain the oxygen regulator and hose. When gas stops flowing and the gauges read zero, close the valve.

(3) When the above operations are performed properly, both high and low pressure gauges on the acetylene and oxygen regulators will register zero.

c.   Release the tension on both regulator screws by turning the screws to the left until they rotate freely.

d.   Coil the hoses without kinking them and suspend them on a suitable holder or hanger. Avoid upsetting the cylinders to which they are attached.

5-14.   REGULATORMALFUNCTIONS AND CORRECTIONS

a.   Leakage of gas between the regulator seat and the nozzle is the principal problem encounter with regulators. It is indicated by a gradual increase in pressure on the working pressure gauge when the adjusting screw is fully released or is in position after adjustment. This defect, called "creeping regulator", is caused by bad valve seats or by foreign matter lodged between the seat and the nozzle.

WARNING
Regulators with leakage of gas between the regulator seat and the nozzle must be replaced immediately to avoid damage to other parts of the regulator or injury to personnel. With acetylene regulators, this leakage is particularity dangerous because acetylene at high pressure in the hose is an explosion hazard.

b.   The leakage of gas, as described. above, can be corrected as outlined below:

(1) Remove and replace the seat if it is worn cracked, or otherwise damaged.

(2) If the malfunction is caused by fouling with dirt or other foreign matter, clean the seat and nozzle thoroughly and blow out any dust or dirt in the valve chamber.

c.   The procedure for removing valve seats and nozzles will vary with the make or design.

d. Broken or buckled gage tubes and distorted or buckled diaphragms are usually caused by backfire at the torch,leaks across the regulator seats, or by failure to release the regulator adjusting screw fully before opening the cylinder valves.

e.   Defective bourdon tubes in the gages are indicated by improper action of the gages or by escaping gas from the gage case. Gages with defective bourdon tubes should be removed and replaced with new gages. Satisfactory repairs cannot be made without special equipment.

f.   Buckled or distorted diaphragms cannot be adjusted properly and should be replaced with new ones. Rubber diaphragms can be replaced easily by removing the spring case with a vise or wrench. Metal diaphragms are sometimes soldered to the valve case and their replacement is a factory or special repair shop job. Such repairs should not be attempted by anyone unfamiliar with the work.

5-15.   TORCH MALFUNCTIONS AND CORRECTIONS

WARNING
Defects in oxyacetylene welding torches which are sources of gas leaks must be corrected immediately, as they may result in flashbacks or backfires, with resultant injury to the operator and/or damage to the welding apparatus.

a.   General. Improved functioning of welding torches is usually due to one or more of the following causes: leaking valves,leaks in the mixing head seat, scored or out-of-round welding tip orifices, clogged tubes or tips, and damaged inlet connection threads. Corrective measures for these common torch defects are described below.

b.   Leaking Valves.

(1) Bent or worn valve stems should be replaced and damaged seats should be refaced.

(2) Loose packing may be corrected by tightening the packing nut or by installing new packing and then tightening the packing nut.

CAUTION
This work should be done by the manufacturer because special reamers are required for trueing these seats.

c. Leaks in the Mixing Heads. These are indicated by popping out of the flame and by emission of sparks from the tips accompanied by a squealing noise. Leaks in the mixing head will cause improper mixing of the oxygen and acetylene causing flashbacks. A flashback causes the torch head and handle to suddenly become very hot. Repair by reaming out and trueing the mixing head seat.

d. Scored or Out-of-Round Tip Orifices. Tips in this condition cause the flame to be irregular and must be replaced.

e. Clogged Tubes and Tips.

(1) Carbon deposits caused by flashbacks or backfire, or the presence of foreign matter that has entered the tubes through the hoses will clog tubes. If the tubes or tips are clogged, greater working pressures will be needed to produce the flame required. The flame produced will be distorted.

(2) The torch should be disassembled so that the tip, mixing head, valves, and hose can be cleaned and cleaned out with compressed air at a pressure of 20 to 30 psi (137.9 to 206.85 kPa).

(3) The tip and mixing head should be cleaned either with a cleaning drill or with soft copper or brass wire and then blown out with compressed air. The cleaning drills should be approximately one drill size smaller than the tip orifice to avoid enlarging the orifice during cleaning.

WARNING
Damages inlet connection threads may cause fires by ignition of the leaking gas, resulting in injury to the welding operator and/or damage to the equipment.

f. Damaged Inlet Connection Threads. Leaks due to damaged inlet connection threads can be detected by opening the cylinder valves and keeping the needle valves closed. Such leaks will cause the regulator pressure to drop. Also, if the threads are damaged, the hose connection at the torch inlet will be difficult or impossible to tighten. To correct this defect, the threads should be recut and the hose connections thoroughly cleaned.

Section II. OXYACETYLENE CUTTING EQUIPMENT

5-16. CUTTING TORCH AND OTHER CUTTING EQUIPMENT

a. The cutting torch (fig. 5-12), like the welding torch, has a tube for oxygen and one for acetylene. In addition, there is a tube for high pressure oxygen, along with a cutting tip or nozzle. The tip (fig. 5-13) is provided with a center hole through which a jet of pure oxygen passes. Mixed oxygen and acetylene pass through holes surrounding the center holes for the preheating flames. The number of orifices for oxyacetlylene flames ranges from 2 to 6, depending on the purpose for which the tip is used. The cutting torch is controlled by a trigger or lever operated valve. The cutting torch is furnished with interchangeable tips for cutting steel from less than 1/4 in. (6.4 mm) to more than 12.0 in. (304.8 mm) in thickness.

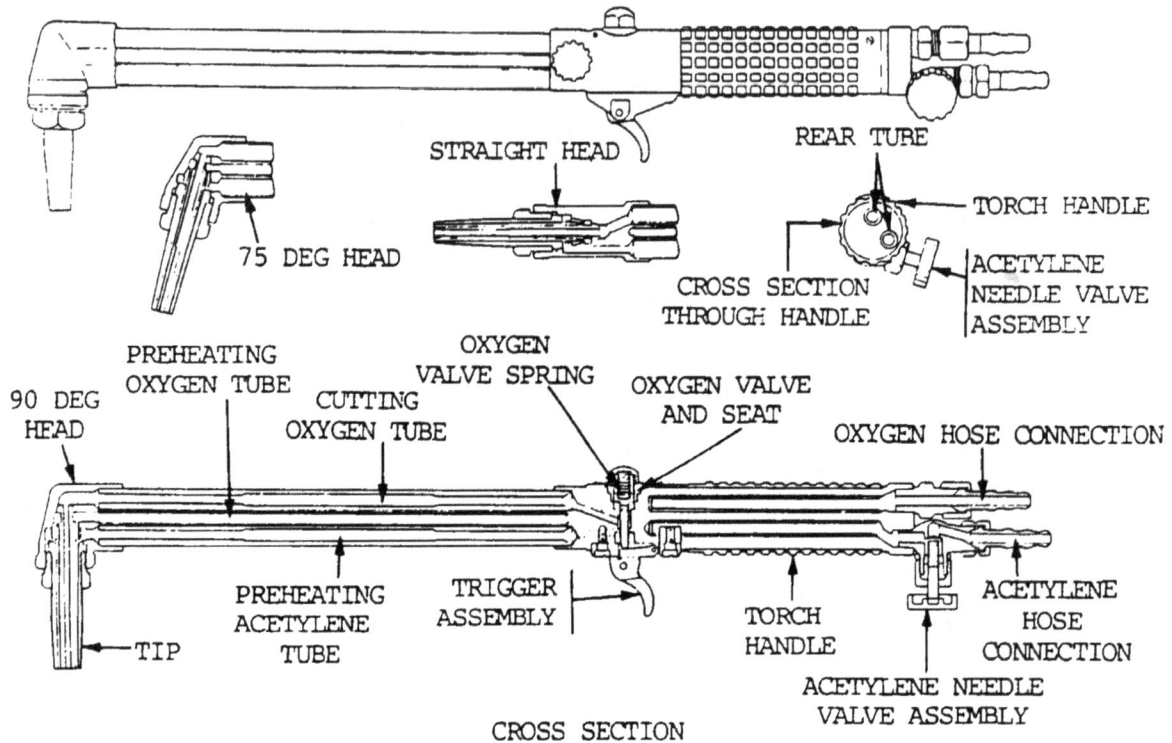

Figure 5-12. Oxyacetylene cutting torch.

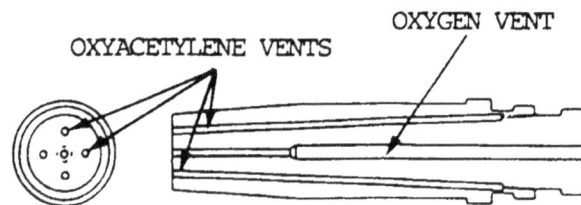

Figure 5-13. Diagram of oxyacetylene cutting tip.

5-16. CUTTING TORCH AND OTHER CUTTING EQUIPMENT (cont)

b. A cutting attachment fitted to a welding torch in place of the welding tip is shown in figure 5-14.

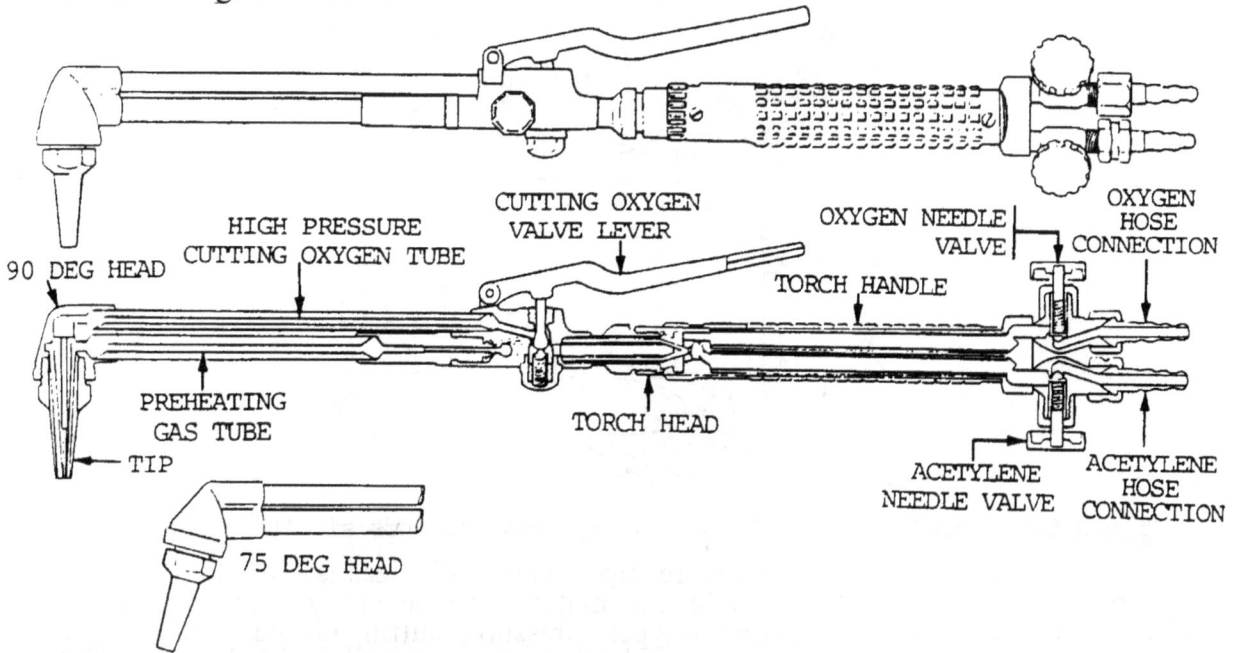

Figure 5-14. Cutting attachment for welding torch.

c. In order to make uniformly clean cuts on steel plate, motor driven cutting machines are used to support and guide the cutting torch. Straight line cutting or beveling is acccomplished by guiding the machine along a straight line on steel tracks. Arcs and circles are cut by guiding the machine with a radius rod pivoted about a central point. Typical cutting machines in operation are shown in figures 5 15 and 5-16.

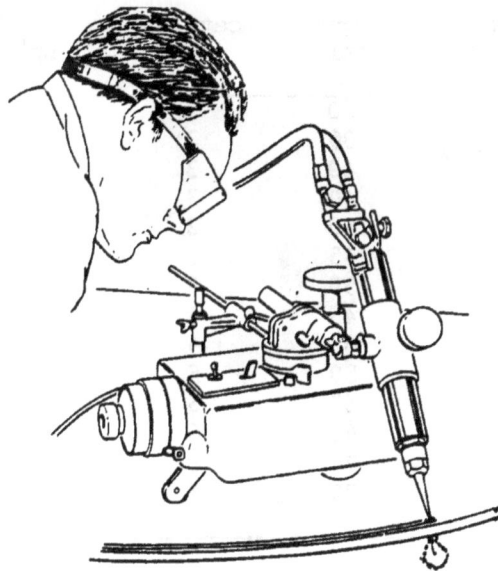

5-15. Making a bevel on a circular path with a cutting machine.

Figure 5-16. Machine for making four oxyacetylene cuts simultaneously.

d. There is a wide variety of cutting tip styles and sizes available to suit various types of work. The thickness of the material to be cut generally governs the selection of the tip. The cutting oxygen pressure, cutting speed, and preheating intensity should be controlled to produce narrow, parallel sided kerfs. Cuts that are improperly made will produce ragged, irregular edges with adhering slag at the bottom of the plates. Table 5-3 identifies cutting tip numbers, gas pressures, and hand-cutting speeds used for cutting mild steel up to 12 in. (304.8 mm) thick.

Table 5-3. Oxyacetylene Cutting Information

Plate thickness (in.)	Cutting tip[1] (size number)	Oxygen (psi)	Acetylene (psi)	Hand-cutting speed (in. per minute)
1/4	0	30	3	16.0 to 18.0
3/8	1	30	3	14.5 to 16.5
1/2	1	40	3	12.0 to 14.5
3/4	2	40	3	12.0 to 14.5
1	2	50	3	8.5 to 11.5
1-1/2	3	45	3	6.0 to 7.5
2	4	50	3	5.5 to 7.0
3	5	45	4	5.0 to 6.5
4	5	60	4	4.0 to 5.0
5	6	50	5	3.5 to 4.5
6	6	55	5	3.0 to 4.0
8	7	60	6	2.5 to 3.5
10	7	70	6	2.0 to 3.0
12	8	70	6	1.5 to 2.0

[1]Various manufacturers do not adhere to the numbering of tips as set forth in this table; therefore, some tips may carry different identification numbers.

5-17.   OPERATION OF CUTTING EQUIPMENT

Attach the required cutting tip to the torch and adjust the oxygen and acetylene pressures in accordance with table 5-3.

NOTE

The oxygen and acetylene gas pressure settings listed are only approximate. In actual use, pressures should be set to effect the best metal cut .

b.   Adjust the preheating flame to neutral.

c.   Hold the torch so that the cutting oxygen lever or trigger can be operated with one hand. Use the other hand to steady and maintain the position of the torch head to the work. Keep the flame at a 90 degree angle to work in the direction of travel.   The inner cones of the preheating flames should be about 1/16 in. (1.6 mm) above the end of the line to be cut. Hold this position until the spot has been raised to a bright red heat, and then slowly open the cutting oxygen valve.

d.   If the cut has been started properly, a shower of sparks will fall from the opposite side of the work. Move the torch at a speed which will allow the cut to continue penetrating the work. A good cut will be clean and narrow.

e.   When cutting billets, round bars, or heavy sections, time and gas are saved if a burr is raised with a chisel at the point where the cut is to start. This small portion will heat quickly and cutting will start immediately. A welding rod can be used to start a cut on heavy sections.When used, it is called a starting rod.

Section III.   ARC WELDING EQUIPMENT AND ACCESSORIES

5-18. GENERAL

In electric welding processes, an arc is produced between an electrode and the work piece (base metal). The arc is formed by passing a current between the electrode and the workpiece across the gap. The current melts the base metal and the electrode (if the electrode is a consumable type), creating a molten pool. On solidifying, the weld is formal. An alternate method employs a nonconsumable electrode, such as a tungsten rod. In this case, the weld is formed by melting and solidifying the base metal at the joint. In some instances, additional metal is required, and is added to the molten pool from a filler rod.

Electrical equipment required for arc welding depends on the source from which the electric power is obtained. If the power is obtained from public utility lines, one or more of the following devices are required: transformers (of which there are several types), rectifiers, motor generators, and control equipment. If public utility power is not available, portable generators driven by gasoline or diesel engines are used.

5-19.   DIRECT CURRENT ARC WELDING MACHINES

a.   The direct current welding machine has a heavy duty direct current generator 5-17).   The generators are made in six standardized ratings for general purposes as described below:

(1) The machines rated 150 and 200 amperes, 30 volts, are used for light shielded metal-arc welding and for gas metal-arc welding. They are also used for general purpose job shop work.

(2) The machines rated 200, 300, and 400 amperes, 40 volts, used for general welding purposes by machine or manual application.

(3) Machines rated 600 amperes, 40 volts, are used for submerged arc welding and for carbon-arc welding.

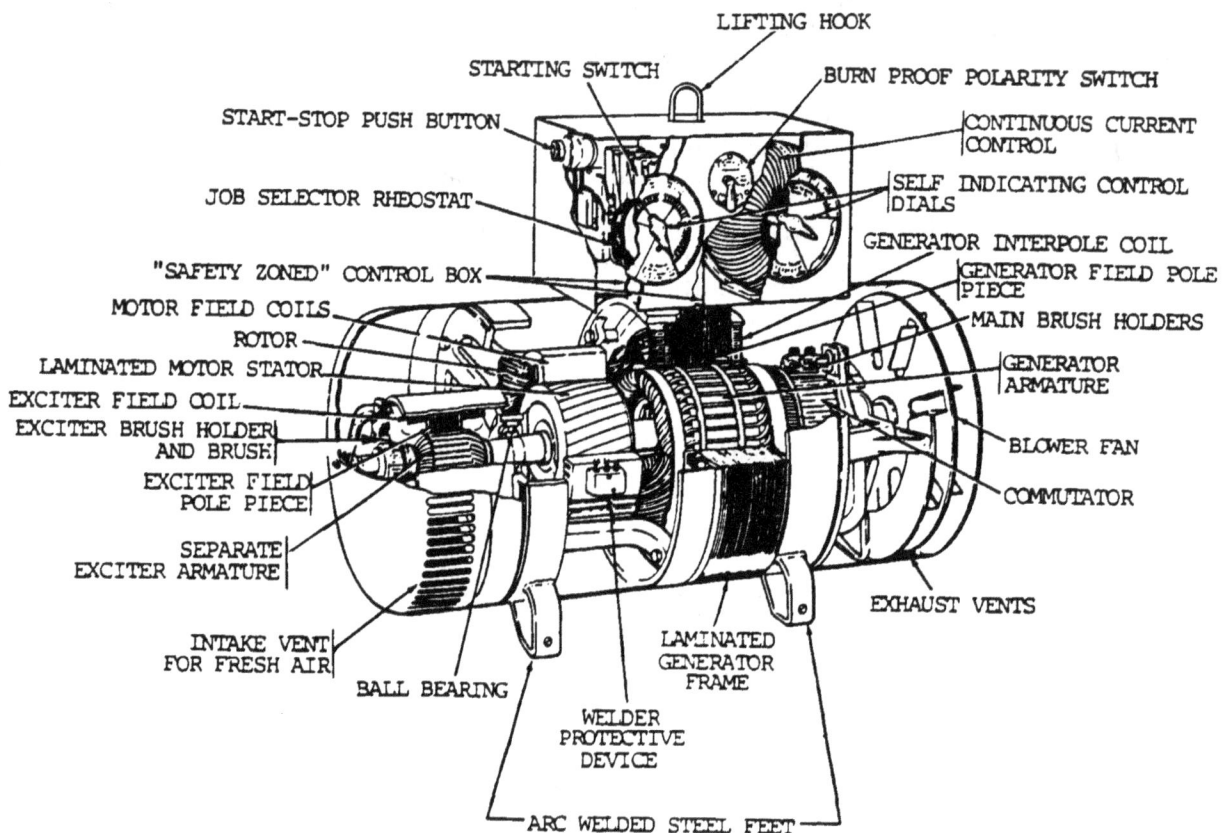

Figure 5-17. Cutaway view of DC welding generator.

b. The electric motors must commonly used to drive the welding generators are 220/440 volts, 3 phase, 60 cycle. The gasoline and diesel engines should have a rated horsepower in excess of the rated output of the generator. This will allow for the rated overload capacity of the generator and for the power required to operate the accessories of the engine. The simple equarion HP = 1.25P/746 can be used; HP is the engine horsepower and P is the generator rating in watts. For exapmle, a 20 horsepower engine would be used to drive a welding generator with a rated 12 kilowatt output.

5-19. DIRECT CURRENT ARC WELDING MACHINES (cont)

c. In most direct current welding machines, the generator is of the variable voltage type, and is arranged so that the voltage is automatically adjusted to the demands of the arc. However, the voltage may be set manually with a rheostat.

d. The welding current amperage is also manually adjustable, and is set by means of a selector switch or series of plug receptacles. In either case, the desired amperage is obtained by tapping into the generator field coils. When both voltage and amperage of the welding machine are adjustable, the machine is known as dual control type. Welding machines are also manufactured in which current controls are maintained by movement of the brush assembly.

e. A direct current welding machine is described in TM 5-3431-221-15, and is illustrated in figure 5-18.

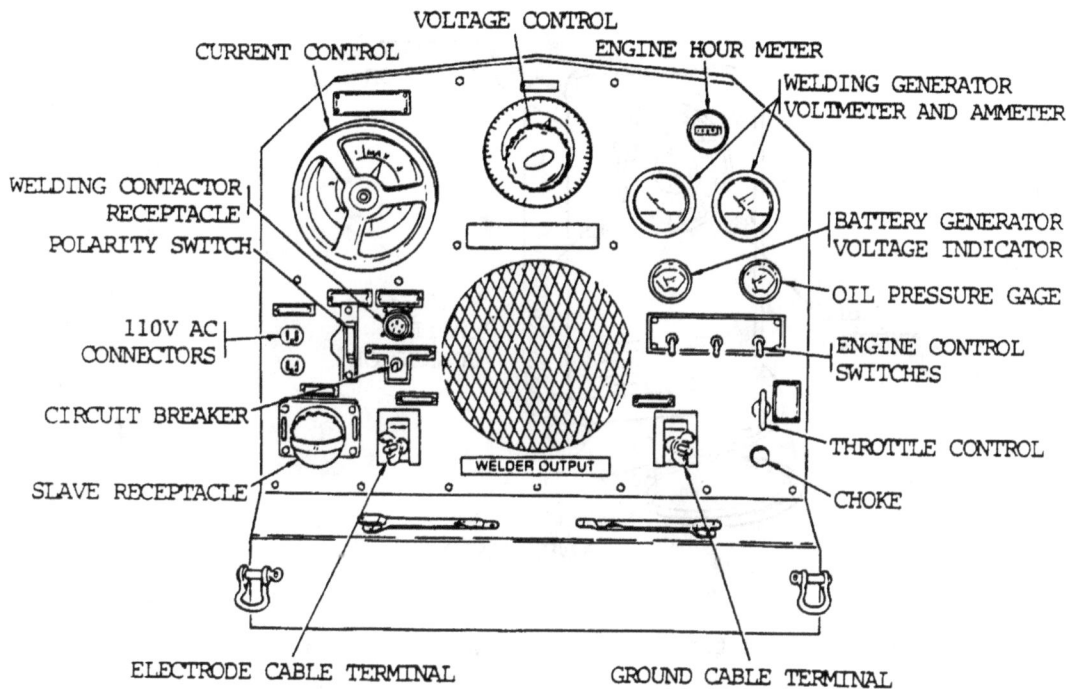

Figure 5-18. Direct current welding machine.

f. A maintenance schedule should be set up to keep the welding machine in good operating condition. The machine should be thoroughly inspected every 3 months and blown free of dust with clean, dry, compressed air. At least once each year, the contacts of the motor starter switches and the rheostat should be cleaned and replaced if necessary. Brushes should be inspected frequently to see if they are making proper contact on the commutator and that they move freely in the brush holders. Clean and true the commutator with sandpaper or a commutator stone if it is burned or roughened. Check the bearings twice a year. Remove all the old grease and replace it with new grease.

g. Direct current rectifier type welding machines have been designed with copper oxide, silicon, or selenium dry plates. These machines usually consist of a transform to reduce the power line voltage to the required 220/440 volts, 3 phase, 60 cycle input current; a reactor for adjustment of the current; and a rectifier to change the alternating current to direct current. Sometimes another reactor is used to reduce ripple in the output current.

5-20. ALTERNATING CURRENT ARC WELDING MACHINES

a. Most of the alternating current arc welding machines in use are of the single operator, static transformer type (fig. 5-19). For manual operation in industrial applications, machines having 200, 300, and 400 amphere ratings are the sizes in general use. Machines with 150 ampere ratings are sometimes used in light industrial, garage and job shop welding.

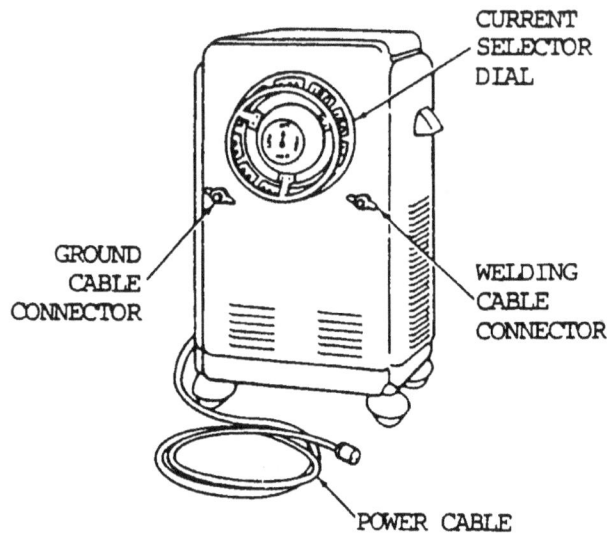

Figure 5-19. Alternating current arc welding machine.

b. The transformers are generally equipped with arc stabilizing capacitors. Current control is provided in several ways. One such method is by means of an adjustable reactor in the output circuit of the transformer. In other types, internal reactions of the transformer are adjustable. A handwheel, usually installed on the front or the top of the machine, makes continuous adjustment of the output current, without steps, possible.

c. The screws and bearings on machines with screw type adjustments should be lubricated every 3 months. The same lubrication schedule applies to chain drives. Contacts, switches, relays, and plug and jack connections should be inspected every 3 months and cleaned or replaced as required. The primary input current at no load should be measured and checked once a year to ensure the power factor connecting capacitors are working, and that input current is as specified on the nameplate or in the manufacturer's instruction book.

5-21.   GAS TUNGSTEN-ARC WELDING (GTAW) EQUIPMENT (TIG)

a.   General.   In tungsten inert gas (TIG) welding, (also known as GTAW), an arc is struck between a virtually nonconsumable tungsten electrode and the workpiece. The heat of the arc causes the edges of the work to melt and flow together. Filler rod is often required to fill the joint. During the welding operation, the weld area is shielded from the atmosphere by a blanket of inert argon gas. A steady stream of argon passes through the torch, which pushes the air away from the welding area and prevents oxidation of the electrode, weld puddle, and heat affected zone.

b.   Equipment.

(1) The basic equipment requirements for manual TIG welding are shown in figure 5-20. Equipment consists of the welding torch plus additional apparatus to supply electrical power, shielding gas, and a water inlet and outlet. Also, personal protective equipment should be worn to protect the operator from the arc rays during welding operations.

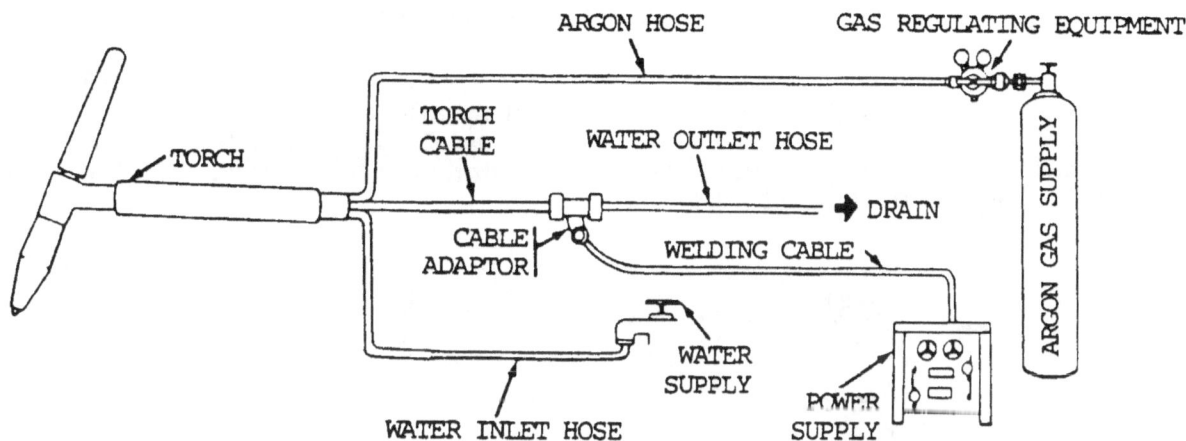

Figure 5-20. Gas tungsten-arc welding setup.

NOTE
Different types of TIG welding equipment are available through normal supply channels. Water-cooled torches and air-cooled torches are both available. Each type carries different amperage ratings. Consult the appropriate manual covering the type torch used.

(2) Argon is supplied in steel cylinders containing approximately 330 cu ft at a pressure to 2000 psi (13,790 kPa). A single or two stage regulator may be used to control the gas flow. A specially designed regulator containing a flowmeter, as shown in figure 5-21, may be used. The flowmeter provides better adjustment via flow control than the single or two stage regulator and is calibrated in cubic feet per hour (cfh). The correct flow of argon to the torch is set by turning the adjusting screw on the regulator.The rate of flow depends on the kind and thickness of the metal to be welded.

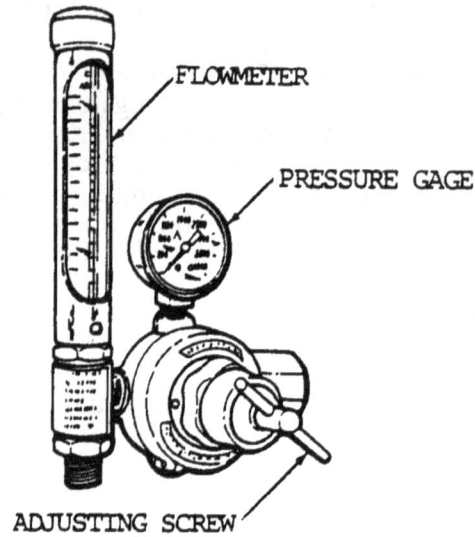

Figure 5-21. Argon regulator with flowmeter.

(3) Blanketing of the weld area is provided by a steady flow of argon gas directed through the welding torch (fig. 5-22). Since argon is slightly more than 1-1/3 times as heavy as air, it pushes the lighter air molecules aside, effectively preventing oxidation of the welding electrode, the molten weld puddle, and the heat affected zone adjacent to the weld bead.

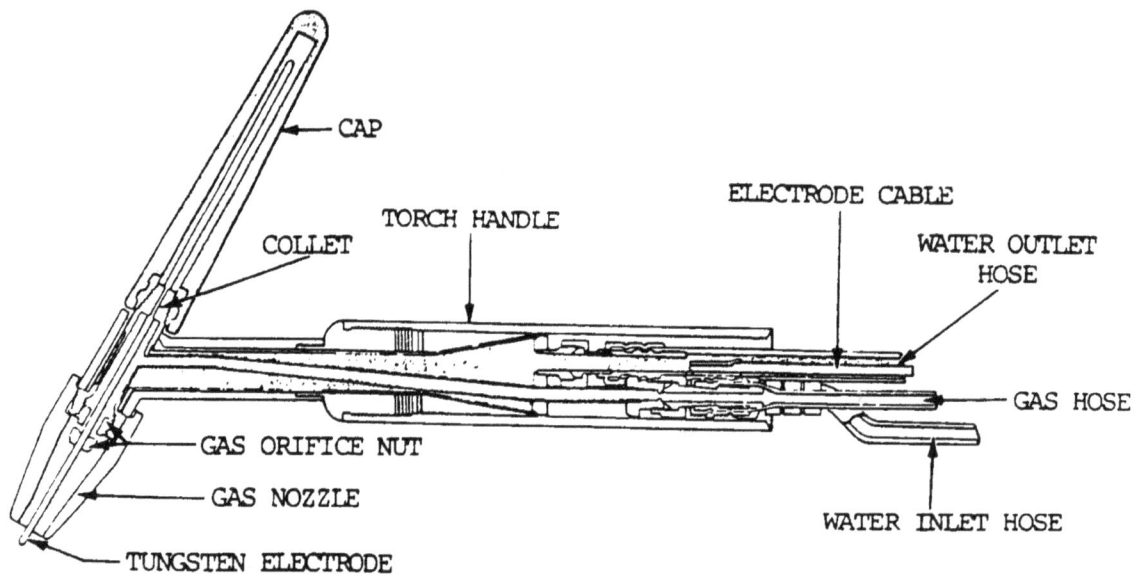

Figure 5-22. TIG welding torch.

5-21. GAS TUNGSTEN-ARC WELDING (GTAW) EQUIPMENT (TIG) (cent)

(4) The tremendous heat of the arc and the high current often used usually necessitate water cooling of the torch and power cable (fig. 5-22). The cooling water must be clean; otherwise, restricted or blocked passages may cause excessive overheating and damage to the equipment. It is advisable to use a suitable water strainer or filter at the water supply source. If a self-contained unit is used, such as the one used in the field (surge tank) where the cooling water is recirculated through a pump, antifreeze is required if the unit is to be used outdoors during the winter months or freezing weather. Some TIG welding torches require less than 55 psi (379 kPa) water pressure and will require a water regulator of some type. Check the operating manual for this information.

c. Nomenclature of Torch (fig. 5-22).

(1) Cap. Prevents the escape of gas from the top of the torch and locks the electrode in place.

(2) Collet. Made of copper; the electrode fits inside and when the cap is tightened, it squeezes against the electrode and leeks it in place.

(3) Gas orifice nut. Allows the gas to escape.

(4) Gas nozzle. Directs the flew of shielding gas onto the weld puddle. Two types of nozzles are used; the one for light duty welding is made of a ceramic material, and the one for heavy duty welding is a copper water-cooled nozzle.

(5) Hoses. Three plastic hoses, connected inside the torch handle, carry water, gas, and the electrode power cable.

5-22. GAS METAL-ARC WELDING (GMAW) EQUIPMENT

a. General. GMAW is most commonly referred to as "MIG" welding, and the following text will use "MIG" or "MIG welding" when referring to GMAW. MIG welding is a process in which a consumable, bare wire electrode is fed into a weld at a controlled rate of speed, while a blanket of inert argon gas shields the weld zone from atmospheric contamination. In addition to the three basic types of metal transfer which characterize the GMAW process, there are several variations of significance.

(1) Pulsed spray welding. Pulsed spray welding is a variation of the MIG welding process that is capable of all–position welding at higher energy levels than short circuiting arc welding. The power source provides two current levels; a steady "background" level, which is too low to produce spray transfer; and a "pulsed peak" current, which is superimposed upon the background current at a regulated interval. The pulse peak is well above the transition current, and usually one drop is transferred during each pulse. The combination of the two levels of current produces a steady arc with axial spray transfer at effective welding currents below those required for conventional spray arc welding. Because the heat input is lower, this variation in operation is capable of welding thinner sections than are practical with the conventional spray transfer.

(2) Arc spot welding. Gas metal arc spot welding is a method of joining similar to resistance spot welding and riveting. A variation of continuous gas

metal arc welding, the process fuses two pieces of sheet metal together by penetrating entirely through one piece into the other.No joint preparation is required other than cleaning of the overlap areas.The welding gun remains stationary while a spot weld is being made. Mild steel, stainless steel, and aluminum are commonly joined by this method.

(3) Electrogas welding. The electrogas (EG) variation of the MIG welding process is a fully automatic, high deposition rate method for the welding of butt, corner, and T-joints in the vertical position. The eletrogas variation essentially combines the mechanical features of electroslag welding (ESW) with the MIG welding process. Water-coded copper shoes span the gap between the pieces being welded to form a cavity for the molten metal.A carriage is mounted on a vertical column; this combination provides both vertical and horizontal movementWelding head, controls, and electrode spools are mounted on the carriage. Both the carriage and the copper shoes move vertically upwards as welding progresses. The welding head may also be oscillated to provide uniform distribution of heat and filler metal. This method is capable of welding metal sections of from 1/2 in. (13 mm) to more than 2 in. (5.08 an)in thickness in a single pass. Deposition rates of 35 to 46 lb (16 to 21 kg) perhour per electrode can be achieved.

b. MIG Equipment.

NOTE
Different types of MIG welding equipment are available through normal supply channels. Manuals for each type must be consulted prior to welding operations.

(1) The MIG welding unit is designed for manual welding with small diameter wire electrodes, using a Spool-on-gun torch. The unit consists of a torch (fig. 5-23), a voltage control box, and a welding contractor (fig. 5-24). The torch handle contains a complete motor and gear reduction unit that pulls the welding wire electrode from a 4 in. (102 mm) diameter spool containing 1 lb (0.5 kg) of wire electrode mounted in the rear of the torch.

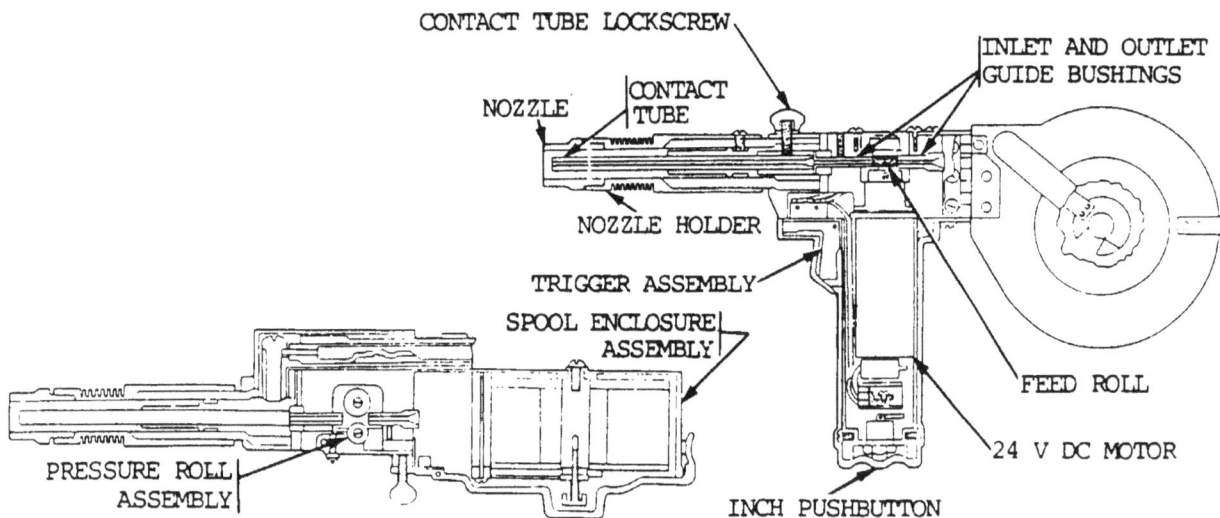

Figure 5-23. MIG welding torch.

5-21. GAS TUNGSTEN-ARC WELDING (GTAW) EQUIPMENT (TIG) (cont)

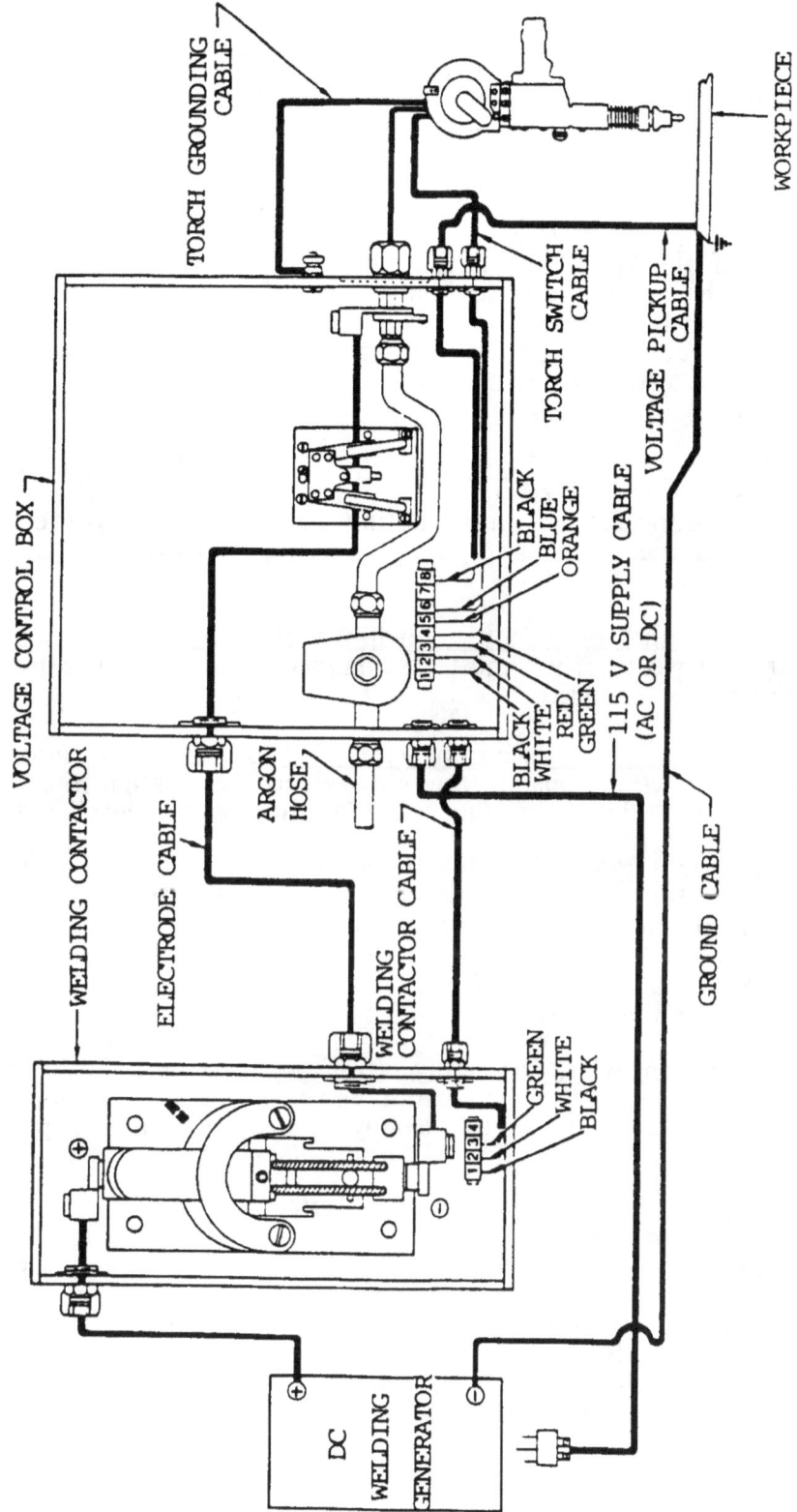

Figure 5-24. Connection diagram for MIG welding.

(2) Three basic sizes of wire electrode maybe used: 3/32 in. (2.38 mm), 3/64 in. (1.19 mm), and 1/16 in. (1.59 mm). Many types of metal may be welded provided the welding wire electrode is of the same composition as the base metal.

(3) The unit is designed for use with an ac-dc conventional, constant-current welding power supply. Gasoline engine-driven arc welding machines issued to field units may be used as both a power source and a welding source.

c. Nomenclature of Torch.

(1) Contact tube (fig. 5-23). This tube is made of copper and has a hole in the center of the tube that is from 0.01 to 0.02 in(0.25 to 0.51 mm) larger than the size of the wire electrode being used. The contact tube and the inlet and outlet guide bushings must be charged when the size of the wire electrode is changed. The contact tube transfers power from the electrode cable to the welding wire electrode. An insulated lock screw is provided which secures the contact tube in the torch.

(2) Nozzle and holder (fig. 5-23). The nozzle is made of copper to dissipate heat and is chrome-plated to reflect the heat. The holder is made of stainless steel and is connected to an insulating material which prevents an arc from being drawn between the nozzle and the ground in case the gun canes in contact with the work.

(3) Inlet and outlet guide bushings (fig. 5-23). The bushings are made of nylon for long wear. They must be changed to suit the wire electrode size when the electrode wire is changed.

(4) Pressure roll assembly (fig. 5-23). This is a smooth roller, under spring tension, which pushes the wire electrode against the feed roll and allows the wire to be pulled from the spool. A thumbscrew applies tension as required.

(5) Motor (fig. 5-23). When the inch button is depressed, the current for running the motor comes from the 110 V ac-dc source and the rotor pulls the wire electrode from the spool before starting the welding operation. When the trigger is depressed, the actual welding operation starts and the motor pulls the electrode from the spool at the required rate of feed. The current for this rotor is supplied by the welding generator.

(6) Spool enclosure assembly (fig. 5-23). This assembly is made of plastic which prevents arc spatter from jamming the wire electrode on the spool. A small window allows the operator to visually check the amount of wire electrode remaining on the spool.

NOTE

If for any reason the wire electrode stops feeding, a burn-back will result. With the trigger depressed, the welding contactor is closed, thereby allowing the welding current to flow through the contact tube. As long as the wire electrode advances through the tube, an arc will be drawn at the end of the wire electrode. Should the wire electrode stop feeding while the trigger is still being depressed, the arc will then form at the end of the contact tube, causing it to melt off. This is called burn-back.

5-21.   GAS TUNGSTEN-ARC WELDING (GTAW) EQUIPMENT (TIG) (cent)

(7) Welding contactor (fig. 5-24). The positive cable from the dc welding generator is connected to a cable coming out of the welding contactor, and the ground cable is connected to the workpiece. The electrode cable and the welding contactor cable are connected between the welding contactor and voltage control box as shown.

(8) Argon gas hose (fig. 5-24). This hose is connected from the voltage control box to the argon gas regulator on the argon cylinder.

(9) Electrode cable (fig. 5-24). The electrode cable enters through the welding current relay and connects into the argon supply line. Both then go out of the voltage control box and into the torch in one line.

(10) Voltage pickup cable (fig. 5-24). This cable must be attached to the ground cable at the workpiece. This supplies the current to the motor during welding when the trigger is depressed.

(11) Torch switch and grounding cables (fig. 5-24). The torch switch cable is connected into the voltage control box, and the torch grounding cable is connected to the case of the voltage control box.

5-23.   OPERATING THE MIG

a.   Starting to Weld.

(1) Press the inch button and allow enough wire electrode to emerge from the nozzle until 1/2 in. (13 mm) protrudes beyond the end of the nozzle. With the main line switch "ON" and the argon gas and power sources adjusted properly, the operator may begin to weld.

(2) When welding in the open air, a protective shield must be installed to prevent the argon gas from being blown away from the weld zone and allowing the weld to become contaminate.

(3) Press the torch trigger. This sends current down the torch switch cable and through the contactor cable, closing the contactor.

(4) When the contactor closes, the welding circuit from the generator to the welding torch is completed.

(5) At the same time the contactor closes, the argon gas solenoid valve opens, allowing a flow of argon gas to pass out of the nozzle to shield the weld zone.

(6) Lower the welding helmet and touch the end of the wire electrode to the workpiece. The gun is held at a 90 degree angle to the work but pointed at a 10 degree angle toward the line of travel.

CAUTION

To prevent overloading the torch motor when stopping the arc, release the trigger; never snap the arc out by raising the torch without first releasing the trigger.

(7) Welding will continue as long as the arc is maintained and the trigger is depressed.

b. Setting the Wire Electrode Feed.

(1) A dial on the front of the voltage control box, labeled WELDING CONTROL, is used to regulate the speed of the wire electrode feed.

(2) To increase the speed of the wire electrode being fed from the spool. turn the dial counterclockwise. This decreases the amount of resistance across the arc and allows the motor to turn faster.Turning the dial clockwise will increase the amount of resistance, thereby decreasing the speed of the wire electrode being fed from the spool.

(3) At the instant that the wire electrode touches the work, between 50 and 100 volts dc is generated. This voltage is picked up by the voltage pickup cable and shunted back through the voltage control box into a resistor.There it is reduced to the correct voltage (24 V dc) and sent to the torch motor.

c. Fuses.

(1) Two 10-ampere fuses, located at the front of the voltage control box, protect and control the electrical circuit within the voltage control box.

(2) A 1-ampere fuse, located on the front of the voltage control box, protects and controls the torch motor.

d. Installing the Wire Electrode.

(1) Open the spool enclosure cover assembly, brake, and pressure roll assembly (fig. 5-23).

(2) Unroll and straighten 6 in.(152 mm) of wire electrode from the top of the spool.

(3) Feed this straightened end of the wire electrode into the inlet and outlet bushings; then place spool onto the mounting shaft.

(4) Close the Pressure roller and secure it in place. Press the inch button. feeding the wire electrode until there is 1/2 in. (13 mm) protruding beyond the end to the nozzle.

e. Setting the Argon Gas Pressure.

(1) Flip the argon switch on the front of the voltage control panel to the MANUAL position.

(2) Turn on the argon gas cylinder valve and set the pressure on the regulator.

(3) When the proper pressure is set on the regulator, flip the argon switch to the AUTOMATIC POSITION.

5-23. OPERATING THE MIG TORCH (cont)

(4) When in the MANUAL position the argon gas continues to flow. When in the AUTOMATIC position the argon gas flows only when the torch trigger is depressed, and stops flowing when the torch trigger is released.

f. Generator Polarity. The generator is set on reverse polarity. When set on straight polarity, the torch motor will run in reverse, withdrawing the wire electrode and causing a severe burn-back.

g. Reclaiming Burned-Back Contact Tubes. When the contact tubes are new, they are 5-3/8 in. (137 mm) long. When burn-backs occur, a maximum of 3/8 in. (9.5 mm) may be filed off. File a flat spot on top of the guide tube, place a drill pilot on the contact tube, then drill out the contact tube. For a 3/64 in. (1.2 mm) contact tube, use a No. 46 or 47 drill bit.

h. Preventive Maintenance.

(1) Keep all weld spatter cleaned out of the inside of the torch. Welding in the vertical or overhead positions will cause spatter to fall down inside the torch nozzle holder and restrict the passage of the argon gas. Keep all hose connections tight.

(2) To replace the feed roll, remove the nameplate on top of the torch, the flathead screw and retainer from the feed roll mounting shaft, and the contact ring and feed roll. Place a new feed roll on the feed roll mounting shaft, making certain that the pins protruding from the shaft engage the slots in the feed roll. Reassemble the contact ring and nameplate.

5-24. OTHER WELDING EQUIPMENT

a. Cables. Two welding cables of sufficient current carrying capacity with heavy, tough, resilient rubber jackets are required. One of the cables should be composed of fine copper strands to permit as much flexibility as the size of the cable will allow. One end of the less flexible cable is attached to the ground lug or positive side of the direct current welding machine; the other end to the work table or other suitable ground. One end of the flexible cable is attached to the electrode holder and the other end to the negative side of a direct current welding machine for straight polarity. Most machines are equipped with a polarity switch which is used to change the polarity without interchanging the welding cables at the terminals of the machine. For those machines not equipped with polarity switches, for reverse polarity, the cables are reversed at the machine.

b. Electrode Holders. An electrode holder is an insulated clamping device for holding the electrode during the welding operation. The design of the holder depends on the welding process for which it is used, as explained below.

(1) Metal-arc electrode holder. This is an insulated clamp in which a metal electrode can be held at any desired angle. The jaws can be opened by means of a lever held in place by a spring (fig. 5-25).

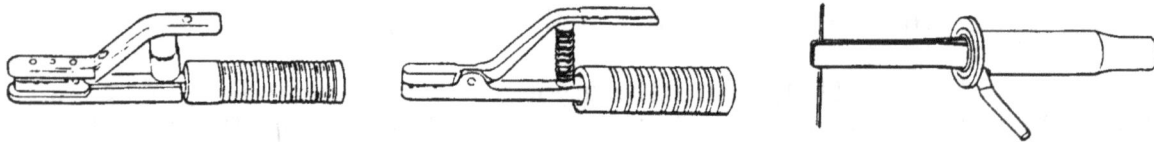

Figure 5-25. Metal-arc welding electrode holders.

(2) Atomic hydrogen torch. This electrode holder or torch consists of two tubes in an insulated handle, through which both hydrogen gas and electric current flow. The hydrogen is supplied to a tube in the rear of the handle from which it is led into the two current carrying tubes by means of a manifold. One of the two electrode holders is movable, and the gap between this and the other holder is adjusted by means of a trigger on the handle (fig. 5-26).

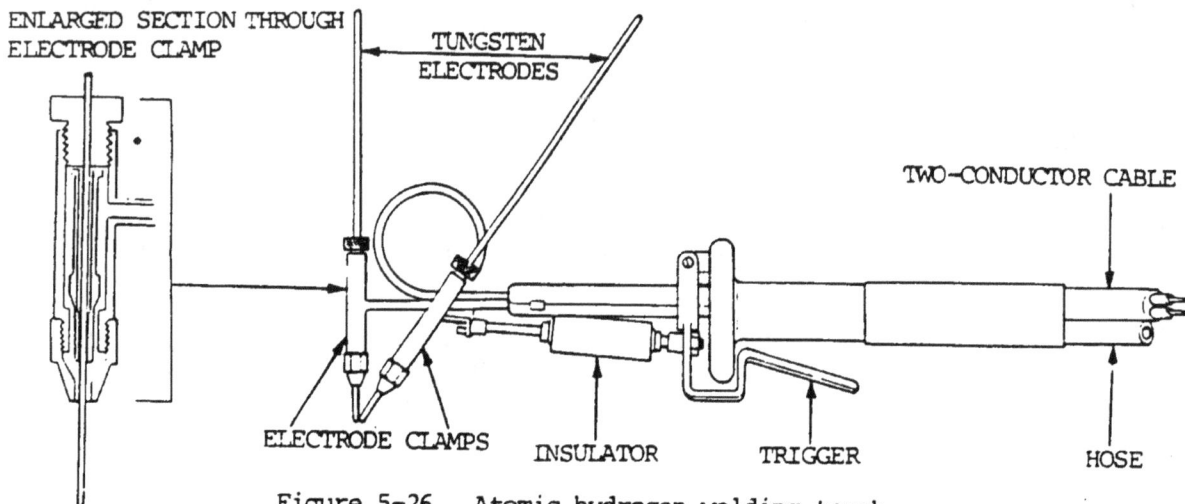

Figure 5-26. Atomic hydrogen welding torch.

(3) Carbon-arc electrode holder. This holder is manufactured in three specific types. One type holds two electrodes and is similar in design to the atomic hydrogen torch, but has no gas tubes; a second type is a single electrode holder equipped with a heat shield; the third type is for high amperage welding and is watercooled.

c. Accessories.

(1) Chipping hammer and wire brush. A chipping hammer is required to loosen scale, oxides and slag. A wire brush is used to clean each weld bead before further welding. Figure 5-27 shows a chipping hammer with an attachable wire brush.

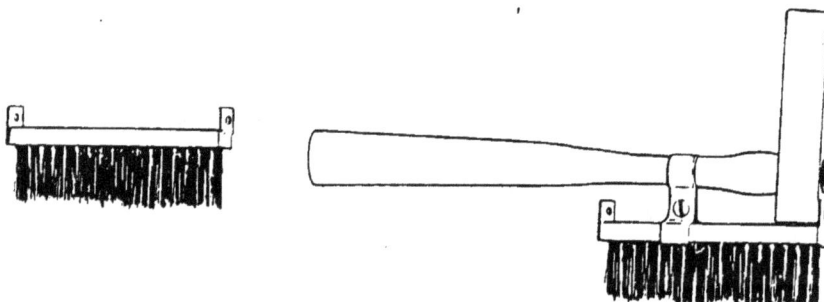

Figure 5-27. Chipping hammer and wire brush.

5-24.   OTHER WELDING EQUIPMENT (cont)

(2) Welding table. A welding table should be of all-steel construction. A container for electrodes with an insulated hook to hold the electrode holder when not in use should be provided. A typical design for a welding table is shown in figure 5-28.

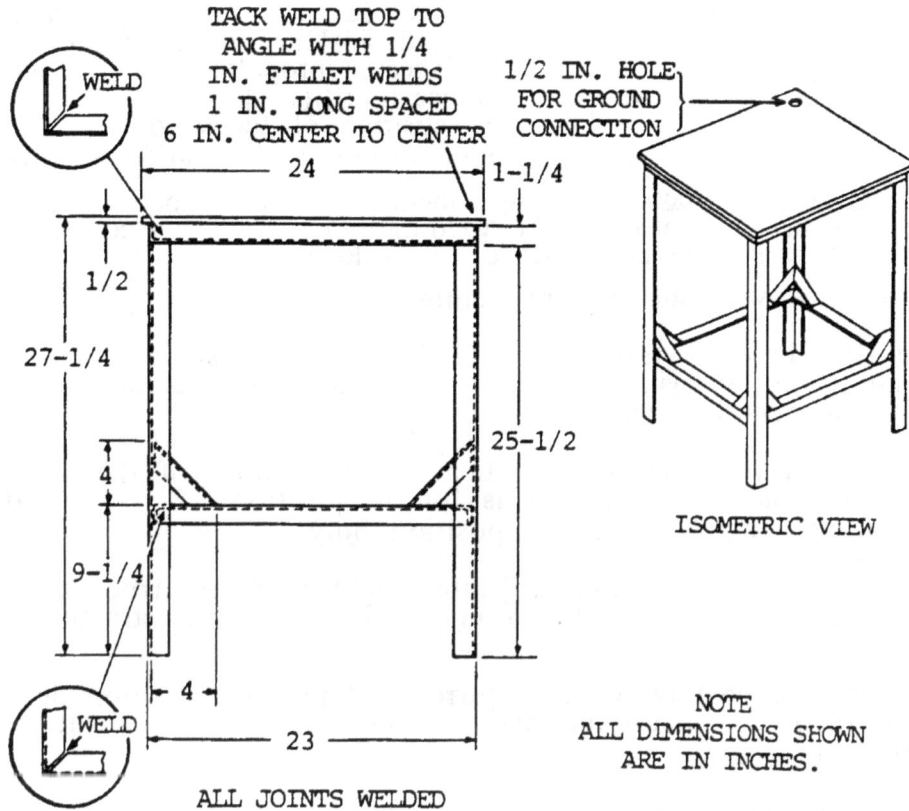

Figure 5-28. Welding table.

(3) Clamps and backup bars. Workpieces for welding should be clamped in position with C-clamps or other clamp brackets. Blocks, strips, or bars of copper or cast iron should be available for use as backup bars in welding light sheet aluminum and in making certain types of joints. Carbon blocks, fire clay, or other fire-resistant material should also be available. These materials are used to form molds which hold molten metal within given limits when building up sections. A mixture of water, glass, and fire clay or carbon powder can be used for making molds .

d. Goggles. Goggles with green lenses shaped to cover the eye orbit should be available to provide glare protection for personnel in and around the vicinity of welding and cutting operations (other than the welder).

NOTE
These goggles should not be used in actual welding operations.

5-25. ELECTRODES AND THEIR USE

a. General. When molten metal is exposed to air, it absorbs oxygen and nitrogen, and becomes brittle or is otherwise adversely affected. A slag cover is needed to protect molten or solidifying weld metal from the atmosphere. This cover can be obtained from the electrode coating, which protects the metal from damage, stabilizes the arc, and improves the weld in the ways described below.

b. Types of Electrodes. The metal-arc electrodes may be grouped and classified as bare electrodes, light coated electrodes and shielding arc or heavy coated electrodes. The type used depends on the specific properties required in the weld deposited. These include corrosion resistance, ductility, high tensile strength, the type of base metal to be welded; the position of the weld (i.e., flat, horizontal, vertical, or overhead); and the type of current and polarity required.

c. Classification of Electrodes. The American Welding Society's classification number series has been adopted by the welding industry. The electrode identification system for steel arc welding is set up as follows:

(1) E indicates electrode for arc welding.

(2) The first two (or three) digits indicate tensile strength (the resistance of the material to forces trying to pull it apart) in thousands of pounds per square inch of the deposited metal.

(3) The third (or fourth) digit indicates the position of the weld. 0 indicates the classification is not used; 1 is for all positions; 2 is for flat and horizontal positions only; 3 is for flat position only.

(4) The fourth (or fifth) digit indicates the type of electrode coating and the type of power supply used; alternating or direct current, straight or reverse polarity.

(5) The types of coating, welding current and polarity position designated by the fourth (or fifth) identifying digit of the electrode classification are as listed in table 5-4.

(6) The number E601O indicates an arc welding electrode with minimum stress relieved tensile strength of 60,000 psi; is used in all positions and reverse polarity direct current is required.

Table 5-4. Coating, Current, and Polarity Types Designated By the Fourth Digit in the Electrode Classification Number.

Digit	Coating	Weld Current
0	*	*
1	Cellulose potassium	ac, dcrp, dcsp
2	Titania sodium	ac, dcsp
3	Titania potassium	ac, dcsp, dcrp
4	Iron powder titania	ac, dcsp, dcrp
5	Low hydrogen sodium	dcrp
6	Low hydrogen potassium	ac, dcrp
7	Iron powder iron oxide	ac, dcsp
8	Iron powder low hydrogen	ac, dcrp, dcsp

*When the fourth (or last) digit is 0, the type of coating and current to be used are determined by the third digit.

5-25. ELECTRODES AND THEIR USE (cont)

(3) The eletrode identification system for stainless steel arc welding is set up as follows:

(a) E indicates electrode for arc welding.

(b) The first three digits indicated the American Iron and Steel Institute type of stainless steel.

(c) The last two digits indicate the current and position used.

(d) The number E-308-16 by this system indicates stainless steel type 308; used in all positions; with alternating or reverse polarity direct current.

d. Bare Electrodes. Bare electrodes are made of wire compositions required for specific applications. These electrodes have no coatings other than those required in wire drawing. These wire drawing coatings have some slight stabilizing effect on the arc but are otherwise of no consequence. Bare electrodes are used for welding manganese steel and other purposes where a coated electrode is not required or is undesirable. A diagram of the transfer of metal across the arc of a bare electrode is shown in figure 5-29.

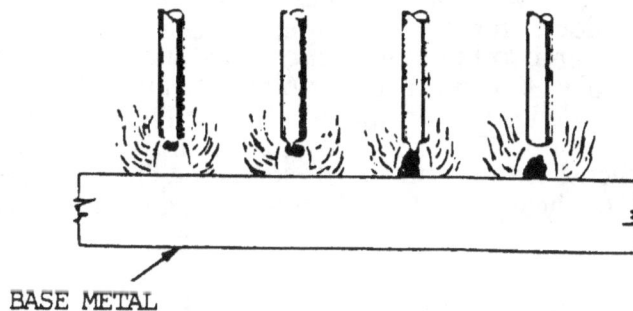

BASE METAL

Figure 5-29. Molten metal transfer with a bare electrode.

e. Light Coated Electrodes.

(1) Light coated electrodes have a definite composition. A light coating has been applied on the surface by washing, dipping, brushing, spraying, tumbling, or wiping to improve the stability and characteristics of the arc stream. They are listed under the E45 series in the electrode identification system.

(2) The coating generally serves the following functions:

(a) It dissolves or reduces impurities such as oxides, sulfur, and phosphorus.

(b) It changes the surface tension of the molten metal so that the globules of metal leaving the end of the electrode are smaller and more frequent, making the flow of molten metal more uniform.

(c) It increases the arc stability by introducing materials readily ionized (i.e., changed into small particles with an electric charge) into the arc stream.

(3) Some of the light coatings may produce a slag, but it is quite thin and does not act in the same manner as the shielded arc electrode type slag. The arc action obtained with light coated electrodes is shown in figure 5-30.

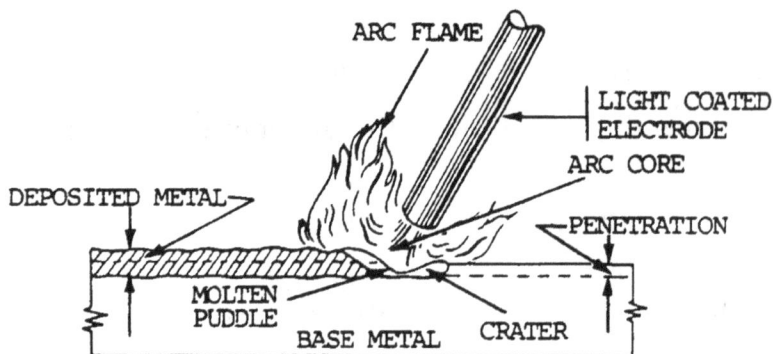

Figure 5-30. Arc action obtained with a light coated electrode.

f. Shielded Arc or HeavyCoated Electrodes. Shielded arc or heavy coated electrodes have a definite composition on which a coating has been applied by dipping or extrusion. The electrodes are manufactured in three general types: those with cellulose coatings; those with mineral coatings; and those with coatings of combinations of mineral and cellulose. The cellulose coatings are composed of soluble cotton or other forms of cellulose with small amounts of potassium, sodium, or titanium, and in some cases added minerals.The mineral coatings consist of sodium silicate, metallic oxides, clay, and other inorganic substances or combinations thereof. Cellulose coated electrodes protect the molten metal with a gaseous zone around the arc as well as slag deposit over the weld zone.The mineral coated electrode forms a slag deposit only. The shielded arc or heavy coated electrodes are used for welding steels, cast iron, and hard surfacing.The arc action obtained with the shielded arc or heavy coated electrode is shown in figure 5-31.

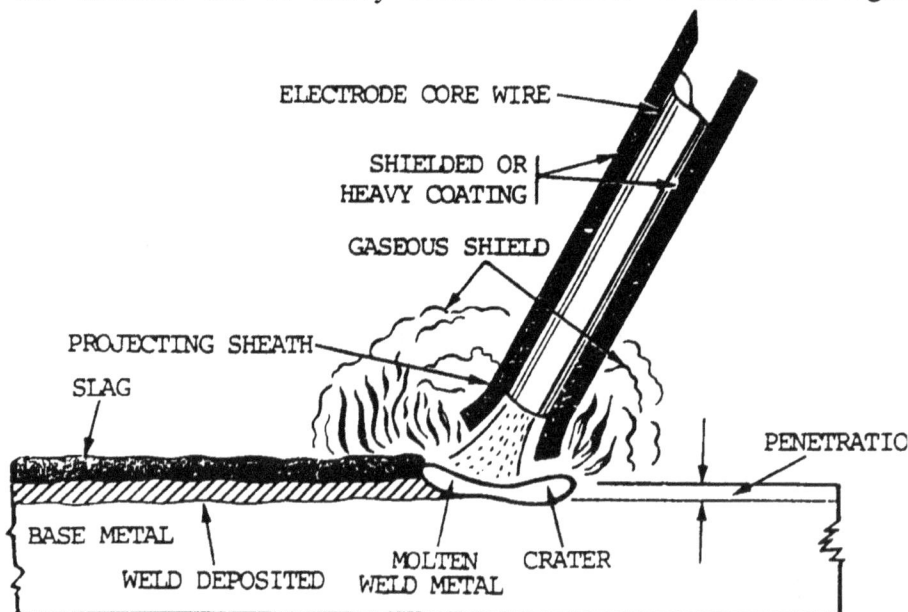

Figure 5-31. Arc action obtained with a shielded arc electrode.

5-25.   ELECTRODES AND THEIR USE (cont)

   g.   Functions of Shielded Arc or Heavy Coated Electrodes.

        (1) These electrodes produce a reducing gas shield around the arc which pre-
vents atmospheric oxygen or nitrogen from contaminating the weld metal. The oxygen
would readily combine with the molten metal,removing alloying elements and causing
porosity. The nitrogen would cause brittleness,low ductility, and in some cases,
low strength and poor resistance to corrosion.

        (2) The electrodes reduce impurities such as oxides, sulfur, and phosphorus
so that these impurities will not impair the weld deposit.

        (3) They provide substances to the arc which increase its stability and elimi-
nate wide fluctuations in the voltage so that the arc can be maintained without
excessive spattering.

        (4) By reducing the attractive force between the molten metal and the end of
the electrode, or by reducing the surface tension of the molten metal, the vaporized and melted coating causes the molten metal at the end of the electrode to
break up into fine, small particles.

        (5) The coatings contain silicates which will form a slag over the molten
weld and base metal. Since the slag solidifies at a relatively slow rate, it holds
the heat and allows the underlying metal to cool and slowly solidify. This slow
solidification of the metal eliminates the entrapment of gases within the weld and
permits solid impurities to float to the surface.Slow cooling also has an annealing effect on the weld deposit.

        (6) The physical characteristics of the weld deposit are modified by incorp-
orating alloying materials in the electrode coating.The fluxing action of the slag
will also produce weld metal of better quality and permit welding at higher speeds.

        (7) The coating insulates the sides of the electrode so that the arc is con-
centrated into a confined area. This facilitates welding in a deep U or V groove.

        (8) The coating produces a cup, cone, or sheath (fig. 5-31) at the tip of the
electrode which acts as a shield, concentrates and directs the arc, reduces heat
losses and increases the temperature at the end of the electrode.

   h.   Storing Electrodes. Electrodes must be kept dry. Moisture destroys the
desirable characteristics of the coating and may cause excessive spattering and
lead to the formation of cracks in the welded area.Electrodes exposed to damp air
for more than two or three hours should be dried by heating in a suitable oven
(fig. 5-32) for two hours at $500^{\circ}F$ ($260^{\circ}C$). After they have dried, they should
be stored in a moisture proof container.Bending the electrode can cause the coating to break loose from the core wire. Electrodes should not be used if the core
wire is exposed.

Figure 5-32. Electrode drying ovens.

i. Tungsten Electrodes.

(1) Nonconsumable electrodes for gas tungsten-arc (TIG) welding are of three types: pure tungsten, tungsten containing 1 or 2 percent thorium, and tungsten containing 0.3 to 0.5 percent zirconium.

(2) Tungsten electrodes can be identified as to type by painted end marks as follows.

(a) Green -- pure tungsten.

(b) Yellow -- 1 percent thorium.

(c) Red -- 2 percent thorium.

(d) Brown – 0.3 to 0.5 percent zirconium.

(3) Pure tungsten (99. 5 percent tungsten) electrodes are generally used on less critical welding operations than the tungstens which are alloyed. This type of electrode has a relatively low current-carrying capacity and a low resistance to contamination.

(4) Thoriated tungsten electrodes (1 or 2 percent thorium) are superior to pure tungsten electrodes because of their higher electron output, better arc-starting and arc stability, high current-carrying capacity, longer life, and greater resistance to contamination.

(5) Tungsten electrodes containing 0.3 to 0.5 percent zirconium generally fall between pure tungsten electrodes and thoriated tungsten electrodes in terms of performance. There is, however, some indication of better performance in certain types of welding using ac power.

(6) Finer arc control can be obtained if the tungsten alloyed electrode is ground to a point (fig. 5-33). When electrodes are not grounded, they must be operated at maximum current density to obtain reasonable arc stability. Tungsten electrode points are difficult to maintain if standard direct current equipment is used as a power source and touch-starting of the arc is standard practice. Maintenance of electrode shape and the reduction of tungsten inclusions in the weld can best be accomplished by superimposing a high-frequency current on the regular welding current. Tungsten electrodes alloyed with thorium and zirconium retain their shape longer when touch-starting is used.

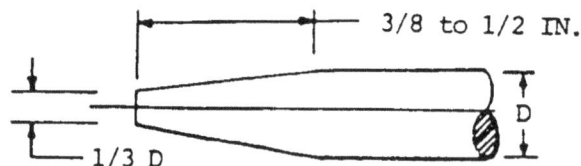

3/8 to 1/2 IN.

D

1/3 D

Figure 5-33. Correct electrode taper.

5-25. ELECTRODES AND THEIR USE (cont)

(7) The electrode extension beyond the gas cup is determined by the type of joint being welded. For example, an extension beyond the gas cup of 1/8 in. (3.2 mm) might be used for butt joints in light gage material, while an extension of approximately 1/4 to 1/2 in. (6.4 to 12.7 mm) might be necessary on some fillet welds. The tungsten electrode of torch should be inclined slightly and the filler metal added carefully to avoid contact with the tungsten. This will prevent contamination of the electrode. If contamination does occur, the electrode must be removed, reground, and replaced in the torch.

j. <u>Direct Current Welding</u>. In direct current welding, the welding current circuit may be hooked up as either straight polarity (dcsp) or reverse polarity (dcrp). The polarity recommended for use with a specific type of electrode is established by the manufacturer.

(1) For dcsp, the welding machine connections are electrode negative and workpiece positive (fig. 5-34); electron flow is from electrode to workpiece. For dcrp, the welding machine connections are electrode positive and workpiece negative; electron flow is from workpiece to electrode.

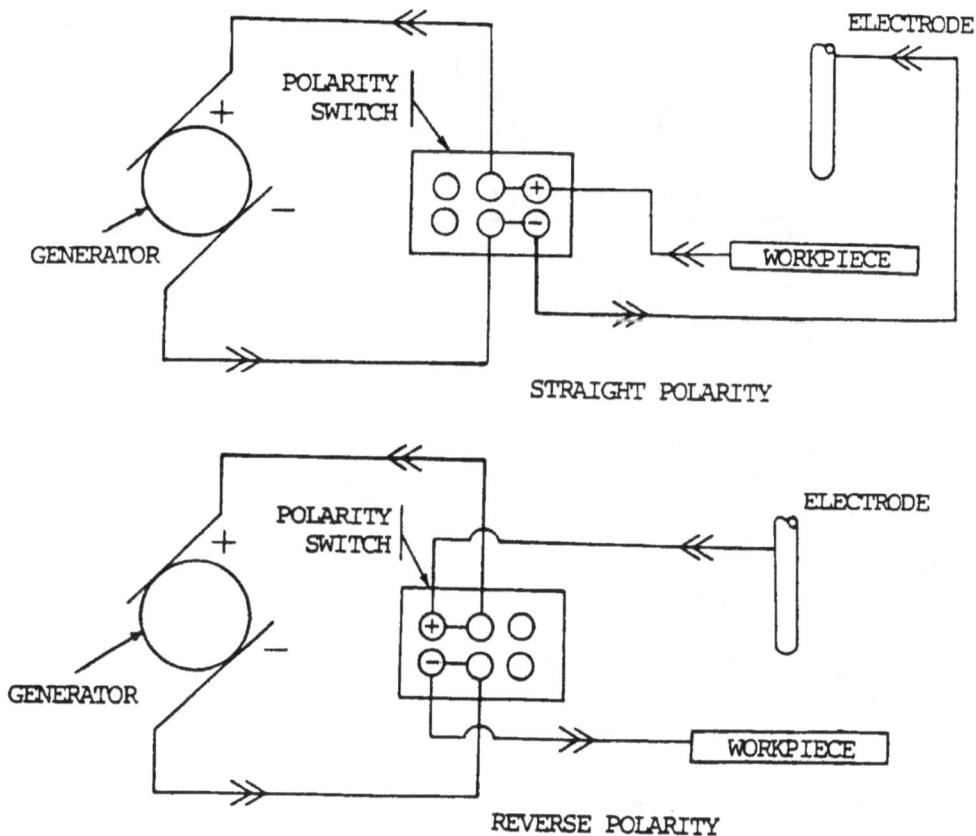

Figure 5-34. Polarity of welding current.

(2) For both current polarities,the greatest part of the heating effect occurs at the positive side of the arc. The workpiece is dcsp and the electrode is dcrp. Thus, for any given welding current, dcrp requires a larger diameter electrode than does dcsp. For example, a 1/16-in. (1.6-mm) diameter pure tungsten electrode can handle 125 amperes of welding current under straight polarity conditions. If the polarity were reversed, however, this amount of current would melt off the electrode and contaminate the weld metal. Hence, a 1/4-in. (6.4-mm) diameter pure tungsten electrode is required to handle 125 amperes dcrp satisfactorily and safely. However, when heavy coated electrodes are used, the composition of the coating and the gases it produces may alter the heat conditions.This will produce greater heat on the negative side of the arc. One type of coating may provide the most desirable heat balance with straight polarity, while another type of coating on the same electrode may provide a more desirable heat balance with reverse polarity.

(3) The different heating effects influence not only the welding action, but also the shape of the weld obtained. DCSP welding will produce a wide, relatively shallow weld (fig. 5-35). DCRP welding, because of the larger electrode diameter and lower currents generally employed, gives a narrow, deep weld.

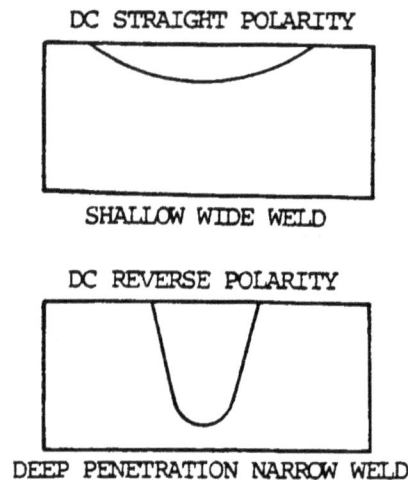

DC STRAIGHT POLARITY

SHALLOW WIDE WELD

DC REVERSE POLARITY

DEEP PENETRATION NARROW WELD

Figure 5-35. Effect of polarity on weld shape.

(4) One other effect of dcrp welding is the so-called plate cleaning effect. This surface cleaning action is caused either by the electrons leaving the plate or by the impact of the gas ions striking the plate, which tends to break up the surface oxides, and dirt usually present.

(5) In general, straight polarity is used with all mild steel, bare, or light coated electrodes. Reverse polarity is used in the welding of non-ferrous metals such as aluminum, bronze, monel, and nickel.Reverse polarity is also used with sane types of electrodes for making vertical and overhead welds.

(6) The proper polarity for a given electrode can be recognized by the sharp, cracking sound of the arc. The wrong polarity will cause the arc to emit a hissing sound, and the welding bead will be difficult to control.

5-25. ELECTRODES AND THEIR USE (cont)

k. <u>Alternating Current Welding</u>.

(1) Alternating current welding, theoretically, is a combination of dcsp and dcrp welding. This can be best explained by showing the three current waves visually. As shown in figure 5-36, half of each complete alternating current (ac) cycle is dcsp, the other half is dcrp.

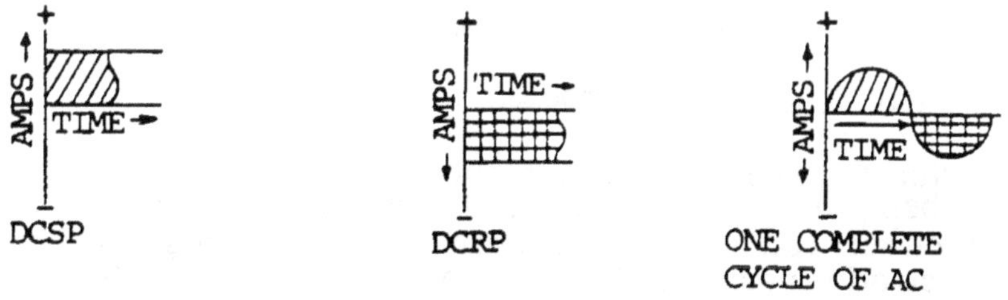

Figure 5-36. AC wave.

(2) Moisture, oxides, scale, etc., on the surface of the plate tend, partially or completely, to prevent the flow of current in the reverse polarity direction. This is called rectification. For example, in no current at all flowed in the reverse polarity direction, the current wave would be similar to figure 5-37.

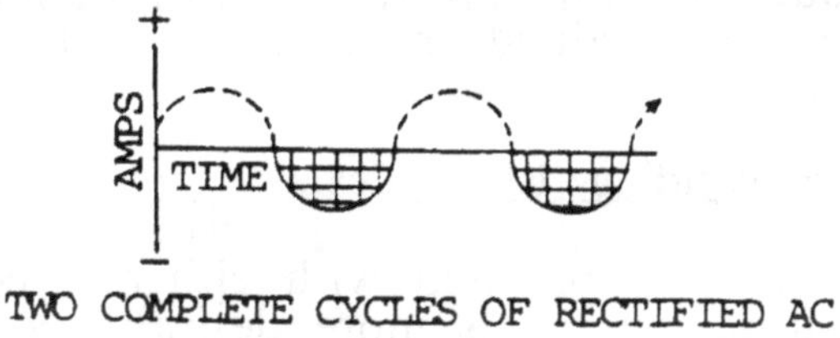

TWO COMPLETE CYCLES OF RECTIFIED AC

Figure 5-37. Rectified ac wave.

(3) To prevent rectification from occurring, it is common practice to introduce into the welding current an additional high-voltage, high-frequency, low-power current. This high-frequency current jumps the gap between the electrode and the workpiece and pierces the oxide film, thereby forming a path for the welding current to follow. Superimposing this high-voltage, high-frequency current on the welding current gives the following advantages:

(a) The arc may be started without touching the electrode to the workpiece.

(b) Better arc stability is obtained.

(c) A longer arc is possible. This is particularly useful in surfacing and hardfacing operations.

(d) Welding electrodes have longer life.

(e) The use of wider current range for a specific diameter electrode is possible.

5-44

(4) A typical weld contour produced with high-frequency stabilized ac is shown in figure 5-38, together with both dcsp and dcrp welds for comparison.

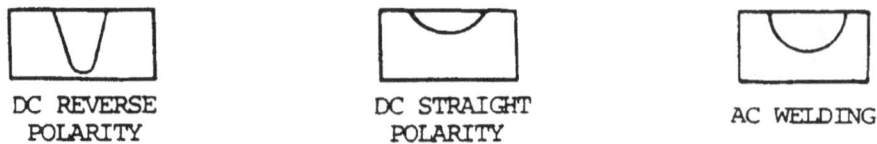

DC REVERSE POLARITY   DC STRAIGHT POLARITY   AC WELDING

Figure 5-38. Comparison of penetration contours.

l. Direct Current Arc Welding Electrodes.

(1) The manufacturer's recommendations should be followed when a specific type of electrode is being used. In general, direct current shielded arc electrodes are designed either for reverse polarity (electrode positive) or for straight polarity (electrode negative), or both. Many, but not all, of the direct current electrodes can be used with alternating current. Direct current is preferred for many types of covered, nonferrous bare and alloy steel electrodes. Recommendations from the manufacturer also include the type of base metal for which given electrodes are suited, corrections for poor fit-ups, and other specific conditions.

(2) In most cases, straight polarity electrodes will provide less penetration than reverse polarity electrodes, and for this reason will permit greater welding speed. Good penetration can be obtained from either with proper welding conditions and arc manipulation.

m. Alternating Current Arc Welding Electrodes.

(1) Coated electrodes which can be used with either direct or alternating current are available. Alternating current is more desirable while welding in restricted areas or when using the high currents required for thick sections because it reduces arc blow. Arc blow causes blowholes, slag inclusions, and lack of fusion in the weld.

(2) Alternating current is used in atomic hydrogen welding and in those carbon arc processes that require the use of two carbon electrodes. It permits a uniform rate of welding and electrode consumption. In carbon-arc processes where one carbon electrode is used, direct current straight poarity is recommended, because the electrode will be consumed at a lower rate.

n. Electrode Defects and Their Effects.

(1) If certain elements or oxides are present in electrode coatings, the arc stability will be affected. In bare electrodes, the composition and uniformity of the wire is an important factor in the control of arc stability. Thin or heavy coatings on the electrodes will riot completely remove the effects of defective wire.

(2) Aluminum or aluminum oxide (even when present in quantities not exceeding 0.01 percent), silicon, silicon dioxide, and iron sulphate cause the arc to be unstable. Iron oxide, manganese oxide, calcium oxide, and iron sulphate tend to stabilize the arc.

5-25. ELECTRODES AND THEIR USE (cont)

(3) When phosphorus or sulfur are present in the electrode in excess of 0.04 percent, they will impair the weld metal because they are transferred from the electrode to the molten metal with very little loss.Phosphorus causes grain growth, brittleness, and "cold shortness" (i.e., brittle when below red heat) in the weld. These defects increase in magnitude as the carbon content of the steel increases. Sulfur acts as a slag, breaks up the soundness of the weld metal, and causes "hot shortness" (i.e., brittle when above red heat).Sulfur is particularly harmful to bare, low-carbon steel electrodes with a low manganese content. Manganese promotes the formation of sound welds.

(4) If the heat treatment, given the wire core of an electrode, is not uniform, the electrode will produce welds inferior to those produced with an electrode of the same composition that has been properly heat treated.

Section IV.   RESISTANCE WELDING EQUIPMENT

5-26. RESISTANCE WELDING

a. General. Resistance welding is a group of welding processes in which the joining of metals is produced by the heat obtained from resistance of the work to the electric current, in a circuit of which the work is a part, and by the application of pressure. The three factors involved in making a resistance weld are the amount of current that passes through the work,the pressure that the electrodes transfer to the work, and the time the current flows through the work.Heat is generated by the passage of electrical current through a resistance current, with the maximum heat being generated at the surfaces being joined. Pressure is required throughout the welding cycle to assure a continuous electrical circuit through the work. The amount of current employed and the time period are related to the heat input required to overcome heat losses and raise the temperature of the metal to the welding temperature. The selection of resistance welding equipment is usually determined by the joint design, construction materials, quality requirements, production schedules, and economic considerations.Standard resistance welding machines are capable of welding a variety of alloys and component sizes. There are seven major resistance welding processes: resistance projection welding, resistance spot welding, resistance flash welding, resistance upset welding, resistance seam welding, resistance percussion welding, and resistance high frequency welding.

b. Principal Elements of Resistance Welding Machines.A resistance welding machine has three principal elements:

(1) An electrical circuit with a welding transformer and a current regulator, and a secondary circuit,including the electrodes which conduct the welding current to the work.

(2) A mechanical system consisting of a machine frame and associated mechanisms to hold the work and apply the welding force.

(3) The control equipment (timing devices) to initiate the time and duration of the current flow. This equipment may also control the current magnitude, as well as the sequence and the time of other parts of the welding cycle.

c. <u>Electrical Operation.</u> Resistance welds are made with either semiautomatic or mechanized machines. With the semiautomatic machine, the welding operator positions the work between the electrodes and pushes a switch to initiate the weld; the weld programmer completes the sequence. In a mechanized setup, parts are automatically fed into a machine, then welded and ejected without welding operator assistance. Resistance welding machines are classified according to their electrical operation into two basic groups: direct energy and stored energy. Machines in both groups may be designed to operate on either single-phase or three-phase power.

d. <u>Spot Welding.</u>

(1) There are several types of spot welding machines, including rocker arm, press, portable, and multiple type. A typical spot welding machine, with its essential operating elements for manual operation, is shown in figure 5-39. In these machines, the electrode jaws are extended in such a manner as to permit a weld to be made at a considerable distance f ran the edge of the base metal sheet. The electrodes are composed of a copper alloy and are assembled in a manner by which considerable force or squeeze may be applied to the metal during the welding process.

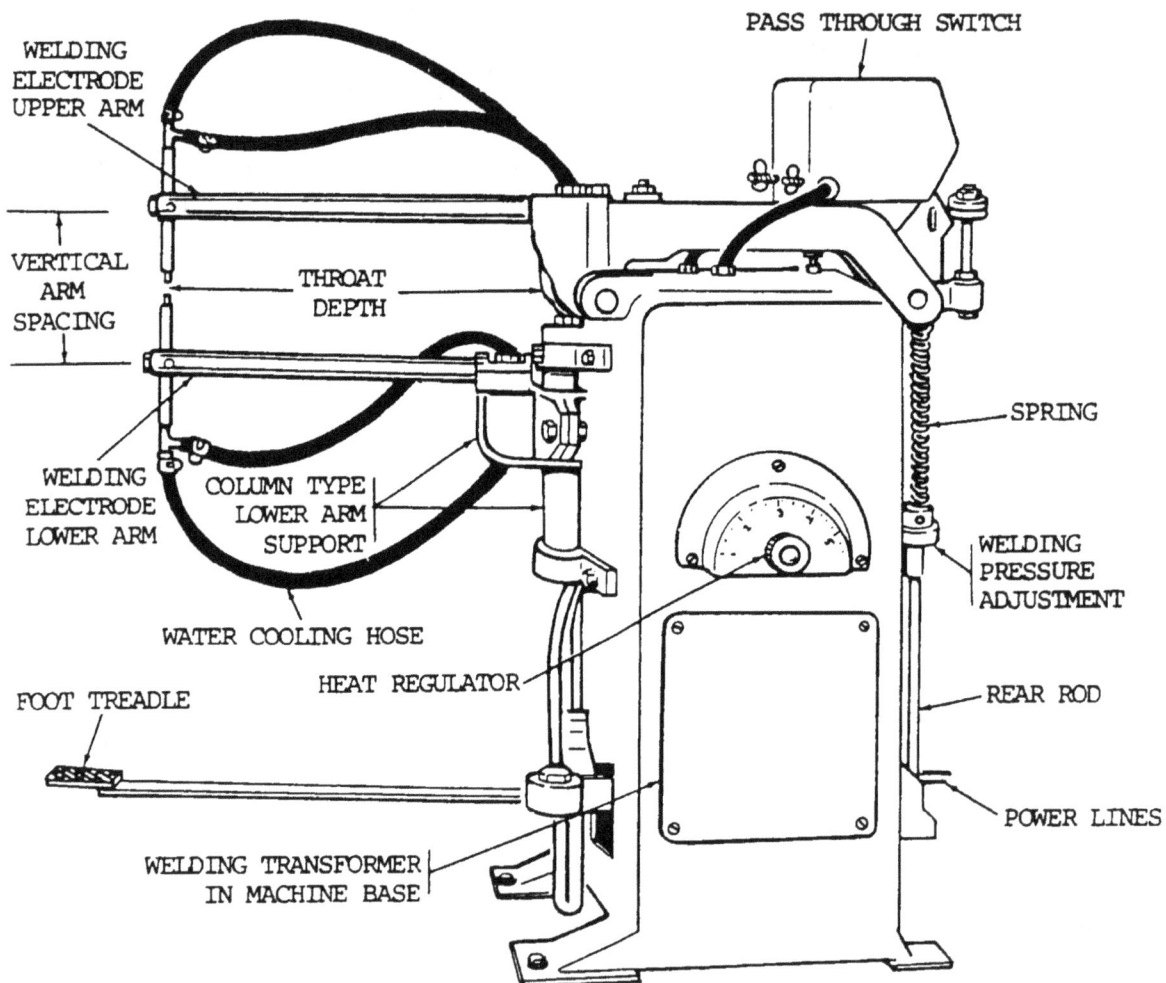

Figure 5-39. Resistance spot welding machine and accessories.

5-26.   RESISTANCE WELDING (cont)

(a) <u>Rocker arm type</u>. These machines consist essentially of a cylindrical arm or extension of an arm which transmits the electrode force and in most cases, the welding current. They are readily adaptable for spot welding of most weldable metals. The travel path of the upper electrode is in an arc about the fulcrum of the upper arm. The electrodes must be positioned so that both are in the plane of the horn axes. Because of the radial motion of the upper electrode, these machines are not recommended for projection welding.

(b) <u>Press type</u>. In this type of machine, the moveable welding head travels in a straight line in guide bearings or ways. Press type machines are classified according to their use and method of force application. They may be designed for spot welding, projection welding, or both. Force may be applied by air or hydraulic cylinders, or manually with small bench units.

(c) <u>Portable type</u>. A typical portable spot welding machine consists of four basic units: a portable welding gun or tool; a welding transformer and, in some cases, a rectifier; an electrical contactor and sequence timer; and a cable and hose unit to carry power and cooling water between the transformer and welding gun. A typical portable welding gun consists of a frame, an air or hydraulic actuating cylinder, hand grips, and an initiating switch. The design of the gun is tailored to the needs of the assembly to be welded.

(d) <u>Multiple spot welding type</u>. These are special-purpose machines designed to weld a specific assembly. They utilize a number of transformers. Force is applied directly to the electrode through a holder by an air or hydraulic cylinder. For most applications, the lower electrode is made of a piece of solid copper alloy with one or more electrode alloy inserts that contact the part to be welded. Equalizing guns are often used where standard electrodes are needed on both sides of the weld to obtain good heat balance, or where variations in parts will not permit consistent contact with a large, solid, lower electrode. The same basic welding gun is used for the designs, but it is mounted on a special "C" frame similar to that for a portable spot welding gun. The entire assembly can move as electrode force is applied to the weld location.

(2) When spot welding aluminum conventional spot welding machines used to weld sheet metal may be used. However, the best results are obtained only if certain refinements are incorporated into these machines. These features include the following:

(a) Ability to handle high current for short welding times.

(b) Precise electronic control of current and length of time it is applied.

(c) Rapid follow up of the electrode force by employing anti-friction bearings and lightweight, low-inertia heads.

(d) High structural rigidity of the welding machine arms, holders, and platens in order to minimize deflection under the high electrode forces used for aluminum, and to reduce magnetic deflections, a variable or dual force cycle to permit forging the weld nugget.

(e) Slope control to permit a gradual buildup and tapering off of the welding current.

(f) Postheat current to allow slower cooling of the weld.

(g) Good cooling of the Class I electrodes to prevent tip pickup or sticking. Refrigerated cooling is often helpful.

e. <u>Projection Welding</u>. The projection welding dies or electrodes have flat surfaces with larger contacting areas than spot welding electrodes. The effectiveness of this type of welding depends on the uniformity of the projections or embossments on the base metal with which the electrodes are in contact (fig. 5-40). The press type resistance welding machine is normally used for projection welding. Flat nose or special electrodes are used.

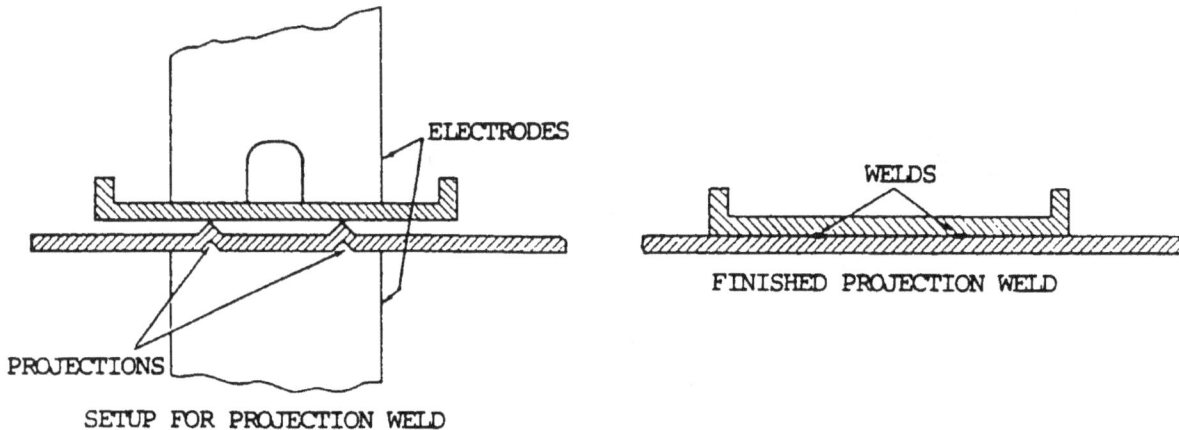

ELECTRODES

WELDS

PROJECTIONS

SETUP FOR PROJECTION WELD

FINISHED PROJECTION WELD

Figure 5-40. Projection welding.

f. <u>Seam Welding</u>. A seam welding machine is similar in principle to a spot welding machine, except that wheel-shaped electrodes are used rather than the electrode tips used in spot welding. Several types of machines are used for seam welding, the type used depending on the sevice requirements. In some machines, the work is held in a fixed position and a wheel type electrode is passed over it. Portable seam welding machines use this principle. In the traveling fixture type seam welding machine, the electrode is stationary and the work is moved. Seam welding machine controls must provide an on-ff sequencing of weld current and a control of wheel rotation. The components of a standard seam welding machine include a main frame that houses the welding transformer and tap switch; a welding head consisting of an air cylinder, a ram, and an upper electrode mounting and drive mechanism; the lower electrode mounting and drive mechanism, if used; the secondary circuit connections; electronic controls and contactor; and wheel electrodes.

g. <u>Upset and Flash Welding</u>. Flash and upset welding machines are similar in construction. The major difference is the motion of the movable platen during welding and the mechanisms used to impart the motion. Flash weld-fig is generally preferred for joining components of equal cross section end-to-end. Upset welding is normally used to weld wire, rod, or bar of small cross section and to join the seam continuously in pipe or tubing. Flash welding machines are generally of much larger capacity than upset welding machines. However, both of these processes can be performed on the same type of machine. The metals that are to be joined serve as electrodes.

5-26.   RESISTANCE WELDING (Cont)

(1) A standard flash welding machine consists of a main frame, stationary platen, movable platen clamping mechanisms and fixtures, a transformer, a tap switch, electrical controls, and a flashing and upsetting mechanismElectrodes that hold the parts and conduct the welding current to them are mounted on the platens.

(2) Upset welding machines consist of a main frame that houses a transform and tap switch, electrodes to hold the parts and conduct the welding current, and means to upset the joint. A primary contactor is used to control the welding current.

h.   Percussion Welding. This process uses heat from an arc produced by a rapid discharge of electrical energy to join metals. Pressure is applied progressively during or immediately following the electrical discharge. The process is similar to flash and upset welding. Two types of welding machines are used in percussion welding: magnetic and capacitor discharge. A unit generally consists of a modified press-type resistance welding machine with specially designed transform, controls, and tooling.

i .   High FrequencyWelding. This process joins metals with the heat generated from the resistance of the work pieces to a high frequency alternating current in the 10,000 to 500,000 hertz range, and the rapid application of an upsetting force after heating is completed. The process is entirely automatic and utilizes equipment designed specifically for this process.

Section V.   THERMIT WELDING EQUIPMENT

5-27.   THERMIT WELDING (TW)

a.   General. .   Thermit material is a mechanical mixture of metallic aluminum and processed iron oxide. Molten steel is produced by the thermit reaction in a magnesite-lined crucible. At the bottom of the crucible, a magnesite stone is burned, into which a magnesite stone thimble is fitted. This thimble provides a passage through which the molten steel is discharged into the mold. The hole through the thimble is plugged with a tapping pin, which is covered with a fire-resistant washer and refractory sand.The crucible is charged by placing the correct quantity of thoroughly mixed thermit material in itIn preparing the joint for thermit welding, the parts to be welded must be cleaned, alined, and held firmly in place. If necessary, metal is removed from the joint to permit a free flow of the thermit metal into the joint.A wax pattern is then made around the joint in the size and shape of the intended weld.A mold made of refractory sand is built around the wax pattern and joint to hold the molten metal after it is poured. The sand mold is then heated to melt out the wax and dry the mold. The mold should be properly vented to permit the escape of gases and to allow the proper distribution of the thermit metal at the jointA thermit welding crucible and mold is shown in figure 5-41.

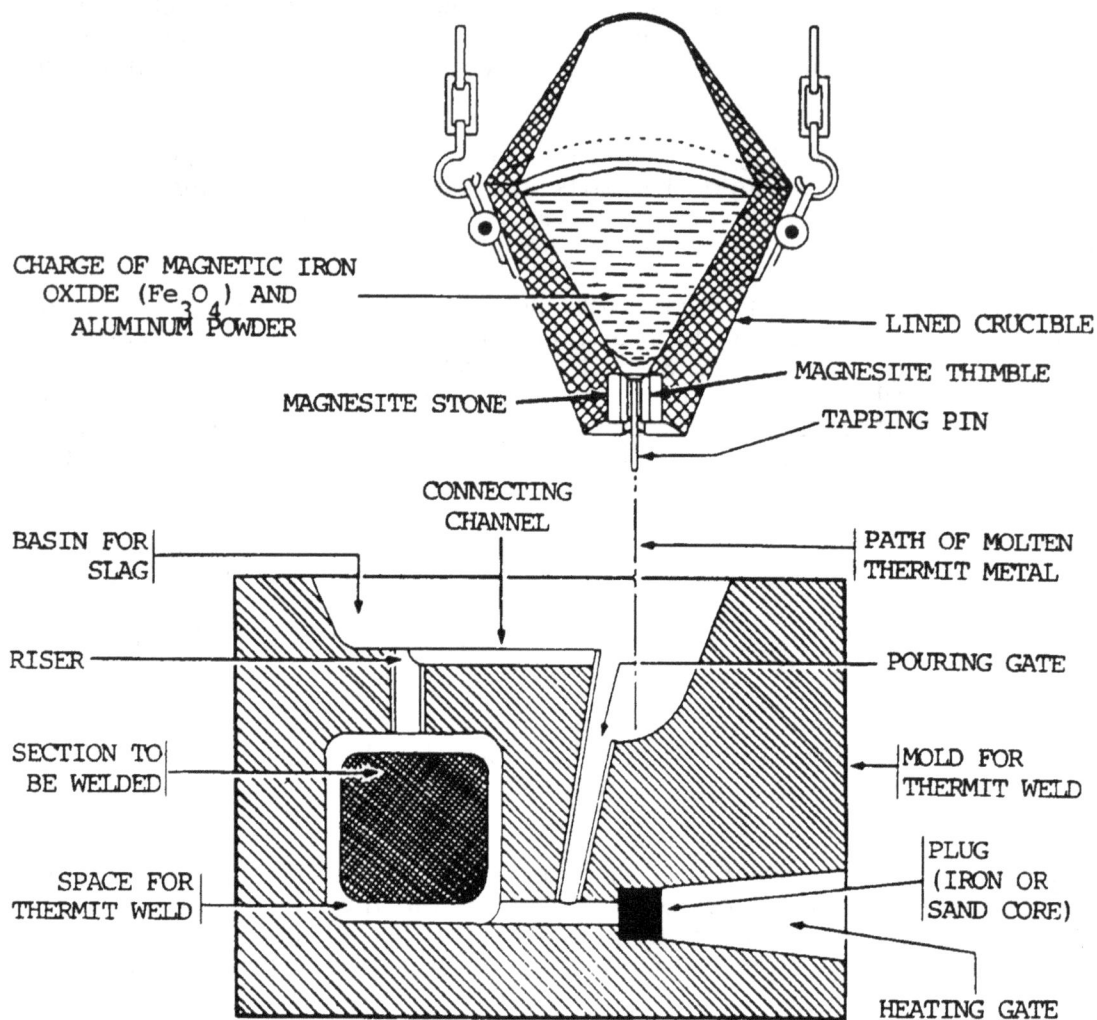

Figure 5-41. Thermit welding crucible and mold.

Labels in figure:

CHARGE OF MAGNETIC IRON OXIDE (Fe$_3$O$_4$) AND ALUMINUM POWDER

LINED CRUCIBLE

MAGNESITE THIMBLE

MAGNESITE STONE

TAPPING PIN

CONNECTING CHANNEL

BASIN FOR SLAG

PATH OF MOLTEN THERMIT METAL

RISER

POURING GATE

SECTION TO BE WELDED

MOLD FOR THERMIT WELD

SPACE FOR THERMIT WELD

PLUG (IRON OR SAND CORE)

HEATING GATE

Section VI. FORGE WELDING TOOLS AND EQUIPMENT

5-28. FORGES

Forge welding is a form of hot pressure welding which joins metals by heating them in an air forge or other furnace, and then applying pressure. The forge, which may be either portable or stationary, is the most important component of forge welding equipment. The two types used in hand forge welding are described below.

5-28. FORGES (cont)

a. Portable Forge. The essential parts of a forge are a hearth, a tuyere, a water tank, and a blower. One type of portable forge is shown in figure 5-42. The tuyere is a valve mechanism designed to direct an air blast into the fire. It is made of cast iron and consists of a fire pot, base with air inlet, blast valve, and ash gate. The air blast passes through the base and is admitted to the fire through the valve. The valve can be set in three different positions to regulate the size and direction of the blast according to the fire required. The valve handle is also used to free the valve from ashes. A portable forge may have a handcrank blower, as shown in figure 5-42, or it may be equipped with an electric blower. The blower produces air blast pressure of about 2 oz per sq in. A hood is provided on the forge for carrying away smoke and fumes.

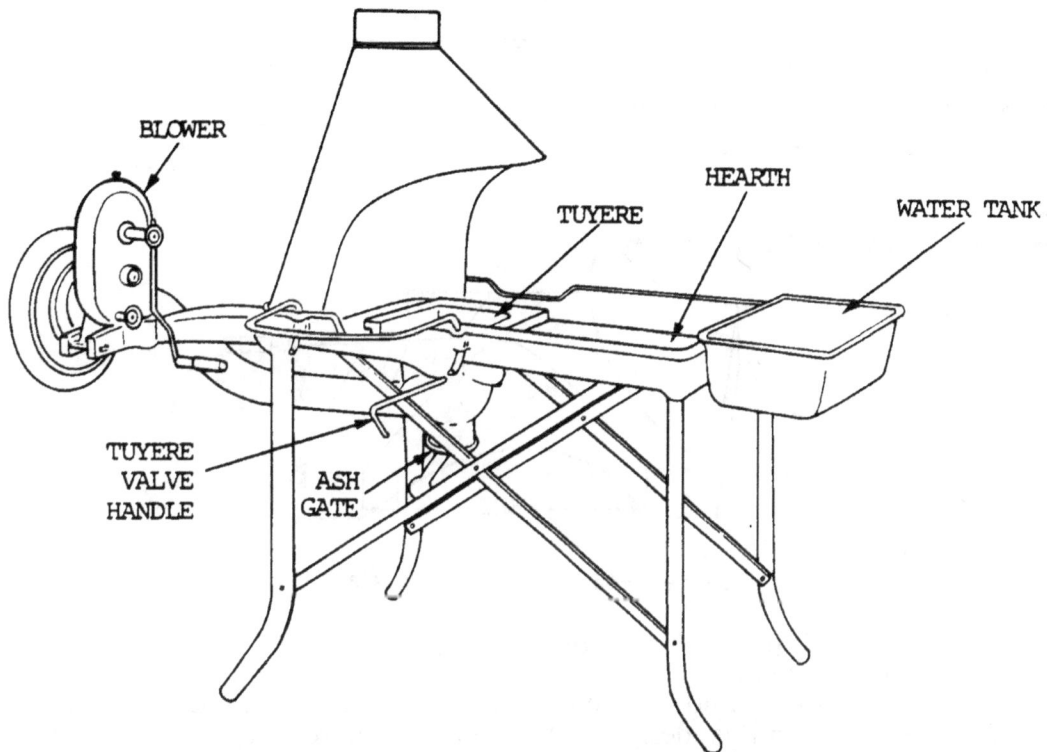

Figure 5-42. Portable forge.

b. Stationary Forge. The stationary forge is similar to the portable forge except that it is usually larger with larger air and exhaust connections. The forge may have an individual blower or there may be a large capacity blower for a group of forges. The air blast valve usually has three slots at the top, the positions of which can be controlled by turning the valve. The opening of these slots can be varied to regulate the volume of the blast and the size of the fire. The stationary forges, like portable forges, are available in both updraft and downdraft types. In the updraft type, the smoke and gases pass up through the hood and chimney by natural draft or are drawn off by an exhaust fan. In the downdraft type, the smoke and fumes are drawn down under an adjustable hood and carried through a duct by an exhaust fan that is entirely separate from the blower. The downdraft forge permits better air circulation and shop ventilation, because the removal of furies and smoke is positive.

5-29. FORGING TOOLS

a. Anvil.

(1) The anvil (fig. 5-43) is usually made of two forgings or steel castings welded together at the waist. The table or cutting block is soft so that cutters and chisels caning in contact with it will not be dulled. The face is made of hardened, tempered tool steel which is welded to the top of the anvil. It cannot be easily damaged by hammering.

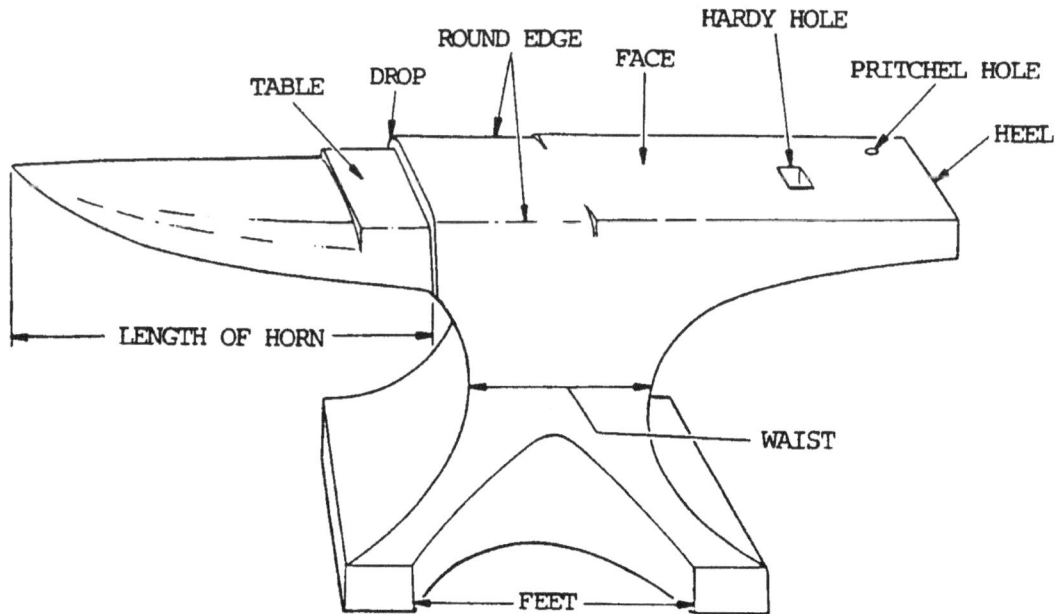

Figure 5-43. Blacksmith's anvil.

(2) The edges of an anvil are rounded for about 4.00 in. (102 mm) back from the table to provide edges where stock can be bent without danger of cutting it. All other edges are sharp and will cut stock when it is hammered against them. The hardy hole is square and is designed to hold the hardy, bottom, swages, fullers, and other special tools. The pritchel hole is round and permits slugs of metal to pass through when holes are punched in the stock.The anvil is usually mounted on a heavy block of wood, although steel pedestals or bolsters are sometimes used. The height of the anvil should be adjusted so that the operator's knuckles will just touch its face when he stands erect with his arms hanging naturally.

(3) Anvils are designated by weight (i.e., No.150 weighs 150 lb), and range in size from No. 100 to No. 300.

b. Other Tools. In addition to the anvil, othetools such as hammers sledges, tongs, fullers, flatters, chisels, swage blocks,punches, and a vise are used in forging operations.

CHAPTER 6
WELDING TECHNIQUES

Section I. DESCRIPTION

6-1. GENERAL

The purpose of this chapter is to outline the various techniques used in welding processes. Welding processes may be broken down into many categories. Various methods and materials may be used to accomplish good welding practices. Common methods of welding used in modern metal fabrication and repair are shown in figure 6-1.

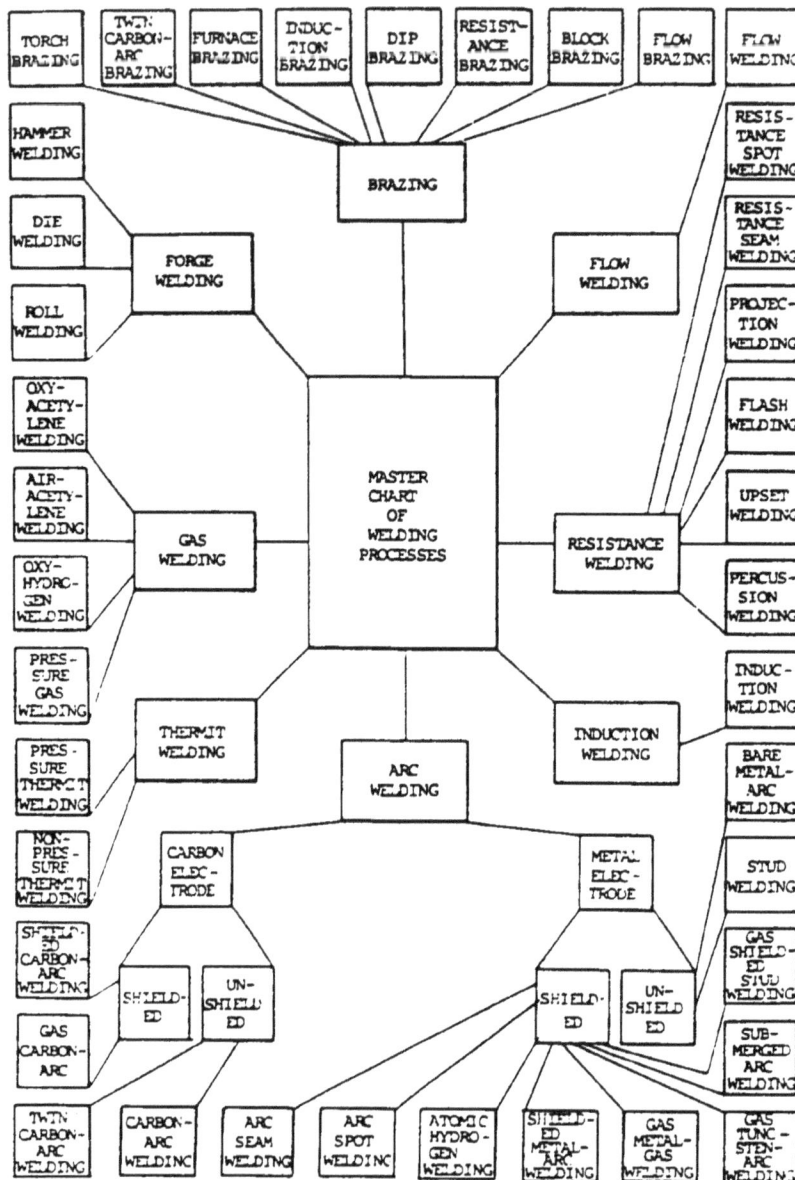

NOTE: SOLDERING NOT INCLUDED

Figure 6-1. Chart of welding processes.

TC 9-237

6-2. ARC WELDING

The term arc welding applies to a large and varied group of processes that use an electric arc as the source of heat to melt and join metals. In arc welding processes, the joining of metals, or weld, is produced by the extreme heat of an electric arc drawn between an electrode and the workpiece or between two electrodes. The formation of a joint between metals being arc welded may or may not require the use of pressure or filler metal. The arc is struck between the workpiece and an electrode that is mechanically or manually moved along the joint, or that remains stationary while the workpiece is roved underneath it. The electrode will be either a consumable wire rod or a nonconsumable carbon or tungsten rod which carries the current and sustains the electric arc between its tip and the workpiece. When a nonconsumable electrode is used, a separate rod or wire can supply filler material, if needed. A consumable electrode is specially prepared so that it not only conducts the current and sustains the arc, but also melts and supplies filler metal to the joint, and may produce a slag covering as well.

a. Metal Electrodes. In bare metal-arc welding, the arc is drawn between a bare or lightly coated consumable electrode and the workpiece. Filler metal is obtained from the electrode, and neither shielding nor pressure is used. This type of welding electrode is rarely used, however, because of its low strength, brittleness, and difficulty in controlling the arc.

(1) Stud welding. The stud welding process produces a joining of metals by heating them with an arc drawn between a metal stud, or similar part, and the workpiece. The molten surfaces to be joined, when properly heated, are forced together under pressure. No shielding gas is used. The most common materials welded with the arc stud weld process are low carbon steel, stainless steel, and aluminum. Figure 6-2 shows a typical equipment setup for arc stud welding.

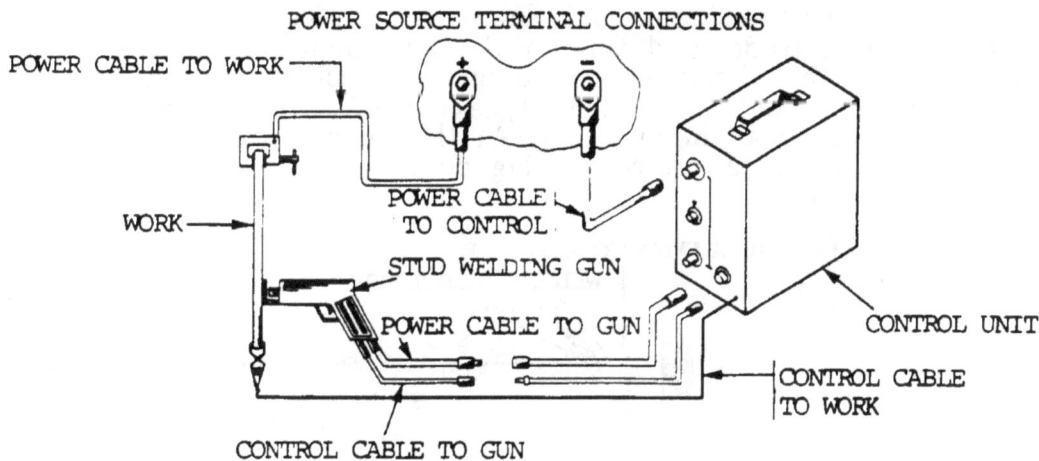

Figure 6-2. Equipment setup for arc stud welding.

(2) Gas shielded stud welding. This process, a variation of stud welding, is basically the same as that used for stud welding, except that an inert gas or flux, such as argon or helium, is used for shielding. Shielding gases and fluxes are

6-2

used when welding nonferrous metals such as aluminum and magnesium. Figure 6-3 shows a typical setup for gas shielded arc stud welding.

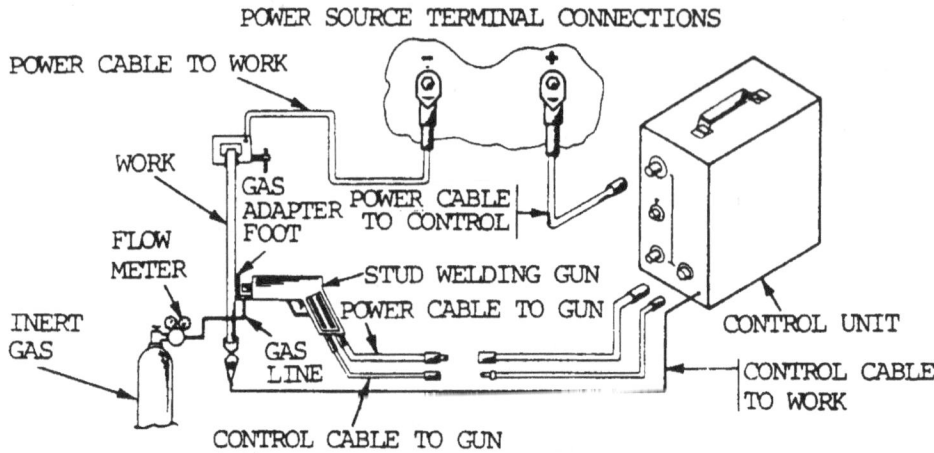

Figure 6-3. Equipment setup for gas shielded arc stud welding.

(3) <u>Submerged arc welding</u>. This process joins metals by heating them with an arc maintained between a bare metal electrode and the workpiece. The arc is shielded by a blanket of granular fusible material and the workpiece. Pressure is not used and filler metal is obtained from the electrode or from a supplementary welding rod. Submerged arc welding is distinguished from other arc welding processes by the granular material that covers the welding area. This granular material is called a flux, although it performs several other important functions. It is responsible for the high deposition rates and weld quality that characterize the submerged arc welding process in joining and surfacing applications. Basically, in submerged arc welding, the end of a continuous bare wire electrode is inserted into a mound of flux that covers the area or joint to be welded. An arc is initiated, causing the base metal, electrode, and flux in the immediate vicinity to melt. The electrode is advanced in the direction of welding and mechanically fed into the arc, while flux is steadily added. The melted base metal and filler metal flow together to form a molten pool in the joint. At the same time, the melted flux floats to the surface to form a protective slag cover. Figure 6-4 shows the submerged arc welding process.

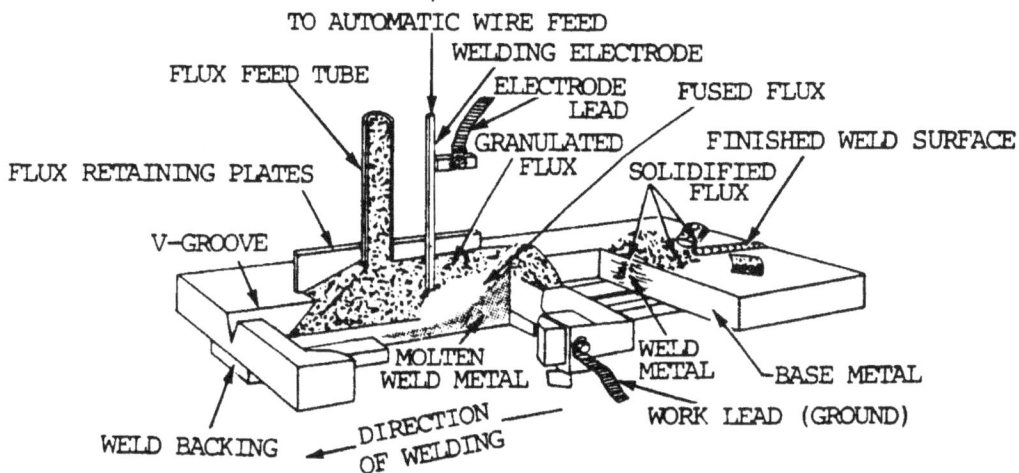

Figure 6-4. Submerged arc welding process.

6-2. ARC WELDING (cont)

(4) <u>Gas tungsten-arc welding (TIG welding or GTAW)</u>.The arc is drawn between a nonconsumable tungsten eletrode and the workpieceShielding is obtained from an inert gas or gas mixture. Pressure and/or filler metal may or may not be used. The arc fuses the metal being welded as well as filler metal, if used. The shield gas protects the electrode and weld pool and provides the required arc characteristics. A variety of tungsten electrodes are used with the process. The electrode is normally ground to a point or truncated cone configuration to minimize arc wandering. The operation of typical gas shielded arc welding machines may be found in TM 5-3431-211-15 and TM 5-3431-313-15. Figure 6-5 shows the relative position of the torch, arc, tungsten electrode, gas shield, and the welding rod (wire) as it is being fed into the arc and weld pool.

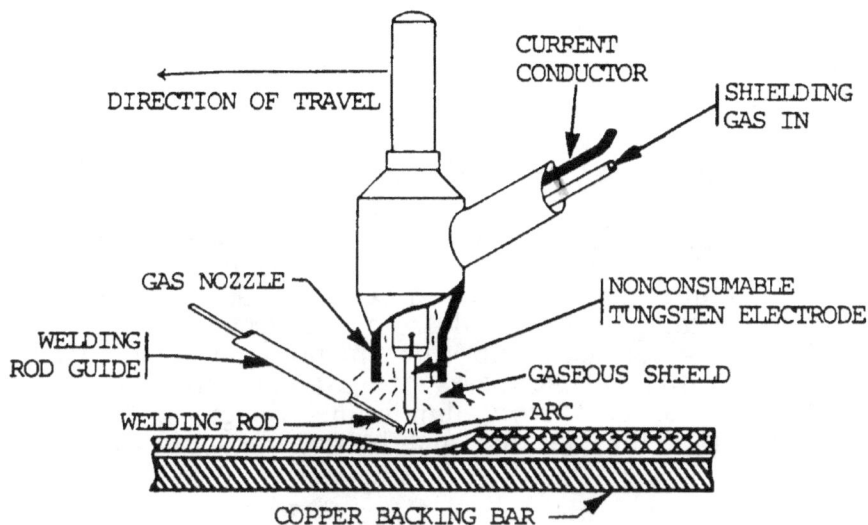

Figure 6-5. Gas tungsten arc welding.

(5) <u>Gas metal-arc Welding (MIG welding or GMAW)</u>In this process, coalescence is produced by heating metals with an arc between a continuous filler metal (consumable) electrode and the workpiece. The arc, electrode tip and molten weld metal are shielded from the atmosphere by a gas.Shielding is obtained entirely from an externally supplied inert gas, gas mixture, or a mixture a gas and a flux. The electrode wire for MIG welding is continuously fed into the arc and deposited as weld metal. Electrodes used for MIG welding are quite small in diameter compared to those used in other types of welding. Wire diameters 0.05 to 0.06 in. (0.13 to 0.15 cm) are average. Because of the small sizes of the electrode and high currents used in MIG welding, the melting rates of the electrodes are very high. Electrodes must always be provided as long, continuous strands of tempered wire that can be fed continuously through the welding equipment. Since the small electrodes have a high surface-to-volume ratio, they should be clean and free of contaminant which may cause weld defects such as porosity and cracking. Figure 6-6 shows the gas metal arc welding process. Allcommercially important metals such as carbon steel, stainless steel, aluminum, and copper can be welded with this process in all positions by choosing the appropriate shielding gas, electrode, and welding conditions.

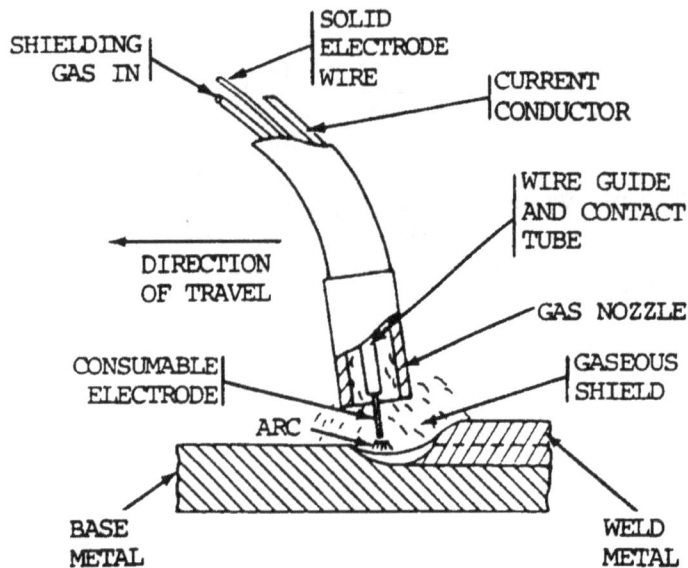

Figure 6-6. Gas metal arc welding.

(6) <u>Shielded metal-arc welding</u>. The arc is drawn between a covered consumable metal electrode and workpiece. The electrode covering is a source of arc stabilizers, gases to exclude air, metals to alloy the weld, and slags to support and protect the weld. Shielding is obtained from the decomposition of the electrode covering. Pressure is not used and filler metal is obtained from the electrode. Shielded metal arc welding electrodes are available to weld carbon and low alloy steels; stainless steels; cast iron; aluminum, copper, and nickel, and their alloys. Figure 6-7 describes the shielded metal arc welding process.

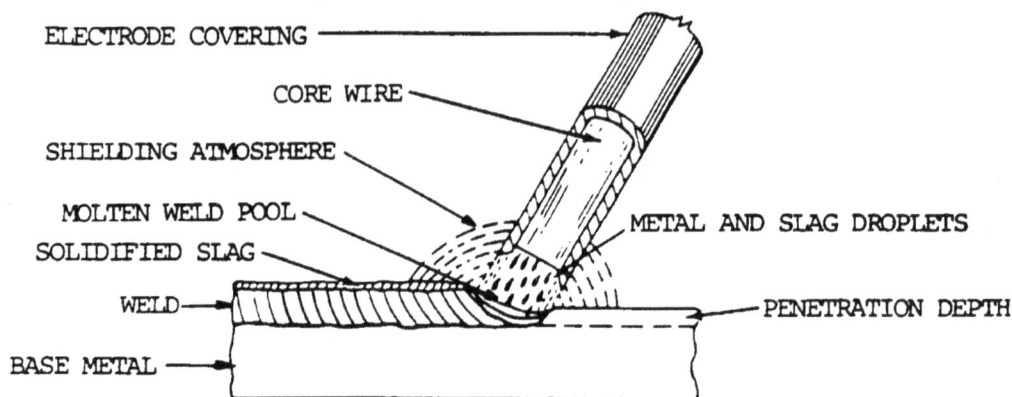

Figure 6-7. Shielded metal arc welding.

(7) <u>Atomic hydrogen welding</u>. The arc is maintained between two metal electrodes in an atmosphere of hydrogen. Shielding is obtained from the hydrogen. Pressure and/or filler metal may or may not be used. Although the process has limited industrial use today, atomic hydrogen welding is used to weld hard-to-weld metals, such as chrome, nickel, molybdenum steels, Inconel, Monel, and stainless steel. Its main application is tool and die repair welding and for the manufacture of steel alloy chain.

6-2. ARC WELDING (cont)

(8) Arc spot welding. An arc spot weld is a spot weld made by an arc welding process. A weld is made in one spot by drawing the arc between the electrode and workpiece. The weld is made without preparing a hole in either member. Filler metal, shielding gas, or flux may or may not be used.Gas tungsten arc welding and gas metal arc welding are the processes mostcommonly used to make arc spot welds. However, flux-cored arc welding and shielded metal arc welding using covered electrodes can be used for making arc spot welds.

(9) Arc seam welding. A continuous weld is made along faying surfaces by drawing the arc between an electrode and workpiece.Filler metal, shielding gas, or flux may or may not be used.

b. Carbon Electrode.

(1) Carbon-arc welding. In this process, the arc is drawnbetween an electrode and the workpiece. No shielding is USA. Pressure and/ofiller metal may or may not he used. Two types of electrodes are used for carbonarc welding: pure qraphite and baked carbon.The pure graphite electrode does notrode away as quickly as the carbon electrode,but is more expensive and more fragile.

(2) Twin carbon-arc welding. In this variation on carbon-arc welding, the arc is drawn between two carbon electrodes.When the two carbon electrodes are brought together,the arc is struck and established between them.The angle of the electrodes provides an arc that forms in front of the apex angle and fans out as a soft source of concentrated heat or arc flame, softer than a single carbon arc. Shielding and pressure are not used. Filler metal may or may not be used. The twin carbon-arc welding process can also be used for brazing.

(3) Gas-carbon arc welding. This process is also a variation of carbon arc welding, except shielding by inert gas or gas mixture is used.The arc is drawn between a carbon electrode and the workpiece.Shielding is obtained from an inert gas or gas mixture. Pressure and/or filler metal may or may not be used.

(4) Shielded carbon-arc welding. In this carbon-arc variation, the arc is drawn between a carbon electrode and the workpieceShielding is obtained from the combustion of a solid material fed into the arc, or from a blanket of flux on the arc, or both. Pressure and/or filler metal may or may not be used.

6-3. GAS WELDING

Gas welding processes are a group of welding processes in which a weld is made by heating with a gas flame or flares. Pressure and/or filler metal may or may not be used. Also referred to as oxyfuel gas welding, the term gas welding is used to describe any welding process that uses a fuel gas combined with oxygen, or in rare cases, with air, to produce a flame having sufficient energy to melt the base metal. The fuel gas and oxygen are mixed in the proper proportions in a chamber, which is generally a part of the welding tip assembly. The torch is designed to give the welder complete control of the welding flare, allowing the welder to regulate the melting of the base metal and the filler metal. The molten metal from the plate edges and the filler metal intermixin a common molten pool and join upon cooling to form one continuous piece.Manual welding methods are generally used.

Acetylene was originally used as the fuel gas in oxyfuel gas welding, but other gases, such as MAPP gas, have also been used. The flames must provide high localized energy to produce and sustain a molten pool. The flames can also supply a protective reducing atmosphere over the molten metal pool which is maintained during welding. Hydrocarbon fuel gases such as propane, butane, and natural gas are not suitable for welding ferrous materials because the heat output of the primary flame is too low for concentrated heat transfer, or the flame atmosphere is too oxidizing. Gas welding processes are outlined below.

a. Pressure Gas Welding. In this process, a weld is made simultaneously over the entire area of abutting surfaces with gas flames obtained from the combustion of a fuel gas with oxygen and the application of pressure. No filler metal is used. Acetylene is normally used as a fuel gas in pressure gas welding. Pressure gas welding has limited uses because of its low flame temperature, but is extensively used for welding lead.

b. Oxy-Hydrogen Welding. In this process, heat is obtained from the combustion of hydrogen with oxygen. No pressure is used, and filler metal may or may not be used. Hydrogen has a maximum flame temperature of 4820°F (2660 °C), but has limited use in oxyfuel gas welding because of its colorless flare, which makes adjustment of the hydrogen-oxygen ratio difficult. This process is used primarily for welding low melting point metals such as lead, light gage sections, and small parts.

c. Air-Acetylene Welding. In this process, heat is obtained from the combustion of acetylene with air. No pressure is used, and filler metal may or may not be used. This process is used extensively for soldering and brazing of copper pipe.

d. Oxy-Acetylene Welding. In this process, heat is obtained from the combustion of acetylene with oxygen. Pressure and/or filler metal may or may not be used. This process produces the hottest flame and is currently the most widely used fuel for gas welding.

e. Gas Welding with MAPP Gas. Standard acetylene gages, torches, and welding tips usually work well with MAPP gas. A neutral MAPP gas flame has a primary cone about 1 1/2 to 2 times as long as the primary acetylene flare. A MAPP gas carburizing flare will look similar to carburizing acetylene flare, and the MAPP gas oxidizing flame will look like the short, intense blue flare of the neutral acetylene flame. The neutral MAPP gas flame is a very deep blue.

6-4. BRAZING

Brazing is a group of welding processes in which materials are joined by heating to a suitable temperature and by using a filler metal with a melting point above 840 °F (449 °C), but helm that of the base metal. The filler metal is distributed to the closely fitted surfaces of the joint by capillary action. The various brazing processes are described below.

6-4. BRAZING (cont)

a. Torch Brazing (TB). Torch brazing is performed by heating the parts to be brazed with an oxyfuel gas torch or torches. Depending upon the temperature and the amount of heat required the fuel gas may be burned with air, compressed air, or oxygen. Brazing filler metal may be preplaced at the joint or fed from hand-held filler metal. Cleaning and fluxing are necessary. Automated TB machines use preplaced fluxes and preplaced filler metal in paste, wire, or shim form. For manual torch brazing, the torch may be equipped with a single tip, either single or multiple flame.

b. Twin Carbon-Arc Brazing. In this process, an arc is maintained between two carbon electrodes to produce the heat necessary for welding.

c. Furnace Brazing. In this process, a furnace produces the heat necessary for welding. In furnace brazing, the flame does not contact the workpiece. Furnace brazing is used extensively where the parts to be brazed can be assembled with the filler metal preplaced near or in the joint. Figure 6-8 illustrates a furnace brazing operation.

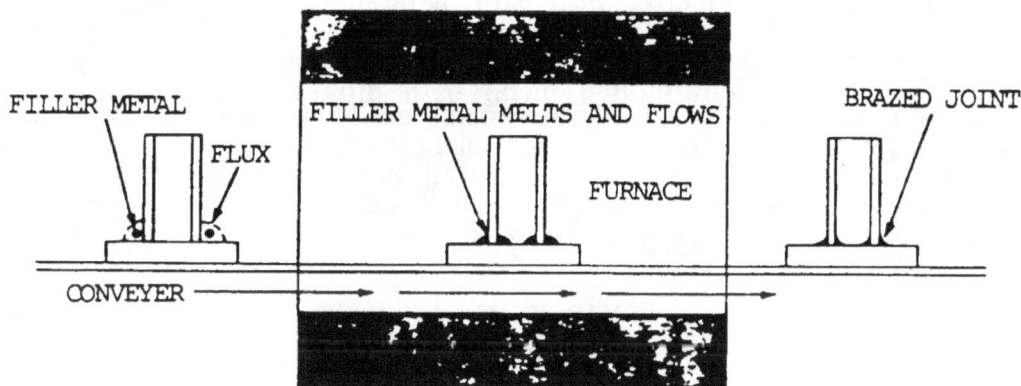

Figure 6-8. Furnace brazing operation.

d. Induction Brazing. In this process, the workpiece acts as a short circuit in the flow of an induced high frequency electrical current. The heat is obtained from the resistance of the workpiece to the current. Once heated in this manner, brazing can begin. Three common sources of high frequency electric current used for induction brazing are the motor-generator resonant spark gap, and vacuum tube oscillator. For induction brazing, the parts are placed in or near a water-cooled coil carrying alternating current. Careful design of the joint and the coil are required to assure the surfaces of all members of the joint reach the brazing temperature at the same time. Typical coil designs are shown in figure 6-9.

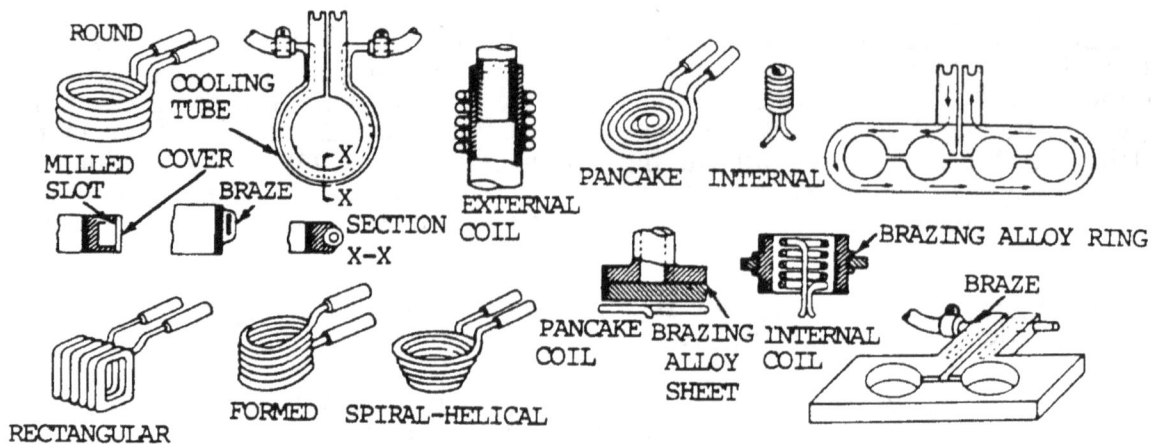

Figure 6-9. Typical induction brazing coils and joints.

e. Dip Brazing. There are two methods of dip brazing—chemcial bath and molten metal bath. In chemical bath dip brazing, the brazing filler metal is preplaced and the assembly is immersed in a bath of molten salt, as shown in figure 6-10. The salt bath furnishes the heat necessary for brazing and usually provides the necessary protection from oxidation. The salt bath is contained in a metal or other suitable pot and heated. In molten metal bath dip brazing, the parts are immersed in a bath of molten brazing filler metal contained in a suitable pot. A cover of flux should be maintained over the molten bath to protect it from oxidation. Dip brazing is mainly used for joining small parts such as wires or narrow strips of metal. The ends of wires or parts must be held firmly together when removed from the bath until the brazing filler metal solidifies.

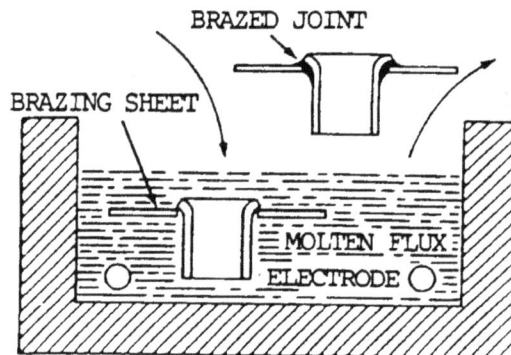

Figure 6-10. Chemical bath dip brazing.

f. Resistance Brazing. The heat necessary for resistance brazing is obtained from the resistance to the flow of an electric current through the electrodes and the joint to be brazed. The parts of the joint are a part of the electrical current. Brazing is done by the use of a low-voltage, high-current transformer. The conductors or electrodes for this process are made of carbon, molybdenum, tungsten, or steel. The parts to be brazed are held between two electrodes and the proper pressure and current are applied. Pressure should be maintained until the joint has solidified.

6-4.   BRAZING (cont)

g.   Block Brazing. In this process, heat is obtained from heated blocks applied to the to be joined.

h.   Flow Brazing. In flow brazing, heat is obtained from molten, nonferrous filler metal poured over the joint until the brazing temperature is obtained.

i.   Infrared Brazing (IRB). Infared brazing uses a high intensity quartz lamp as a heat source. The process is suited to the brazing of very thin materials and is normally not used on sheets thicker than 0.50 in. (1.27 cm) Assembies to be brazed are supported in a position which enables radiant energy to be focused on the joint. The assembly and the lamps can be placed in an evacuated or controlled atmosphere. Figure 6-11 illustratesthe equipment used for infrared brazing.

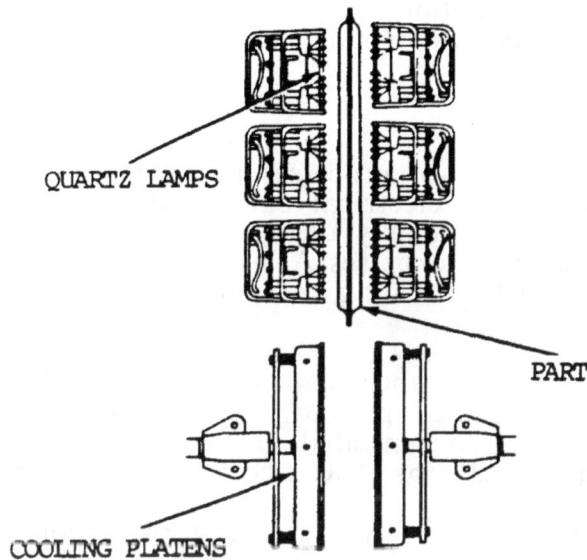

Figure 6-11. Infrared brazing apparatus.

j.   Diffusion Brazing (DFB). Unlike all of the previous brazing processes, diffusion brazing is not defined by its heat source, but by the mechanism in-volved. A joint is formed by holding the brazement at a suitable temperature for a sufficient time to allow mutual diffusion of the base and filler metal.The joint produced has a composition considerably different than either the filler metal or base metal, and no filler metal should be discernible in the finished micro-structure. The DFB process produces stronger joints than the normal brazed joint. Also, the DFB joint remelts at temperatures approaching that of the base metal. The typical thickness of the base metals that are diffusion brazed range from very thin foil up to 1 to 2 in. (2.5 to 5.1 cm). Much heavier parts can also be brazed since thickness has very little bearing on the process. Many parts that are difficult to braze by other processes can be diffusion brazed. Both butt and lap joints having superior mechanical properties can be produced, and the parts are usually fixtured mechanically or tack welded together.Although DFB requires a relatively long period of time (30minutes to as long as 24 hours) to complete, it can produce many parts at the same time at a reasonable cost. Furnances are most frequently used for this method ofprocessing.

k. Special Processes.

(1) Blanket brazing is another process used for brazing. A blanket is resistance heated, and most of the heat is transferred to the parts by conduction and radiation. Radiation is responsible for the majority of the heat transfer.

(2) Exothemic brazing is another special process, by which the heat required to melt and flow a commercial filler metal is generated by a solid state exothermic chemical reaction. An exothermic chemical reaction is any reaction between two or more reactants in which heat is given off due to the free energy of the system. Exothermic brazing uses simple tooling and equipment. The process uses the reaction heat in bringing adjoining or nearby metal interfaces to a temperature where preplaced brazing filler metal will melt and wet the metal interface surfaces. The brazing filler metal can be a commercially available one having suitable melting and flow temperatures. The only limitations may be the thickness of the metal that must be heated through and the effects of this heat, or any previous heat treatment, on the metal properties.

6-5. RESISTANCE WELDING

Resistance welding consists of a group of processes in which the heat for welding is generated by the resistance to the electrical current flow through the parts being joined, using pressure. It is commonly used to weld two overlapping sheets or plates which may have different thicknesses. A pair of electrodes conducts electrical current through the sheets, forming a weld. The various resistance processes are outlined below.

a. Resistance Spot Welding. In resistance spot welding, the size and shape of the individually formed welds are limited primarily by the size and contour of the electrodes. The welding current is concentrated at the point of joining using cylindrical electrodes with spherical tips. The electrodes apply pressure.

b. Resistance Seam Welding. This weld is a series of overlapping spot welds made progressively along a joint by rotating the circular electrodesSuch welds are leaktight. A variation of this process is the roll spot weld, in which the spot spacing is increased so that the spots do not-overlap and the weld is not leaktight. In both processes, the electrodes apply pressure.

c. Projection Welding. These welds are localized at points predetermced by the design of the parts to be welded. The localization is usually accomplished by projections, embossments, or intersections. The electrodes apply pressure.

d. Flash Welding. In this process, heat is created at the joint by its resistance to the flow of the electric current, and the metal is heated above its melting point. Heat is also created by arcs at the interface. A force applied immediately following heating produces an expulsion of metal and the formation of a flash. The weld is made simultaneously over the entire area of abutting surfaces by the application of pressure after the heating is substantially completed.

6-5. RESISTANCE WELDING (cont)

    e. Upset Welding. In this process, the weld is made either simultaneously over the entire area of two abutting surfaces, or progressively along a joint. Heat for welding is obtained from the resistance to the flow of electric current through the metal at the joint. Force is applied to upset the joint and start a weld when the metal reaches welding temperature. In some cases, force is applied before heating starts to bring the faying surfaces in contact. Pressure is maintained throughout the heating period.

    f. Percussion Welding. This weld is made simultaneously over the entire area of abutting surfaces by the heat obtained from an arc. The arc is produced by a rapid discharge of electrical energy. It is extinguished by pressure applied percussively during the discharge.

    g. High–Frequency Welding. High frequency welding includes those processes in which the joining of metals is produced by the heat generated from the electrical resistance of the workpiece to the flow of high-frequency current, with or without the application of an upsetting force. The two processes that utilize high-frequency current to produce the heat for welding are high-frequency resistance welding and high-frequency induction welding, sometimes called induction resistance welding. Almost all high-frequency welding techniques apply sane force to bring the heated metals into close contact. During the application or force, an upset or bulging of metal occurs in the weld area.

6-6. THERMIT WELDING

    a. Thermit welding (TW) is a process which joins metals by heating them with superheated liquid metal from a chemical reaction between a metal oxide and aluminum or other reducing agent, with or without the application of pressure. Filler metal is obtained from the liquid metal.

    b. The heat for welding is obtained from an exothermic reaction or chemical change between iron oxide and aluminum. This reaction is shown by the following formula:

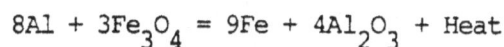

$$8Al + 3Fe_3O_4 = 9Fe + 4Al_2O_3 + Heat$$

The temperature resulting from this reaction is approximately 4500 (2482 °C).

    c. The superheated steel is contained in a crucible located immediately above the weld joint. The exothermic reaction is relatively slow and requires 20 to 30 seconds, regardless of the amount of chemicals involved. The parts to be welded are alined with a gap between them. The superheated steel runs into a mold which is built around the parts to be welded. Since it is almost twice as hot as the melting temperature of the base metal, melting occurs at the edges of the joint and alloys with the molten steel from the crucible. Normal heat losses cause the mass of molten metal to solidify, coalescence occurs, and the weld is completed. If the parts to be welded are large, preheating withtin the mold cavity may be necessary to bring the pats to welding temperature and to dry out the mold. If the parts are small, preheating is often eliminated. The thermit welding process is applied only in the automatic mode. Once the reaction is started, it continues until completion.

d. Themit welding utilizes gravity, which causes the molten metal to fill the cavity between the parts being welded. It is very similar to the foundry practice of pouring a casting. The difference is the extremely high temperature of the molten metal. The making of a thermit weld is shown in figure 6-12. When the filler metal has cooled, allunwanted excess metal may be removed by oxygen cutting, machining, or grinding. The surface of the completed weld is usually sufficiently smooth and contoured so that it does not require additional metal finishing. Information on thermit welding equipment may be found on p 5-50.

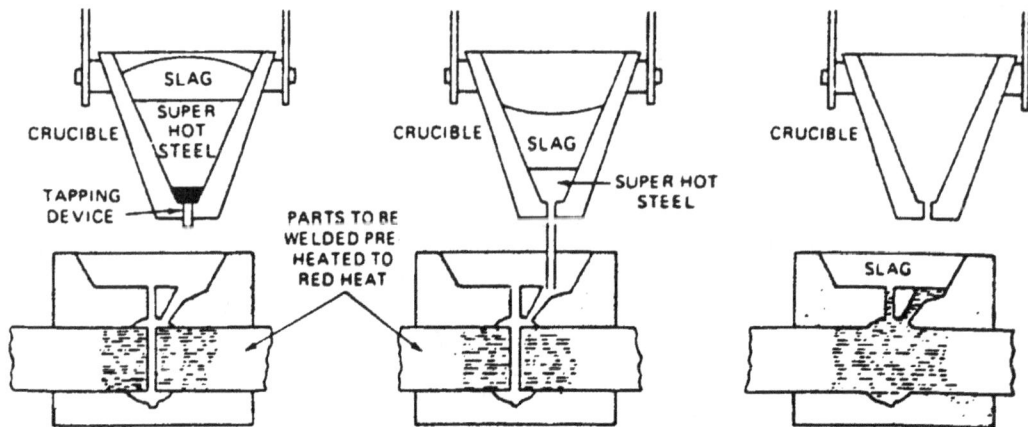

Figure 6-12. Steps in making a thermit weld.

e. The amount of thermit is calculated to provide sufficient metal to produce the weld. The amount of steel produced by the reaction is approximately one-half the original quantity of thermit material by weight and one-third by volume.

f. The deposited weld metal is homgenous and quality is relatively high. Distortion is minimized since the weld is accomplished in one pass and since cooling is uniform across the entire weld cross section. There is normally shrinkage across the joint, but little or no angular distortion.

g. Welds can be made with the parts to be joined in almost any position as long as the cavity has vertical sides. If the cross-sectional area or thicknesses of the parts to be joined are quite large, the primary problem is to provide sufficient thermit metal to fill the cavity.

h. Thermit welds can also be used for welding nonferrous materials. The most popular uses of nonferrous thermit welding are the joining of copper and aluminum conductors for the electrical industry. In these cases, the exothermic reaction is a reduction of copper oxide by aluminum, which produces molten superheated copper. The high-temperature molten copper flows into the mold, melts the ends of the parts to be welded, and, as the metal cools, a solid homgenous weld results In welding copper and aluminum cables, the molds are made of graphite and can be used over and over. When welding nonferrous materials, the parts to be joined must be extremely clean. A flux is normally applied toc the joint prior to welding. Special kits are available that provide the molds for different sizes of cable and the premixed thermit material. This material also includes enough of the igniting material so that the exothermic reaction is started by means of a special lighter.

Section II. NOMENCLATURE OF THE WELD

6-7. GENERAL

Common terms used to describe the various facets of the weld are explained in paragraphs 6-8 and 6-9 and are illustrated in figure 6-13.

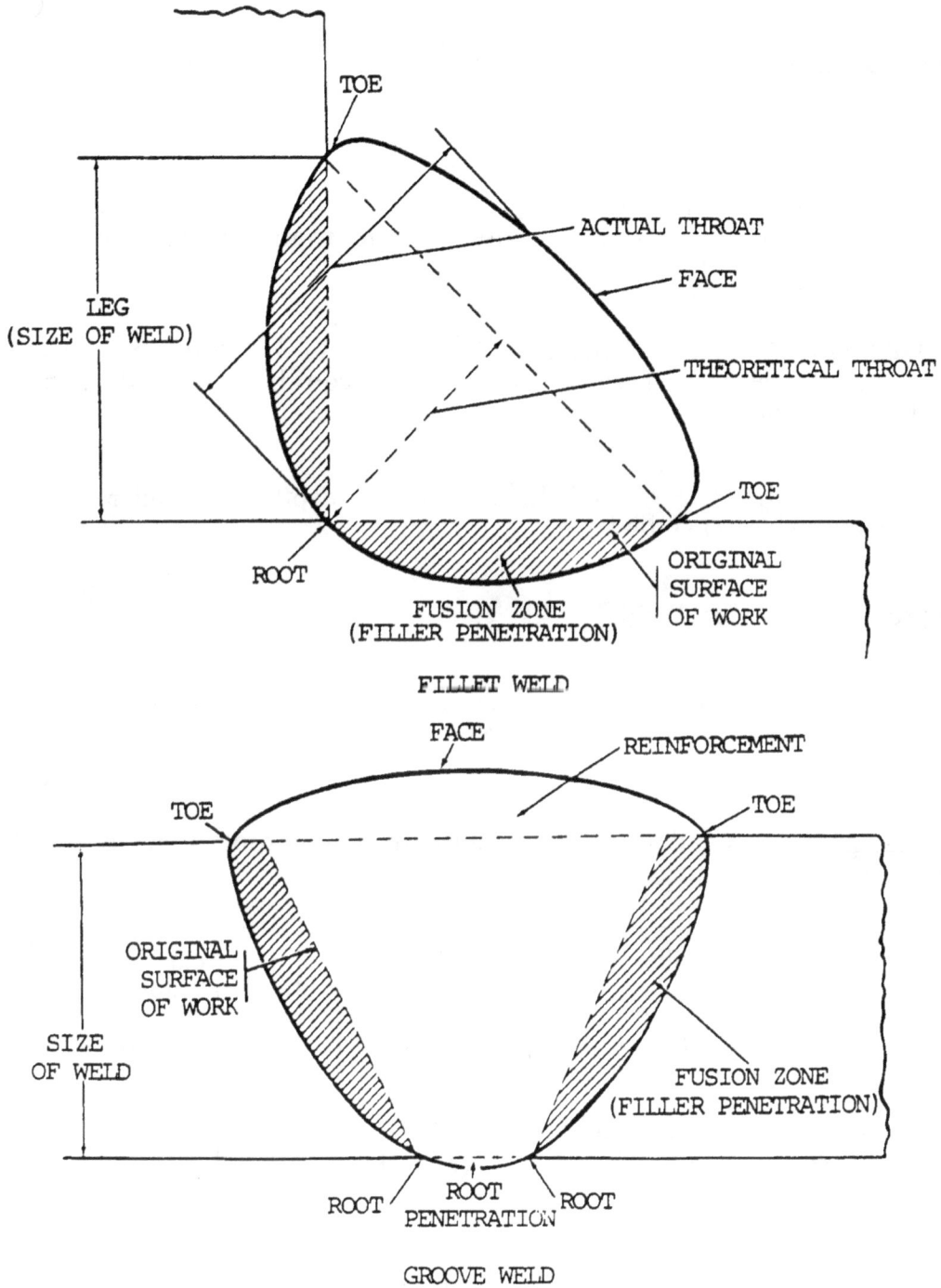

Figure 6-13. Nomenclature of welds.

6-8. SECTIONS OF A WELD

a. _Fusion Zone (Filler Penetration)_. The fusion zone is the area of base metal melted as determined in the cross section of a weld.

b. _Leg of a Fillet Weld_. The leg of a fillet weld is the distance from the root of the joint to the toe of the fillet weld. There are two legs in a fillet weld.

c. _Root of the Weld_. This is the point at which the bottom of the weld intersects the base metal surface, as shown in the cross section of weld.

d. _Size of the Weld_.

(1) _Equal leg-length fillet welds_. The size of the weld is designated by the leg-length of the largest isosceles right triangle that can be inscribed within the fillet weld cross section.

(2) _Unequal leg-length fillet welds_. The size of the weld is designated by the leg-length of the largest right triangle that can be inscribed within the fillet weld cross section.

(3) _Groove weld_. The size of the weld is the depth of chamfering plus the root penetration when specified.

e. _Throat of a Fillet Weld_.

(1) _Theoretical throat_. This is the perpendicular distance between the root of the weld and the hypotenuse of the largest right triangle that can be inscribed within tie fillet weld cross section.

(2) _Actual throat_. This is the distance from the root of a fillet weld to the center of its face.

f. _Face of the Weld_. This is the exposed surface of the weld, made by an arc or gas welding process, on the side from which the welding was done.

g. _Toe of the Weld_. This is the junction between the face of the weld and the base metal.

h. _Reinforcement of the Weld_. This is the weld metal on the face of a groove weld in excess of the metal necessary for the specified weld size.

6-9. MULTIPASS WELDS

a. The nomenclature of the weld, the zones affected by the welding heat when a butt weld is made by more than one pass or layer, and the nomenclature applying to the grooves used in butt welding are shown in figure 6-14. Figure 6-15 is based on weld type and position.

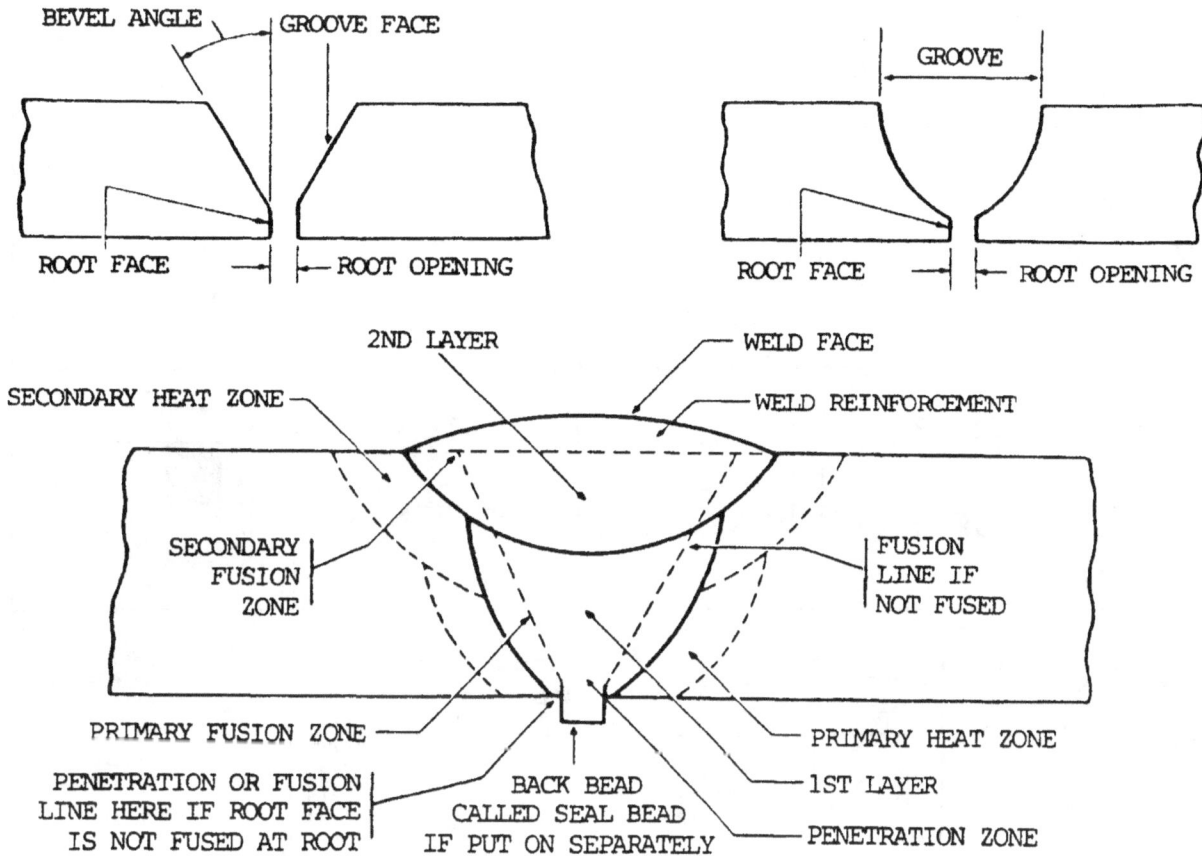

Figure 6-14. Heat affected zones in a multipass weld.

b. The primary heat zone is the area fused or affected by heat in the first pass or application of weld metal. The secondary heat zone is the area affected in the second pass and overlaps the primary heat zone. The portion of base metal that hardens or changes its properties as a result of the welding heat in the primary zone is partly annealed or softened by the welding heat in the secondary zone. The weld metal in the first layer is also refined in structure by the welding heat of the second layer. The two heating conditions are important in determing the order or sequence in depositing weld metal in a particular joint design.

FILLET SIZE	WELDING POSITION			
	FLAT 1F	HORIZONTAL 1F	VERTICAL UP 3F (U)	OVERHEAD 4F
1/4	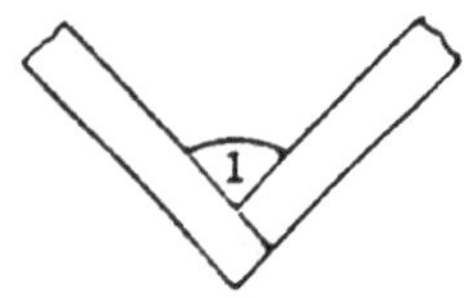	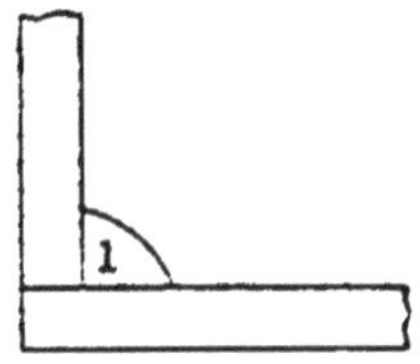	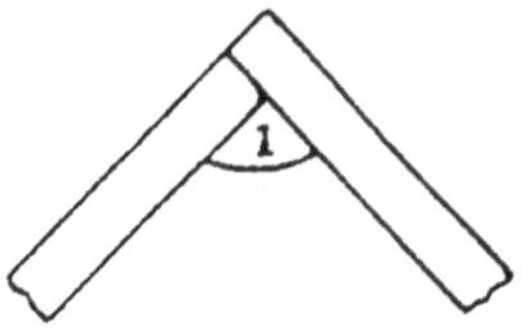	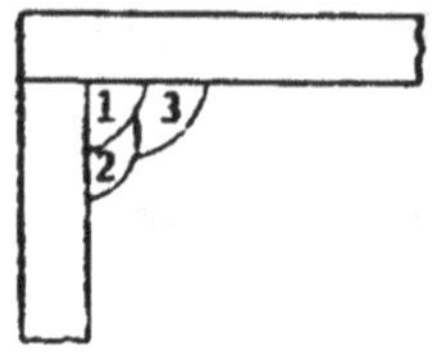
1/2	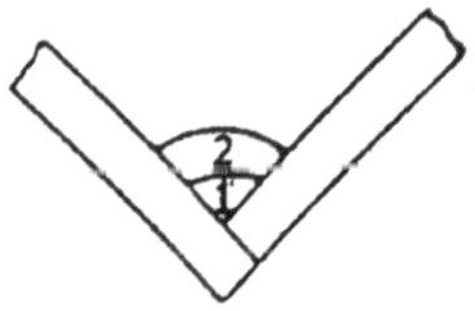	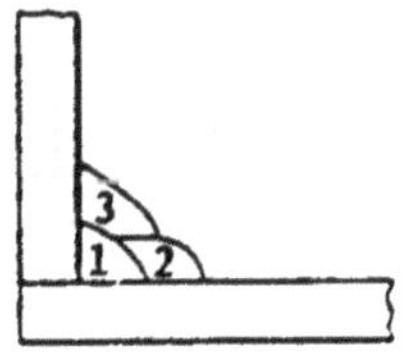	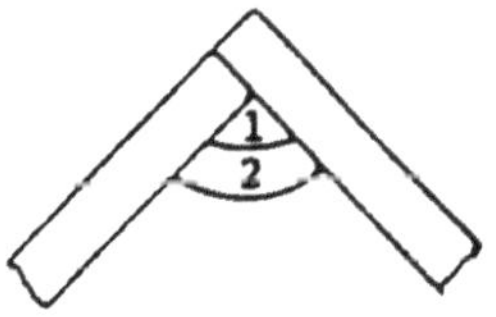	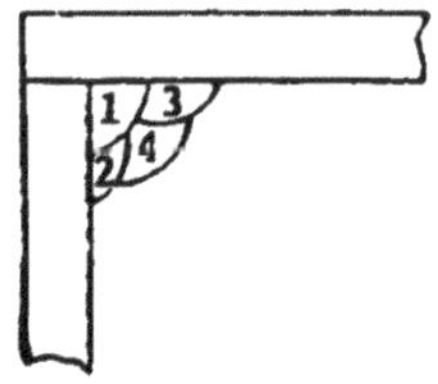
3/4	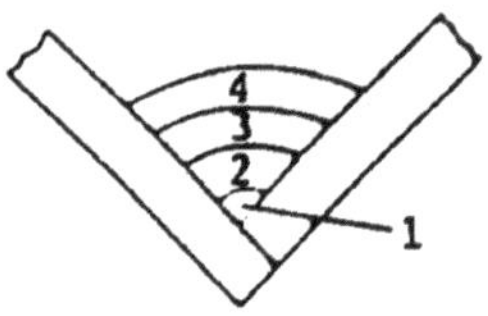	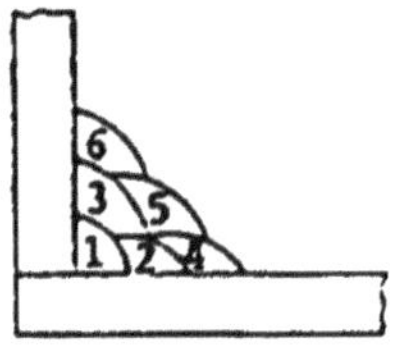	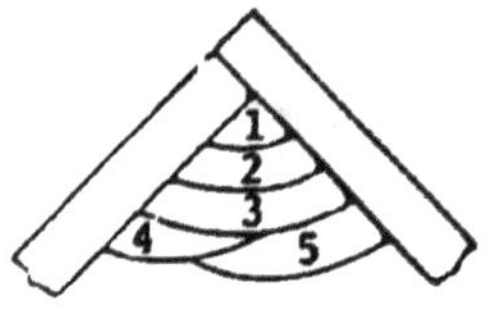	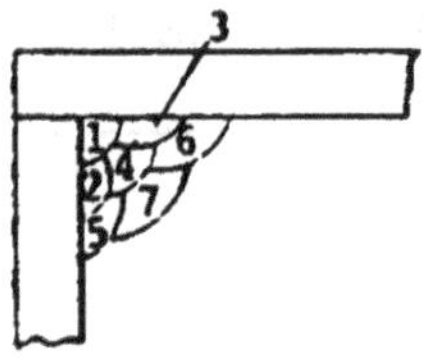

WELDING POSITION	SUGGESTED ELECTRODE TYPE	ELECTRODE DIAMETER FOR MATERIAL THICKNESS (IN.)		
		1/4	1/2	3/4
1F, 2F	E7024	1/4	1/4	1/4
3F (U)	E7018	5/32	5/32	5/32
4F	E6010	3/16	3/16	3/16
	E7018	5/32	5/32	5/32

MATERIAL THICKNESS (IN.)	WELDING POSITION
1/8	ALL POSITIONS
3/16	
1/4	

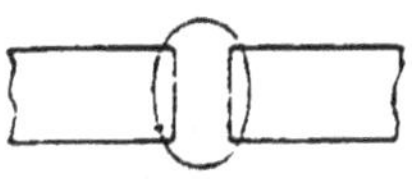

WELDING POSITION	SUGGESTED ELECTRODE TYPE	ELECTRODE DIAMETER FOR MATERIAL THICKNESS (IN.)		
		1/8	3/16	1/4
1G	E6010	3/32	1/8	5/32
2G, 3G(U)	E6010, E6012			
3G(D), 4G	E6014, E6013	3/32	1/8	5/32

Figure 6-15. Welding procedure schedule--various welds (sheet 1 of 3).

6-9. MULTIPASS WELDS (cont)

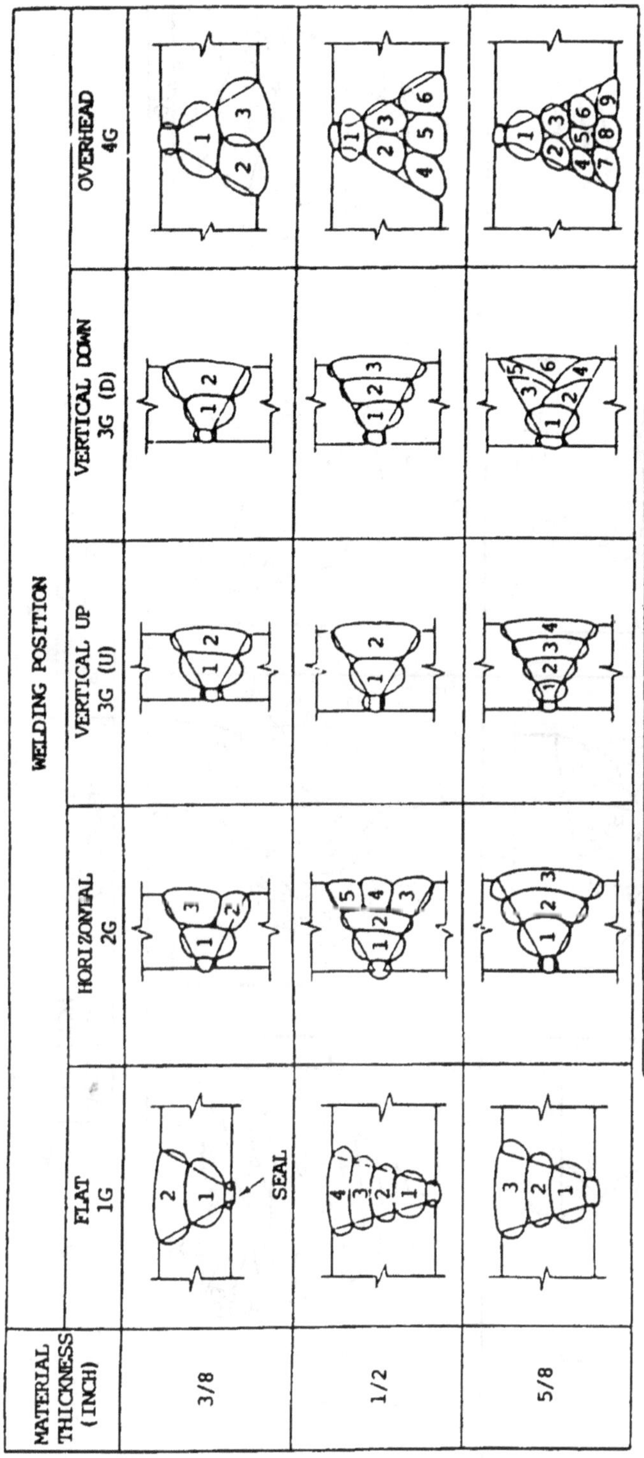

Figure 6-15. Welding procedure schedule--various welds (sheet 2 of 3).

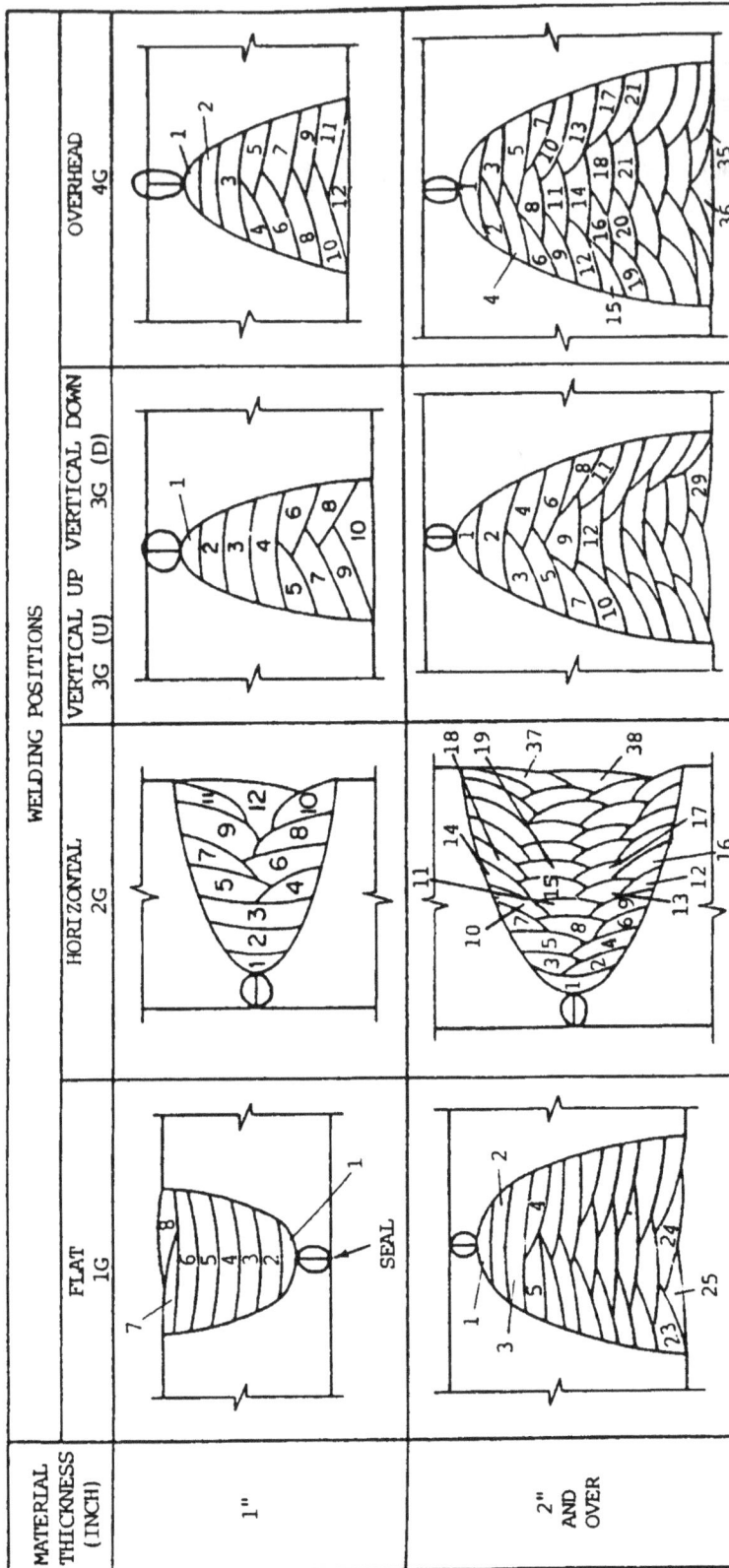

Figure 6-15. Welding procedure schedule--various welds (sheet 3 of 3).

Section III. TYPES OF WELDS AND WELDED JOINTS

6-10. GENERAL

a. <u>Welding</u> is a materials joining process used in making welds. A <u>weld</u> is a localized coalescence of metals or nonmetals produced either by heating the materials to a suitable temperate with or without the application of pressure, or by the application of pressure alone, with or without the use of filler metal. <u>Coalescence</u> is a growing together or a growing into one body, and is used in all of the welding process definitions. A <u>weldment</u> is an assembly of component parts joined by welding, which can be made of many or few metal parts. A weldment may contain metals of different compositions, and the pieces may be in the form of rolled shapes, sheet, plate, pipe, forgings, or castings. To produce a usable structure or weldnent, there must be weld joints between the various pieces that make the weldment. The <u>joint</u> is the junction of members or the edges of members which are to be joined or have been joined. <u>Filler metal</u> is the material to be added in making a welded, brazed, or soldered joint. <u>Base metal</u> is the material to be welded, soldered, or cut.

b. The properties of a welded joint depend partly on the correct preparation of the edges being welded. All mill scale, rust, oxides, and other impurities must be removed from the joint edges or surfaces to prevent their inclusion in the weld metal. The edges should be prepared to permit fusion without excessive melting. Care must be taken to keep heat loss due to radiation into the base metal from the weld to a minimum. A properly prepared joint will keep both expansion on heating and contraction on coaling to a minimum.

c. Preparation of the metal for welding depends upon the form, thickness, and kind of metal, the load the weld will be required to support, and the available means for preparing the edges to be joined.

d. There are five basic types of joints for bringing two members together for welding. These joint types or designs are also used by other skilled trades. The five basic types of joints are described below and shown in figure 6-16.

(1) <u>B. Butt joint</u> - parts in approximately the same plane.

(2) <u>C. Corner joint</u> - parts at approximately right angles and at the edge of both parts.

(3) <u>E. Edge joint</u> - an edge of two or more parallel parts.

(4) <u>L. Lap joint</u> - between overlapping parts.

(5) <u>T. T joint</u> - parts at approximately right angles, not at the edge of one part.

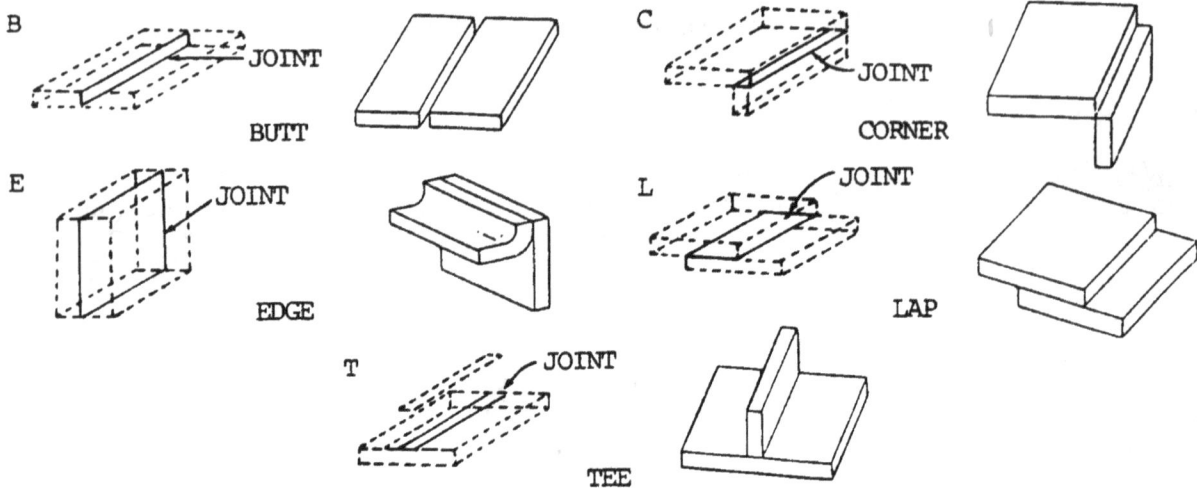

Figure 6-16. Basic joint types.

6-11. BUTT JOINT

a. This type of joint is used to join the edges of two plates or surfaces located in approximately the same plane. Plane square butt joints in light sections are shown in figure 6-17. Grooved butt joints for heavy sections with several types of edge preparation are shown in figure 6-18. These edges can be prepared by flame cutting, shearing, flame grooving, machining, chipping, or carbon arc air cutting or gouging. The edge surfaces in each case must be free of oxides, scales, dirt, grease, or other foreign matter.

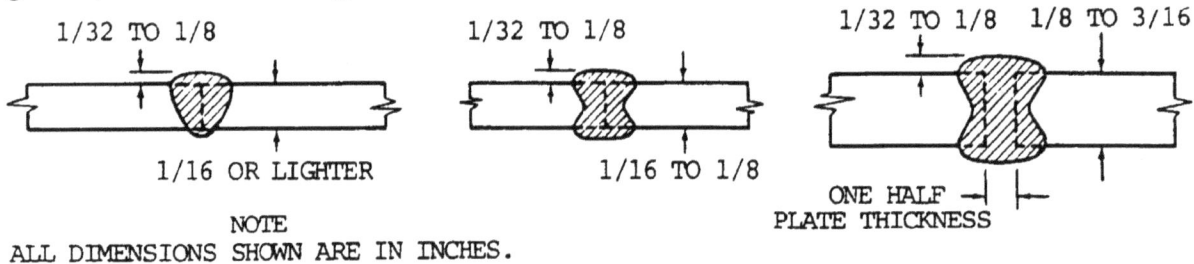

NOTE
ALL DIMENSIONS SHOWN ARE IN INCHES.

Figure 6-17. Butt joints in light sections.

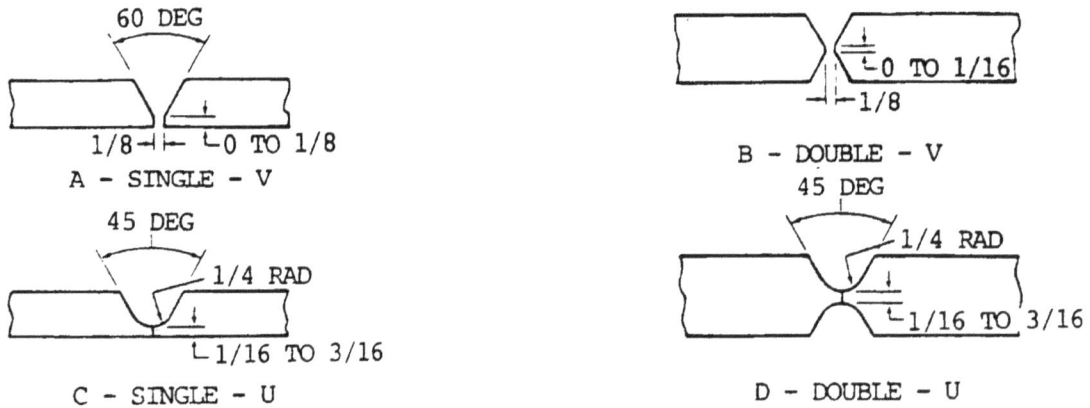

Figure 6-18. Butt joints in heavy sections.

6-11.   BUTT JOINT (cont)

b.   The square butt joints shown in figure 6-16 are used for butt welding light sheet metal. Plate thicknesses 3/8 to 1/2 in. (0.95 to 1.27 an) can be welded using the single V or single U joints as shown in views A and C, figure 6-18, p. 6-21.   The edges of heavier sections (1/2 to 2 in. (1.27 to 5.08 an)) are prepared as shown in view B, figure 6-18, p 6-21. Thicknesses of 3/4 in. (1.91 cm) and up are prepared as shown in view D, figure 6-18, p 6-21.The edges of heavier sections should be prepared as shown in views B and D, figure 6-18, p 6-21. The single U groove (view C, fig. 6-18, p 6-21) is more satisfactory and requires less filler metal than the single V groove when welding heavy sections and when welding in deep grooves. The double V groove joint requires approximately one-half the amount of filler metal used to produce the single V groove joint for the same plate thickness.   In general, butt joints prepared from both sides permit easier welding, produce less distortion, and insure better weld metal qualities in heavy sections than joints prepared from one side only.

6-12.   CORNER JOINT

a.   The common corner joints are classified as flush or closed, half open, and full open.

b.   This type of joint is used to join two members located at approximately right angles to each other in the form of an L.The fillet weld corner joint (view A, fig. 6-19) is used in the construction of boxes, box frames, tanks, and similar fabrications.

c.   The closed corner joint (view B, fig. 6-19) is used on light sheet metal, usually 20 gage or less, and on lighter sheets when high strength is not required at the joint. In making the joint by oxyacetylene welding, the overlapping edge is melted dawn, and little or no filler metal is added.In arc welding, only a very light bead is required to make the joint.When the closed joint is used for heavy sections, the lapped plate is V beveled or U grooved to permit penetration to the root of the joint.

d.   Half open comer jointsare suitable for material 12 gageand heavier. This joint is used when welding canonly be performed on one side andwhen loads will not be severe.

e.   The open corner joint (view C,fig. 6-19) is used on heavier sheets and plates.   The two edges are melted down and filler metal is added to fill up the corner.   This type of joint is the strongest of the corner joints.

f.   Corner joints on heavy plates are welded from both sides as shown in view D, figure 6-19.   The joint is first welded from the outside, then reinforced fromthe back side with a seal bead.

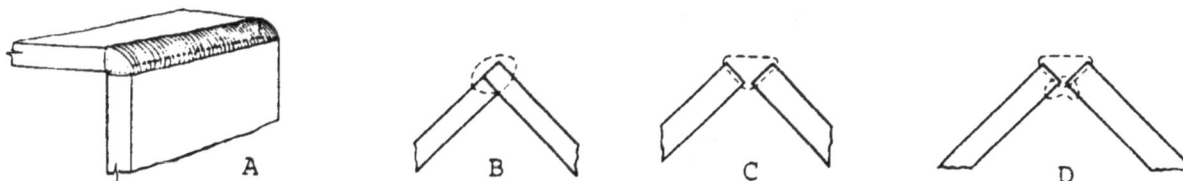

Figure 6-19.   Corner joints for sheets and plates.

6-13.   EDGE JOINT

This type of joint is used to join two or more parallel or nearly parallel members. It is not very strong and is used to join edges of sheet metal, reinforcing plates in flanges of I beams, edges of angles, mufflers, tanks for liquids, housing, etc. Two parallel plates are joined together as shown in view A, figure 6-20. On heavy plates, sufficient filler metal is added to fuse or melt each plate edge completely and to reinforce the joint.

b.   Light sheets are welded as shown in view B, figure 6-20. No preparation is necessary other than to clean the edges and tack weld them in position. The edges are fused together so no filler metal is required. The heavy plate joint as shown in view C, figure 6-20, requires that the edges be beveled in order to secure good penetration and fusion of the side walls. Filler metal is used in this joint.

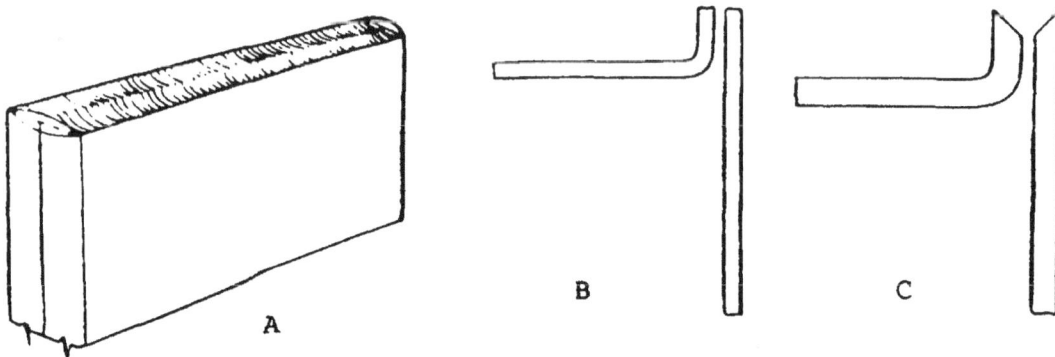

Figure 6-20.   Edge joints for light sheets and plates.

6-14.   LAP JOINT

This type of joint is used to join two overlapping members. A single lap joint where welding must be done from one side is shown in view A, figure 6-21. The double lap joint is welded on both sides and develops the full strength of the welded members (view B, fig. 6-21). An offset lap joint (view C, fig. 6-21) is used where two overlapping plates must be joined and welded in the same plane. This type of joint is stronger than the single lap type, but is more difficult to prepare.

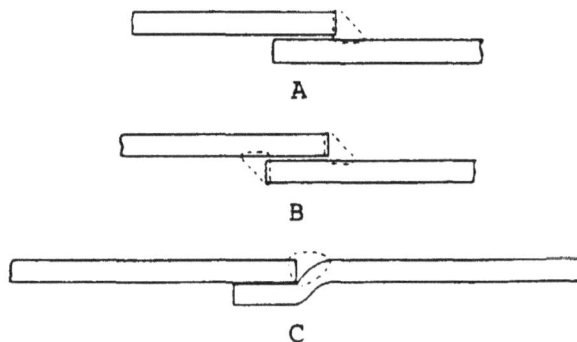

Figure 6-21. Lap joints.

6-15. TEE JOINT

a. Tee joints are used to weld two plates or sections with surfaces located approximately 90 degrees to each other at the joint, but the surface of one plate or section is not in the same plane as the end of the other surface. A plain tee joint welded from both sides is shown in view B, figure 6-22. The included angle of bevel in the preparation of tee joints is approximately half that required for butt joints.

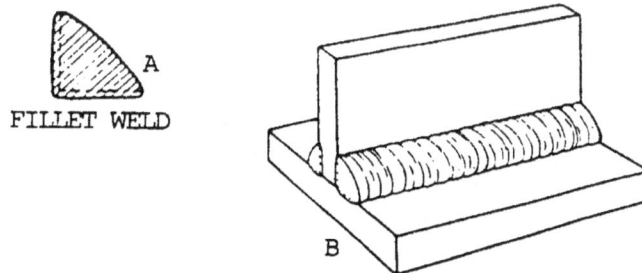

Figure 6-22. Tee joint - single pass fillet weld.

b. Other edge preparations used in tee joints are shown in figure 6-23. A plain tee joint, which requires no preparation other than cleaning the end of the vertical plate and the surface of the horizontal plate, is shown in view A, figure 6-23. The single beveled joint (view B, fig. 6-23) is used in plates and sections Up to 1/2 in. (1.27 cm) thick. The double beveled joint (view C, fig. 6-23) is used on heavy plates that can be welded from both sides. The single J joint (view) D, fig. 6-23) is used for welding plates 1 in. thick or heavier where welding is done from one side. The double J joint (view E, fig. 6-23) is used for welding very heavy plates from both sides.

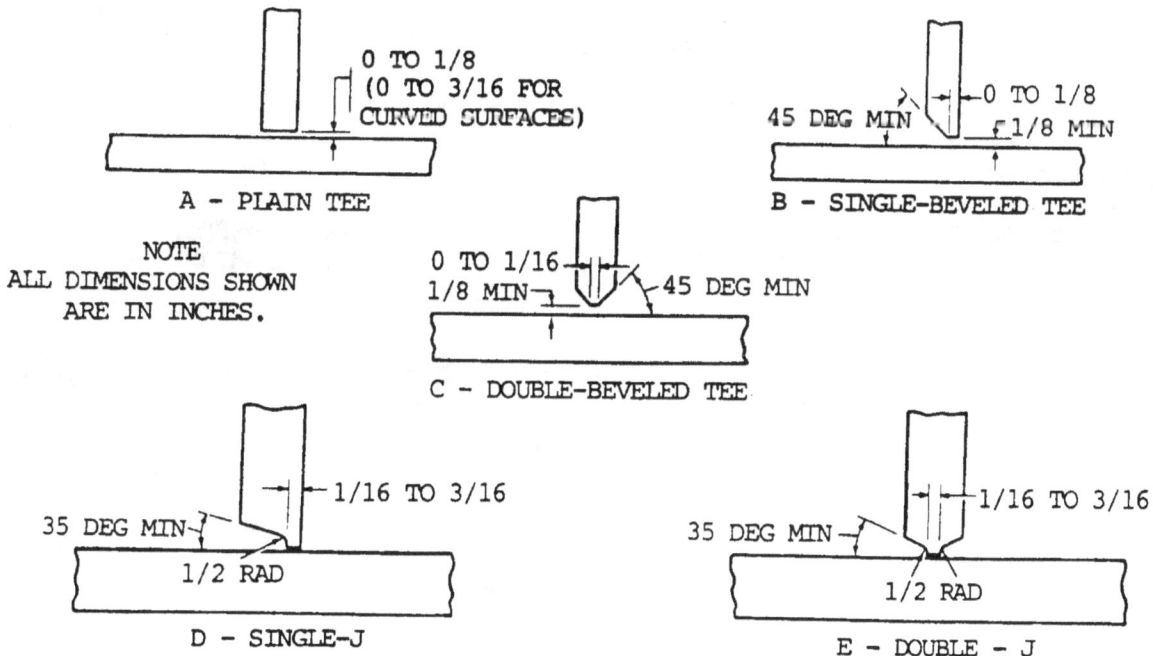

Figure 6-23. Edge preparation for tee joints.

c. Care must be taken to insure penetration into the root of the weld. This penetration is promoted by root openings between the ends of the vertical members and the horizontal surfaces.

6-16. TYPES OF WELDS

a. General. It is important to distinguish between the joint and the weld. Each must be described to completely describe the weld joint. There are many different types of welds, which are best described by their shape when shown in cross section. The most popular weld is the fillet weld, named after its cross-sectional shape. Fillet welds are shown by figure 6-24. The second nest popular is the groove weld. There are seven basic types of groove welds, which are shown in figure 6-25. Other types of welds include flange welds, plug welds, slot welds, seam welds, surfacing welds, and backing welds. Joints are combined with welds to make weld joints. Examples are shown in figure 6-26, p 6-26. The type of weld used will determine the manner in which the seam, joint, or surface is prepared.

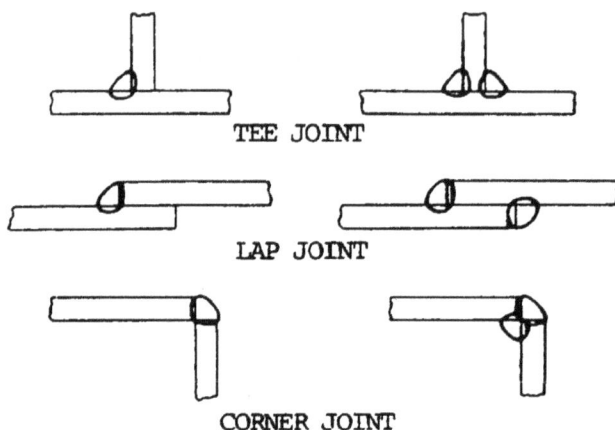

TEE JOINT

LAP JOINT

CORNER JOINT

Figure 6-24. Applications of fillet welds--single and double.

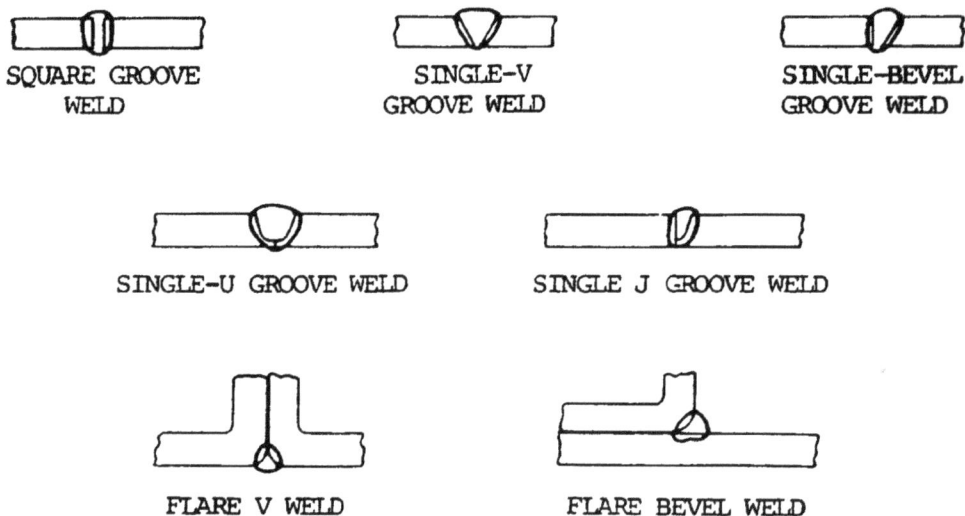

SQUARE GROOVE WELD

SINGLE-V GROOVE WELD

SINGLE-BEVEL GROOVE WELD

SINGLE-U GROOVE WELD

SINGLE J GROOVE WELD

FLARE V WELD

FLARE BEVEL WELD

Figure 6-25. Basic groove welds.

6-16.   TYPES OF WELDS (cont)

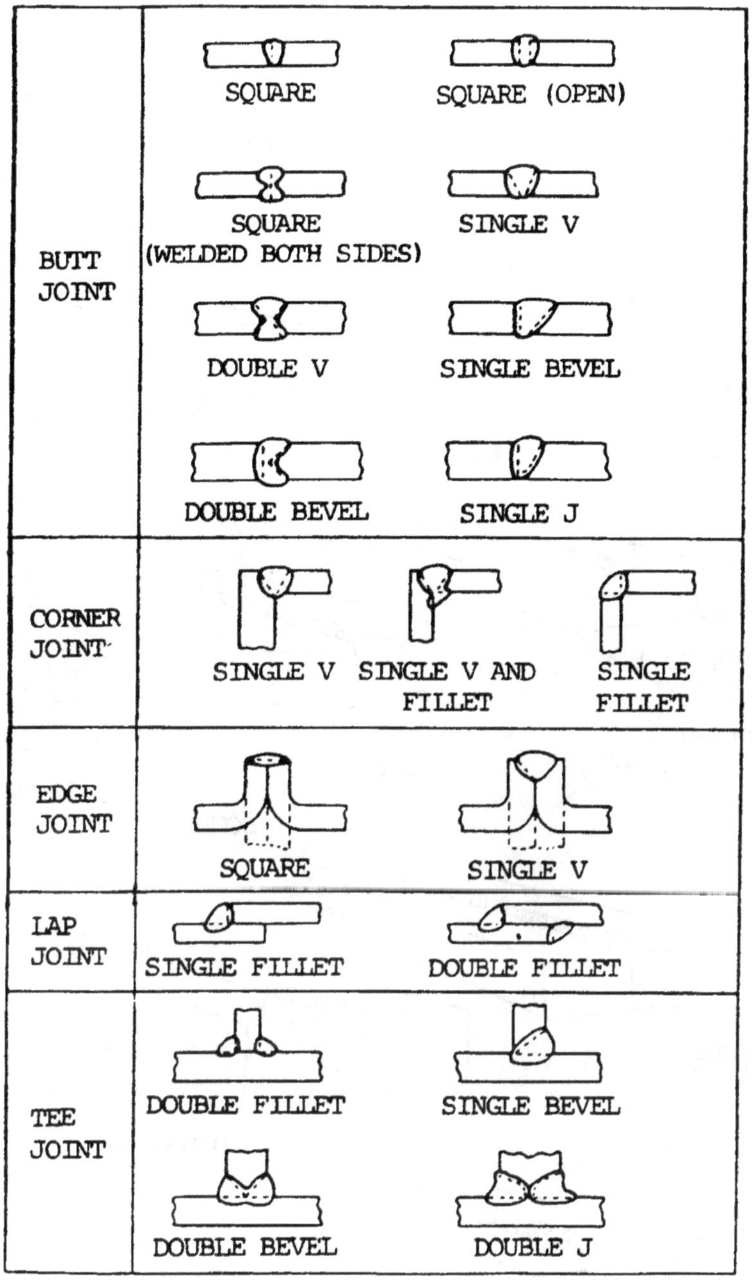

BUTT JOINT	SQUARE · SQUARE (OPEN) · SQUARE (WELDED BOTH SIDES) · SINGLE V · DOUBLE V · SINGLE BEVEL · DOUBLE BEVEL · SINGLE J
CORNER JOINT	SINGLE V · SINGLE V AND FILLET · SINGLE FILLET
EDGE JOINT	SQUARE · SINGLE V
LAP JOINT	SINGLE FILLET · DOUBLE FILLET
TEE JOINT	DOUBLE FILLET · SINGLE BEVEL · DOUBLE BEVEL · DOUBLE J

Figure 6-26. Typical weld joints

   b.   Groove Weld. These are beads deposited in a groove between two members to be joined. See figure 6-27 for the standard types of groove welds.

LESS THAN
1/8 THICK

SQUARED FOR ONE-SIDE WELD   SQUARED FOR WELDING BOTH SIDES

SINGLE-BEVEL

DOUBLE-BEVEL

90 TO
120 DEG

1/16

3/32 TO 1/8   (1/8 TO 1/4
THICK)

SINGLE-V

(OVER 1/4
THICK)

90 TO
120 DEG

1/16

3/32 TO 1/8

DOUBLE V

SINGLE-J

DOUBLE-J

SINGLE-U

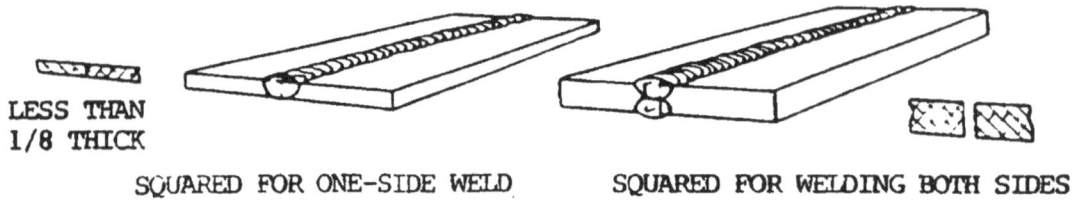

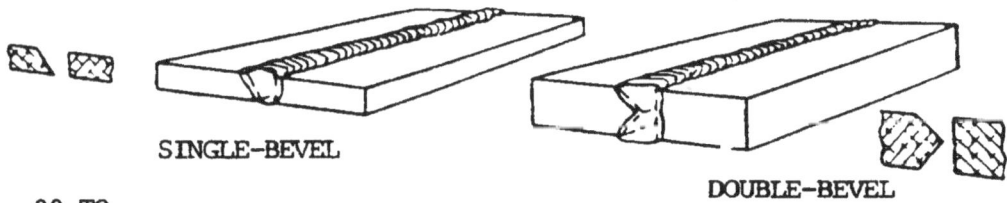

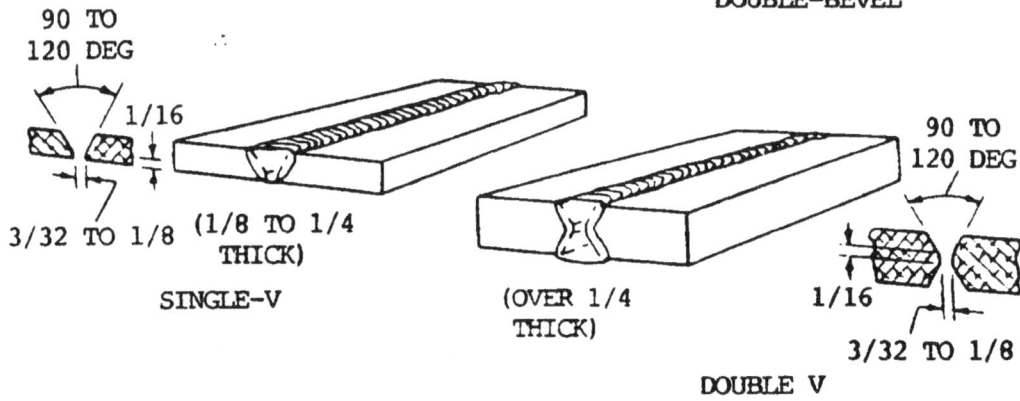

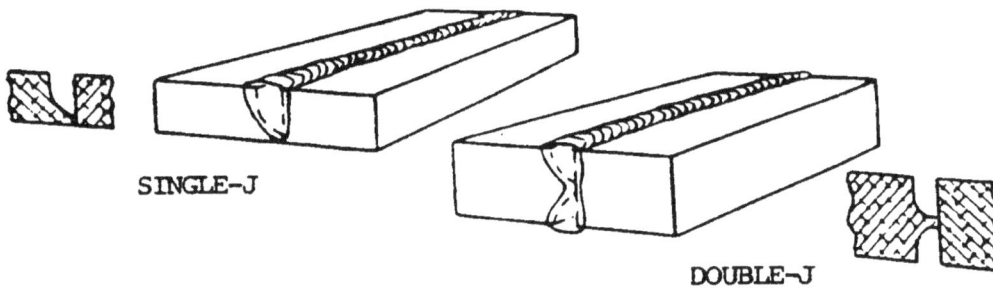

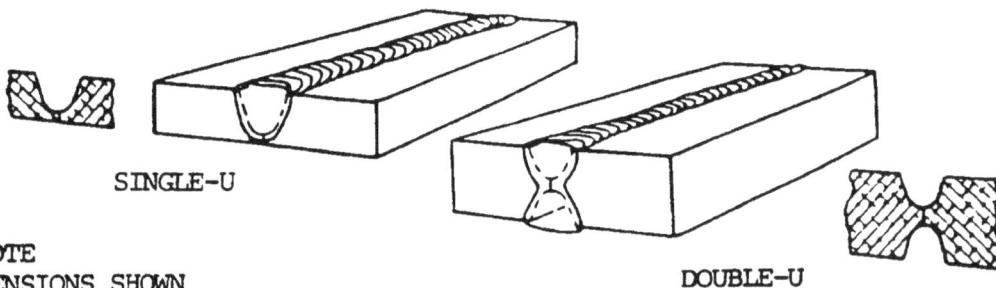

NOTE
ALL DIMENSIONS SHOWN
ARE IN INCHES.

DOUBLE-U

Figure 6-27. Types of groove welds.

6-16. TYPES OF WELDS (cont)

c. <u>Surfacing weld (fig. 6-28)</u>. These are welds composed of one or more strings or weave beads deposited on an unbroken surface to obtain desired properties or dimensions. This type of weld is used to build up surfaces or replace metal on worn surfaces. It is also used with square butt joints.

d. <u>Plug Weld (fig. 6-28)</u>. Plug welds are circular welds made through one member of a lap or tee joint joining that member to the other. The weld may or may not be made through a hole in the first member; if a hole is used, the walls may or may not be parallel and the hole may be partially or completely filled with weld metal. Such welds are often used in place of rivets.

NOTE
A fillet welded hole or a spot weld does not conform to this definition.

e. <u>Slot Weld (fig. 6-28)</u>. This is a weld made in an elongated hole in one member of a lap or tee joint joining that member to the surface of the other member that is exposed through the hole. This hole may be open at one end and may be partially or completely filled with weld metal.

NOTE
A fillet welded slot does not conform to this definition.

f. <u>Fillet Weld (top, fig. 6-28)</u>. This is a weld of approximately triangular cross section joining two surfaces at approximately right angles to each other, as in a lap or tee joint.

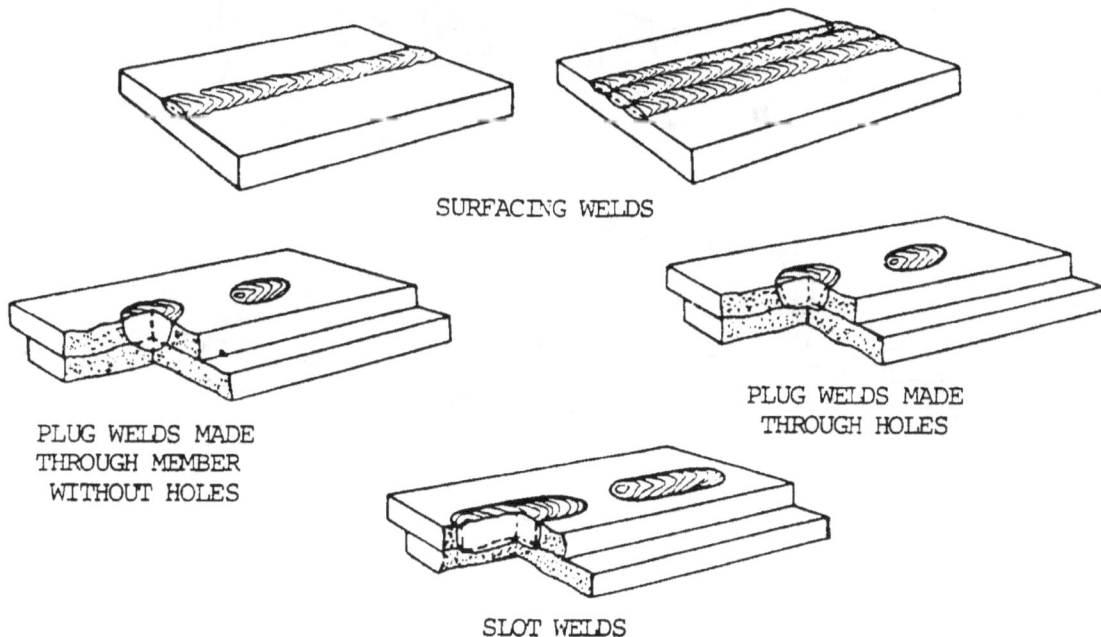

SURFACING WELDS

PLUG WELDS MADE
THROUGH MEMBER
WITHOUT HOLES

PLUG WELDS MADE
THROUGH HOLES

SLOT WELDS

Figure 6-28. Surfacing, plug, and slot welds.

g. <u>Flash Weld (fig. 6-29)</u>. A weld made by flash welding (p 6-11).

h. <u>Seam Weld (fig. 6-29)</u>. A weld made by arc seam or resistance seam welding (p 6-11). Where the welding process is not specified, this term infers resistance seam welding.

i. <u>Spot Weld (fig. 6-29)</u>. A weld made by arc spot or resistance spot welding (p 6-11). Where the welding process is not specified, this term infers a resistance spot weld.

j. <u>Upset Weld (fig. 6-29)</u>. A weld made by upset welding (para 6-12).

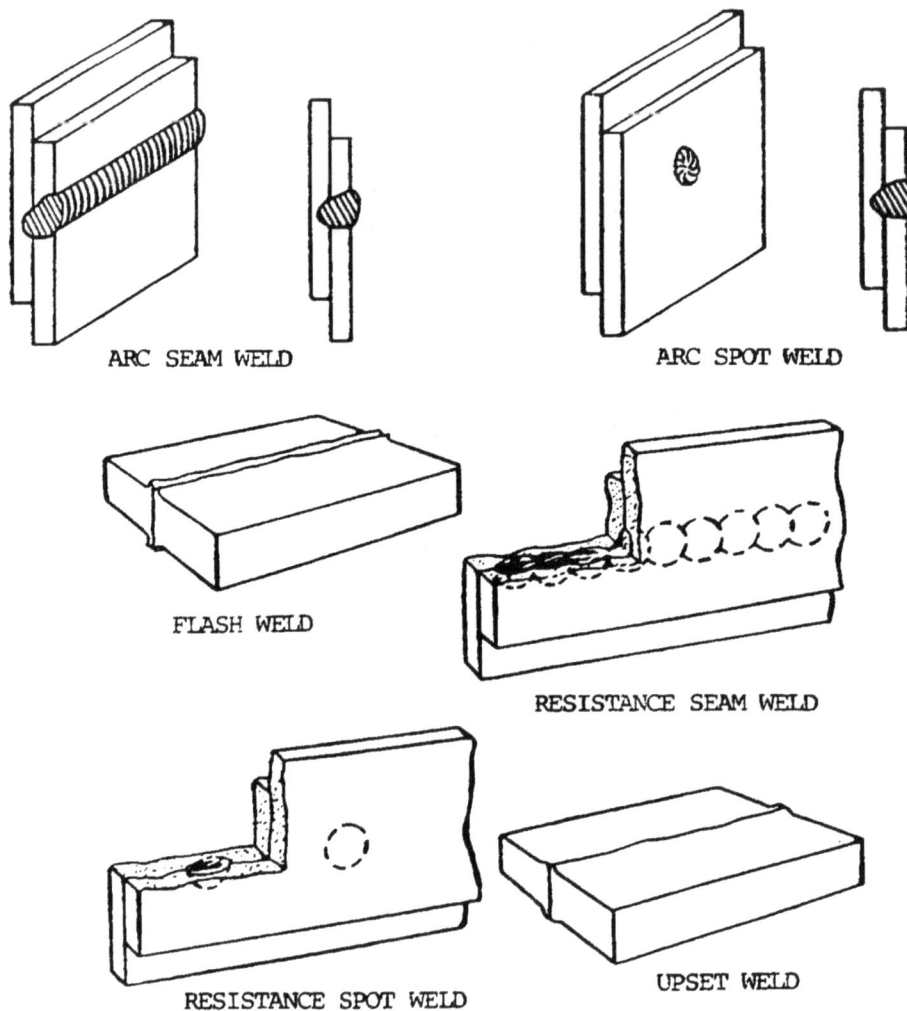

ARC SEAM WELD

ARC SPOT WELD

FLASH WELD

RESISTANCE SEAM WELD

RESISTANCE SPOT WELD

UPSET WELD

Figure 6-29. Flash, seam, spot, and upset welds.

Section IV. WELDING POSITIONS

6-17. GENERAL

Welding is often done on structures in the position in which they are found. Techniques have been developed to allow welding in any position. Sane welding processes have all-position capabilities, while others may be used in only one or two positions. All welding can be classified according to the position of the workpiece or the position of the welded joint on the plates or sections being welded. There are four basic welding positions, which are illustrated in figures 6-30 and 6-31. Pipe welding positions are shown in figure 6-32. Fillet, groove, and surface welds may be made in all of the following positions.

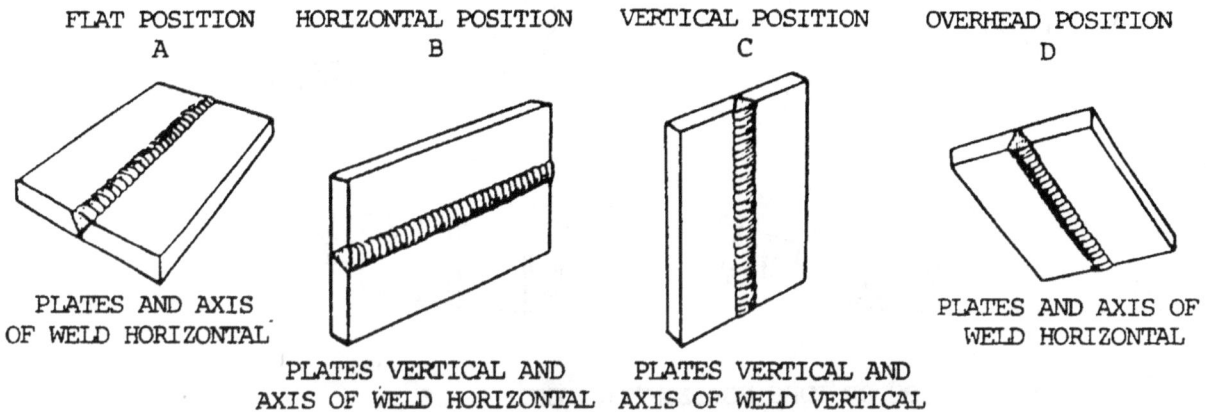

FLAT POSITION
A

HORIZONTAL POSITION
B

VERTICAL POSITION
C

OVERHEAD POSITION
D

PLATES AND AXIS
OF WELD HORIZONTAL

PLATES VERTICAL AND
AXIS OF WELD HORIZONTAL

PLATES VERTICAL AND
AXIS OF WELD VERTICAL

PLATES AND AXIS OF
WELD HORIZONTAL

Figure 6-30. Welding positions--groove welds--plate.

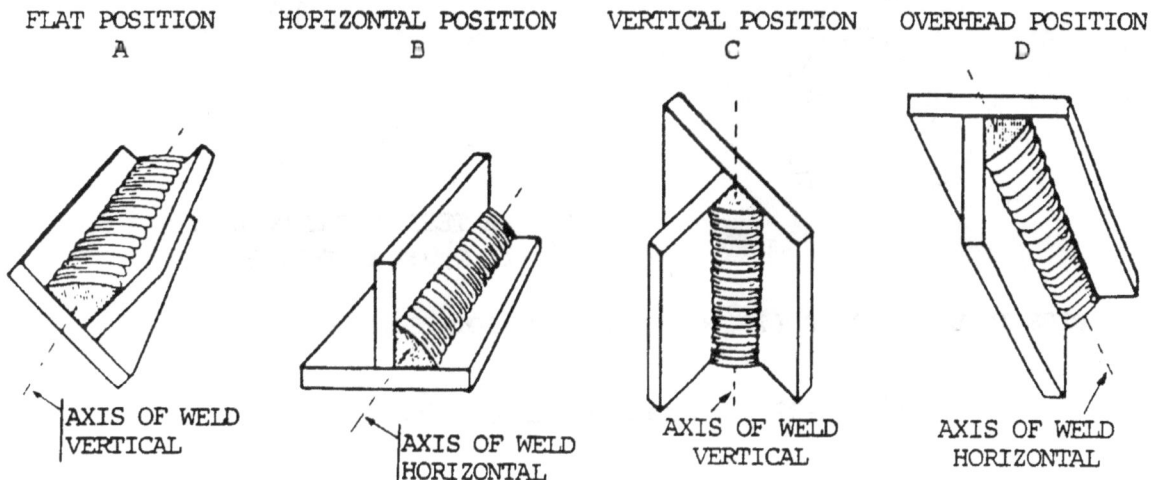

FLAT POSITION
A

HOPIZONTAL POSITION
B

VERTICAL POSITION
C

OVERHEAD POSITION
D

AXIS OF WELD
VERTICAL

AXIS OF WELD
HORIZONTAL

AXIS OF WELD
VERTICAL

AXIS OF WELD
HORIZONTAL

Figure 6-31. Welding positions--fillet welds--plate.

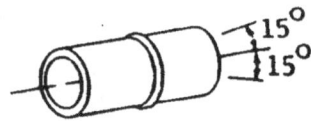

PIPE HORIZONTAL AND ROTATED.
WELD FLAT ($\pm$ 15°). DEPOSIT FILLER
METAL AT OR NEAR THE TOP.

PIPE OR TUBE VERTICAL
AND NOT ROTATED DURING
WELDING. WELD HORIZONTAL
($\pm$ 15°).

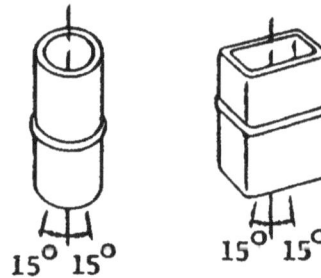

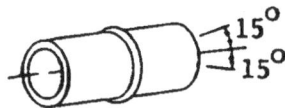

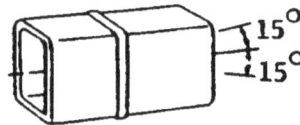

PIPE OR TUBE HORIZONTAL FIXED ($\pm$ 15°).
WELD FLAT, VERTICAL, OVERHEAD

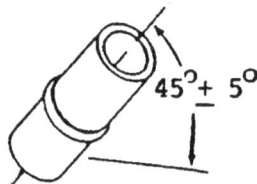

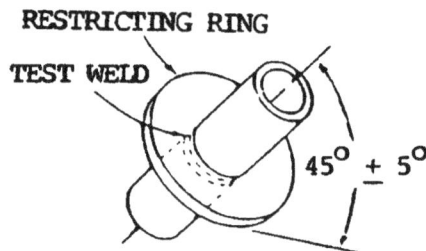

RESTRICTING RING

TEST WELD

E TEST POSITION 6GR
(T, K, OR Y CONNECTIONS)

PIPE INCLINED FIXED (45° $\pm$ 5°) AND NOT ROTATED DURING WELDING.

Figure 6-32. Welding position--pipe welds.

6-18. FLAT POSITION WELDING

In this position, the welding is performed from the upper side of the joint, and the face of the weld is approximately horizontal. Flat welding is the preferred term; however, the same position is sometimes called downhand.(See view A, figure 6-30 and view A, figure 6-31 for examples of flat position welding for fillet and groove welds).

TC 9-237

6-19. HORIZONTAL POSITION WELDING

NOTE
The axis of a weld is a line through the length of the weld, perpendicular to the cross section at its center of gravity.

a. Fillet Weld. In this position, welding is performed on the upper side of an approximately horizontal surface and against an approximately vertical surface. View B, figure 6-31, p 6-30 illustrates a horizontal fillet weld.

b. Groove Weld. In this position, the axis of the weld lies in an approximately horizontal plane and the face of the weld lies in an approximately vertical plane. View B, figure 6-30, p 6-30 illustrates a horizontal groove weld.

c. Horizontal Fixed Weld. In this pipe welding position, the axis of the pipe is approximately horizontal and the pipe is not rotated during welding. Pipe welding positions are shown in figure 6-32, p 6-31.

d. Horizontal Rolled Weld. In this pipe welding position, welding is performed in the flat position by rotating the pipe. Pipe welding positions are shown in figure 6-32, p 6-31.

6-20. VERTICAL POSITION WELDING

a. In this position, the axis of the weld is approximately vertical. Vertical welding positions are shown in view C, figures 6-30 and 6-31, p 6-30.

b. In vertical position pipe welding, the axis of the pipe is vertical, and the welding is performed in the horizontal position. The pipe may or may not be rotated. Pipe welding positions are shown in figure 6-32, p 6-31.

6-21. OVERHEAD POSITION WELDING

In this welding position, the welding is performed from the underside of a joint. Overhead position welds are illustrated in view D, figures 6-30 and 6-31, p 6-30.

6-22. POSITIONS FOR PIPE WELDING

Pipe welds are made under many different requirements and in different welding situations. The welding position is dictated by the job. In general, the position is fixed, but in sane cases can be rolled for flat-position work. Positions and procedures for welding pipe are outlined below.

a. Horizontal Pipe Rolled

(1) Align the joint and tack weld or hold in position with steel bridge clamps with the pipe mounted on suitable rollers (fig. 6-33). Start welding at point C, figure 6-33, progressing upward to point B. When point B is reached, rotate the pipe clockwise until the stopping point of the weld is at point C and again weld upward to point B. When the pipe is being rotated, the torch should be held between points B and C and the pipe rotated past it.

(2) The position of the torch at point A (fig. 6-33) is similar to that for a vertical weld. As point B is approached, the weld assumes a nearly flat position and the angles of application of the torch and rod are altered slightly to compensate for this change.

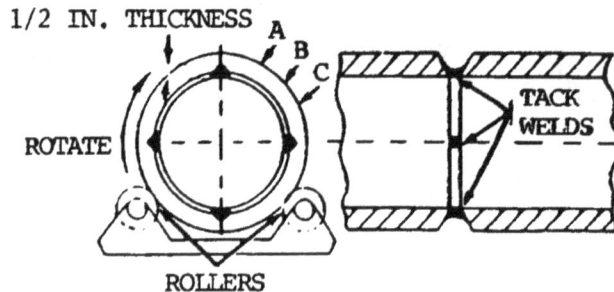

Figure 6-33. Diagram of tack welded pipe on rollers.

(3) The weld should be stopped just before the root of the starting point so that a small opening remains. The starting point is then reheated, so that the area surrounding the junction point is at a uniform temperature. This will insure a complete fusion of the advancing weld with the starting point.

(4) If the side wall of the pipe is more than 1/4 in. (0.64 cm) in thickness, a multipass weld should be made.

b. Horizontal Pipe Fixed Position Weld.

(1) After tack welding, the pipe is set up so that the tack welds are oriented approximately as shown in figure 6-34. After welding has been started, the pipe must not be moved in any direction.

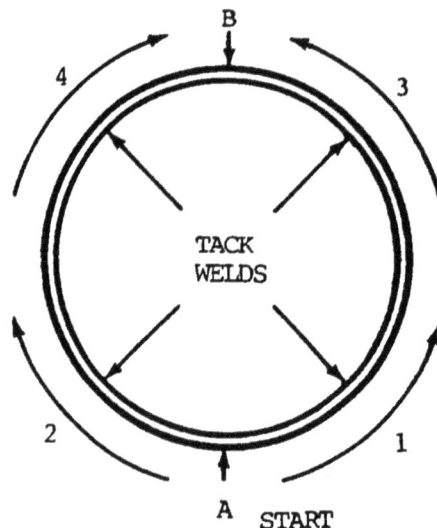

Figure 6-34. Diagram of horizontal pipe weld with uphand method.

6-22. POSITIONS FOR PIPE WELDING (cont)

(2) When welding in the horizontal fixed position, the pipe is welded in four steps as described below.

Step 1. Starting at the bottom or 6'clock position, weld upward to the 3 o'clock position.

Step 2. Starting back at the bottom, weld upward to the 9 o'clock position.

Step 3. Starting back at the 3 o'clock position, weld to the top

Step 4. Starting back at the 9 o'clock position, weld upward to the top. overlapping the bead.

(3) When welding downward, the weld is made in two stages. Start at the top (fig. 6-35) and work down one side (1, fig. 6-35) to the bottom, then return to the top and work down the other side (2, fig. 6-35) to join with the previous weld at the bottom. The welding downward method is particularly effective with arc welding, since the higher temperature of the electric arc makes possible the use of greater welding speeds. With arc welding, the speed is approximately three times that of the upward welding method.

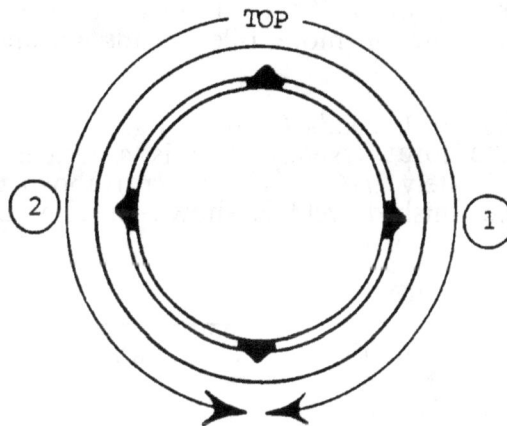

Figure 6-35. Diagram of horizontal pipe weld with downhand method.

(4) Welding by the backhand method is used for joints in low carbon or low alloy steel piping that can be rolled or are in horizontal position. One pass is used for wall thicknesses not exceeding 3/8 in. (0.95 cm), two passes for wall thicknesses 3/8 to 5/8 in. (0.95 to 1.59 cm), three passes for wall thicknesses 5/8 to 7/8 in. (1.59 to 2.22 cm), and four passes for wall thicknesses 7/8 to 1-1/8 in. (2.22 to 2.87 cm).

c. Vertical Pipe Fixed Position Weld. Pipe in this position, wherein the joint is horizontal, is most frequently welded by the backhand method (fig. 6-36). The weld is started at the tack and carried continuously around the pipe.

Figure 6-36. Vertical pipe fixed position weld with backhand method.

d. Multipass Arc Welding.

(1) Root beads. If a lineup clamp is used, the root bead (view A, fig. 6-37) is started at the bottom of the groove while the clamp is in position. When no backing ring is used, care should be taken to build up a slight bead on the inside of the pipe. If a backing ring is used, the root bead should be carefully fused to it. As much root bead as the bars of the lineup clamp will permit should be applied before the clamp is removed. Complete the bead after the clamp is removed.

(2) Filler beads. Care should be taken that the filler beads (view B, fig. 6-37) are fused into the root bead, in order to remove any undercut causal by the deposition of the root bead. One or more filler beads around the pipe usually will be required.

(3) Finish beads. The finish beads (view C, fig. 6-37) are applied over the filler beads to complete the joint. Usually, this is a weave bead about 5/8 in. (1. 59 cm) wide and approximately 1/16 in. (O. 16 cm) above the outside surface of the pipe when complete. The finished weld is shown in view D, figure 6-37.

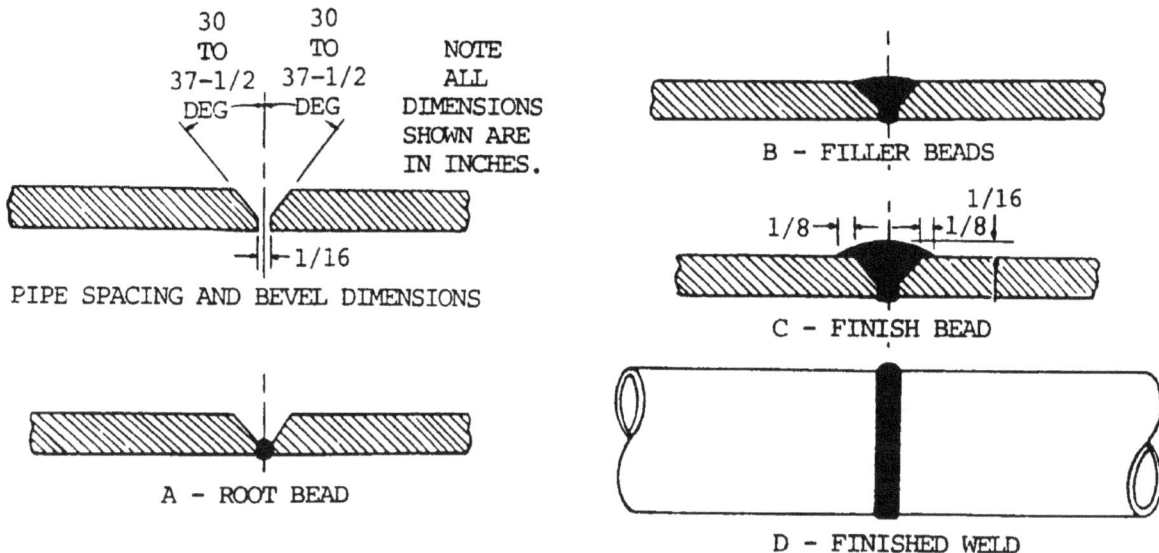

Figure 6-37. Deposition of root, filler, and finish weld beads.

6-22.   POSITIONS FOR PIPE WELDING (cont)

e.. <u>Aluminum pipe welding</u>.For aluminum pipe, special joint details have been developed and are normally associated with combination-type procedures. A backing ring is not used in most cases. The rectangular backing ring is rarely used when fluids are transmitted through the piping system. It may be used for structural applications in which pipe and tubular members are used to transmit loads rather than materials.

6-23.   FOREHAND WELDING

a.   <u>Work angle</u> is the angle that the electrode, or centerline of the welding gun, makes with the referenced plane or surface of the base metal in a plane perpendicular to the axis of a weld. Figure 6-38 shows the work angle for a fillet weld and a groove weld. For pipe welding, the work angle is the angle that the electrode, or centerline of the welding gun, makes with the referenced plane or surface of the pipe in a plane extending from the center of the pipe through the puddle. <u>Travel angle</u> is the angle that the electrode, or centerline of the welding gun, makes with a reference line perpendicular to the axis of the weld in the plane of the weld axis. Figure 6-39 illustrates the travel angle for fillet and groove welds. For pipe welding, the travel angle is the angle that the electrode, or centerline of the welding gun, makes with a reference line extending from the center of the pipe through the arc in the plane of the weld axis. The travel angle is further described as a <u>drag angle</u> or a <u>push angle</u>. Figure 6-39 shows both drag angles and push angles. The push angle, which points forward in the direction of travel, is also known as <u>forehand welding</u>.

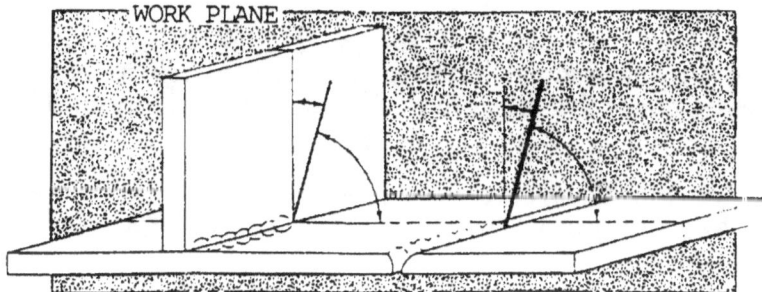

Figure 6-38. Work angle--fillet and groove weld.

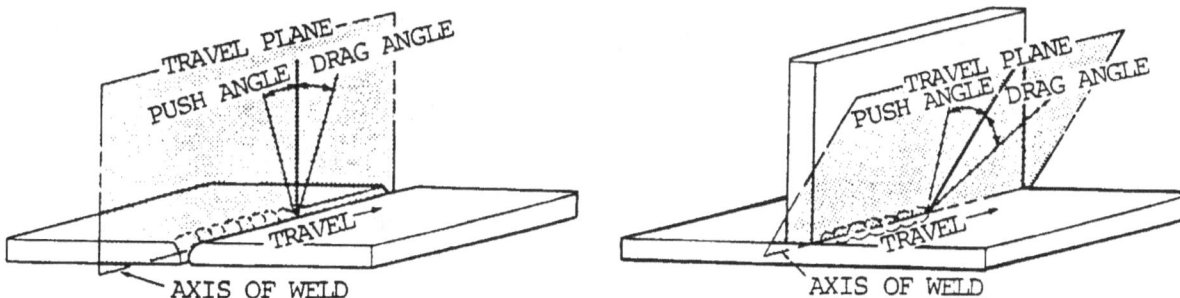

Figure 6-39. Travel angle--fillet and groove weld.

b. In forehand welding, the welding rod precedes the torch.The torch is held at an approximately 30 degree angle from vertical in the direction of welding as shown in figure 6-40. The flame is pointed in the direction of welding and directed between the rod and the molten puddle.This position permits uniform preheating of the plate edges immediately ahead of the molten puddle. By moving the torch and the rod in opposite semicircular paths, the heat can be carefully balanced to melt the end of the rod and the side walls of the plate into a uniformly distributed molten puddle. The rod is dipped into the leading edge of the puddle so that enough filler metal is melted to produce an even weld joint.The heat reflected backwards from the rod keeps the metal molten.The metal is distributed evenly to both edges being welded by the motion of the tip and rod.

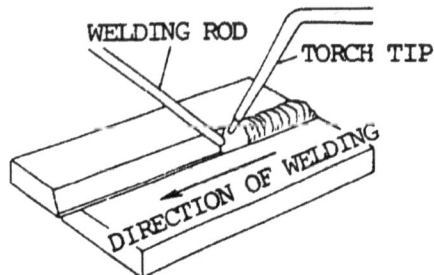

WELDING ROD
TORCH TIP
DIRECTION OF WELDING

NOTE
TORCH AND ROD ANGLES ARE 45 DEG AS VIEWED BY THE OPERATOR AND PERPENDICULAR (90 DEG) TO THE WORK SURFACE AS VIEWED FROM THE END OF THE WORKPIECE.

Figure 6-40. Forehand welding.

c. This method is satisfactory for welding sheets and light plates in all positions. Some difficulties are encountered in welding heavier plates for the reasons given below:

(1) In forehand welding, the edges of the plate must be beveled to provide a wide V with a 90 degree included angle.This edge preparation is necessary to insure satisfactory melting of the plate edges, good penetration, and fusion of the weld metal to the base metal.

(2) Because of this wide V, a relatively large molten puddle is required. It is difficult to obtain a good joint when the puddle is too large.

6-24. BACKHAND WELDING

a. Backhand welding, also known as drag angle, is illustrated in figure 6-41. The drag angle points backward from the direction of travel.

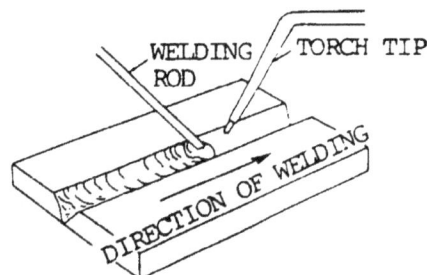

WELDING ROD
TORCH TIP
DIRECTION OF WELDING

NOTE
TORCH AND ROD ANGLES ARE AS VIEWED BY THE OPERATOR AND PERPENDICULAR (90 DEG) TO THE WORK SURFACE AS VIEWED FROM THE END OF THE WORKPIECE.

Figure 6-41. Backhand welding.

6-24. BACKHAND WELDING (cont)

b. In this method, the torch precedes the welding rod, as shown in figure 6-41, p 37. The torch is held at an angle approximately 30 degrees from the vertical, away from the direction of welding, with the flame directed at the molten puddle. The welding rod is between the flame and the molten puddle. This position requires less transverse motion than is used in forehand welding.

c. Backhand welding is used principally for welding heavy sections because it permits the use of narrower V's at the joint. A 60 degree included angle of bevel. is sufficient for a good weld. In general, there is less puddling, and less welding rod is used with this method than with the forehand method.

Section V. EXPANSION AND CONTRACTION IN WELDING OPERATIONS

6-25. GENERAL

a. Most of the welding processes involve heat. High-temperature heat is responsible for much of the welding warpages and stresses that occur. When metal is heated, it expands in all directions. When metal cools, it contracts in all directions. Some distortions caused by weld shrinkage are shown in figure 6-42.

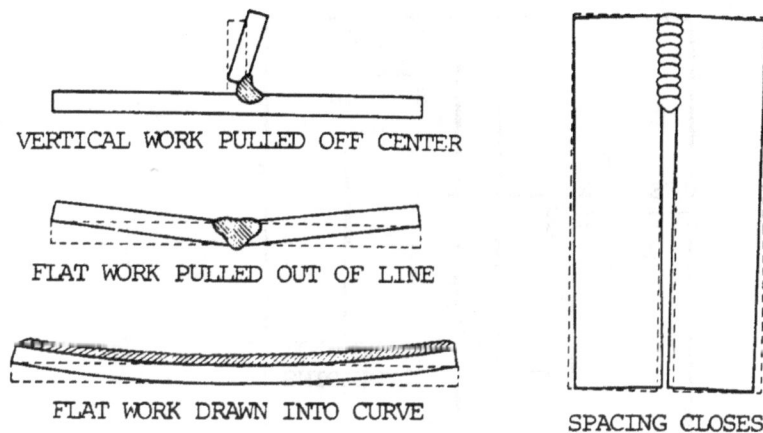

VERTICAL WORK PULLED OFF CENTER

FLAT WORK PULLED OUT OF LINE

FLAT WORK DRAWN INTO CURVE

SPACING CLOSES

Figure 6-42. Results of weld metal shrinkage.

b. There is a direct relationship between the amount of temperature change and change in dimension. This is based on the coefficient of thermal expansion. Thermal expansion is a measure of the linear increase in unit length based on the change in temperature of the material. The coefficient of expansion is different for the various metals. Aluminum has one of the highest coefficient of expansion ratios, and changes in dimension almost twice as much as steel for the same temperature change.

c. A metal expands or contracts by the same amount when heated or cooled the same temperature if it is not restrained. If the expansion of the part being welded is restrained, buckling or warping may occur. If contraction is restrained, the parts may be cracked or distorted because of the shrinkage stresses.

d. when welding, the metals that are heated and cooled are not unrestrained since they are a part of a larger piece of metal which is not heated to the same temperature. Parts not heated or not heated as much tend to restrain that portion of the same piece of metal that is heated to a higher temperature. This non-uniform heating always occurs in welding. The restraint caused by the part being non-uniformly heated is the principal cause for the thermal distortion and warpages that occur in welding.

e. Residual stresses that occur when metal is subjected to non-uniform temperature change are called thermal stresses. These stresses in weldments have two major effects: they produce distortion, and may cause premature failure in weldments.

6-26. CONTROLLING CONTRACTION IN SHEET METAL

a. The welding procedure should be devised so that contraction stresses will be held to a minimum order to keep the desired shape and strength of the welded part. Some of the methods used for controlling contraction are described below.

b. The backstep method as shown in view A, figure 6-43, may be used. With the backstep method, each small weld increment has its own shrinkage pattern, which then becomes insignificant to the total pattern of the entire weldment.

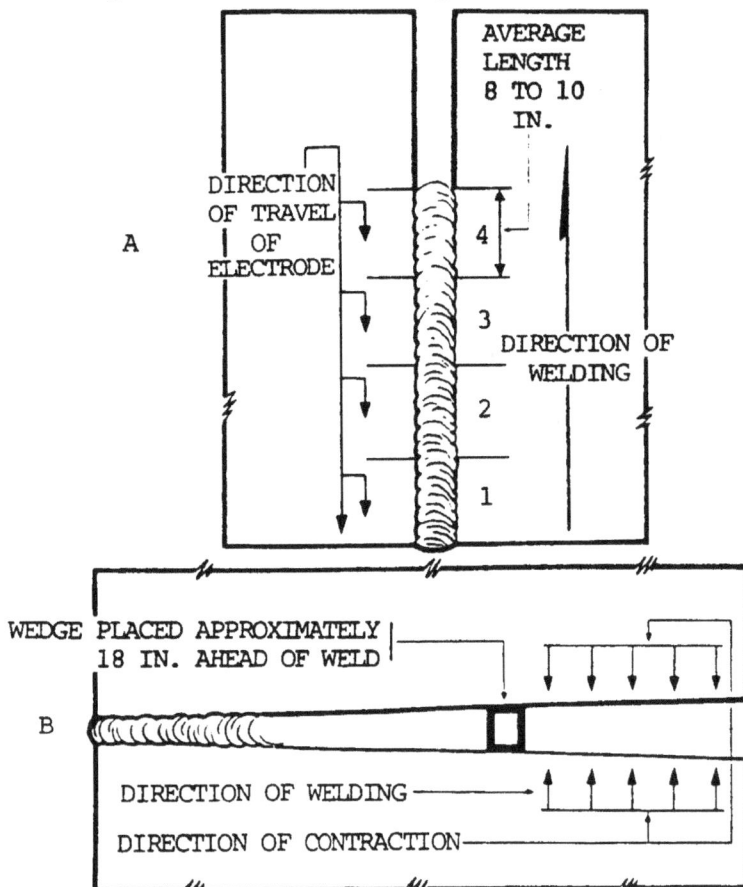

Figure 6-43. Methods of counteracting contractions.

6-26. CONTROLLING CONTRACTION IN SHEET METAL (cont)

c. In welding long seams, the contraction of the metal deposited at the joint will cause the edges being welded to draw together and possibly overlap. This action should be offset by wedging the edges apart as shown in view B, figure 6-43, p 6-39. The wedge should be moved forward as the weld progresses. The spacing of the wedge depends on the type of metal and its thickness. Spacing for metals more than 1/8 in. (3.2 mm) thick is approximately as follows:

Metal	In. per ft
Steel	1/4 to 3/8
Brass and Bronze	3/16
Aluminum	1/4
Copper	3/16
Lead	5/16

d. Sheet metal under 1/16 in. (0. 16 cm) thick may be welded by flanging the edges as shown in figure 6-20, p 6-23, and tacking at internals along the seam before welding. A weld can be produced in this manner without the addition of filler metal.

e. Buckling and warping can be prevented by the use of quench plates as shown in figure 6-44. The quench plates are heavy pieces of metal clamped parallel to the seam being welded with sufficient space between to permit the welding operation. These quench plates absorb the heat of welding, thereby decreasing the stresses due to expansion and contraction.

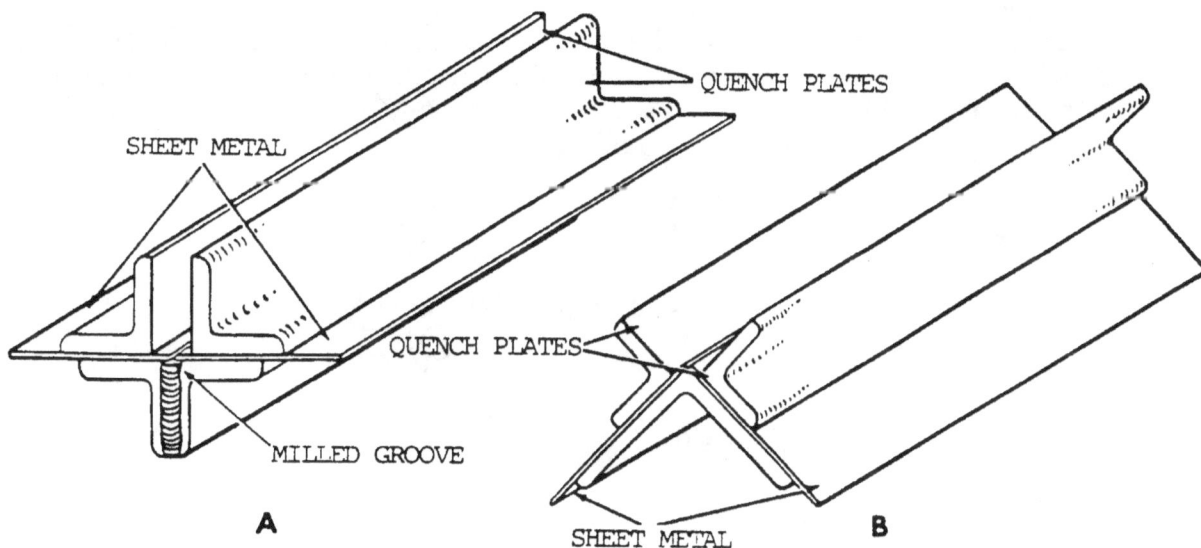

Figure 6-44. Quench plates used in the welding of sheet metal.

f. Jigs and fixtures may be used to hold members in place for welding. These are usually heavy sections in the vicinity of the seam (fig. 6-45). The heavy sections cool the plate beyond the area of the weld.

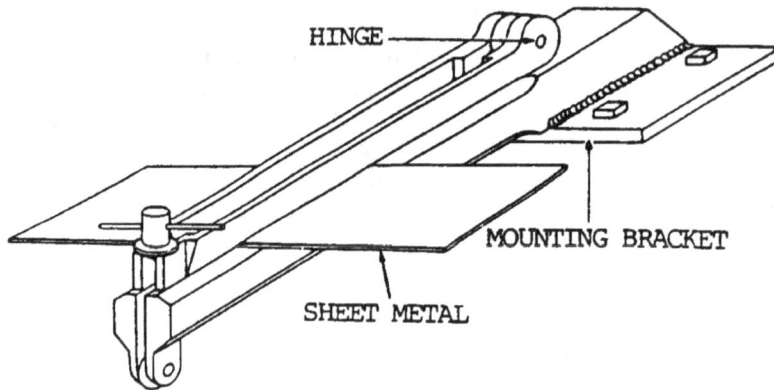

Figure 6-45. Fixture used in the welding of sheet metal.

g. In pipe welding, spacing as illustrated in figure 6-43, p 6-39, is not practical. Proper alignment of pipe can be best obtained by tack welding to hold pieces in place. The pipes should be separated by a gap of 1/8 to 1/4 in. (0.32 to) 0.64 cm), depending on the size of the pipe being welded.

6-27. CONTROLLING CONTRACTION AND EXPANSION IN CASTINGS

a. Prior to welding gray iron castings, expansion and contraction are provided for by preheating. Before welding, small castings can be preheated by means of a torch to a very dull red heat, visible in a darkened room. After welding, a reheating and controlled slow cooling or annealing will relieve internal stresses and assure a proper gray iron structure.

b. For larger castings, temporary charcoal-fired furnaces built of fire brick and covered with fire resistant material are often used. Only local preheating of parts adjacent to the weld is usually necessary (fig. 6-46). Such local preheating can be done with a gasoline, kerosene, or welding torch.

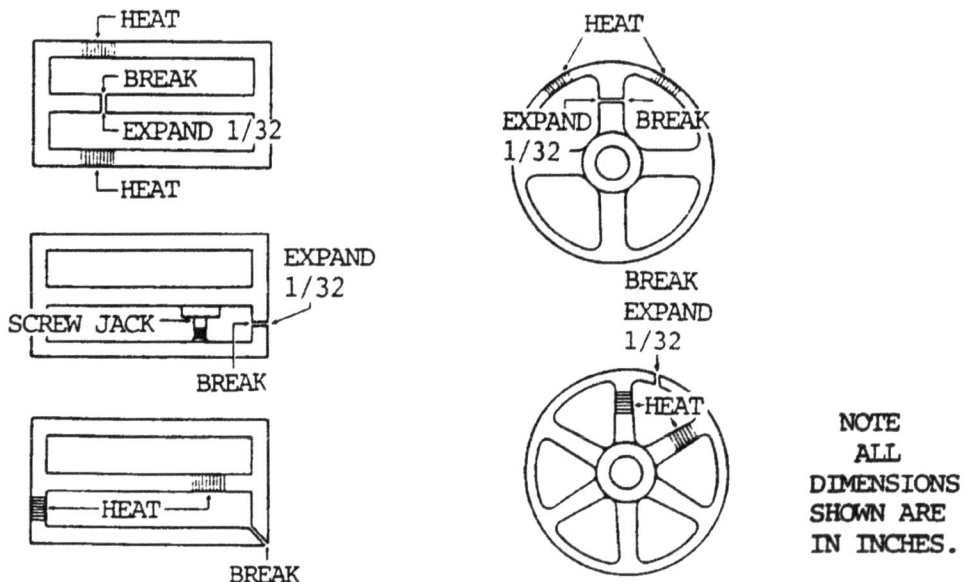

Figure 6-46. Controlling expansion and contraction of castings by preheating.

6-27. CONTROLLING CONTRACTION AND EXPANSION IN CASTINGS (cont)

c. Before welding a crack that extends from the edge of a casting, it is advisable to drill a small hole 1/2 to 1 in. (1.27 to 2.54 cm) beyond the visible end of the crack. If the applied heat causes the crack to run, it will only extend to the drill hole.

d. If a crack does not extend to the end of a casting, it is advisable to drill a small hole 1/2 to 1 in. (1.27 to 2.54 cm) beyond each end of the visible crack.

e. The above procedures apply to gray iron castings, as well as bronze welded castings, except that less preheat is required for bronze welded castings.

6-28. WELDING DISTORTION AND WARPAGE

a. General. The high temperature heat involved in most welding processes is largely responsible for the distortion, warpage and stresses that occur. When heated, metal expands in all directions and when it cools, it contracts in all directions. As described in paragraph 6-25, there is a direct relationship between the amount of temperature change and the change in dimension of the metal. A metal expands or contracts by the same amount when heated or cooled the same temperature, if it is not restrained. However, in welding, the metals that are heated and cooled are not unrestrained, because they are a part of a larger piece of metal which is not heated to the same temperature. This non-uniform heating and partial restraint is the main cause of thermal distortion and warpage that occur in welding. Figure 6-47 shows the effects of expansion on a cube of metal. When the cube of metal is exposed to a temperature increase, it will expand in the x, y, and z directions. When it cools, if unrestricted, it will contract by the same amount as it expanded.

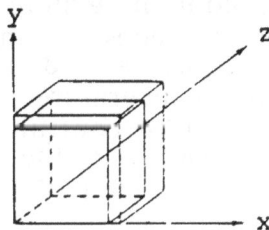

Figure 6-47. Cube of metal showing expansion.

b. A weld is usually made progressively, which causes the solidified portions of the weld to resist the shrinkage of later portions of the weld bead. The portions welded first are forced in tension down the length of the weld bead (longitudinal to the weld) as shown in figure 6-48. In the case of a butt weld, little motion of the weld is permitted in the direction across the material face (transverse direction) because of the weld joint preparation or stiffening effect of underlying passes. In these welds, as shown in figure 6-48, there will also be transverse residual stresses. For fillet welds, as shown in figure 6-49, the shrinkage stresses are rigid down the length of the weld and across its face.

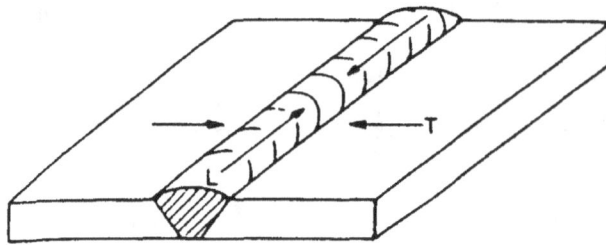

Figure 6-48. Longitudinal (L) and transverse (T) shrinkage stresses in a butt weld.

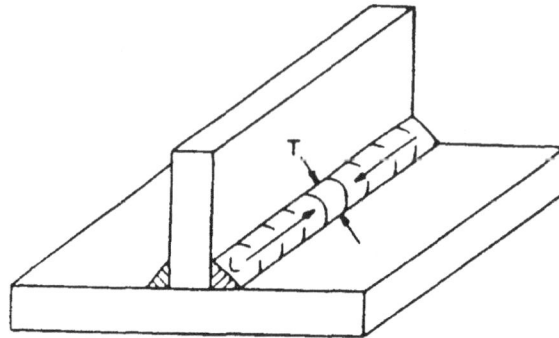

Figure 6-49. Longitudinal (L) and transverse (T)
shrinkage stresses in a fillet weld.

c. At the point of solidification, the molten metal has little or no strength. As it cools, it squires strength. It is also in its expanded form because of its high temperature. The weld metal is now fused to the base metal, and they work together. As the metal continues to cool, it acquires higher strength and is now contracting in three directions. The arc depositing molten metal is a moving source of heat and the cooling differential is also a moving factor, but tends to follow the travel of the arc. With the temperature still declining and each small increment of heated metal tending to contractcontracting stresses will occur, and there will be movement in the metal adjacent to the weld. The unheated metal tends to resist the cooling dimension changes of the previously molten metalTempera- ture differential has an effect on this.

d. The temperate differential is determined by thermal conductivity. The higher the thermal conductivity of the metal, the less effect differential heating will have. For example, the thermal conductivity of copper is the highest, alumi- num is half that amount, and steel about one-fifth that of copperHeat would move more quickly through a copper bar than through a steel bar, and the temperature differential would not be so greatThis physical property must be consiered when welding, along with the fact that arc temperatures are very similar but the metal melting points are somewhat different.

e. Another factor is the travel speed of the heat source or arc. If the travel speed is relatively fast, the effect of the heat of the arc will cause expansion of the edges of the plates, and they will bow outward and open up the joint. This is the same as running a bead on the edge of the plateIn either case, it is a momen- tary situation which continues to change as the weld progresses. By adjusting the current and travel speed, the exact speed can be determined for a specific joint design so that the root will neither open up nor close together.

6-28. WELDING DISTORTION AND WARPAGE (cont)

f. Residual stresses in weldments produce distortion and may be the cause of premature failure in weldments. Distortion is caused when the heated weld region contracts non-uniformly, causing shrinkage in one part of the weld to exert eccentric forces on the weld cross section. The weldment strains elastically in response to these stresses, and this non-uniform strain is seen in macroscopic distortion. The distortion may appear in butt joints as both longitudinal and transverse shrinkage or contraction and as angular change (rotation) when the face of the weld shrinks more than the root. The angular change produces transverse bending in the plates along the weld length. These effects are shown in figure 6-50.

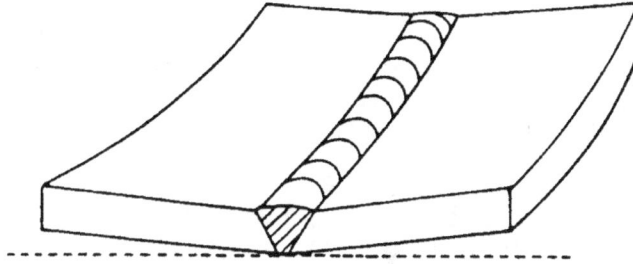

Figure 6-50. Distortion in a butt weld.

g. Distortion in fillet welds is similar to that in butt welds. Transverse and longitudinal shrinkage as well as angular distortion result from the unbalanced nature of the stresses in these welds (fig. 6-51). Since fillet welds are often used in combination with other welds in a weldment, the distortion may be complex.

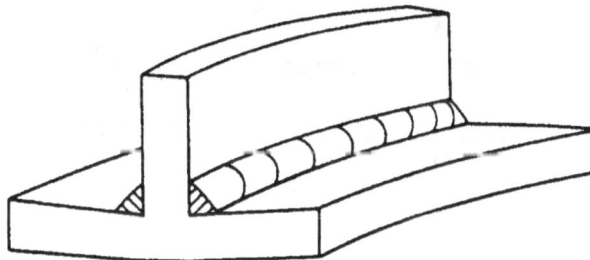

Figure 6-51. Distortion in a fillet weld.

h. Residual stresses and distortion affect materials by contributing to buckling, curling, and fracturing at low applied stress levels. When residual stresses and their accompanying distortion are present buckling may occur at liner compressive loads than would be predicted otherwise. In tension, residual stresses may lead to high local stresses in weld regions of low toughness and may result in running brittle cracks which can spread to low overall stress areas. Residual stresses may also contribute to fatigue or corrosion failures.

i. Control of distortion can be achieved by several methods. Commonly used methods include those which control the geometry of the weld joint, either before or during welding. These methods include prepositioning the workplaces before welding so that weld distortion leaves them in the desired final geometry, or restraining the workplaces so they cannot move and distort during welding. Designing the joint so that weld deposits are balanced on each side of the center line is another useful technique. Welding process selection and weld sequence also influence distortion and residual stress. Some distorted weldments can be straightened mechanically after welding, and thermal or flame straightening can also be applied.

j. Residual stresses may be eliminated by both thermal and mechanical means. During thermal stress relief, the weldment is heated to a temperature at which the yield point of the metal is low enough for plastic flow to occur and allow relaxation of stress. The mechanical properties of the weldment may also change, but not always toward a more uniform distribution across the joint. For example, the brittle fracture resistance of many steel weld-rents is improved by thermal stress relief not only because the residual stresses in the weld are reduced, but also because hard weld heat-affected zones are tempered and made tougher by this procedure. Mechanical stress relief treatments will also reduce residual stresses, but will not change the microstructure or hardness of the weld or heat-affected zone. Peening, proof stressing, and other techniques are applied to weldments to acccomplish these ends.

k. The welder must consider not only reducing the effects of residual stresses and distortion, but also the reduction of cracks, porosity, and other discontinuities; material degradation due to thermal effects during welding; the extent of nondestructive testing; and fabrication cost. A process or procedure which produces less distortion may also produce more porosity and cracking in the weld zone. Warping and distortion can be minimized by several methods. General methods include:

(1) Reducing residual stresses and distortion prior to welding by selecting proper processes and procedures.

(2) Developing better means for stress relieving and removing distortion.

(3) Changing the structural design and the material so that the effects of residual stresses and distortion can be minimized.

The following factors should be taken into consideration when welding in order to reduce welding warpage:

(1) The location of the neutral axis and its relationship in both directions.

(2) The location of welds, size of welds, and distance from the neutral axis in both directions.

(3) The time factor for welding and cooling rates when making the various welds.

(4) The opportunity for balancing welding around the neutral axis.

(5) Repetitive identical structure and varying the welding techniques based on measurable warpage.

(6) The use or procedures and sequences to minimize weldment distortion.

6-28. WELDING DISTORTION AND WARPAGE (cont)

When welding large structures and weldments, it is important to establish a procedure to minimize warpage. The order of joining plates in a deck or on a tank will
affect stresses and distortion. As a general rule, transverse welds should be made
before longitudinal welds. Figure 6-52 shows the order in which the joints should
be welded.

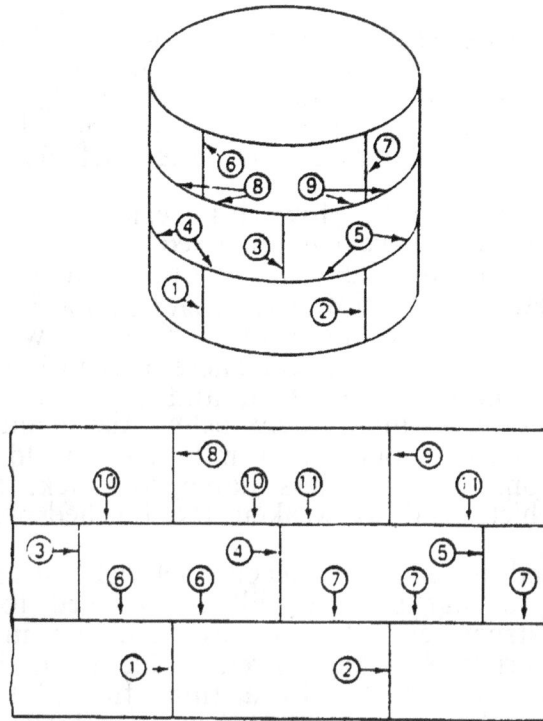

Figure 6-52. The order in which to make weld joints.

Warpage can be minimized in smaller structures by different techniques, which include the following:

(1) The use of restraining fixtures, strong backs, or many tack welds.

(2) The use of heat sinks or the fast cooling of welds.

(3) The predistortion or prebending of parts prior to welding.

(4) Balancing welds about the weldment neutral axis or using wandering sequences or backstep welding.

(5) The use of intermittent welding to reduce the volume of weld metal.

(6) The use of proper joint design selection and minimum size.

(7) As a last resort, use preheat or peening.

Section VI. WELDING PROBLEMS AND SOLUTIONS

6-29. STRESSES AND CRACKING

a. In this section, welding stresses and their effect on weld cracking is explained. Factors related to weldment failure include weld stresses, cracking, weld distortion, lamellar tearing, brittle fracture, fatigue cracking, weld design, and weld defects.

b. When weld metal is added to the metal being welded, it is essentially cast metal. Upon cooling, the weld metal shrinks to a greater extent than the base metal in contact with the weld, and because it is firmly fused, exerts a drawing action. This drawing action produces stresses in and about the weld which may cause warping, buckling, residual stresses, or other defects.

c. Stress relieving is a process for lowering residual stresses or decreasing their intensity. Where parts being welded are fixed too firmly to permit movement, or are not heated uniformly during the welding operation, stresses develop by the shrinking of the weld metal at the joint. Parts that cannot move to allow expansion and contraction must be heated uniformly during the welding operation. Stress must be relieved after the weld is completed. These precautions are important in welding aluminum, cast iron, high carbon steel, and other brittle metals, or metals with low strength at temperatures immediately below the malting point. Ductile materials such as bronze, brass, copper, and mild steel yield or stretch while in the plastic or soft conditions, and are less liable to crack. However, they may have undesirable stresses which tend to weaken the finished weld.

d. When stresses applied to a joint exceed the yield strength, the joint will yield in a plastic fashion so that stresses will be reduced to the yield point. This is normal in simple structures with stresses occurring in one direction on parts made of ductile materials. Shrinkage stresses due to normal heating and cooling do occur in all three dimensions. In a thin, flat plate, there will be tension stresses at right angles. As the plate becomes thicker, or in extremely thick materials, the stresses occur in three directions.

e. When simple stresses are imposed on thin, brittle materials, the material will fail in tension in a brittle manner and the fracture will exhibit little or no pliability. In such cases, there is no yield point for the material, since the yield strength and the ultimate strength are nearly the same. The failures that occur without plastic deformation are known as brittle failures. When two or more stresses occur in a ductile material, and particularly when stresses occur in three directions in a thick material, brittle fracture may occur.

f. Residual stresses also occur in castings, forgings, and hot rolled shapes. In forgings and castings, residual stresses occur as a result of the differential cooling that occurs. The outer portion of the part cools first, and the thicker and inner portion cools considerably faster. As the parts cool, they contract and pick up strength so that the portions that cool earlier go into a compressive load, and the portions that cool later go into a tensile stress mode. In complicated parts, the stresses may cause warpage.

6-29.   STRESSES AND CRACKING (cont)

g.   Residual stresses are not always detrimental. They may have no effect or may have a beneficial effect on the service life of parts. Normally, the outer fibers of a part are subject to tensile loading and thus, with residual compression loading, there is a tendency to neutralize stress in the outer fibers of the part. An example of the use of residual stress is in the shrink fit of parts. A typical example is the cooling of sleeve bearings to insert them into machined holes, and allow them to expand to their normal dimension to retain then in the proper location.   Sleeve bearings are used for heavy,slow machinery, and are subject to compressive residual loading, keeping them within the hole. Large roller bearings are usually assembled to shafts by heating to expand them slightly so they will fit on the shaft, then allowing them to cool, to produce a tight assembly.

h.   Residual stresses occur in all arc welds. The most common method of measuring stress is to produce weld specimens and then machine away specific amounts of metal, which are resisting the tensile stress in and adjacent to the weld.The movement that occurs is then measured.Another method is the use of grid marks or data points on the surface of weldments that can be measured in multiple directions.   Cuts are made to reduce or release residual stresses from certain parts of the weld joint, and the measurements are taken again. The amount of the movement relates to the magnitude of the stresses. A third method utilizes extremely small strain gauges.   The weldment is gradually and mechanically cut from adjoining portions to determine the change in internal stresses. With these methods, it is possible to establish patterns and actually determine amounts of stress within parts that were caused by the thermal effects of welds.

i.   Figure 6-53 shows residual stresses in an edge weld. The metal close to the weld tends to expand in all directions when heated by the welding arc.This metal is restrained by adjacent cold metal and is slightly upset, or its thickness slightly increased, during this heating period. When the weld metal starts to cool, the upset area attempts to contract, but is again restrained by cooler metal.This results in the heated zone becoming stressed in tension.When the weld has cooled to room temperature, the weld metal and the adjacent base metal are under tensile stresses close to the yield strength. Therefore, there is a portion that is compressive, and beyond this, another tensile stress area. The two edges are in tensile residual stress with the center in compressive residual stress, as illustrated.

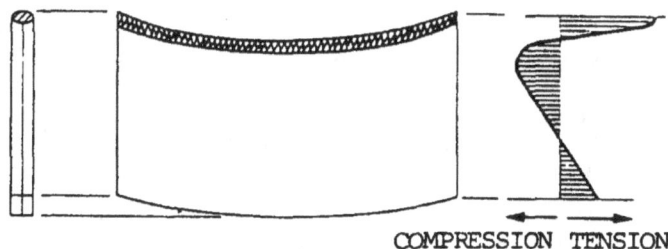

COMPRESSION TENSION

Figure 6-53.   Edge welded joint -- residual stress pattern.

j. The residual stresses in a butt weld joint made of relatively thin plate are more difficult to analyze. This is because the stresses occur in the longitudinal direction of the weld and perpendicular to the axis of the weld. The residual stresses within the weld are tensile in the longitudinal direction of the weld and the magnitude is at the yield strength of the metal. The base metal adjacent to the weld is also at yield stress, parallel to the weld and along most of the length of the weld. When moving away from the weld into the base metal, the residual stresses quickly fall to zero, and in order to maintain balance, change to compression. This is shown in figure 6-54. The residual stresses in the weld at right angles to the axis of the weld are tensile at the center of the plate and compressive at the ends. For thicker materials when the welds are made with multipasses, the relationship is different because of the many passes of the heat source. Except for single-pass, simple joint designs, the compressive and tensile residual stresses can only be estimated.

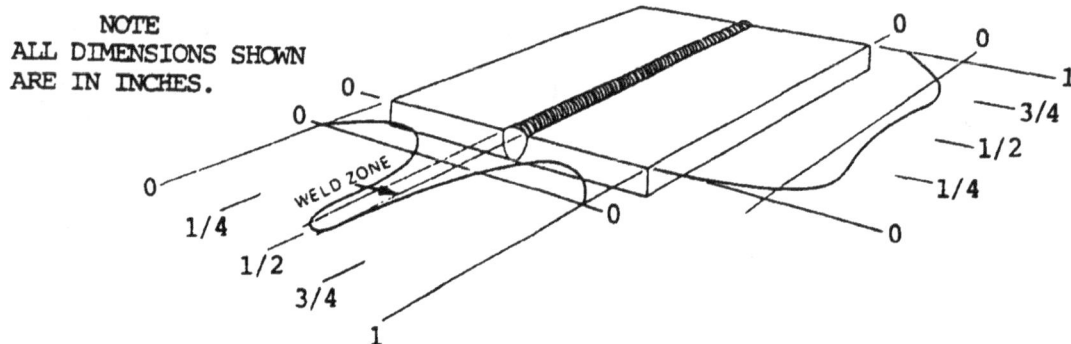

Figure 6-54. Butt welded joint -- residual stress pattern.

k. As each weld is made, it will contract as it solidifies and gain strength as the metal cools. As it contracts, it tends to pull, and this creates tensile stresses at and adjacent to the weld. Further from the weld or bead, the metal must remain in equilibrium, and therefore compressive stresses occur. In heavier weldments when restraint is involved, movement is not possible, and residual stresses are of a higher magnitude. In a multipass single-groove weld, the first weld or root pass originally creates a tensile stress. The second, third, and fourth passes contract and cause a compressive load in the root passAs passes are made until the weld is finished, the top passes will be in tensile load, the center of the plate in compression, and the root pass will have tensile residual stress.

l. Residual stresses can be decreased in several ways, as described below:

(1) If the weld is stressed by a load beyond its yield, strength plastic deformation will occur and the stresses will be more uniform, but are still located at the yield petit of the metal. This will not eliminate residual stresses, but will create a more uniform stress pattern. Another way to reduce high or peak residual stresses is by means of loading or stretching the weld by heating adjacent areas, causing them to expand. The heat reduces the yield strength of the weld metal and the expansion will tend to reduce peak residual stresses within the weld. This method also makes the stress pattern at the weld area more uniform.

6-29.   STRESSES AND CRACKING (cont)

        (2) High residual stresses can be reduced by stress relief heat treatment. With heat treatment, the weldment is uniformly heated to an elevated temperature, at which the yield strength of the metal is greatly reduced. The weldment is then allowed to cool slowly and uniformly so that the temperature differential between parts is minor. The cooling will be uniform and a uniform low stress pattern will develop within the weldment.

        (3) High-temperature preheating can also reduce residual stress, since the entire weldment is at a relatively high temperature, and will cool more or less uniformly from that temperature and so reduce peak residual stresses.

    m.   Residual stresses also contribute to weld cracking. Weld cracking sometimes occurs during the manufacture of the weldment or shortly after the weldment is completed. Cracking occurs due to many reasons and may occur years after the weldment is completed. Cracks are the most serious defects that occur in welds or weld joints in weldments. Cracks are not permitted in most weldments, particularly those subject to low-temperature service, impact loading, reversing stresses, or when the failure of the weldment will endanger life.

    n.   Weld cracking that occurs during or shortly after the fabrication of the weldment can be classified as hot cracking or cold cracking. In addition, welds may crack in the weld metal or in the base metal adjacent to the weld metal, usually in the heat-affected zone. Welds crack for many reasons, including the following:

        (1) Insufficient weld metal cross section to sustain the loads involved.

        (2) Insufficient ductility of weld metal to yield under stresses involved.

        (3) Under-bead cracking due to hydrogen pickup in a hardenable type of base material.

    o.   Restraint and residual stresses are the main causes of weld cracking during the fabrication of a weldment. Weld restraint can come from several factors, including the stiffness or rigidity of the weldment itself. Weld metal shrinks as it cools, and if the parts being welded cannot move with respect to one another and the weld metal has insufficient ductility, a crack will result. Movement of welds may impose high loads on other welds and cause them to crack during fabrication. A more ductile filler material should be used, or the weld should be made with sufficient cross-sectional area so that as it cools it will have enough strength to withstand cracking tendencies. Typical weld cracks occur in the root pass when the parts are unable to move.

    p.   Rapid cooling of the weld deposit is also responsible for weld cracking. If the base metal being joined is cold and the weld is small, it will cool quickly.

Shrinkage will also occur quickly, and cracking can occur. If the parts being joined are preheated even slightly, the cooling rate will be lower and cracking can be eliminated.

q. Alloy or carbon content of base material can also affect cracking. When a weld is made with higher-carbon or higher-alloy base material, a certain amount of the base material is melted and mixed with the electrode to produce the weld metal. The resulting weld metal has higher carbon and alloy content. It may have higher strength, but it has less ductility. As it shrinks, it may not have enough ductility to cause plastic deformation, and cracking may occur.

r. Hydrogen pickup in the weld metal and in the heat-affected zone can also cause cracking. When using cellulose-covered electrodes or when hydrogen is present because of damp gas, damp flux, or hydrocarbon surface materials, the hydrogen in the arc atmosphere will be absorbed in the molten weld metal and in adjoining high-temperature base metal. As the metal cools, it will reject the hydrogen, and if there is enough restraint, cracking can occur. This type of cracking can be reduced by increasing preheat, reducing restraint, and eliminating hydrogen from the arc atmosphere.

s. When cracking is in the heat-affected zone or if cracking is delayed, the cause is usually hydrogen pickup in the weld metal and the heat-affected zone of the base metal. The presence of higher-carbon materials or high alloy in the base metal can also be a cause. When welding high-alloy or high-carbon steels, the buttering technique can be used to prevent cracking. This involves surfacing the weld face of the joint with a weld metal that is much lower in carbon or alloy content than the base metal. The weld is then made between the deposited surfacing material and avoids the carbon and alloy pickup in the weld metal, so a more ductile weld deposit is made. Total joint strength must still be great enough to meet design requirements. Underbead cracking can be reduced by the use of low-hydrogen processes and filler metals. The use of preheat reduces the rate of cooling, which tends to decrease the possibility of cracking.

t. Stress Relieving Methods.

(1) Stress relieving in steel welds may be accomplished by preheating between 800 and 1450 °F (427 and 788 °C), depending on the material, and then slowly cooling. Cooling under some conditions may take 10 to 12 hours. Small pieces, such as butt welded high speed tool tips, may be annealed by putting them in a box of fire resistant material and cooling for 24 hours. In stress relieving mild steel, heating the completed weld for 1 hour per 1.00 in. (2.54 cm) of thickness is common practice. On this basis, steel 1/4 in. (0.64 cm) thick should be preheated for 15 minutes at the stress relieving temperature.

(2) Peening is another method of relieving stress on a finished weld, usually with compressed air and a roughing or peening tool. However, excessive peening may cause brittleness or hardening of the finished weld and may actually cause cracking.

6-29. STRESSES AND CRACKING (cont)

(3) Preheating facilitates welding in many cases. It prevents cracking in the heat affected zone, particularly on the first passes of the weld metal. If proper preheating times and temperatures are used, the cooling rate is slowed sufficiently to prevent the formation of hard martensite, which causes cracking. Table 6-1 lists preheating temperatures of specific metals.

Table 6-1. Preheating Temperatures*

Metal	Temperature	
	°F	°C
Low carbon steels (up to 0.30 percent carbon)	200 to 300	93 to 149
Medium carbon steels (0.30 to 0.55 percent carbon)	300 to 500	149 to 260
High carbon steels (0.55 to 0.83 percent carbon)	500 to 800	260 to 427
Carbon molybdenum steels (0.10 to 0.30 percent carbon)	300 to 600	149 to 316
Carbon molybdenum steels (0.30 to 0.35 percent carbon)	500 to 800	260 to 427
High strength constructional alloy	100 to 400	38 to 204
Manganese steels (up to 1.75 percent carbon)	300 to 900	149 to 482
Manganese steels (up to 15.0 percent manganese)	Usually not required	
Nickel steels (up to 3.50 percent nickel)	200 to 700	93 to 371
Chromium steels	300 to 500	149 to 260
Nickel and chromium steels	200 to 1100	93 to 593
Stainless steels	Usually not required	
Cast iron	700 to 900	371 to 482
Aluminum	500 to 700	260 to 371
Copper	500 to 800	260 to 427
Nickel	200 to 300	93 to 149
Monel	200 to 300	93 to 149
Brass and bronze	300 to 500	149 to 260

*The preheating temperatures for alloy steels are governed by the carbon as well as the alloy content of the steel.

(4) The need for preheating steels and other metals is increased under the following conditions:

(a) When the temperature of the part or the surrounding atmosphere is at or below freezing.

(b) When the diameter of the welding rod is small in comparison to the thickness of the metal being joined.

(c) When the welding speed is high.

(d) When the shape and design of the parts being welded are complicated.

(e) When there is a great difference in mass of the parts being welded.

(f) When welding steels with a high carbon, low manganese, or other alloy content.

(g) When the steel being welded tends to harden when cooled in air from the welding temperature.

u. The following general procedures can be used to relieve stress and to reduce cracking:

(1) Use ductile weld metal.

(2) Avoid extremely high restraint or residual stresses.

(3) Revise welding procedures to reduce restraint.

(4) Utilize low-alloy and low-carbon materials.

(5) Reduce the cooling rate by use of preheat.

(6) Utilize low-hydrogen welding processes and filler metals.

(7) When welds are too small for the service intended, they will probably crack. The welder should ensure that the size of the welds are not smaller than the minimum weld size designated for different thicknesses of steel sections.

6-30. IN-SERVICE CRACKING

Weldments must be designed and built to perform adequately in service. The risk of failure of a weldment is relatively small, but failure can occur in structures such as bridges, pressure vessels, storage tanks, ships, and penstocks. Welding has sometimes been blamed for the failure of large engineering structures, but it should be noted that failures have occurred in riveted and bolted structures and in castings, forgings, hot rolled plate and shapes, as well as other types of construction. Failures of these types of structures occurred before welding was widely used and still occur in unwelded structures today. However, it is still important to make weldments and welded structures as safe against premature failure of any type as possible. There are four specific types of failures, including brittle fracture, fatigue fracture, lamellar tearing, and stress corrosion cracking.

a. Brittle Fracture. Fracture can be classified into two general categories, ductile and brittle.

(1) Ductile fracture occurs by deformation of the crystals and slip relative to each other. There is a definite stretching or yielding and a reduction of cross-sectional area at the fracture (fig. 6-55) .

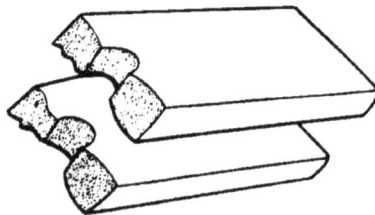

Figure 6-55. Ductile fracture surface.

6-30.   IN-SERVICE CRACKING (cont)

(2) <u>Brittle fracture</u> occurs by cleavage across individual crystals. The fracture exposes the granular structure and there is little or no stretching or yielding.   There is no reduction of area at the fracture (fig. 6-56).

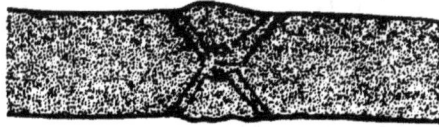

Figure 6-56. Brittle fracture surface.

(3) It is possible that a broken surface will display both ductile and brittle fracture over different areas of the surface. This means that the fracture which propagated across the section changed its mode of fracture.

(4) There are four factors that should be reviewed when analyzing a fractured surface. They are growth marking, fracture mode, fracture surface texture and appearance, and amount of yielding or plastic deformation at the fracture surface.

(5) Growth markings are one way to identify the type of failure. Fatigue failures are characterized by a fine texture surface with distinct markings produced by erratic growth of the crack as it progresses. The chevron or herringbone pattern occurs with brittle or impact failures. The apex of the chevron appearing on the fractured surface always points toward the origin of the fracture and is an indicator of the direction of crack propagation.

(6) Fracture mode is the second factor. Ductile fractures have a shear mode of crystalline failure. The surface texture is silky or fibrous in appearance. Ductile fractures often appear to have failed in shear as evidenced by all parts of the fracture surface assinning an angle of approximately 45 degrees with respect to the axis of the load stress.

(7) The third factor is fracture surface and texture. Brittle or cleavage fractures have either a granular or a crystalline appearance. Brittle fractures usually have a point of origin. The chevron pattern will help locate this point.

(8) An indication of the amount of plastic deformation is the necking down of the surface. There is little or no deformation for a brittle fracture and usually a considerable necked down area in the case of a ductile fracture.

(9) One characteristic of brittle fracture is that the steel breaks quickly and without warning. The fractures increase at very high speeds, and the steels fracture at stresses below the normal yield strength for steel. Mild steels, which show a normal degree of ductility when tested in tension as a normal test bar, may fail in a brittle manner. In fact, mild steel may exhibit good toughness characteristics at roan temperature. Brittle fracture is therefore more similar to the fracture of glass than fracture of normal ductile materials. A combination of conditions must be present at the same time for brittle fracture to occur. Some of

these factors can be eliminated and thus reduce the possibility of brittle fracture. The following conditions must be present for brittle fracture to occur:w lo temperature, a notch or defect, a relatively high rate of loading, and triaxial stresses normally due to thickness of residual stresses. The microstructure of the metal also has an effect.

(10) Temperature is an important factor which must be considered in conjunction with microstructure of the material and the presence of a notch.Impact testing of steels using a standard notched bar specimen at different temperatures shows a transition from a ductile type failure to a brittle type failure based on a lowered temperature, which is known as the transition temperature.

(11) The notch that can result from faulty workmanship or from improper design produces an extremely high stress concentration which prohibits yielding. A crack will not carry stress across it,and the load is transmitted to the end of the crack. It is concentrated at this point and little or no yielding will occur. Metal adjacent to the end of the crack which does not carry load will not undergo a reduction of area since it is not stressed. It is, in effect, a restraint which helps set up triaxial stresses at the base of the notch or the end of the crack. Stress levels much higher than normal occur at this point and contribute to starting the fracture.

(12) The rate of loading is the time versus strain rate. The high rate of strain, which is a result of impact or shock loading, does not allow sufficient time for the normal slip process to occur.The material under load behaves elastically, allowing a stress level beyond the normal yield point. When the rate of loading, from impact or shock stressesoccurs near a notch in heavy thick material, the material at the base of the notch is subjected very suddenly to very high stresses. The effect of this is often complete and rapid failure of a structure and is what makes brittle fracture so dangerous.

(13) Triaxial stresses are more likely to occur in thicker material than in thin material. The z direction acts as a restraint at the base of the notch, and for thicker material, the degree of restraint in the through direction is higher. This is why brittle fracture is more likely to occur in thick plates or complex sections than in thinner materials.Thicker plates also usually have less mechanical working in their manufacture than thinner plates and are more susceptible to lower ductility in the z axis. The microstructure and chemistry of the material in the center of thicker plates have poorer properties than the thinner material, which receives more mechanical working.

(14) The microstructure of the material is of major importance to the fracture behavior and transition temperature range.Microstructure of a steel depends on the chemical composition and production processes used in manufacturing it. A steel in the as-rolled condition will have a higher transition temperature or liner toughness than the same steel in a normalized condition. Normalizing, or heating to the proper temperature and cooling slowly, produces a grain refinement which provides for higher toughness. Unfortunately, fabrication operations on steel, such as hot and cold forming, punching, and flame cutting, affect the original microstructure. This raises the transition temperature of the steel.

6-30. IN-SERVICE CRACKING (cont)

(15) Welding tends to accentuate some of the undesirable characteristics that contribute to brittle fracture. The thermal treatment resulting from welding tends to reduce the toughness of the steel or to raise its transition temperature in the heat–affected zone. The monolithic structure of a weldment means that more energy is locked up and there is the possibility of residual stresses which may be at yield point levels. The monolithic structure also causes stresses and strains to be transmitted throughout the entire weldment, and defects in weld joints can be the nucleus for the notch or crack that will initiate fracture.

(16) Brittle fractures can be reduced in weldments by selecting steels that have sufficient toughness at the service temperatures. The transition temperature should be below the service temperature to which the weldment will be subjected. Heat treatment, normalizing, or any method of reducing locked-up stresses will reduce the triaxial yield strength stresses within the weldment. Design notches must be eliminated and notches resulting from poor workmanship must not occur. Internal cracks within the welds and unfused root areas must be eliminated.

b. Fatigue Failure. Structures sometimes fail at nominal stresses considerably below the tensile strength of the materials involved. The materials involved were ductile in the normal tensile tests, but the failures generally exhibited little or no ductility. Most of these failures development after the structure had been subjected to a large number of cycles of loading. This type of failure is called a fatigue failure.

(1) Fatigue failure is the formation and development of a crack by repeated or fluctuating loading. when sudden failure occurs, it is because the crack has increased enough to reduce the load-carrying capacity of the part. Fatigue cracks may exist in some weldments, but they will not fail until the load-carrying area is sufficiently reduced. Repeated loading causes progressive enlargement of the fatigue cracks through the material. The rate at which the fatigue crack increases depends upon the type and intensity of stress as well as other factors involving the design, the rate of loading, and type of material.

(2) The fracture surface of a fatigue failure is generally smooth and frequently shins concentric rings or areas spreading from the point where the crack initiated These rings show the propagation of the crack, which might be related to periods of high stress followed by periods of inactivity. The fracture surface also tends to become rougher as the rate of propagation of the crack increases. Figure 6-57 shows the characteristic fatigue failure surface.

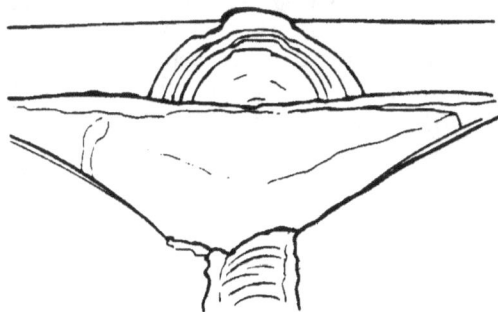

Figure 6-57. Fatigue fracture surface.

(3) Many structures are designed to a permissible static stress based on the yield point of the material in use and the safety factor that has been selected. This is based on statically loaded structuresthe stress of which remains relatively constant. Many structures, however, are subject to other than static loads in service. These changes may range from simple cyclic fluctuations to completely random variations. In this type of loading, the structure must be designed for dynamic loading and considered with respect to fatigue stresses.

(4) The varying loads involved with fatigue stresses can be categorized in different manners. These can be alternating cycles from tension to compression, or pulsating loads with pulses from zero load to a maximum tensile load, or from a zero load to a compressive load, or loads can be high and rise higher, either tensile or compressive. In addition to the loadings, it is important to consider the number of times the weldment is subjected to the cyclic loading.For practical purposes, loading is considers in millions of cycles. Fatigue is a cumulative process and its effect is in no way healed during periods of inactivity. Testing machines are available for loading metal specimens to millions of cycles.The results are plotted on stress vs.cycle curves, which show the relationship between the stress range and the number of cycles for the particular stress used. Fatigue test specimens are machined and polished, and the results obtained on such a specimen may not correlate with actual service life of a weldment.It is therefore important to determine those factors which adversely affect the fatigue life of a weldment.

(5) The possibility of a fatigue failure depends on four factors: the material used, the number of loading cycles,the stress level and nature of stress variations, and total design and design details.The last factor is controllable in the design and manufacture of the weldment.Weld joints can be designed for uniform stress distribution utilizing a full-penetration weld, but in other cases, joints may not have full penetration because of an unfused roofThis prohibits uniform stress distribution. Even with a full-penetration weld, if the reinforcement is excessive, a portion of the stress will flow through the reinforced area and will not be uniformly distributed.Welds designed for full penetration might not have complete penetration because of workmanship factors such as cracks, slag inclusions, and incomplete penetration, and therefore contain a stress concentration. One reason fatigue failures in welded structures occur is because the welded design can introduce more severe stress concentrations than other types of desigfThe weld defects mentioned previously,including excessive reinforcement, undercut, or negative reinforcement, will contribute to the stress concentration factoA weld also forms an integral part of the structureand when parts are attached by welding, they may produce sudden changes of section which contribute to stress concentrations under normal types of loading.Anything that can be done to smooth out the stress flow in the weldment will reduce stress concentrations and make the weldment less subject to fatigue failure. Total design with this in mind and careful workmanship will help to eliminate this type of problem.

6-30.   IN-SERVICE CRACKING (cont)

c.   <u>Lamellar Tearing</u>. Lamellar tearing is a cracking which occurs beneath welds, and is found in rolled steel plate weldments. The tearing always lies within the base metal, usually outside the heat-affected zone and generally parallel to the weld fusion boundary. This type of cracking has been found in corner joints where the shrinkage across the weld tended to open up in a manner similar to lamination of plate steel. In these cases, the lamination type crack is removed and replaced with weld metal. Before the advent of ultrasonic testing, this type of failure was probably occuring and was not foundIt is only when welds subjected the base metal to tensile loads in the z or through, direction of the rolled steel that the problem is encountered. For many years, the lower strength of rolled steel in the through direction was recognized and the structural code prohibited z-directional tensile loads on steel spacer plates. Figure 6-58 shows how lamellar tearing will come to the surface of the metal. Figure 6-59, shining a tee joint, is a more common type of lamellar tearing, which is much more difficult to find. In this case, the crack does not cane to the surface and is under the weld. This type of crack can only be found with ultrasonic testing or if failure occurs, the section can actually come out and separate from the main piece of metal.

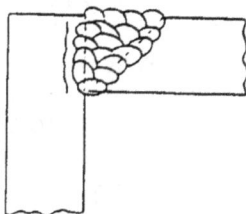

Figure 6-58.   Corner joint.

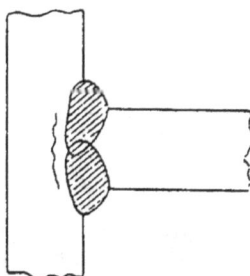

Figure 6-59.   Tee joint.

(1) Three conditions must occur to cause lamellar tearing. These are strains in the through direction of the plate caused by weld metal shrinkage in the joint and increased by residual stresses and by loading; stress through the joint across the plate thickness or in the z direction due to weld orientation in which the fusion line beneath the weld is roughly parallel to the lamellar separation; and poor ductility of the material in the z, or through, direction.

(2) Lamellar tearing can occur during flame-cutting operations and also in cold-shearing operations. It is primarily the low strength of the material in the z, or through, direction that contributes to the problem. A stress placed in the z

direction triggers the tearing. The thermal heating and stresses resulting from weld shrinkage create the fracture. Lamellar tearing is not associated with the under-bead hydrogen cracking problem.It can occur soon after the weld has been made, but on occasion will occur at a period months later.Also, the tears are under the heat-affected zone, and are more apt to occur in thicker materials and in higher-strength materials.

(3) Only a very small percentage of steel plates are susceptible to lamellar tearing. There are only certain plates where the concentration of inclusions are coupled with the unfavorable shape and type that present the risk of tearing. These conditions rarely occur with the other two factors mentioned previously. In general, three situations must occur in combination: structural restraint, joint design, and the condition of the steel.

(4) Joint details can be changed to avoid the possibility of lamellar tearing. In tee joints, double-fillet weld joints are less susceptible than full-penetration welds. Balanced welds on both sides of the joint present less risk of lamelalar tearing than large single-sided welds. corner joints are common in box columns. Lamellar tearing at the corner joints is readily detected on the exposed edge of the plate. Lamellar tearing can be overcome in corner joints by placing the bevel for the joint on the edge of the plate that would exhibit the tearing rather than on the other plate. This is shown by figure 6-60. Butt joints rarely are a problem with respect to lamellar tearing since the shrinkage of the weld does not set up a tensile stress in the thickness direction of the plates.

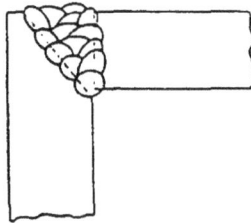

Figure 6-60. Redesigned corner joint to avoid lamellar tearing.

(5) Arc welding processes having higher heat input are less likely to create lamellar tearing. This may be because of the fewer number of applications of heat and the lesser number of shrinkage cycles involved in making a weld. Deposited filler metal with lower yield strength and high ductility also reduces the possibility of lamellar tearing. Preheat and stress relief heat treatment are not specifically advantageous with respect to lamellar tearing. The buttering technique of laying one or more layers of low strength, high-ductility weld metal deposit on the surface of the plate stressed in the z direction will reduce the possibility of lamellar tearing. This is an extreme solution and should only be used as a last resort. By observing the design factors mentioned above, the lamellar tearing problem is reduced.

d. Stress Corrosion Cracking. Stress corrosion cracking and delayed cracking due to hydrogen embrittlement can both occur when the weldment is subjected to the type of environment that accentuates this problem.

6-30.   IN-SERVICE CRACKING (cont)

(1) Delayed cracking is caused by hydrogen absorbed in the base metal or weld metal at high temperatures. Liquid or molten steel will absorb large quantities of hydrogen. As the metal solidifies, it cannot retain all of the hydrogen and is forced out of solution. The hydrogen coming out of the solution sets up high stresses, and if enough hydrogen is present, it will cause cracking in the weld or the heat-affected zone. These cracks develop over a period of time after the weld is completed. The concentration of hydrogen and the stresses resulting from it when coupled with residual stresses promote cracking. Cracking will be accelerated if the weldment is subjected to thermal stresses due to repeated heating and cooling.

(2) Stress corrosion cracking in steels is sometimes called caustic embrittlement. This type of cracking takes place when hot concentrated caustic solutions are in contact with steel that is stressed in tension to a relatively high level. The high level of tension stresses can be created by loading or by high residual stresses. Stress corrosion cracking will occur if the concentration of the caustic solution in contact with the steel is sufficiently high and if the stress level in the weldment is sufficiently high. This situation can be reduced by reducing the stress level and the concentration of the caustic solution. Various inhibitors can be added to the solution to reduce the concentration. Close inspection must be maintained on highly stressed areas.

(3) Graphitization is another type of cracking, caused by long service life exposed to thermal cycling or repeated heating and cooling. This may cause a breakdown of carbides in the steel into small areas of graphite and iron. This formation of graphite in the edge of the heat-affected area exposed to the thermal cycling causes cracking. It will often occur in carbon steels deoxidized with aluminum. The addition of molybdenum to the steel tends to restrict graphitization, and for this reason, carbon molybdenum steels are normally used in high-temperature power plant service. These steels must be welded with filler metals of the same composition.

6-31.   ARC BLOW

a. General. Arc blow is the deflection of an electric arc from its normal path due to magnetic forces. It is mainly encountered with dc welding of magnetic materials, such as steel, iron, and nickel, but can also be encountered when welding nonmagnetic materials. It will usually adversely affect appearance of the weld, cause excessive spatter, and can also impair the quality of the weld. It is often encountered when using the shielded metal arc welding process with covered electrodes. It is also a factor in semiautomatic and fully automatic arc welding processes. Direct current, flowing through the electrode and the base metal, sets up magnetic fields around the electrode, which deflect the arc from its intended path. The welding arc is usually deflected forward or backward of the direction of travel; however, it may be deflected from one side to the other. Back blow is encountered when welding toward the ground near the end of a joint or into a corner. Forward blow is encountered when welding away from the ground at the start of a joint. Arc blow can become so severe that it is impossible to make a satisfactory weld. Figure 6-61 shows the effect of ground location on magnetic arc blow.

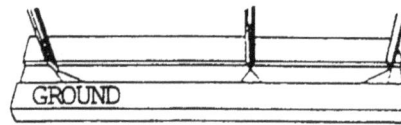

Figure 6-61. Effect of ground location magnetic arc blow.

b. When an electric current passes through an electrical conductor, it produces a magnetic flux in circles around the conductor in planes perpendicular to the conductor and with their centers in the conductor.The right-hand rule is used to determine the direction of the magnetic flux. It states that when the thumb of the right hand points in the direction in which the current flows (conventional flow) in the conductor, the fingers point in the direction of the flux. The direction of the magnetic flux produces polarity in the magnetic field, the same as the north and south poles of a permanent magnet.This magnetic field is the same as that produced by an electromagnet. The rules of magnetism,which state that like poles repel and opposite poles attract, apply in this situation.Welding current is much higher than the electrical current normally encountered. Likewise, the magnetic fields are also much stronger.

c. The welding arc is an electrical conductor and the magnetic flux is set up surrounding it in accordance with the right-hand rule.The magnetic field in the vicinity of the welding arc is the field produced by the welding current which passes through it from the electrode and to the base metal or work. This is a self-induced circular magnetic field which surrounds the arc and exerts a force on it from all sides according to the electrical-magnetic rule.As long as the magnetic field is symmetrical, there is no unbalanced magnetic force and no arc deflection. Under these conditions, the arc is parallel or in line with the centerline of the electrode and takes the shortest path to the base plate.If the symmetry of this magnetic field is disturbed, the forces on the arc are no longer equal and the arc is deflected by the strongest force.

d. The electrical-magnetic relationship is used in welding applications for magnetically moving, or oscillating, the welding arc.The gas tungsten arc is deflected by means of magnetic flux. It can be oscillated by transverse magnetic fields or be made to deflect in the direction of travel. Moving the flux field surrounding the arc and introducing an external-like polarity field roves the arc magnetically. Oscillation is obtained by reversing the external transverse field to cause it to attract the field surrounding the arc. As the self-induced field around the arc is attracted and repelled, it tends to move the arc column, which tries to maintain symmetry within its own self-induced magnetic field. Magnetic oscillation of the gas tungsten welding arc is used to widen the deposition.Arcs can also be made to rotate around the periphery of abutting pipes by means of rotating magnetic fields. Longer arcs are moved more easily than short arcs. The amount of magnetic flux to create the movement must be of the same order as the flux field surrounding the arc column.Whenever the symmetry of the field is disturbed by some other magnetic force,it will tend to move the self-induced field surrounding the arc and thus deflect the arc itself.

6-31.   ARC BLOW (cont)

e.   Except under the most simple conditions the self-induced magnetic field is not symmetrical throughout the entire electric circuit and changes direction at the arc.   There is always an unbalance of the magnetic field around the arc because the arc is roving and the current flow pattern through the base material is not constant.   The magnetic flux will pass through a magnetic material such as steel much easier than it will pass through air, and the magnetic flux path will tend to stay within the steel and be more concentrated and stronger than in air. Welding current passes through the electrode lead the electrode holder to the welding electrode, then through the arc into the base metal. At this point the current changes direction to pass to the work lead connection then through the work lead back to the welding machine. This is shown by figure 6-62. At the point the arc is in contact with the work, the change of direction is relatively abrupt, and the fact that the lines of force are perpendicular to the path of the welding current creates a magnetic unbalance. The lines of force are concentrated together on the inside of the angle of the current path through the electrode and the work, and are spread out on the outside angle of this path. Consequently, the magnetic field is much stronger on the side of the arc toward the work lead connection than on the other side, which produces a force on the stronger side and deflects the arc to the left.   This is toward the weaker force and is opposite the direction of the current path.   The direction of this force is the same regardless of the direction of the current.   If the welding current is reversed, the magnetic field is also reversed, but the direction of the magnetic force acting on the arc is always in the same direction, away from the path of the current through the work.

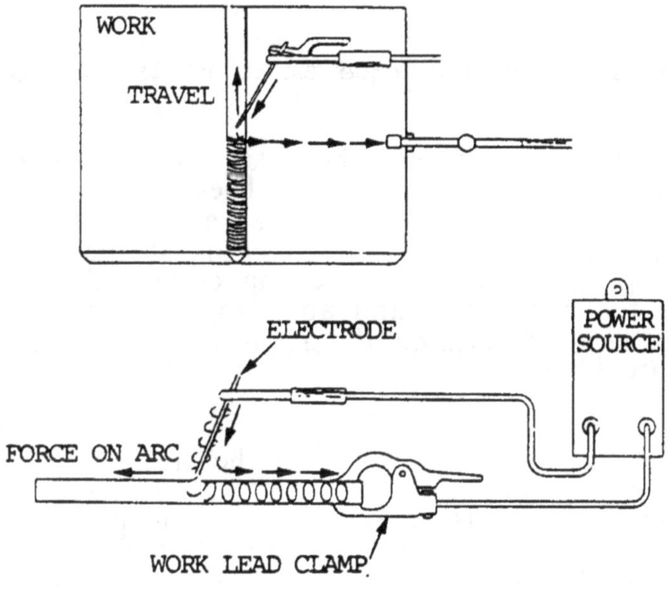

Figure 6-62. Unbalanced magnetic force due to current direction change.

f. The second factor that keeps the magnetic field from being symmetrical is the fact that the arc is moving and depositing weld metal. As a weld is made joining two plates, the arc moves from one end of the joint to the other and the magnetic field in the plates will constantly change. Since the work lead is immediately under the arc and moving with the arc, the magnetic path in the work will not be concentric about the point of the arc, because the lines of force take the easiest path rather than the shortest path.Near the start end of the joint the lines of force are crowded together and will tend to stay within the steel. Toward the finish end of the joint, the lines of force will be separated since there is more area. This is shown by figure 6-63. In addition, where the weld has been made the lines of force go through steel. Where the weld is not made, the lines of force must cross the air gap or root opening.The magnetic field is more intense on the short end and the unbalance produces a force which deflects the arc to the right or toward the long end.

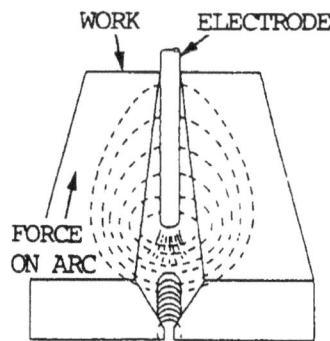

Figure 6-63. Unbalanced magnetic force due to unbalanced magnetic path.

g. When welding with direct current, the total force tending to cause the arc to deflect is a combination of these two forces.These forces may add or subtract from each other, and at times may meet at right angles. The polarity or direction of flow of the current does not affect the direction of these forces nor the resultant force. By analyzing the path of the welding current through the electrode and into the base metal to the work lead, and analyzing the magnetic field within the base metal, it is possible to determine the resultant forces and predict the resulting arc deflection or arc blow.

h. Forward blow exists for a short time at the start of a weld, then diminishes. This is because the flux soon finds an easy path through the weld metal. Once the magnetic flux behind the arc is concentrated in the plate and the weld, the arc is influenced mainly by the flux in front of it as this flux crosses the root opening. At this point, back blow may be encountered.Back blow can occur right up to the end of the joint. As the weld approaches the end, the flux ahead of the arc becomes more crowded, increasing the back blow.Back blow can become extremely severe right at the very end of the joint.

6-31. ARC BLOW (cont)

i. The use of alternating current for welding greatly reduces the magnitude of deflection or arc blow; however, ac welding does not completely eliminate arc blow. Reduction of arc blow is reduced because the alternating current sets up other currents that tend to either neutralize the magnetic field or greatly reduce its strength. Alternating current varies between maximum value of one polarity and the maximum value of the opposite polarity.The magnetic field surrounding the alternating current conductor does the same thingThe alternating magnetic field is a roving field which induces current in any conductor through which it passes, according to the induction principle.Currents are induced in nearby conductors in a direction opposite that of the inducing currentThese induced currents are called eddy currents. They produce a magnetic field of their own which tends to neutralize the magnetic field of the arc current.These currents are alternating currents of the same frequency as the arc current and are in the part of the work nearest the arc. They always flow from the opposite direction as shown by figure 6-64. When alternating current is used for welding, eddy currents are induced in the workpiece, which produce magnetic fields and reduce the intensity of the field acting on the arc. Alternating current cannot be used for all welding applications and for this reason changing from direct current to alternating current may not always be possible to eliminate or reduce arc blow.

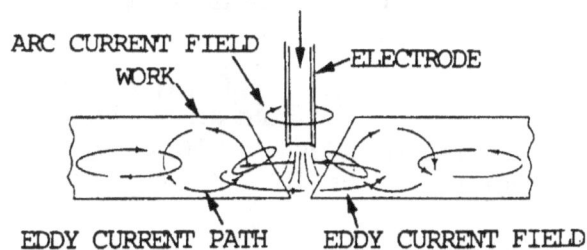

Figure 6-64. Reduction of magnetic force due to induced fields.

j. Summary of Factors Causing Arc Blow.

(1) Arc blow is caused by magnetic forces. The induced magnetic forces are not symmetrical about the magnetic field surrounding the path of the welding current. The location of magnetic material with respect to the arc creates a magnetic force on the arc which acts toward the easiest magnetic path and is independent of electrode polarity. The location of the easiest magnetic path changes constantly as welding progresses; therefore, the intensity and the direction of the force changes.

(2) Welding current will take the easiest path but not always the most direct path through the work to the work lead connectionThe resultant magnetic force is opposite in direction to the current from the arc the work lead connection, and is independent of welding current polarity.

(3) Arc blow is not as severe with alternatingcurrent because the induction principle creates current flow within the base metalwhich creates magnetic fields that tend to neutralize the magnetic field affectithge arc.

(4) The greatest magnetic force on the arc is caused by the difference in resistance of the magnetic path in the base metal around the arc. The location of the work lead connection is of secondary importance, but may have an effect on reducing the total magnetic force on the arc. It is best to have the work lead connection at the starting point of the weld. This is particularly true in electroslag welding where the work lead should be connected to the starting sump. On occasion, the work lead can be changed to the opposite end of the joint. In sane cases, leads can be connected to both ends.

k. <u>Minimizing Arc Blow.</u>

(1) The magnetic forces acting on the arc can be modified by changing the magnetic path across the joint. This can be accomplished by runoff tabs, starting plates, large tack welds, and backing strips, as well as the welding sequence.

(2) An external magnetic field produced by an electromagnet may be effective. This can be accomplished by wrapping several turns of welding lead around the workpiece.

(3) Arc blow is usually more pronounced at the start of the weld seam. In this case, a magnetic shunt or runoff tab will reduce the blow.

(4) Use as short an arc as possible so that there is less of an arc for the magnetic forces to control.

(5). The welding fixture can be a source of arc blow; therefore, an analysis with respect to fixturing is important. The hold-down clamps and backing bars must fit closely and tightly to the work. In general, copper or nonferrous metals should be used. Magnetic structure of the fixture can affect the magnetic forces controlling the arc.

(6) Place ground connections as far as possible from the joints to be welded.

(7) If back blow is the problem, place the ground connection the start of welding, and weld toward a heavy tack weld.

(8) If forward blow causes trouble, place the ground connection the end of the joint to be welded.

(9) Position the electrode so that the arc force counteracts the arc blow.

(10) Reduce the welding current.

(11) Use the backstep sequence of welding.

(12) Change to ac, which may require a change in electrode classification.

(13) Wrap the ground cable around the workpiece in a direction such that the magnetic field it sets up will counteract the magnetic field causing the arc blow.

(14) Another major problem can result from magnetic fields already in the base metal, particularly when the base metal has been handled by magnet lifting cranes. Residual magnetism in heavy thick plates handled by magnets can be of such magnitude that it is almost impossible to make a weld. Attempt to demagnetize the parts, wrap the part with welding leads to help overcome their effect, or stress relieve or anneal the parts.

6-32. WELD FAILURE ANALYSIS

a. General. Only rarely are there failures of welded structures, but failures of large engineered structures do occur occasionally. Catastrophic failures of major structures are usually reported whenever they occur. The results of investigations of these failures are usually reported and these reports often provide information that is helpful in avoiding future similar problems. In the same manner, there are occasional failures of noncritical welds and weldments that should also be investigated. Once the reason is determined it can then be avoided. An objective study must be made of any failure of parts or structures to determine the cause of the failure. This is done by investigating the service life, the conditions that led up to the failure, and the actual mode of the failure. An objective study of failure should utilize every bit of information available, investigate all factors that could remotely be considered and evaluate all this information to find the reason for the failure. Failure investigation often uncovers facts that lead to changes in design, manufacturing, or operating practice, that will eliminate similar failures in the future. Failures of insignificant parts can also lead to advances in knowledge and should be done objectively, as with a large structure. Each failure and subsequent investigation will lead to changes that will assure a more reliable product in the future.

b. The following four areas of interest should be investigated to determine the cause of weld failure and the interplay of factors involved:

(1) Initial observation. The detailed study by visual inspection of the actual component that failed should be made at the failure site as quickly as possible. Photographs should be taken, preferably in color, of all parts, structures, failure surfaces, fracture texture appearance, final location of component debris, and all other factors. Witnesses to the failure should all be interviewed and all information determined from them should be recorded.

(2) Background data. Investigators should gather all information concerning specifications, drawings, component design, fabrication methods, welding procedures, weld schedules, repairs in and during manufacturing and in service, maintenance, and service use. Efforts should be made to obtain facts pertinent to all possible failure modes. Particular attention should be given to environmental details, including operating temperatures, normal service loads, overloads, cyclic loading, and abuse.

(3) Laboratory studies. Investigators should make tests to verify that the material in the failed parts actually possesses the specified composition, mechanical properties, and dimensions. Studies should also be made microscopically in those situations in which it would lead to additional information. Each failed part should be thoroughly investigated to determine what bits of information can be added to the total picture. Fracture surfaces can be extremely important. Original drawings should be obtained and marked showing failure locations, along with design stress data originally used in designing the product. Any other defects in the structure that are apparent, even though they might not have contributed to the failure, should also be noted and investigated.

(4) Failure assumptions. The investigator should list not only all positive facts and evidence that may have contributed to the failure, but also all negative responses that may be learned about the failure. It is sometimes important to know what specific things did not happen or what evidence did not appear to

help determine what happened. The data should be tabulated and the actual failure should be synthesized to include all available evidence.

c. Failure cause can usually be classified in one of the following three classifications:

(1) Failure due to faulty design or misapplication of material.

(2) Failure due to improper processing or improper workmanship.

(3) Failure due to deterioration during service.

d. The following is a summary of the above three situations:

(1) Failure due to faulty design or misapplication of the material involves failure due to inadequate stress analysis, or a mistake in design such as incorrect calculations on the basis of static loading instead of dynamic or fatigue loading. Ductile failure can be caused by a load too great for the section area or the strength of the material. Brittle fracture may occur from stress risers inherent in the design, or the wrong material may have been specified for producing the part.

(2) Failures can be caused by faulty processing or poor workmanship that may be related to the design of the weld joint, or the weld joint design can be proper but the quality of the weld is substandard. The poor quality weld might include such defects as undercut, lack of fusion, or cracks. Failures can be attributed to poor fabrication practice such as the elimination of a root opening, which will contribute to incomplete penetration. There is also the possibility that the incorrect filler metal was used for welding the part that failed.

(3) Failure due to deterioration during service can cause overload, which may be difficult to determine. Normal wear and abuse to the equipment may have resulted in reducing sections to the degree that they no longer can support the load. Corrosion due to environmental conditions and accentuated by stress concentrations will contribute to failure. In addition, there may be other types of situations such as poor maintenance, poor repair techniques involved with maintenance, and accidental conditions beyond the user's control. The product might be exposed to an environment for which it was not designed.

e. Conclusion. Examination of catastrophic and major failures has led the welding industry to appreciate the following facts:

(1) Weldments are monolithic in character.

(2) Anything welded onto a structure will carry part of the load whether intended or not.

(3) Abrupt changes in section, either because of adding a deckhouse or removing a portion of the deck for a hatch opening, create stress concentration. Under normal loading, if the steel at the point of stress concentration is notch sensitive at the service temperature, failure can result.

6-33. OTHER WELDING PROBLEMS

a. There are two other welding problems that require some explanation and solutions. These are welding over painted surfaces and painting of welds.

CAUTION

Cutting painted surfaces with arc or flame processes should be done with caution. Demolition of old structural steel work that had been painted many times with flame-cutting or arc-cutting techniques can create health problems. Cutting through many layers of lead paint will cause an abnormally high lead concentration in the immediate area and will require special precautions such as extra ventilation or personnel protection.

b. Welding over paint is discouraged. In every code or specification, it is specifically stated that welding should be done on clean metal. In some industries, however, welds are made over paint and in others flame cutting is done on painted base metal.

(1) In the shipbuilding industry and in several other industries, steel when it is received from tie steel mill, is shot blasted, given a coating of prime paint, and then stored outdoors. Painting is done to preserve the steel during storage, and to identify it. In sane shipyards a different color paint is used for different classes of steel. When this practice is used, every effort should be made to obtain a prime paint that is compatible with welding.

(2) There are at least three factors involved with the success of the weld when welding over painted surfaces: the compatibility of the paint with welding; the dryness of the paint; and the paint film thickness.

(3) Paint compatibility varies according to the composition of the paint. Certain paints contain large amounts of aluminum or titanium dioxide, which are usually compatible with welding. Other paints may contain zinc, lead, vinyls, and other hydrocarbons, and are not compatible with welding. The paint manufacturer or supplier should be consulted. Anything that contributes to deoxidizing the weld such as aluminum, silicon, or titanium will generally be compatible. Anything that is a harmful ingredient such as lead, zinc and hydrocarbons will be detrimental. The fillet break test can be used to determine compatibility. The surfaces should be painted with the paint under consideration. The normal paint film thickness should be used, and the paint must be dry.

(4) The fillet break test should be run using the proposed welding procedure over the painted surface. It should be broken and the weld examined. If the weld breaks at the interface of the plate with the paint it is obvious that the paint is not compatible with the weld.

(5) The dryness of the paint should be considered. Many paints employ an oil base which is a hydrocarbon. These paints dry slowly, since it takes a considerable length of time for the hydrocarbons to evaporate. If welding is done before the paint is dry, hydrogen will be in the arc atmosphere and can contribute to underbead cracking. The paint will also cause porosity if there is sufficient oil present. Water based paints should also be dry prior to welding.

(6) The thickness of the paint film is another important factor. Some paints may be compatible if the thickness of the film is a maximum of to 4 mm. If the paint film thicknesses are double that amount such as occurs at an overlap area, there is the passability of weld porosity. Paint films that are to be welded over should be of the minimum thickness possible.

(7) Tests should be run with the dry maximum film thickness to be used with the various types of paints to determine which paint has the least harmful effects on the weld deposit.

c. Painting over welds is also a problem. The success of any paint film depends on its adherence to the base metal and the weld, which is influenced by surface deposits left on the weld and adjacent to it. The metallurgical factors of the weld bead and the smoothness of the weld are of minor importance with regard to the success of the paint. Paint failure occurs when the weld and the immediate area are not properly cleaned prior to painting. Deterioration of the paint over the weld also seems to be dependent upon the amount of spatter present. Spatter on or adjacent to the weld leads to rusting of the base material under the paint. It seems that the paint does not completely adhere to spatter and some spatter does fall off in time, leaving bare metal spots in the paint coating.

CAUTION

Aluminum and aluminum alloys should not be cleaned with caustic soda or cleaners having a pH above 10, as they may react chemically. Other nonferrous metals should be investigated for reactivity prior to cleaning.

(1) The success of the paint job can be insured by observing both preweld and postweld treatment. Preweld treatment found most effective is to use antispatter compounds, as well as cleaning the weld area, before welding. The antispatter compound extends the paint life because of the reduction of spatter. The antispatter compound must be compatible with the paint to be used.

(2) Postweld treatment for insuring paint film success consists of mechanical and chemical cleaning. Mechanical cleaning methods can consist of hand chipping and wire brushing, power wire brushing, or sand or grit blasting. Sand or grit blasting is the most effective mechanical cleaning method. If the weldment is furnace stress relieved and then grit blasted, it is prepared for painting. When sand or grit blasting cannot be used, power wire brushing is the next most effective method. In addition to mechanical cleaning, chemical bath washing is also recommended. Slag coverings on weld depsits must be thoroughly removed from the surface of the weld and from the adjacent base metal. Different types of coatings create more or less problems in their removal and also with respect to paint adherence. Weld slag of many electrodes is alkaline in nature and for this reason must be neutralized to avoid chemical reactions with the paint, which will cause the paint to loosen and deteriorate. For this reason, the weld should be scrubbed with water, which will usually remove the residual coating slag and smoke film from the weld. If a small amount of phosphoric acid up to a 5% solution is used, it will be more effective in neutralizing and removing the slag. It must be followed by a water rinse. If water only is used, it is advisable to add small amounts of phosphate or chromate inhibitors to the water to avoid rusting, which might otherwise occur.

6-33.   OTHER WELDING PROBLEMS (cont)

(3) It has been found that the method of applying paint is not an important factor in determining the life of the paint over welds. The type of paint employed must be suitable for coating metals and for the service intended.

(4) Successful paint jobs over welds can be obtained by observing the following: minimize weld spatter using a compatible anti-spatter compound; mechanically clean the weld and adjacent area; and wash the weld area with a neutralizing bath and rinse.

CHAPTER 7
METALS IDENTIFICATION

Section I. CHARACTERISTICS

7-1. GENERAL

Most of the metals and alloys used in Army materiel can be welded by one or more of the processes described in this manual. This section describes the characteristics of metals and their alloys, with particular reference to their significance in welding operations.

7-2. PROPERTIES OF METALS

a. <u>Definitions</u>. All metals fall within two categories, ferrous or nonferrous.

(1) <u>Ferrous metals</u> are metals that contain iron. Ferrous metals appear in the form of cast iron, carbon steel, and tool steel. The various alloys of iron, after undergoing cetain processes, are pig iron, gray cast iron, white iron, white cast iron, malleable cast iron, wrought iron, alloy steel, and carbon steel. All these types of iron are mixtures of iron and carbon, manganese, sulfur, silicon, and phosphorous. Other elements are also present, but in amounts that do not appreciably affect the characteristics of the metal.

(2) <u>Nonferrous metals</u> are those which do not contain iron. Aluminum, copper, magnesium, and titanium alloys are among those metals which belong to this group .

b. <u>Physical Properties</u>. Many of the physical properties of metals determine if and how they can be welded and how they will perform in service. Physical properties of various metals are shown in table 7-1, p 7-2.

7-2. PROPERTIES OF METALS (cont)

Table 7-1. Physical Properties of Metals

Properties Base Metal Or Alloy	Specific Gravity	Density lb/ft³	Melting Point (Liquidus)		Boiling Point		Relative Thermal Conductivity Copper = 1	Co-efficient of linear Expansion × 10⁻⁶ per degree	
			°F	°C	°F	°C		°F	°C
Aluminum and alloys	2.70	166	1218	659	3270	2480	0.52	13.8	24.8
Brass, navy	8.60	532	1650	900	NA	NA	0.28	11.8	21.2
Bronze, alum (90Cu-9Al)	7.69	480	1905	1040	NA	NA	0.15	16.6	29.9
Bronze, phosphor (90Cu-10Sn)	8.78	551	1830	1000	NA	NA	0.12	10.2	18.4
Bronze, silicon (96Cu-3Si)	8.72	542	1880	1025	NA	NA	0.10	10.0	18.0
Copper (deoxidized)	8.89	556	1981	1081	4700	2600	1.00	9.8	17.6
Copper nickel (70Cu-30Ni)	8.81	557	2140	1172	NA	NA	0.07	9.0	16.2
Everdur (96Cu-3Si-1Mn)	8.37	523	1866	1019	NA	NA	0.09	10.0	18.0
Gold	19.30	1205	1945	1061	5380	2950	0.76	7.8	14.0
Inconel (72Ni-16Cr-8Fe)	8.25	530	2600	1425	NA	NA	0.04	6.4	11.5
Iron, cast	7.50	450	2300	1260	NA	NA	0.12	6.0	10.8
Iron, wrought	7.80	485	2750	1510	5500	3000	0.16	6.7	12.1
Lead	11.34	708	621	328	3100	1740	0.08	16.4	29.5
Magnesium	1.74	108	1202	650	2010	1100	0.40	14.3	25.7
Monel (67Ni-30Cu)	8.47	551	2400	1318	NA	NA	0.07	7.8	14.0
Nickel	8.80	556	2650	1452	5250	3000	0.16	7.4	13.3
Nickel silver	8.44	546	2030	1110	NA	NA	0.09	9.0	16.2
Silver	10.45	656	1764	962	4010	2210	1.07	10.6	19.1
Steel, low alloy	7.85	490	2600	1430	NA	NA	0.12	6.7	12.1
Steel, high carbon	7.85	490	2500	1374	NA	NA	0.17	6.7	12.1
Steel, low carbon	7.84	490	2700	1483	NA	NA	0.17	6.7	12.1
Steel, manganese (14Mn)	7.81	490	2450	1342	NA	NA	0.04	6.7	12.1
Steel, medium carbon	7.84	490	2600	1430	NA	NA	0.17	6.7	12.1
Steel, stainless (austentic)	7.90	495	2550	1395	NA	NA	0.12	9.6	17.3
Steel, stainless (martensitic)	7.70	485	2600	1430	NA	NA	0.17	9.5	17.1
Steel, stainless (ferritic)	7.70	485	2750	1507	NA	NA	0.17	9.5	17.1
Tantalum	16.60	1035	5162	2996	7410	5430	0.13	3.6	6.5
Tin	7.29	455	449	232	4100	2270	0.15	12.8	23.0
Titanium	4.50	281	3031	1668	5900	3200	0.04	4.0	7.2
Tungsten	18.80	1190	6170	3420	10,600	5600	0.42	2.5	4.5
Zinc	7.13	442	788	419	1660	907	0.27	22.1	39.8

(1) <u>Color.</u> Color relates to the quality of light reflected from the metal.

(2) <u>Mass or density.</u> Mass or density relates to mass with respect to volume. Commonly known as <u>specific gravity,</u> this property is the ratio of the mass of a given volume of the metal to the mass of the same volume of water at a specified temperature, usually $39°F$ ($4°C$). For example, the ratio of weight of one cubic foot of water to one cubic foot of cast iron is the specific gravity of cast iron. This property is measured by grams per cubic millimeter or centimeter in the metric system.

(3) <u>Melting point.</u> The melting point of a metal is important with regard to welding. A metal's fusibility is related to its melting point, the temperature at which the metal changes from a solid to a molten state. Pure substances have a sharp melting point and pass from a solid state to a liquid without a change in temperature. During this process, however, there is an absorption of heat during melting and a liberation of heat during freezing. The absorption or release of thermal energy when a substance changes state is called its <u>latent heat</u>. Mercury is the only common metal that is in its molten state at normal room temperature. Metals having low melting temperatures can be welded with lower temperature heat sources. The soldering and brazing processes utilize low-temperature metals to join metals having higher melting temperatures.

(4) <u>Boiling point.</u> Boiling point is also an important factor in welding. The boiling point is the temperature at which the metal changes from the liquid state to the vapor state. Some metals, when exposed to the heat of an arc, will vaporize.

(5) <u>Conductivity.</u> Thermal and electrical conductivity relate to the metal's ability to conduct or transfer heat and electricity. <u>Thermal conductivity,</u> the ability of a metal to transmit heat throughout its mass, is of vital importance in welding, since one metal may transmit heat from the welding area much more quickly than another. The thermal conductivity of a metal indicates the need for preheating and the size of heat source required. Thermal conductivity is usually related to copper. Copper has the highest thermal conductivity of the common metals, exceeded only by silver. Aluminum has approximately half the thermal conductivity of copper, and steels have abut one-tenth the conductivity of copper. Thermal conductivity is measured in calories per square centimeter per second per degree Celsius. <u>Electrical conductivity</u> is the capacity of metal to conduct an electric current. A measure of electrical conductivity is provided by the ability of a metal to conduct the passage of electrical current. Its opposite is resistivity, which is measured in micro-ohms per cubic centimeter at a standardize temperature, usually $20°C$. Electrical conductivity is usually considered as a percentage and is related to copper or silver. Temperature bears an important part in this property. As temperature of a metal increases, its conductivity decreases. This property is particularly important to resistance welding and to electrical circuits.

7-2. PROPERTIESOF METALS (cont)

(6) <u>Coefficient of linear thermal expansion</u>.With few exceptions, solids expand when they are heated and contract when they are cooled. The coefficient of linear thermal expansion is a measure of the linear increase per unit length based on the change in temperature of the metal.Expansion is the increase on the dimension of a metal caused by heat. The expansion of a metal in a longitudinal direction is known as the linear expansion.The coefficient of linear expansion is expressed as the linear expansion per unit length for one degree of temperature increase. When metals increase in size, they increase not only in length but also in breadth and thickness. This is called <u>volumetric expansion</u>. The coefficient of linear and volumetric expansion varies over a wide range for different metals. Aluminum has the greatest coefficient of expansion, expanding almost twice as much as steel for the same temperaturechange. This is important for welding with respect to warpage, wapage controhnd fixturing, and for welding together dissimilar metals.

(7) <u>Corrosion resistance</u>. Corrosion resistance is the resistance to eating or wearing away by air, moisturepr other agents.

c. <u>Mechanical Properties</u>. The mechanical properties of metals determine the range of usefulness of the metal and establish the service that can be expected. Mechanical properties are also used to help specify and identify the metalsThey are important in welding because the weld must provide the same mechanical properties as the base metals being joined. The adequacy of a weld depends on whether or not it provides properties equal to or exceeding those of the metals being joined. The most common mechanical properties considered are strength, hardness, ductility, and impact resistance. Mechanical properties of various metals are shown in table 7-2.

Table 7-2. Mechanical Properties of Metals

Properties Base Metal Or Alloy	YIELD STRENGTH			TENSILE STRENGTH			Elongation % in 2 in. (50mm)	Hardness BHN
	lb/in.^{2}	MPa	kg/mm^2	lb/in.^{2}	MPa	kg/mm^2		
Aluminum and alloys	5,000	34.5	3.5	13,000	89.60	9.1	35.0	23.0
Brass, navy	20,000	206.8	21.0	62,000	427.40	43.6	47.0	89.0
Bronze, alum. (90Cu-9A1)	30,000	206.8	21.0	76,000	523.90	53.4	10.0	125.0
Bronze, phosphor (90Cu-10Sn)	28,000	193.0	19.7	66,000	455.00	46.4	35.0	148.0
Bronze, silicon (96Cu-3Si)	15,000	103.4	10.5	40,000	275.80	28.1	52.0	119.0
Copper (deoxidized)	10,000	68.9	7.0	33,000	227.50	23.2	40.0	30.0
Copper nickel (70Cu-30Ni)	20,000	137.9	14.0	55,000	379.20	38.6	45.0	95.0
Everdur (96Cu-3Si-1Mn)	20,000	137.9	14.0	55,000	379.20	38.6	60.0	75.0
Gold	-	-	-	17,000	117.20	11.9	45.0	25.0
Inconel (76Ni-16Cr-8Fe)	35,000	241.3	24.6	85,000	586.00	59.7	45.0	150.0
Iron, cast	-	-	-	25,000	172.40	17.5	0.5	180.0
Iron, wrought	27,000	186.1	19.0	40,000	275.80	28.1	25.0	100.0
Lead	19,000	131.0	13.4	2,500	17.20	1.7	45.0	6.0
Magnesium	13,000	89.6	9.1	25,000	172.40	17.5	4.0	40.0
Monel (67Ni-30Cu)	35,000	241.3	24.6	75,000	517.10	52.7	45.0	125.0
Nickel	8,500	58.6	6.0	46,000	317.10	32.3	40.0	85.0
Nickel silver	20,000	137.9	14.0	58,000	399.80	40.7	35.0	90.0
Silver	8,000	55.2	5.6	23,000	158.60	16.2	35.0	90.0
Steel, low alloy	50,000	344.7	35.1	75,000	517.10	52.7	28.0	170.0
Steel, high carbon	90,000	620.5	63.2	140,000	965.20	98.4	20.0	201.0
Steel, low carbon	36,000	248.2	25.3	60,000	413.60	42.2	35.0	310.0
Steel, manganese (14Mn)	75,000	517.1	52.7	118,000	813.50	82.9	22.0	200.0
Steel, medium carbon	52,000	358.5	36.5	87,000	599.80	61.2	24.0	170.0
Steel, stainless (austentic)	40,000	275.8	28.1	90,000	620.50	63.2	23.0	160.0
Steel, stainless (matensitic)	80,000	551.5	56.2	100,000	68.90	70.3	26.0	250.0
Steel, stainless (ferritic)	45,000	310.2	31.6	75,000	517.10	52.7	30.0	155.0
Tantalum	-	-	-	50,000	344.70	35.1	40.0	300.0

Table 7-2. Mechanical Properties of Metals (cont)

Base Metal Or Alloy	YIELD STRENGTH lb/in.^{2}	MPa	kg/mm^2	TENSILE STRENGTH lb/in.^{2}	MPa	kg/mm^2	Elongation % in 2 in. (50mm)	Hardness BHN
Tin	1,710	11.8	1.2	3,130	21.60	2.2	50.0	5.3
Titanium	40,000	275.8	28.1	60,000	413.60	42.2	28.0	-
Tungsten	-	-	-	500,000	3447.00	351.5	15.0	230.0
Zinc	18,000	124.1	12.6	25,000	172.35	17.5	20.0	38.0

NOTE Values depend on heat treatment or mechanical condition or mass of the metal.

(1) Tensile strength. Tensile strength is defined as the maximum load in tension a material will withstand before fracturing, or the ability of a material to resist being pulled apart by opposing forces. Also known as ultimate strength, it is the maximum strength developed in a metal in a tension test. The tension test is a method for determing the behavior of a metal under an actual stretch loading. This test provides the elastic limit, elongation, yield point, yield strength, tensile strength, and the reduction in area. The tensile strength is the value most commonly given for the strength of a material and is given in pounds per square inch (psi) (kiloPascals (kPa)). The tensile strength is the number of pounds of force required to pull apart a bar of material 1.0 in. (25.4 mm) wide and 1.00 in. (25.4 mm) thick (fig. 7-1).

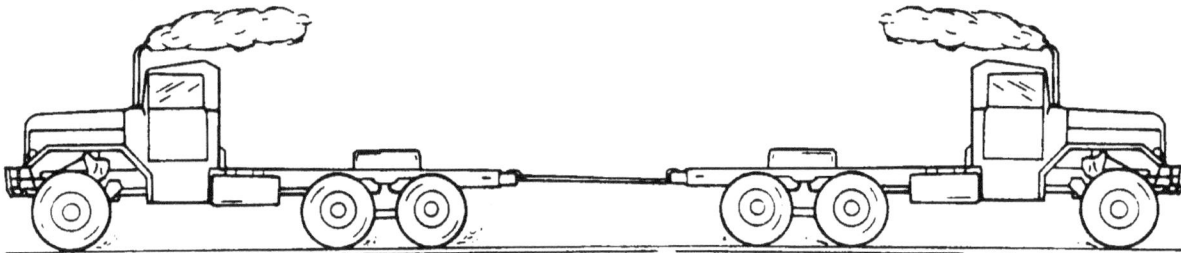

Figure 7-1. Tensile strength.

(2) Shear strength. Shear strength is the ability of a material to resist being fractured by opposing forces acting of a straight line but not in the same plane, or the ability of a metal to resist being fractured by opposing forces not acting in a straight line (fig. 7-2).

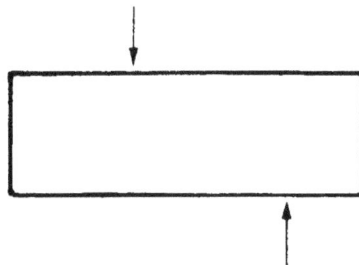

Figure 7-2. Shear strength.

7-2. PROPERTIES OF METALS (cont)

(3) <u>Fatigue strength</u>. Fatigue strength is the maximum load a material can withstand without failure during a large number of reversals of load. For example, a rotating shaft which supports a weight has tensile forces on the top portion of the shaft and compressive forces on the bottom. As the shaft is rotated, there is a repeated cyclic change in tensile and compressive strength. Fatigue strength values are used in the design of aircraft wings and other structures subject to rapidly fluctuating loads. Fatigue strength is influenced by microstructure, surface condition, corrosive environment, and cold work.

(4) <u>Compressive strength</u>. Compressive strength is the maximum load in compression a material will withstand before a predetermined amount of deformation, or the ability of a material to withstand pressures acting in a given plane (fig. 7-3). The compressive strength of both cast iron and concrete are greater than their tensile strength. For most materials, the reverse is true.

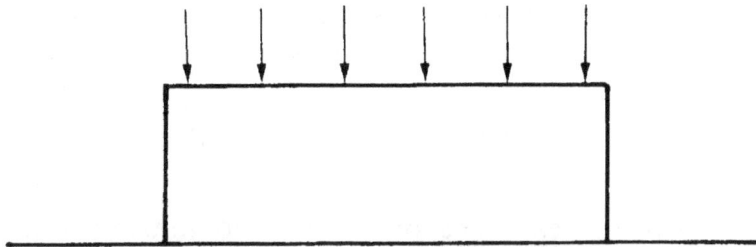

Figure 7-3. Compressive strength.

(5) <u>Elasticity.</u> Elasticity is the ability of metal to return to its original size, shape, and dimensions after being deformed, stretched, or pulled out of shape. The elastic limit is the point at which permanent damage starts. The yield point is the point at which definite damage occurs with little or no increase in load. The yield strength is the number of pounnds per square inch (kiloPascals) it takes to produce damage or deformation to the yield point.

(6) <u>Modulus of elasticity</u>. The modulus of elasticity is the ratio of the internal stress to the strain produced.

(7) <u>Ductility.</u> The ductility of a metal is that property which allows it to be stretched or otherwise changed in shape without breaking, and to retain the changed shape after the load has been ramoved. It is the ability of a material, such as copper, to be drawn or stretched permanently without fracture. The ductility of a metal can be determied by the tensile test by determiing the percentage of elongation. The lack of ductility is brittleness or the lack of showing any permanent damage before the metal cracks or breaks (such as with cast iron).

(8) <u>Plasticity.</u> Plasticity is the ability of a metal to be deformed extensively without rupture. Plasticity is similar to ductility.

(9) <u>Malleability.</u> Malleability is another form of plasticity, and is the ability of a material to deform permanently under compression without rupture. It is this property which allows the hammering and rolling of metals into thin sheets. Gold, silver, tin, and lead are examples of metals exhibiting high malleability. Gold has exceptional malleability and can be rolled into sheets thin enough to transmit light.

(10) <u>Reduction of area</u>. This is a measure of ductility and is obtained from the tensile test by measuring the original cross-sectional area of a specimen to a cross-sectional area after failure.

(11) <u>Brittleness</u>. Brittleness is the property opposite of plasticity or ductility. A brittle metal is one than cannot be visibly deformed permanently, or one that lacks plasticity.

(12) <u>Toughness</u>. Toughness is a combination of high strength and medium ductility. It is the ability of a material or metal to resist fracture, plus the ability to resist failure after the damage has begun. A tough metal, such as cold chisel, is one that can withstand considerable stress, slowly or suddenly applied, and which will deform before failure. Toughness is the ability of a material to resist the start of permanent distortion plus the ability to resist shock or absorb energy.

(13) <u>Machinability and weldability</u>. The property of machinability and weldability is the ease or difficulty with which a material can be machined or welded.

(14) <u>Abrasion resistance</u>. Abrasion resistance is the resistance to wearing by friction.

(15) <u>Impact resistance</u>. Resistance of a metal to impacts is evaluated in terms of impact strength. A metal may possess satisfactory ductility under static loads, but may fail under dynamic loads or impact. The impact strength of a metal is determined by measuring the energy absorbed in the fracture.

(16) <u>Hardness</u>. Hardness is the ability of a metal to resist penetration and wear by another metal or material. It takes a combination of hardness and toughness to withstand heavy pounding. The hardness of a metal limits the ease with which it can be machined, since toughness decreases as hardness increases. Table 7-3, p 7-8, illustrates hardness of various metals.

(a) <u>Brinell hardness test</u>. In this test, a hardened steel ball is pressed slowly by a known force against the surface of the metal to be tested. The diameter of the dent in the surface is then measured, and the Brinell hardness number (bhn) is determined by from standard tables (table 7-3, p 7-8).

(b) <u>Rockwell hardness test</u>. This test is based upon the difference between the depth to which a test point is driven into a metal by a light load and the depth to which it is driven in by a heavy load. The light load is first applied and then, without moving the piece, the heavy load is applied. The hardness number is automatically indicated on a dial. The letter designations on the Rockwell scale, such as B and C, indicate the type of penetrator used and the amount of heavy load (table 7-3, p 7-8). The same light load is always used.

(c) <u>Scleroscope hardness test</u>. This test measures hardness by letting a diamond-tipped hammer fall by its own weight from a fixed height and rebound from the surface; the rebound is measured on a scale. It is used on smooth surfaces where dents are not desired.

7-2. PROPERTIES OF METALS (cont)

Table 7-3. Hardness Conversion Table

BRINELL			ROCKWELL			Approximate
			C	B		
Diameter in mm, 8000 kg Load 10 mm Ball	Hardness No.	Vickers or Firth Hardness No.	150 kg Load 120° Diamond Cone	100 kg Load 1/16 in. dia Ball	Scleroscope No.	Tensile Strength 1000 psi
2.05	898					440
2.10	857					420
2.15	817					401
2.20	780	1150	70		106	384
2.25	745	1050	68		100	368
2.30	712	960	66		95	352
2.35	682	885	64		91	337
2.40	653	820	62		87	324
2.45	627	765	60		84	311
2.50	601	717	58		81	298
2.55	578	675	57		78	287
2.60	555	633	55	120	75	276
2.65	534	598	53	119	72	266
2.70	514	567	52	119	70	256
2.75	495	540	50	117	67	247
2.80	477	515	49	117	65	238
2.85	461	494	47	116	63	229
2.90	444	472	46	115	61	220
2.95	429	454	45	115	59	212
3.00	415	437	44	114	57	204
3.05	401	420	42	113	55	196
3.10	388	404	41	112	54	189
3.15	375	389	40	112	52	182
3.20	363	375	38	110	51	176
3.25	352	363	37	110	49	170
3.30	341	350	36	109	48	165
3.35	331	339	35	109	46	160
3.40	321	327	34	108	45	155
3.45	311	316	33	108	44	150
3.50	302	305	32	107	43	146
3.55	293	296	31	106	42	142
3.60	285	287	30	105	40	138
3.65	277	279	29	104	39	134
3.70	269	270	28	104	38	131
3.75	262	263	26	103	37	128
3.80	255	256	25	102	37	125
3.85	248	248	24	102	36	122
3.90	241	241	23	100	35	119
3.95	235	235	22	99	34	116
4.00	229	229	21	98	33	113
4.05	223	223	20	97	32	110
4.10	217	217	18	96	31	107
4.15	212	212	17	96	31	104
4.20	207	207	16	95	30	101
4.25	202	202	15	94	30	99
4.30	197	197	13	93	29	97
4.35	192	192	12	92	28	95
4.40	187	187	10	91	28	93

Table 7-3. Hardness Conversion Table (cont)

BRINELL			ROCKWELL			
			C	B		Approximate
Diameter in mm, 8000 kg Load 10 mm Ball	Hardness No.	Vickers or Firth Hardness No.	150 kg Load 120° Diamond Cone	100 kg Load 1/16 in. dia Ball	Scleroscope No.	Tensile Strength 1000 psi
4.45	183	183	9	90	27	91
4.50	179	179	8	89	27	89
4.55	174	174	7	88	26	87
4.60	170	170	6	87	26	85
4.65	166	166	4	86	25	83
4.70	163	163	3	85	25	82
4.75	159	159	2	84	24	80
4.80	156	156	1	83	24	78
4.85	153	153		82	23	76
4.90	149	149		81	23	75
4.95	146	146		80	22	74
5.00	143	143		79	22	72
5.05	140	140		78	21	71
5.10	137	137		77	21	70
5.15	134	134		76	21	68
5.20	131	131		74	20	66
5.25	128	128		73	20	65
5.30	126	126		72		64
5.35	124	124		71		63
5.40	121	121		70		62
5.45	118	118		69		61
5.50	116	116		68		60
5.55	114	114		67		59
5.60	112	112		66		58
5.65	109	109		65		56
5.70	107	107		64		56
5.75	105	105		62		54
5.80	103	103		61		53
5.85	101	101		60		52
5.90	99	99		59		51
5.95	97	97		57		50
6.00	95	95		56		49

7-3. CATEGORIES OF METALS (cont)

a. <u>General</u>. It is necessary to know the composition of the metal being welded in order to produce a successful weld. Welders and metal workers must be able to identify various metal products so that proper work methods may be applied. For Army equipment, drawings (MWOs) should be available. They must be examined in order to determine the metal to be used and its heat treatment, if required. After some practice, the welder will learn that certain parts of machines or equipment are always cast iron, other parts are usually forgings, and so on.

b. <u>Tests.</u> There are seven tests that can be performed in the shop to identify metals. Six of the different tests are summarized in table 7-4, p 7-10. These should be supplemented by tables 7-1 and 7-2 (p 7-2 and 7-4) which present physical and mechanical properties of metal, and table 7-3, which presents hardness data. These tests are as follows:

Table 7-4. Summary of Identification Tests of Metals

Base Metal or Alloy	Color	Properties Magnet	Chisel	Fracture	Flame or Torch	Spark
Aluminum and alloys	bluish-white	non-magnetic	easily cut	white	melts wo/col	non-spark
Brass, navy	yellow or reddish	non-magnetic	easily cut	not used	not used	non-spark
Bronze, alum. (90Cu-9Al)	reddish yellow	non-magnetic	easily cut	not used	not used	non-spark
Bronze, phosphor (90Cu-10Sn)	reddish yellow	non-magnetic	easily cut	not used	not used	non-spark
Bronze, silicon (96Cu-3Si)	reddish yellow	non-magnetic	easily cut	not used	not used	non-spark
Copper (deoxidized)	red; 1 cent piece	non-magnetic	easily cut	red	not used	non-spark
Copper nickel (70Cu-30 Ni)	white; 5 cent piece	non-magnetic	easily cut	not used	not used	non-spark
Everdur (96Cu-3Si-1 Mn)	gold	non-magnetic	easily cut	not used	not used	non-spark
Gold	yellow	non-magnetic	easily cut	not used	not used	non-spark
Inconel (76Ni-16Cr-8Fe)	white	non-magnetic	easily cut	not used	not used	non-spark
Iron, cast	dull gray	magnetic	not easily chipped	brittle	melts slowly	see text
Iron, wrought	light gray	magnetic	easily cut	bright gray fibers	melts fast	see text
Lead	dark gray	non-magnetic	very soft	white; crystal	melts quick	non-spark
Magnesium	silvery white	slightly magnetic	soft	not used	burns in air	non-spark
Monel (67Ni-30Cu)	light gray	magnetic	tough	light gray	not used	see text
Nickel	white	non-magnetic	easily cut	almost white	not used	non-spark
Nickel silver	white	non-magnetic	easily chipped	not used	not used	non-spark
Silver	white; pre-1965 10¢ pc	non-magnetic	not used	not used	not used	non-spark
Steel, low alloy	blue-gray	magnetic	depends on comp	medium gray	shows color	see text
Steel, high carbon	dark gray	magnetic	hard to chip	very lgt gray	shows color	see text
Steel, low carbon	dark gray	magnetic	continuous chip	bright gray	shows color	see text
Steel, manganese (14Mn)	dull	non-magnetic	work hardens	coarse grained	shows color	see text
Steel, medium carbon	dark gray	magnetic	easily cut	very lgt gray	shows color	see text
Steel, stainless (austentic)	bright silvery	see text	continuous chip	deps on type	melts fast	see text
Steel, stainless (matensitic)	gray	slightly magnetic	continuous chip	deps on type	melts fast	see text
Steel, stainless (ferritic)	bright silvery	slightly-magnetic	-	deps on type	-	see text
Tantalum	gray	non-magnetic	hard to chip	-	high temp	-
Tin	silvery white	non-magnetic	usually as plating	usually as plating	melts quick	non-spark
Titanium	steel gray	non-magnetic	hard	not used	not used	see text
Tungsten	steel gray	non-magnetic	hardest metal	brittle	highest temp	non-spark
Zinc	dark gray	non-magnetic	usually as plating	at R.T.	melts quick	non-sprak

(1) <u>Appearance test</u>. The appearance test includes such things as color and appearance of machined as well as unmachined surfaces. Form and shape give definite clues as to the identity of the metal. The shape can be descriptive; for example, shape includes such things as cast engine blocks, automobile bumpers, reinforcing rods, I beams or angle irons, pipes, and pipe fittings. Form should be considered and may show how the part was rode, such as a casting with its obvious surface appearance and parting mold lines, or hot rolled wrought material, extruded or cold rolled with a smooth surface. For example, pipe can be cast, in which case it would be cast iron, or wrought, which would normally be steel. Color provides a very strong clue in metal identification. It can distinguish many metals such as copper, brass, aluminum, magnesium, and the precious metals. If metals are oxidized, the oxidation can be scraped off to determine the color of the unoxidized metal. This helps to identify lead, magnesium, and even copper. The oxidation on steel, or rust, is usually a clue that can be used to separate plain carbon steels from the corrosion-resisting steels.

(2) <u>Fracture test</u>. Some metal can be quickly identified by looking at the surface of the broken part or by studying the chips produced with a hammer and chisel. The surface will show the color of the base metal without oxidation. This will be true of copper, lead, and magnesium. In other cases, the coarseness or roughness of the broken surface is an indication of its structure. The ease of breaking the part is also an indication of its ductility of lack of ductility. If the piece bends easily without breaking, it is one of the more ductile metals. If it breaks easily with little or no bending, it is one of the brittle metals.

(3) <u>Spark test</u>. The spark test is a method of classifying steels and iron according to their composition by observing the sparks formed when the metal is held against a high speed grinding wheel. This test does not replace chemical analysis, but is a very convenient and fast method of sorting mixed steels whose spark characteristics are known. When held lightly against a grinding wheel, the different kinds of iron and steel produce sparks that vary in length, shape, and color. The grinding wheel should be run to give a surface speed of at least 5000 ft (1525 m) per minute to get a good spark stream. Grinding wheels should be hard enough to wear for a reasonable length of time, yet soft enough to keep a free-cutting edge. Spark testing should be done in subdued light, since the color of the spark is important. In all cases, it is best to use standard samples of metal for the purpose of comparing their sparks with that of the test sample.

(a) Spark testing is not of much use on nonferrous metals such as coppers, aluminums, and nickel-base alloys, since they do not exhibit spark streams of any significance. However, this is one way to separate ferrous and nonferrous metals.

7-3.   CATEGORIES OF METALS (cont)

(b) The spark resulting from the test should be directed downward and studied. The color, shape, length, and activity of the sparks relate to characteristics of the material being tested. The spark stream has specific items which can be identified. The straight lines are called carrier lines. They are usually solid and continuous. At the end of the carrier line, they may divide into three short lines, or forks. If the spark stream divides into more lines at the end, it is called a sprig. Sprigs also occur at different places along the carrier line. These are called either star or fan bursts. In some cases, the carrier line will enlarge slightly for a very short length, continue, and perhaps enlarge again for a short length. When these heavier portions occur at the end of the carrier line, they are called spear points or buds. High sulfur creates these thicker spots in carrier lines and the spearheads. Cast irons have extremely short streams, whereas low–carbon steels and most alloy steels have relatively long streams. Steels usually have white to yellow color sparks, while cast irons are reddish to straw yellow. A 0.15 percent carbon steel shins sparks in long streaks with some tendency to burst with a sparkler effect; a carbon tool steel exhibits pronounced bursting; and a steel with 1.00 percent cabon shows brilliant and minute explosions or sparklers. As the carbon content increases, the intensity of bursting increases.

(c) One big advantage of this test is that it can be applied to metal in all stages, bar stock in racks, machined forgings or finished parts. The spark test is best conducted by holding the steel stationary and touching a high speed portable grinder to the specimen with sufficient pressure to throw a horizontal spark stream about 12.00 in. (30.48 cm) long and at right angles to the line of vision. Wheel pressure against the work is important because increasing pressure will raise the temperature of the spark stream and give the appearance of higher carbon content. The sparks near and around the wheel, the middle of the spark stream, and the reaction of incandescent particles at the end of the spark stream should be observed. Sparks produced by various metals are shown in figure 7-4.

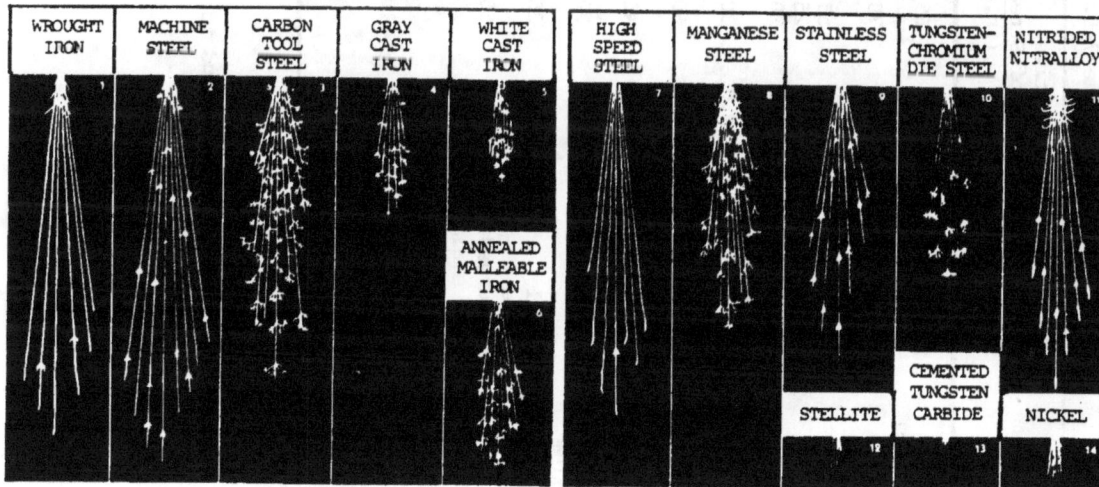

Figure 7-4.   Characteristics of sparks generated by the grinding of metals.

Table 7-5. Summary of Spark Test

Metal	Volume of Stream	Relative Length of Stream (mm)	(in.)	Color of Stream Close to Wheel	Color of Stream Near End of Stream	Quantity of Spurts	Nature of Spurts
1. Wrought iron	Large	1651.0	65	Straw	White	Very few	Forked
2. Machine steel (AISI 1020)	Large	1778.0	70	White	White	Few	Forked
3. Carbon tool steel	Moderately large	1397.0	55	White	White	Very many	Fine, repeating
4. Gray cast iron	Small	635.0	25	Red	Straw	Many	Fine, repeating
5. White cast iron	Very small	508.0	20	Red	Straw	Few	Fine, repeating
6. Annealed malleable iron	Moderate	762.0	30	Red	Straw	Many	Fine, repeating
7. High-speed steel (18-4-1)	Small	1524.0	60	Red	Straw	Extremely few	Forked
8. Austenitic manganese steel	Moderately large	1143.0	45	White	White	Many	Fine, repeating
9. Stainless steel (Type 410)	Moderate	1270.0	50	Straw	White	Moderate	Forked
10. Tungsten-chromium die steel	Small	889.0	35	Red	Straw	Many	Fine, repeating
11. Nitrided nitralloy	Large (curved)	1397.0	55	White	White	Moderate	Forked
12. Stellite	Very small	254.0	10	Orange	Orange	None	- -
13. Cemented tungsten carbide	Extremely small	50.8	2	Light orange	Light orange	None	- -
14. Nickel	Very small	254.0	10	Orange	Orange	None	- -
15. Copper, brass aluminum	None	- -	- -	- -	- -	None	- -

NOTE

The numbers on the left correspond to illustrations of spark streams shown in figure 7-4.

7-3. CATEGORIESOF METALS (cont)

CAUTION

The torch test should be used with discretion, as it may damage the part being tested. Additionally, magnesium may ignite when heated in the open atmosphere.

(4) Torch test. With the oxyacetylene torch, the welder can identify various metals by studying how fast the metal melts and how the puddle of molten metal and slag looks, as well as color changes during heating. When a sharp corner of a white metal part is heated, the rate of melting can be an indication of its identity. If the material is aluminum, it will not melt until sufficient heat has been used because its high conductivity. If the part is zinc, the sharp corner will melt quickly, since zinc is not a good conductor.In the case of copper, if the sharp comer melts, it is normally deoxidized copper. If it does not melt until much heat has been applied, it is electrolytic copper.Copper alloys, if composed of lead, will boil. To distinguish aluminum from magnesium, apply the torch to filings. Magnesium will burn with a sparkling white flame. Steel will show characteristic colors before melting.

(5) Magnetic test. The magnetic test can be quickly performed using a small pocket magnet. With experience, it is possible to judge a strongly magnetic material from a slightly magnetic material.The nonmagnetic materials are easily recognized. Strongly magnetic materials include the carbon and low-alloy steels, iron alloys, pure nickel, and martensitic stainless steels. A slightly magnetic reaction is obtained from Monel and high-nickel alloys and the stainless steel of the 18 chrome 8 nickel type when cold worked, such as in a seamless tube. Nonmagnetic materials include copper-base alloys, aluminum-base alloys, zinc-base alloys, annealed 18 chrome 8 nickel stainless, the magnesium, and the precious metals.

(6) Chisel test. The chip test or chisel test may also be used to identify metals. The only tools required are a banner and a cold chisel. Use the cold chisel to hammer on the edge or corner of the material being examined.The ease of producing a chip is an indication of the hardness of the metal. If the chip is continuous, it is indicative of a ductile metal, whereas if chips break apart, it indicates a brittle material. On such materials as aluminum, mild steel and malleable iron, the chips are continuous. They are easily chipped and the chips do not tend to break apart. The chips for gray cast iron are so brittle that they become small, broken fragments. On high-carbon steel, the chips are hard to obtain because of the hardness of the material, but can be continuous.

(7) Hardness test. Refer to table 7-3, p 7-8, for hardness values of the various metals, and to p 7-7 for information on the three hardness tests that are commonly used. A less precise hardness test is the file test. A summary of the reaction to filing, the approximate Brinell hardness, and the possible type of steel is shown in table 7-6. A sharp mill file must be used. It is assumed that the part is steel and the file test will help identify the type of steel.

Table 7-6. Approximate Hardness of Steel by the File Test

File Reaction	Brinell Hardness	Type Steel
File bites easily into metal	100 BHN	Mild steel
File bites into metal with pressure	200 BHN	Medium carbon steel
File does not bite into metal except with extreme pressure	300 BHN	High alloy steel-high carbon steel
Metal can only be filed with difficulty	400 BHN	Unhardened tool steel
File will mark metal but metal is nearly as hard as the file and filing is impractical	500 BHN	Hardened tool steel
Metal is harder than file	600 + BHN	

(8). Chemical test. There are numerous chemical test than can be made in the shop to identify some material. Monel can be distinguished form Inconel by one drop of nitric acid applied to the surface. It will turn blue-green on Monel, but will show no reaction on Inconel. A few drops of a 45 percent phosphoric acid will bubble on low-chomium stainless steels. Magnesium can be distinguished from aluminum using silver nitrate, which will leave a black deposit on magnesium, but not on aluminum. These tests can become complicated, and for this reason are not detailed further here.

c. Color Code for Marking Steel Bars. The Bureau of Standards of the United States Department of Commerce has a color code for making steel bars. The color markings provided in the code may be applied by painting the ends of bars. Solid colors usually mean carbon steel, while twin colors designate alloy and free-cutting steel.

d. Ferrous Metal The basic substance used to make both steel and cast iron (gray and malleable) is iron. It is used in the form of pig iron. Iron is produced form iron ore that occurs chiefly in nature as an oxide, the two most important oxides being hematite and magetite. Iron ore is reduced to pig iron in a blast furnace, and the impurities are removed in the form of slag (fig. 7-5, p 7-16). Raw materials charged into the furnace include iron ore, coke, and limestone. The pig iron produced is udes to manufacture steel or cast iron

Plain carbon steel consists of iron and carbon. Carbon is the hardening element. Tougher alloy steel contains other elements such as chromium, nickel, and molybdenum. Cast iron is nothing more than basic carbon steel with more carbon added, along with silicon. The carbon content range for steel is 0.03 to 1.7 percent, and 4.5 percent for cast iron.

7-3. CATEGORIES OF METALS (cont)

Steel is produced in a variety of melting furnaces, such as open-hearth, Bessemer converter, crucible, electric-arc, and induction. Most carbon steel is made in open-hearth furnaces, while alloy steel is melted in electric-arc and induction furnaces. Raw materials charged into the furnace include mixtures of iron ore, pig iron, limestone, and scrap. After melting has been completed, the steel is tapped from the furnace into a ladle and then poured into ingots or patterned molds. The ingots are used to make large rectangular bars, which are reduced further by rolling operations. The molds are used for castings of any design.

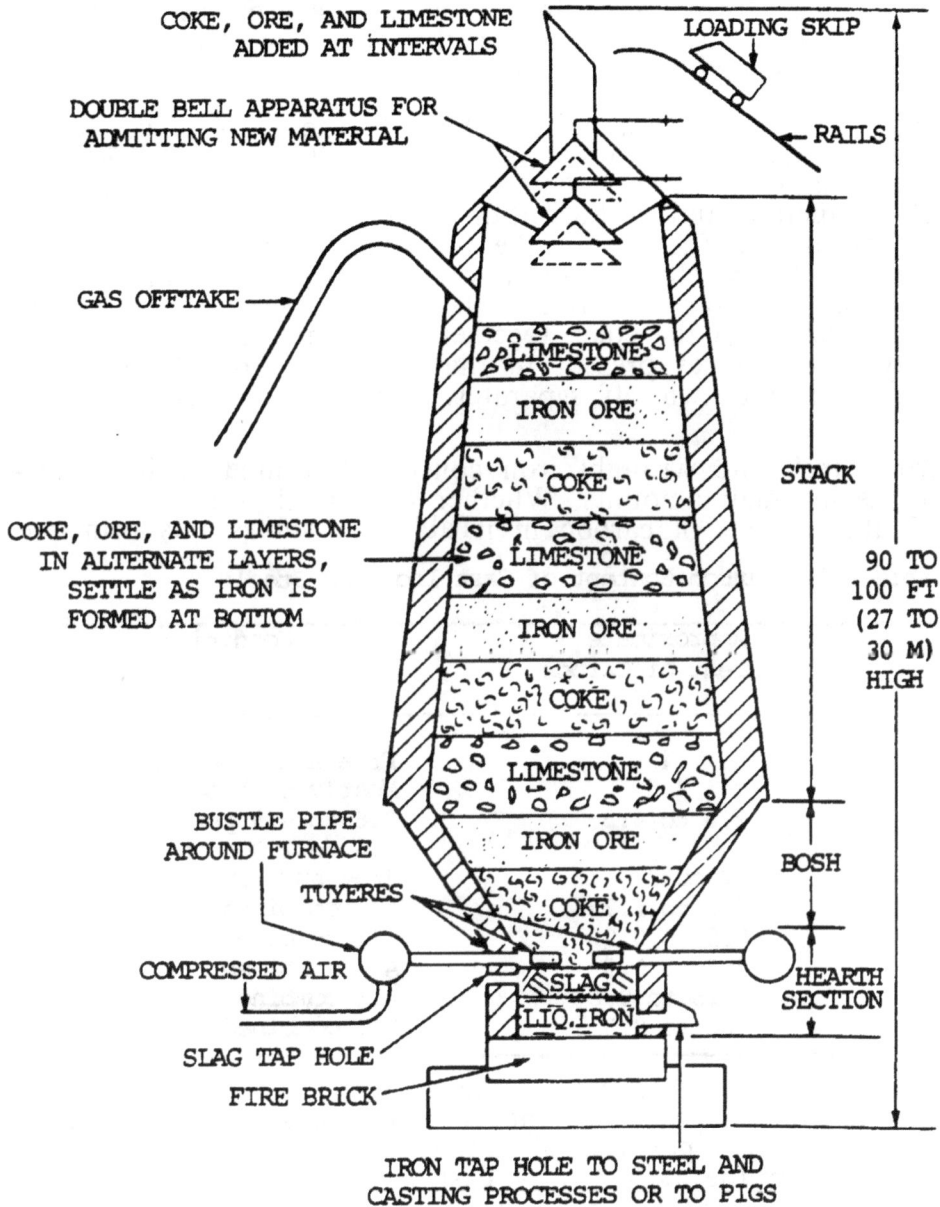

Figure 7-5. Blast furnace.

Cast iron is produced by melting a charge of pig iron, limestone, and coke in a cupola furnace. It is then poured into sand or alloy steel molds. When making gray cast iron castings, the molten metal in the mold is allowed to become solid and cool to room temperature in open air. Malleable cast iron, on the other hand, is made from white cast iron, which is similar in content to gray cast iron except that malleable iron contains less carbon and silicon. White cast iron is annealed for more than 150 hours at temperatures ranging from 1500 to 1700 (815 to 927 °C). The result is a product called malleable cast iron. The desirable properties of cast iron are less than those of carbon steel because of the difference in chemical makeup and structure. The carbon present in hardened steel is in solid solution, while cast iron contains free carbon known as graphite. In gray cast iron, the graphite is in flake form, while in malleable cast iron the graphite is in nodular (rounded) form. This also accounts for the higher mechanical properties of malleable cast iron as compared with gray cast iron.

Iron ore is smelted with coke and limestone in a blast furnace to remove the oxygen (the process of reduction) and earth foreign matter from it. Limestone is used to combined with the earth matter to form a liquid slag. Coke is used to supply the carbon needed for the reduction and carburization of the ore. The iron ore, limestone, and coke are charged into the top of the furnace. Rapid combustion with a blast of preheated air into the smelter causes a chemical reaction, during which the oxygen is removed from the iron. The iron melts, and the molten slag consisting of limestone flux and ash from the coke, together with compounds formed by reaction of the flux with substances present in the ore, floats on the heavier iron liquid. Each material is then drawn off separately (fig. 7-6, p 7-18).

All forms of cast iron, steel, and wrought iron consist of a mixture of iron, carbon, and other elements in small amounts. Whether the metal is cast iron or steel depends entirely upon the amount of carbon in it. Table 7-7 shows this principle.

Table 7-7. Carbon Content of Cast Iron and Steel

Item	Approximate Percent of Carbon	Condition of Incorporated Carbon
Pig iron	4	Free and combined
White cast iron	3.5	Mostly combined
Gray cast iron	2.5 to 4.5	0.6 to 0.9 percent free 2.6 to 2.9 percent combined
Malleable cast iron	2 to 3.5	Free and combined
Tool steel	0.9 to 1.7	All combined
High-carbon steel	0.5 to 0.9	All combined
Medium-carbon steel	0.3 to 0.5	All combined
Cast steel	0.15 to 0.6	All combined
Low-carbon steel	up to 0.3	All combined

Cast iron differs from steel mainly because its excess of carbon (more than 1.7 percent) is distributed throughout as flakes of graphite, causing most of the remaining carbon to separate. These particles of graphite form the paths through which failures occur, and are the reason why cast iron is brittle. By carefully controlling the silicon content and the rate of cooling, it is possible to cause any definite amount of the carbon to separate as graphite or to remain combined. Thus, white, gray, and malleable cast iron are all produced from a similar base.

7-3. CATEGORIES OF METALS (cont)

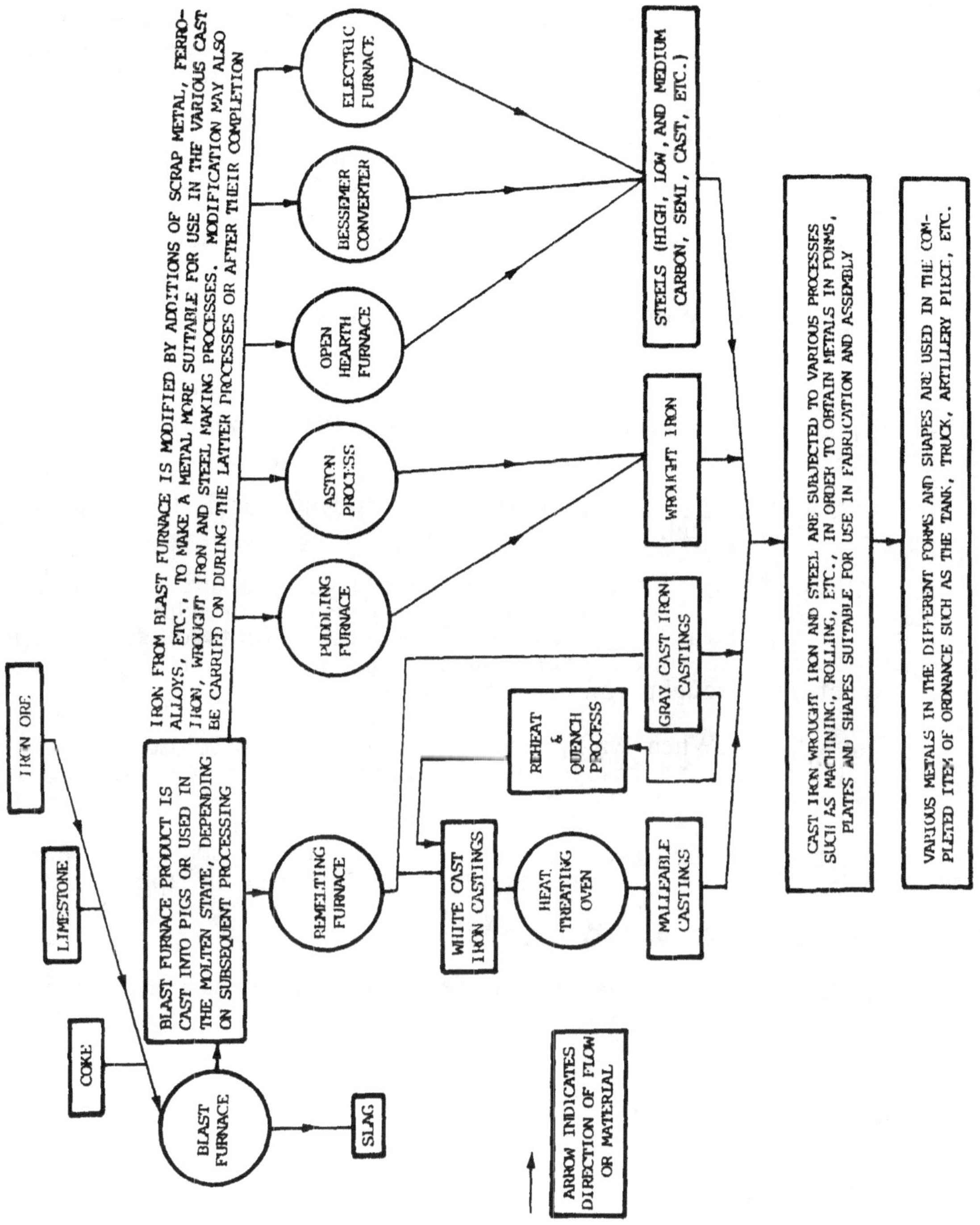

Figure 7-6. Conversion of iron ore into cast iron, wrought iron, and steel.

(1) <u>Wrought iron</u>.

(a) <u>General.</u> Wrought iron is almost pure iron. It is made from pig iron in a puddling Furnace and has a carbon content of less than 0.08 percent. Carbon and other elements present in pig iron are taken out, leaving almost pure iron. In the process of manufacture, some slag is mixed with iron to form a fibrous structure in which long stringers of slag running lengthwise, are mixed with long threads of iron. Because of the presence of slag, wrought iron resists corrosion and oxidation, which cause rusting .

(b) <u>Uses.</u> Wrought iron is used for porch railings, fencing, farm implements, nails, barbed wire, chains, modern household furniture, and decorations.

(c) <u>Capabilities</u> Wrought iron can be gas and arc welded, machined, plated, and is easily formed.

(d) <u>Limitations</u>. Wrought iron has low hardness and low fatigue strength.

(e) <u>Properties</u> Wrought iron has Brinell hardness number of 105; tensile strength of 35,000 psi; specific gravity of 7.7; melting point of 2750F (1510 ^{0}C); and is ductile and corrosion resistant.

(f) <u>Appearance test</u>. The appearance of wrought iron is the same as that of rolled, low-carbon steel.

(g) <u>Fracture test</u>. Wrought iron has a fibrous structure due to threads of slag. As a result, it can be split in the direction in which the fibers run. The metal is soft and easily cut with a chisel, and is quite ductile. When nicked and bent, it acts like rolled steel. However, the break is very jagged due to its fibrous structure. Wrought iron cannot be hardened.

(h) <u>Spark test.</u> When wrought iron is ground, straw-colored sparks form near the grinding wheel, and change to white forked sparklers near the end of the stream.

(i) <u>Torch test</u> Wrought iron melts quietly without sparking. It has a peculiar slag coating with white lines that are oily or greasy in appearance.

(2) <u>Cast iron (gray, white, and malleable</u>).

(a) <u>General</u> Cast iron is a manmade alloy of iron, carbon, and silicon. A portion of the carbon exists as free carbon or graphite. Total carbon content is between 1.7 and 4.5 percent.

(b) <u>Uses.</u> Cast iron is used for water pipes, machine tool castings, transmission housing, engine blocks, pistons, stove castings, etc.

(c) <u>Capabilities.</u> Cast iron may be brazed or bronze welded, gas and arc welded, hardened, or machined.

(d) <u>Limitations</u>. Cast iron must be preheated prior to welding. It cannot be worked cold.

7-3. CATEGORIES OF METALS (cont)

(e) <u>Properties</u> Cast iron has a Brinell hardness number of 150 to 220 (no alloys) and 300 to 600 (alloyed);tensile strength of 25,000 to 50,000 psi (172,375 to 344,750 kPa) (no alloys) and 50,000 to 100,OOO psi (344,750 to 689,500 kPa) (alloyed); specific gravity of 7.6; high compressive strength that is four times its tensile strength; high rigidity; good wear resistance; and fair corrosion resistance.

(f) <u>Gray cast iron</u>. If the molten pig iron is permitted to cool slowly, the chemical compund of iron and carbon breaks up to a certain extent. Much of the carbon separates as tiny flakes of graphite scattered throughout the metal. This graphite-like carbon, as distinguish from combined carbon, causes the gray appearance of the fracture, which characterizes ordinary gray cast iron.Since graphite is an excellent lubricant and the metal is shot throughout with tiny, flaky cleavages, gray cast iron is easy to machine but cannot withstand a heavy shock. Gray cast iron consists of 90 to 94 percent metallic iron with a mixture of carbon, manganese, phosphorus, sulfur, and silicon.Special high-strength grades of this metal also contain 0.75 to 1.50 percent nickel and 0.25 to 0.50 percent chromium or 0.25 to 1.25 percent molybdenum.Commercial gray iron has 2.50 to 4.50 percent carbon. About 1 percent of the carbon is combined with the iron, while about 2.75 percent remains in the free or graphitic state. In making gray cast iron, the silicon content is usually increased,since this allows the formation of graphitic carbon. The combined carbon (iron carbide), which is a small percentage of the total carbon present in cast iron, is known as cementite.In general, the more free carbon (graphitic carbn) present in cast iron, the lower the combined carbon content and the softer the iron.

<u>1. Appearance test</u>. The unmachined surface of gray cast iron castings is a very dull gray in color and may be somewhat roughened by the sand mold used in casting the part. Cast iron castings are rarely machined all over. Unmachined castings may be ground in places to remove rough edges.

<u>2. Fracture test</u>. Nick a corner all around with a chisel or hacksaw and strike the corner with a sharp blow of the hammer. The dark gray color of the broken surface is caused by fine black specks of carbon present in the form of graphite. Cast iron breaks short when fractured. Small, brittle chips made with a chisel break off as soon as they are formed.

<u>3. Spark test</u>. A small volume of dull-red sparks that follow a straight line close to the wheel are given off when this metal is spark tested. These break up into many fine, repeated spurts that change to a straw color.

<u>4. Torch test</u>. The torch test results in a puddle of molten metal that is quiet and has a jelly like consistency. When the torch flame is raised, the depression in the surface of the molts-puddle disappears instantly. A heavy, tough film forms on the surface as it melts.The molten puddle takes time to harden and gives off no sparks.

(g) <u>White cast iron</u>. When gray cast iron is heated to the molten state, the carbon completely dissolves in the iron, probably combining chemically with it. If this molten metal is cooled quickly, the two elements remain in the combined state, and white cast iron is formed.The carbon in this type of iron measures above 2.5 to 4.5 percent by weight, and is referred to as combined carbon. White cast iron is very hard and brittle, often impossible to machine, and has a silvery white fracture.

(h) <u>Malleable cast iron</u>. Malleable cast iron is made by heating white cast iron froman 1400 to 1700F (760 and 927 °C) for abut 150 hours in boxes containing hematite ore or iron scale. This heating causes a part of the combined carbon to change into the free or uncombined state. This free carbon separates in a different way from carbon in gray cast iron and is called temper carbon. It exists in the form of small, rounded particles of carbon which give malleable iron castings the ability to bend before breaking and to withstand shock better than gray cast iron. The castings have properties more like those of pure iron: high strength, ductility, toughness, and ability to resist shock. Malleable cast iron can be welded and brazed. Any welded part should be annealed after welding.

1. <u>Appearance test</u>. The surface of malleable cast iron is very much like gray cast iron, but is generally free from sand.It is dull gray and somewhat lighter in color than gray cast iron.

2. <u>Fracture test</u>. When malleable cast iron is fractured, the central portion of the broken surface is dark gray with a bright, steel-like band at the edges. The appearance of the fracture may best be described as a picture frame. When of good quality, malleable cast iron is much tougher than other cast iron and does not break short when nicked.

3. <u>Spark test</u>. When malleable cast iron is ground, the outer, bright layer gives off bright sparks like steel. As the interior is reached, the sparks quickly change to a dull-red color near the wheel.These sparks from the interior section are very much like those of cast iron; however, they are somewhat longer and are present in large volume.

4. <u>Torch test</u>. Molten malleable cast iron boils under the torch flame. After the-flame has been withdrawn, the surface will be full of blowholes. When fractured, the melted parts are very hard and brittle, having the appearance of white cast iron (they have been changed to white or chilled iron by melting and fairly rapid cooling). The outside, bright, steel-like band gives off sparks, but the center does not.

(3) <u>Steel</u>.

(a) <u>General</u>. A form of iron, steel contains less carbon than cast iron, but considerably more than wrought ironThe carbon content is from 0.03 to 1.7 percent. Basic carbon steels are alloyed with other elements, such as chromium and nickel, to increase certain physical properties of the metal.

(b) <u>Uses.</u> Steel is used to make nails, rivets, gears, structural steel, roles, desks, hoods, fenders, chisels, hammers, etc.

(c) <u>Capabilities.</u> Steel can be machined, welded, and forged, all to varying degrees, depending on the type of steel.

(d) <u>Limitations</u>. Highly alloyed steel is difficult to produce.

7-3. CATEGORIES OF METALS (cont)

(e) <u>Properties.</u> Steel has tensile strength of 45,000 psi (310,275 kPa) for low-carbon steel, 80,000 psi (551,600 kPa) for medium-carbon steel, 99,000 psi (692,605 kPa) for high-carbon steel, and 150,000 psi (1,034,250 kPa) for alloyed steel; and a melting point of 2800°F (1538 °C).

(f) <u>Low-carbon steel (carbon content up to 0.30 percent</u>. This steel is soft and ductile, and can be rolled, punched, sheared, and worked when either hot or cold. It is easily machined and can readily be welded by all methods. It does not harden to any great amount; however, it can easily be case hardened.

<u>1. Appearance test</u>. The appearance of the steel depends upon the method of preparation rather than upon composition. Cast steel has a relatively rough, dark-gray surface, except where it has been machined. Rolled steel has fine surface lines running in one direction. Forged steel is usually recognizable by its shape, hammer marks, or fins.

<u>2. Fracture test</u>. When low-carbon steel is fractured, the color is bright crystalline gray. It is tough to chip or nick. Low carbon steel, wrought iron, and steel castings cannot be hardened.

<u>3. Spark test</u>. The steel gives off sparks in long yellow-orange streaks, brighter than cast iron, that show some tendency to burst into white, forked sparklers.

<u>4. Torch test</u>. The steel gives off sparks when melted, and hardens almost instantly.

(g) <u>Medium-carbon steel (carbon content ranging from 0.30 to 0.50 percent)</u>. This steel may be heat-treated after fabrication. It is used for general machining and forging of parts that require surface hardness and strength. It is made in bar form in the cold-rolled or the normalized and annealed condition. During welding, the weld zone will become hardened if cooled rapidly and must be stress-relieved after welding.

(h) <u>High-carbon steel (carbon content ranging from 0.50 to 0.90 percent)</u>. This steel is used for the manufacture of drills, taps, dies, springs, and other machine tools and hand tools that are heat treated after fabrication to develop the hard structure necessary to withstand high shear stress and wear. It is manufactured in bar, sheet, and wire forms, and in the annealed or normalized condition in order to be suitable for machining before heat treatment. This steel is difficult to weld because of the hardening effect of heat at the welded joint.

<u>1. Appearance test</u>. The unfinished surface of high-carbon steel is dark gray and similar to other steel. It is more expensive, and is usually worked to produce a smooth surface finish.

<u>2. Fracture test</u>. High-carbon steel usually produces a very fine-grained fracture, whiter than low-carbon steel. Tool steel is harder and more brittle than plate steel or other low-carbon material. High-carbon steel can be hardened by heating to a good red and quenching in water.

3. Spark test. High-carbon steel gives off a large volume of bright yellow-orange sparks.

4. Torch test. Molten high-carbon steel is brighter than low-carbon steel, and the melting surface has a porous appearance. It sparks more freely than low-carbon (mild) steels, and the sparks are whiter.

(i) High carbon tool steel. Tool steel (carbon content ranging from 0.90 to 1.55 percent) is used in the manufacture of chisels, shear blades, cutters, large taps, wood-turning tools, blacksmith's tools, razors, and similar parts where high hardness is required to maintain a sharp cutting edge. It is difficult to weld due to the high carbon content. A spark test shows a moderately large volume of white sparks having many fine, repeating bursts.

(4) Cast steel.

(a) General. Welding is difficult on steel castings containing over 0.30 percent carbon and 0.20 percent silicon. Alloy steel castings containing nickel, molybdenum, or both of these metals are easily welded if the carbon content is low. Those containing chromium or vanadium are more difficult to weld. Since manganese steel is nearly always used in the form of castings, it is also considered with cast steel. Its high resistance to wear is its most valuable property.

(b) Appearance test. The surface of cast steel is brighter than cast or malleable iron and sometimes contains small bubble-like depressions.

(c) Fracture test The color of a fracture in cast steel is bright crystalline gray. This steel is tough and does not break short. Steel castings are tougher than malleable iron and chips made with a chisel curl up more. Manganese steel, however, is so tough that is cannot be cut with a chisel nor can it be machined.

(d) Spark test. The sparks created from cast steel are much brighter than those from cast iron. Manganese steel gives off marks that explode, throwing off brilliant sparklers at right angles to the original-path of the spark:

(e) Torch test. When melted, cast steel sparks and hardens quickly.

(5) Steel forgings.

(a) General Steel forgings may be of carbon or alloy steels. Alloy steel forgings are harder and more brittle than low carbon steels.

(b) Appearance test. The surface of steel forgings is smooth. Where the surface of drop forgings has not been finished, there will be evidence of the fin that results from the metal squeezing out between the two forging dies. This fin is removed by the trimming dies, but enough of the sheared surface remains for identification. All forgings are covered with reddish brown or black scale, unless they have been purposely cleaned.

7-3. CATEGORIES OF METALS (cont)

(c) Fracture test. The color of a fracture in a steel forging varies from bright crystalline to silky gray. Chips are tough; and when a sample is nicked, it is harder to break than cast steel and has a finer grain.Forgings may be of low- or high-carbon steel or of alloy steel. Tool steel is harder and more brittle than plate steel or other low-carbon material. The fracture is usually whiter and finer grained. Tool steel can be hardened by heating to a good red and then quenching in water. Low-carbon steel, wrought iron, and steel castings cannot be usefully hardened.

(d) Spark test. The sparks given off are long, yellow-orange stream-ers and are typical steel sparks. Sparks from high-carbon steel (machinery and tool steel) are much brighter than those from low-carbon steel.

(e) Torch test. Steel forgings spark when melted, and the sparks increase in number and brightness as the carbon content becomes greater.

(6) Alloy steel.

(a) General. Alloy steel is frequently recognizable by its use. There are many varieties of alloy steel used in the manufacture of Army equipment. They have greater strength and durability than carbon steel, and a given strength is secured with less material weight. Manganese steel is a special alloy steel that is always used in the cast condition (see cast steel, p 7-23).

Nickel, Chromium,vanadium, tungsten, molybdenum, and silicon are the most common elements used in alloy steel.

1. Chromium is used as an alloying element in carbon steels to increase hardenability, corrosion resistance, and shock resistance.It imparts high strength with little loss in ductility.

2. Nickel increases the toughness, strength, and ductility of steels, and lowers the hardening temperatures so than an oil quench, rather than a water quench, is used for hardening.

3. Manganese is used in steel to produce greater toughness, wear resistance, easier hot rolling, and forging. An increase in manganese content decreases the weldability of steel.

4. Molybdenum increases hardenability, which is the depth of hardening possible through heat treatment.The impact fatigue property of the steel is improved with up to 0.60 percent molybdenum.Above 0.60 percent molybde-num, the impact fatigue property is impaired.Wear resistance is improved with molybdenum content above 0.75 percent.Molybdenum is sometimes combined with chro-mium, tungsten, or vanadium to obtain desired properties.

5. Titanium and columbium (niobium) are used as additional alloy-ing agents in low-carbon content, corrosion resistant steels.They support resis-tance to intergranular corrosion after the metal is subjected to high temperatures for a prolonged time period.

<u>6.</u> Turgsten, as an alloying element in tool steel, produces a fine, dense grain-when used insmall quantities. When used in larger quantities, from 17 to 20 percent, and in combination with other alloys, it produces a steel that retains its hardness at high temperatures.

<u>7.</u> Vanadium is used to help control grain size. It tends to increase hardenability and causes marked secondary hardness, yet resists tempering. It is also added to steel during manufacture to remove oxygen.

<u>8.</u> Silicon is added to steel to obtain greater hardenability and corrosion resistance, and is often used with manganese to obtain a strong, tough steel. High speed tool steels are usually special alloy compositions designed for cutttig tools. The carbon content ranges from 0.70 to 0.80 percent. They are difficult to weld except by the furnace induction method.

<u>9.</u> High yield strength, low alloy structural steels (often referred to as constructional alloy steels) are special low carbon steels containing specific small amounts of alloying elements. These steels are quenched and tempered to obtain a yield strength of 90,000 to 100,000 psi (620,550 to 689,500 kPa) and a tensile strength of 100,000 to 140,000 psi (689,500 to 965,300 kPa), depending upon size and shape. Structural members fabricated of these high strength steels may have smaller cross sectional areas than common structural steels, and still have equal strength. In addition, these steels are more corrosion and abrasion resistant. In a spark test, this alloy appears very similar to the low carbon steels.

NOTE

This type of steel is much tougher than low carbon steels, and shearing machines must have twice the capacity required for low carbon steels.

(b) <u>Apperance test</u> Alloy steel appear the same as drop-forged steel.

(c) <u>Fracture test</u> Alloy steel is usually very close grained; at times the fracture appears velvety.

(d) <u>Spark test</u>. Alloy steel produces characteristic sparks both in color and shape. some of the more common alloys used in steel and their effects on the spark stream are as follows:

<u>1.</u> <u>Chromium</u>. Steels containing 1 to 2 percent chromium have no outstanding features inthe spark test. Chromium in large amounts shortens the spark stream length to one-half that of the same steel without chromium, but does not appreciably affect the stream's brightness. Other elements shorten the stream to the same extent and also make it duller. An 18 percent chromium, 8 percent nickel stainless steel produces a spark similar to that of wrought iron, but only half as long. Steel containing 14 percent chromium and no nickel produces a shorter version of the low-carbon spark. An 18 percent chromium, 2 percent carbon steel (chromium die steel) produces a spark similar to that of carbon tool steel, but one-third as long.

7-3. CATEGORIES OF METALS (cont)

2. <u>Nickel.</u> The nickel spark has a short, sharply defined dash of brilliant light just before the fork. In the amounts found in S.A.E. steels, nickel can be recognized only when the carbon content is so low that the bursts are not too noticeable.

3. <u>High chromium-nickel alloy(stainless) steels.</u> The sparks given off during a spark test are straw colored near the grinding wheel and white near the end of the streak. There is a medium volume of streaks having a moderate number of forked bursts.

4. <u>Manganese</u>. Steel containing this element produces a spark similar to a carbon steel spark. A moderate increase in manganese increases the volume of the spark stream and the force of the bursts. Steel containing more than the normal amount of manganese will spark in a manner similar to high-carbon steel with low manganese content.

5. <u>Molybdenum</u>. Steel containing this element produces a characteristic spark with a detached arrowhead similar to that of wrought iron. It can be seen even in fairly strong carbon bursts. Molybdenum alloy steel contains nickel, chromium, or both.

6. <u>Molybdenum with other elements</u>. When molybdenum and other elements are substituted for some of the tungsten in high-speed steel, the spark stream turns orange. Although other elements give off a red spark, there is enough difference in their color to tell them from a tungsten spark.

7. <u>Tungsten</u>. Tungsten will inpart a dull red color to the spark stream near the wheel. It also shortens the spark stream, decreases the size, or completely eliminates the carbon burst. Steel containing 10 percent tungsten causes short, curved, orange spear points at the end of the carrier line. Still lower tungsten content causes small white bursts to appear at the end of the spear pint. Carrier lines may be anything from dull red to orange in color, depending on the other elements present, if the tungsten content is not too high.

8. <u>Vanadium</u>. Alloy steels containing vanadium produce sparks with a detached arrowhead at the end of the carrier line similar to those arising from molybdenum steels. The spark test is not positive for vanadium steels.

9. <u>High speed tool steels</u>. A spark test in these steels will impart a few long; forked sparks which are red near the wheel, and straw-colored near the end of the spark stream.

(7) <u>Special steel.</u> Plate steel is used in the manufacture of built-up welded structures such as gun carriages. In using nickel plate steel, it has been found that commericial grades of low-alloy structural steel of not over 0.25 percent carbon, and several containing no nickel at all, are better suited to welding than those with a maximum carbon content of 0.30 percent. Armorplate, a low carbon alloyed steel, is an example of this kind of plate. Such plate is normally used in the "as rolled" condition. Electric arc welding with a covered electrode may require preheating of the metal, followed by a proper stress-relieving heat treatment (post heating), to produce a structure in which the welded joint has properties equal to those of the plate metal.

e. Nonferrous metal.

(1) Aluminum (Al).

(a) General Aluminum is a lightweight, soft, low strength metal which can easily be cast, forged, machined, formed, and welded.It is suitable only in low temperature applications, except when alloyed with specific elements. Commercial aluminum alloys are classified into two groups, wrought alloys and cast alloys. The wrought alloy group includes those alloys which are designed for mill products whose final physical forms are obtained by working the metal mechanically. The casting alloy group includes those alloys whose final shapes are obtained by allowing the molten metal to solidify in a mold.

(b) Uses. Aluminum is used as a deoxidizer and alloying agent in the manufacture of steel. Castings, pistons, torque converter pump housings, aircraft structures, kitchen utensils, railways cars, and transmission lines are made of aluminum.

(c) Capabilities Aluminum can be cast, forged, machined, formed and welded.

(d) Limitations. Direct metal contact of aluminum with copper and copper alloys should be avoided. Aluminum should be used in low-temperature applications.

(e) Properties Pure aluminum has a Brinell hardness number of 17 to 27; tensile strength of 6000 to 16,000 psi (41,370 to 110,320 kPa); specific gravity of 2.7; and a melting point of 1220°F (660 °C). Aluminum alloys have a Brinell hardness number of 100 to 130, and tensile strength of 30,000 to 75,000 psi (206,850 to 517,125 kPa). Generally, aluminum and aluminum alloys have excellent heat conductivity; high electrical conductivity (60 percent that of copper, volume for volume; high strength/weight ratio at room temperature; and unfairly corrosion resistant.

(f) Appearance test. Aluminum is light gray to silver in color, very bright when polished, dull when oxidized, and light in weight. Rolled and sheet aluminum materials are usually pure metal.Castings are alloys of aluminum with other metals, usually zinc, copper, silicon and sometimes iron and magnesium. Wrought aluminum alloys may contain chromium silicon, magnesium, or manganese. Aluminum strongly resembles magnesium in appearance. Aluminum is distinguished from magnesium by the application of a drop of silver nitrate solution on each surface. The silver nitrate will not react with the aluminum, but leaves a black deposit of silver on the magnesium.

(g) Fracture test. A fracture in rolled aluminum sections shows a smooth, bright structure. A fracture in an aluminum casting shins a bright crystalline structure.

(h) Spark test. No sparks are given off from aluminum.

(i) Torch test Aluminum does not turn red before melting.It holds its shape until almost molten, then collapses (hot shorts) suddenly. A heavy film of white oxide forms instantly on the molten surface.

7-3. CATEGORIESOF METALS (cont)

(2) Chromium (Cr).

(a) General Chromium is an alloying agent used in steel, cast iron, and nonferrous alloys of nickel, copper, aluminum, and cobalt. It is hard, brittle, corrosion resistant, can be welded, machined, forged, and is widely used in electroplating. Chromium is not resistant to hydrochloric acid and cannot be used in its pure state because of its difficulty to work.

(b) Uses. Chromium is one of the most widely used alloys. It is used as an alloying agent-in steel and cast iron (0.25 to 0.35 percent) and in nonferrous alloys of nickel, copper, aluminum, and cobalt.It is also used in electroplating for appearance and wear, in powder metallurgy, and to make mirrors and stainless steel.

(c) Capabilities. Chromium alloys can be welded, machined, and forged. Chromium is never used in its pure state.

(d) Limitations. Chromium is not resistant to hydrochloric acid, and cannot be used in the pure state because of its brittleness and difficulty to work.

(e) Properties (pure). Chromium has a specific gravity of 7.19; a melting point of 3300 °F (1816 °C); Brinell hardness number of 110 to 170; is resistant to acids other than hydrochloric; and is weak, heat, and corrosion resistant.

(3) Cobalt (Co).

(a) General Cobalt is a hard, white metal similar to nickel in appearance, but has a slightly bluish cast.

(b) Uses. Cobalt is mainly used as an alloying element in permanent and soft magnetic materials, high-speed tool bits and cutters, high-temperature, creep-resisting alloys, and cemented carbide tools, bits, and cutters.It is also used in making insoluble paint pigmnts and blue ceramic glazesIn the metallic form, cobalt does not have many-uses. However, when combined with other elements, it is used for hard-facing materials.

(c) Capabilities Cobalt can be welded, machined (limited), and cold-drawn.

(d) Limitations. Cobalt must be machined with cemented carbide cutters. Welding high carbon cobalt steel often causes cracking.

(e) Properties. Pure cobalt has a tensile strength of 34,000 psi (234,430 kPa); Brinell hardness number of 125; specific gravity of 8.9; and a melting point of 2720 °F (1493 °C). Cobalt alloy (Stellite 21) has a tensile strength of 101,000 psi (696,395 kPa) and is heat and corrosion resistant.

(4) Copper (Cu).

(a) General Copper is a reddish metal,is very ductile and malleable, and has high electrical and heat conductivity. It is used as a major element in hundreds of alloys. Commercially pure copper is not suitable for welding.

Though it is very soft, it is very difficult to machine due to its high ductility. Beryllium copper contains from 1.50 to 2.75 percent beryllium. It is ductile when soft, but loses ductility and gains tensile strength when hardened. Nickel copper contains either 10, 20, or 30 percent nickel. Nickel alloys have moderately high to high tensile strength, which increases with the nickel content. They are moderately hard, quite tough, and ductile. They are very resistant to the erosive and corrosive effects of high velocity sea water, stress corrosion, and corrosion fatigue. Nickel is added to copper zinc alloys (brasses) to lighten their color; the resultant alloys are called nickel silver. These alloys are of two general types, one type containing 65 percent or more copper and nickel combined, the other containing 55 to 60 percent copper and nickel combined. The first type can be cold worked by such operations as deep drawing, stamping, and spinning. The second type is much harder end is not processed by any of the cold working methods. Gas welding is the preferred process for joining copper and copper alloys.

(b) Uses. The principal use of commericially pure copper is in the electrical industry where it is made into wire or other such conductors. It is also used in the manufacture of nonferrous alloys such as brass, bronze, and Monel metal. Typical copper products are sheet roofing, cartridge cases, bushings, wire, bearings, and statues.

(c) Capababilities. Copper can be forged, cast, and cold worked. It can also be welded, but its machinability is only fair. Copper alloys can be welded.

(d) Limitations. Electrolytic tough pitch copper cannot be welded satisfactorily. Pure copper is not suitable for welding and is difficult to machine due to its ductility.

(e) Properties Pure copper is nonmagnetic; has a Brinell hardness number of 60 to 110; a tensile strength of 32,000 to 60,000 psi (220,640 to 413,700 kPa); specific gravity of 8.9; melting point of $1980\,^{\circ}F$ ($1082\,^{\circ}C$); and is corrosion resistant. Copper alloys have a tensile strength of 50,000 to 90,000 psi (344,750 to 620,550 kPa) and a Brinell hardness number of 100 to 185.

(f) Appearance test. Copper is red in color when polished, and oxidizes to various shades of green.

(g) Fracture test Copper presents a smooth surface when fractured, which is free from crystalline appearance.

(h) Spark test. Copper gives off no sparks.

(i) Torch test Because copper conducts heat rapidly, a larger flame is required to produce fusion of copper than is needed for the same size piece of steel. Copper melts suddenly and solidifies instantly. Copper alloy, containing small amounts of other metals, melts more easily and solidifies more slowly than pure copper.

7-3. CATEGORIES OF METALS (cont)

(j) Brass and bronze. Brass, an alloy of copper and zinc (60 to 68 percent copper and 32 to 40 percent zinc) , has a low melting point and high heat conductivity. There are several types of brass, such as naval, red, admiralty, yellow, and commercial. All differ in copper and zinc content; may be alloyed with other elements such as lead, tin, manganese, or iron; have good machinability; and can be welded. Bronze is an alloy of copper and tin and may contain lead, zinc, nickel, manganese, or phosphorus. It has high strength, is rust or corrosion resistant, has good machinability, and can be welded.

1. Appearance test. The color of polished brass and bronze varies with the composition from red, almost like copper, to yellow brass. They oxidize to various shades of green, brown, or yellow.

2. Fracture test. The surface of fractured brass or bronze ranges from smooth to crystalline, depending upon composition and method of preparation; i.e., cast, rolled, or forged.

3. Spark test. Brass and bronze give off no sparks.

4. Torch test. Brass contains zinc, which gives off white fumes when it is melted. Bronze contains tin. Even a slight amount of tin makes the alloy flow very freely, like water. Due to the small amount of zinc or tin that is usually present, bronze may fume slightly, but never as much as brass.

(k) Aluminum bronze.

1. Appearance test. When polished, aluminum bronze appears a darker yellow than brass.

2. Fracture test. Aluminum bronze presents a smooth surface when fractured.

3. Spark test. Aluminum bronze gives off no sparks.

4. Torch test. Welding aluminum bronze is very difficult. The surface is quickly covered with a heavy stun that tends to mix with the metal and is difficult to remove.

(5) Lead (Pb).

CAUTION
Lead dust and fumes are poisonous. Exercise extreme care when welding lead, and use personal protective equipment as described in chapter 2.

(a) General. Lead is a heavy, soft, malleable metal with low melting point, low tensile strength, and low creep strength. It is resistant to corrosion from ordering atmosphere, moisture, and water, and is effective against many acids. Lead is well suited for cold working and casting. The low melting point of lead makes the correct welding rod selection very important.

(b) Uses. Lead is used mainly in the manufacture of electrical equipment such as lead-coated power and telephone cables, and storage batteries. It is also used in building construction in both pipe and sheet form, and in solder. Zinc alloys are used in the manufacture of lead weights, bearings, gaskets, seals, bullets, and shot. Many types of chemical compounds are produced from lead; among these are lead carbonate (paint pigment) and tetraethyl lead (antiknock gasoline). Lead is also used for X-ray protection (radiation shields). Lead has more fields of application than any other metal.

(c) Capabilities. Lead can be cast, cold worked, welded, and machined . It is corrosion, atmosphere, moisture, and water resistant, and is resistant to many acids.

(d) Limitations. Lead has low strength with heavy weight. Lead dust and fumes are very poisonous.

(e) Properties Pure lead has tensile strength of 2500 to 3000 psi (17,237.5 to 20,685 kPa); specific gravity of 11.3; and a melting point of 620°F (327 C). Alloy lead B32-467 has tensile strength of 5800 psi (39,991 kPa). Generally, lead has low electrical conductivity; is self-lubricating; is malleable; and is corrosion resistant.

(6) Magnesium (Mg).

(a) General Magnesium is an extremely light metal, is white in color, has a low melting point, excellent machinability, and is weldable. Welding by either the arc or gas process requires the use of a gaseous shield. Magnesium is moderately resistant to atmospheric exposure, many chemicals such as alkalies, chromic and hydrofluoric acids, hydrocarbons, and most alcohols, phenols, esters, and oils. It is nonmagnetic. Galvanic corrosion is an important factor in any assembly with magnesium.

(b) Uses. Magnesium is used as a deoxidizer for brass, bronze, nickel, and silver. Because of its light weight, it is used in many weight-saving applications, particularly in the aircraft industry. It is also used in the manufacture and use of fireworks for railroad flares and signals, and for military purposes. Magnesium castings are used for engine housings, blowers, hose pieces, landing wheels, and certain parts of the fuselage of aircraft. Magnesium alloy materials are used in sewing machines, typewriters, and textile machines.

(c) Capabilities. Magnesium can be forged, cast, welded, and machined.

(d) Limitations. Magnesium in fine chip form will ignite at low temperatures (800 to 1200°F (427 to 649 °C)). The flame can be mothered with suitable materials such as carbon dioxide (CO_2) foam, and sand.

(e) Properties Pure magnesium has tensile strength of 12,000 psi (82,740 kPa) (cast) and tensile strength of 37,000 psi (255,115 kPa) (rolled); Brinell hardness number of 30 (cast) and 50 (rolled); specific gravity of 1.7; and a melting point of 1202°F (650 °C). Magnesium alloy has Brinell hardness number of 72 (hard) and 50 (forged); and tensile strength of 42,000 psi (289,590 kPa) (hard) and 32,000 psi (220,640 kPa) (forged).

7-3. CATEGORIES OF MATERIALS (cont)

    (f) <u>Appearance test</u>. Magnesium resembles aluminum in appearance. The polished surface is silver-white, but quickly oxidizes to a grayish film. Like aluminum, it is highly corrosion resistant and has a good strength-to-weight ratio, but is lighter in weight than aluminum. It has a very low kindling point and is not very weldable, except when it is alloyed with manganese and aluminum. Magnesium is distinguished from aluminum by the use of a silver nitrate solution. The solution does not react with aluminum, but leaves a black deposit of silver on magnesium. Magnesium is produced in large quantities from sea water. It has excellent machinability, but special care must be used when machining because of its low kindling point.

    (g) <u>Fracture test</u>. Magnesium has a rough surface with a fine grain structure.

    (h) <u>Spark test.</u> No sparks are given off.

CAUTION
Magnesium may ignite and burn when heated in the open atmosphere.

    (i) <u>Torch test</u>. Magnesium oxidizes rapidly when heated in open air, producing an oxide film which is insoluble in the liquid metal. A fire may result when magnesium is heated in the open atmosphere. As a safety precaution, magnesium should be melted in an atmosphere of inert gas.

    (7) <u>Manganese (Mn)</u>.

    (a) <u>General</u>. Pure manganese has a relatively high tensile strength, but is very brittle. Manganese is used as an alloying agent in steel to deoxidize and desulfurize the metal. In metals other than steel, percentages of 1 to 15 percent manganese will increase the toughness and the hardenability of the metal involved.

    (b) <u>Uses.</u> Manganese is used mainly as an alloying agent in making steel to increase tensile strength. It is also added during the steel-making process to remove sulfur as a slag. Austenitic manganese steels are used for railroad track work, power shovel buckets, and rock crushers. Medium-carbon manganese steels are used to make car axles and gears.

    (c) <u>Capabilities.</u> Manganese can be welded, machined, and cold-worked.

    (d) <u>Limitations</u>. Austenitic manganese steels are best machined with cemented carbide, cobalt, and high-speed steel cutters.

    (e) <u>Properties</u>. Pure manganese has tensile strength of 72,000 psi (496,440 kPa) (quenched) Brinell hardness number of 330; specific gravity of 7.43: a melting point of 2270°F (1243°C); and is brittle. Manganese alloy has a tensile strength of 110,000 psi (758,450 kPa). Generally, manganese is highly polishable and brittle.

(8) <u>Molybdenum (Mo)</u>.

(a) <u>General</u> Pure molybdenum has a high tensile strength and is very resistant to heat. It is principally used as an alloying agent in steel to increase strength, hardenability, and resistance to heat.

(b) <u>Uses.</u> Molybdenum is used mainly as an alloy. Heating elements, switches, contacts, thermocouplers, welding electrodes, and cathode ray tubes are made of molybdenum.

(c) <u>Capabilities</u> Molybdenum can be swaged, rolled, drawn, or machined.

(d) <u>Limitations</u>. Molybdenum can only be welded by atomic hydrogen arc, or butt welded by resistance heating in vacuum.It is attacked by nitric acid, hot sulfuric acid, and hot hydrochloric acid.

(e) <u>Properties</u> Pure molybdenum has a tensile strength of 100,000 psi (689,500 kPa) (sheet) and 30,000 Psi (206,850 kPa) (wire); Brinell hardness number of 160 to 185; specific gravity of 10.2; meting point of 4800F (2649 °C); retains hardness and strength at high temperatures; and is corrosion resistant.

(9) <u>Nickel (Ni)</u>.

(a) <u>General.</u> Nickel is a hard, malleable, ductile metal. As an alloy, it will increase ductility, has no effect on grain size, lowers the critical point for heat treatment, aids fatigue strength, and increases impact values in low temperature operations. Both nickel and nickel alloys are machinable and are readily welded by gas and arc methods.

(b) <u>Uses.</u> Nickel is used in making alloys of both ferrous and nonferrous metal. Chemical and food processing equipment, electrical resistance heating elements, ornamental trim, and parts that must withstand elevated temperatures are all produced from nickel-containing metal.Alloyed with chromium, it is used in the making of stainless steel.

(c) <u>Capabilities</u> Nickel alloys are readily welded by either the gas or arc methods. Nickel alloys can be machined, forged, cast, and easily formed.

(d). <u>Limitations.</u> Nickel oxidizes very slowly in the presence of moisture or corrosive gases.

(e) <u>Properties.</u> Pure nickel has tensile strength of 46,000 psi (317,170 kPa); Brinell hardness number 220; specific gravity of 8.9; and melting point of 2650 °F (1454 °C). Nickel alloys have Brinell hardness number of 140 to 230. Monel-forged nickel has tensile strength of 100,000 psi (689,500 kPa), and high strength and toughness at high temperatures.

(f) <u>Appearance.</u> Pure nickel has a grayish white color.

(g) <u>Fracture</u> The fracture surface of nickel is smooth and fine grained.

7-3. CATEGORIESOF METALS (cont)

(h) Spark test. In a spark test, nickel produces a very small amount of short, orangestreaks which are generally wavy.

(i) Monel metal. Monel metal is a nickel alloy of silver-white color containing about67.00 percent nickel, 29.00 to 80.00 percent copper, 1.40 percent iron, 1.00 percent manganese, 0.10 percent silicon, and 0.15 percent carbon. In appearance, it resembles untarnished nickel.After use, or after contact with chemical solutions, the silver-white color takes on a yellow tinge, and some of the luster is lost. It has a very high resistance to corrosion and can be welded.

(10) Tin (Sn).

(a) General. Tin is a very soft, malleable, somewhat ductile, corrosion resistant metal having low tensile strength and highrystalline structure. It is used in coating metals to prevent corrosion.

(b) Uses. The major application of tin is in coating steel. It serves as the best container for preserving perishable focal. Tin, in the form of foil, is often used in wrapping food products.A second major use of tin is as an alloying element. Tin is alloyed with copper to produce tin brass and bronze, with lead to produce solder and with antimony and lead to form babbitt.

(c) Capabilities. Tin can be die cast, cold worked (extruded), machined, and soldered.

(d) Limitations. Tin is not weldable.

(e) Properties. Pure tin has tensile strength of 2800 psi (19,306 kPa); specific gravity of 7.29; melting point of 450°F (232 °C); and is corrosion resistant Babbitt alloy tin has tensile strength of10,000 psi' (68,950 kPa) and Brinell hardness number of 30.

(f) Appearance. Tin is silvery white in color.

(g) Fracture test. The fracture surfaceof tin is silvery white and fairly smooth.

(h) Spark test. Tin gives off no sparksin a spark test.

(i) Torch test. Tin melts at 450F (232 °C), and will boil under the torch.

(11) Titanium (Ti).

(a) General Titanium is a very soft, silvery white, medium-strength metal having very good corrosion resistance.It has a high strength to weight ratio, and its tensile strength increases as the temperature decreasesTitanium has low impact and creep strengths, as well as seizing tendencies, at temperatures above 80°F (427 °C).

(b) <u>Uses.</u> Titanium is a metal of the tin group which occurs naturally as titanium oxide or in other oxide forms. The free element is separated by heating the oxide with aluminum or by the electrolysis of the solution in calcium chloride. Its most important compound is titanium dioxide, which is used widely in welding electrode coatings. It is used as a stabilizer in stainless steel so that carbon will not be separated during the welding operation It is also used as an additive in alloying aluminum, copper, magnesium, steel, and nickel; making powder for fireworks; and in the manufacture of turbine blades, aircraft firewalls, engine nacelles, frame assemblies, ammunition tracks, and mortar base plates.

(c) <u>Capabilities</u> Titanium can be machined at low speeds and fast feeds; formal; spot- and seam-welded, and fusion welded using inert gas.

(d) <u>Limitations</u>. Titanium has low impact strength, and low creep strength at high temperatures (above 800F (427 °C)). It can only be cast into simple shapes, and it cannot be welded by any gas welding process because of its high attraction for oxygen. Oxidation causes this metal to become quite brittle. The inert gas welding process is reommended to reduce contamination of the weld metal.

(e) <u>Properties.</u> Pure titanium has a tensile strength of 100,000 psi; Brinell hardness number of 200; specific gravity of 4.5; melting point of 3300F (1851 °C); and good corrosion resistance. Alloy titanium has a Brinell hardness number of 340; tensile strength of 150,000 psi; and a high strength/weight ratio (twice that of aluminum alloy at 400F (204 °C)).

(f) <u>Appearance test</u>. Titanium is a soft, shiny, silvery-white metal burns in air and is the only element that burns in nitrogen. Titanium alloys look like steel, and can be distinguished from steel by a copper sulfate solution. The solution will not react with titanium, but will leave a coating of copper on steel.

(g) <u>Spark test</u>. The sparks given off are large, brilliant white, and of medium length.

(12) <u>Turgsten (W)</u>.

(a) <u>General</u> Tungsten is a hard, heavy, nonmagnetic metal which will melt at approximately 6150°F (3400 °C).

(b) <u>Uses.</u> Tungsten is used in making light bulb filaments, phonograph needles, and as an alloying agent in production of high-speed steel, armorplate, and projectiles. It is also used as an alloying agent in nonconsumable welding electrodes, armor plate, die and tool steels, and hard metal carbide cutting tools.

(c) <u>Capabilities.</u> Tungsten can be cold and hot drawn.

(d) <u>Limitations</u>. Tungsten is hard to machine, requires high temperatures for melting, and is produced by powered metallurgy (sintering process).

(e) <u>Properties</u> Tungsten has a melting point of 6170 ± 35F (3410 ± 19 °C); is ductile; has tensile strength of 105,OOO psi (723,975 kPa); a specific gravity of 19.32; thermal conductivity of 0.397 a Brinell hardness number of 38; and is a dull white color.

7-3. CATEGORIES OF METALS (cont)

(f) <u>Appearance</u>. Tungsten is steel gray in color.

(g) <u>Spark test.</u> Tungsten produces a very small volume of short, straight, orange streaks in a spark test.

(13) <u>Zinc (Zn).</u>

(a) <u>General</u> Zinc is a medium low strength metal having a very low melting point. It is easy to machine, but coarse grain zinc should be heated to approximately 180 F (82°C) to avoid cleavage of crystals. Zinc can be soldered or welded if it is properly cleaned and the heat input closely controlled.

(b) <u>Uses.</u>

<u>1.</u> Galvanizing metal is the largest use of zinc and is done by dipping the part in molten zinc or by electroplating it. Examples of items made in this way are galvanized pipe, tubing, sheet metal, wire, nails, and bolts. Zinc is also used as an alloying element in producing alloys such as brass and bronze. Those alloys that are made up primarily of zinc itself.

<u>2.</u> Typical parts made with zinc alloy are die castings, toys, ornaments, building equipment, carburetor and fuel pump bodies, instrument panels, wet and dry batteries, fuse plugs, pipe organ pipes, munitions, cooking utensils, and flux. Other forms of zinc include zinc oxide and zinc sulfide, widely used in paint and rubber, and zinc dust, which is used in the manufacture of explosives and chemical agents.

(c) <u>Capabilities.</u> Zinc can be cast, cold worked (extruded), machined, and welded.

(d) <u>Limitations</u>. DO not use zinc die castings in continuous contact with steam.

(e) <u>Properties.</u> Zinc has a tensile strength of 12,000 psi (82,740 kPa) (cast) and 27,000 psi (186,165 kPa) (rolled); a specific gravity of 7.1; a melting point of 790°F (421°C); is corrosion resistant; and is brittle at 220F (104°C).

(f) <u>Appearance.</u> Both zinc and zinc alloys are blue-white in color when polished, and oxidize to gray.

(g) <u>Fracture test</u> Zinc fractures appear somewhat granular.

(h) <u>Spark test.</u> Zinc and zinc alloys give off no sparks in a spark test.

(i) <u>Zinc die castings.</u>

1. Appearance test. Die castings are usually alloys of zinc, aluminum, magnesium, lead, and tin. They are light in weight, generally silvery white in color (like aluminum), and sometimes of intricate design. A die-cast surface is much smoother than that of a casting made in sand, and is almost as smooth as a machined surface. Sometimes, die castings darkened by use may be mistaken for malleable iron when judged simply by looks, but the die casting is lighter in weight and softer.

2. Fracture test. The surface of a zinc die casting is white and has a slight granular structure.

3. Spark test. Zinc die castings give off no sparks.

4. Torch test. Zinc die castings can be recognized by their low melting temperatures. The metal boils when heated with the oxyacetylene flame. A die casting, after thorough cleaning, can be welded with a carburizing flare using tin or aluminum solders as filler metal. If necessary, the die-cast part can be used as a pattern to make a new brass casting.

(14) White metal die castings.

(a) General. These are usually made with alloys of aluminum, lead, magnesium, or tin. Except for those made of lead and tin, they are generally light in weight and white in color.

(b) Appearance. The surface is much smoother than that produced by castings made in sand.

(c) Fracture test. Fractured surface is white and somewhat granular.

(d) Spark test. No sparks given off in a spark test.

(e) Torch test. Melting points are low, and the metal boils under the torch.

Section II. STANDARD METAL DESIGNATIONS

7-4. GENERAL

The numerical index system for the classification of metals and their alloys has been generally adopted by industry for use on drawings and specifications. In this system, the class to which the metal belongs, the predominant alloying agent, and the average carbon content percentage are given.

7-5. STANDARD DESIGNATION SYSTEM FOR STEEL

a. Numbers are used to designate different chemical compositions. A four-digit number series designates carbon and alloying steels according to the type and classes shown in table 7-8, p 7-38. This system has been expanded, and in some cases five digits are used to designate certain alloy steels.

b. Two letters are often used as a prefix to the numerals. The letter C indicates basic open hearth carbon steels, and E indicates electric furnace carbon and alloy steels. The letter H is sometimes used as a suffix to denote steels manufactured to meet hardenability limits.

7-5. STANDARD DESIGNATION SYSTEM FOR STEEL (cont)

c. The first two digits indicate the major alloying metals in a steel, such as manganese, nickel-chromium, and chrome-molybdenum.

d. The last digits indicate the approximate middle of the carbon content range in percent. For example, 0.21 indicates a range of 0.18 to 0.23 percent carbon. In a few cases, the system deviates from this rule, and some carbon ranges relate to the ranges of manganese, sulfur, phosphorous, chromium, and other elements.

e. The system designates the major elements of a steel and the approximate carbon range of the steel. It also indicates the manufacturing process used to produce the steel. The complete designation system is shown in table 7-9, p 7-40.

Table 7-8. Standard Steel and Steel Alloy Number Designations

Series Designation	Types and Classes
10xx	Non-resulfurized carbon steel grades (plain carbon steel)
11xx	Resulfurized carbon steel grades (free cutting carbon steel)
13xx	Manganese 1.75%
20xx	Nickel steels
23xx	Nickel 3.50%
25xx	Nickel 5.00%
30xx	Nickel-chromium steels*
31xx	Nickel 1.25%-chromium 0.65 or 0.80%
33xx	Nickel 3.50%-chromium 1.55%
40xx	Molybdenum 0.25%
41xx	Chromium 0.50-0.95%-molybdenum 0.12 or 0.20%
43xx	Nickel 1.80%-chromium 0.50 or 0.80%-molybdenum 0.25%*
46xx	Nickel 1.55 or 1.80%-molybdenum 0.20 or 0.25%
47xx	Nickel 1.05%-chromium 0.45%-molybdenum 0.25%*
48xx	Nickel 3.50%-molybdenum 0.25%
50xx	Chromium 0.28 or 0.40%
51xx	Chromium 0.80, 0.90, 0.95, 1.00 or 1.05%
5xxxx	Carbon 1.00%-chromium 0.50, 1.00, or 1.45%
60xx	Chrome-vanadium steels
61xx	Chromium 0.80 or 0.95%-vanadium 0.10 or 0.15% min
70xx	Heat resisting casting alloys
80xx	Nickel-chrome-molybdenum steels*
86xx	Nickel 0.55%-chromium 0.50 or 0.65%-molybdenum 0.20%
87xx	Nickel 0.55%-chromium 0.50%-molybdenum 0.25%
90xx	Silicon-manganese steels
92xx	Manganese 0.85%-silicon 2.00%
93xx	Nickel 3.25%-chromium 1.20%-molybdenum 0.12%
94xx	Manganese 1.00%-nickel 0.45%-chromium 0.40%-molybdenum 0.12%
97xx	Nickel 0.55%-chromium 0.17%-molybdenum 0.20%
98xx	Nickel 1.00%-chromium 0.80%-molybdenum 0.25%*

*Stainless steels always have a high chromium content, often considerable amounts of nickel, and sometimes contain molybdenum and other elements. Stainless steels are identified by a three-digit number beginning with 2, 3, 4, or 5.

f. The number 2340 by this system indicates a nickel steel with approximately 3 percent nickel and 0.40 percent carbon. The number 4340 indicates a nickel-chrome-molybdenum metal with 0.40 percent carbon.

S.A.E. Steel Specifications
The following numerical system for identifying carbon and alloy steels of various specifications has been adopted by the Society of Automotive Engineers.

COMPARISION
A.I.S.I.--S.A.E. Steel Specifications

The ever-growing variety of chemical compositions and quality requirements of steel specifications have resulted in several thousand different combinations of chemical elements being specified to meet individual demands of purchasers of steel products

The S.A.E. developed a system of nomenclature for identification of various chemical compositions which symbolize certain standards as to machining, heat treating, and carburizing performance. The American Iron and Steel Institute has now gone further in this regard with a new standardization setup with similar nomenclature, but with restricted carbon ranges and combinations of other elements which have been accepted as standard by all manufacturers of bar steel in the steel industry. The Society of Automotive Engineers have, as a result, revised most of their specifications to coincide with those set up by the American Iron and Steel Institute.

PREFIX LETTERS

    No prefix for basic open-hearth alloy steel.
    (B) Indicates acid Bessemer carbon steel.
    (C) Indicates basic open-hearth carbon steel.
    (E) Indicates electric furnace steel.

NUMBER DESIGNATIONS
    (10XX series) Basic open-hearth and acid Bessemer carbon steel grades, non
sulfurized and non-phosphorized.
    (11XX series) Basic open-hearth and acid Bessemer carbon steel grades,
sulfurized but not phosphorized.
    (1300 series) Manganese 1.60 to 1.90%
    (23XX series) Nickel 3.50%
    (25XX series) Nickel 5.0%
    (31XX series) Nickel 1.25%-chromium 0.60%
    (33XX series) Nickel 3.50%-chromium 1.60%
    (40XX series) Molybdenum
    (41XX series) Chromium molybdenum
    (43XX series) Nickel-chromium-molybdenum
    (46XX series) Nickel 1.65%-molybdenum 0.25%
    (48XX series) Nickel 3.25%-molybdenum 0.25%
    (51XX series) Chromium
    (52XX series) Chromium and high carbon
    (61XX series) Chromium vanadium
    (86XX series) Chrome nickel molybdenum
    (87XX series) Chrome nickel molybdenum
    (92XX series) Silicon 2.0%-chromium
    (93XX series) Nickel 3.0%-chromium-molybdenum
    (94XX series) Nickel-chromium-molybdenum
    (97XX series) Nickel-chromium-molybdenum
    (98XX series) Nickel-chromium-molybdenum

7-5. STANDARD DESIGNATION SYSTEM FOR STEEL (Cont)

Table 7-9. AISI-SAE Numerical Designation of Carbon and Alloy Steels

		Carbon Steels			
SAE No.	C	Mn	P Max	S Max	AISI Numbe
-	0.06 max	0.35 max	0.040	0.050	C100
1006	0.08 max	0.25-0.40	0.040	0.050	C100
1008	0.10 max	0.25-0.50	0.040	0.050	C100
1010	0.08-0.13	0.30-0.60	0.040	0.050	C101
-	0.10-0.15	0.30-0.60	0.040	0.050	C101
-	0.11-0.16	0.50-0.80	0.040	0.050	C101
1015	0.13-0.18	0.30-0.60	0.040	0.050	C101
1016	0.13-0.18	0.60-0.90	0.040	0.050	C101
1017	0.15-0.20	0.30-0.60	0.040	0.050	C101
1018	0.15-0.20	0.60-0.90	0.040	0.050	C101

7-6. STANDARD DESIGNATION SYSTEM FOR ALUMINUM AND ALUMINUM ALLOYS

a. Currently, there is no standard designation system for aluminum castings. Wrought aluminum and aluminum alloys have a standard four-digit numbering system.

b. The first digit represents the major alloying element.

c. The second digit identifies alloy modifications (a zero means the original alloy).

d. The last two digits seine only to identify different aluminum alloys which are in common commercial use, except in the 1XXX class. In the 1XXX class, the last two digits indicate the aluminum content above 99 percent, in hundredths of one percent.

e. In number 1017, the 1 indicates a minimum aluminum composition of 99 percent; the O indicates it is the original composition; and the 17 indicates the hundredths of one percent of aluminum above the 99 percent minimum composition. In this example, the aluminum content is 99.17 percent.

f. In number 3217, the 3 indicates a manganese aluminum alloy; the 2 indicates the second modification of this particular alloy; and the 17 indicatescommonly used commercial alloy.

g. The various classes of aluminum and aluminum alloys are identified by numbers as shown in table 7-10.

Table 7-10. Standard Aluminum and Aluminum Alloy Number Designations

Major alloying element	Number
Aluminum (99% minimum)	1XXX
Copper	2XXX
Manganese	3XXX
Silicon	4XXX
Magnesium	5XXX
Magnesium-silicon	6XXX
Zinc	7XXX
Other element	8XXX
Unused class	9XXX

7-7. STANDARD DESIGNATION SYSTEM FOR MAGNESIUM AND MAGNESIUM ALLOYS

a. Wrought magnesium and magnesium alloys are identified by a combination of letters and numbers. The letters identify which alloying elements were used in the magnesium alloy (table 7-11). Numbers, which may follow the letters, designate the percentage of the elements in the magnesium alloy. There may be an additional letter following the percentage designators which indicates the alloy modifications. For example, the letter A means 1; B means 2; and C means 3.

b. In the identification number AZ93C, the A indicates aluminum; the Z indicates zinc; the 9 indicates there is 9 percent aluminum in the alloy; the 3 indicates there is 3 percent zinc in the alloy; and the C indicates the third modification to the alloy. The first digit, 9 in this example, always indicates the percentage of the first letter, A in this example. The second digit gives the percentage of the second letter (table 7-12, p 7-42).

c. Temper designations may be added to the basic magnesium designation, the two being separated by a dash. The temper designations are the same as those used for aluminum (see Heat Treatment of Steel, p 12-72).

Table 7-11. Letters Used to Identify Alloying Elements in Magnesium Alloys

Letter	Alloying Element
A	Aluminum
B	Bismuth
C	Copper
D	Cadmium
E	Rare earth
F	Iron
H	Thorium
K	Zirconium
L	Beryllium
M	Manganese
N	Nickel
P	Lead
Q	Silver
R	Chromium
S	Silicon
T	Tin
Z	Zinc

7-8. STANDARD DESIGNATION SYSTEM FOR COPPER AND COPPER ALLOYS

Table 7-12. Composition of Magnesium Alloys

Alloy	Aluminum	Manganese	Zinc	Zirconium	Rare earths	Thorium	Magnesium
				NOMINAL COMPOSITION--PERCENT			
Sand and permanent mold castings							
AZ92A	9.0	0.15	2.0	–	–	–	Balance
AZ63A	6.9	0.25	3.0	–	–	–	Balance
AZ81A	7.6	0.13 min.	0.7	–	–	–	Balance
AZ91C	8.7	0:20	0.7	–	–	–	Balance
EK30A	–	–	–	0.35	3.0	–	Balance
EK41A	–	–	–	0.6	4.0	–	Balance
EZ33A	–	–	2.7	0.7	3.0	–	Balance
HK31A	–	–	–	0.7	–	3.0	Balance
HZ32A	–	–	2.1	0.7	–	3.0	Balance
Die castings							
AZ91A AZ91B	9.0	0.20	0.6	–	–	–	Balance
Extrusions							
AZ31B AZ31C	3.0	0.45	1.0	–	–	–	Balance
AZ61A	6.5	0.30	1.0	–	–	–	Balance
M1A	–	1.50	–	–	–	–	Balance
AZ80A	8.5	0.25	0.5	–	–	–	Balance
ZK60A	–	–	5.7	0.55	–	–	Balance
Sheet and plate							
AZ31B	3.0	0.45	1.0	–	–	–	Balance
HK31A	–	–	–	0.7	–	3.0	Balance

Per ASTM B275 magnesium alloys (abridged).

a. There are over 300 different wrought copper and copper alloys commercially available. The Copper Development Association, Inc., has established an alloy designation system that is widely accepted in North America. It is not a specification system but rather a method of identifying and grouping different coppers and copper alloys. This system has been updated so that it now fits the unified numbering system (UNS). It provides one unified numbering ring system which includes all of the commercially available metals and alloys. The UNS designation consists of the prefix letter C followed by a space, three digits, another space, and, finally, two zeros.

b. The information shown by table 7-13 is a grouping of these copper alloys by common names which normally include the constituent alloys. Welding information for those alloy groupings is provided. There may be those alloys within a grouping that may have a composition sufficiently different to create welding problems. These are the exception, however, and the data presented will provide starting point guidelines. There are two categories, wrought materials and cast materials. The welding information is the same whether the material is cast or rolled.

Table 7-13. Copper and Copper Alloy Designation System

Copper Number	Wrought Alloys-Groups
C11X00	Oxygen free-high conductivity copper (99.95 + %)
C11X00 C12X00 C13X00	Tough pitch copper (99.88 + %)
C19X00	High copper alloys (96 + % copper)
C2XX00	Copper-zinc-alloys (brasses)
C3XX00	Copper-zinc-lead alloys (leaded brasses)
C4XX00	Copper-zinc-tin alloys (tin brasses)
C50X00 C51X00 C52X00	Copper-tin alloys (phosphor bronzes)
C53X00 C54X00	Copper-tin-lead alloys (leaded phosphor bronzes)
C61X00 C62X00 C63X00	Copper-aluminum alloys (aluminum bronzes)
C64X00 C65X00	Copper-silicon alloys (silicon bronzes)
C66X00 C67X00 C68X00 C69X00	Copper-zinc alloys (misc. brasses & bronzes)

7-8. STANDARD DESIGNATION SYSTEM FOR COPPER AND COPPER ALLOYS (cent)

Table 7-13. Copper and Copper Alloy Designation System (cont)

C70X00 C71X00 C72X00	Copper-nickel alloys
C73X00 C74X00 C75X00 C76X00 C77X00 C78X00 C79X00	Copper-nickel-zinc alloys (nickel silvers)
	Cast Alloys--Groups
C80X00	Copper alloys (99 + % copper)
C81X00 C82X00	High copper alloys (beryllium copper)
C83X00	Copper-tin-zinc + copper-tin-zinc-lead alloys (red brasses and leaded RB)
C84X00	Semi-red brasses and leaded semi-red brasses
C85X00	Yellow brasses and leaded yellow brasses
C86X00	Manganese and leaded manganese bronze alloys
C87X00	Copper-zinc-silicon alloys (silicon bronzes and brasses)
C90X00 C91X00	Copper-tin alloys (tin bronzes)
C92X00	Copper-tin-lead alloy (leaded tin bronze)
C93X00	Copper-tin-lead alloy (high leaded tin bronze)

7-9. STANDARD DESIGNATION SYSTEM FOR TITANIUM

There is no recognized standard designation system for titanium and titanium alloys. However, these compositions are generally designated by using the chemical
symbol for titanium, Ti, followed by the percentage number(s) and the chemical sym-
bols(s) of the alloying element(s). For example, Ti-5 A1-2.5 Sn would indicate
that 5 percent aluminum and 2-1/2 percent tin alloying elements are present in the
titanium metal.

Section III. GENERAL DESCRIPTION AND WELDABILITY OF FERROUS METALS

7-10. LOW CARBON STEELS

a. General. The low carbon (mild) steels include those with a carbon content
of up to 0.30 percent (fig. 7-7). In most low carbon steels, carbon ranges from
0.10 to 0.25 percent, manganese from 0.25 to 0.50 percent, phosphorous O.40 percent
maximum, and sulfur 0.50 percent maximum. Steels in this range are most widely
used for industrial fabrication and construction These low carbon steels do not
harden appreciably when welded, and therefore do not require preheating or
postheating except in special cases, such as when heavy sections are to be welded.
In general, no difficulties are encountered when welding low carbon steels. Properly made low carbon steel welds will equal or exceed the base metal in strength.
Low carbon steels are soft, ductile, can be rolled, punched, sheared, and worked
when either hot or cold. They can be machined and are readily welded. Cast steel
has a rough, dark gray surface except where machined. Rolled steel has fine surface lines running in one direction. Forged steel is usually recognizable by its
shape, hammer marks, or fins. The fracture color is bright crystalline gray, and
the spark test yields sparks with long, yellow-orange streaks that have a tendency
to burst into white, forked sparklers. Steel gives off sparks when melted and
solidifies almost instantly. Low carbon steels can be easily welded with any of
the arc, gas, and resistance welding processes.

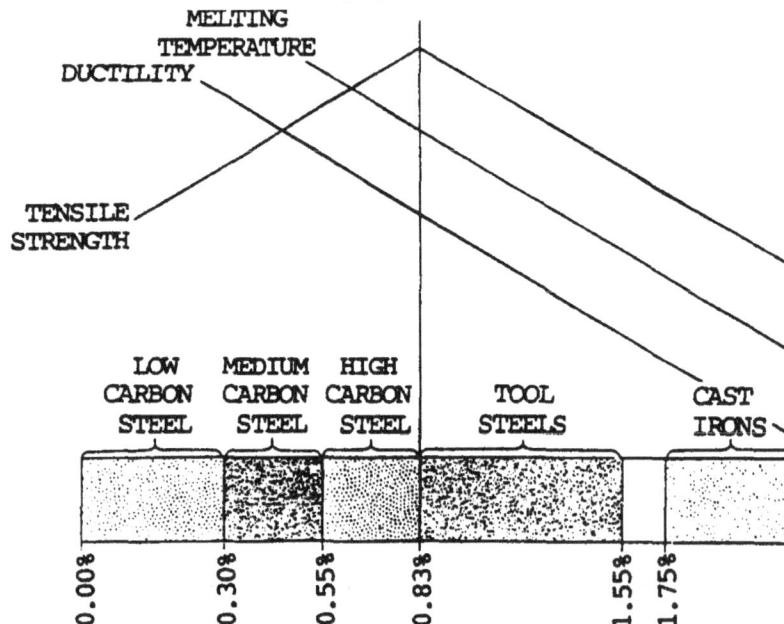

Figure 7-7. How steel qualities change as carbon is added.

TC 9-237

7-10.   LOW CARBON STEELS (cont)

   b.   Copper coated low carbon rods should be used for welding low carbon steel.
The rod sizes for various plate thicknesses are as follows:

Plate thickness	Rod diameter
1/16 to 1/8 in. (1.6 to 3.2 mm)	1/16 in. (1.6 mm)
1/8 to 3/8 in. (3.2 to 9.5 mm)	1/8 in. (3.2 mm)
3/8 to 1/2 in. (9.5 to 12.7 mm)	3/16 in. (4.8 mm)
1/2 in. (12.7 mm) and heavier	1/4 in. (6.4 mm)

NOTE

Rods from 5/16 to 3/8 in. (7.9 to 9.5 mm) are available for heavy
welding. However, heavy welds can be made with the 3/16 or 1/4 in.
(4.8 or 6.4 mm) rods by properly controlling the puddle and melting
rate of the rod.

   c.   The joints may be prepared by flame cutting or machining. The type of preparation (fig. 7-8) is determined by the plate thickness and the welding position.

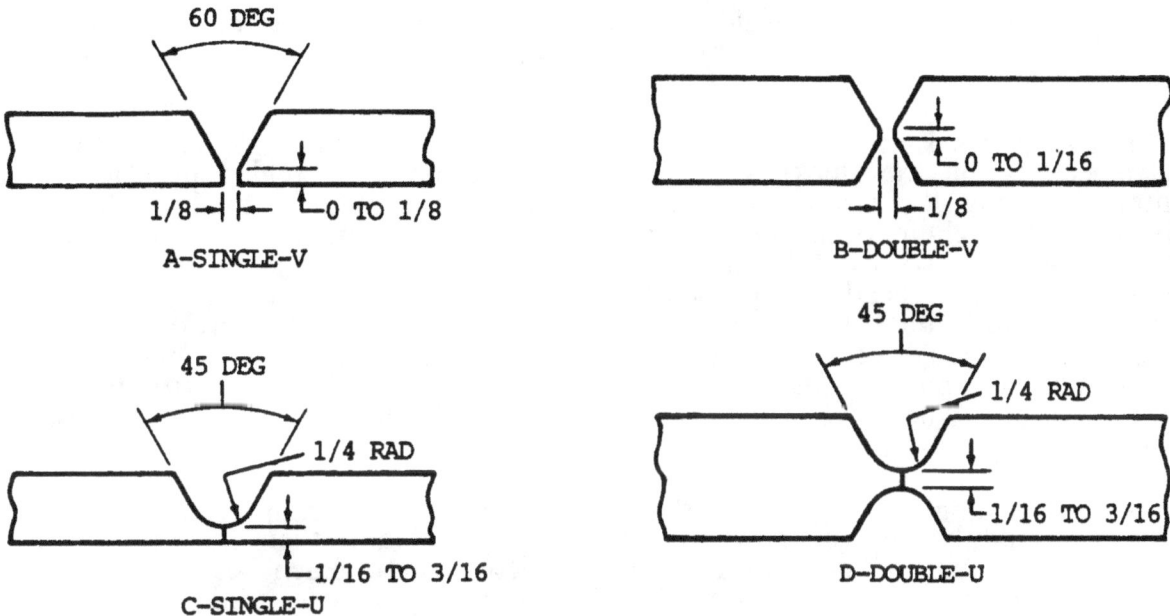

NOTE: ALL DIMENSIONS SHOWN ARE IN INCHES.

Figure 7-8. Weld preparation.

   d. The flame should be adjusted to neutral. Either the forehand or backhand
welding method may be used (p 6-36), depending on the thickness of the plates being
welded .

   e.   The molten metal should not be overheated, because this will cause the metal
to boil and spark excessively. The resultant grain structure of the weld metal
will be large, the strength lowered, and the weld badly scarred.

7-46

f. The low carbon steels do not harden in the fusion zone as a result of welding.

g. <u>Metal-Arc Welding</u>.

(1) When metal-arc welding low carbon steels,the bare, thin coated or heavy coated shielded arc types of electrodes may be used. These electrodes are ofwlo carbon type (0.10 to 0.14 percent).

(2) Low carbon sheet or plate materials that have been exposed to low temperatures should be preheated slightly to room temperature before welding.

(3) In welding sheet metal up to 1/8 in. (3.2 mm) in thickness, the plain square butt joint type of edge preparation may be used. When long seams are to be welded in these materials, the edges should be spaced to allow for shrinkage, because the deposited metal tends to pull the plates togetherThis shrinkage is less severe in arc welding than in gas welding, and spacing of approximately 1/8 in. (3.2 mm) will be sufficient.

(4) The backstep, or skip, welding technique should be used for short seams that are fixed in place. This will prevent warpage or distortion, and will minimize residual stresses.

(5) Heavy plates should be beveled to provide an included angle of up to 60 degrees, depending on the thickness. The parts should be tack welded in place at short intervals along the seam. The first, or root, bead should be made with an electrode small enough in diameter to obtain good penetration and fusion at the base of the joint. A 1/8 or 5/32 in. (3.2 or 4.0 mm) electrode is suitable for this purpose. The first bead should be thoroughly cleaned by chipping and wire brushing before additional layers of weld metal are deposited. Additional passes of the filler metal should be made with a 5/32 or 3/16 in. (4.0 or 4.8 mm) electrode. The passes should be made with a weaving motion for flat, horizontal, or vertical positions. When overhead welding, the best results are obtained by using string beads throughout the weld.

(6) When welding heavy sections that have been beveled from both sides, the weave beads should be deposited alternately on one side and then the other. This will reduce the amount of distortion in the welded structureEach bead should be cleaned thoroughly to remove all scale, oxides, and slag before additional metal is deposited. The motion of the electrode should be controlled so as to make the bead uniform in thickness and to prevent undercutttig and overlap at the edges of the weld. All slag and oxides must be removed from the surface of the completed weld to prevent rusting.

h. <u>Carbon-Arc Welding</u>. Low carbon sheet and plate up to 3/4 in. (19.0 mm) in thickness can be welded using the carbon-arc welding process. The arc is struck against the plate edges, which are prepared in a manner similar to that required for metal-arc welding. A flux should be used on the joint and filler metal should be added as in oxyacetylene welding. A gaseous shield should be provided around the molten base. Filler metal, by means of a flux coated welding rod, should also be provided. Welding must be done without overheating the molten metal. Failure to observe these precautions can cause the weld metal to absorb an excessive amount of carbon from the electrode and oxygen and nitrogen from the air, and cause brittleness in the welded joint.

7-11. MEDIUM CARBON STEELS

a. General. Medium carbon steels are non-alloy steels which contain from 0.30 to 0.55 percent carbon. These steels may be heat treated after fabrication and used for general machining and forging of parts which require surface hardness and strength. They are manufactured in bar form and in the cold rolled or the normalized and annealed condition. When heat treated steels are welded, they should be preheated from 300 to 500°F (149 to 260°C), depending on the carbon content (O.25 to 0.45 percent) and the thickness of the steel. The preheating temperature may be checked by applying a stick of 50-50 solder (melting point 450°F (232°C)) to the plate at the joint and noting when the solder begins to melt. During welding, the weld zone will become hardened if cooled rapidly, and must be stress relieved after welding. Medium carbon steels may be welded with any of the arc, gas, and resistance welding processes.

b. With higher carbon and manganese content the low-hydrogen type electrodes should be used, particularly in thicker sections. Electrodes of the low-carbon, heavy coated, straight or reverse polarity type, similar to those used for metal-arc welding of low carbon steels, are satisfactory for welding medium carbon steels.

c. Small parts should be annealed to induce softness before welding. The parts should be preheated at the joint and welded with a filler rod that produces heat treatable welds. After welding, the entire piece should be heat treated to restore its original properties.

d. Either a low carbon or high strength rod can be used for welding medium carbon steels. The welding flame should be adjusted to slightly carburizing, and the puddle of metal kept as small as possible to make a sound joint. Welding with a carburizing flame causes the metal to heat quickly, because heat is given off when steel absorbs carbon. This permits welding at higher speeds.

e. Care should be taken to slowly cool the parts after welding to prevent cracking of the weld. The entire welded part should be stress relieved by heating to between 1100 and 1250°F (593 and 677°C) for one hour per inch (25.4 mm) of thickness, and then slowly cooling. Cooling can be accomplished by covering the parts with fire resistant material or sand.

f. Medium carbon steels can be brazed by using a preheat of 200 to 400°F (93 to 204°C), a good bronze rod, and a brazing flux. However, these steels are better welded by the metal-arc process with mild steel shielded arc electrodes.

g. When welding mild steels, keep the following general techniques in mind:

(1) The plates should be prepared for welding in a manner similar to that used for welding low carbon steels. When welding with low carbon steel electrodes, the welding heat should be carefully controlled to avoid overheating the weld metal and excessive penetration into the side walls of the joint. This control is accomplished by directing the electrode more toward the previously deposited filler metal adjacent to the side walls than toward the side walls directly. By using this procedure, the weld metal is caused to wash up against the side of the joint and fuse with it without deep or excessive penetration.

(2) High welding heats will cause large areas of the base metal in the fusion zone adjacent to the welds to become hard and brittle. The area of these hard zones in the base metal can be kept to a minimum by making the weld with a series of small string or weave beads, which will limit the heat input. Each bead or layer of weld metal will refine the grain in the weld immediately beneath it, and will anneal and lessen the hardness produced in the base metal by the previous bead.

(3) When possible, the finished joint should be heat treated after welding. Stress relieving is normally used when joining mild steel, and high carbon alloys should be annealed.

(4) In welding medium carbon steels with stainless steel electrodes, the metal should be deposited in string beads in order to prevent cracking of the weld metal in the fusion zone. When depositing weld metal in the upper layers of welds made on heavy sections, the weaving motion of the electrode should not exceed three electrode diameters.

(5) Each successive bead of weld should be chipped, brushed, and cleaned prior to the laying of another bead.

7-12. HIGH CARBON STEELS

a. General. High carbon steels include those with a carbon content exceeding 0.55 percent. The unfinished surface of high carbon steels is dark gray and similar to other steels. High carbon steels usually produce a very fine grained fracture, whiter than low carbon steels. Tool steel is harder and more brittle than plate steel or other low carbon material. High carbon steel can be hardened by heating to a good red and quenching in water. Low carbon steel, wrought iron, and steel castings cannot be hardened. Molten high carbon steel is brighter than low carbon steel, and the melting surface has a cellular appearance. It sparks more freely than low carbon (mild) steel, and the sparks are whiter. These steels are used to manufacture tools which are heat treated after fabrication to develop the hard structure necessary to withstand high shear stress and wear. They are manufactured in bar, sheet, and wire forms, and in the annealed or normalized and annealed condition in order to be suitable for machining before heat treatment. The high carbon steels are difficult to weld because of the hardening effect of heat at the welded joint. Because of the high carbon content and the heat treatment usually given to these steels, their basic properties are impaired by arc welding.

b. The welding heat changes the properties of high carbon steel in the vicinity of the weld. To restore the original properties, heat treatment is necessary.

c. High carbon steels should be preheated from 500 to 800°F (260 to 427 °C) before welding. The preheating temperature can be checked with a pine stick, which will char at these temperatures.

d. Since high carbon steels melt at lower temperatures than low and medium carbon steels, care should be taken not to overheat the weld or base metal. Overheating is indicated by excessive sparking of the molten metal. Welding should be completed as soon as possible and the amount of sparking should be used as a check on the welding heat. The flame should be adjusted to carburizing. This type of flame tends to produce sound welds.

7-12. HIGH CARBON STEELS (cont)

e. Either a medium or high carbon welding rod should be used to make the weld. After welding, the entire piece should be stress relieved by heating to between 1200 and 1450 °F (649 and 788 °C) for one hour per inch (25.4 mm) of thickness, and then slowly cooling. If the parts can easily be softened before welding, a high carbon welding rod should be used to make the joint. The entire piece should then be heat treated to restore the original properties of the base metal.

f. In some cases, minor repairs to these steels can be made by brazing. This process does not require temperatures as high as those used for welding, so the properties of the base metal are not seriously affected. Brazing should only be used in special cases, because the strength of the joint is not as high as the original base metal.

g. Either mild or stainless steel electrodes can be used with high carbon steels.

h. Metal-arc welding in high carbon steels requires critical control of the weld heat. The following techniques should be kept in mind:

(1) The welding heat should be adjusted to provide good fusion at the side walls and root of the joint without excessive penetration. Control of the welding heat can be accomplished by depositing the weld metal in small string beads. Excessive puddling of the metal should be avoided, because this can cause carbon to be picked up from the base metal, which in turn will make the weld metal hard and brittle. Fusion between the filler metal and the side walls should be confined to a narrow zone. Use the surface fusion procedure prescribed for medium carbon steels (para 7-11, p 7-48).

(2) The same procedure for edge preparation, cleaning of the welds, and sequence of welding beads as prescribed for low and medium carbon steels also applies to high carbon steels.

(3) Small, high carbon steel parts are sometimes repaired by building up worn surfaces. When this is done, the piece should be annealed or softened by heating to a red heat and cooling slowly. The piece should then be welded or built up with medium carbon or high strength electrodes, and heat treated after welding to restore its original properties.

7-13. TOOL STEELS

a. General. Steels used for making tools, punches, and dies are perhaps the hardest, strongest, and toughest steels used in industry. In general, tool steels are medium to high carbon steels with specific elements included in different amounts to provide special characteristics. A spark test shows a moderately large volume of white sparks having many fine, repeating bursts.

b. Carbon is provided in tool steel to help harden the steel for cutting and wear resistance. Other elements are added to provide greater toughness or strength. In some cases, elements are added to retain the size and shape of the tool during its heat treat hardening operation, or to make the hardening operation safer and to provide red hardness so that the tool retains its hardness and strength when it becomes extremely hot. Iron is the predominant element in the composition of tool steels. Other elements added include chromium, cobalt, manganese, molybdenum, nickel, tungsten, and vanadium. The tool or die steels are designed for special purpose that are dependent upon composition. Certain tool steels are made for producing die blocks; some are made for producing molds, other for hot working, and others for high-speed cutting application.

c. Another way to classify tool steels is according to the type of quench required to harden the steel. The most severe quench after heating is the water quench (water-hardening steels). A less severe quench is the oil quench, obtained by cooling the tool steel in oil baths (oil-hardening steels). The least drastic quench is cooling in air (air-hardening steels).

d. Tool steels and dies can also be classified according to the work that is to be done by the tool. This is based on class numbers.

(1) Class I steels are used to make tools that work by a shearing or cutting actions, such as cutoff dies, shearing dies, blanking dies, and trimming dies.

(2) Class II steels are used to make tools that produce the desired shape of the part by causing the material being worked, either hot or cold, to flow under tension. This includes drawing dies, forming dies, reducing dies, forging dies, plastic molds, and die cast molding dies.

(3) Class III steels are used to make teds that act upon the material being worked by partially or wholly reforming it without changing the actual dimensions. This includes bending dies, folding dies, and twisting dies.

(4) Class IV steels are used to make dies that work under heavy pressure and that produce a flow of metal or other material caressing it into the desired form. This includes crimping dies, embossing dies, heading dies, extrusion dies, and staking dies.

e. Steels in the tool steels group have a carbon content ranging from 0.83 to 1.55 percent. They are rarely welded by arc welding because of the excessive hardness produced in the fusion zone of the base metal. If arc welding must be done, either mild steel or stainless steel electrodes can be used.

f. Uniformly high preheating temperatures (up to 1000°F (583 °C)) must be used when welding tool steels.

g. In general, the same precautions should be taken as those required for welding high carbon steels (para 6-12, p 6-22). The welding flare should be adjusted to carburizing to prevent the burning out of carbon in the weld metal. The welding should be done as quickly as possible, taking care not to overheat the molten metal. After welding, the steel should be heat treated to restore its original properties.

h. Drill rods can be used as filler rods because their high carbon content compares closely with that of tool steels.

i. A flux suitable for welding cast iron should be used in small quantities to protect the puddle of high carbon steel and to remove oxides in the weld metal.

7-13. TOOL STEEL (cont)

Welding Technique. When welding tool steels, the following techniques should be kept in mind:

(1) If the parts to be welded are small, they should be annealed or softened before welding. The edges should then be preheated up to 1000F (538 °C), depending on the carbon content and thickness of the plate. Welding should be done with either a mild steel or high strength electrode.

(2) High carbon electrodes should not be used for welding tool steels. The carbon picked up from the base metal by the filler metal will cause the weld to become glass hard, whereas the mild steel weld metal can absorb additional carbon without becoming excessively hard. The welded part should then be heat treated to restore its original properties.

(3) When welding with stainless steel electrodes, the edge of the plate should be preheated to prevent the formation of hard zones in the base metal. The weld metal should be deposited in small string beads to keep the heat input to a minimum. In general, the application procedure is the same as that required for medium and high carbon steels.

k. There are four types of die steels that are weld repairable. These are water-hardening dies, oil-hardening dies, air-hardening dies, and hot work tools. High-speed tools can also be repaired.

7-14. HIGH HARDNESS ALLOY STEELS

a. General. A large number and variety of alloy steels have been developed to obtain high strength, high hardness, corrosion resistance, and other special properties. Most of these steels depend on a special heat treatment process ordered to develop the desired characteristic in the finished state. Alloy steels have greater strength and durability than other carbon steels, and a given strength is secured with less material weight.

b. High hardness alloy steels include the following:

(1) Chromium alloy steels. Chromium is used as an alloying element in carbon steels to increase hardenability, corrosion resistance, and shock resistance, and gives high strength with little loss in ductility. Chromium in large amounts shortens the spark stream to one half that of the same steel without chromium, but does not affect the stream's brightness.

(2) Nickel alloy steels. Nickel increases the toughness, strength, and ductility of steels, and liners the hardening temperature so that an oil quench, rather than a water quench, is used for hardening. The nickel spark has a short, sharply defined dash of brilliant light just before the fork.

(3) High chromium-nickel alloy (stainless) steels. These high alloy steels cover a wide range of compositions. Their stainless, corrosion, and heat resistant properties vary with the alloy content, and are due to the formation of a very thin oxide film which forms on the surface of the metal. Sparks are straw colored near

the grinding wheel, and white near the end of the streak. There is a medium volume of streaks which have a moderate number of forked bursts.

(4) Manganese alloy steels. Manganese is used in steel to produce greater toughness, wear resistance, easier hot rolling, and forging. An increase in manganese content decreases the weldability of steel. Steels containing manganese produce a spark similar to a carbon spark. A moderate increase in manganese increases the volume of the spark stream and the intensity of the bursts. A steel containing more than a normal amount of manganese will produce a spark similar to a high carbon steel with a lower manganese content.

(5) Molybdenum alloy steels. Molybdenum increases hardenability, which is the depth of hardening possible through heat treatment. The impact fatigue property of the steel is improved with up to 0.60 percent molybdenum. Above 0.60 percent molybdenum, the impact fatigue proper is impaired. Wear resistance is improved with molybdemnn content above about 0.75 percent. Molybdenum is sometimes cabined with chromium, tungsten, or vanadium to obtain desired properties. Steels containing this element produce a charcteristic spark with a detached arrowhead similar to that of wrought iron, which can be seen even in fairly strong carbon bursts. Molybdenum alloy steels contain either nickel and/or chromium.

(6) Titanium and columbium (niobium) alloy steels. These elements are used as additional alloying agents in low carbon content, corrosion resistant steels. They support resistance to intergranular corrosion after the metal is subjected to high temperatures for a prolonged period of time.

(7) Tungsten alloy steels. Tungsten, as an alloying element in tool steel, tends to produce a fine, dense grain when used in relatively small quantities. when used in larger quantities, from 17 to 20 percent, and in combination with other alloys, tungsten produces a steel that retains its hardness at high temperatures. This element is usually used in combination with chromium or other alloying agents. In a spark test, tungsten will show a dull red color in the spark stream near the wheel. It also shortens the spark stream and decreases the size of or completely eliminates the carbon burst. A tungsten steel containing about 10 percent tungsten causes short, curved, orange spear points at the end of the carrier lines. Still lower tungsten content causes small, white bursts to appear at the end of the spear petit. Carrier lines may be from dull red to orange, depending on the other elements present, providing the tunsten content is not too high.

(8) Vanadium alloy steels. Vanadium is used to help control grain size. It tends to increase hardenability and causes marked secondary hardness, yet resists tempering. It is added to steel during manufacture to remove oxygen. Alloy steels containing vanadium produce sparks with detached arrowheads at the end of the carrier line similar to those produced by molybdenum steels.

(9) Silicon alloy steels. Silicon is added to steel to obtain greater hardenability and corrosion resistance. It is often used with manganese to obtain a strong, tough steel.

(10) High speed tool steels. These steels are usually special alloy compositions designed for cutting tools. The carbon content ranges from 0.70 to 0.80 percent. They are difficult to weld, except by the furnace induction method. A spark test will show a few long, forked spades which are red near the wheel, and straw colored near the end of the spark stream.

7-14.   HIGH HARDNESS ALLOY STEELS (cont)

c.   Many of these steels can be welded with a heavy coated electrode of the shielded arc type, whose composition is similar to that of the base metal. Low carbon electrodes can also be used with some steels.Stainless steel electrodes are effective where preheating is not feasible or desirable.Heat treated steels should be preheated, if possible, in order to minimize the formation of hard zones, or layers, in the base metal adjacent to the weld.The molten metal should not be overheated, and the welding heat should be controlled by depositing the metal in narrow string beads. In many cases, the procedures for welding medium carbon steels (para 7-11, p 7-48) and high carbon steels (para 7-12, p 7-49) can be used in the welding of alloy steels.

7-15.   HIGH YIELD STRENGTH, LOW ALLOY STRUCTURAL STEELS

a.   General. High yield strength, low alloy structural steels (constructional alloy steels) are special steels that are tempered to obtain extreme toughness and durability. The special alloys and general makeup of these steels require special treatment to obtain satisfactory weldments. These steels are special, low-carbon steels containing specific, small amounts of alloying elements.They are quenched and tempered to obtain a yield strength of 90,000 to 100,000 psi (620,550 to 689,500 kPa) and a tensile strength of 100,000 to 140,000 psi (689,500 to 965,300 kPa), depending upon size and shape.Structural members fabricated from these high strength steels may have smaller cross-sectional areas than common structural steels and still have equal strength. These steels are also more corrosion and abrasion resistant than other steels.In a spark test, these alloys produce a spark very similar to low carbon steels.

b.   Welding Technique. Reliable welding of high yield strength, low alloy structural steels can be performed by using the following guidelines:

CAUTION
To prevent underbead cracking, only low hydrogen electrodes should be used when welding high yield strength, low alloy structural steels.

(1)   Correct electrodes.Hydrogen is the number one enemy of sound welds in alloy steels; therefore, use only low hydrogen (MIL-E-18038 or MIL-E-22200/1) electrodes to prevent underbead cracking.Underbead cracking is caused by hydrogen picked up in the electrode coating, released into the arc, and absorbed by the molten metal.

(2)   Moisture control of electrodes. If the electrodes are in an airtight container, place them, immediately upon opening the container, in a ventilated holding oven set at 250 to 300°F (121 to 149 °C) . In the event that the electrodes are not in an airtight container, put them in a ventilated baking oven and bake for 1-1/4 hours at 800°F (427 °C). Baked electrodes should, while still warm, be placed in the holding oven until used.Electrodes must be kept dry to eliminate absorption of hydrogen.Testing for moisture should be in accordance with MIL-E-22200.

NOTE

Moisture stabilizer NSN 3439-00-400-0090 is an ideal holding oven for field use (MIL-M-45558).

c. Low Hydrogen Electrode Selection. Electrodes are identified by classification numbers which are always marked on the electrode containers. For low hydrogen coatings, the last two nunbers of the classification should be 15, 16, or 18. Electrodes of 5/32 and 1/8 in. (4.0 and 3.2 mm) in diameter are the most commonly used, since they are more adaptable to all types of welding of this type steel. Table 7-14 lists electrodes used to weld high yield strength, low alloy structural steels. Table 7-15 is a list of electrodes currently established in the Army supply system

Table 7-14. Electrode Numbers

E8015$_1$	E9015^2	E10015	E11015	E12015
E8016^2	E9016	E10016	E11016	E12016
E8018	E9018	E10018	E11018	E12018

[1]The E indicates electrode; the first two or three digits indicate tensile strength; the last two digits indicate covering. The numbers 15, 16, and 18 all indicate a low hydrogen covering.
[2]Low hydrogen electrodes E80 and E90 are recommended for fillet welds, since they are more ductile than the higher strength electrodes, which are desirable for butt welds.

Table 7-15. Electrodes in the Army Supply System

Electrode Number	Size (in.)	NSN
E9018	1/8 dia x 14 lg	3439-00-853-2716
E9018	5/32 dia x 14 lg	3439-00-853-2718
E11018	1/8 dia x 14 lg	3439-00-587-2412
E11018	5/32 dia x 14 lg	3439-00-587-2413
E11018	3/16 dia x 14 lg	3439-00-878-2158

d. Selecting Wire-Flux and Wire-Gas Combinations. Wire electrodes for submerged arc and gas-shielded arc welding are not classified according to strength. Welding wire and wire-flux combinations used for steels to be stress relieved should contain no more than 0.05 recent vanadium. Weld metal with more than 0.05 percent vanadium may brittle if stress relieved. When using either the submerged arc or gas metal-arc welding processes to weld high yield strength, wlo alloy structural steels to lower strength steels the wire-flux and wire-gas combination should be the same as that recommend for the lower strength steels.

7-15. HIGH YIELD STRENGTH, LOW ALLOY STRUCTURAL STEELS (cont)

e. _Preheating._ For welding plates under 1.0 in. (25.4 mm) thick preheating above 50 °F (10 °C) is not required except to remove surface moisture from the base metal. Table 7-16 contains suggested preheating temperatures.

Table 7-16. Suggested Preheat Temperatures

Plate Thickness (in.)	Shielded Metal-Arc (Manual Arc) Welding[2]	Gas Metal-Arc Welding[3]	Submerged arc welding	
			Carbon Steel or Alloy Wire Neutral Flux[4]	Carbon Steel Wire, Alloy Flux[5]
Up to 1/2, inclusive	50 °F (10 °C)	50 °F (10 °C)	50 °F (10 °C)	50 °F (10 °C)
Over 1/2 to 1, inclusive	50 °F (10 °C)	50 °F (10 °C)	50 °F (10 °C)	200 °F (93 °C)
Over 1 to 2, inclusive	150 °F (66 °C)	150 °F (66 °C)	200 °F 93 °C)	300 °F (149 °C)
Over 2	200 °F (93 °C)	200 °F (93 °C)	300 °F (149 °C)	400 °F (204 °C)

[1] Preheated temperatures above the minimum shown may be necessary for highly restrained welds. However, preheat or interpass temperatures should never exceed 400 °F (204 °C) for thicknesses up to and including 1-1/2 in. (38.1 mm) or 450 °F (232 °C) for thicknesses over 1-1/2 in. (38.1 mm).
[2] Electrode E11018 is normal for this type steel. However, E12015, 16 or 18 may be necessary for thin sections, depending on design stress. Lower strength low hydrogen electrodes E1OOXX may also be used.
[3] Example: A-632 wire (Airco) and argon with 1 percent oxygen.
[4] Example: Oxweld 100 wire (Linde) and 709-5 flux.
[5] Example: L61 wire (Lincoln) and A0905 X 10 flux.

f. Welding Heat.

(1) _General._ It is important to avoid excessive heat concentration in order to allow the weld area to cool quickly. Either the heat input nomograph or the heat input calculator can be used to detmine the heat input into the weld.

(2) _Heat input nomograph._ To use the heat input nomograph (fig. 7-9), find the volts value in column 1 and draw a line to the amps value in column 3. Fran the point where this line intersects Colunm 2, draw another line to the in./min value in column 5. Read the heat units at the point where this second line intersects column 4. The heat units represent thousands of joules per inch. For example, at 20 volts and 300 amps, the line intersects column 2 at the value 6. At 12 in./min, the heat input is detmined as 30 heat units, or 30,000 joules/in.

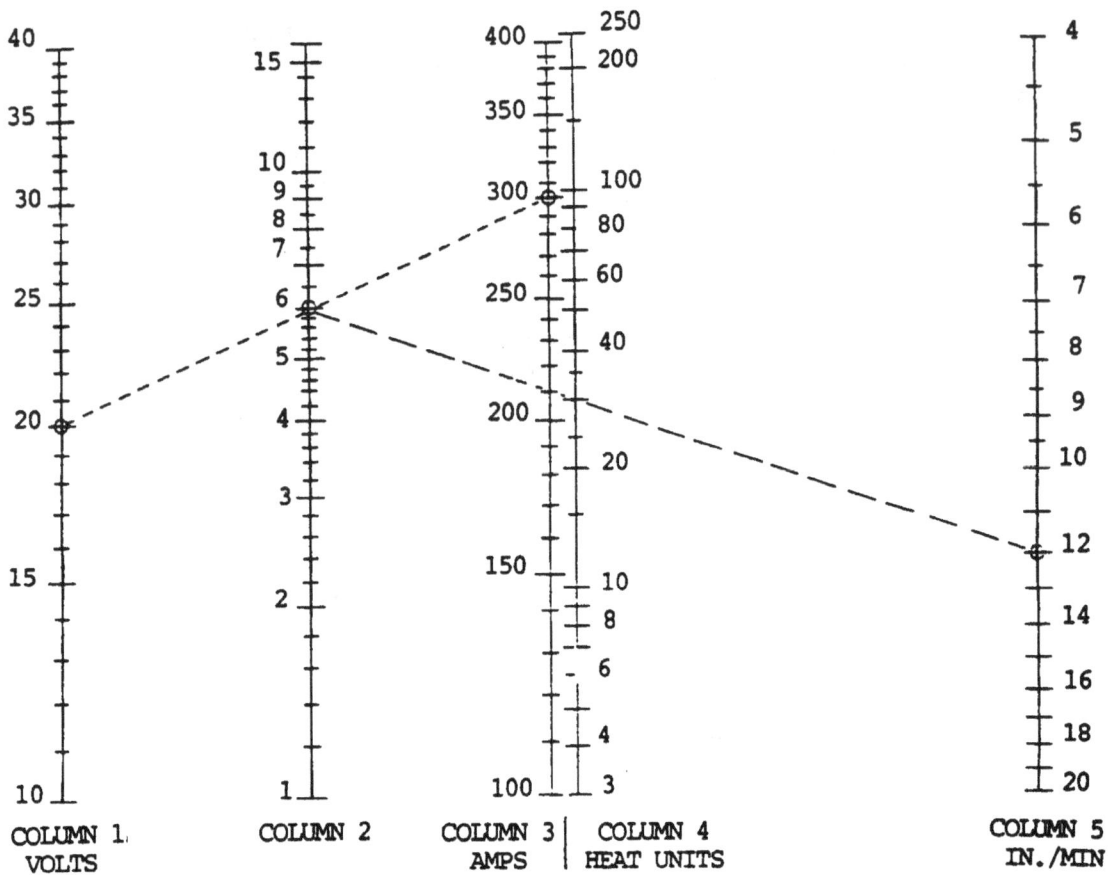

Figure 7-9. Heat input nomograph.

(3) Heat input calculator. The heat input calculator can be made by copying the pattern printed on the inside of the back cover of this manual onto plastic, light cardboard, or other suitable material and cutting out the piece. If no suitable material is available, the calculator may be assembled by cutting the pattern out of the back cover. After the two pieces are cut out, a hole is punched in the center of each. They are then assembled using a paper fastener, or some similar device, which will allow the pieces to rotate. To determine welding heat input using the calculator, rotate until the value on the volts scale is aligned directly opposite the value on the speed (in./min) scale. The value on the amps scale will then be aligned directly opposite the calculated value for heat units. As with the nomograph, heat units represent thousands of joules per inch.

7-15. HIGH YIELD STRENGTH, LOW ALLOY STRUCTURAL STEELS (cont)

(4) <u>Maximum heat input</u>. Check the heat input value obtained from the nomograph or calculator against the suggested maximums in tables 7-17 and 7-18. If the calculated value is too high, adjust the amperes, travel speed, or preheat temperature until the calculated heat input is within the proper range. The tables are applicable only to single-arc, shielded metal-arc, submerged arc, gas tungsten-arc, flux-cored arc, and gas metal-arc processes. They are not applicable to multiple-arc or electroslag welding, or other high heat input vertical-welding processes, since welds made by these in the "T-1" steels should be heat treated by quenching and tempering.) For welding conditions exceeding the range of the nomograph or calculator, the heat input can be calulated using the following formula:

$$\text{Heat Input (1,000 Joules/in.)} = \frac{\text{Amps} \times \text{Volts} \times 60}{\text{speed (in./min)}}$$

Table 7-17. Maximum Heat Inputs for T1 Steel[1]

Thickness, In.	Preheat and Interpass Temperature				
	70 °F (21 °C)	150 °F (60 °C)	200 °F (93 °C)	300 °F (149 °C)	400 °F (204 °C)
3/16	27	23	21	17	13
1/4	36	32	29	24	19
1/2	70	62	56	47	40
3/4	121	107	99	82	65
1	any	188	173	126	93
1-1/4	any	any	any	175	127
1-1/2	any	any	any	any	165
2	any	any	any	any	any

[1]Maximum heat inputs are based on minimum Charpy V-notch impact value of 10 ft-lb at -50 °F (-46 °C) in the heat-affected zone.

Table 7-18. Maximum Heat Inputs for T1 Type A and Type B Steels[1]

Thickness, In.	Preheat and Interpass Temperature				
	70 °F (21 °C)	150 °F (66 °C)	200 °F (93 °C)	300 °F (149 °C)	400 °F (204 °C)
3/16	17.5	15.3	14.0	11.5	9.0
1/4	23.7	20.9	19.2	15.8	12.3
3/8	35.0	30.7	28.0	23.5	18.5
1/2	47.4	41.9	38.5	31.9	25.9
5/8	64.5	57.4	53.0	42.5	33.5
3/4	88.6	77.4	69.9	55.7	41.9
1	any	120.0	110.3	86.0	65.6
1-1/4	any	any	154.0	120.0	94.0

heat inputs are based on a minimum Charpy V-notch impact value of 10 ft-lb at 0°F (-18 °C) in the heat-affected zone.

g. Welding Process. Reliable welding of high yield strength, low alloy structural steel can be per formal by choosing an electrode with low hydrogen content or selecting the proper wire-flux or wire gas combination when using the submerged arc or gas metal arc processes. Use a straight stringer bead whenever possible. Avoid using the weave pattern; however, if needed, it must be restricted to a partial weave pattern. Best results are obtained by a slight circular motion of the electrode with the weave area never exceeding two elect-rode diameters. Never use a full weave pattern. The partial weave pattern should not exceed twice the diameter of the electrode. Skip weld as practical. Peening of the weld is sometimes recommended to relieve stresses while cooling larger pieces. Fillet welds should be smooth and correctly contoured.Avoid toe cracks and undercutting. Electrodes used for fillet welds should be of lower strength than those used for butt welding. Air-hammerpeening of fillet welds can help to prevent cracks, especially if the welds are to be stress relieved. A soft steel wire pedestal can help to absorb shrinkage forces. Butter welding in the toe area before actual fillct welding strengths the area where a toe crack may startA bead is laid in the toe area, then ground off prior to the actual fillet welding. This butter weld bead must be located so that the toe of the fillet will be laid directly over it during actual fillet welding. Because of the additional mateial involved in fillet welding, the cooling rate is increased and heat inputs may be extended about 25 percent.

7-16. CAST IRON

a. General. A cast iron is an alloy of iron, carbon, and silicon, in which the amount of carbon is usually more than 1.7 percent and less than 4.5 percent.

(1) The most widely used type of cast iron is known as gray iron. Gray iron has a variety of compositions, but is usually such that it is primarily perlite with many graphite flakes dispersed throughout.

(2) There are also alloy cast irons which contain small amounts of chromium, nickel, molybdenum, copper, or other elements added to provide specific properties.

(3) Another alloy iron is austenitic cast iron, which is modified by additions of nickel and other elements to reduce the transformation temperature so tha the structure is austenitic at room or normal temperatures. Austenitic cast irons have a high degree of corrosion resistance.

(4) In white cast iron almost all the carbon is in the combined form. This provides a cast iron with higher hardness, which is used for abrasion resistance.

(5) Malleable cast iron is made by giving white cast iron a special annealing heat treatment to change the structure of the carbon in the iron. The structure i: changed to perlitic or ferritic, which increases its ductility.

(6) Nodular iron and ductile cast iron are made by the addition of magnesium or aluminum which will either tie up the carbon in a combined state or will give the free carbon a spherical or nodular shape, rather than the normal flake shape i gray cast iron. This structure provides a greater degree of ductility or malleability of the casting.

7-16. CAST IRON (cont)

(7) Cast irons are widely used in agricultural equipment; on machine tools as bases, brackets, and covers;for pipe fittings and cast iron pipe; and for automobile engine blocks, heads,manifolds, and water preps. Cast iron is rarely used in structural work except for compression members.It is widely used in construction machinery for counterweights and in other applications for which weight is required.

b. Gray cast iron has low ductility and therefore will not expand or stretch to any considerable extent before breaking or cracking.Because of this characteristic, preheating is necessary when cast iron is welded by the oxyacetylene welding process. It can, however, be welded with the metal-arc process without preheating if the welding heat is carefully controlled. This can be accomplished by welding only short lengths of the joint at a time and allowing these sections to cool. By this procedure, the heat of welding is confined to a small area, and the danger of cracking the casting is eliminated. Large castings with complicated sections, such as motor blocks, can be welded without dismantling or preheating.Special electrodes designed for this purpose are usually desirable.Ductile cast irons, such as malleable iron, ductile iron, and nodular iron, can be successfully welded. For best results, these types of cast irons should be welded in the annealed condition.

c. Welding is used to salvage new iron castings, to repair castings that have failed in service, and to join castings to each other or to steel parts in manufacturing operations. Table 7-19 shows the welding processes that can be used for welding cast, malleable, and nodular irons.The selection of the welding process and the welding filler metals depends on the type of weld properties desired and the service life that is expected. For example, when using the shielded metal arc welding process, different types of filler metal can be used. The filler metal will have an effect on the color match of the weld compared to the base material. The color match can be a determing factor, specifically in the salvage or repair of castings, where a difference of color would not be acceptable.

Table 7-19. Welding Processes and Filler Metals for Cast Iron

Welding Process & Filler Metal Type	Filler Metal Spec	Filler Metal Type	Color Match	Machineable Deposit
SMAW (Stick)				
Cast iron	E-C1	Cast iron	Good	Yes
Copper-tin[2]	ECuSn A & C	Copper-5 or 8% tin	No	Yes
Copper-aluminum[2]	ECuAl-A2	Copper-10% aluminum	No	Yes
Mild steel	E-St	Mild steel	Fair	No
Nickel	ENi-CI	High nickel alloy	No	Yes
Nickel-iron	ENiFe-CI	50% Nickel plus iron	No	Yes
Nickel-copper	ENiCu-A & B	55 or 65% Ni + 40 or 30% W	No	Yes
Oxy Fuel Gas				
Cast iron	RCI & A & B	Cast iron-with minor alloys	Good	Yes
Copper zinc[2]	RCuZn B & C	58% Copper-zinc	No	Yes

Table 7-19. Welding Processes and Filler Metals for Cast Iron (cont)

Welding Process & Filler Metal Type	Filler Metal Spec[1]	Filler Metal Type[1]	Color Match	Machinable Deposit
Brazing[3]				
Copper zinc	RBCuZn A & D	Copper-zinc & copper-Zinc-nickel	No	Yes
GMAW (MIG)				
Mild steel	E60S-3	Mild steel	Fair	No
Copper base[2]	ECuZn-C	Silicon bronze	No	Yes
Nickel-copper	ENiCu-B	High nickel	No	Yes
FCAW				
Mild steel	E70T-7	Mild steel	Fair	No
Nickel type	No spec	50% nickel plus iron	No	Yes

NOTE 1 See AWS Specification for Welding Rods and Covered Electrode for Welding Cast Iron.
2 Would be considered a brass weld.
3 Heat source any for brazing also carbon arc, twin carbon arc, gas tungsten arc, or plasma arc.

d. No matter which of the welding processes is selected, certain preparatory steps should be made. It is important to determine the exact type of cast iron to be welded, whether it is gray cast iron or a malleable or ductile type. If exact information is not known, it is best to assume that it is gray cast iron with little or no ductility. In general, it is not recommended to weld repair gray iron castings that are subject to heating and cooling in normal service, especially when heating and cooling vary over a range of temperatures exceeding 400°F (204 °C). Unless cast iron is used as the filler material, the weld metal and base metal may have different coefficients of expansion and contraction. This will contribute to internal stresses which cannot be withstood by gray cast iron. Repair of these types of castings can be made, but the reliability and service life on such repairs cannot be predicted with accuracy.

e. Preparation for Welding.

(1) In preparing the casting for welding, it is necessary to remove all surface materials to completely clean the casting in the area of the weld. This means removing paint, grease, oil, and other foreign material from the weld zone. It is desirable to heat the weld area for a short time to remove entrapped gas from the weld zone of the base metal. The skin or high silicon surface should also be removed adjacent to the weld area on both the face and root side. The edges of a joint should be chipped out or ground to form a 60° angle or bevel. Where grooves are involved, a V groove from a 60-90° included angle should be used. The V should extend approximately 1/8 in. (3.2 mm) from the bottom of the crack. A small hole should be drilled at each end of the crack to keep it from spreading. Complete penetration welds should always be used, since a crack or defect not completely removed may quickly reappear under service conditions.

7-16.   CAST IRON (cont)

(2) Preheating is desirable for welding cast irons with any of the welding processes. It can be reduced when using extremely ductile filler metal. Preheating will reduce the thermal gradient between the weld and the remainder of the cast iron. Preheat temperatures should be related to the welding process, the filler metal type, the mass, and the complexity of the casting.Preheating can be done by any of the normal methods. Torch heating is normally used for relatively small castings weighing 30.0 lb (13.6 kg) or less. Larger parts may be furnace preheated, and in some cases, temporary furnaces are built around the part rather than taking the part to a furnace. In this way, the parts can be maintained at a high interpass temperature in the temporary furnace during welding. Preheating should be general, since it helps to improve the ductility of the material and will spread shrinkage stresses over a large area to avoid critical stresses at any one point. Preheating tends to help soften the area adjacent to the weld; it assists in degassing the casting, and this in turn reduces the possibility of porosity of the deposited weld metal; and it increases welding speed.

(3) Slow cooling or postheating improves the machinability of the heat-affected zone in the cast ironadjacent to the weld. The post cooling should be as slow as possible. This can bedone by covering the casting with insulating materials to keep the air or breezesfrom it.

f. Welding Technique.

(1) Electrodes.

(a) Cast iron can be welded with a coated steel electrode, but this method should be used as an emergency measure only.When using a steel electrode, the contraction of the steel weld metal, the carbon picked up from the cast iron by the weld metal, and the hardness of the weld metal caused by rapid cooling must be considered. Steel shrinks more than cast iron when ceded from a molten to a solid state. When a steel electrode is used, this uneven shrinkage will cause strains at the joint after welding. When a large quantity of filler metal is applied to the joint, the cast iron may crack just back of the line of fusion unless preventive steps are taken. To overcome these difficulties, the prepared joint should be welded by depositing the weld metal in short string beads, 0.75 to 1.0 in. long (19.0 to 25.4 mm). These are made intermittently and, in some cases, by the backstep and skip procedure. To avoid hard spots, the arc should be struck in the V, and not on the surface of the base metal.Each short length of weld metal applied to the joint should be lightly peened while hot with a small ball peen hammer, and allowed to cool before additional weld metal is applied. The peening action forges the metal and relieves the cooling strains.

(b) The electrodes used should be 1/8 in. (3.2 mm) in diameter to prevent excessive welding heat. Welding should be done with reverse polarity. Weaving of the electrode should be held to a minimum.Each weld metal deposit should be thoroughly cleaned before additional metal is added.

(c) Cast iron electrodes must be used where subsequent machining of the welded joint is required. Stainless steel electrodes are used when machining of the weld is not required. The procedure for making welds with these electrodes is the same as that outlined for welding with mild steel electrodes. Stainless steel electrodes provide excellent fusion between the filler and base metals. Great care must be taken to avoid cracking in the weld because stainless steel expands and contracts approximately 50 percent more than mild steel in equal changes of temperature.

(2) Arc Welding.

(a) The shielded metal arc welding process can be utilized for welding cast iron. There are four types of filler metals that may be used: cast iron covered electrodes; covered copper base alloy electrodes; covered nickel base alloy electrodes; and mild steel covered electrodes. There are reasons for using each of the different specific types of electrodes, which include the machinability of the deposit, the color match of the deposit, the strength of the deposit, and the ductility of the final weld.

(b) When arc welding with the cast iron electrodes (ECI), preheat to between 250 and 800 °F (121 and 425 °C), depending on the size and complexity of the casting and the need to machine the deposit and adjacent areas. The higher degree of heating, the easier it will be to machine the weld deposit. In general, it is best to use small-size electrodes and a relatively low current setting. A medium arc length should be used, and, if at all possible, welding should be done in the flat position. Wandering or skip welding procedure should be used, and peening will help reduce stresses and will minimize distortion. Slow cooling after welding is recommended. These electrodes provide an excellment color match cm gray iron. The strength of the weld will equal the strength of the base metal. There are two types of copper-base electrodes: the copper tin alloy and the copper aluminum types. The copper zinc alloys cannot be used for arc welding electrodes because of the low boiling temperature of zinc. Zinc will volatilize in arc and will cause weld metal porosity.

(c) When the copper base electrodes are used, a preheat of 250 to 400 °F (121 to 204 °C) is recommended. Small electrodes and low current should be used. The arc should be directed against the deposited metal or puddle to avoid penetration and mixing the base metal with the weld metal. Slow cooling is recommended after welding. The copper-base electrodes do not provide a good color match.

(d) There are three types of nickel electrodes used for welding cast iron. These electrodes can be used without preheat; however, heating to 100 °F (38 °C) is recommended. These electrodes can be used in all positions; however, the flat position is recommended. The welding slag should be removed between passes. The nickel and nickel iron deposits are extremely ductile and will not become brittle with the carbon pickup. The hardness of the heat-affected zone can be minimized by reducing penetration into the cast iron base metal. The technique mentioned above, playing the arc on the puddle rather than on the base metal, will help minimize dilution. Slow cooling and, if necessary, postheating will improve machinability of the heat-affected zone. The nickel-base electrodes do not provide a close color match.

(e) Copper nickel type electrodes cane in two grades. Either of these electrodes can be used in the sames manner as the nickel or nickel iron electrode with about the same technique and results. The deposits of these electrodes do not provide a color match.

7-16.   CAST IRON (cont)

(f) Mild steel electrodes are not recommded for welding cast iron if the deposit is to be machined. The mild steel deposit will pick up sufficient carbon to make a high-carbon deposit, which is impossible to machine. Additionally, the mild steel deposit will have a reduced level of ductility as a result of increased carbon content. This type of electrode should be used only for small repairs and should not be used when machining is required.Minimum preheat is possible for small repair jobs. Small electrodes at low current are recommded to minimize dilution and to avoid the concentration of shrinkage stresses. Short welds using a wandering sequence should be used, and the weld should be peened as quickly as possible after welding. The mild steel electrode deposit provides a fair color match.

(3) Carbon-arc welding of cast iron. Iron castings may be welded with a carbon arc, a cast iron rod, and a cast iron welding flux.The joint should be preheated by moving the carbon electrodes along the surface. This prevents too-rapid cooling after welding. The molten puddle of metal can be worked with the carbon electrode so as to move any slag or oxides that are formed to the surface. Welds made with the carbon arc cool more slowly and are not as hard as those made with the metal arc and a cast iron electrode.The welds are machinable.

(4) Oxyfuel gas welding. The oxyfuel gas process is often used for welding cast iron. Most of the fuel gases can be used. The flame should be neutral to slightly reducing. Flux should be used. Two types of filler metals are available: the cast iron rods and the copper zinc rods. Welds made with the proper cast iron electrode will be as strong as the base metal.Good color match is provided by all of these welding reds. The optimum welding procedure should be used with regard to joint preparation, preheat, and post heat.The copper zinc rods produce braze welds. There are two classifications: a managanese bronze and a low-fuming bronze. The deposited bronze has relatively high ductility but will not provide a color match.

(5) Brazing and braze welding.

(a) Brazing is used for joining cast iron to cast iron and steels. In these cases, the joint design must be selected for brazing so that capillary attraction causes the filler metal to flow between closely fitting parts. The torch method is normally used. In addition, the carbon arc, the twin carbon arc, the gas tungsten arc, and the plasma arc can all be used as sources of heat. Two brazing filler metal alloys are normally used; both are copper zinc alloys. Braze welding can also be used to join cast iron. In braze welding, the filler metal is not drawn into the joint by capillary attraction. This is sometimes called bronze welding. The filler material having a liquidous above 850°F (454 °C) should be used. Braze welding will not provide a color match.

(b) Braze welding can also be acccmpolished by the shielded metal arc and the gas metal arc welding processes. High temperature preheating is not usually required for braze welding unless the part is extremely heavy or complex in geometry. The bronze weld metal deposit has extremely high ductility, which compensates for the lack of ductility of the cast iron. The heat of the arc is sufficient to bring the surface of the cast iron up to a temperature at which the copper base filler metal alloy will make a bond to the cast iron. Since there is little or no intermixing of the materials,the zone adjacent to the weld in the base metal is not appreciably hardened. The weld and adjacent area are machinable after the weld is completed. In general, a 200 °F (93 °C) preheat is sufficient for most application. The cooling rate is not extremely critical and a stress relief heat treatment is not usually required. This type of welding is commonly used for repair welding of automotive parts, agricultural implement parts, and even automotive engine blocks and heads. It can only be used when the absence of color match is not objectionable.

(6) Gas metal arc welding. The gas metal arc welding process can be used for making welds between malleable iron and carbon steels. Several types of electrode wires can be used, including:

(a) Mild steel using 75% argon + 25% CO_2 or shielding.
(b) Nickel copper using 100% argon for shielding.
(c) Silicon bronze using 50% argon + 50% helium for shielding.

In all cases, small diameter electrode wire should be used at low current. With the mild steel electrode wire, the Argon-CO_2 shielding gas mixture issued to minimize penetration. In the case of the nickel base filler metal and the Copper base filler metal, the deposited filler metal is extremely ductile. The mild steel provides a fair color match. A higher preheat is usually required to reduce residual stresses and cracking tendencies.

(7) Flux-cored arc welding. This process has recently been used for welding cast irons. The more successful application has been using a nickel base flux-cored wire. This electrode wire is normally operated with CO_2 shielding gas, but when lower mechanical properties are not objectionable, can be operated without external shielding gas. The minimum preheat temperatures can be used. The technique should minimize penetration into the cast iron base metal. Postheating is normally not required. A color match is not obtained.

(8) Studding. Cracks in large castings are somtimes repaired by studding (fig. 7-10). In this process, the fracture is removed by grinding a V groove. Holes are drilled and tapped at an angle on each side of the groove, and studs are screwed into these holes for a distance equal to the diameter of the studs, with the upper ends projecting approximately 1/4 in(6.4 mm) above the cast iron surface. The studs should be seal welded in place by one or two beads around each stud, and then tied together by weld metal beads. Welds should be made in short lengths, and each length peened while hot to prevent high stresses or cracking upon cooling. Each bead should be allowed to cool and be thoroughly cleaned before additional metal is deposited. If the studding method cannot be applied, the edges of the joint should be chipped out or machined with a round-nosed tool to form a U groove into which the weld metal should be deposited.

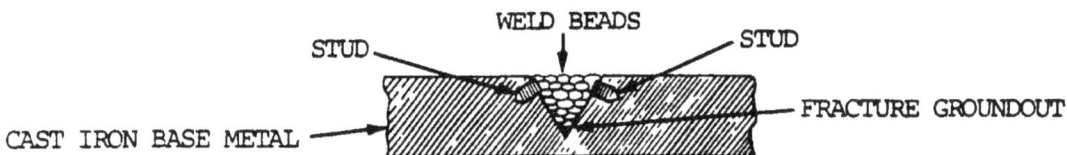

Figure 7-10. Studding method for cast iron repair.

(9) Other welding processes can be used for cast iron. Thermit welding has been used for repairing certain types of cast iron machine tool parts. Soldering can be used for joining cast iron, and is sometimes used for repairing small defects in small castings. Flash welding can also be used for welding cast iron.

Section IV. GENERAL DESCRIPTION AND WELDABILITY OF NONFERROUS METALS

7-17. ALUMINUMWELDING

a. General. Aluminum is a lightweight, soft, low strength metal which can easily be cast, forged, machined, formed and welded. Unless alloyed with specific elements, it is suitable only in low temperature applications. Aluminum is light gray to silver in color, very bright when polished, and dull when oxidized. A fracture in aluminum sections shins a smooth, bright structure. Aluminum gives off no sparks in a spark test, and does not show red prior to melting. A heavy film of white oxide forms instantly on the molten surface. Its combination of light weight and high strength make aluminum the second most popular metal that is welded. Aluminum and aluminum alloys can be satisfactorily welded by metal-arc, carbon-arc, and other arc welding processes. The principal advantage of using arc welding processes is that a highly concentrated heating zone is obtained with the arc. For this reason, excessive expansion and distortion of the metal are prevented.

b. Alloys. Many alloys of aluminum have been developed. It is important to know which alloy is to be welded. A system of four-digit numbers has been developed by the Aluminum Association, Inc., to designate the various wrought aluminum alloy types. This system of alloy groups, shown by table 7-20, is as follows:

(1) 1XXX series. These are aluminums of 99 percent or higher purity which are used primarily in the electrical and chemical industries.

(2) 2XXX series. Copper is the principal alloy in this group, which provides extremely high strength when properly heat treated. These alloys do not produce as good corrosion resistance and are often clad with pure aluminum or special-alloy aluminum. These alloys are used in the aircraft industry.

(3) 3XXX series. Manganese is the major alloying element in this group, which is non-heat-treatable. Manganese content is limited to about 1.5 percent. These alloys have moderate strength and are easily worked.

(4) 4XXX series. Silicon is the major alloying element in this group. It can be added in sufficient quantities to substantially reduce the melting point and is used for brazing alloys and welding electrodes. Most of the alloys in this group are non-heat-treatable.

(5) 5XXX series. Magnesium is the major alloying element of this group, which are alloys of medium strength. They possess good welding characteristics and good resistance to corrosion, but the amount of cold work should be limited.

(6). 6XXX series. Alloys in this group contain silicon and magnesium, which make them heat treatable. These alloys possess medium strength and good corrosion resistance.

(7) 7XXX series. Zinc is the major alloying element in this group. Magnesium is also included in most of these alloys. Together, they form a heat-treatable alloy of very high strength, which is used for aircraft frames.

Table 7-20. Designation of Aluminum Alloy Groups

Designation	Major Alloying Element
1xxx	99.0% minimum aluminum and over
2xxx	Copper
3xxx	Manganese
4xxx	Silicon
5xxx	Magnesium
6xxx	Magnesium and silicon
7xxx	Zinc
8xxx	Other element

c. Welding Aluminum Alloys. Aluminum posesses a number of properties that make welding it different than the welding of steels. These are: aluminum oxide surface coating; high thermal conductivity; high thermal expansion coefficient; low melting temperature; and the absence of color change as temperature approaches the melting point. The normal metallurgical factors that apply to other metals apply to aluminum as well.

(1) Aluminum is an active metal which reacts with oxygen in the air to produce a hard, thin film of aluminum oxide on the surface. The melting point of aluminum oxide is appoximately 3600 °F (1982 °C) which is almost three times the melting point of pure aluminum (1220F (660 °C)). In addition, this aluminum oxide film absorbs moisture from the air, particularly as it becomes thicker. Moisture is a source of hydrogen, which causes porosity in aluminum welds. Hydrogen may also come from oil, paint, and dirt in the weld area. It also comes from the oxide and foreign materials on the electrode or filler wire, as well as from the base metal. Hydrogen will enter the weld pool and is soluble in molten aluminum. As the aluminum solidifies, it will retain much less hydrogen. The hydrogen is rejected during solidification. With a rapid cooling rate, free hydrogen is retained within the weld and will cause porosity. Porosity will decrease weld strength and ductility, depending on the amount.

CAUTION
Aluminum and aluminum alloys should not be cleaned with caustic soda or cleaners with a pH above 10 as they may react chemically.

(a) The aluminum oxide film must be removed prior to welding. If it is not completely removed, small particles of unmelted oxide will be trapped in the weld pool and will cause a reduction in ductility, lack of fusion, and possibly weld cracking.

(b) The aluminum oxide can be removed by mechanical, chemical, or electrical means. Mechanical removal involves scrapting with a sharp tool, sandpaper, wire brush (stainless steel), filing, or any other mechanical method. Chemical removal can be done in two ways. One is by use of cleaning solutions, either the etching types or the nonetching types. The nonetching types should be used only when starting with relatively clean parts, and are used in conjunction with other solvent cleaners. For better cleaning, the etching type solutions are recommended, but must be used with care. When dipping is employed, hot and cold rinsing is highly recommended. The etching type solutions are alkaline solutions. The time in the solution must be controlled so that too much etching does not occur.

7-17. ALUMINUM WELDING (cont)

(c) Chemical cleaning includes the use of welding fluxes. Fluxes are used for gas welding, brazing, and soldering. The coating on covered aluminum electrodes also maintains fluxes for cleaning the base metal. Whenever etch cleaning or flux cleaning is used, the flux and alkaline etching materials must be completely removed from the weld area to avoid future corrosion.

(d) The electrical oxide removal system uses cathodic bombardment. Cathodic bombardment occurs during the half cycle of alternating current gas tungsten arc welding when the electrode is positive (reverse polarity). This is an electrical phenomenon that actually blasts away the oxide coating to produce a clean surface. This is one of the reasons why AC gas tungsten arc welding is so popular for welding aluminum.

(e) Since aluminum is so active chemically, the oxide film will immediately start to reform. The time of buildup is not extremely fast, but welds should be made after aluminum is cleaned within at least 8 hours for quality welding. If a longer time period occurs, the quality of the weld will decrease.

(2) Aluminum has a high thermal conductivity and low melting temperature. It conducts heat three to five times as fast as steel, depending on the specific alloy. More heat must be put into the aluminum, even though the melting temperature of aluminum is less than half that of steel. Because of the high thermal conductivity, preheat is often used for welding thicker sections. If the temperature is too high or the time period is too long, weld joint strength in both heat-treated and work-hardend alloys may be diminished. The preheat for aluminum should not exceed 400 °F (204 °C), and the parts should not be held at that temperature longer than necessary. Because of the high heat conductivity, procedures should utilize higher speed welding processes using high heat input. Both the gas tungsten arc and the gas metal arc processes supply this requirement. The high heat conductivity of aluminum can be helpful, since the weld will solidify very quickly if heat is conducted away from the weld extremely fast. Along with surface tension, this helps hold the weld metal in position and makes all-position welding with gas tungsten arc and gas metal arc welding practical.

(3) The thermal expansion of aluminum is twice that of steel. In addition, aluminum welds decrease about 6 percent in volume when solidifying from the molten state. This change in dimension may cause distortion and cracking.

(4) The final reason aluminum is different from steels when welding is that it does not exhibit color as it approaches its melting temperature until it is raised above the melting point, at which time it will glow a dull red. When soldering or brazing aluminum with a torch, flux is used. The flux will melt as the temperature of the base metal approaches the temperature required. The flux dries out first, and melts as the base metal reaches the correct working temperature. When torch welding with oxyacetylene or oxyhydrogen, the surface of the base metal will melt first and assume a characteristic wet and shiny appearance. (This aids in knowing when welding temperatures are reached.) When welding with gas tungsten arc or gas metal arc color is not as important, because the weld is completed before the adjoining area melts.

d. Metal-Arc Welding of Aluminum.

(1) Plate welding. Because of the difficulty of controlling the arc, butt and fillet welds are difficult to produce in plates less than 1/8 in. (3.2 mm) thick. when welding plate heavier than 1/8 in. (3.2 mm), a joint prepared with a 20 degree bevel will have strength equal to a weld made by the oxyacetylene process. This weld may be porous and unsuitable for liquid- or gas-tight joints. Metal-arc welding is, however, particularly suitable for heavy material and is used on plates up to 2-1/2 in. (63.5 mm) thick.

(2) Current and polarity settings. The current and polarity settings will vary with each manufacturer's type of electrodes.The polarity to be used should be determined by trial on the joints to be made.

(3) Plate edge preparation. In general, the design of welded joints for aluminum is quite consistent with that for steel joints.However, because of the higher fluidity of aluminum under the welding arc, some important general principles should be kept in mind. With the lighter gauges of aluminum sheet, less groove spacing is advantageous when weld dilution is not a factor. The controlling factor is joint preparation. A specially designed V groove that is applicable to aluminum is shown in A, figure 7-11. This type of joint is excellent where welding can be done from one side only and where a smooth, penetrating bead is desired. The effectiveness of this particular design depends upon surface tension, and should be applied on all material over 1/8 in. (3.2 mm) thick. The bottom of the special V groove must be wide enough to contain the root pass completely. This requires adding a relatively large amount of filler alloy to fill the groove. Excellent control of the penetration and sound root pass welds are obtained. This edge preparation can be employed for welding in all positions. It eliminates difficulties due to burn-through or over-penetration in the overheat and horizontal welding positions. It is applicable to all weldable base alloys and all filler alloys.

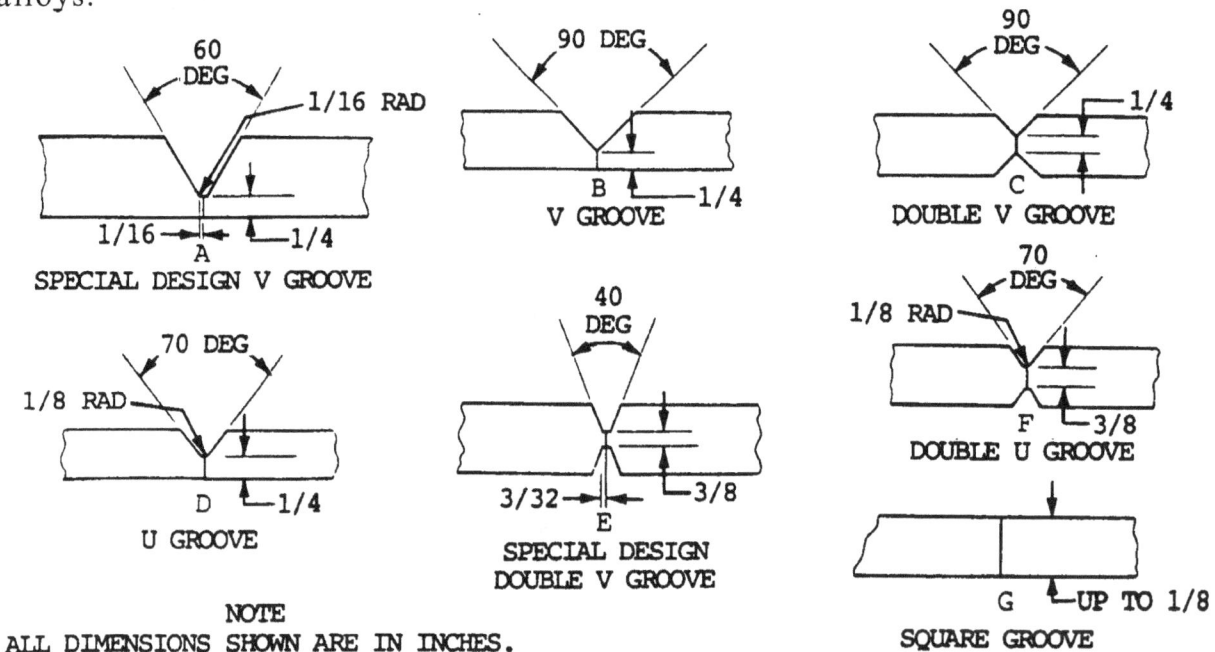

Figure 7-11. Joint design for aluminum plates.

7-17.   ALUMINUM WELDING (cont)

  e.   Gas Metal-Arc (MIG) Welding (GMAW).

      (1) General. This fast, adaptable process is used with direct current re-
verse polarity and an inert gas to weld heavier thicknesses of aluminum alloys, in
any position, from 1/16 in. (1.6 mm) to several inches thick. TM 5-3431-211-15
describes the operation of a typical MIG welding set.

      (2) Shielding gas. Precautions should be taken to ensure the gas shield is
extremely efficient. Welding grade argon, helium, or a mixture of these gases is
used for aluminum welding. Argon produces a smother and more stable arc than
helium. At a specific current and arc length, helium provides deeper penetration
and a hotter arc than argon.Arc voltage is higher with helium, and a given change
in arc length results in a greater change in arc voltage.The bead profile and
penetration pattern of aluminum welds made with argon and helium differ. With
argon, the bead profile is narrower and more convex than helium. The penetration
pattern shows a deep central section.Helium results in a flatter, wider bead, and
has a broader under-bead penetration patternA mixture of approximately 75 percent helium and 25 percent argon provides the advantages of both shielding gases
with none of the undesirable characteristics of either.Penetration pattern and
bead contour show the characteristics of both gasesArc stability is comparable
to argon. The angle of the gun or torch is more critical when welding aluminum
with inert shielding gas. A 30° leading travel angle is recommeded . The electrode wire tip should be oversize for aluminum.Table 7-21 provides welding procedure schedules for gas metal-arc welding of aluminum.

      (3) Welding technique. The electrode wire must be clean. The arc is struck
with the electrode wire protruding about 1/2 in. (12.7 mm) from the cup. A frequently used technique is to strike the arc approximately 1.0 in. (25.4 mm) ahead
of the beginning of the weld and then quickly bring the arc to the weld starting
point, reverse the direction of travel, and proceed with normal welding. Alterna-
atively, the arc may be struck outside the weld groove on a starting tab. When
finishing or terminatingweld, a similar practice may be followed by reversin
the direction of welding, and simultaneously increasing the speed of welding to
taper the width Of the molten pool prior to breaking the arc. This helps to avert
craters and crater cracking. Runoff tabs are commonly used. Having established
the arc, the welder moves the electrode along the joint while maintaining a 70 to
85 degree forehand angle relative to the work.A string bead technique is normally
preferred. Care should be taken that the forehand angle is not changed or increased as the end of the weld is approached.Arc travel speed controls the bead
size. When welding aluminum with this process, it is must important that high travel speeds be maintained. When welding uniform thicknesses, the electrode to work
angle should be equal on both sides of the weld.When welding in the horizontal
position, best results are obtained by pointing the gun slightly upward. When
welding thick plates to thin plates, it is helpful to direct the arc toward the
heavier section. A slight backhand angle is sometimes helpful when welding thin
sections to thick sections. The root pass of a joint usually requires a short arc
to provide the desired penetration.Slightly longer arcs and higher arc voltages
may be used on subsequent passes.

Table 7-21. Welding Procedure Schedules for Gas Metal-Arc Welding (GMAW) of Aluminum (MIG Welding)

Material Thickness (or Fillet Size)			Type of Weld Fillet or Groove	Electrode Diameter		WELDING POWER		Wire Feed Speed ipm	Shielding Gas Flow cfh	No. of Passes	Travel Speed (per pass) ipm
ga	in.	mm		in.	mm	Current Amps DC	Arc Volt EP				
—	0.050	—	Sq. groove & fillet	0.030	0.8	50	12-14	268-308	30	1	17-25
—	0.062	1.6	Sq. groove & fillet	0.030	0.8	55-60	12-14	295-320	30	1	17-25
—	0.062	1.6	Sq. groove & fillet	3/64	1.2	110-125	19-21	175-185	30	1	20-27
—	0.093	2.4	Sq. groove & fillet	0.030	0.8	90-100	14-18	330-370	30	1	24-36
—	0.125	3.2	Fillet	0.030	0.8	110-125	19-22	410-460	30	1	20-24
11	0.125	3.2	Sq. groove	3/64	1.2	110-125	20-24	175-190	40	1	20-24
3/16	0.187	4.7	Sq. groove & fillet	3/64	1.2	160-195	20-24	215-225	40	1	20-25
1/4	0.250	6.4	Fillet	3/64	1.2	160-195	20-24	215-225	40	1	20-25
1/4	0.250	6.4	Vee groove	1/16	1.6	175-225	22-26	150-195	40	3	20-25
3/8	0.375	9.5	Vee groove & fillet	1/16	1.6	200-300	22-26	170-275	40	2-5	25-30
1/2	0.500	12.7	Vee groove & fillet	1/16	1.6	220-230	22-27	195-205	40	3-8	12-18
1/2	0.500	12.7	Double vee groove	3/32	2.4	320-340	22-29	140-150	45	2-5	15-17
3/4	0.750	19.0	Double vee groove	1/16	1.6	255-275	22-27	230-250	50	4-10	8-18
3/4	0.750	19.0	Double vee groove	3/32	2.4	355-375	22-29	155-160	50	4-10	14-16
1	1.000	25.4	Double vee groove	1/16	1.6	255-290	22-27	230-265	50	4-14	6-18
1	1.000	25.4	Double vee groove	3/32	2.4	405-425	22-27	175-180	50	4-8	8-12

NOTE

For groove and fillet welds—material thickness also indicates fillet weld size. Use vee groove for 3/16" and thicker. Use argon for thin and medium material; use 50% argon and 50% helium for thick material. Increase gas flow rate 10% for overhead position. Increase amperage 10-20% when backup is used. Decrease amperage 10-20% when welding out of position.

7-17. ALUMINUM WELDING (cont)

The wire feeding equipment for aluminum welding must be in good adjustment for efficient wire feeding. Use nylon type liners in cable assemblies. Proper drive rolls must be selected for the aluminum wire and for the size of the electrode wire. It is more difficult to push extremely small diameter aluminum wires through long gun cable assemblies than steel wires. For this reason, the spool gun or the newly developed guns which contain a linear feed motor are used for the small diameter electrode wires. Water-cooled guns are required except for low-current welding. Both the constant current (CC) power source with matching voltage sensing wire feeder and the constant voltage (CV) power source with constant speed wire feeder are used for welding aluminum. In addition, the constant speed wire feeder is somtimes used with the constant current power source. In general, the CV system is preferred when welding on thin material and using all diameter electrode wire. It provides better arc starting and regulation. The CC system is preferred when welding thick material using linger electrode wires. The weld quality seems better with this system. The constant current power source with a moderate drop of 15 to 20 volts per 100 amperes and a constant speed wire feeder provide the most stable power input to the weld and the highest weld quality.

(4) Joint design. Edges may be prepare for welding by sawing, machining, rotary planing, routing or arc cutting. Acceptable joint designs are shown in figure 7-12.

f. Gas Tungsten-Arc (TIG) Welding (GTAW).

(1) The gas tungsten arc welding process is used for welding the thinner sections of aluminum and aluminum alloys. There are several precautions that should be mentioned with respect to using this process.

(a) Alternating current is recommeded for general-purpose work since it provides the half-cycle of cleaning action. Table 7-22, p 7-74, provides welding procedure schedules for using the process on different thicknesses to produce different welds. AC welding, usually with high frequency, is widely used with manual and automatic applications. Procedures should be followed closely and special attention given to the type of tungsten electrode size of welding nozzle, gas type, and gas flow rates. When manual welding, the arc length should be kept short and equal to the diameter of the electrode. The tungsten electrode should not protrude too far beyond the end of the nozzle. The tungsten electrode should be kept clean. If it does accidentally touch the molten metal, it must be redressed.

(b) Welding power sources designed for the gas tungsten arc welding process should be used. The newer equipment provides for programing, pre- and post-flow of shielding gas, and pulsing.

(c) For automatic or machine welding, direct current electrode negative (straight polarity) can be used. Cleaning must be extremely efficient, since there is no cathodic bombardment to assist. When dc electrode negative is used, extremely deep penetration and high speeds can be obtained. Table 7-23, p 7-75 lists welding procedure schedules for dc electrode negative welding.

(d) The shielding gases are argon, helium, or a mixture of the two. Argon is used at a lower flow rate. Helium increases penetration, but a higher flow rate is required. When filler wire is used, it must be clean. Oxide not removed from the filler wire may include moisture that will produce porosity the weld deposit.

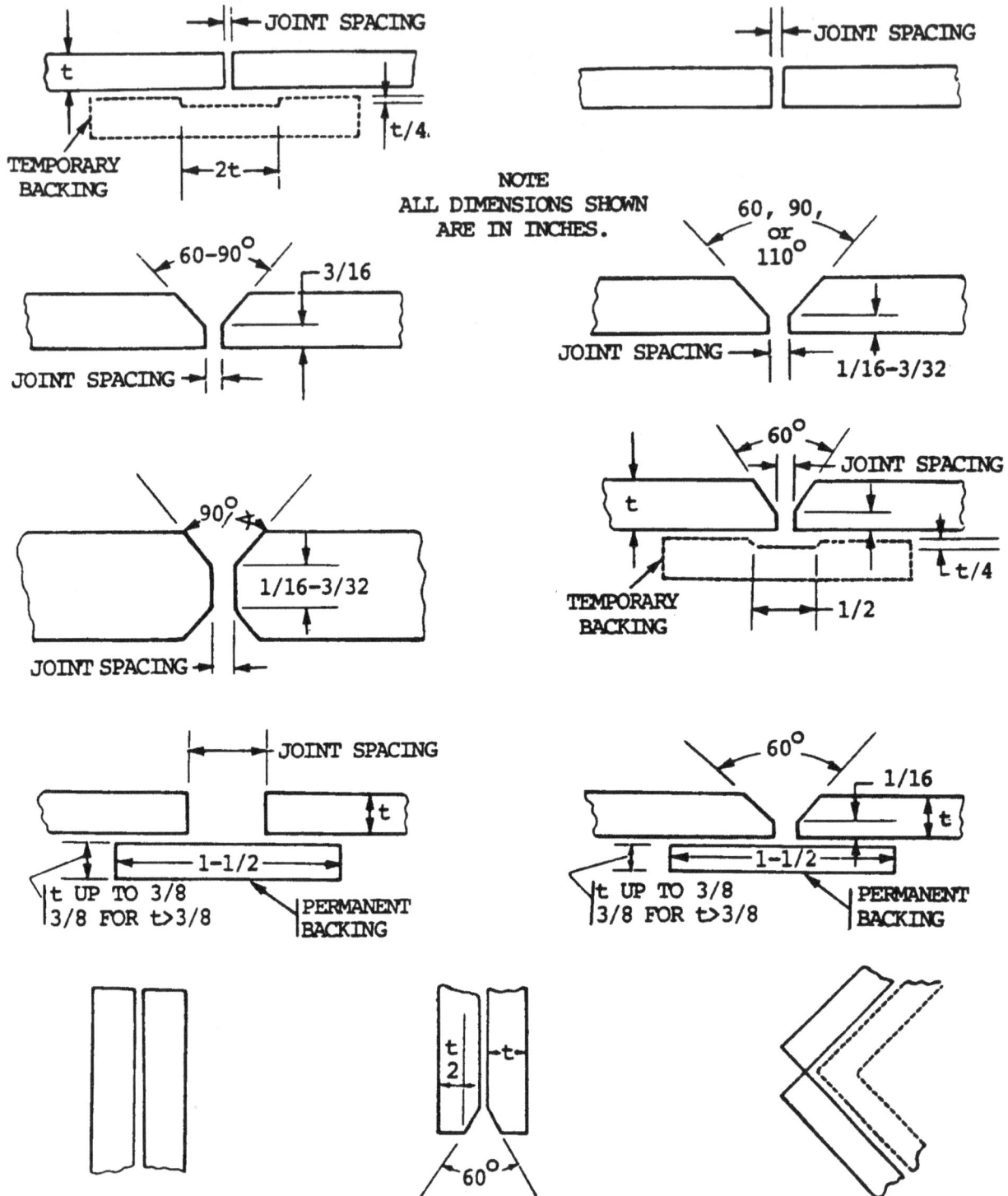

Figure 7-12. Aluminum joint designs for gas metal-arc welding processes.

7-17. ALUMINUM WELDING (cont)

Table 7-22. Welding Procedure Schedules for AC-GTAW Welding of Aluminum (TIG Welding)

| Material Thickness (or Fillet Size) | | | Type of Weld Fillet or Groove | Tungsten Electrode Diameter | | Filler Rod Diameter | | Nozzle Size Inside Dia. in. | Shielding Gas Flow cfh | Welding Current Amps AC | No. of Passes | Travel Speed (per pass) ipm |
ga	in.	mm		in.	mm	in.	mm					
3/64	0.046	1.2	Sq. Groove & Fillet	1/16	1.6	1/16	1.6	1/4-3/8	20	40-60	1	14-18
1/16	0.063	1.6	Sq. Groove & Fillet	3/32	2.4	3/32	2.4	5/16-3/8	20	70-90	1	8-12
3/32	0.094	2.4	Sq. Groove & Fillet	3/32	2.4	3/32	2.4	5/16-3/8	20	95-115	1	10-12
1/8	0.125	3.2	Sq. Groove & Fillet	1/8	3.2	1/8	3.2	3/8	20	120-140	1	9-12
3/16	0.187	4.7	Fillet	5/32	3.9	5/32	3.9	7/16-1/2	25	160-200	1	9-12
3/16	0.187	4.7	Vee Groove	5/32	3.9	5/32	3.9	7/16-1/2	25	160-180	2	10-12
1/4	0.250	6.4	Fillet	3/16	4.8	3/16	4.8	7/16-1/2	30	230-250	1	8-11
1/4	0.250	6.4	Vee Groove	3/16	4.8	3/16	4.8	7/16-1/2	30	200-220	2	8-11
3/8	0.375	9.5	Vee Groove	3/16	4.8	3/16	4.8	1/2	35	250-310	2-3	9-11
1/2	0.500	12.7	Vee or U Groove	1/4	6.4	1/4	6.4	5/8	35	400-470	3-4	6

NOTE

Increase amperage when backup is used. Data is for all welding positions. Use low side of range for out of position welding. For tungsten electrodes-- 1st choice--pure tungsten EWP; 2nd choice-- zirconated EWZr. Normally argon is used for shielding, however, mixtures of 10% or more helium with argon are sometimes used for increased penetration in aluminum 1/4 in. (64 mm) thick and over. The gas flow should be increased when helium is added. A mixture of 75% He + 25% argon is popular. When 100% helium is used, gas flow rates are about twice those used for argon.

Table 7-23. Welding Procedure Schedules for DC-GTAW Welding of Aluminum (TIG Welding)

| Material Thickness (or Fillet Size) | | | Type of Weld Fillet or Groove | Tungsten Electrode Diameter | | Filler Rod Diameter | | Nozzle Size Inside Dia. | Shielding Gas Flow | Welding Current Amps | No. of Passes | Travel Speed (per pass) |
ga	in.	mm		in.	mm	in.	mm	in.	cfh	DCEN		ipm
20	0.032	0.8	Sq. groove & fillet	3/32	2.4	None		3/8	30	65-70	1	52
18	0.046	1.2	Sq. groove & fillet	3/64	1.2	3/64	1.2	3/8	30	35-95	1	45
16	0.063	1.6	Sq. groove & fillet	3/64	1.2	3/64	1.2	3/8	30	45-120	1	36
13	0.094	2.4	Sq. groove & fillet	1/16	1.6	1/16	1.6	3/8	30	90-185	1	32
11	1/8	3.2	Sq. groove & fillet	1/8	3.2	1/8	3.2	3/8	30	120-220	1	20
11	1/8	3.2	Sq. groove & fillet	1/8	3.2	None		3/8	30	180-200	1	24
--	1/4	6.4	Sq. groove & fillet	1/8	3.2	1/8	3.2	1/2	40	230-340	1	22
--	1/4	6.4	Sq. groove & fillet	1/8	3.2	None		1/2	40	220-240	1	22
--	1/2	12.7	Vee groove	3/16	4.8	1/8	3.2	1/2	40	300-450	1	20
--	1/2	12.7	Sq. groove	5/32	3.9	None		1/2	40	260-300	2	20
--	3/4	19.1	Vee groove	3/16	4.8	1/8	3.2	1/2	40	300-450	2	6
--	3/4	19.1	Sq. groove	3/16	4.8	None		1/2	40	450-470	2	6
--	1	25.4	Vee groove	3/16	4.8	1/8	3.2	5/8	40	300-450	2	5

NOTE

Normally for automatic travel. Use Helium or 75% helium 25% argon.

7-17. ALUMINUM WELDING (cont)

(2) Alternating current.

(a) Characteristics of process. The welding of aluminum by the gas tung-sten-arc welding process using alternating current produces an oxide cleaning action. Argon shielding gas is used. Better results are obtained when welding aluminum with alternating current by using equipment designed to produce a balanced wave or equal current in both directions.Unbalance will result in loss of power and a reduction in the cleaning action of the arc.Characteristics of a stable arc are the absence of snapping or cracking, smooth arc starting, and attraction of added filler metal to the weld puddle rather than a tendency to repulsion. A stable arc results in fewer tungsten inclusions.

(b) Welding technique. For manual welding of aluminum with ac, the electrode holder is held in one hand and filler rod, if used, in the other. An initial arc is struck on a starting block to heat the electrodeThe arc is then broken and reignited in the joint. This technique reduces the tendency for tungsten inclusions at the start of the weld. The arc is held at the starting point until the metal liquifies and a weld pool is established. The establishment and maintenance of a suitable weld pool is important, and welding must not proceed ahead of the puddle. If filler metal is required, it may be added to the front or leading edge of the pool but to one side of the center line.Both hands are moved in unison with a slight backward and forward motion along the jointThe tungsten electrode should not touch the filler rod. The hot end of the filler rod should not be withdrawn from the argon shield. A short arc length must be maintained to obtain sufficient penetration and avoid undercutting, excessive width of the weld bead, and consequent loss of penetration control and weld contourOne rule is to use an arc length approximately equal to the diameter of the tungsten electrode. When the arc is broken, shrinkage cracks may occur in the weld crater, resulting in a defective weld. This defect can be prevented by gradually lengthening the arc while adding filler metal to the crater. Then, quickly break and restrike the arc several times while adding additional filler metal to the crater, or use a foot control to reduce the current at the end of the weld. Tacking before welding is helpful in controlling distortion. Tack welds should be of ample size and strength and should be chipped out or tapered at the ends before welding over.

(c) Joint design. The joint designs shown in figure 7-11, p 7-69 are applicable to the gas tungsten-arc welding process with minor exceptionsInexperienced welders who cannot maintain a very short arc may require a wider edge preparation, included angle, or joint spacing. Joints may be fused with this process without the addition of filler metal if the base metal alloy also makes a satisfactory filler alloy. Edge and corner welds are rapidly made without addition of filler metal and have a good appearance, but a very close fit is essential.

(3) Direct current straigh polarity.

(a) Charcteristics of process.This process, using helium and thoriated tungsten electrodes is advantageous for many automatic welding operations, especially in the welding of heavy sections. Since there is less tendency to heat the electrode, smiler electrodes can be used for a given welding current. This will contribute to keeping the weld bead narrow.The use of direct current straight polarity (dcsp) provides a greater heat input than can be obtained with ac current. Greater heat is developed in the weld pool, which is consequently deeper and narrower.

(b) Welding techniques. A high frequency current should be used to initiate the arc. Touch starting will contaminate the tungsten electrode. It is not necessary to form a puddle as in ac welding, since melting occurs the instant the arc is struck. Care should be taken to strike the arc within the weld area to prevent undesirable marking of the material. Standard techniques such as runoff tabs and foot operated heat controls are used. These are helpful in preventing or filling craters, for adjusting the current as the work heats, and to adjust for a change in section thickness. In dcsp welding, the torch is moved steadily forward. The filler wire is fed evenly into the leading edge of the weld puddle, or laid on the joint and melted as the arc roves forward. In all cases, the crater should be filled to a point above the weld bead to eliminate crater cracks. The fillet size can be controlled by varying filler wire size. DCSP is adaptable to repair work. Preheat is not required even for heavy sections, and the heat affected zone will be smaller with less distortion.

(c) Joint designs. The joint designs shown in figure 7-11, p 7-69, are applicable to the automatic gas tungsten-arc dcsp welding process with minor exceptions. For manual dcsp, the concentrated heat of the arc gives excellent root fusion. Root face can be thicker, grooves narrower, and build up can be easily controlled by varying filler wire size and travel speed.

g. Square Wave Alternating Current Welding (TIG).

(1) General. Square wave gas tungsten-arc welding with alternating current differs frozen conventional balanced wave gas tungsten-arc welding in the type of wave from used. With a square wave, the time of current flow in either direction is adjustable from 20 to 1. In square wave gas tungsten-arc welding, there are the advantages of surface cleaning produced by positive ion bombardment during the reversed polarity cycle, along with greater weld depth to width ratio produced by the straight polarity cycle. Sufficient aluminum surface cleaning action has been obtained with a setting of approximately 10 percent dcrp. Penetration equal to regular dcsp welding can be obtained with 90 percent dcsp current.

(2) Welding technique. It is necessary to have either superimposed high frequency or high open circuit voltage, because the arc is extinguished every half cycle as the current decays toward zero, and must be restarted each tire. Precision shaped thoriated tungsten electrodes should be used with this process. Argon, helium, or a combination of the two should be used as shielding gas, depending on the application to be used.

(3) Joint design. Square wave alternating current welding offers substantial savings over conventional alternating current balanced wave gas tungsten arc welding in weld joint preparation. Smaller V grooves, U grooves, and a thicker root face can be used. A greater depth to width weld ratio is conducive to less weldment distortion, along with favorable welding residual stress distribution and less use of filler wire. With some slight modification, the same joint designs can be used as in dcsp gas tungsten-arc welding (fig. 7-11, p 7-69).

h. Shielded Metal-Arc Welding. In the shielded metal-arc welding process, a heavy dipped or extruded flux coated electrode is used with dcrp. The electrodes are covered similarly to conventional steel electrodes. The flux coating provides a gaseous shield around the arc and molten aluminum puddle, and chemically combines and removes the aluminum oxide, forming a slag. When welding aluminum, the process is rather limited due to arc spatter, erratic arc control, limitations on thin material, and the corrosive action of the flux if it is not removed properly.

7-17. ALUMINUM WELDING (cont)

i. Shielded Carbon-Arc Welding. The shielded carbon-arc welding process can be used in joining aluminum. It requires flux and produces welds of the same appearance, soundness, and structure as those produced by either oxyacetylene or oxyhydrogen welding. Shielded carbon-arc welding is done both manually and automatically. A carbon arc is used as a source of heat while filler metal is supplied from a separate filler rod. Flux must be removed after welding; otherwise severe corrosion will result. Manual shielded carbon-arc welding is usually limited to a thickness of less than 3/8 in. (9.5 mm), accomplished by the same method used for manual carbon arc welding of other material.Joint preparation is similar to that used for gas welding. A flux covered rod is used.

j. Atomic Hydrogen Welding. This welding process consists of maintaining an arc between two tungsten electrodes in an atmosphere of hydrogen gas.The process can be either manual or automatic with procedures and techniques closely related to those used in oxyacetylene welding. Since the hydrogen shield surrounding the base metal excludes oxygen, smaller amounts of flux are required to combine or remove aluminum oxide. Visibility is increased, there are fewer flux inclusions, and a very sound metal is deposited.

k. Stud Welding.

(1) Aluminum stud welding may be accomplished with conventional arc stud welding equipment, using either the capacitor discharge or drawn arc capacitor discharge techniques. The conventional arc stud welding process may be used to weld aluminum studs 3/16 to 3/4 in. (4.7 to 19.0 mm) diameter.The aluminum stud welding gun is modified slightly by the addition of a special adapter for the control of the high purity shielding gases used during the welding cycle. An added accessory control for controlling the plunging of the stud at the completion of the weld cycle adds materially to the quality of weld and reduces spatter loss. Reverse polarity is used, with the electrode gun positive and the workpiece negative. A small cylindrical or cone shaped projection on the end of the aluminum stud initiates the arc and helps establish the longer arc length required for aluminum welding.

(2) The unshielded capacitor discharge or drawn arc capacitor discharge stud welding processes are used with aluminum studs 1/16 to 1/4 in. (1.6 to 6.4 mm) diameter. Capacitor discharge welding uses a low voltage electrostatic storage system, in which the weld energy is stored at a low voltage in capacitors with high capacitance as a power source. In the capacitor discharge stud welding process, a small tip or projection on the end of the stud is used for arc initiation. The drawn arc capacitor discharge stud welding process uses a stud with a pointed or slightly rounded end. It does not require a serrated tip or projection on the end of the stud for arc initiation. In both cases, the weld cycle is similar to the conventional stud welding process. However, use of the projection on the base of the stud provides the most consistent welding.The short arcing time of the capacitor discharge process limits the melting so that shallow penetration of the workpiece results. The minimum aluminum work thickness considered practical is 0.032 in. (0.800 mm).

l. Electron Beam Welding. Electron beam welding is a fusion joining process in which the workpiece is bombarded with a dense stream of high velocity electrons, and virtually all of the kinetic energy of the electrons is transformed into heat upon impact. Electron beam welding usually takes place in an evacuated chamber. The chamber size is the limiting factor on the weldment size. Conventional arc and gas heating melt little more than the surface.Further penetration comes solely by conduction of heat in all directions from this molten surface spot. The fusion zone widens as it depends. The electron beam is capable of such intense local heating that it almost instantly vaporizes a hole through the entire joint thickness. The walls of this hole are molten, and as the hole is moved along the joint, more metal on the advancing side of the hole is melted. This flaws around the bore of the hole and solidifies along the rear side of the hole to make the weld. The intensity of the beam can be diminished to give a partial penetration with the same narrow configuration. Electron beam welding is generally applicable to edge, butt, fillet, melt-thru lap, and spot welds. Filler metal is rarely used except for surfacing.

    m.   Resistance Welding.

      (1) General. The resistance welding processes (spot, seam, and flash welding) are important in fabricating aluminum alloysThese processes are especially useful in joining the high strength heat treatable alloys, which are difficult to join by fusion welding, but can be joined by the resistance welding process with practically no loss in strength. The natural oxide coating on aluminum has a rather high and erratic electrical resistance. To obtain spot or seam welds of the highest strength and consistency, it is usually necessary to reduce this oxide coating prior to welding.

      (2) Spot welding. Welds of uniformly high strength and good appearance depend upon a consistently low surface resistance between the workplaces. For most applications, some cleaning operations are necessary before spot or seam welding aluminum. Surface preparation for welding generally consists of removal of grease, oil, dirt, or identification markings, and reduction and improvement of consistency of the oxide film on the aluminum surface. Satisfactory performance of spot welds in service depends to a great extent upon joint design. Spot welds should always be designed to carry shear loads. However, when tension or combined loadings may be expected, special tests should be conducted to determine the actual strength of the joint under service loading. The strength of spot welds in direct tension may vary from 20 to 90 percent of the shear strength.

      (3) Seam welding. Seam welding of aluminum and its alloys is very similar to spot welding, except that the electrodes are replaced by wheels. The spots made by a seam welding machine can be overlapped to form a gas or liquid tight joint. By adjusting the timing, the seam welding machine can produce uniformly spaced spot welds equal in quality to those produced on a regular spot welding machine, and at a faster rate. This procedure is called roll spot or intermittent seam welding.

      (4) Flash welding. All aluminum alloys may be joined by the flash welding process. This process is particularly adapted to making butt or miter joints between two parts of similar cross section. It has been adapted to joining aluminum to copper in the form of bars and tubing.The joints so produced fail outside of the weld area when tension loads are applied.

7-17. ALUMINUM WELDING (cont)

n. <u>Gas welding</u>. Gas welding has been done on aluminum using both oxyacetylene and oxyhydrogen flames. In either case, an absolutely neutral flame is required. Flux is used as well as a filler rod. The process also is not too popular because of low heat input and the need to remove flux.

o. <u>Electroslag welding</u>. Electroslag welding is used for joining pure aluminum, but is not successful for welding the aluminum alloys. Submerged arc welding has been used in some countries where inert gas is not available.

p. <u>Other processes</u>. Most of the solid state welding processes, including friction welding, ultrasonic welding, and cold welding are used for aluminums. Aluminum can also be joined by soldering and brazing. Brazing can be accomplished by most brazing methods. A high silicon alloy filler material is used.

7-18. BRASS AND BRONZE WELDING

a. <u>General</u>. Brass and bronze are alloys of copper. Brass has zinc, and bronze has tin as the major alloying elements. However, some bronze metals contain more zinc than tin, and some contain zinc and no tin at all. High brasses contain from 20 to 45 percent zinc. Tensile strength, hardness and ductility increase as the percentage of zinc increases. These metals are suitable for both hot and cold working.

b. <u>Metal-Arc Welding</u>. Brasses and bronzes can be successfully welded by the metal-arc process. The electrode used should be of the shielded arc type with straight polarity (electrode positive). Brasses can be welded with phosphor bronze, aluminum bronze, or silicon bronze electrodes depending on the base metal composition and the service required. Backing plates of matching metal or copper should be used. High welding current should not be used for welding copper-zinc alloys (brasses), otherwise the zinc content will be volatilized. All welding should be done in the flat position. If possible, the weld metal should be deposited with a weave approximately three times the width of the electrode.

c. <u>Carbon-Arc Welding</u>. This method can be used to weld brasses and bronzes with filler reds of approximately the same composition as the base metal. In this process, welding is accomplished in much the same way the bronze is bonded to steel. The metal in the carbon arc is superheated, and this very hot metal is alloyed to the base metal in the joint.

d. <u>Oxyacetylene Welding</u>. The low brasses are readily jointed by oxyacetylene welding. This process is particularly suited for piping because it can be done in all welding positions. Silicon copper welding rods or one of the brass welding rods may be used. For oxyacetylene welding of the high brasses, low-fuming welding rods are used. These low-fuming rods have composition similar to many of the high brasses. A flux is required, and the torch flame should be adjusted to a slightly oxidizing flame to assist in controlling fuming. Preheating and an auxiliary heat source may also be necessary. The welding procedures for copper are also suitable for the brasses.

e. <u>Gas Metal Arc Welding</u>. Gas metal arc welding is recommended for joining large phosphor bronze fabrications and thick sections. Direct current, electrode positive, and argon shielding are normally used.The molten weld pool should be kept small and the travel speed rather high.Stringer beads should be used. Hot peening of each layer will reduce welding stresses and the likelihood of cracking.

f. <u>Gas Tungsten Arc Welding</u>. Gas tungsten arc welding is used primarily for repair of castings and joining of phosphor bronze sheet. As with gas metal arc welding, hot peening of each layer of weld metal is beneficial. Either stabilized ac or direct current, electrode negative can be used with helium or argon shielding. The metal should be preheated to the 350 to 400°F (177 to 204 °C) range, and the travel speed should be as fast as practical.

g. <u>Shielded Metal Arc Welding</u>. Phosphor bronze covered electrodes are available for joining bronzes of similar compositions.These electrodes are designed for use with direct current, electrode positive.Filler metal should be deposited as stringer beads for best weld joint mechanical properties. Postweld annealing at 900 °F (482 °C) is not always necessary, but is desirable for maximum ductility, particularly if the weld metal is to be cold worked. Moisture, both on the work and in the electrode coverings, must be strictly avoided. Baking the electrodes at 250 to 300 °F (121 to 149 °C) before use may be necessary to reduce moisture in the covering to an acceptable level.

7-19. COPPER WELDING

a. <u>General</u>. Copper and copper-base alloys have specific properties which make them widely used. Their high electrical conductivity makes them widely used in the electrical industries, and corrosion resistance of certain alloys makes them very useful in the process industries. Copper alloys are also widely used for friction or bearing applications. Copper can be welded satisfactorily with either bare or coated electrodes. The oxygen free copper can be welded with more uniform results than the oxygen bearing copper, which tends to become brittle when welded. Due to the high thermal conductivity of copper, the welding currents are higher than those required for steel, and preheating of the base metal is necessary. Copper shares some of the characteristics of aluminum, but is weldable. Attention should be given to its properties that make the welding of copper and copper alloys different from the welding of carbon steels. Copper alloys possess properties that require special attention when welding. These are:

(1) High thermal conductivity.

(2) High thermal expansion coefficient.

(3) Relatively low melting point.

(4) Hot short or brittle at elevated temperatures.

(5) Very fluid molten metal.

(6) High electrical conductivity.

(7) Strength due to cold working.

7-19.   COPPER WELDING (cont)

Copper has the highest thermal conductivity of all commercial metals, and the comments made concerning thermal conductivity of aluminum apply to copper, to an even greater degree.

Copper has a relatively high coefficient of thermal expansion, approximately 50 percent higher than carbon steel, but lower than aluminum.

The melting point of the different copper alloys varies over a relatively wide range but is at least $1000\,°F$ ($538\,°C$) lower than carbon steel. Some of the copper alloys are hot short. This means that they become brittle at high temperatures, because some of the alloying elements form oxides and other compounds at the grain boundaries, embrittling the material.

Copper does not exhibit heat colors like steel and when it melts it is relatively fluid. This is essentially the result of the high preheat normally used for heavier sections. Copper has the highest electrical conductivity of any of the commercial metals. This is a definite problem in the resistance welding processes.

All of the copper alloys derive their strength from cold working. The heat of welding will anneal the copper in the heat–affected area adjacent to the weld, and reduce the strength provided by cold working. This must be considered when welding high-strength joints.

There are three basic groups of copper designations. The first is the oxygen-free type which has a copper analysis of 99.95 percent or higher. The second subgroup are the tough pitch coppers which have a copper composition of 99.88 percent or higher and some high copper alloys which have 96.00 percent or more copper.

The oxygen-free high-conductivity copper contains no oxygen and is not subject to grain boundary migration. Adequate gas coverage should he used to avoid oxygen of the air caning into contact with the molten metal. Welds should be made as quickly as possible, since too much heat or slow welding can contribute to oxidation. The deoxidized coppers are preferred because of their freedom from embrittlement by hydrogen. Hydrogen enbrittlement occurs when copper oxide is exposed to a reducing gas at high temperature. The hydrogen reduces the copper oxide to copper and water vapor. The entrapped high temperature water vapor or steam can create sufficient pressure to cause cracking. In common with all copper welding, preheat should be used and can run from 250 to 1000°F (121 to 538 °C), depending on the mass involved .

The tough pitch electrolytic copper is difficult to weld because of the presence of copper oxide within the material. During welding, the copper oxide will migrate to the grain boundaries at high temperatures, which reduces ductility and tensile strength. The gas-shielded processes are recommended since the welding area is more localized and the copper oxide is less able to migrate in appreciable quantities.

The third copper subgroup is the high-copper alloys which may contain deoxidizers such as phosphorus. The copper silicon filler wires are used with this material. The preheat temperatures needed to make the weld quickly apply to all three grades.

c. <u>Gas Metal-Arc (MIG) Welding (GMAW)</u>.

(1) The gas metal arc welding process is used for welding thicker materials. It is faster, has a higher deposition rate, and usually results in less distortion. It can produce high-quality welds in all positions. It uses direct current, electrode positive. The CV type power source is recommended.

(2) Metal-arc welding of copper differs from steel welding as indicated below:

(a) Greater root openings are required.

(b) Tight joints should be avoided in light sections.

(c) Larger groove angles are required, particularly in heavy sections, in order to avoid excessive undercutting, slag inclusions, and porosity. More frequent tack welds should be used.

(d) Higher preheat and interpass temperatures are required (800 °F (427 °C) for copper, 700 °F (371 °C) for beryllium copper).

(e) Higher currents are required for a given size electrode or plate thickness.

(3) Most copper and copper alloy coated electrodes are designed for use with reverse (electrode positive) polarity. Electrodes for use with alternating currents are available.

(4) Peening is used to reduce stresses in the joints. Flat-nosed tools are used for this purpose. Numerous moderate blows should be used, because vigorous blows could cause crystallization or other defects in the joint.

d. <u>Gas Tungsten-Arc (TIG) Welding (GTAW)</u>.

CAUTION
Never use a flux containing fluoride when welding copper or copper alloys.

(1) Copper can be successfully welded by the gas tungsten-arc welding process. The weldability of each copper alloy group by this process depends upon the alloying elements used. For this reason, no one set of welding conditions will cover all groups.

(2) Direct current straight polarity is generally used for welding most copper alloys. However, high frequency alternating current or direct current reverse polarity is used for beryllium copper or copper alloy sheets less than 0.05 in. (0.13 cm) thick.

(3) For some copper alloys, a flux is recommended. However, a flux containing fluoride should never be used since the arc will vaporize the fluoride and irritate the lungs of the operator.

7-19. COPPER WELDING (cont)

e. <u>Carbon-Arc Welding</u>.

(1) This process for copper welding is most satisfactory for oxygen-free copper, although it can be used for welding oxygen-bearing copper up to 3/8 in. (9.5 mm) in thickness. The root opening for thinner material should be 3/16 in. (4.8 mm), and 3/8 in. (9.5 mm) for heavier material. The electrode should be graphite type carbon, sharpened to a long tapered point at least equal to the size of the welding rod. Phosphor bronze welding rods are used most frequently in this process.

(2) The arc should be sharp and directed entirely on the weld metal, even at the start. If possible, all carbon-arc welding should be done in the flat welding position or on a moderate slope.

7-20. MAGNESIUM WELDING

a. <u>General</u>. Magnesium is a white, very lightweight, machinable, corrosion resistant, high strength metal. It can be alloyed with small quantities of other metals, such as aluminum, manganese, zinc and zirconium, to obtain desired properties. It can be welded by most of the welding processes used in the metal working trades. Because this metal oxidizes rapidly when heated to its melting point in air, a protective inert gas shield must be provided in arc welding to prevent destructive oxidation.

b. Magnesium possesses properties that make welding it different from the welding of steels. Many of these are the same as for aluminum. These are:

(1) Magnesium oxide surface coating which increases with an increase in temperature.

(2) High thermal conductivity.

(3) Relatively high thermal expansion coefficient.

(4) Relatively low melting temperature.

(5) The absence of color change as temperature approaches the melting point.

The normal metallurgical factors that apply to other metals apply to magnesium as well.

c. The welds produced between similar alloys will develop the full strength of the base metals; however, the strength of the heat-affected zone may be reduced slightly. In all magnesium alloys, the solidification range increases and the melting point and the thermal expansion decrease as the alloy content increases. Aluminum added as an alloy up to 10 percent improves weldability, since it tends to refine the weld grain structure. Zinc of more than 1 percent increases hot shortness, which can result in weld cracking. The high zinc alloys are not recommeded for arc welding because of their cracking tendencies. Magnesium, containing small amounts of thorium, possesses excellent welding qualities and freedom from cracking Weldments of these alloys do not require stress relieving. Certain magnesium alloys are subject to stress corrosion Weldments subjected to corrosive attack over a Period of time may crack adjacent to welds if the residual stresses are not removed. Stress relieving is required for weldments intended for this type of service.

d. <u>Cleaning.</u> An oil coating or chrome pickle finish is usually provided on magnesium alloys for surface protection during shipment and storageThis oil, along with other foreign matter and metallic oxides, must be removed from the surface prior to welding. Chemical cleaning is preferred, because it is faster ah more uniform in its action. Mechanical cleaning can be utilized if chemical cleaning facilities are not available A final bright chrome pickle finish is recommended for parts that are to be arc welded. The various methods for cleaning magnesium are described below.

WARNING

The vapors from some chlorinated solvents (e.g., carbon tetrachloride, trichloroethylene, and perchloroethylene) break down under the ultraviolet radiation of an electric arc and form a toxic gas. Avoid welding where such vapors are present. These solvents vaporize easily, and prolonged inhalation of the vapor can be hazardous. These organic vapors should be removed from the work area before welding begins.

Dry cleaning solvent and mineral spirits paint thinner are highly flammable. Do not clean parts near an open flame or in a smoking area. Dry cleaning solvent and mineral spirits paint thinner evaporate quickly and have a defatting effect on the skin. When used without protective gloves, these chemicals may cause irritation or cracking of the skin. Cleaning operations should be performed only in well ventilated areas.

(1) Grease should be removed by the vapor degreasing system in which trichloroethylene is utilized or with a hot alkaline cleaning compound. Grease may also be removed by dipping small parts in dry cleaning solvent or mineral spirits paint thinner

(2) Mechanical cleaning can be done satisfactorily with 160 and 240 grit aluminum oxide abrasive cloth, stainless steel wool, or by wire brushing.

WARNING

Precleaning and postcleaning acids used in magnesium welding and brazing are highly toxic and corrosive. Goggles, rubber gloves, and rubber aprons should be worn when handling the acids and solutions. D o not inhale fumes and mists. When spilled on the body or clothing, wash immediately with large quantities of cold water, and seek medical attention. Never pour water into acid when preparing solution: instead, pour acid into water. Always mix acid and water slowly. Cleaning operations should be performed only in well ventilated areas.

TC 9-237

7-20. MAGNESIUM WELDING (cont)

(3) Immediately after the grease, oil, and other foreign materials have been removed from the surface, the metal should be dipped for 3 minutes in a hot solution with the following composition:

Chromic acid (CrO) -- 24 oz (680 g)

Sodium nitiate (NaNO) -- 4 oz (113 g)

Calcium or magnesium fluoride -- 1/8 oz (3.5 g)

Water -------------------------- to make 1 gal. (3.8 1)

The bath should be operated at 70F (21 °C). The work should be removed from the solution, thoroughly rinsed with hot water, and air dried. The welding rod should also be cleaned to obtain the best results.

e. Joint Preparation. Edges that are to be welded must be smooth and free of loose pieces and cavities that might contain contaminating agents, such as oil or oxides. Joint preparations for arc welding various gauges of magnesium are shown in figure 7-13.

f. Safety Precautions

CAUTION
Magnesium can ignite and burn when heated in the open atmosphere.

(1) Goggles, gloves, and other equipment designed to protect the eyes and skin of the welder must be worn.

(2) The possibility of fire caused by welding magnesium metal is very remote. The temperature of initial fusion must be reached before solid magnesium metal ignites. Sustained burning occurs only if this temperature is maintained. Finely divided magnesium particles such as grinding dust, filings, shavings, borings, and chips present a fire hazard. They ignite readily if proper precautions are not taken. Magnesium scrap of this type is not common to welding operations. If a magnesium fire does start, it can be extinguished with dry sand, dry powdered soapstone, or dry cast iron chips. The preferred extinguishing agents for magnesium fires are graphite base powders.

g. Gas Tungsten-Arc (TIG) Welding (GTAW) of Magnesium.

(1) Because of its rapid oxidation when magnesium is heated to its melting point, an inert gas (argon or helium) is used to shield metal during arc welding. This process requires no flux and permits high welding speeds, with sound welds of high strength.

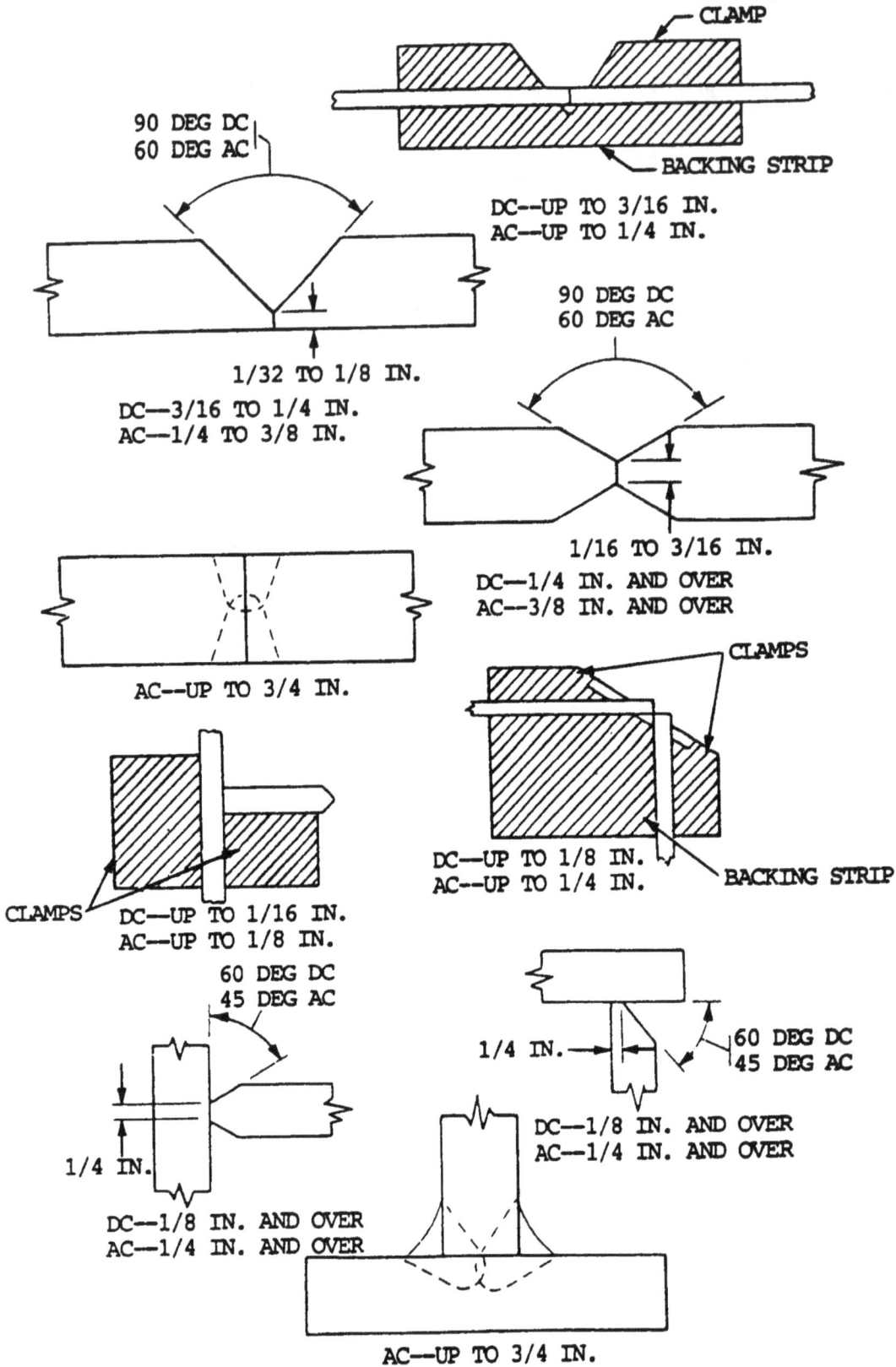

Figure 7-13. Joint preparation for arc welding magnesium.

7-20. MAGNESIUM WELDING (cont)

(2) Direct current machines of the constant current type operating on straight polarity (electrode positive) and alternating current machines are used with a high frequency current superimposed on the welding current. Both alternating and direct current machines are used for thin gauge material. However, because of better penetrating power, alternating current machines are used on material over 3/16 in. (4.8 mm) thick. Helium is considered more practical than argon for use with direct current reverse polarity. However, three times as much helium by volume as argon is required for a given amount of welding. Argon is used with alternating current.

(3) The tungsten electrodes are held in a water cooled torch equipped with required electrical cables and an inlet and nozzle for the inert gas.

(4) The two magnesium alloys, in the form of sheet, plate, and extrusion, that are most commonly used for applications involving welding are ASTM-1A (Fed Spec QQ-M-54), which is alloyed with manganese, and ASTM-AZ31A (Fed Spec QQ-44), which is alloyed with aluminum, manganese, and zinc.

(5) In general, less preparation is required for welding with alternating current than welding with direct current because of the greater penetration obtained. Sheets up to 1/4 in. (6. 4 mm) thickness may be welded from one side with a square butt joint. Sheets over 1/4 in. (6. 4 mm) thickness should be welded from both sides whenever the nature of the structure permits, as sounder welds may be obtained with less warpages. For a double V joint, the included angle should extend from both sides to leave a minimum 1/16 in. (1.6 mm) root face in the center of the sheets. When welding a double V joint, the back of the first bead should be chipped out using a chipping hammer fitted with a cape chisel. Remove oxide film, dirt, and incompletely fused areas before the second bead is added. In this manner, maximum soundness is obtained.

(6) The gas should start flowing a fraction of a second before the arc is struck. The arc is struck by brushing the electrode over the surface. With alternating current, the arc should be started and stopped by means of a remote control switch. The average arc length should be about 1/8 in. (3.2 mm) when using helium and 1/16 in. (1.6 mm) when using argon. Current data and rod diameter are shown in table 7-24.

(7) When welding with alternating current, maximum penetration is obtained when the end of the electrode is held flush with or slightly below the surface of the work. The torch should be held nearly perpendicular to the surface of the work, and the welding rod added from a position as neatly parallel with the work as possible (fig. 7-14). The torch should have a slightly leading travel angle.

Table 7-24. Magnesium Weld Data

Sheet Thickness (in.)[1]	Current (amps)[2]	Rod Diameter (in.)[1]
0.030	20	1/16
0.040	30	1/16
0.050	35	3/32
0.060	45	3/32
0.070	55	1/8
0.080	60	1/8
0.090	65	1/8
0.100	70	1/8
0.125	75	1/8
0.150	80	5/32
0.200[3]	90	5/32
0.250	100	5/32
0.500	115	5/32
1.000	130	5/32

[1]Dimensions are given in inches.
[2]Currents shown are for all alloys except alloy Ml, which requires 5 to 10 amperes more current for materials up to 0.05 in.(1.27 mm) thick and 15 to 30 amperes more current for thicker materials. Currents given are for welding speeds of 12 in. (304.8 mm) per minute.
[3]Sheets thicker than 0.15 in. (3.81 mm) should be welded in more than one pass. A current of about 60 amperes is used on the first pass and the currents given in the table are used for subsequent passes.

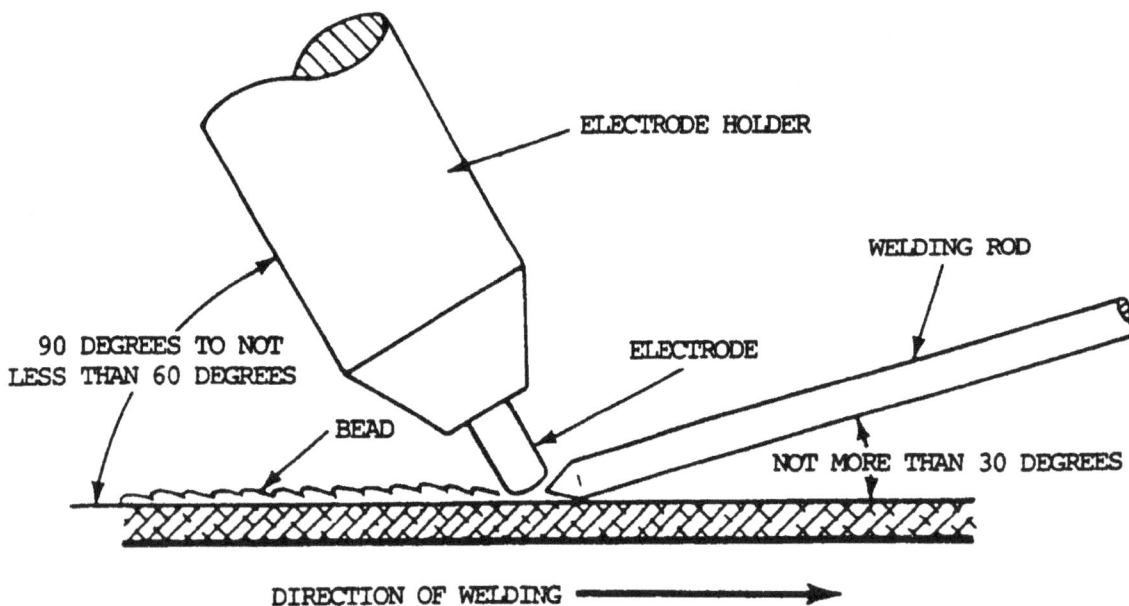

Figure 7-14. Position of torch and welding rod.

7-20. MAGNESIUM WELDING (cont)

(8) Welding should progress in a straight line at a uniform speed. There should be no rotary or weaving motion of the rod or torch, except for larger corner joints or fillet welds. The welding rod can be fed either continuously or intermittently. Care should be taken to avoid withdrawing the heated end from the protective gaseous atmosphere during the welding operation.The cold wire filler metal should be brought in as near to horizontal as possible (on flat work). The filler wire is added to the leading edge of the weld puddle. Runoff tabs are recommended for welding any except the thinner metals. Forehand welding, in which the welding rod precedes the torch in the direction of welding, is preferredIf stops are necessary, the weld should be started about 1/2 in(12.7 mm) back from the end of the weld when welding is resumed.

(9) Because of the high coefficient of thermal expansion and conductivity, control of distortion in the welding of magnesium requires jigging, small beads, and a properly selected welding sequence to helpminimize distortion. Magnesium parts can be straightened by holding them in position with clamps and heating to 300 to 400 °F (149 to 204 °C). If this heating is done by local torch application, care must be taken not to overheat the metal and destroy its properties.

(10) If cracking is encountered during the welding of certain magnesium alloys, starting and stopping plates may be used to overcome this difficult. These plates consist of scrap pieces of magnesium stock butted against opposite ends of the joint to be welded as shown in A, figure 7-15. The weld is started on one of the abutting plates, continued across the junction along the joint to be welded, and stopped on the opposite abutting plateIf a V groove is used, the abutting plates should also be grooved.An alternate method is to start the weld in the middle of the joint and weld to each edge (B, fig. 7-15). Cracking may also be miniimized by preheating the plate and holding the jig to 200 to 400 (93 to 204 °C) by increasing the speed of the weld.

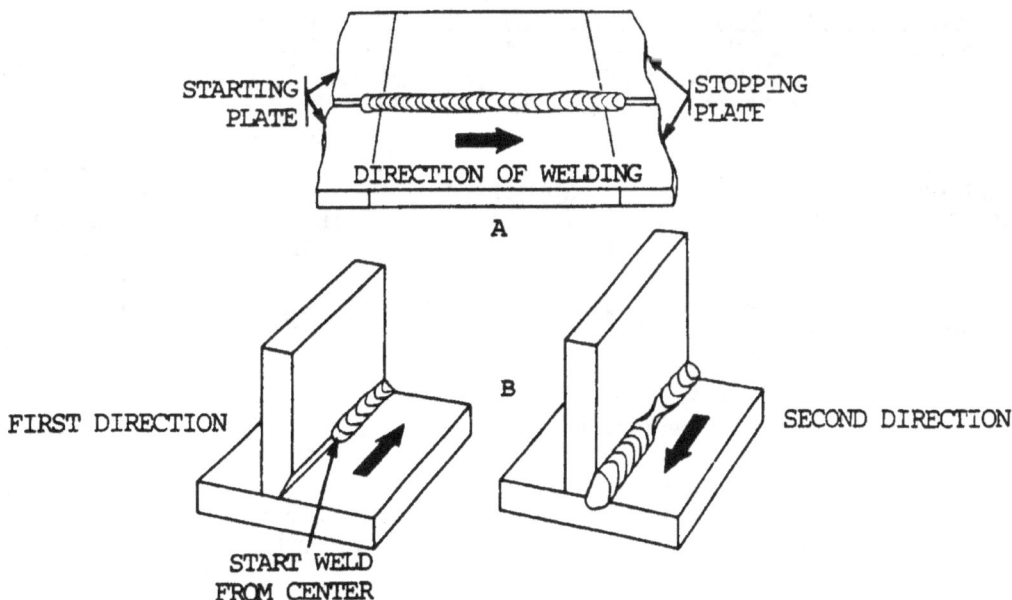

Figure 7-15. Minimizing cracking during welding.

(11) Filler reds must be of the same composition as the alloy being joined when arc welding. One exception is when welding AZ31B. In this case, grade C rod (MIL-R-6944), which produces a stronger weld metal, is used to reduce cracking.

(12) Residual stress should be relieved through heat treatment. Stress relief is essential so that lockup stresses will not cause stress corrosion cracking. The recommended stress relieving treatment for arc welding magnesium sheet is shown in table 7-25.

Table 7-25. Magnesium Stress Relief Data

Alloy	Temperature °F	°C	Time at Temperature (hour)
AZ31B (annealed)	500	260	0.25
AZ31B (hard rolled)	265	129	1.00
M1 (annealed)	500	260	0.25
M1 (hard rolled)	400	204	1.00

(13) The only cleaning required after arc welding of magnesium alloys is wire brushing to remove the slight oxide deposit on the surface. Brushing may leave traces of iron, which may cause galvanic corrosion. If necessary, clean as in b above. Arc welding smoke can be removed by immersing the parts for 1/2 to 2 minutes at 180 to 212°F (82 to 100°C), in a solution composed of 16 oz (453 g) tetrasodium pyrophosphate ($Na_4P_2O_7$), 16 oz (453 g) sodium metaborate ($NaBO_2$) and enough water to make 1 gallon (3.8 l).

(14) Welding procedure schedules for GTAW of magnesium (TIG welding) are shown in table 7-26, p 7-92.

h. Gas Metal-Arc (MIG) Welding of Magnesium (GMAW). The gas metal arc welding process is used for the medium to thicker sections. It is considerably faster than gas tungsten arc welding. Special high-speed gear ratios are usually required in the wire feeders since the magnesium electrode wire has an extremely high meltoff rate. The normal wire feeder and power supply used for aluminum welding is suitable for welding magnesium. Different types of arc transfer can be obtained when welding magnesium. This is primarily a matter of current level or current density and voltage setting. The short-circuiting transfer and the spray transfer are recommended. Argon is usually used for gas metal arc welding of magnesium; however, argon-helium mixtures can be used. In general, the spray transfer should be used on material 3/16 in. (4.8 mm) and thicker and the short-circuiting arc used for thinner metals. Welding procedure schedules for GMAW of magnesium (MIG welding) are shown in table 7-27, p 7-93.

i. Other Welding Processes. Magnesium can be welded using the resistance welding processes, including spot welding, seam welding, and flash welding. Magnesium can also be joined by brazing. Most of the different brazing techniques can be used. In all cases, brazing flux is required and the flux residue must be completely removed from the finish part. Soldering is not as effective, since the strength of the joint is relatively low. Magnesium can also be stud welded, gas welded, and plasma-arc welded.

Table 7-26. Welding Procedure Schedules for Gas Tungsten Arc Welding (GTAW) of Magnesium (TIG Welding)

| Material Thickness (or Fillet Size) | | | Type of Weld Fillet or Groove | Tungsten Electrode Diameter | | Filler Rod Diameter | | Nozzle Size Inside Dia. | Shielding Gas Flow | Welding Current Amps | No. of Passes | Travel Speed (per pass) |
ga	in.	mm		in.	mm	in.	mm	in.	cfh	DCEN		ipm
20	0.038	0.9	Square groove	1/16	1.5	3/32	2.4	1/4	15	25-40	1	20
20	0.038	0.9	Fillet	1/16	1.5	3/32	2.4	1/4	15	30-45	1	20
16	0.063	1.6	Square groove	1/16	1.5	3/32	2.4	1/4	15	45-60	1	20
16	0.063	1.6	Fillet	1/16	1.6	3/32	2.4	1/4	15	45-60	1	20
14	0.078	1.9	Square groove	1/16	1.6	3/32	2.4	1/4	15	60-75	1	17
14	0.078	1.9	Fillet	1/16	1.6	3/32	2.4	1/4	15	60-75	1	17
12	0.109	2.8	Square groove	3/32	2.4	1/8	3.2	5/16	15	80-100	1	17
12	0.109	2.8	Fillet	3/32	2.4	1/8	3.2	5/16	15	80-100	1	17
11	0.125	3.2	Square groove	3/32	2.4	1/8	3.2	5/16	25	95-115	1	17
11	0.125	3.2	Fillet	3/32	2.4	1/8	3.2	5/16	25	95-115	1	17
3/16	0.187	4.7	Vee groove	1/8	3.2	1/8	3.2	3/8	25	95-115	2	26
1/4	0.250	6.4	Vee groove	1/8	3.2	3/16	4.8	1/2	25	110-130	2	24
3/8	0.375	9.5	Vee groove	1/8	3.2	3/16	4.8	1/2	30	135-165	2	20

NOTE

Increase amperage when backup is used. Data is for flat position. Reduce amperage 10% to 20% when welding in horizontal, vertical or overhead positions. Tungsten electrode. Select filler metal in accordance with selection chart. Shielding gas is normally argon. A mixture of 75% helium + 25% argon is used for heavier thickness. For heavy thickness, 100% helium is used. Gas flow rates for helium are approximately twice those used for argon.

Table 7-27. Welding procedure Schedules for Gas Metal Arc Welding (GMAW) of Magnesium (MIG Welding)

Material Thickness (or Fillet Size) ga	in.	mm	Type of Weld Fillet or Groove	Electrode Diameter in.	mm	WELDING POWER Current Amps DC	Arc Volt EP	Wire Feed Speed ipm	Shielding Gas Flow cfh	No. of Passes	Travel Speed (per pass) ipm
0.025	—	—	Sq. groove & fillet	0.040	1.0	26-27	13-16	180	40-60	1	24-36
0.040	—	—	Sq. groove & fillet	0.040	1.0	35-50	13-16	250-340	40-60	1	24-36
0.063	1/16	1.6	Sq. groove & fillet	0.063	1.6	60-75	13-16	140-170	40-60	1	24-36
0.090	3/32	2.4	Sq. groove & fillet	0.063	1.6	95-125	13-16	210-280	40-60	1	24-36
0.125	1/8	3.2	Sq. groove & fillet	0.094	2.4	110-135	13-16	100-130	40-60	1	24-36
0.160	5/32	3.9	Sq. groove & fillet	0.094	2.4	135-140	13-16	130-140	40-60	1	24-36
0.190	3/16	4.8	Vee groove & fillet	0.094	2.4	175-205	13-16	160-190	40-60	2	24-36
0.250	1/4	6.4	Vee groove & fillet	0.063	1.6	240-290	24-30	550-660	50-80	2	24-36
0.375	3/8	9.5	Vee groove & fillet	0.094	2.4	320-350	24-30	350-385	50-80	2	24-36
0.500	1/2	12.7	Vee groove & fillet	0.094	2.4	350-420	24-30	385-415	50-80	2	24-36
1.000	1	25.4	Vee groove & fillet	0.094	2.4	350-420	24-30	385-415	50-80	4	24-36

NOTE

Values are for flat position welding. For groove and fillet welds--material thickness also indicates fillet weld size. Use vee groove for 1/4 in. (6.4 mm) and thicker. Shielding gas is argon. For heavier thicknesses, use helium-argon mixtures. Above 200 amps and 200 volts, metal transfer is spray type. Below 200 amps and 20 volts, metal transfer is short circuiting type.

7-21. TITANIUMWELDING

a. General.

(1) Titanium is a soft, silvery white, medium strength metal with very good corrosion resistance. It has a high strength to weight ratio and its tensile strength increases as the temperature decreases. Titanium has low impact and creep strengths. It has seizing tendencies at temperatures above 800°F (427 °C).

(2) Titanium has a high affinity for oxygen and other gases at elevated temperatures, and for this reason, cannot be welded with any process that utilizes fluxes, or where heated metal is exposed to the atmosphere. Minor amounts of impurities cause titanium to become brittle.

(3) Titanium has the characteristic known as the ductile-brittle transition. This refers to a temperature at which the metal breaks in a brittle manner, rather than in a ductile fashion. The recrystallization of the metal during welding can raise the transition temperature. Contamination during the high temperate period and impurities can raise the transition temperature period and impurities can raise the transition temperature so that the material is brittle at room temperatures. If contamination occurs so that transition temperature is raised sufficiently, it will make the welding worthless. Gas contamination can occur at temperatures below the melting point of the metal. These temperatures range from 700°F (371 °C) up to 1000 °F (538 °C).

(4) At room temperature, titanium has an impervious oxide coating that resists further reaction with air. The oxide coating melts at temperatures considerably higher than the melting point of the base metal and creates problems. The oxidized coating may enter molten weld metal and create discontinuities which greatly reduce the strength and ductility of the weld.

(5) The procedures for welding titanium and titanium alloys are similar to other metals. Some processes, such as oxyacetylene or arc welding processes using active gases, cannot be used due to the high chemical activity of titanium and its sensitivity to embrittlement by contamination. Processes that are satisfactory for welding titanium and titanium alloys include gas shielded metal-arc welding, gas tungsten arc welding, and spot, seam, flash, and pressure welding. Special procedures must be employed when using the gas shielded welding processes. These special procedures include the use of large gas nozzles and trailing shields to shield the face of the weld from air. Backing bars that provide inert gas to shield the back of the welds from air are also used. Not only the molten weld metal, but the material heated above 1000°F (538 °C) by the weld must be adequately shielded in order to prevent embrittlement. All of these processes provide for shielding of the molten weld metal and heat affected zones. Prior to welding, titanium and its alloys must be free of all scale and other material that might cause weld contamination.

b. Surface Preparation.

WARNING

The nitric acid used to preclean titanium for inert gas shield arc welding is highly toxic and corrosive. Goggles, rubber gloves, an d rubber aprons must be worn when handling acid solutions. Do not inhale gases and mists. When spilled on the body or clothing, wash immediately with large quantities of cold water, and seek medical help. Never pour water into acid when preparing the solution; instead, pour acid into water. Always mix acid and water slowly. Perform cleaning operations only in well ventilated places .

The caustic chemicals (including hydrode) used to preclean titanium for inert gas shielded arc welding are highly toxic and corro— sive. Goggles, rubber gloves, and rubber aprons must be worn when handling these chemicals. Do not inhale gases or mists. When spilled on the body or clothing, wash immediately with large quantities of cold water and seek medical help. Special care should be taken at all times to prevent any water from coming in contact with the molten bath or any other large amount of sodium hydride, as this will cause the fomation of highly explosive hydrogen gas.

(1) Surface cleaning is important in preparing titanium and its alloys for welding. Proper surface cleaning prior to welding reduces contamination of the weld due to surface scale or other foreign materials. Small amounts of contamination can render titanium completely brittle.

(2) Several cleaning procedures are used, depending on the surface condition of the base and filler metals. Surface conditions most often encountered are as follows:

(a) Scale free (as received from the mill).

(b) Light scale (after hot forming or annealing at intermediate temperature; ie., less than 1300 ^{0}F (704 ^{0}C).

(c) Heavy scale (after hot forming, annealing, or forging at high temperature).

(3) Metals that are scale free can be cleaned by simple decreasing.

(4) Metals with light oxide scale should be cleaned by acid pickling. In order to minimize hydrogen pickup, pickling solutions for this operation should have a nitric acid concentration greater than 20 percent. Metals to be welded should be pickled for 1 to 20 minutes at a bath temperature from 80 to 160 (27 to 71 ^{0}C). After pickling, the parts are rinsed in hot water.

(5) Metals with a heavy scale should be cleaned with sand, grit, or vaporblasting, molten sodium hydride salt baths, or molten caustic baths. Sand, grit, or vaporblasting is preferred where applicable. Hydrogen pickup may occur with molten bath treatments, but it can be minimized by controlling the bath temperature and pickling time. Bath temperature should be held at about 750 to 850 (399 to 454 ^{0}C). Parts should not be pickled any longer than necessary to remove scale. After heavy scale is removed, the metal should be pickled as described in (4) above.

7-21. TITANIUM WELDING (cont)

(6) Surfaces of metals that have undergone oxyacetylene flame cutting operations have a very heavy scale, and may contain microscopic cracks due to excessive contamination of the metallurgical characteristics of the alloys.The best cleaning method for flame cut surfaces is to remove the contaminated layer and any cracks by machining operations. Certain alloys can be stress relieved immediately after cutting to prevent the propagation of these cracks.This stress relief is usually made in conjunction with the cutting operation.

c. MIG or TIG Welding of titanium.

(1) General. Both the MIG and TIG welding processes are used to weld titanium and titanium alloys. They are satisfactory for manual and automatic installations. With these processes, contamination of the molten weld metals and adjacent heated zones is minimized by shielding the arc and the root of the weld with inert gases (see (2)(b)) or special backing bars (see (2)(c)). In some cases, inert gas filler welding chambers (see (3)) are used to provide the required shielding. When using the TIG welding process, a thoriated tungsten electrode should be used.The electrode size should be the smallest diameter that will carry the welding current. The electrode should be ground to a point.The electrode may extend 1-1/2 times its diameter beyond the end of the nozzle.Welding is done with direct current, electrode negative (straight polarity).Welding procedure for TIG welding titanium are shown in table 7-28. Selection of the filler metal will depend upon the titanium alloys being joined. When welding pure titanium, a pure titanium wire should be used. When welding a titanium alloy, the next lowest strength alloy should be used as a filler wire. Due to the dilution which will take place dining welding, the weld deposit will pick up the required strength. The same considerations are true when MIG welding titaniun.

(2) Shielding.

(a) General Very good shielding conditions ue necessary to produce arc welded joints with maximum ductility and toughness.To obtain these conditions, the amount of air or other active gases which contact the molten weld metals and. adjacent heated zones must be very low.Argon is normally used with the gas-shielded process. For thicker metal, use helium or a mixture of argon and helium. Welding grade shielding gases are generally free from contamination; however, tests can be made before welding. A simple test is to make a bead on a piece of clean scrap titanium, and notice its color.The bead should be shiny. Any discoloration of the surface indicates a contamination.Extra gas shielding provides protection for the heated solid metal next to the weld metal.This shielding is provided by special trailing gas nozzles, or by chill bars laid immediately next to the weld. Backup gas shielding should be provided to protect the underside of the weld joint. Protection of the back side of the joint can also be provided by placing chill bars in intimate contact with the backing strips.If the contact is close enough, backup shielding gas is not required.For critical applications, use an inert gas welding chamber. These can be flexible, rigid, or vacuum-purge chambers.

Table 7-28. Welding Procedure Schedule for Metal-Arc Welding (GMAW) of Titanium (MIG Welding)

Material Thickness (or Fillet Size) ga	in.	mm	Type of Weld Fillet or Groove	Tungsten Electrode Diameter in.	mm	Filler Rod Diameter in.	mm	Nozzle Size Inside Dia. in.	Shielding Gas Flow cfh	Welding Current Amps DCEN	No. of Passes	Travel Speed (per pass) ipm
24	0.024	0.6	Sq. groove & fillet	1/16	1.6	None		3/8	18	20-35	1	6
16	0.063	1.6	Sq. groove & fillet	1/16	1.6	None		5/8	18	85-140	1	6
3/32	0.093	1.6	Sq. groove & fillet	3/32	2.4	1/16	1.6	5/8	25	170-215	1	8
1/8	0.125	3.2	Sq. groove & fillet	3/32	2.4	1/16	1.6	5/8	25	190-235	1	8
3/16	0.188	4.8	Sq. groove & fillet	3/32	2.4	1/8	3.2	5/8	25	220-280	2	8
1/4	0.250	6.4	Vee groove & fillet	1/8	3.2	1/8	3.2	5/8	30	275-320	2	8
3/8	0.375	9.5	Vee groove & fillet	1/8	3.2	1/8	3.2	3/4	35	300-350	2	6
1/2	0.500	12.7	Vee groove & fillet	1/8	3.2	5/32	3.9	3/4	40	325-425	3	6

NOTE

Tungsten used, 1st choice 2% thoriated EWTh2-2nd choice 1% thoriated EWTh1. Use filler metal one or two grades lower in strength than the base metal. Adequate gas shielding is a must not only for the arc but also heated metal. Backing gas is recommended at all times. A trailing gas shield is also recommended. Argon is preferred. For high heat input on thicker material use argon-helium mixture. Without backup or chill bar, decrease current 20%.

7-21. TITANIUM WELDING (cont)

(b) Inert gases. Both helium and argon are used as the shielding gases. With helium as the shielding gas high welding speeds and better penetration are obtained than with argon, but the arc is more stable in argon. For open air welding operations, most welders prefer argon as the shielding gas because its density is greater than that of air. Mixtures of argon and helium are also used. With mixtures, the arc characteristics of both helium and argon are obtained. The mixtures usually vary in composition from about 20 to 80 percent argon. They are often used with the consumable electrode process. To provide adequate shielding for the face and root sides of welds, special precautions often are taken. The precautions include the use of screens and baffles (see (c) 3), trailing shields (see (c) 7), and special backing fixtures (see (c) 10) in open air welding, and the use of inert gas filler welding chambers.

(c) Open air welding.

1. In open air welding operations, the methods used to shield the face of the weld vary with joint design, welding conditions, and the thickness of the materials being joined. The most critical area in regard to the shielding is the molten weld puddle. Impurities diffuse into the molten metal very rapidly and remain in solution. The gas flowing through a standard welding torch is sufficient to shield the molten zone. Because of the low thermal conductivity of titanium, however, the molten puddle tends to be larger than most metals. For this reason and because of shielding conditions required in welding titanium, larger nozzles are used on the welding torch, with proportionally higher gas flows that are required for other metals. Chill bars often are used to limit the size of the puddle.

2. The primary sources of contamination in the molten weld puddle are turbulence in the gas flow, oxidation of hot filler reds, insufficient gas flow, small nozzles on the welding torch, and impure shielding gases. The latter three sources are easily controlled.

3. If turbulence occurs in the gas flowing from the torch, air will be inspired and contamination will result. Turbulence is generally caused by excessive amounts of gas flowing through the torch, long arc lengths, air currents blowing across the weld, and joint design. Contamination from this source can be minimized by adjusting gas flows and arc lengths, and by placing baffles alongside the welds. Baffles protect the weld from drafts and tend to retard the flow of shielding gas from the joint area. Chill bars or the clamping toes of the welding jig can serve as baffles (fig. 7-16). Baffles are especially important for making corner type welds. Additional precautions can be taken to protect the operation from drafts and turbulence. This can be achieved by erecting a canvas (or other suitable material) screen around the work area.

4. In manual welding operations with the tungsten-arc process, oxidation of the hot filler metal is a very important source of contamination. To control it, the hot end of the filler wire must be kept within the gas shield of the welding torch. Welding operators must be trained to keep the filler wire shielded when welding titanium and its alloys. Even with proper manipulation, however, contamination from this source probably cannot be eliminated completely.

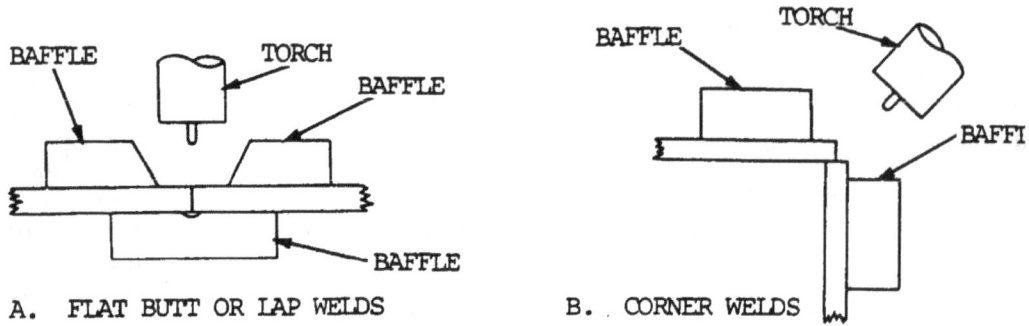

Figure 7-16. Baffle arrangements to improve shielding.

5. Weld contamination which occurs in the molten weld puddle is especially hazardous. The impurities go into solution, and do not cause discoloration. Although discolored welds may have been improperly shielded while molten, weld discoloration is usually caused by contamination which occurs after the weld has solidified.

6. Most of the auxiliary equipment used on torches to weld titanium is designed to improve shielding conditions for the welds as they solidify and cool. However, if the welding heat input is low and the weld cools to temperatures below about 1200 to 1300°F (649 to 704°C) while shielded, auxiliary shielding equipment is not required. If the weld is at an excessively high temperature after it is no longer shielded by the welding torch, auxiliary shielding must be supplied.

7. Trailing shields often are used to supply auxiliary shielding. These shields extend behind the welding torch and vary considerably in size, shape, and design. They are incorporated into special cups which are used on the welding torch, or may consist only of tubes or hoses attached to the torch or manipulated by hand to direct a stream of inert gas on the welds. Figure 7-17 shows a drawing of one type of trailing shield currently in use. Important features of this shield are that the porous diffusion plate allows an even flow of gas over the shielded area. This will prevent turbulence in the gas stream The shield fits on the torch so that a continuous gas stream between the torch and shield is obtained.

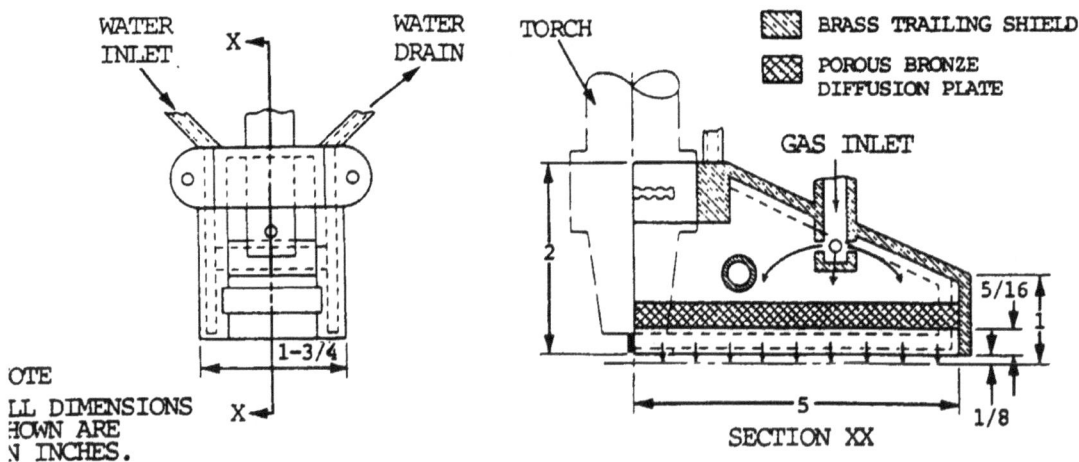

Figure 7-17. Trailing shield.

7-21. TITANIUM WELDING (cont)

      8. Baffles are also beneficial in improving shielding conditions for welds by retarding the flow of shielding gas from the joint area. Baffles may be placed alongside the weld, over the top, or at the ends of the weld. In some instances, they may actually form a chamber around the arc and molten weld puddle. Also, chill bars may be used to increase weld cooling rates and may make auxiliary

      9. Very little difficulty has been encountered in shielding the face of welds in automatic welding operations. However, considerable difficulty has been encountered in manual operations.

      10. In open air welding operations, means must be provided for shielding the root or back of the welds. Backing fixtures are often used for this purpose. In one type, an auxiliary supply of inert gas is provided to shield the back of the weld. In the other, a solid or grooved backing bar fits tightly against the back of the weld and provides the required shilding. Fixtures which provide an inert gas shield are preferred, especiallly in manual welding operations with low welding speeds. Figure 7-18 shows backing fixtures used in butt welding heavy plate and thin sheet, respectively. Similar types of fixtures are used for other joint designs. However, the design of the fixtures varies with the design of the joints. For fillet welds on tee joints, shielding should be supplied for two sides of the weld in addition to shielding the face of the weld

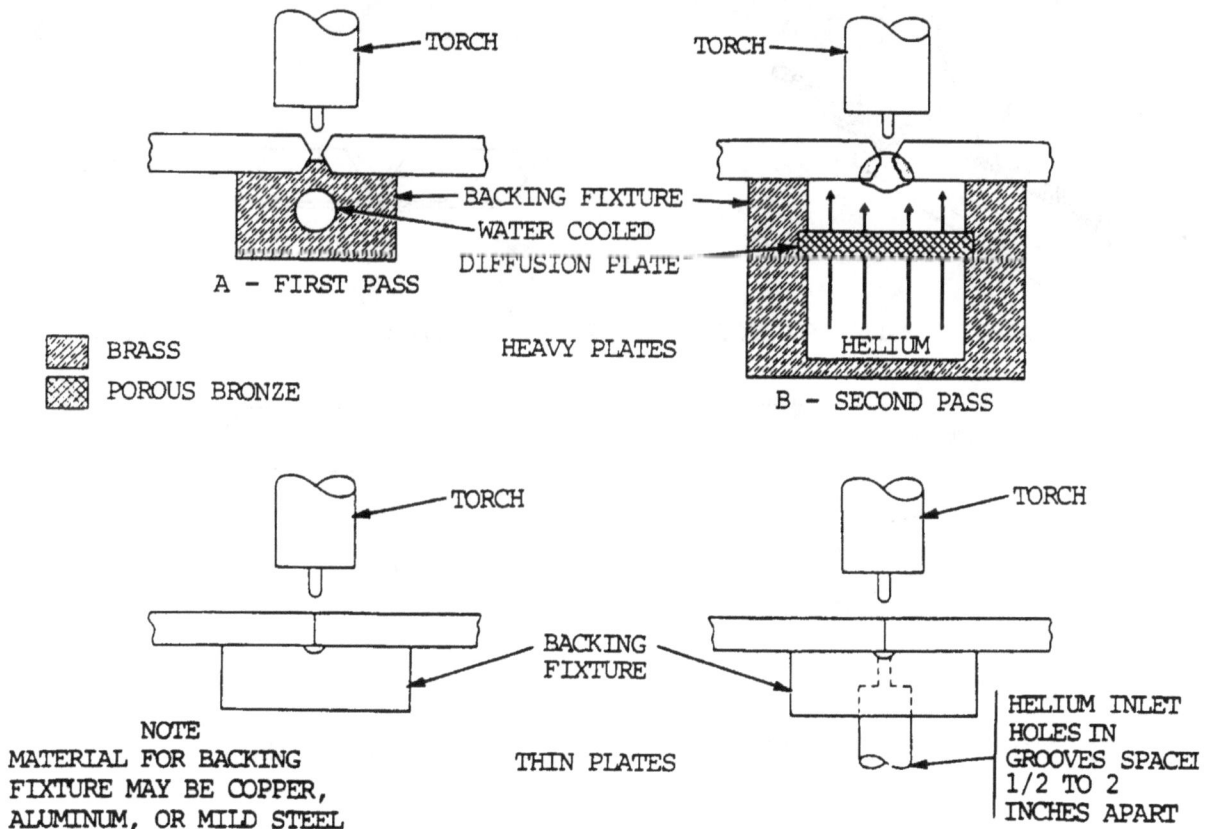

Figure 7-18. Backing fixtures for butt welding heavy plate and thin sheet.

11. For some applications, it may be easier to enclose the back of the weld, as in a tank, and supply inert gas for shielding purposes. This method is necessary in welding tanks, tubes, or other enclosed structures where access to the back of the weld is not possible. In some weldments, it may be necessary to machine holes or grooves in the structures in order to provide shielding gas for the back or root of the welds.

WARNING
When using weld backup tape, the weld must be allowed to cool for several minutes before attempting to remove the tape from the workpiece.

12. Use of backing fixtures such as shown in figure 7-18 can be eliminated in many cases by the use of weld backup tape. This tape consists of a center strip of heat resistant fiberglass adhered to a wider strip of aluminum foil, along with a strip of adhesive on each side of the center strip that is used to hold tape to the underside of the tack welded joint. During the welding, the fiberglass portion of the tape is in direct contact with the molten metal, preventing excessive penetration. Contamination or oxidation of the underside of the weld is prevented by the airtight seal created by the aluminum foil strip. The tape can be used on butt or corner joints (fig. 7-19) or, because of its flexibility, on curved or irregularly shaped surfaces. The surface to which the tape is applied must be clean and dry. Best results are obtained by using a root gap wide enough to allow full penetration.

WELDING SHEET METAL-MIG OR TIG PROCESSES   JOINT PREPARATION FOR BUTT WELDING PLATE

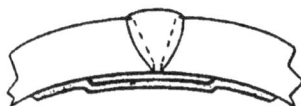

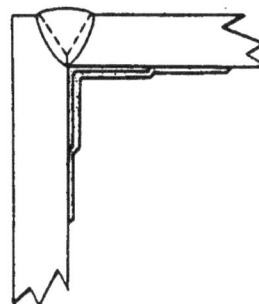

BUTT WELDED JOINT ON CURVED SURFACE   JOINT PREPARATION FOR CORNER WELD

Figure 7-19. Use of weld backup tape.

7-21. TITANIUM WELDING (cont)

13. Bend or notch toughness tests are the best methods for evaluating shielding conditions, but visual inspection of the weld surface, which is not an infallible method, is the only nondestructive mean for evaluating weld quality at the present time. With this method, the presence of a heavy gray scale with a nonmetallic luster on the weld bead indicates that the weld has been contaminate badly and has low ductility. Also, the weld surface may be shiny but have different colors, ranging from grayish blue to violet to brown. This type of discoloration may be found on severely contaminated welds or may be due only to surface contamination, while the weld itself may be satisfactory. However, the quality of the weld cannot be determined without a destructive test. With good shielding procedures, weld surfaces are shiny and show no discoloration.

(3) Welding chambers.

(a) For some applications, inert gas filled welding chambers are used. The advantage of using such chambers is that good shielding may be obtained for the root and face of the weld without the use of special fixtures. Also, the surface appearance of such welds is a fairly reliable measure of shielding conditions. The use of chambers is especially advantageous when complex joints are being welded. However, chambers are not required for many applications, and their use may be limited.

(b) Welding chambers vary in size and shape, depending on their use and the size of assemblies to be welded. The inert atmospheres maybe obtained by evacuating the chamber and filling it with helium or argon, purging the chamber with inert gas, or collapsing the chamber to expel air and refilling it with an inert gas. Plastic bags have been used in this latter manner. when the atmospheres are obtained by purging or collapsing the chambers, inert gas usually is supplied through the welding torch to insure complete protection of the welds.

(4) Joint designs. Joint designs for titanium are similar to those used for other metals. For welding a thin sheet, the tungsten-arc process generally is used. With this process, butt welds may be made with or without filler rod, depending on the thickness of the joint and fitup. Special shearing procedures somtimes are used so that the root opening does not exceed 8 percent of the sheet thickness. If fitup is this good, filler rod is not required. If fitup is not this good, filler metal is added to obtain full thickness joints. In welding thicker sheets (greater than 0.09 in. (2.3 mm)), both the tungsten-arc and consumable electrode processes are used with a root opening. For welding titanium plates, bars, or forgings, both the tungsten-arc and consumable electrode processes also are used with single and double V joints. In all cases, good weld penetration may be obtained with excessive drop through. However, penetration and dropthrough are controlled more easily by the use of proper backing fixtures.

NOTE

Because of the low thermal conductivity of titanium, weld beads tend to be wider than normal. However, the width of the beads is generally controlled by using short arc lengths, or by placing chill bars or the clamping toes of the jig close to the sides of the joints.

(5) <u>Welding variables.</u>

(a) Welding speed and current for titanium alloys depend on the process used, shielding gas, thickness of the material being welded, design of the backing fixtures, along with the spacing of chill bars or clamping bars in the welding jig. Welding speeds vary from about 3.0 to 40.0 in. (76.2 to 1016.0 mm) per minute. The highest welding speeds are obtained with the consumable electrode process. In most cases, direct current is used with straight polarity for the tungsten-arc process. Reverse polarity is used for the consumable electrode process.

(b) Arc wander has proven troublesome in some automatic welding operations. With arc wander, the arc from the tungsten or consumable electrode moves from one side of the weld joint to the other side. A straight, uniform weld bead will not be produced. Arc wander is believed to be caused by magnetic disturbances, bends in the filler wire, coatings on the filler wire, or a combination of these. Special metal shields and wire straighteners have been used to overcome arc wander, but have not been completely satisfactory. Also, constant voltage welding machines have been used in an attempt to overcome this problem. These machines also have not been completely satisfactory.

(c) In setting up arc welding operations for titanium, the welding conditions should be evaluated on the basis of weld joint properties and appearance. Radiographs will show if porosity or cracking is present in the weld joint. A simple bend test or notch toughness test will show whether or not the shielding conditions are adequate. A visual examination of the weld will show if the weld penetration and contour are satisfactory. After adequate procedures are established, careful controls are desirable to ensure that the shielding conditions are not changed.

(6) <u>Weld defects.</u>

(a) <u>General</u> Defects in arc welded joints in titanium alloys consist mainly of porosity (see (b)) and cold cracks (see (c)). Weld penetration can be controlled by adjusting welding conditions.

(b) <u>Porosity.</u> Weld porosity is a major problem in arc welding titanium alloys. Although acceptable limits for porosity in arc welded joints have not been establish, porosity has been observed in tungsten-arc welds in practically all of the alloys which appear suitable for welding operations. It does not extend to the surface of the weld, but has been detected in radiographs. It usually occurs close to the fusion line of the welds. Weld porosity may be reduced by agitating the molten weld puddle and adjusting welding speeds. Also, remelting the weld will eliminate some of the porosity present after the first pass. However, the latter method of reducing weld porosity tends to increase weld contamination.

7-21. TITANIUMWELDING (cont)

(c) Cracks.

1. With adequate shielding procedures and suitable alloys, cracks should not be a problem. However, cracks have been troublesome in welding some alloys. Weld cracks are attributed to a number of causes. In commercially pure titanium, weld metal cracks are believed to be caused by excessive oxygen or nitrogen contamination. These cracks are usually observed in weld craters. In some of the alpha-beta alloys, transverse cracks in the weld metal and heat affected zones are believed to be due to the low ductility of the weld zones. However, cracks in these alloys also may be due to contamination. Cracks also have been observed in alpha-beta welds made under restraint and with high external stresses. These cracks are sometimes attributed to the hydrogen content of the alloys.

NOTE
If weld cracking is due to contamination, it may be controlled by improving shielding conditions. However, repair welding on excessively contaminated welds is not practical in many cases.

2. Cracks which are caused by the low ductility of welds in alpha-beta alloys can be prevented by heat treating or stress relieving the weldment in a furnance immediately after welding. Oxyacetylene torches also have been used for this purpose. However, care must be taken so that the weldment is not overheated or excessively contaminated by the torch heating operation.

3. Cracks due to hydrogen may be prevented by vacuum annealing treatments prior to welding.

(7) Availability of welding filler wire. Most of the titanium alloys which are being used in arc welding applications are available as wire for use as welding filler metal. These alloys are listed below:

(a) Commercially pure titanium— commercially available as wire.

(b) Ti-5Al-2-1/2Sn alloy -- available as wire in experimental quantities.

(c) Ti-1-1/2Al-3Mn alloy -- available as wire in experimental quantities.

(d) Ti-6Al-4V alloy -- available as wire in experimental quantities.

(e) There has not been a great deal of need for the other alloys as welding filler wires. However, if such a need occurs, most of these alloys also could be reduced to wire. In fact, the Ti-8Mn alloy has been furnished as welding wire to met some requests.

d. Pressure Welding. Solid phase or pressure welding has been used to join titanium and titanium alloys. In these Processes, the surfaces to be jointed are not melted. They are held-together under pressure and heated to elevated temperatures (900 to 2000 °F (482 to 1093 °C)). One methd of heating used in pressure welding is the oxyacetylene flame. With suitable pressure and upset, good welds are obtainable in the high strength alpha-beta titanium alloys. The contaminated area on the surface of the weld is displaced from the joint area by the upset, which occurs during welding. This contaminated surface is machined off after welding. Another method of heating is by heated dies. Strong lap joints are obtained with this method in commercially pure titanium and a high strength alpha-beta alloy. By heating in this manner, welds may be made in very short periods of time, and inert gas shielding may be supplied to the joint. With all of the heating methods, less than 2 minutes is required to complete the welding operation. With solid phase or pressure welding processes, it is possible to produce ductile welds in the high strength alpha-beta alloys by using temperatures which do not cause embrittlement in these alloys.

7-22. NICKEL AND MONTEL WELDING

a. Generel. Nickel is a hard, malleable, ductile metal. Nickel and its alloys are commonly used when corrosion resistance is required. Nickel and nickel alloys such as Monel can, in general, be welded by metal-arc and gas welding methods. Some nickel alloys are more difficult to weld due to different compositions. The operator should make trial welds with reverse polarity at several current values and select the one best suited for the work. Generally, the oxyacetylene welding methods are preferred for smiler plates. However, small plates can be welded by the metal-arc and carbon-arc processes, and large plates are most satisfactorily joined, especially if the plate is nickel clad steel.

When welding, the nickel alloys can be treated much in the same manner as austenitic stainless steels with a few exceptions. These exceptions are:

(1) The nickel alloys will acquire a surface or coating which melts at a temperature approximately 1000°F (538 °C) above the melting point of the base metal.

(2) The nickel alloys are susceptible to embrittlement at welding temperatures by lead, sulfur, phosphorus, and some low-temperature metals and alloys.

(3) Weld penetration is less than expected with other metals.

When compensation is made for these three factors, the welding procedures used for the nickel alloys can he the same as those used for stainless steel. This is because the melting point, the coefficient of thermal expansion, and the thermal conductivity are similar to austenitic stainless steel.

It is necessary that each of these precautions be considered. The surface oxide should be completely removed from the joint area by grinding, abrasive blasting, machining, or by chemical means. When chemical etches are used, they must be completely removed by rinsing prior to welding. The oxide which melts at temperatures above the melting point of the base metal may enter the weld as a foreign material, or impurity, and will greatly reduce the strength and ductility of the weld. The problem of embrittlement at welding temperatures also means that the weld surface must be absolutely clean. paints, crayon markings, grease, oil, machining lubricants, and cutting oils may all contain the ingredients which will cause embrittlement. They must be completely removed for the weld area to avoid embrittlement. It is necessary to increase the opening of groove angles and to provide adequate root openings when full-penetration welds are used. The bevel or groove angles should be increased to approximately 40 percent over those used for carbon steel.

7-22. NICKEL AND MONEL WELDING (cont)

b. <u>Joint Design</u>. Butt joints are preferred but corner and lap joints can be effectively welded. Beveling is not required on plates 1/16 to 1/8 in. (1.6 to 3.2 mm) thick. With thicker materials, a bevel angle of 35 to 37-1/2 degrees should be made. When welding lap joints, the weld should be made entirely with nickel electrodes if water or air tightness is required.

c. <u>Welding Techniques</u>.

(1) Clean all surfaces to be welded either mechanically by machine, sandblasting, grinding, or with abrasive cloth; or chemically by pickling.

(2) Plates having U or V joints should be assembled, and if nickel clad steel, should be tacked on the steel side to prevent warping and distortion. After it is determined that the joint is even and flat, complete the weld on the steel side. Chip out and clean the nickel side and weld. If the base metal on both sides is nickel, clean out the groove on the unwelded side prior to beginning the weld on that side.

(3) If desired, the nickel side maybe completed first. However, the steel side must be tacked and thoroughly cleaned and beveled (or gouged) down to the root of the nickel weld prior to welding.

(4) Lap and corner joints are successfully welded by depositing a bead of nickel metal into the root and then weaving successive beads over the root weld.

(5) The arc drawn for nickel or nickel alloy welding should be slightly shorter than that used in normal metal-arc welding. A 1/16 to 1/8 in. (1.6 to 3.2 mm) arc is a necessity.

(6) Any position weld can be accomplished that can be satisfactorily welded by normal metal-arc welding of steel.

d. <u>Welding Methods</u>.

(1) Almost all the welding processes can be used for welding the nickel alloys. In addition, they can be joined by brazing and soldering.

(2) <u>Welding nickel alloys</u>. The most popular processes for welding nickel alloys are the shielded metal arc welding process, the gas tungsten arc welding process, and the gas metal arc welding process. Process selection depends on the normal factors. When shielded metal arc welding is used the procedures are essentially the same as those used for stainless steel welding.

The welding procedure schedule for using gas tungsten arc welding (TIG) is shown by table 7-29. The Welding procedure schedule for gas metal arc welding (MIG) is shown by table 7-30, p 7-108. The procedure information set forth on these tables will provide starting points for developing the welding procedures.

(3) No postweld heat treatment is required to maintain or restore corrosion resistance of the nickel alloys. Heat treatment is required for precipitating hardening alloys. Stress relief may be required to meet certain specifications to avoid stress corrosion cracking in applications involving hydrofluoric acid vapors or caustic solutions.

Table 7-29. Welding Procedure Schedules for Gas Tungsten Arc Welding (GTAW) Nickel Alloys (TIG Welding)

Material Thickness (or Fillet Size) ga	in.	mm	Type of Weld Fillet or Groove	Tungsten Electrode Diameter in.	mm	Filler Rod Diameter in.	mm	Nozzle Size Inside Dia. in.	Shielding Gas Flow cfh	Welding Current Amps DCEN	No. of Passes	Travel Speed (per pass) ipm
24	0.024	0.6	Sq. groove & fillet	1/16	1.6	None		3/8	15	8-10	1	8
16	0.063	1.6	Sq. groove & fillet	3/32	2.4	1/16	1.6	1/2	18	25-45	1	8
1/8	0.125	3.2	Sq. groove & fillet	1/8	3.2	3/32	2.4	1/2	25	125-175	1	11
1/4	0.250	6.4	Vee groove & fillet	1/8	3.2	1/8	3.2	1/2	30	125-175	2	8

NOTE

Tungsten used; 1st choice 2% thoriated EWTh2-2nd choice 1% thoriated EWTh1. Adequate gas shielding is required not only for the arc but also for the heated metal. Backing gas is recommended at all times. A trailing gas shield is also recommended. Argon is preferred, but for higher heat input on thicker material, use argon-helium mixture. Data is for flat position. Reduce amperage 10% to 20% when welding in horizontal, vertical, or overhead position.

7-22. NICKEL AND MONEL WELDING (cont)

Table 7-30. Welding Procedure Schedules for Gas Metal Arc Welding (GMAW) Nickel Alloys (MIG Welding)

Material Thickness (or Fillet Size)			Type of Weld Fillet or Groove	Electrode Diameter		WELDING POWER			Wire Feed Speed ipm	Shielding Gas Flow cfh	No. of Passes	Travel Speed (per pass) ipm
ga	in.	mm		in.	mm	Current Amps DC	Arc Volt EP					
1/16	0.062	1.6	Sq. groove & fillet	3/36	1.2	200-250	23-27		200-250	50	1	55-65
1/8	0.125	3.2	Sq. groove & fillet	1/16	1.6	290-340	25-35		150-175	60	1	30-35
1/4	0.250	6.4	Double vee & fillet	1/16	1.6	300-350	28-38		170-200	80	3	20-35

NOTE

Use 50% helium and 50% argon for thin metal and 100% helium for thick--higher voltage is for helium. Increase amperage 10-20% when backup is used. Data is for flat position. Reduce current 10-20% for other positions.

CHAPTER 8

ELECTRODES AND FILLER METALS

Section I. TYPES OF ELECTRODES

8-1. COVERED ELECTRODES

a. General. When molten metal is exposed to air, it absorbs oxygen and nitrogen, and becomes brittle or is otherwise adversely affected. A slag cover is needed to protect molten or solidifying weld metal from the atmosphere. This cover can be obtained from the electrode coating. The composition of the electrode coating determines its usability, as well as the composition of the deposited weld metal and the electrode specification. The formulation of electrode coatings is based on well-established principles of metallurgy, chemistry, and physics. The coating protects the metal from damage, stabilizes the arc, and improves the weld in other ways, which include:

(1) Smooth weld metal surface with even edges.

(2) Minimum spatter adjacent to the weld.

(3) A stable welding arc.

(4) Penetration control.

(5) A strong, tough coating.

(6) Easier slag removal.

(7) Improved deposition rate.

The metal-arc electrodes may be grouped and classified as bare or thinly coated electrodes, and shielded arc or heavy coated electrodes. The covered electrode is the most popular type of filler metal used in arc welding. The composition of the electrode covering determines the usability of the electrode, the composition of the deposited weld metal, and the specification of the electrode. The type of electrode used depends on the specific properties required in the weld deposited. These include corrosion resistance, ductility, high tensile strength, the type of base metal to be welded, the position of the weld (flat, horizontal, vertical, or overhead); and the type of current and polarity required.

b. Types of Electrodes. The coatings of electrodes for welding mild and low alloy steels may have from 6 to 12 ingredients, which include cellulose to provide a gaseous shield with a reducing agent in which the gas shield surrounding the arc is produced by the disintegration of cellulose; metal carbonates to adjust the basicity of the slag and to provide a reducing atmosphere; titanium dioxide to help form a highly fluid, but quick-freezing slag and to provide ionization for the arc; ferromanganese and ferrosilicon to help deoxidize the molten weld metal and to supplement the manganese content and silicon content of the deposited weld metal; clays and gums to provide elasticity for extruding the plastic coating material and to help provide strength to the coating; calcium fluoride to provide shielding gas to protect the arc, adjust the basicity of the slag, and provide fluidity and solubility of the metal oxides; mineral silicates to provide slag and give strength to the electrode covering; alloying metals including nickel, molybdenum, and chromium to provide alloy content to the deposited weld metal; iron or manganese oxide to adjust the fluidity and properties of the slag and to help stabilize the arc; and iron powder to increase the productivity by providing extra metal to be deposited in the weld.

The principal types of electrode coatings for mild steel and are described below.

(1) Cellulose-sodium (EXX10). Electrodes of this type cellulosic material in the form of wood flour or reprocessed low alloy electrodes have up to 30 percent paper. The gas shield contains carbon dioxide and hydrogen, which are reducing agents. These gases tend to produce a digging arc that provides deep penetration. The weld deposit is somewhat rough, and the spatter is at a higher level than other electrodes. It does provide extremely good mechanical properties, particularly after aging. This is one of the earliest types of electrodes developed, and is widely used for cross country pipe lines using the downhill welding technique. It is normally used with direct current with the electrode positive (reverse polarity).

(2) Cellulose-potassium (EXX11). This electrode is very similar to the cellulose-sodium electrode, except more potassium is used than sodium. This provides ionization of the arc and makes the electrode suitable for welding with alternating current. The arc action, the penetration, and the weld results are very similar. In both E6010 and E6011 electrodes, small amounts of iron powder may be added. This assists in arc stabilization and will slightly increase the deposition rate.

(3) Rutile-sodium (EXX12). When rutile or titanium dioxide content is relatively high with respect to the other components, the electrode will be especially appealing to the welder. Electrodes with this coating have a quiet arc, an easily controlled slag, and a low level of spatter. The weld deposit will have a smooth surface and the penetration will be less than with the cellulose electrode. The weld metal properties will be slightly lower than the cellulosic types. This type of electrode provides a fairly high rate of deposition. It has a relatively low arc voltage, and can be used with alternating current or with direct current with electrode negative (straight polarity).

(4) Rutile-potassium (EXX13). This electrode coating is very similar to the rutile-sodium type, except that potassium is used to provide for arc ionization. This makes it more suitable for welding with alternating current. It can also be used

with direct current with either polarity. It produces a very quiet, smooth running arc.

(5) Rutile-iron powder (EXXX4). This coating is very similar to the rutile coatings mentioned above, except that iron powder is added. If iron content is 25 to 40 percent, the electrode is EXX14. If iron content is 50 percent or more, the electrode is EXX24. With the lower percentage of iron powder, the electrode can be used in all positions. With the higher percentage of iron paler, it can only be used in the flat position or for making horizontal fillet welds. In both cases, the deposition rate is increased, based on the amount of iron powder in the coating.

(6) Low hydrogen-sodium (EXXX5). Coatings that contain a high proportion of calcium carbonate or calcium fluoride are called low hydrogen, lime ferritic, or basic type electrodes. In this class of coating, cellulose, clays, asbestos, and other minerals that contain combined water are not used. This is to ensure the lowest possible hydrogen content in the arc atmosphere. These electrode coatings are baked at a higher temperature. The low hydrogen electrode family has superior weld metal properties. They provide the highest ductility of any of the deposits. These electrodes have a medium arc with medium or moderate penetration. They have a medium speed of deposition, but require special welding techniques for best results. Low hydrogen electrodes must be stored under controlled conditions. This type is normally used with direct current with electrode positive (reverse polarity).

(7) Low hydrogen-potassium (EXXX6). This type of coating is similar to the low hydrogen-sodium, except for the substitution of potassium for sodium to provide arc ionization. This electrode is used with alternating current and can be used with direct current, electrode positive (reverse polarity). The arc action is smother, but the penetration of the two electrodes is similar.

(8) Low hydrogen-potassium (EXXX6). The coatings in this class of electrodes are similar to the low-hydrogen type mentioned above. However, iron powder is added to the electrode, and if the content is higher than 35 to 40 percent, the electrode is classified as an EXX18.

(9) Low hydrogen-iron powder (EXX28). This electrode is similar to the EXX18, but has 50 percent or more iron powder in the coating. It is usable only when welding in the flat position or for making horizontal fillet welds. The deposition rate is higher than EXX18. Low hydrogen coatings are used for all of the higher-alloy electrodes. By additions of specific metals in the coatings, these electrodes become the alloy types where suffix letters are used to indicate weld metal compositions. Electrodes for welding stainless steel are also the low-hydrogen type.

(10) Iron oxide-sodium (EXX20). Coatings with high iron oxide content produce a weld deposit with a large amount of slag. This can be difficult to control. This

coating type produces high-speed deposition, and provides medium penetration with low spatter level. The resulting weld has a very smooth finish. The electrode is usable only with flat position welding and for making horizontal fillet welds. The electrode can be used with alternating current or direct current with either polarity.

(11) Iron-oxide-iron power (EXX27). This type of electrode is very similar to the iron oxide-sodium type, except it contains 50 percent or more iron power. The increased amount of iron power greatly increases the deposition rate. It may be used with alternating direct current of either polarity.

(12) There are many types of coatings other than those mentioned here, most of which are usually combinations of these types but for special applications such as hard surfacing, cast iron welding, and for nonferrous metals.

c. Classification and Storage of Electrodes. Refer to paragraph 5-25 for classification and storage of electrodes.

d. Deposition Rates. The different types of electrodes have different deposition rates due to the composition of the coating. The electrodes containing iron power in the coating have the highest deposition rates. In the United States, the percentage of iron power in a coating is in the 10 to 50 percent range. This is based on the amount of iron power in the coating versus the coating weight. This is shown in the formula:

$$\% \text{ Iron powder} = \frac{\text{Weight of iron powder}}{\text{Total weight of coating}} \times 100$$

These percentages are related to the requirements of the American Welding Society (AWS) specifications. The European method of specifying iron power is based on the weight of deposited weld metal versus the weight of the bare core wire consumed. This is shown as follows:

$$\% \text{ Iron powder} = \frac{\text{Weight of deposited metal}}{\text{Weight of bare core wire}} \times 100$$

Thus, if the weight of the deposit were double the weight of the core wire, it would indicate a 200 percent deposition efficiency, even though the amount of the iron power in the coating represented only half of the total deposit. The 30 percent iron power formula used in the United States would produce a 100 to 110 percent deposition efficiency using the European formula. The 50 percent iron power electrode figured on United States standards would produce an efficiency of approximately 150 percent using the European formula.

e. Light Coated Electrodes.

(1) Light coated electrodes have a definite composition. A light coating has been applied on the surface by washing, dipping, brushing, spraying, tumbling, or wiping. The coatings improve the characteristics of the arc stream. They are listed under the E45 series in the electrode identification system, refer to paragraph 5-25.

(2) The coating generally serves the functions described below:

(a) It dissolves or reduces impurities such as oxides, sulfur, and phosphorus.

(b) It changes the surface tension of the molten metal so that the globules of metal leaving the end of the electrode are smaller and more frequent. This helps make flow of molten metal more uniform.

(c) It increases the arc stability by introducing materials readily ionized (i.e., changed into small particles with an electric charge) into the arc stream.

(3) Some of the light coatings may produce a slag. The slag is quite thin and does not act in the same manner as the shielded arc electrode type slag.

f. Shielded Arc or Heavy Coated Electrodes. Shielded arc or heavy coated electrodes have a definite composition on which a coating has been applied by dipping or extrusion. The electrodes are manufactured in three general types: those with cellulose coatings; those with mineral coatings; and those whose coatings are combinations of mineral and cellulose. The cellulose coatings are composed of soluble cotton or other forms of cellulose with small amounts of potassium, sodium, or titanium, and in some cases added minerals. The mineral coatings consist of sodium silicate, metallic oxides clay, and other inorganic substances or combinations thereof. Cellulose coated electrodes protect the molten metal with a gaseous zone around the arc as well as the weld zone. The mineral coated electrode forms a slag deposit. The shielded arc or heavy coated electrodes are used for welding steels, cast iron, and hard surfacing.

g. Functions of Shielded Arc or Heavy Coated Electrodes.

(1) These electrodes produce a reducing gas shield around the arc. This prevents atmospheric oxygen or nitrogen from contaminating the weld metal. The oxygen readily combines with the molten metal, removing alloying elements and causing porosity. Nitrogen causes brittleness, low ductility, and in Some cases low strength and poor resistance to corrosion.

(2) They reduce impurities such as oxides, sulfur, and phosphorus so that these impurities will not impair the weld deposit.

(3) They provide substances to the arc which increase its stability. This eliminates wide fluctuations in the voltage so that the arc can be maintained without excessive spattering.

(4) By reducing the attractive force between the molten metal and the end of the electrodes, or by reducing the surface tension of the molten metal, the vaporized and melted coating causes the molten metal at the end of the electrode to break up into fine, small particles.

(5) The coatings contain silicates which will form a slag over the molten weld and base metal. Since the slag solidifies at a relatively slow rate, it holds the heat and allows the underlying metal to cool and solidify slowly. This slow solidification of the metal eliminates the entrapment of gases within the weld and permits solid impurities to float to the surface. Slow cooling also has an annealing effect on the weld deposit.

(6) The physical characteristics of the weld deposit are modified by incorporating alloying materials in the electrode coating. The fluxing action of the slag will also produce weld metal of better quality and permit welding at higher speeds.

h. Direct Current Arc Welding Electrodes.

(1) The manufacturer's recommendations should be followed when a specific type of electrode is being used. In general, direct current shielded arc electrodes are designed either for reverse polarity (electrode positive) or for straight polarity (electrode negative), or both. Many, but not all, of the direct current electrodes can be used with alternating current. Direct current is preferred for many types of covered nonferrous, bare and alloy steel electrodes. Recommendations from the manufacturer also include the type of base metal for which given electrodes are suitable, corrections for poor fit-ups, and other specific conditions.

(2) In most cases, reverse polarity electrodes will provide more penetration than straight polarity electrodes. Good penetration can be obtained from either type with proper welding conditions and arc manipulation.

i. Alternating Current Arc Welding Electrodes.

(1) Coated electrodes which can be used with either direct or alternating current are available. Alternating current is more desirable while welding in restricted areas or when using the high currents required for thick sections because it reduces arc blow. Arc blow causes blowholes, slag inclusions, and lack of fusion in the weld.

(2) Alternating current is used in atomic hydrogen welding and in those carbon arc processes that require the use of two carbon electrodes. It permits a uniform rate of welding and electrode consumption ion. In carbon-arc processes where one

carbon electrode is used, direct current straight polarity is recommended, because the electrode will be consumed at a lower rate.

j. Electrode Defects and Their Effect.

(1) If certain elements or oxides are present in electrode coatings, the arc stability will be affected. In bare electrodes, the composition and uniformity of the wire is an important factor in the control of arc stability. Thin or heavy coatings on the electrodes will not completely remove the effects of defective wire.

(2) Aluminum or aluminum oxide (even when present in quantities not exceeding 0.01 percent), silicon, silicon dioxide, and iron sulfate cause the arc to be unstable. Iron oxide, manganese oxide, calcium oxide, and iron sulfide tend to stabilize the arc.

(3) When phosphorus or sulfur are present in the electrode in excess of 0.04 percent, they will impair the weld metal. They are transferred from the electrode to the molten metal with very little loss. Phosphorus causes grain growth, brittleness, and "cold shortness" (i.e., brittle when below red heat) in the weld. These defects increase in magnitude as the carbon content of the steel increases. Sulfur acts as a slag, breaks up the soundness of the weld metal, and causes "hot shortness" (i.e., brittle when above red heat). Sulfur is particularly harmful to bare low carbon steel electrodes with a low manganese content. Manganese promotes the formation of sound welds.

(4) If the heat treatment given the wire core of an electrode is not uniform, the electrode will produce welds inferior to those produced with an electrode of the same composition that has been properly heat treated.

8-2. SOLID ELECTRODE WIRES

a. General. Bare or solid wire electrodes are made of wire compositions required for specific applications, and have no coatings other than those required in wire drawing. These wire drawing coatings have a slight stabilizing effect on the arc, but are otherwise of no consequence. Bare electrodes are used for welding manganese steels and for other purposes where a covered electrode is not required or is undesirable. A sketch of the transfer of metal across the arc of a bare electrode is shown in figure 8-1.

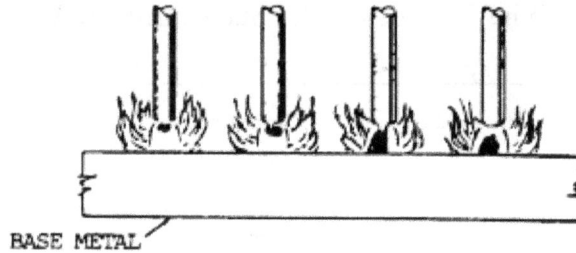

Figure 8-1. Transfer of metal across the arc of a bare electrode.

b. Solid steel electrode wires may not be bare. Many have a very thin copper coating on the wire. The copper coating improves the current pickup between contact tip and the electrode, aids drawing, and helps prevent rusting of the wire when it is exposed to the atmosphere. Solid electrode wires are also made of various stainless steels, aluminum alloys, nickel alloys, magnesium alloys, titanium alloys, copper alloys, and other metals.

c. When the wire is cut and straightened, it is called a welding rod, which is a form of filler metal used for welding or brazing and does not conduct the electrical current. If the wire is used in the electrical circuit, it is called a welding electrode, and is defined as a component of the welding circuit through which current is conducted. A bare electrode is normally a wire; however, it can take other forms.

d. Several different systems are used to identify the classification of a particular electrode or welding rod. In all cases a prefix letter is used.

    (1) <u>Prefix R</u>. Indicates a welding rod.

    (2) <u>Prefix E</u>. Indicates a welding electrode.

    (3) <u>Prefix RB</u>. Indicates use as either a welding rod or for brazing filler metal.

    (4) <u>Prefix ER</u>. Indicates wither an electrode or welding rod.

e. The system for identifying bare carbon steel electrodes and rods for gas shielded arc welding is as follows:

    (1) <u>ER</u> indicates an electrode or welding rod.

    (2) <u>70</u> indicates the required minimum as-welded tensile strength in thousands of pounds per square inch (psi).

    (3) <u>S</u> indicates solid electrode or rod.

    (4) <u>C</u> indicates composite metal cored or stranded electrode or rod.

    (5) <u>1</u> suffix number indicates a particular analysis and usability factor.

Table 8-1. Mild Steel Electrode Wire Composition for Submerged Arc Welding

AWS* Classification	C	Mn	Si	S	S	P	Total Other Elements
Chemical Composition-Percent							
Low manganese classes							
EL8	0.10	0.30 to 0.55	0.05	0.035	0.03	0.30	0.50
EL8K	0.10	0.30 to 0.55	0.10 to 0.20	0.035	0.03	0.30	0.50
EL12	0.07 to 0.15	0.35 to 0.60	0.05	0.035	0.03	0.30	0.50
Medium manganese classes							
EM5K	0.06	0.90 to 1.40	0.40 to 0.70	0.035	0.03	0.30	0.50
EM12	0.07 to 0.15	0.85 to 1.25	0.05	0.035	0.03	0.30	0.50
EM12K	0.07 to 0.15	0.85 to 1.25	0.15 to 0.35	0.035	0.03	0.30	0.50
EM13K	0.07 to 0.19	0.90 to 1.40	0.45 to 0.70	0.035	0.03	0.30	0.50
EM15K	0.12 to 0.20	0.85 to 1.25	0.15 to 0.35	0.035	0.03	0.30	0.50
High manganese class							
EH14	0.10 to 0.18	1.75 to 2.25	0.05	0.035	0.03	0.30	0.50

*American Welding Society

f. Submerged Arc Electrodes. The system for identifying solid bare carbon steel for submerged arc is as follows:

(1) The prefix letter E is used to indicate an electrode. This is followed by a letter which indicates the level of manganese, i.e., L for low, M for medium, and H for high manganese. This is followed by a number which is the average amount of carbon in points or hundredths of a percent. The composition of some of these wires is almost identical with some of the wires in the gas metal arc welding specification.

(2) The electrode wires used for submerged arc welding are given in American Welding Society specification, "Bare Mild Steel Electrodes and Fluxes for Submerged Arc Welding." This specification provides both the wire composition and the weld deposit chemistry based on the flux used. The specification does give composition of the electrode wires. This information is given in table 8-1. When these electrodes are used with specific submerged arc fluxes and welded with proper procedures, the deposited weld metal will meet mechanical properties required by the specification.

(3) In the case of the filler reds used for oxyfuel gas welding, the prefix letter is R, followed by a G indicating that the rod is used expressly for gas welding. These letters are followed by two digits which will be 45, 60, or 65. These designate the approximate tensile strength in 1000 psi (6895 kPa).

(4) In the case of nonferrous filler metals, the prefix E, R, or RB is used, followed by the chemical symbol of the principal metals in the wire. The initials for one or

two elements will follow. If there is more than one alloy containing the same elements, a suffix letter or number may be added.

(5) The American Welding Society's specifications are most widely used for specifying bare welding rod and electrode wires. There are also military specifications such as the MIL-E or -R types and federal specifications, normally the QQ-R type and AMS specifications. The particular specification involved should be used for specifying filler metals.

g. The most important aspect of solid electrode wires and rods in their composition, which is given by the specification. The specifications provide the limits of composition for the different wires and mechanical property requirements.

h. Occasionally, on copper-plated solid wires, the copper may flake off in the feed roll mechanism and create problems. It may plug liners, or contact tips. A light copper coating is desirable. The electrode wire surface should be reasonably free of dirt and drawing compounds. This can be checked by using a white cleaning tissue and pulling a length of wire through it. Too much dirt will clog the liners, reduce current pickup in the tip, and may create erratic welding operation.

i. Temper or strength of the wire can be checked in a testing machine. Wire of a higher strength will feed through guns and cables better. The minimum tensile strength recommended by the specification is 140,000 psi (965,300 kPa).

j. The continuous electrode wire is available in many different packages. They range from extremely small spools that are used on spool guns, through medium-size spools for fine-wire gas metal arc welding. Coils of electrode wire are available which can be placed on reels that are a part of the welding equipment. There are also extremely large reels weighing many hundreds of pounds. The electrode wire is also available in drums or payoff packs where the wire is laid in the round container and pulled from the container by an automatic wire feeder.

8-3. FLUX-CORED OR TUBULAR ELECTRODES

a. General. The flux-cored arc welding process is made possible by the design of the electrode. This inside-outside electrode consists of a metal sheath surrounding a core of fluxing and alloying compounds. The compounds contained in the electrode perform essentially the same functions as the coating on a covered electrode, i.e., deoxidizers, slag formers, arc stabilizers, alloying elements, and may provide shielding gas. There are three reasons why cored wires are developed to supplement solid electrode wires of the same or similar analysis.

(1) There is an economic advantage. Solid wires are drawn from steel billets of the specified analyses. These billets are not readily available and are expensive. A single billet might also provide more solid electrode wire than needed. In the case

of cored wires, the special alloying elements are introduced in the core material to provide the proper deposit analysis.

(2) Tubular wire production method provides versatility of composition and is not limited to the analysis of available steel billets.

(3) Tubular electrode wires are easier for the welder to use than solid wires of the same deposit analysis, especially for welding pipe in the fixed position.

b. Flux-Cored Electrode Design. The sheath or steel portion of the flux-cored wire comprises 75 to 90 percent of the weight of the electrode, and the core material represents 10 to 25 percent of the weight of the electrode.

For a covered electrode, the steel represents 75 percent of the weight and the flux 25 percent. This is shown in more detail below:

	Flux Cored Electrode Wire (E70T-1)			Covered Electrode (E7016)	
By area	Flux	25%	By area	Flux	55%
	steel	75%		steel	45%
By weight	Flux	15%	By weight	Flux	24%
	steel	85%		steel	76%

More flux is used on covered electrodes than in a flux-cored wire to do the same job. This is because the covered electrode coating contains binders to keep the coating intact and also contains agents to allow the coating to be extruded.

c. Self-Shielding Flux-Cored Electrodes. The self-shielding type flux-cored electrode wires include additional gas forming elements in the core. These are necessary to prohibit the oxygen and nitrogen of the air from contacting the metal transferring across the arc and the molten weld puddle. Self-shielding electrodes also include extra deoxidizing and denigrating elements to compensate for oxygen and nitrogen which may contact the molten metal. Self-shielding electrodes are usually more voltage-sensitive and require electrical stickout for smooth operation. The properties of the weld metal deposited by the self-shielding wires are sometimes inferior to those produced by the externally shielded electrode wires because of the extra amount of deoxidizers included. It is possible for these elements to build up in multipass welds, lower the ductility, and reduce the impact values of the deposit. Some codes prohibit the use of self-shielding wires on steels with yield strength exceeding 42,000 psi (289,590 kPa). Other codes prohibit the self-shielding wires from being used on dynamically loaded structures.

d. <u>Metal Transfer</u>. Metal transfer from consumable electrodes across an arc has been classified into three general modes. These are spray transfer, globular transfer, and short circuiting transfer. The metal transfer of flux-cored electrodes resembles a fine globular transfer. On cored electrodes in a carbon dioxide shielding atmosphere, the molten droplets build up around the outer sheath of the electrode. The core material appears to transfer independently to the surface of the weld puddle. At low currents, the droplets tend to be larger than when the current density is increased. Transfer is more frequent with smaller drops when the current is increased. The larger droplets at the lower currents cause a certain amount of splashing action when they enter the weld puddle. This action decreases with the smaller droplet size. This explains why there is less visible spatter, the arc appears smoother to the welder, and the deposition efficiency is higher when the electrode is used at high current rather than at the low end of its current range.

e. <u>Mild Steel Electrodes</u>. Carbon steel electrodes are classified by the American Welding Society specification, "Carbon Steel Electrodes for Flux-cored-Arc Welding". This specification includes electrodes having no appreciable alloy content for welding mild and low alloy steels. The system for identifying flux-cored electrodes follows the same pattern as electrodes for gas metal arc welding, but is specific for tubular electrodes. For example, in E70T-1, the E indicates an electrode; 70 indicates the required minimum as-welded tensile strength in thousands of pounds per square inch (psi); T indicates tubular, fabricated, or flux-cored electrode; and 1 indicates the chemistry of the deposited weld metal, gas type, and usability factor.

f. <u>Classification of Flux-Cored Electrodes</u>.

(1) <u>E60T-7 electrode classification</u>. Electrodes of this classification are used without externally applied gas shielding and may be used for single-and multiple-pass applications in the flat and horizontal positions. Due to low penetration and to other properties, the weld deposits have a low sensitivity to cracking.

(2) <u>E60T-8 electrode classifications</u>. Electrodes of this classification are used without externally applied gas shielding and may be used for single-and multiple-pass applications in the flat and horizontal positions. Due to low penetration and to other properties, the weld deposits have a low sensitivity to cracking.

(3) <u>E70T-1 electrode classification</u>. Electrodes of this classification are designed to be used with carbon dioxide shielding gas for single-and multiple-pass welding in the flat position and for horizontal fillets. A quiet arc, high-deposition rate, low spatter loss, flat-to-slightly convex bead configuration, and easily controlled and removed slag are characteristics of this class.

(4) <u>E70T-2 electrode classification</u>. Electrodes of this classification are used with carbon dioxide shielding gas and are designed primarily for single-pass welding in the flat position and for horizontal fillets. However, multiple-pass welds can be made when the weld beads are heavy and an appreciable amount of mixture of the base and filler metals occurs.

(5) <u>E70T-3 electrode classification</u>. Electrodes of this classification are used without externally applied gas shielding and are intended primarily for depositing single-pass, high-speed welds in the flat and horizontal positions on light plate and gauge thickness base metals. They should not be used on heavy sections or for multiple-pass applications.

(6) <u>E70T-4 electrode classification</u>. Electrodes of this classification are used without externally applied gas shielding and may be used for single-and multiple-pass applications in the flat and horizontal positions. Due to low penetration, and to other properties, the weld deposits have a low sensitivity to cracking.

(7) <u>E70T-5 electrode classification</u>. This classification covers electrodes primarily designed for flat fillet or groove welds with or without externally applied shielding gas. Welds made using-carbon dioxide shielding gas have better quality than those made with no shielding gas. These electrodes have a globular transfer, low penetration, slightly convex bead configuration, and a thin, easily removed slag.

(8) <u>E70T-6 electrode classification</u>. Electrodes of this classification are similar to those of the E70T-5 classification, but are designed for use without an externally applied shielding gas.

(9) <u>E70T-G electrode classification</u>. This classification includes those composite electrodes that are not included in the preceding classes. They may be used with or without gas shielding and may be used for multiple-pass work or may be limited to single-pass applications. The E70T-G electrodes are not required to meet chemical, radiographic, bend test, or impact requirements; however, they are required to meet tension test requirements. Welding current type is not specified.

g. The flux-cored electrode wires are considered to be low hydrogen, since the materials used in the core do not contain hydrogen. However, some of these materials are hydroscopic and thus tend to absorb moisture when exposed to a high-humidity atmosphere. Electrode wires are packaged in special containers to prevent this. These electrode wires must be stored in a dry room.

h. <u>Stainless Steel Tubular Wires</u>. Flux-cored tubular electrode wires are available which deposit stainless steel weld metal corresponding to the A.I.S.I. compositions. These electrodes are covered by the A.W.S specification, "Flux-Cored Corrosion Resisting Chromium and Chromium-Nickel Steel Electrodes." These electrodes are identified by the prefix E followed by the standard A.I.S.I. code number. This is followed by the letter T indicating a tubular electrode. Following this and a dash are four-possible suffixes as follows:

(1) -1 indicates the use of CO_2 (carbon dioxide) gas for shielding and DCEP.

(2) -2 indicates the use of argon plus 2 percent oxygen for shielding and DCEP.

(3) -3 indicates no external gas shielding and DCEP.

(4) -G indicates that gas shielding and polarity are not specified.

Tubular or flux-cored electrode wires are also used for surfacing and submerged arc welding applications.

i. Deposition Rates and Weld Quality. The deposition rates for flux-cored electrodes are shown in figure 8-2. These curves show deposition rates when welding with mild and low-alloy steel using direct current electrode positive. Two type of of covered electrodes are shown for comparison. Deposition rates of the smaller size flux-cored wires exceed that of the covered electrodes. The metal utilization of the flux-cored electrode is higher. Flux-cored electrodes have a much broader current range than covered electrodes, which increases the flexibility of the process. The quality of the deposited weld metal produced by the flux-cored arc welding process depends primarily on the flux-cored electrode wire that is used. It can be expected that the deposited weld metal will match or exceed the properties shown for the electrode used. This assures the proper matching of base metal, flux-cored electrode type and shielding gas. Quality depends on the efficiency of the gas shielding envelope, on the joint detail, on the cleanliness of the joint, and on the skill of the welder. The quality level of of weld metal deposited by the self-shielding type electrode wires is usually lower than that produced by electrodes that utilize external gas shielding.

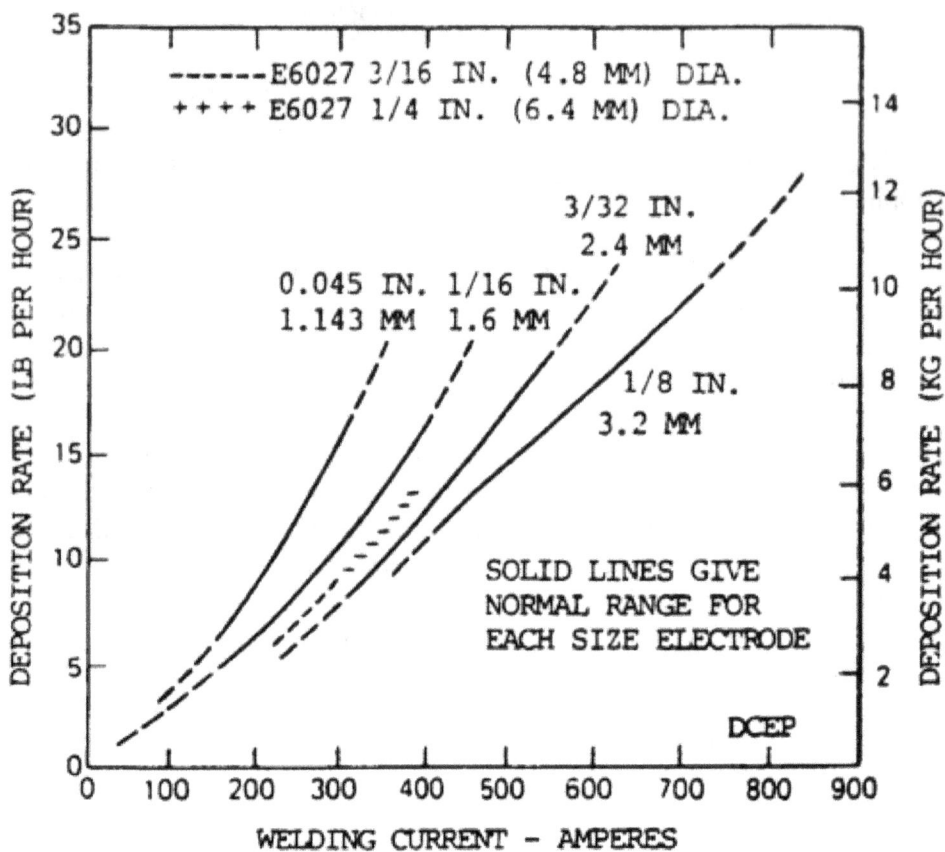

Figure 8-2. Deposition rates of steel flux-cored electrodes.

Section II. OTHER FILLER METALS

8-4. GENERAL

There are other filler metals and special items normally used in making welds. These include the nonconsumable electrodes (tungsten and carbon), and other materials, including backing tapes, backing devices, flux additives, solders, and brazing alloys. Another type of material consumed in making a weld are the consumable rings used for root pass welding of pipe. There are also ferrules used for stud welding and the guide tubes in the consumable guide electroslag welding method. Other filler materials are solders and brazing alloys.

8-5. NONCONSUMABLE ELECTRODES

a. Types of Nonconsumable Electrodes. There are two types of nonconsumable electrodes. The carbon electrode is a non-filler metal electrode used in arc welding or cutting, consisting of a carbon graphite rod which may or may not be coated with copper or other coatings. The second nonconsumable electrode is the tungsten electrode, defined as a non-filler metal electrode used in arc welding or cutting, made principally of tungsten.

b. Carbon Electrodes. The American Welding Society does not provide specification for carbon electrodes but there is a military specification, no. MIL-E-17777C, entitled, "Electrodes Cutting and Welding Carbon-Graphite Uncoated and Copper Coated". This specification provides a classification system based on three grades: plain, uncoated, and copper coated. It provides diameter information, length information, and requirements for size tolerances, quality assurance, sampling, and various tests. Applications include carbon arc welding, twin carbon arc welding, carbon cutting, and air carbon arc cutting and gouging.

c. Tungsten Electrodes.

(1) Nonconsumable electrodes for gas types: pure tungsten, tungsten containing tungsten arc (TIG) welding are of four 1.0 percent thorium, tungsten containing 2.0 percent thorium, and tungsten containing 0.3 to 0.5 percent zirconium. They are also used for plasma-arc and atomic hydrogen arc welding.

(2) Tungsten electrodes can be identified by painted end marks:

(a) Green - pure tungsten.

(b) Yellow - 1.0 percent thorium.

(c) Red - 2.0 percent thorium.

(d) Brown - 0.3 to 0.5 percent zirconium.

(3) Pure tungsten (99. 5 percent tungsten) electrodes are generally used on less critical welding operations than the tungstens which are alloyed. This type of electrode has a relatively low current carrying capacity and a low resistance to contamination.

(4) Thoriated tungsten electrodes (1.0 or 2.0 percent thorium) are superior to pure tungsten electrodes because of their higher electron output, better arc starting and arc stability, high current-carrying capacity, longer life, and greater resistance to contamination.

(5) Tungsten electrodes containing 0.3 to 0.5 percent zirconium generally fall between pure tungsten electrodes and thoriated tungsten electrodes in terms of

performance. There is, however, some indication of better performance in certain types of welding using ac power.

(6) Finer arc control can be obtained if the tungsten alloyed electrode is ground to a point (fig. 8-3). When electrodes are not grounded, they must be operated at maximum current density to obtain reasonable arc stability. Tungsten electrode points are difficult to maintain if standard direct current equipment is used as a power source and touch--starting arc is standard practice. Maintenance of electrode shape and the reduction of tungsten inclusions in the weld can best be ground by superimposing a high-frequency current on the regular welding current. Tungsten electrodes alloyed with thorium retain their shape longer when touch-starting is used. Unless high frequency alternating current is available, touch-starting must be used with thorium electrodes.

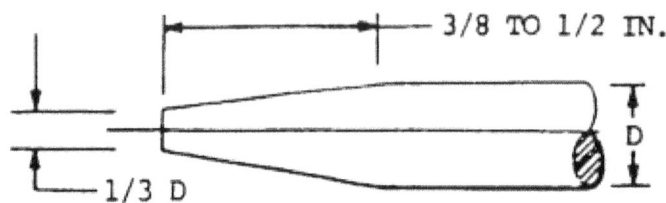

Figure 8-3. Correct electrode taper.

(7) The electrode extension beyond the gas cup is determined by the type of joint being welded. For example, an extension beyond the gas cup of 1/8 in. (0.32 cm) might be used for butt joints in light gauge material, while an extension of approximately 1/4 to 1/2 in. (0.64 to 1.27 cm) might be necessary on some fillet welds. The tungsten electrode or torch should be inclined slightly and the filler metal added carefully to avoid contact with the tungsten to prevent contamination of the electrode. If contamination does occur, the electrode must be removed, reground, and replaced in the torch.

d. Backing Materials. Backing materials are being used more frequently for welding. Special tapes exist, some of which include small amounts of flux, which can be used for backing the roots of joints. There are also different composite backing materials, for one-side welding. Consumable rings are used for making butt welds in pipe and tubing. These are rings made of metal that are tack welded in the root of the weld joint and are fused into the joint by the gas tungsten arc. There are three basic types of rings called consumable inert rings which are available in different analyses of metal based on normal specifications.

8-6. SUBMERGED ARC FLUX ADDITIVES

Specially processed metal powder is sometimes added to the flux used for the submerged arc welding process. Additives are provided to increase productivity or enrich the alloy composition of the deposited weld metal. In both cases, the additives are of a proprietary

TC 9-237

nature and are described by their manufacturers, indicating the benefit derived by using the particular additive. Since there are no specifications covering these types of materials, the manufacturer's information must be used.

8-7. SOLDERING

a. <u>General</u>. Soldering is the process of using fusible alloys for joining metals. The kind of solder used depends on the metals being joined. Hard solders are called spelter, and hard soldering is called silver solder brazing. This process gives greater strength and will withstand more heat than soft solder. Soft soldering is used for joining most common metals with an alloy that melts at a temperature below that of the base metal, and always below 800°F (427°C). In many respects, this is similar to brazing, in that the base is not melted, but merely tinned on the surface by the solder filler metal. For its strength, the soldered joint depends on the penetration of the solder into the pores of the base metal and. the formation of a base metal-alloy solder.

b. Solders of the tin-lead alloy system constitute the largest portion of all solders in use. They are used for joining most metals and have good corrosion resistance to most materials. Most cleaning and soldering processes may be used with the tin-lead solders. Other solders are: tin-antimony; tin-antimony-lead; tin-silver; tin-lead-silver; tin-zinc; cadmium-silver; cadmium-zinc; zinc-aluminum; bismuth (fusible) solder; and indium solders. These are described below. Fluxes of all types can also be used; the choice depends on the base metal to be joined.

(1) <u>Tin-antinmony solder</u>. The 95 percent tin-5 percent antimony solder provides a narrow melting range at a temperature higher than the tin-lead eutectic. the solder is used many plumbing, refrigeration, and air conditioning applications because of its good creep strength.

(2) <u>Tin-antimony-lead solders</u>. Antimony may be added to a tin-lead solder as a substitute for some of the tin. The addition of antimony up to 6 percent of the tin content increases the mechanical properties of the solder with only slight impairment to the soldering characteristics. All standard methods of cleaning, fluxing, and heating may be used.

(3) <u>Tin-silver and tin-lead-silver solders</u>. The 96 percent tin-4 percent silver solder is free of lead and is often used to join stainless steel for food handling equipment. It has good shape and creep strengths, and excellent flow characteristics. The 62 percent tin-38 percent lead-2 percent silver solder is used when soldering silver-coated surfaces for electronic applications. The silver addition retards the dissolution of the silver coating during the soldering operation. The addition of silver also increases creep strength. The high lead solders containing tin and silver provide higher temperature solders or many applications. They exhibit good tensile, shear, and creep strengths and are recommended for cryogenic applications. Because of their high melting range, only inorganic fluxes are recommended for use with these solders.

(4) <u>Tin-zinc solders</u>. A large number of tin-zinc solders have come into use for joining aluminum. Galvanic corrosion of soldered joints in aluminum is minimized if the metals in the joint are close to each other in the electrochemical series. Alloys containing 70 to 80 percent tin with the balance zinc are recommended for soldering aluminum. The addition of 1 to 2 percent aluminum, or an increase of the zinc content to as high as 40 percent, improves corrosion resistance. However, the liquidus temperature rises correspondingly, and these solders are therefore more difficult to apply. The 91/9 and 60/40 tin-zinc solders may be used for high temperature applications (above 300°F (149°C)), while the 80/20 and the 70/30 tin-zinc solders are generally used to coat parts before soldering.

CAUTION

Cadmium fumes can be health hazards. Improper use of solders containing cadmium can be hazardous to personnel.

(5) <u>Cadmium-silver solder</u>. The 95 percent cadmium-5 percent silver solder is in applications where service temperatures will be higher than permissible with lower melting solders. At room temperature, butt joints in copper can be made to produce tensile strengths of 170 MPa (25,000 psi). At 425°F (218°C), a tensile strength of 18 MPa (2600 psi) can be obtained. Joining aluminum to itself or to other metals is possible with this solder. Improper use of solders containing cadmium may lead to health hazards. Therefore, care should be taken in their application, particularly with respect to fume inhalation.

(6) <u>Cadmium-zinc solders</u>. These solders are also useful for soldering aluminum. The cadmium-zinc solders develop joints with intermediate strength and corrosion resistance when used with the proper flux. The 40 percent cadmium-60 percent zinc solder has found considerable use in the soldering of aluminum lamp bases. Improper use of this solder may lead to health hazards, particularly with respect to fume inhalation.

(7) <u>Zinc-aluminum solder</u>. This solder is specifically for use on aluminum. It develops joints with high strength and good corrosion resistance. The solidus temperature is high, which limits its use to applications where soldering temperature is in excess of 700°F (371°C) can be tolerated. A major application is in dip soldering the return bends of aluminum air conditioner coils. Ultrasonic solder pots are employed without the use of flux. In manual operations, the heated aluminum surface is rubbed with the solder stick to promote wetting without a flux.

(8) <u>Fusible alloys</u>. Bismuth-containing solders, the fusible alloys, are useful for soldering operations where soldering temperatures helm 361°F (183°C) are required. The low melting temperature solders have applications in cases such as soldering heat treated surfaces where higher soldering temperatures would result

in the softening of the part; soldering joints where adjacent material is very sensitive to temperature and would deteriorate at higher soldering temperatures; step soldering operations where a low soldering temperature is necessary to avoid destroying a nearby joint that has been made with a higher melting temperature solder; and on temperature-sensing devices, such as fire sprinkler systems, where the device is activated when the fusible alloy melts at relatively low temperature. Many of these solders, particularly those containing a high percentage of bismuth, are very difficult to use successfully in high-speed soldering operations. Particular attention must be paid to the cleanliness of metal surfaces. Strong, corrosive fluxes must be used to make satisfactory joints on uncoated surfaces of metals, such as copper or steel. If the surface can be plated for soldering with such metals as tin or tin-lead, noncorrosive rosin fluxes may be satisfactory; however, they are not effective below 350°F (177°C).

(9) Indium solders. These solders possess certain properties which make them valuable for some special applications. Their usefulness for any particular application should be checked with the supplier. A 50 percent indium-50 percent tin alloy adheres to glass readily and may be used for glass-to-metal and glass-to-glass soldering. The low vapor pressure of this alloy makes it useful for seals in vacuum systems. Iridium solders do not require special techniques during use. All of the soldering methods, fluxes, and techniques used with the tin-lead solders are applicable to iridium solders.

8-8. BRAZING ALLOYS

a. General.

(1) Brazing is similar to the soldering processes in that a filler rod with a melting point lower than that of the base metal, but stove 800°F (427°C) is used. A groove, fillet, plug, or slot weld is made and the filler metal is distributed by capillary attraction. In brazing, a nonferrous filler rod, strip, or wire is used for repairing or joining cast iron, malleable iron, wrought iron, steel, copper, nickel, and high melting point brasses and bronzes. Some of these brasses and bronzes, however, melt at a temperature so near to that of the filler rod that fusion welding rather than brazing is required.

(2) Besides a welding torch with a proper tip size, a filler metal of the required composition and a proper flux are important to the success of any brazing operation.

(3) The choice of the filler metal depends on the types of metals to be joined. Copper-silicon (silicon-bronze) rods are used for brazing copper and copper alloys. Copper-tin (phosphor-bronze) rods are used for brazing similar copper alloys and for brazing steel and cast iron. Other compositions are used for brazing specific metals.

(4) Fluxes are used to prevent oxidation of the filler metal and the base metal surface, and to promote the free flowing of the filler metal. They should be chemically active and fluid at the brazing temperature. After the joint members have been fitted and thoroughly cleaned, an even coating of flux should be brushed over the adjacent surfaces of the joint, taking care that no spots are left uncovered. The proper flux is a good temperate indicator for torch brazing because the joint should be heated until the flux remains fluid when the torch flame is momentarily removed.

b. Characteristics. For satisfactory use in brazing applications, brazing filler metals must possess the following properties:

(1) The ability to form brazed joints possessing suitable mechanical and physical properties for the intended service application.

(2) A melting point or melting range compatible with the base metals being joined and sufficient fluidity at brazing temperature to flow and distribute into properly prepared joints by capillary action.

(3) A composition of sufficient homogeneity and stability to minimize separation of constituents (liquation) under the brazing conditions encountered.

(4) The ability to wet the surfaces of the base metals being joined and form a strong, sound bond.

(5) Depending on the requirements, ability to produce or avoid base metal-filler metal interactions.

c. Filler Metal Selection. The following factors should be considered when selecting a brazing filler metal:

(1) Compatibility with base metal and joint design.

(2) Service requirements for the brazed assembly. Compositions should be selected to suit operating requirements, such as service temperature (high or cryogenic), thermal cycling, life expectancy, stress loading, corrosive conditions, radiation stability, and vacuum operation.

(3) Brazing temperature required. Low brazing temperatures are usually preferred to economize on heat energy; minimize heat effects on base metal (annealing, grain growth, warpage, etc.); minimize base metal-filler metal interaction; and increase the life of fixtures and other teals. High brazing temperatures are preferred in order to take advantage of a higher melting, but more economical, brazing filler metal; to combine annealing, stress relief, or heat treatment of the base metal with brazing; to permit subsequent processing at elevated temperatures; to promote base metal-filler metal interactions to increase the joint

remelt temperature; or to promote removal of certain refractory oxides by vacuum or an atmosphere.

(4) Method of heating. Filler metals with narrow melting ranges (less than 50°F (28°C) between solidus and liquidus) can be used with any heating method, and the brazing filler metal may be preplaced in the joint area in the form of rings, washers, formed wires, shims, powder, or paste. Such alloys may also be manually or automatically face fed into the joint after the base metal is heated. Filler metals that tend to liquate should be used with heating methods that bring the joint to brazing temperature quickly, or allow the introduction of the brazing filler metal after the base metal reaches the brazing temperature.

d. Aluminum-Silicon Filler Metals. This group is used for joining aluminum and aluminum alloys. They are suited for furnace and dip brazing, while some types are also suited for torch brazing using lap joints rather than butt joints. Flux should be used in all cases and removed after brazing, except when vacuum brazing. Use brazing sheet or tubing that consists of a core of aluminum alloy and a coating of lower melting filler metal to supply aluminum filler metal. The coatings are aluminum-silicon alloys and may be applied to one or both sides of sheet. Brazing sheet or tubing is frequently used as one member of an assembly with the mating piece made of an unclad brazeable alloy. The coating on the brazing sheet or tubing melts at brazing temperature and flows by capillary attraction and gravity to fill the joints.

e. Magnesium Filler Metals. Because of its higher melting range, one magnesium filler metal (BMg-1) is used for joining AZ10A, KIA, and MIA magnesium alloys, while the other alloy (BMg-2a), with a lower melting range, is used for the AZ31B and ZE10A compositions. Both filler metals are suited for torch, dip, or furnace brazing processes. Heating must be closely controlled with both filler metals to prevent melting of the base metal.

f. Copper and Copper-Zinc Filler Metals. These brazing filler metals are used for joining various ferrous metals and nonferrous metals. They are commonly used for lap and butt joints with various brazing processes. However, the corrosion resistance of the copper-zinc alloy filler metals is generally inadequate for joining copper, silicon bronze, copper-nickel alloys, or stainless steel.

(1) The essentially pure copper brazing filler metals are used for joining ferrous metals, nickel base, and copper-nickel alloys. They are very free flowing and are often used in furnace brazing with a combusted gas, hydrogen, or dissociated ammonia atmosphere without flux. However, with metals that have components with difficult-to-reduce oxides (chromium, manganese, silicon, titanium, vanadium, and aluminum), a higher quality atmosphere or mineral flux may be required. copper filler metals are available in wrought and powder forms.

(2) Copper-zinc alloy filler metals are used on most common base metals. A mineral flux is commonly used with the filler metals.

(3) Copper-zinc filler metals are used on steel, copper, copper alloys, nickel and nickel base alloys, and stainless steel where corrosion resistance is not a requirement. They are used with the torch, furnace, and induction brazing processes. Fluxing is required, and a borax-boric acid flux is commonly used.

g. Copper-Phosphorus Filler Metals. These filler metals are primarily used for joining copper and copper alloys and have some limited use for joining silver, tungsten, and molybdenum. They should not be used on ferrous or nickel base alloys, or on copper-nickel alloys with more than 10 percent nickel. These filler metals are suited for all brazing processes and have self fluxing properties when used on copper. However, flux is recommended with all other metals, including copper alloys.

h. Silver Filler Metals.

(1) These filler metals are used for joining most ferrous and nonferrous metals, except aluminum and magnesium, with all methods of heating. They may be prep laced in the joint or fed into the joint area after heating. Fluxes are generally required, but fluxless brazing with filler metals free of cadmium and zinc can be done on most metals in an inert or reducing atmosphere (such as dry hydrogen, dry argon, vacuum, and combusted fuel gas).

CAUTION

Do not overheat filler metals containing cadmium. Cadmium oxide fumes are hazardous.

(2) The addition of cadmium to the silver-copper-zinc alloy system lowers the melting and flew temperatures of the filler metal. Cadmium also increases the fluidity and wetting action of the filler metal on a variety of base metals. Cadmium bearing filler metals should be used with caution. If they are improperly used and subjected to overheating, cadmium oxide frees can be generated. Cadmium oxide fumes are a health hazard, and excessive inhalation of these fumes must be avoided.

(3) Of the elements that are commonly used to lower the melting and flow temperatures of copper-silver alloys, zinc is by far the most helpful wetting agent when joining alloys based on iron, cobalt, or nickel. Alone or in combination with cadmium or tin, zinc produces alloys that wet the iron group metals but do not alloy with them to any appreciable depth.

(4) Tin has a low vapor pressure at normal brazing temperatures. It is used in silver brazing filler metals in place of zinc or cadmium when volatile constituents are objectionable, such as when brazing is done without flux in atmosphere or vacuum furnaces, or when the brazed assemblies will be used in high vacuum at elevated temperatures. Tin additions to silver-copper alloys produce filler metals with wide melting ranges. Alloys containing zinc wet ferrous metals more

effectively than those containing tin, and where zinc is tolerable, it is preferred to tin.

(5) Stellites, cemented carbides, and other molybdenum and tungsten rich refractory alloys are difficult to wet with the alloys previously mentioned. Manganese, nickel, and infrequently, cobalt, are often added as wetting agents in brazing filler metals for joining these materials. An important characteristic of silver brazing filler metals containing small additions of nickel is improved resistance to corrosion under certain conditions. They are particularly recommended where joints in stainless steel are to be exposed to salt water corrosion.

(6) When stainless steels and other alloys that form refractory oxides are to be brazed in reducing or inert atmospheres without flux, silver brazing filler metals containing lithium as the wetting agent are quite effective. Lithium is capable of reducing the adherent oxides on the base metal. The resultant lithium oxide is readily displaced by the brazing alloy. Lithium bearing alloys are advantageously used in very pure dry hydrogen or inert atmospheres.

i. Gold Filler Metals. These filler metals are used for joining parts in electron tube assemblies where volatile components are undesirable; and the brazing of iron, nickel, and cobalt base metals where resistance to oxidation or corrosion is required. Because of their low rate of interaction with the base metal, they are commonly used on thin sections, usually with induction, furnace, or resistance heating in a reducing atmosphere or in vacuum without flux. For certain applications, a borax-boric acid flux may be used.

j. Nickel Filler Metals.

(1) These brazing filler metals are generally used for their corrosion resistance and heat resistant properties up to 1800°F (982°C) continuous service, and 2200°F (1204°C) short time service, depending on the specific filler metals and operating environment. They are generally used on 300 and 400 series stainless steels and nickel and cobalt base alloys. Other base metals such as carbon steel, low alloy steels, and copper are also brazed when specific properties are desired. The filler metals also exhibit satisfactory room temperature and cryogenic temperature properties down to the liquid point of helium. The filler metals are normally applied as powders, pastes, or in the form of sheet or rod with plastic binders.

(2) The phosphorus containing filler metals exhibit the lowest ductility because of the presence of nickel phosphides. The boron containing filler metals should not be used for brazing thin sections because of their erosive action. The quantity of filler metal and time at brazing temperatures should be controlled because of the high solubility of some base metals in these filler metals.

k. Cobalt Filler Metal. This filler metal is generally used for its high temperature properties and its compatibility with cobalt base metals. For optimum results, brazing should be performed in a high quality atmosphere. Special high temperature fluxes are available.

1. Filler Metals for Refractory Metals.

(1) Brazing is an attractive means for fabricating many assemblies of refractory metals, in particular those involving thin sections. The use of brazing to join these materials is somewhat restricted by the lack of filler metals specifically designed for brazing them. Although several references to brazing are present, the reported filler metals that are suitable for applications involving both high temperature and high corrosion are very limited.

(2) Low melting filler metals, such as silver-copper-zinc, copper-phosphorus, and copper, are used to join tungsten for electrical contact applications. These filler metals are limited in their applications, however, because they cannot operate at very high temperatures. The use of higher melting metals, such as tantalum and columbium, is warranted in those cases. Nickel base and precious-metal base filler metals may be used for joining tungsten.

(3) A wide variety of brazing filler metals may be used to join molybdenum. The brazing temperature range is the same as that for tungsten. Each filler metal should be evaluated for its particular applicability. The service temperature requirement in many cases dictates the brazing filler metal selection. However, consideration must -be given to the effect of brazing temperature on the base metal properties, specifically recrystallization. When brazing above the recrystallization temperature, time should be kept as short as possible. When high temperature service is not required, copper and silver base filler metals may be used. For electronic parts and other nonstructural applications requiring higher temperatures, gold-copper, gold-nickel, and copper-nickel filler metals can be used. Higher melting metals and alloys may be used as brazing filler metals at still higher temperatures.

(4) Copper-gold alloys containing less than 40 percent gold can also be used as filler metals, but gold content between 46 and 90 percent tends to form age hardening compounds which are brittle. Although silver base filler metals have been used to join tantalum and columbium, they are not recommended because of a tendency to embrittle the base metals.

m. Filler metal specifications and welding processes are shown in table 8-2.

Table 8-2. A.W.S.* filter metal specification and welding processes.

AWS Specification	Specification Title	FOR PROCESS SHOWN					
		OAW	SMAW	GTAW	GMAW	SAW	Other
A5.1	Carbon steel covered arc-welding electrodes		X				
A5.2	Iron & steel gas welding rods	X					
A5.3	Aluminum & aluminum alloy arc welding electrodes		X				
A5.4	Corrosion-resisting chromium & chromium-nickel steel covered welding electrodes		X				
A5.5	Low-alloy steel covered arc welding electrodes		X				
A5.6	Copper & copper alloy covered electrodes		X				
A5.7	Copper & copper alloy welding rods	X		X			PAW
A5.8	Brazing filler metal						BR
A5.9	Corrosion-resisting chromium & chromium-nickel steel bare & composite metal cored & standard arc welding electrodes & rods			X	X	X	PAW

*American Welding Society

Table 8-2. A.W.S.* Filler Metal Specification and Welding Processes (cont)

AWS Specification	Specification Title	FOR PROCESS SHOWN					
		OAW	SMAW	GTAW	GMAW	SAW	Other
A5.10	Aluminum & aluminum alloy welding rods & bare electrodes	X		X	X		PAW
A5.11	Nickel & nickel alloy covered welding electrodes		X				
A5.12	Tungsten arc welding electrodes			X			PAW
A5.13	Surfacing welding rods & electrodes	X		X			CAW
A5.14	Nickel & nickel alloy bare welding rods and electrodes	X		X	X	X	PAW
A5.15	Welding rods & covered electrodes for welding cast iron	X	X				CAW
A5.16	Titanium & titanium alloy bare welding rods & electrodes			X	X		
A5.17	Bare carbon steel electrodes & fluxes for submerged-arc welding					X	
A5.18	Carbon steel filler metals for gas shielded arc welding			X	X		PAW
A5.19	Magnesium alloy welding rods & bare electrodes	X		X	X		PAW
A5.20	Carbon steel electrodes for flux cored arc welding						FCAW
A5.21	Composite surfacing welding rods & electrodes	X	X	X			
A5.22	Flux cored corrosion-resisting chromium & chromium-nickel steel electrodes						FCAW
A5.23	Bare low-alloy steel electrodes and fluxes for submerged arc welding					X	
A5.24	Zirconium & zirconium alloy bare welding rods and electrodes			X	X		PAW
A5.25	Consumables used for electro-slag welding of carbon & high strength low alloy steels						ES
A5.26	Consumables used for electrogas welding of carbon and high strength low-alloy steels				X (EG)		FCAW (EG)
A5.27	Copper and copper alloy gas welding rods	X					
A5.28	Low-alloy steel filler metals for gas shielding arc welding			X	X		PAW

Note: If GTAW is shown, the specification will also apply to PAW even though not stated.

*American Welding Society

CHAPTER 9
MAINTENANCE WELDING OPERATIONS
FOR MILITARY EQUIPMENT

9-1. SCOPE

a. This chapter contains information necessary to determine the size of the welding job and proper welding procedures for military items.

b. Appendix A contains references to formal DA publications covering additional equipment used by military item and other equipment not covered by standard welding procedures as set forth in other chapters of this manual. Appendix A also contains references to formal DA publications covering additional equipment used by military personnel which are not included in this chapter.

c. Welding techniques for equipment containing high yield strength, low alloy structural steels (such as TI) used for bulldozer blades, armor, and heavy structural work are covered in chapter 12, section VII of this circular.

9-2. SIZING UP THE JOB

a. General   All of the materials used in the manufacture of military materiel, as well as the assembled equipment are thoroughly tested before the material is issued to the using services in the field. Therefore, most of the damage to and failures of the equipment are due to accidents, overloading, or unusual shocks for which the equipment was not designed to withstand.It is in this class of repair work that field service welding is utilized most frequently.

b. Determination of Weldability. Before repairing any damaged materiel, it must be determined whether or not the materiel can be satisfactorily welded. This determination is based upon the factors listed below.

(1) Determine the nature and extent of the damage and the amount of straightening and fitting of the metal that will be required.

(2) Determine the possibility of restoring the structure to usable condition without the use of welding.

(3) Determine the type of metal used in the damaged part, whether it was heat treated, and if so, what heat treatment was used.

(4) Determine if the welding heat will distort the shape or in any manner impair the physical properties of the part to be repaired.

(5) Determine if heat treating or other equipment or materials will be required in order to make the repair by welding.

9-2.   SIZING UP THE JOB (cont)

c.   Repairing Heat Treated Parts.

(1) In emergency cases, some heat treated parts can be repaired in the heat treated condition by welding with stainless steel electrodes containing 25 percent chromium and 20 percent nickel, or an 18 percent chromium-8 nickel electrode containing manganese or molybdenum. These electrodes will produce a satisfactory weld, although a narrow zone in the base metal in the vicinity of the weld will be affected by the heat of welding.

(2) Minor defects on the surface of heat treated parts may be repaired by either hard surfacing or brazing, depending on their application in service. In any of these repairs, the heat treated part will lose some of its strength, hardness, or toughness, even though the weld metal deposited has good properties.

(3) The preferred metal of repairing heat treated steels, when practicable, requires the annealing of the broken part and welding with a high strength rod. This method produces a welded joint that can be heat treated. The entire part should be heat treated after welding to obtain the properties originally found in the welded parts. This method should not be attempted unless proper heat treating equipment is available.

9-3.   IDENTIFYING THE METAL

Welding repairs should not be made until the type of metal used for the components or sections to be repaired has been determined. This information can be obtained by previous experience with similar materiel by test procedures as described in chapter 7, or from assembly drawings of the components These drawings should be carried by maintenance companies in the field and should show the type of material and the heat treatment of the parts.

9-4.   DETERMINING THE WELDABLE PARK

a.   Welding operations on ordnance materiel are restricted largely to those parts whose essential physical properties are not impaired by the welding heat.

b.   Successful welded repairs cannot be made on machined parts that carry a dynamic load . This applies particularly to high alloy steels that are heat treated for hardness or toughness, or both.

c.   Gears, shafts, antifriction bearings, springs, connecting rods, piston rods, pistons, valves, and cam are considered to be unsuitable for field welding because welding heat alters or destroys the heat treatment of these parts.

9-5.   SELECTING THE PROPER WELDING PROCDURES

The use of welding equipment and the application of welding processes to different metals is covered in other chapters of this manual. A thorough working knowledge of these processes and metals is necessary before a welding procedure for any given job can be selected. When it has been decided by competent authority that the repair can be made by welding, the factors outlined below must be considered.

a. The proper type and size of electrode, together with the current and polarity setting, must be determined if an arc welding process is used. If a gas welding process is used, the proper type of welding rod, correct gas pressure, tip size, flux, and flame adjustment must be determined.

b. In preparing the edges of plates or parts to be welded, the proper cleaning and beveling of the parts to be joined must be considered. The need for backing strips, quench plates, tack welding, and preheating must be determined.

c. Reducing warping and internal stresses requires the use of the proper sequence for welding, control and proper distribution of the welding heat, spacing of the parts to permit some movment, control of the size and location of the deposited weld metal beads, and proper cooling procedure.

d. Military materiel is designed for lightness and the safety factors are, of necessity, low in some cases. This necessitates sane reinforcement at the joint to compensate for the strength lost in the welded part due to the welding heat. A reinforcement must be designed that will provide the required strength without producing high local rigidity or excessive weight.

9-6. PRELIMINARY PRECAUTIONS

Before beginning any welding or cutting operations on the equipment, the safety precautions listed below must be considered.

a. Remove all ammunition from, on, or about the vehicle or materiel.

b. Drain the fuel tank and close the fuel and oil tank shut off valves. If welding or cutting is to be done on the tanks, prepare them for welding in accordance with the instructions in chapter 2, section V.

c. Have a fire extinguisher nearby.

d. Keep heat away from optical elements.

e. Be familiar with and observe the safety precautions prescribed in chapter 2 of this circular.

CHAPTER 10

ARC WELDING AND CUTTING PROCESS

Section I. GENERAL

10-1. DEFINITION OF ARC WELDING

a. Definition. In the arc welding process, the weld is produced by the extreme heat of an electric arc drawn between an electrode and the workpiece, or in some cases, between two electrodes. Welds are made with or without the application of pressure and with or without filler metals. Arc welding processes may be divided into two classes based on the type of electrode used: metal electrodes and carbon electrodes. Detailed descriptions of the various processes may be found in chapter 6, paragraph 6-2.

(1) Metal electrodes. Arc welding processes that fall into this category include bare metal-arc welding, stud welding, gas shielded stud welding, submerged arc welding, gas tungsten arc welding, gas metal-arc welding, shielded metal-arc welding, atomic hydrogen welding, arc spot welding, and arc seam welding.

(2) Carbon electrodes. Arc welding processes that fall into this category include carbon-arc welding, twin carbon-arc welding, gas carbon-arc welding, and shielded carbon-arc welding.

b. Weld Metal Deposition.

(1) General. In metal-arc welding, a number of separate forces are responsible for the transfer of molten filler metal and molten slag to the base metal. These forces are described in (2) through (7) helm.

(2) Vaporization and condensation. A small part of the metal passing through the arc, especially the metal in the intense heat at the end of the electrode, is vaporized. some of this vaporized metal escapes as spatter, but most of it is condensed in the weld crater, which is at a much lower temperature. This occurs with all types of electrodes and in all welding positions.

(3) Gravity. Gravity affects the transfer of metal in flat position welding. In other positions, small electrodes must be used to avoid excessive loss of weld metal, as the surface tension is unable to retain a large amount of molten metal in the weld crater.

(4) Pinch effect. The high current passing through the molten metal at the tip of the electrode sets up a radial compressive magnetic force that tends to pinch the molten globule and detach it from the electrode.

(5) Surface tension. This force holds filler metal and the slag globules in contact with the molten base or weld metal in the crater. It has little to do with the transfer of metal across the arc but is an important factor in retaining the molten weld metal in place and in the shaping of weld contours.

10-1. DEFINITION OF ARC WELDING (cont)

(6) Gas stream from electrode coatings. Gases are produced by the burning and volatilization of the electrode covering and are expanded by the heat of the boiling electrode tip. The velocity and movement of this gas stream give the small particles in the arc a movement away from the electrode tip and into the molten crater on the work.

(7) Carbon monoxide evolution from electrode. According to this theory of metal movement in the welding arc, carbon monoxide is evolved within the molten metal at the electrode tip, causing miniature explosions which expel molten metal away from the electrode and toward the work. This theory is substantiated by the fact that bare wire electrodes made of high purity iron or "killed steel" (i.e., steel that has been almost completely deoxidized in casting) cannot he used successfully in the overhead position. The metal transfer from electrode to the work, the spatter, and the crater formation are, in this theory, caused by the decarburizing action in molten steel.

c. Arc Crater. Arc craters are formal by the pressure of expanding gases from the electrode tip (arc blast), forcing the liquid metal towards the edges of the crater. The higher temperature of the center, as compared with that of the sides of the crater, causes the edges to cool first. Metal is thus drawn from the center to the edges, forming a low spot.

10-2. WELDING WITH CONSTANT CURRENT

The power source is the heart of all arc welding process. Two basic types of power sources are expressed by their voltage-ampere output characteristics. The constant current machine is considered in this paragraph. The other power source, the constant voltage machine, is discussed in paragraph 10-3. The static output characteristic curve produced by both sources is shown in figure 10-1. The characteristic curve of a welding machine is obtained by measuring and plotting the output voltage and the output current while statically loading the machine.

a. The conventional machine is known as the constant current (CC) machine, or the variable voltage type. The CC machine has the characteristic drooping volt-ampere curve, (fig. 10-1), and has been used for many years for the shielded metal arc welding process. A constant-current arc-welding machine is one which has means for adjusting the arc current. It also has a static volt-ampere curve that tends to produce a relatively constant output current. The arc voltage, at a given welding current, is responsive to the rate at which a consumable electrode is fed into the arc. When a nonconsumable electrode is used, the arc voltage is responsive to the electrode-to-work distance. A constant-current arc-welding machine is usually used with welding processes which use manually held eletrodes, continuously fed consumable electrodes, or nonconsumable electrodes. If the arc length varies because of external influences, and slight changes in the arc voltage result, the welding current remains constant.

b. The conventional or constant current (CC) type power source may have direct current or alternating current output. It is used for the shielded metal-arc welding process, carbon arc welding and gouging, gas tungsten arc welding, and plasma arc welding. It is used for stud welding and can be used for the continuous wire processes when relatively large electrode wires are used.

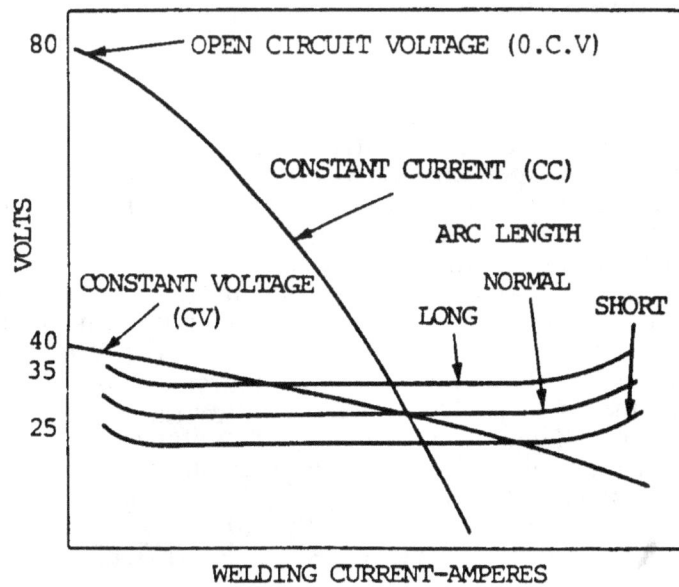

Figure 10-1. Characteristic curve for welding power source.

c. There are two control systems for constant current welding machines: the single-control machine and the dual-control machine.

(1) The single-control machine has one adjustment which changes the current output from minimum to maximum which is usually greater than the rated output of the machine. The characteristic volt-ampere curve is shown by figure 10-2. The shaded area is the normal arc voltage range. By adjusting the current control, a large number of output curves can be obtained. The dotted lines show intermediate adjustments of the machine. With tap or plug-in machines, the number of covers will correspond to the number of taps or plug-in combinations available. Most transformer and transformer-rectifier machines are single-control welding machines.

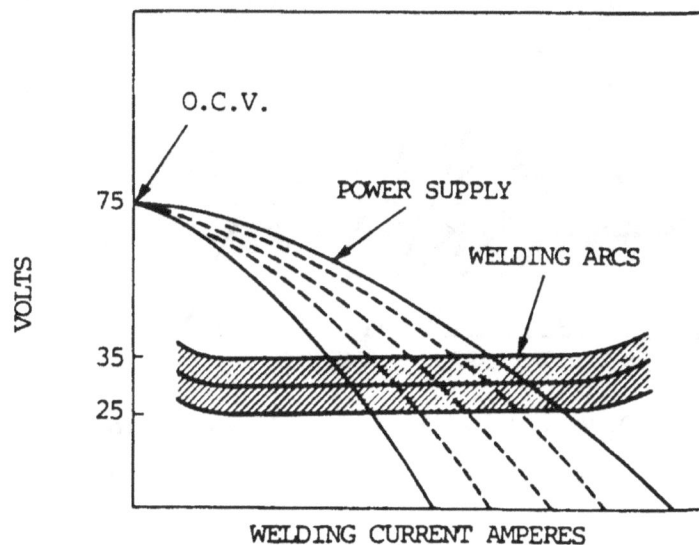

Figure 10-2. Curve for single control welding machine.

TC 9-237

10-2. WELDING WITH CONSTANT CURRENT(Cont)

(2) Dual control machines have both current and voltage controls. They have two adjustments one for coarse-current control and the other for fine-current control, which also acts as an open-circuit voltage adjustment. Generator welding machines usually have dual controls. They offer the welder the most flexibility for different welding requirements. These machines inherently have slope control. The slope of the characteristic curve can be changed from a shallow to a steep slope according to welding requirements. Figure 10-3 shows some of the different curves that can be obtained. Other tunes are obtained with intermediate open-circuit voltage settings. The slope is changed by changing the open-circuit voltage with the fine—curent control adjustment bob. The coarse adjustment sets the current output of the machine in steps from the minimum to the maximum current. The fine—current control will change the open-circuit voltage from approximately 55 volts to 85 volts. However, when welding, this adjustment does not change arc voltage. Arc voltage is controlled by the welder by changing the length of the welding arc. The open-circuit voltage affects the ability to stike an arc. If the open-circuit voltage is much below 60 volts, it is difficult to strike an arc.

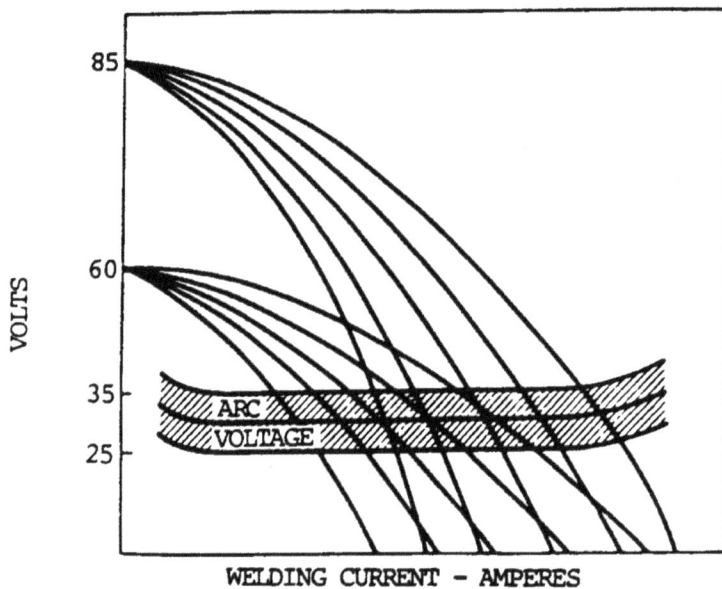

Figure 10-3. Curve for dual control welding machines.

(a) The different slopes possible with a dual-control machine have an important effect on the welding characteristic of the arc. The arc length can vary, depending on the welding technique. A short arc has lower voltage and the long arc has higher voltage. With a short arc (lower voltage), the power source produces more current, and with a longer arc (higher voltage), the power source provides less welding current. This is illustrated by figure 10-4, which shows three curves of arcs and two characteristic curves of a dual-control welding machine. The three arc curves are for a long arc, a normal arc, and the lower curve is for a short arc. The intersection of a curve of an arc and a characteristic curve of a welding machine is known as an operating point. The operating point changes continuously during welding. While welding, and without changing the control on the machine, the welder can lengthen or shorten the arc and change the arc voltage from 35 to 25 volts. With the same machine setting, the short arc (lower voltage) is a high-current arc. Conversely, the long arc (high voltage) is a lower current arc. This allows the welder to control the size of the molten puddle while welding. When the welder purposely and briefly lengthens the arc, the current is reduced, the arc spreads out, and the puddle freezes quicker. The amount of molten metal is reduced, which provides the control needed for out-of-position work. This type of control is built into conventional constant current type of machine, single- or dual-control, ac or dc.

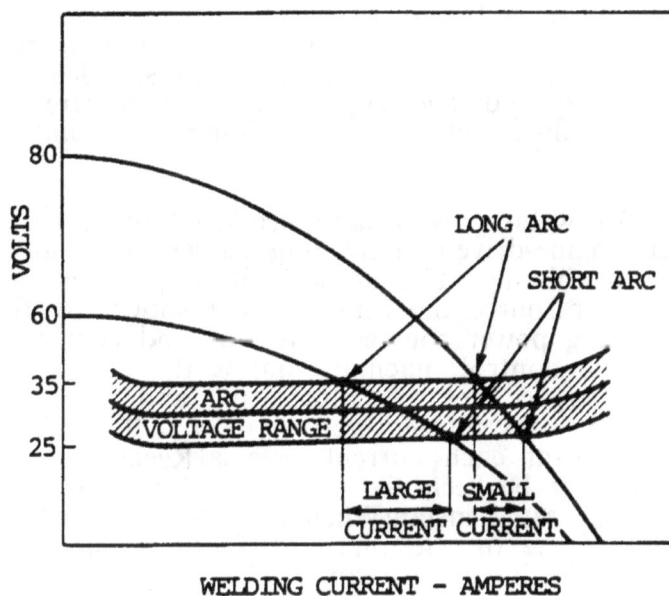

Figure 10-4. Volt ampere slope vs welding operation.

(b) With the dual-control machine, the welder can adjust the machine for more or less change of current for a given change of arc voltage. Both curves in figure 10-4 are obtained on a dud-control machine by adjusting the fine control knob. The top curve shows an 80-volt open-circuit voltage and the bottom curve shins a 60-volt open-circuit voltage. With either adjustment, the voltage and current relationship will stay on the same curve or line. Consider first the 80-volt open-circuit curve which produces the steeper slope. When the arc is long

TC 9-237

10-2. WELDING WITH CONSTANT CURRENT (cont)

with 35 volts and is shortened to 25 volts, the current increases. This is done without touching the machine control. The welder manipulates the arc. With the flatter, 60-volt open-circuit curve, when the arc is shortened from 35 volts to 25 volts, the welding current will increase almost twice as much as it did when following the 80-volt open-circuit curve. The flatter slope curve provides a digging arc where an equal change in arc voltage produces a greater change in arc current. The steeper slope curve has less current change for the same change in arc length and provides a softer arc. There are many characteristic curves between the 80 and 60 open circuit voltage curves, and each allows a different current change for the same arc voltage change. This is the advantage of a dual-control welding machine over a single-control type, since the slope of the curve through the arc voltage range is adjustable only on a dual-control machine. The dual-control generator welding machine is the most flexible of all types of welding power sources, since it allows the welder to change to a higher current arc for deep penetration or to a lower-current, less penetrating arc by changing the arc length. This ability to control the current in the arc over a fairly wide range is extremely useful for making pipe welds.

d. The rectifier welding machine, technically known as the transformer-rectifier, produces direct current for welding. These machines are essentially single-control machines and have a static volt ampere output characteristic curve similar to that shown by figure 10-4, p 10-5. These machines, though not as flexible as the dual-control motor generator can be used for all types of shielded metal arc welding where direct current is required. The slope of the volt-ampere curve through the welding range is generally midway between the maximum and minimum of a dual-control machine.

e. Alternating current for welding is usually produced by a transformer type welding machine, although engine-driven alternating current generator welding machines are available for portable use. The static volt ampere characteristic curve of an alternating current power source the same as that shown by figure 10-4, p 10-5. some transformer welding power sources have fine and coarse adjustment knobs, but these are not dual control machines unless the open-circuit voltage is changed appreciably. The difference between alternating and direct current welding is that the voltage and current pass through zero 100 or 120 times per second, according to line frequency or at each current reversal. Reactance designed into the machine causes a phase shift between the voltage and current so that they both do not go through zero at the same instant. When the current goes through zero, the arc is extinguished, but because of the phase difference, there is voltage present which helps to re-establish the arc quickly. The degree of ionization in the arc stream affects the voltage required to re-establish the arc and the overall stability of the arc. Arc stabilizers (ionizers) are included in the coatings of electrodes designed for ac welding to provide a stable arc.

f. The constant-current type welding machine can be used for some automatic welding processes. The wire feeder and control must duplicate the motions of the welder to start and maintain an arc. This requires a complex system with feedback from the arc voltage to compensate for changes in the arc length. The constant-current power supplies are rarely used for very small electrode wire welding processes.

g. Arc welding machines have been developed with true constant-current volt-ampere static characteristics, within the arc voltage range, as shown by figure 10-5. A welder using this type of machine has little or no control over welding

10-6

current by shortening or lengthening the arc, since the welding current remains the same whether the arc is short or long. This is a great advantage for gas tungsten
current by shortening or lengthening the arc, since the welding current remains the same whether the arc is short or long. This is a great advantage for gas tungsten arc welding, since the working arc length of the tungsten arc is limited. In shield metal-arc welding, to obtain weld puddle control, it is necessary to be able to change the current level while welding. This is done by the machine, which can be programmed to change from a high current (HC) to a low current (LC) on a repetitive basis, known as pulsed welding. In pulsed current welding there are two current levels, the high current and low current, sometimes called background cur- By programming a control circuit, the output of the machine continuously switches from the high to the low current as shown by figure 10-6, p 10-8. The level of both high and low current is adjustable. In addition, the length of time for the high and low current pulses is adjustable. This gives the welder the necessary control over the arc and weld puddle. Pulsed current welding is useful for shielded metal-arc welding of pipe when using certain types of electrodes. Pulsed arc is very useful when welding with the gas tungsten arc welding process.

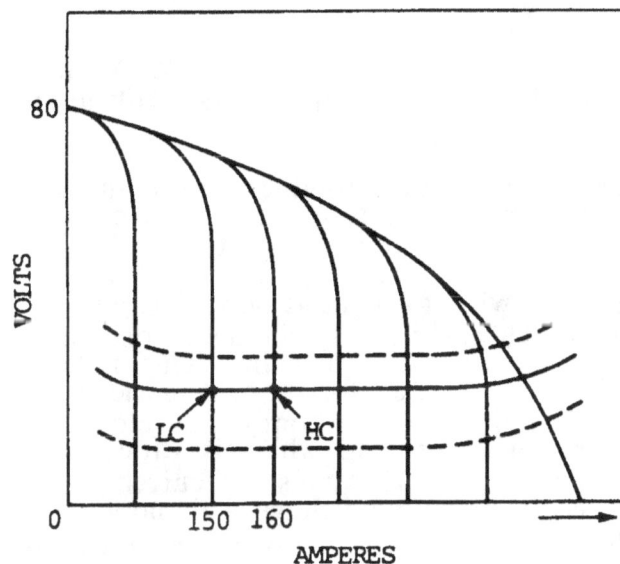

Figure 10-5. Volt ampere curve for true constant current machine.

10-2. WELDING WITH CONSTANT CURRENT (cont)

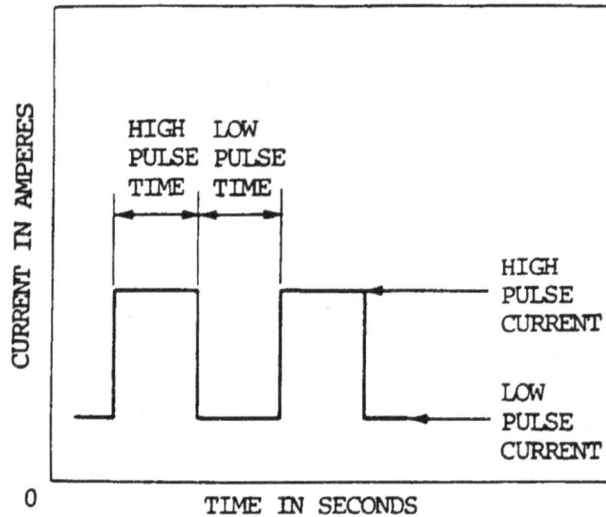

Figure 10-6. Pulsed current welding.

10-3. WELDING WITH CONSTANT VOLTAGE

The second type of power source is the constant voltage (CV) machine or the constant potential (CP) machine. It has a relatively flat volt-ampere characteristic curve.

The static output characteristic tune produced by both the CV and CC machine is shown by figure 10-1, p 10-3. The characteristic curve of a welding machine is obtained by measuring and plotting the output voltage and the output current while statically loading the machine. The constant voltage (CV) characteristic curve is essentially flat but with a slight droop. The curve may be adjusted up and down to change the voltage however, it will never rise to as high an open-circuit voltage as a constant current (CC) machine. This is one reason that the constant voltage (CV) machine is not used for manual shielded metal arc welding with covered electrodes. It is only used for continuous electrode wire welding. The circuit consists of a pure resistance load which is varied from the minimum or no load to the maximum or short circuit. The constant current (CC) curve shows that the machine produces maximum output voltage with no load, and as the load increases, the output voltage decreases. The no-load or open-circuit voltage is usually about 80 volts.

b. The CV electrical system is the basis of operation of the entire commercial electric power system. The electric power delivered to homes and available at every receptacle has a constant voltage. The same voltage is maintained continuously at each outlet whether a small light bulb, with a very low wattage rating, or a heavy-duty electric heater with a high wattage rating, is connected. The current that flows through each of these circuits will be different based on the resistance of the particular item or appliance in accordance with Ohm's law. For example, the small light bulb will draw less than 0.01 amperes of current while the electric heater may draw over 10 amperes. The voltage throughout the system remains constant, but the current flowing through each appliance depends on its resistance or electrical load. The same principle is utilized by the CV welding system.

c. When a higher current is used when welding, the electrode is melted off more rapidly. With low current, the electrode melts off slower. This relationship between melt-off rate and welding current applies to all of the arc welding processes that use a continuously fed electrode. This is a physical relationship that depends upon the size of the electrode, the metal composition, the atmosphere that surrounds the arc, and welding current. Figure 10-7 shows the melt-off rate curves for different sizes of steel electrode wires in a CO_2 atmosphere. Note that these curves are nearly linear, at least in the upper portion of the curve. Similar curves are available for all sizes of electrode wires of different compositions and in different shielding atmospheres. This relationship is definite and fixed, but some variations can occur. This relationship is the basis of the simplified control for wire feeding using constant voltage. Instead of regulating the electrode wire feed rate to maintain the constant arc length, as is done when using a constant current power source, the electrode wire is fed into the arc at a fixed speed. The power source is designed to provide the necessary current to melt off the electrode wire at this same rate. This concept prompted the development of the constant voltage welding power source.

d. The volt-ampere characteristics of the constant voltage power source shown by figure 10-8, p 10-10, was designed to produce substantially the same voltage at no load and at rated or full load. It has characteristics similar to a standard commercial electric power generator. If the load in the circuit changes, the power source automaticaly adjusts its current output to satisfy this requirement, and maintains essentially the same voltage across the output terminals. This ensures a self-regulating voltage power source.

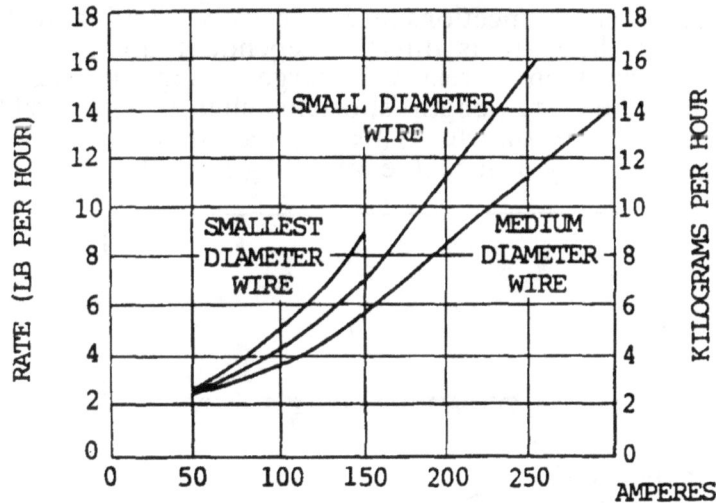

Figure 10-7. Burn-off rates of wire vs current.

10-3.   WELDING WITH CONSTANT VOLTAGE (cont)

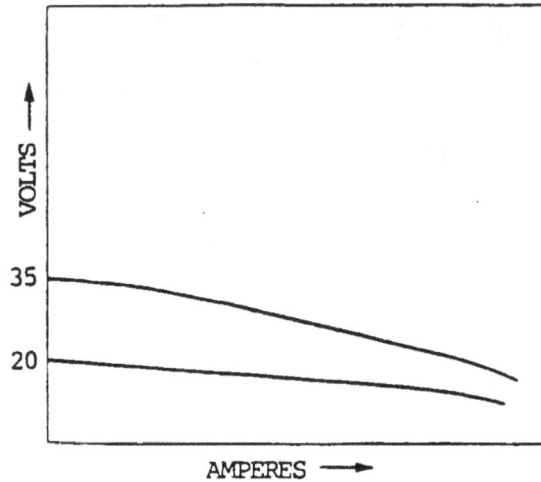

Figure 10-8.   Static volt amp characteristic curve of CV machine.

e.   Resistances or voltage drops occur in the welding arc and in the welding cables and connecters, in the welding gun, and in the electrode length beyond the current pickup tip. These voltage drops add up to the output voltage of the welding machine, and represent the electrical resistance load on the welding power source.   When the resistance of any component in the external circuit changes, the voltage balance will be achieved by changing the welding current in the system. The greatest voltage drop occurs across the welding arc. The other voltage drops in the welding cables and connections are relatively small and constant. The voltage drop across the welding arc is directly dependent upon the arc length. A small change in arc volts results in a relatively large change in welding current. Figure 10-9 shins that if the arc length shortens slightly, the welding current increases by approximately 100 amperes. This change in arc length greatly increases the melt-off rate and quickly brings the arc length back to normal.

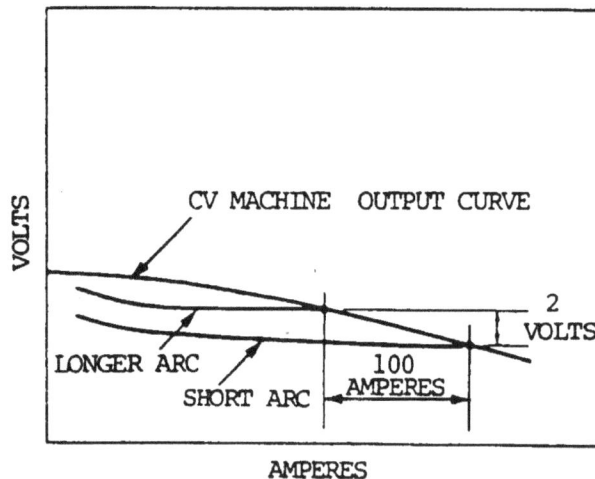

Figure 10-9.   Static volt amp curve with arc range.

f. The constant voltage power source is continually changing its current output in order to maintain the voltage drop in the external portion of the welding circuit. Changes in wire feed speed which might occur when the welder moves the gun toward or away from the work are compensated for by changing the current and the melt-off rate briefly until equilibrium is re-established. The same corrective action occurs if the wire feeder has a temporary reduction in speed. The CV power source and fixed wire feed speed system is self-regulating. Movemment of the cable assembly often changes the drag or feed rate of the electrode wireThe CV welding power source provides the proper current so that the malt-off is equal to the wire feed rate. The arc length is controlled by setting the voltage on the power source. The welding current is controlled by adjusting the wire feed speed.

g. The characteristics of the welding power source must be designed to provide a stable arc when gas metal arc welding with different electrode sizes and metals and in different atmospheres. Most constant voltage power sources have taps or a means of adjusting the slope of the volt-ampere curve.A curve having a slope of 1-1/2 to 2 volts per hundred amperes is best for gas metal arc welding with nonferrous electrodes in inert gas, for submerged arc welding, and for flux-cored arc welding with larger-diameter electrode wires. A curve having a medium slope of 2 to 3 volts per hundred amperes is preferred for CO_2 gas shielded metal arc welding and for small flux-cored electrode wires. A steeper slope of 3 to 4 volts per hundred amperes is recomended for short circuiting arc transfer. These three slopes are shown in figure 10-10. The flatter the curve, the more the current changes for an equal change in arc voltage.

h. The dynamic characteristics of the power source must be carefully engineered. Refer again to figure 10-9. If the voltage changes abruptly with a short circuit, the current will tend to increase quickly to a very high value. This is an advantage in starting the arc but will create unwanted spatter if not controlled. It is controlled by adding reactance or inductance in the circuit. This changes the time factor or response time and provides for a stable arcIn most machines, a different amount of inductance is included in the circuit for the different slopes.

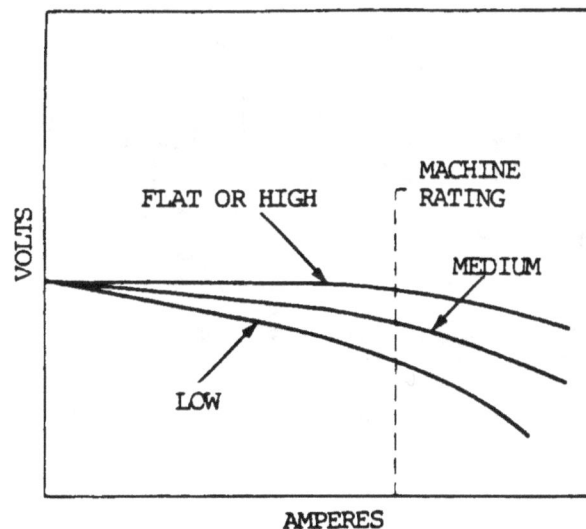

Figure 10-10. Various slopes of characteristic curves.

10-3. WELDING WITH CONSTANT VOLTAGE (cont)

i. The constant voltage welding power system has its greatest advantage when the current density of the electrode wire is high. The current density (amperes/sq in.) relationship for different electrode wire sizes and different currents is shown by figure 10-11. There is a vast difference between the current density employed for gas metal arc welding with a fine electrode wire compared with conventional shielded metal arc welding with a covered electrode.

j. Direct current electrode positive (DCEP) is used for gas metal arc welding. When dc electrode negative (DCEN) is used, the arc is erratic and produces an inferior weld. Direct current electrode negative (DCEN) can be used for submerged arc welding and flux-cored arc welding.

k. Constant voltage welding with alternating current is normally not used. It can be used for submerged arc welding and for electroslag welding.

l. The constant voltage power system should not be used for shielded metal-arc welding. It may overload and damage the power source by drawing too much current too long. It can be used for carbon arc cutting and gouging with small electrodes and the arc welding processes.

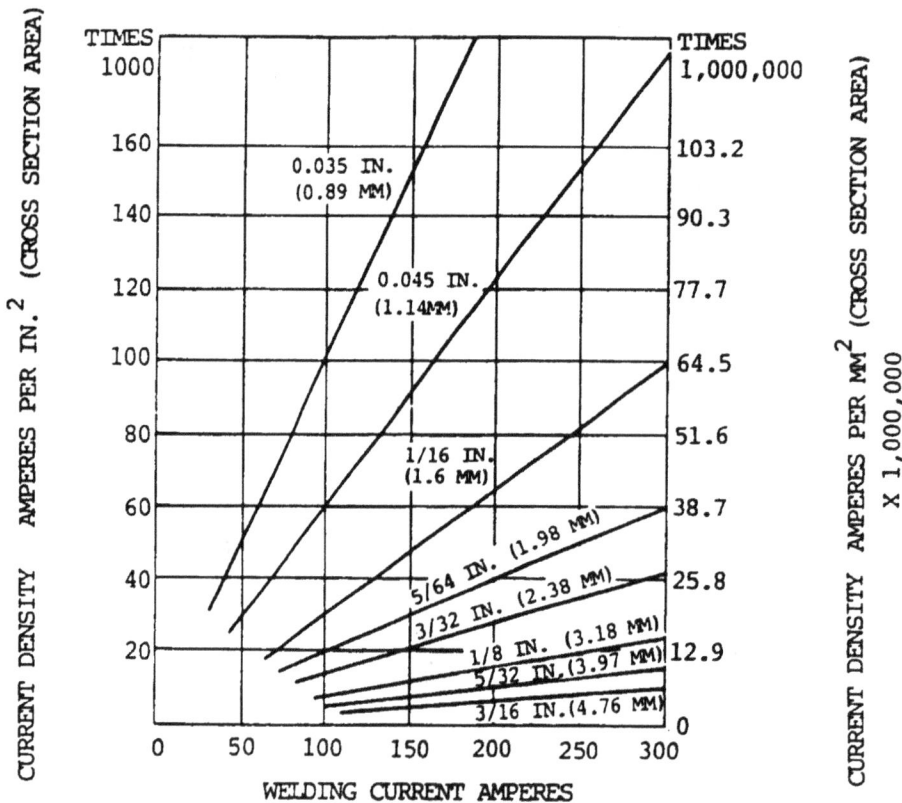

Figure 10-11. Current density—various electrode signs.

10-4. DC STRAIGHT AND REVERSE POLARITY WELDING

a. <u>General</u>. The electrical arc welding circuit is the same as any electrical circuit. In the simplest electrical circuits, there are three factors: current, or the flow of electricity; pressure, or the force required to cause the current to flow; and resistance, or the force required to regulate the flow of current.

(1) Current is a rate of flow and is measured by the amount of electricity that flows through a wire in one second. The term ampere denotes the amount of current per second that flows in a circuit. The letter I is used to designate current amperes.

(2) Pressure is the force that causes a current to flow. The measure of electrical pressure is the volt. The voltage between two points in an electrical circuit is called the difference in potential. This force or potential is called electromotive force or EMF. The difference of potential or voltage causes current to flow in an electrical circuit. The letter E is used to designate voltage or EMF.

(3) Resistance is the restriction to current flow in an electrical circuit. Every component in the circuit,including the conductor, has some resistance to current flow. Current flows easier through some conductors than others; that is, the resistance of some conductors is less than others.Resistance depends on the material, the cross-sectional area,and the temperature of the conductor.The unit of electrical resistance is the ohm. It is designated by the letter R.

b. <u>Electrical circuits</u>. A simple electrical circuit is shown by figure 10-12. This circuit includes two meters for electrical measurement: a voltmeter, and an ammeter. It also shows a symbol for a battery. The longer line of the symbol represents the positive terminal. Outside of a device that sets up the EMF, such as a generator or a battery, the current flows from the negative (-) to the positive (+). The arrow shows the direction of current flow. The ammeter is a low resistance meter shown by the round circle and arrow adjacent to the letter I. The pressure or voltage across the battery can be measured by a voltmeter. The voltmeter is a high resistance meter shown by the round circle and arrow adjacent to the letter E. The resistance in the circuit is shown by a zigzag symbol. The resistance of a resistor can be measured by an ohmmeterAn ohmmeter must never be used to measure resistance in a circuit when current is flowing.

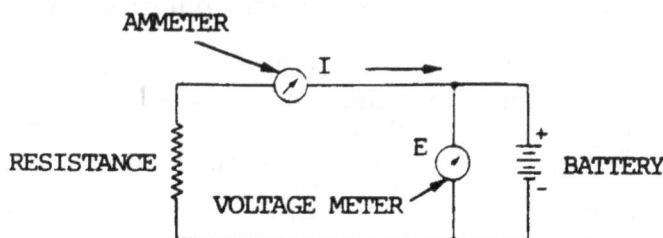

Figure 10-12. Electrical circuit.

10-4.   DC STRAIGHT AND REVERSE POLARITY WELDING (cont)

c. <u>Arc Welding Circuit</u>. A few changes to the circuit shown by figure 10-12, p 10-13, can be made to represent an arc welding circuit. Replace the battery with a welding generator, since they are both a source of EMF (or voltage), and replace the resistor with a welding arc which is also a resistance to current flow. The arc welding circuit is shown by figure 10-13. The current will flow from the negative terminal through the resistance of the arc to the positive terminal.

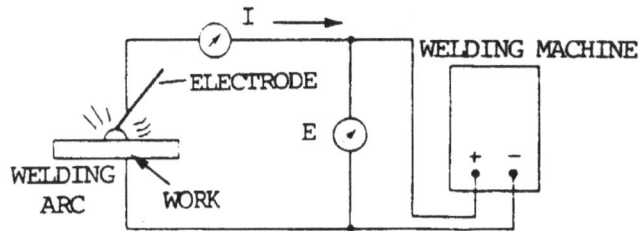

Figure 10-13. Welding electrical circuit.

d. <u>Reverse and Straight Polarity</u>. In the early days of arc welding, when welding was done with bare metal electrodes on steel, it was normal to connect the positive side of the generator to the work and the negative side to the electrode. This provided 65 to 75 percent of the heat to the work side of the circuit to increase penetration. When welding with the electrode negative, the polarity of the welding current was termed straight. When conditions such as welding cast iron or nonferrous metals made it advisable to minimize the heat in the base metal, the work was made negative and the electrode positive and the welding current polarity was said to be reverse. In order to change the polarity of the welding current, it was necessary to remove the cables from the machine terminals and replace them in the reverse position. The early coated electrodes for welding steel gave best results with the electrode positive or reverse polarity; however, bare electrodes were still used. It was necessary to change polarity frequently when using both bare and covered electrodes. Welding machines were equipped with switches that changed the polarity of the terminals and with dual reading meters. The welder could quickly change the polarity of the welding current. In marking welding machines and polarity switches, these old terms were used and indicated the polarity as straight when the electrode was negative, and reverse when the electrode was positive. Thus, electrode negative (DCEN) is the same as straight polarity (dcsp), and electrode positive (DCEP) is the same as reverse polarity (dcrp).

e. The ammeter used in a welding circuit is a millivoltmeter calibrated in amperes connected across a high current shunt in the welding circuit. The shunt is a calibrated, very low resistance conductor. The voltmeter shown in figure 10-12 will measure the welding machine output and the voltage across the arc, which are essentially the same. Before the arc is struck or if the arc is broken, the voltmeter will read the voltage across the machine with no current flowing in the circuit. This is known as the open circuit voltage, and is higher than the arc voltage or voltage across the machine when current is flowing.

f. Another unit in an electrical circuit is the unit of power. The rate of producing or using energy is called power, and is measured in watts. Power in a circuit is the product of the current in amperes multiplied by the pressure in volts. Power is measured by a wattmeter, which is a combination of an ammmeter and a voltmeter.

g. In addition to power, it is necessary to know the amount of work involved. Electrical work or energy is the product of power multiplied by time, and is expressed as watt seconds, joules, or kilowatt hours.

10-5. WELDING ARCS

a. General. The arc is used as a concentrated source of high temperature heat that can be moved and manipulated to melt the base metal and filler metal produce welds.

b. Types of Welding Arcs. There are two basic types of welding arcs. One uses the nonconsumable electrode and the other uses the consumable electrode.

(1) The nonconsumable electrode does not melt in the arc and filler metal is not carried across the arc stream. The welding processes that use the nonconsumable electrode arc are carbon arc welding, gas tungsten arc welding, and plasma arc welding.

(2) The consumable electrode melts in the arc and is carried across the arc in a stream to become the deposited filler metal. The welding processes that use the consumable electrode arc are shielded metal arc welding, gas metal arc welding, flux-cored arc welding, and submerged arc welding.

c. Function of the Welding Arc.

(1) The main function of the arc is to procduce heat. At the same time, it produces a bright light, noise, and in a special case, bombardment that removes surface films from the base metal.

(2) A welding arc is a sustained electrical discharge through a high conducting plasma. It produces sufficient thermal energy which is useful for joining metals by fusion. The welding arc is a steady-state condition maintained at the gap between an electrode and workpiece that can carry current ranging from as low as 5 amperes to as high as 2000 amperes and a voltage as low as 10 volts to the highest voltages used on large plasma units. The welding arc is somewhat different from other electrical arcs since it has a point-to-plane geometric configuration, the point being the arcing end of the electrode and the plane being the arcing area of the workpiece. Whether the electrode is positive or negative, the arc is restricted at the electrode and spreads out toward the workpiece.

(3) The length of the arc is proportional to the voltage across the arc. If the arc length is increased beyond a certain point, the arc will suddenly go out. This means that there is a certain current necessary to sustain an arc of different lengths. If a higher current is used, a longer arc can be maintained.

(4) The arc column is normally round in cross section and is made up of an inner core of plasma and an outer flame. The plasma carries most of the current. The plasma of a high—current arc can reach a temperature of 5000 to 50,000 °Kelvin. The outer flame of the arc is much cooler and tends to keep the plasma in the center. The temperature and the diameter of the central plasma depend on the amount of current passing through the arc, the shielding atmosphere, and the electrode size and type.

10-5. WELDING ARCS (cont)

(5) The curve of an arc, shown by figure 10-14, takes on a nonlinear form which in one area has a negative slope. The arc voltage increases slightly as the current increases. This is true except for the very low—cuent arc which has a higher arc voltage. This is because the low—current plasma has a fairly small cross-sectional area, and as the current increases the cross section of the plasma increases and the resistance is reduced. The conductivity of the arc increases at a greater rate than simple proportionality to current.

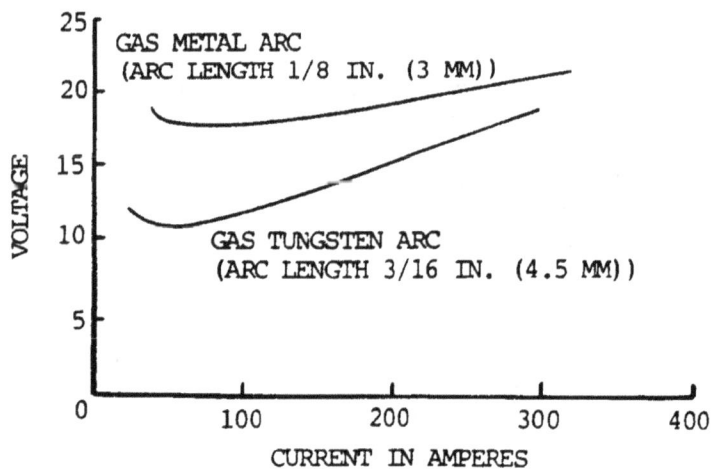

Figure 10-14. Arc characteristic volt amp curve.

(6) The arc is maintained when electrons are emitted or evaporated from the surface of the negative pole (cathode) and flow across a region of hot electrically charged gas to the positive pole (anode), where they are absorbed Cathode and. anode are electrical terms for the negative and positive poles.

(7) Arc action can best be explained by considering the dc tungsten electrode arc in an inert gas atmosphere as shown by figure 10-15. On the left, the tungsten arc is connected for direct current electrode negative (DCEN). When the arc is started, the electrode becomes hot and emits electrons.The emitted electrons are attracted to the positive pole,travel through the arc gap, and raise the temperature of the argon shielding gas atoms by colliding with themThe collisions of electrons with atoms and molecules produce thermal ionization of some of the atoms of the shielding gas. The positively charged gaseous atoms are attracted to the negative electrode where their kinetic (motion) energy is converted to heatThis heat keeps the tungsten electrode hot enough for electron emission. Emission of electrons from the surface of the tungsten cathode is known as thermionic emission. Positive ions also cross the arc. They travel from the positive pole, or the work, to the negative pole, or the electrode. Positive ions are much heavier than the electrons,but help carry the current flow of the relatively low voltage welding arc. The largest portion of the current flow, approximately 99 percent, is via electron flew rather than through the flow of positive ions. The continuous feeding of electrons into the welding circuit from the power source accounts for the continuing balance between electrons and ions in the arc. The electrons colliding with the work create the intense localized heat which provides melting and deep penetration of the base metals.

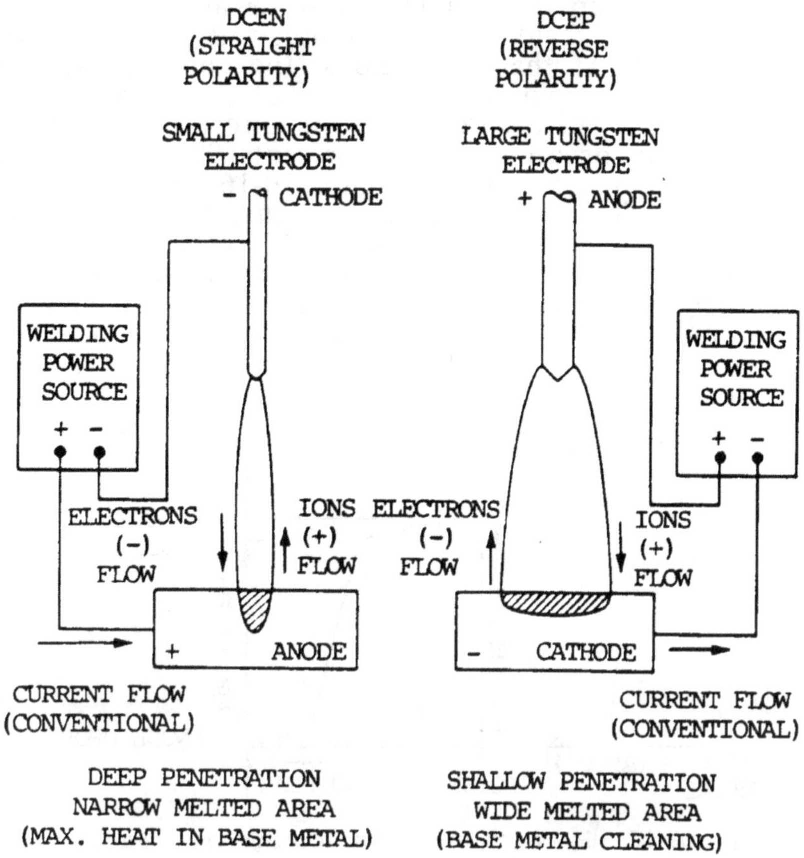

DCEN
(STRAIGHT
POLARITY)

DCEP
(REVERSE
POLARITY)

SMALL TUNGSTEN
ELECTRODE

LARGE TUNGSTEN
ELECTRODE

Figure 10-15. The dc tungsten arc.

(8) In the dc tungsten to base metal arc in an inert gas atmosphere, the maximum heat occurs at the positive pole (anode).When the electrode is positive (anode) and the work is negative (cathode)as shown by figure 10-15, the electrons flow from the work to the electrode where they create intense heat. The electrode tends to overheat. A larger electrode with more heat-absorbing capacity is used for DCEP (dcsp) than for DCEN (dcrp) for the same welding current. In addition, since less heat is generated at the work, the penetration is not so great. One result of DCEP welding is the cleaning effect on the base metal adjacent to the arc area. This appears as an etched surface and is known as catholic etching. It results from positive ion bombardment. This positive ion bombardment also occurs during the reverse polarity half-cycle when using alternating current for welding.

(9) Constriction occurs in a plasma arc torch by making the arc pass through a small hole in a water-cooled copper nozzle. It is a characteristic of the arc that the more it is cooled the hotter it gets; however, it requires a higher voltage. By flowing additional gas through the small hole, the arc is further constricted and a high velocity, high temperature gas jet or plasma emerges. This plasma is used for welding, cutting, and metal spraying.

10-5. WELDINGARCS (cont)

(10) The arc length or gap between the electrode and the work can be divided into three regions: a central region, a region adjacent to the electrode, and a region adjacent to the work. At the end regions, the cooling effect of the electrode and the work causes a rapid drop in potential These two regions are known as the anode and cathode drop, according to the direction of current flow. The length of the central region or arc column represents 99 percent of the arc length and is linear with respect to arc voltage. Figure 10-16 shows the distribution of heat in the arc, which varies in these three regions. In the central region, a circular magnetic field surrounds the arc. This field, produced by the current flow, tends to constrict the plasma and is known as the magnetic pinch effect. The constriction causes high pressures in the arc plasma and extremely high velocities. This, in turn, produces a plasma jet. The speed of the plasma jet approaches sonic speed.

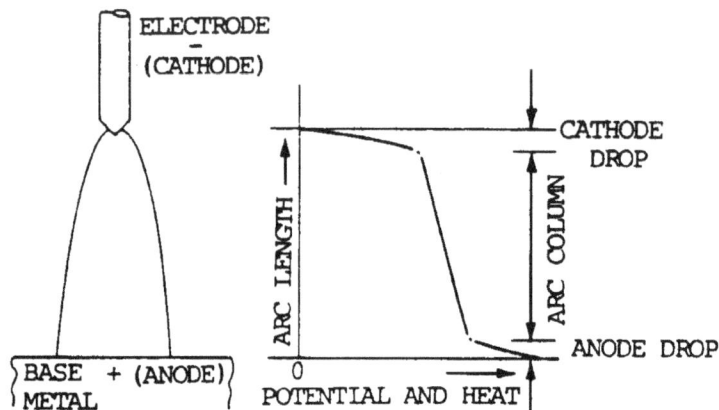

Figure 10-16. Arc length vs voltage and heat.

(11) The cathode drop is the electrical connection between the arc column and the negative pole (cathode). There is a relatively large temperature and potential drop at this point. The electrons are emitted by the cathode and given to the arc column at this point. The stability of an arc depends on the smoothness of the flow of electrons at this point. Tungsten and carbon provide thermic emissions, since both are good emitters of electrons. They have high melting temperatures, are practically nonconsumable and are therefore used for welding electrodes. Since tungsten has the highest melting point of any metal, it is preferred.

(12) The anode drop occurs at the other end of the arc and is the electrical connection between the positive pole (anode) and the arc column The temperature changes from that of the arc column to that of the anode, which is considerably lower. The reduction in temperature occurs because there are fewer ions in this region. The heat liberated at the anode and at the cathode is greater than that from the arc column.

d. Carbon Arc. In the carbon arc, a stable dc arc is obtained when the carbon is negative. In this condition, about 1/3 of the heat occurs at the negative pole (cathode), or the electrode, and about 2/3 of the heat occurs at the positive pole (anode), or the workpiece.

e. <u>Consumable Electrode Arc</u>. In the consumable electrode welding arc, the electrode is melted and molten metal is carried across the arc. A uniform arc length is maintained between the electrode and the base metal by feeding the electrode into the arc as fast as it melts. The arc atmosphere has a great effect on the polarity of maximum heat. In shielded metal arc welding, the arc atmosphere depends on the composition of the coating on the electrode. Usually the maximum heat occurs at the negative ple (cathode). When straight polarity welding with an E6012 electrode, the electrode is the negative pole (DCEN) and the melt-off rate is high. Penetration is minimum. When reverse polarity welding with an E6010 electrode (DCEP), the maximum heat still occurs at the negative pole (cathode), but this is now the base metal, which provides deep penetration. This is shown by figure 10-17. With a bare steel electrode on steel, the polarity of maximum heat is the positive pole (anode). Bare electrodes are operated on straight polarity (DCEN) so that maximum heat is at the base metal (anode) to ensure enough penetration. When coated electrodes are operated on ac, the same amount of heat is produced on each polarity of the arc.

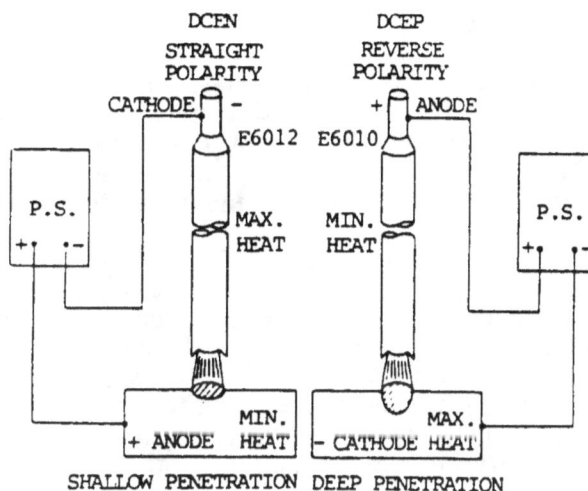

Figure 10-17. The dc shielded metal arc.

f. <u>Consumable Electrode Arc</u>.

(1) The forces that cause metal to transfer across the arc are similar for all the consumable electrode arc welding processes. The type of metal transfer dictates the usefulness of the welding process. It affects the welding position that can be used, the depth of weld penetration, the stability of the welding pool, the surface contour of the weld, and the amount of spatter loss. The metal being transferred ranges from small droplets, smaller than the diameter of the electrode, to droplets larger in diameter than the electrode. The type of transfer depends on the current density, the polarity of the electrode, the arc atmosphere, the electrode size, and the electrode composition.

(2) Several forces affect the transfer of liquid metal across an arc. These are surface tension, the plasma jet, gravity in flat position welding, and electromagnetic force.

10-5. WELDING ARCS (cont)

(a) Surface tension of a liquid causes the surface of the liquid to contract to the smallest possible area. This tension tends to hold the liquid drops on the end of a melting electrode without regard to welding position. This force works against the transfer of metal across the arc and helps keep molten metal in the weld pool when welding in the overhead position.

(b) The welding arc is constricted at the electrode and spreads or flares out at the workpiece. The current density and the arc temperature are the highest where the arc is most constricted, at the end of the electrode. An arc operating in a gaseous atmosphere contains a plasma jet which flows along the center of the arc column between the electrode and the base metal. Molten metal drops in the process of detachment from the end of the electrode, or in flight, are accelerated towards the work piece by the plasma jet.

(c) Earth gravity detaches the liquid drop when the electrode is pointed downward and is a restraining force when the electrode is pointing upward. Gravity has a noticeable effect only at low currents. The difference between the mass of the molten metal droplet and the mass of the workpiece has a gravitational effect which tends to pull the droplet to the workpiece. An arc between two electrodes will not deposit metal on either.

(d) Electromagnetic force also helps transfer metal across the arc. When the welding current flows through the electrode, a magnetic field is set up around it. The electromagnetic force acts on the liquid metal drop when it is about to detach from the electrode. As the metal melts, the cross-sectional area of the electrode changes at the molten tip. The electromagnetic force depends upon whether the cross section is increasing or decreasing. There are two ways in which the electromagnetic force acts to detach a drop at the tip of the electrode. When a drop is larger in diameter than the electrode and the electrode is positive (DCEP), the magnetic force tends to detach the drop. When there is a constriction or necking down which occurs when the drop is about to detach, the magnetic force acts away from the point of constriction in both directions. The drop that has started to separate will be given a push which increases the rate of separation. Figure 10-18 illustrates these two points. Magnetic force also sets up a pressure within the liquid drop. The maximum pressure is radial to the axis of the electrode and at high currents causes the drop to lengthen. It gives the drop stiffness and causes it to project in line with the electrode regardless of the welding position.

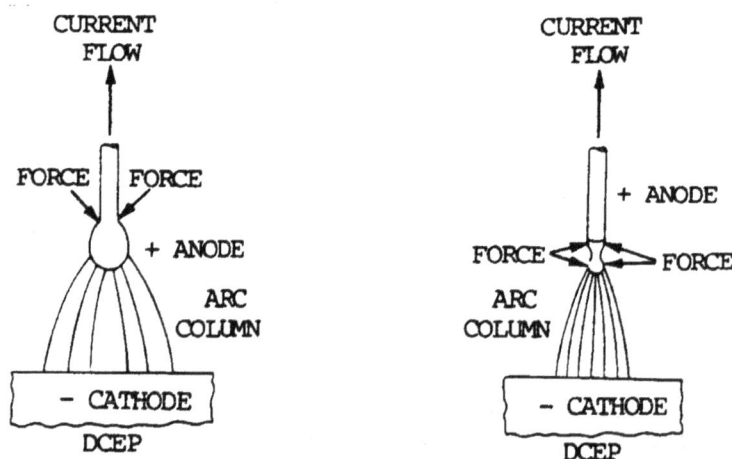

Figure 10-18. The dc consumable electrode metal arc.

10-6. AC WELDING

a. General. Alternating current is an electrical current which flows back and forth at regular intervals in a circuit. When the current rises from zero to a maximum, returns to zero,increases to a maximum in the opposite direction, and finally returns to zero again, it is said to have completed one cycle.

(1) A cycle is divided into 360 degrees. Figure 10-19 is a graphical representation of a cycle and is called a sine wave. It is generated by one revolution of a single loop coil armature in a two-pole alternating current generator. The maximum value in one direction is reached at the 90° position, and in the other direction at the 270° position.

(2) The number of times this cycle is repeated in one second is called the frequency, measured in hertz.

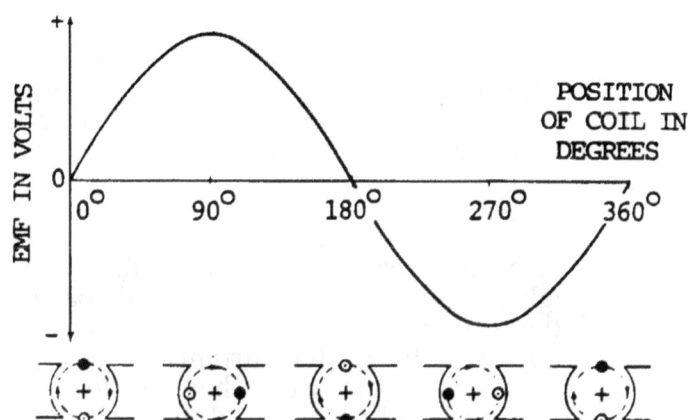

Figure 10-19. Sine wave generation.

b. Alternating current for arc welding normally has the same frequency as the line current. The voltage and current in the ac welding arc follow the sine wave and return to zero twice each cycle. The frequency is so fast that the arc appears continuous and steady. The sine wave is the simplest form of alternating current.

c. Alternating current and voltage are measured with ac meters. An ac voltmeter measures the value of both the positive and negative parts of the sine wave. It reads the effective, or root-mean-square (RMS) voltage. The effective direct current value of an alternating current or voltage is the product of 0.707 multiplied by the maximum value.

d. An alternating current has no unit of its own, but is measured in terms of direct current, the ampere. The ampere is defined as a steady rate of flow, but an alternating current is not a steady currentAn alternating current is said to be equivalent to a direct current when it produces the same average heating effect under exactly similar conditions. This is used since the heating effect of a negative current is the same as that of a positive current.Therefore, an ac ammeter will measure a value, called the effective value, of an alternating current which is shown in amperes. All ac meters, unless otherwise marked, read effective values of current and voltage.

10-6. AC WELDING (cont)

e. Electrical power for arc welding is obtained in two different ways. It is either generated at the point of use or converted from available power from the utility line. There are two variations of electrical power conversion.

(1) In the first variation, a transformer converts the relatively high voltages from the utility line to a liner voltage for ac welding.

(2) The second variation is similar in that it includes the transformer to lower the voltage, but it is followed by a rectifier which changes alternating current to direct current for dc welding.

f. With an alternating flew of current, the arc is extinguished during each half-cycle as the current reduces to zero, requiring reignition as the voltage rises again. After reignition, it passes, with increasing current, through the usual falling volts-amperes characteristic. As the current decreases again, the arc potential is lower because the temperature and degree of ionization of the arc path correspond to the heated condition of the plasma, anode, and cathode during the time of increasing current.

g. The greater the arc length, the less the arc gas will be heated by the hot electrode terminals, and a higher reignition potential will be required. Depending upon the thermal inertia of the hot electrode rterminals and plasma, the cathode emitter may cool enough during the fall of the current to zero to stop the arc completely. When the electrode and welding work have different thermal inertia ability to emit electrons, the current will flow by different amounts during each half-cycle. This causes rectification to a lesser or greater degree. Complete rectification has been experienced in arcs with a hot tungsten electrode and a cold copper opposing terminal. Partial rectification of one half-cycle is common when using the TIG welding process with ac power.

10-7. MULTILAYER WELDING

a. Multiple layer welding is used when maximum ductility of a steel weld is desired or several layers are required in welding thick metal. Multiple layer welding is accomplished by depositing filler metal in successive passes along the joint until it is filled (fig. 10-20). Since the area covered with each pass is small, the weld puddle is reduced in size. This procedure enables the welder to obtain complete joint penetration without excessive penetration and overheating while the first few passes are being deposited. The smaller puddle is more easily controlled, and the welder can avoid oxides, slag inclusions, and incomplete fusion with the base metal.

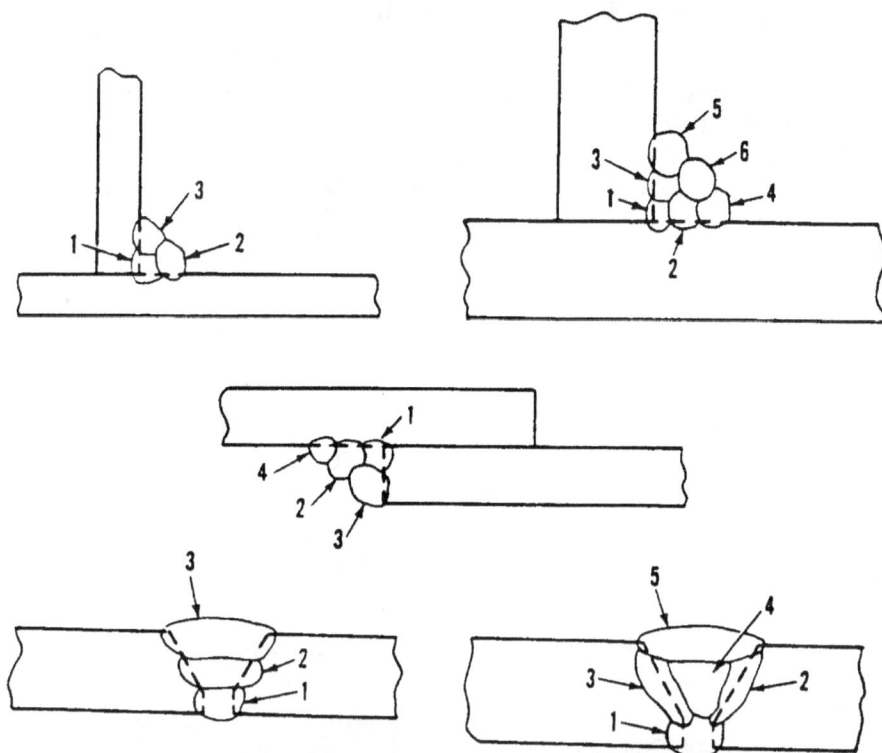

Figure 10-20. Sequences in multilayer welding.

b. The multilayer method allows the welder to concentrate on getting good penetration at the root of the V in the first pass or layer. The final layer is easily controlled to obtain a good smooth surface.

This method permits the metal deposited in a given layer to be partly or wholly refined by the succeeding layers, and therefore improved in ductility. The lower layer of weld metal, after cooling, is reheated by the upper layer and then cooled again. In effect, the weld area is being heat treated. In work where this added quality is desired in the top layer of the welded joint, an excess of weld metal is deposited on the finshed weld and then machined off. The purpose of this last layer is simply to provide welding heat to refine layer of weld metal.

Section II. ARC PROCESSES

10-8. SHIELDED METAL-ARC WELDING (SMAW)

a. General. This is the most widely used method for general welding applications. It is also refereed to as metallic arc, manual metal-arc, or stick-electrode welding. It is an arc welding process in which the joining of metals is produced by heat from an electric arc that is maintained between the tip of a covered electrode and the base metal surface of the joint being welded.

10-8.   SHIELDEDMETAL-ARC WELDING (SMAW) (cont)

b.   Advantages. The SMAW process can be used for welding most structural and alloy steels. These include low-carbon or mild steels; low-alloy, heat-treatable steels; and high-alloy steels such as stainless steels. SMAW is used for joining common nickel alloys and can be used for copper and aluminum alloys. This welding process can be used in all positions--flat, vertical, horizontal, or overhead--and requires only the simplest equipment.Thus, SMAW lends itself very well to field work (fig. 10-21).

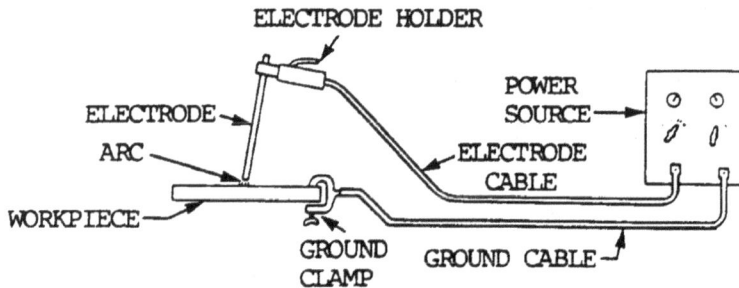

Figure 10-21. Schematic drawing of SMAW equipment.

c.   Disadvantages. Slag removal, unused electrode stubs, and spatter add to the cost of SMAW. Unused electrode stubs and spatter account for about 44 percent of the consumed electrodes. Another cost is the entrapment of slag in the form of inclusions, which may have to be removed.

d.   Processes.

(1) The core of the covered electrode consists of either a solid metal rod of drawn or cast material, or one fabricated by encasing metal powders in a metallic sheath. The core rod conducts the electric current to the arc and provides filler metal for the joint. The electrode covering shields the molten metal from the atmosphere as it is transferred across the arc and improves the smoothness or stability of the arc.

(2) Arc shielding is obtained from gases which form as a result of the decomposition of certain ingredients in the covering.The shielding ingredients vary according to the type of electrode. The shielding and other ingredients in the covering and core wire control the mechanical properties, chemical composition, and metallurgical structure of the weld metal, as well as arc characteristics of the electrode.

(3) Shielded metal arc welding employs the heat of the arc to melt the base metal and the tip of a consumable covered electrode.The electrode and the work are part of an electric circuit known as the welding circuit, as shown in figure 10-22. This circuit begins with the electric power source and includes the welding cables, an electrode holder, a ground clamp, the work, and an arc welding electrode. One of the two cables from the power source is attached to the work. The other is attached to the electrode holder.

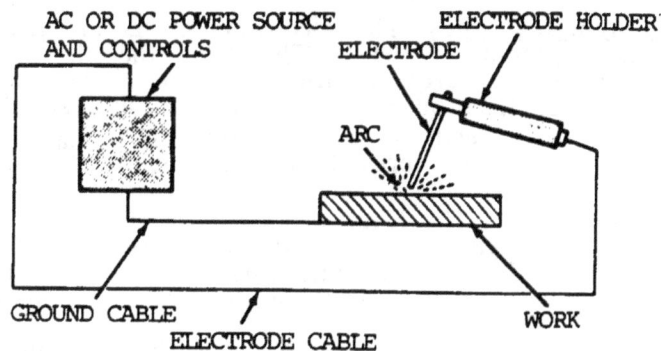

Figure 10-22. Elements of a typical welding circuit for shielded metal arc welding.

(4) Welding begins when an electric arc is struck between the tip of the electrode and the work. The intense heat of the arc melts the tip of the electrode and the surface of the work beneath the arc. Tiny globules of molten metal rapidly form on the tip of the electrode, then transfer through the arc stream into the molten weld pool. In this manner, filler metal is deposited as the electrode is progressively consumed. The arc is moved over the work at an appropriate arc length and travel speed, melting and fusing a portion of the base metal and adding filler metal as the arc progresses. Since the arc is one of the hottest of the commercial sources of heat (temperatures above 9000 °F (5000 °C) have been measured at its center), melting takes place almost instantaneously as the arc contacts the metal. If welds are made in either the flat or the horizontal position, metal transfer is induced by the force of gravity, gas expansion, electric and electromagnetic forces, and surface tension. For welds in other positions, gravity works against the other forces.

(a) Gravity. Gravity is the principal force which accounts for the transfer of filler metal in flat position welding. In other positions, the surface tension is unable to retain much molten metal and slag in the crater. Therefore, smaller electrodes must be used to avoid excessive loss of weld metal and slag. See figure 10-23.

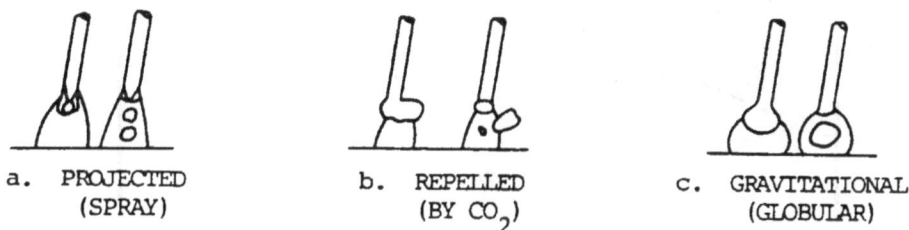

a. PROJECTED (SPRAY)   b. REPELLED (BY CO_2)   c. GRAVITATIONAL (GLOBULAR)

Figure 10-23. Three types of free-flight metal transfer in a welding arc.

(b) Gas expansion. Gases are produced by the burning and volatilization of the electrode coating, and are expanded by the heat of the boiling electrode tip. The coating extending beyond the metal tip of the electrode controls the direction of the rapid gas expansion and directs the molten metal globule into the weld metal pool fomed in the base metal.

10-8. SHIELDED METAL-ARC WELDING (SMAW) (cont)

(c) Electromagnetic forces. The electrode tip is an electrical conductor, as is the moliten metal globule at the tip. Therefore, the globule is affected by magnetic forces acting at 90 degrees to the direction of the current flow. These forces produce a pinching effect on the metal globules and speed up the separation of the molten metal from the end of the electrode. This is particularly helpful in transferring metal in horizontal, vertical, and overhead position welding.

(d) Electrical forces. The force produced by the voltage across the arc pulls the small, pinched-off globule of metal, regardless of the position of welding. This force is especially helpful when using direct-current, straight-polarity, mineral-coated electrodes, which do not produce large volumes of gas.

(e) Surface tension. The force which keeps the filler metal and slag globules in contact with molten base or weld metal in the crater is known as surface tension. It helps to retain the molten metal in horizontal, vertical, and overhead welding, and to determine the shape of weld countours.

e. Equipment. The equipment needed for shielded metal-arc welding is much less complex than that needed for other arc welding processes. Manual welding equipment includes a power source (transformer, dc generator, or dc rectifier) electrode holder, cables, connectors, chipping hammer, wire brush, and electrodes.

f. Welding Parameters.

(1) Welding voltage, current, and travel speed are very important to the quality of the deposited SMAW bead. Figures 10-24 thru 10-30 show the travel speed limits for the electrodes listed in table 10-1, p 10-30. Table 10-1 shows voltage limits for some SMAW electrodes.

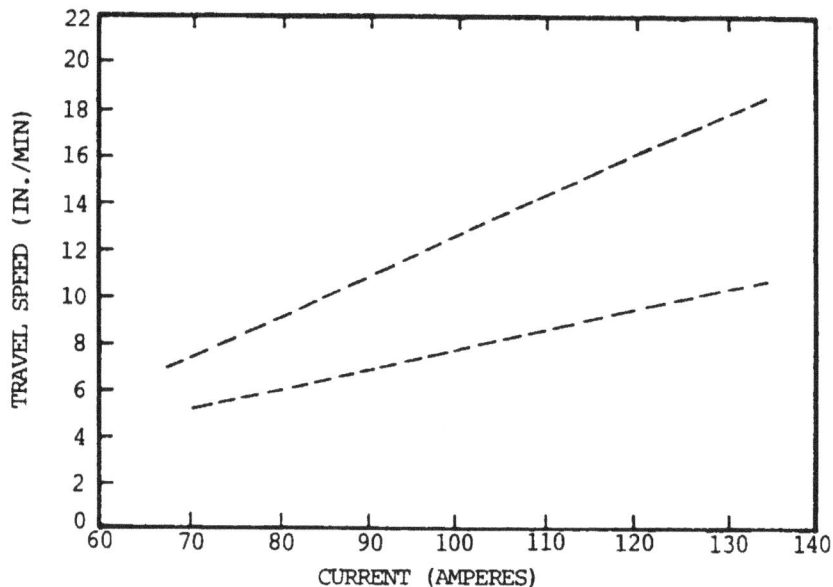

Figure 10-24. Travel speed limits for current levels used for 1/8-inch-diameter E6010 SMAW electrode. Dashed lines show travel speed limits as determined by amount of undercut and bead shape.

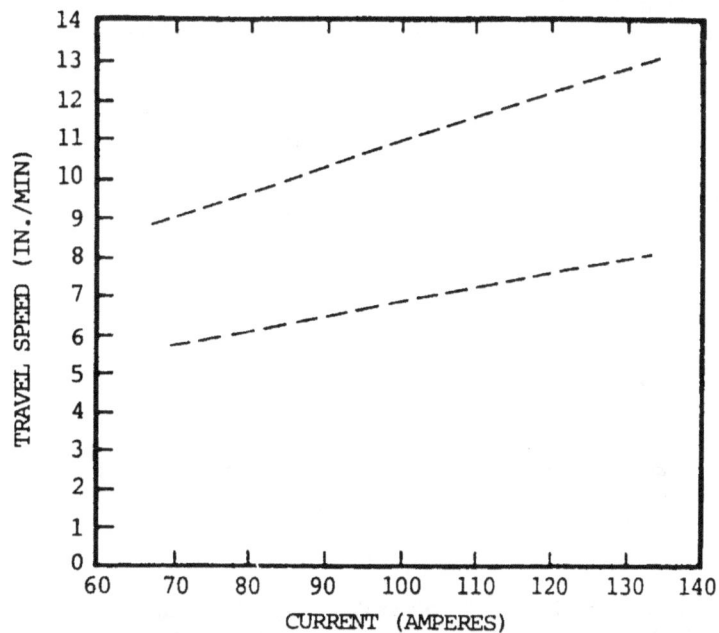

Figure 10-25. Travel speed limits for current levels used for 1/8-inch-diameter
E6011 SMAW electrode. Dashed lines show travel speed limits as
determined by amount of undercut and bead shape.

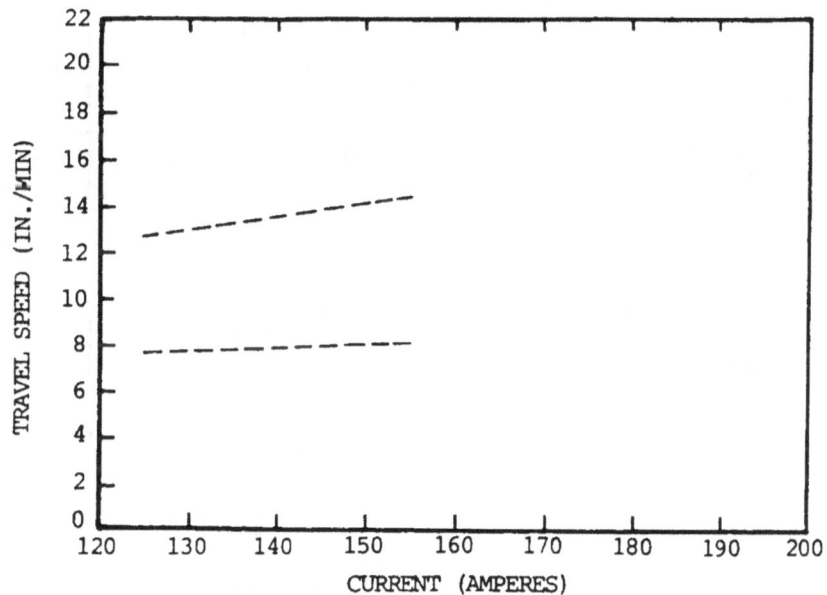

Figure 10-26. Travel speed limits for current levels used for 1/8-inch-diameter
E6013 SMAW electrode. Dashed lines show travel speed limits as
determined by amount of undercut and bead shape.

10-8. SHIELDED METAL-ARC WELDING (SMAW) (cont)

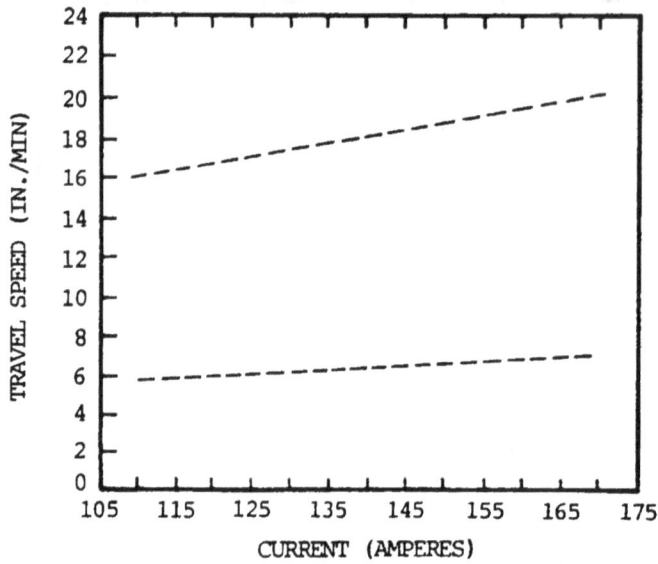

Figure 10-27. Travel speed limits for current levels used for 1/8-inch-diameter E7018 SMAW electrode. Dashed lines show travel speed limits as determined by amount of undercut and bead shape.

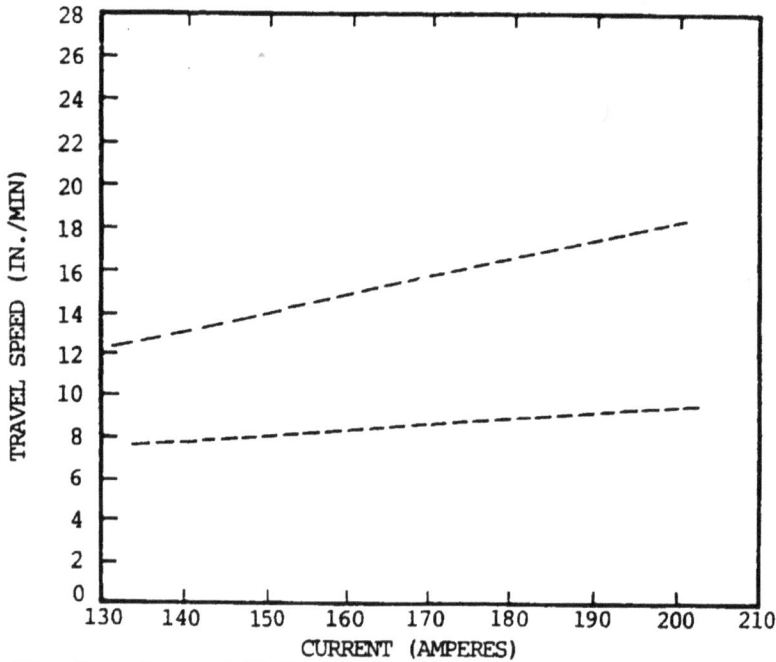

Figure 10-28. Travel speed limits for current levels used for 1/8-inch-diameter E7024 SMAW electrode. Dashed lines show travel speed limits as determined by amount of undercut and bead shape.

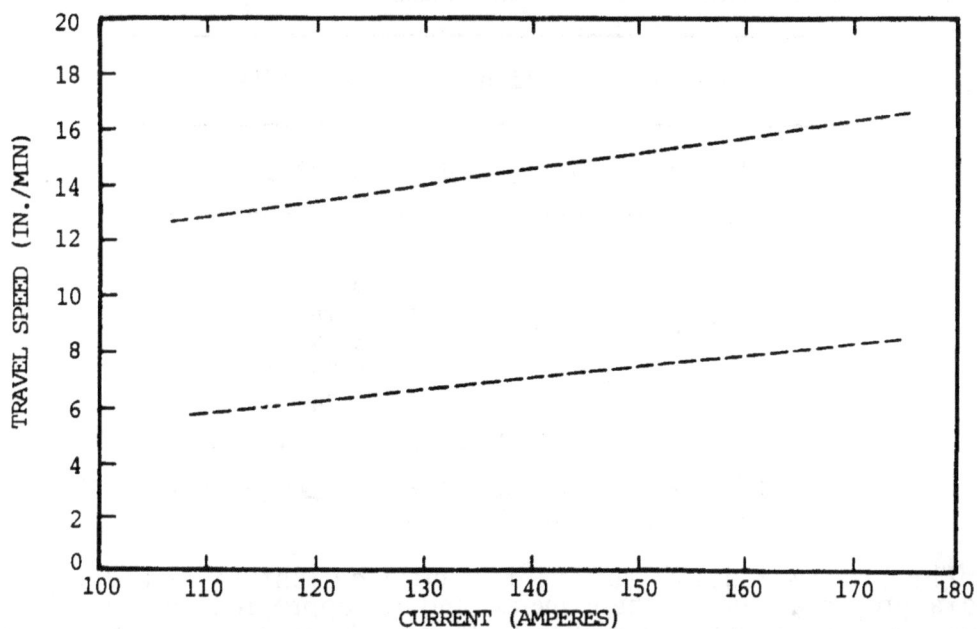

Figure 10-29. Travel speed limits for current levels used for 5/32-inch-diameter
E8018 SMAW electrode. Dashed lines show travel speed limits as
determined by amount of undercut and bead shape.

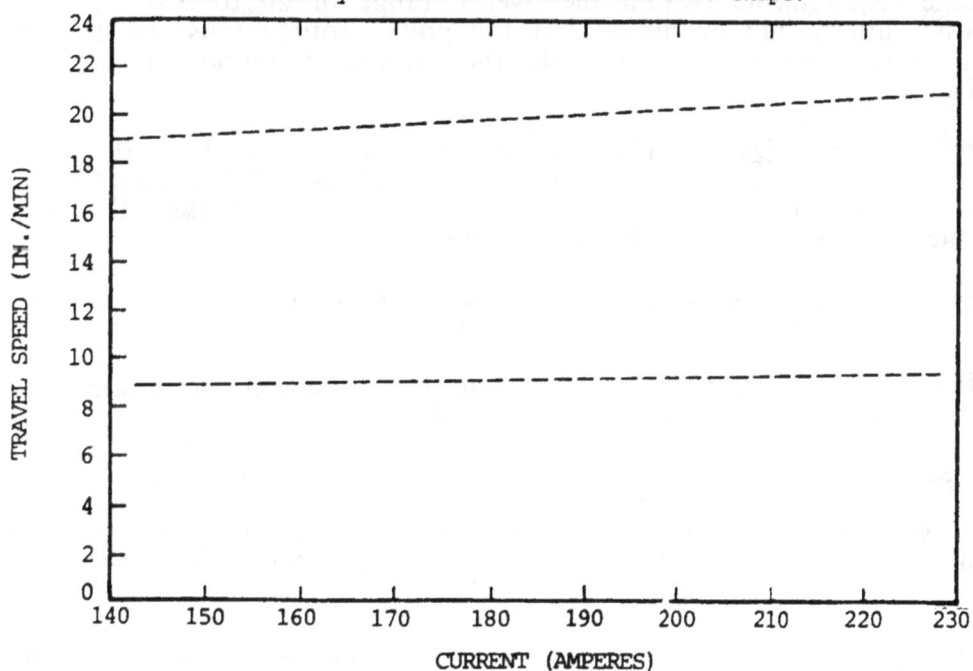

Figure 10-30. Travel speed limits for current levels used for 1/8-inch-diameter
E11018 SMAW electrode. Dashed lines show travel speed limits as
determined by amount of undercut and bead shape.

10-8.   SHIELDED METAL-ARC WELDING (SMAW) (cont)

Table 10-1.   Established Voltage Limits

Electrode*	Voltage limits, V
E6010	28 to 32
E6011	28 to 32
E6013	22 to 26
E7018	25 to 28
E7024	26 to 32
E8018	22 to 28
E11018	25 to 30

*Note all electrodes 1/8-inch diameter except E8018, which is 5/32-inch (4-mm) diameter.

(2) The process requires sufficient electric current to melt both the electrode and a proper amount of base metal, and an appropriate gap between the tip of the electrode and base metal or molten weld pool.These requirements are necessary for coalescence. The sizes and types of electrodes for shielded metal arc welding define the arc voltage requirements (within the overall range of 16 to 40 V) and the amperage requirements (within the overall range of 20 to 550 A). The current may be either alternating or direct, but the power source must be able to control the current level in order to respond to the complex variables of the welding process itself.

g. Covered Electrodes. In addition to establishing the arc and supplying filler metal for the weld deposit,the electrode introduces other materials into or around the arc. Depending upon the type of electrode being used, the covering performs one or more of the following functions:

(1) Provides a gas to shield the arc and prevent excessive atmospheric contamination of the molten filler metal as it travels across the arc.

(2) Provides scavengers, deoxidizers, and fluxing agents to cleanse the weld and prevent excessive grain growth in the weld metal.

(3) Establishes the electrical characteristics of the electrode.

(4) Provides a slag blanket to protect the hot weld metal from the air and enhance the mechanical properties, bead shape, and surface cleanliness of the weld metal.

(5) Provides a means of adding alloying elements to change the mechanical properties of the weld metal.

Functions 1 and 4 prevent the pick-up of oxygen and nitrogen from the air by the molten filler metal in the arc stream and by the weld metal as it solidifies and cools.

The covering on shielded metal arc electrodes is applied by either the extrusion or the dipping process. Extrusion is much more widely used. The dipping process is used primarily for cast and some fabricated core rods.In either case, the covering contains most of the shielding,scavenging, and deoxidizing materials. Most SMAW electrodes have a solid metal core. Some are made with a fabricated or composite core consisting of metal powders encased in a metallic sheath.In this latter case, the purpose of some or even all of the metal powders is to produce an alloy weld deposit.

In addition to improving the mechanical properties of the weld metal, the covering on the electrode can be designed for welding with alternating current.With ac, the welding arc goes out and is reestablished each time the current reverses its direction. For good arc stability,it is necessary to have a gas in the arc stream that will remain ionized during each reversal of the current. This ionized gas makes possible the reignition of the arc. Gases that readily ionize are available from a variety of compunds,including those that contain potassium.It is the incorporation of these compounds in the electrode covering that enables the electrode to operate on ac.

To increase the deposition rate,the coverings of some carbon and low alloy steel electrodes contain iron powder. The iron powder is another source of metal available for deposition, in addition to that obtained from the core of the electrode. The presence of iron powder in the covering also makes more efficient use of the arc energy. Metal powders other than iron are frequently used to alter the mechanical properties of the weld metal.

The thick coverings on electrodes with relatively large amounts of iron powder increase the depth of the crucible at the tip of the electrode.This deep crucible helps contain the heat of the arc and maintains a constant arc length by using the "drag" technique. When iron or other metal powders are added in relatively large amounts, the deposition rate and welding speed usually increase. Iron powder electrodes with thick coverings reduce the level of skill needed to weld.The tip of the electrode can be dragged along the surface of the work while maintaining a welding arc. For this reason,heavy iron powder electrodes frequently are called "drag electrodes." Deposition rates are high;but because slag solidification is slow, these electrodes are not suitable for out-of-position use.

    h. Electrode Classification System. The SMAW electrode classification cede contains an E and three numbers, followed by a dash and either "15" or "16" (EXXX-15). The E designates that the material is an electrode and the three digits indicate composition. Sometimes there are letters following the three digits; these letters indicate a modification of the standard composition. The "15" or "16" specifies the type of current with which these electrodes may be used.Both designations indicate that the electrode is usable in all positions:flat, horizontal, vertical and overhead.

    (1) The "15" indicates that the covering of this electrode is a lime type, which contains a large proportion of calcium or alkaline earth materials. These electrodes are usable with dc reverse-polarity only.

    (2) The designation "16" indicates electrodes that have a lime- or titania-type covering with a large proportion of titanium-bearing minerals.The coverings of these electrodes also contain readily ionizing elements, such as potassium, to stabilize the arc for ac welding.

10-8.   SHIELDED METAL-ARC WELDING (SMAW) (cont)

i.   Chemical Requirements. The AWS divides SMAW electrodes into two groups: mild steel and low-alloy steel. The E60XX and E70XX electrodes are in the mild steel specification. The chemical requirements for E70XX electrodes are listed in AWS A5.1 and allow for wide variations of composition of the deposited weld metal. There are no specified chemical requirements for the E60XX electrodes. The low-alloy specification contains electrode classifications E70XX through E120XX. These codes have a suffix indicating the chemical requirements of the class of electrodes (e.g., E7010-Al or E8018-Cl). The composition of low-alloy E70XX electrodes is controlled much more closely than that of mild steel E70XX electrodes. Low-alloy electrodes of the low-hydrogen classification (EXX15, EXX16, EXX18) require special handling to keep the coatings from picking up water. Manufacturers' recommendations abut storage and rebaking must be followed for these electrodes. AWS A5.5 provides a specific listing of chemical requirements.

Weld Metal Mechanical Properties. The AWS requires the deposited weld metal to have a minimum tensile strength of 60,000 to 100,000 psi (413,700 to 689,500 kPa), with minimum elongations of 20 to 35 percent.

k.   Arc Shielding.

(1) The arc shielding action, illustrated in figure 10-31, is essentially the same for the different types of electrodes, but the specific method of shielding and the volume of slag produced vary from type to type. The bulk of the covering materials in some electrodes is converted to gas by the heat of the arc, and only a small amount of slag is produced. This type of electrode depends largely upon a gaseous shield to prevent atmospheric contamination. Weld metal from such electrodes can be identified by the incomplete or light layer of slag which covers the bead.

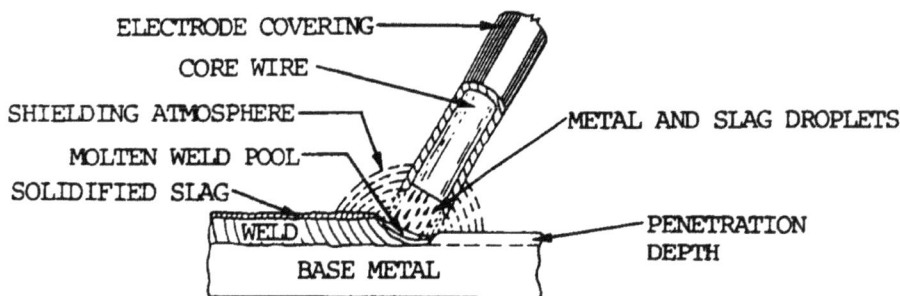

Figure 10-31.   Shielded metal arc welding.

(2) For electrodes at the other extreme, the bulk of the covering is converted to slag by the arc heat, and only a small volume of shielding gas is produced. The tiny globules of metal transferred across the arc are entirely coated with a thin film of molten slag. This slag floats to the weld puddle surface because it is lighter than the metal. It solidifies after the weld metal has solidified. Welds made with these electrodes are identified by the heavy slag deposit that completely cover the weld beads. Between these extremes is a wide variety of electrode types, each with a different combination of gas and slag shielding.

(3) The variations in the amount of slag and gas shielding also influence the welding characteristics of the different types of covered electrodes.Electrodes that have a heavy slag carry high amperage and have high deposition rates. These electrodes are ideal for making large beads in the flat positionElectrodes that develop a gaseous arc shield and have a light layer of slag carry lower amperage and have lower deposition rates. These electrodes produce a smaller weld pool and are better suited for making welds in the vertical and overhead positionBecause of the differences in their welding characteristics, one type of covered electrode will usually be best suited for a given application.

10-9. GAS TUNGSTEN ARC (TIG) WELDING (GTAW)

a. General. Gas tungsten arc welding (TIG welding or GTAW) is a process in which the joining of metals is produced by heating therewith an arc between a tungsten (nonconsumable) electrode and the work.A shielding gas is used, normally argon. TIG welding is normally done with a pure tungsten or tungsten alloy rod, but multiple electrodes are sometimes used. The heated weld zone, molten metal, and tungsten electrode are shielded from the atmosphere by a covering of inert gas fed through the electrode holder. Filler metal may or may not be added. A weld is made by applying the arc so that the touching workpiece and filler metal are melted and joined as the weld metal solidifies. This process is similar to other arc welding processes in that the heat is generated by an arc between a nonconsumable electrode and the workpiece, but the equipment and electrode type distinguish TIG from other arc welding processes. See figure 10-32.

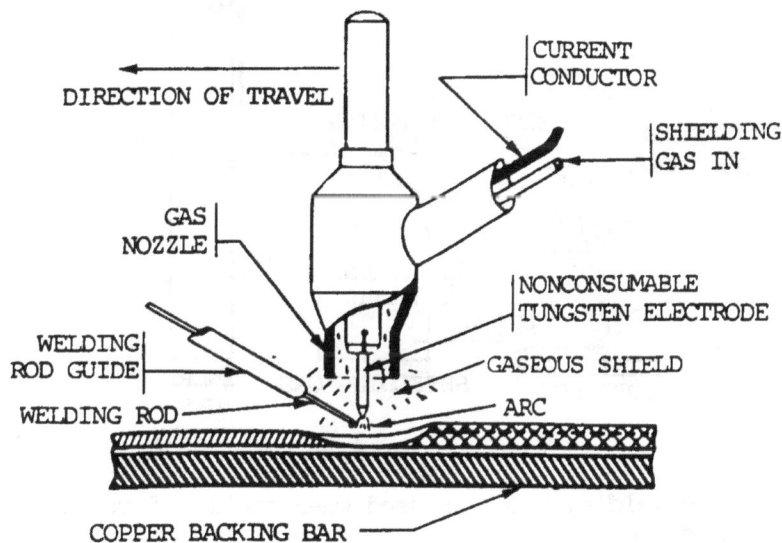

Figure 10-32. Gas tungsten arc (TIG) welding (GTAW).

10-9.   GAS TUNGSTEN ARC (TIG) WELDING (GTAW) (cont)

b. Equipment. The basic features of the equipment used in TIG welding are shown in figure 10-33. The major components required for TIG welding are:

(1) the welding machine, or power source

(2) the welding electrode holder and the tungsten electrode

(3) the shielding gas supply and controls

(4) Several optional accessories are available, which include a foot rheostat to control the current while welding, water circulating systems to cool the electrode holders, and arc timers.

NOTE

There are ac and dc power units with built-in high frequency generators designed specifically for TIG welding. These automatically control gas and water flow when welding begins and ends. If the electrode holder (torch) is water-cooled, a supply of cooling water is necessary. Electrode holders are made so that electrodes and gas nozzles can readily be changed. Mechanized TIG welding equipment may include devices for checking and adjusting the welding torch level, equipment for work handling, provisions for initiating the arc and controlling gas and water flow, and filler metal feed mechanisms.

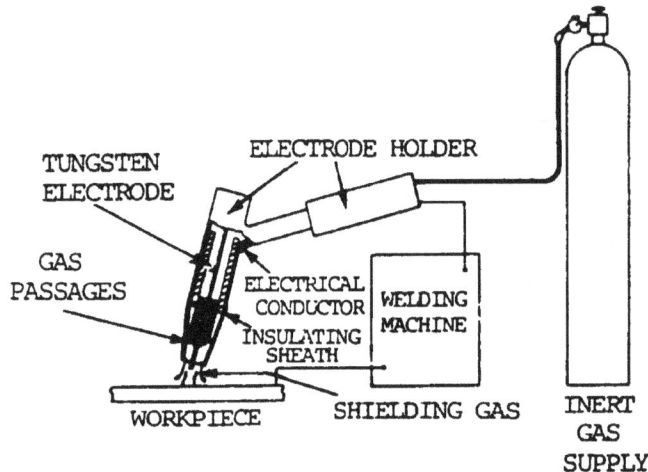

NOTE
A water-cooled welding torch is used when cooling from the inert gas shield is inadequate.

Figure 10-33. Gas tungsten arc welding equipment arrangement.

c.   Advantages. Gas tungsten arc welding is the most popular method for welding aluminum stainless steels, and nickel-base alloys. It produces top quality welds in almost all metals and alloys used by industry. The process provides more precise control of the weld than any other arc welding process, because the arc heat and filler metal are independently controlled. Visibility is excellent because no smoke or fumes are produced during welding, and there is no slag or spatter that must be cleaned between passes or on a completed weld. TIG welding also has reduced distortion in the weld joint because of the concentrated heat source. The gas tungsten arc welding process is very good for joining thin base metals because of excellent control of heat input. As in oxyacetylene welding, the heat source and the addition of filler metal can be separately controlled. Because the electrode is nonconsumable, the process can be used to weld by fusion alone without the addition of filler metal. It can be used on almost all metals, but it is generally not used for the very low melting metals such as solders, or lead, tin, or zinc alloys. It is especially useful for joining aluminum and magnesium which form refractory oxides, and also for the reactive metals like titanium and zirconium, which dissolve oxygen and nitrogen and become embrittled if exposed to air while melting. In very critical service applications or for very expensive metals or parts, the materials should be carefully cleaned of surface dirt, grease, and oxides before welding.

   d. Disadvantages. TIG welding is expensive because the arc travel speed and weld metal deposition rates are lower than with some other methods. Some limitations of the gas tungsten arc process are:

   (1) The process is slower than consumable electrode arc welding processes.

   (2) Transfer of molten tungsten from the electrode to the weld causes contamination. The resulting tungsten inclusion is hard and brittle.

   (3) Exposure of the hot filler rod to air using improper welding techniques causes weld metal contamination.

   (4) Inert gases for shielding and tungsten electrode costs add to the total cost of welding compared to other processes. Argon and helium used for shielding the arc are relatively expensive.

   (S) Equipment costs are greater than that for other processes, such as shielded metal arc welding, which require less precise controls.

For these reasons, the gas tungsten arc welding process is generally not commercially competitive with other processes for welding the heavier gauges of metal if they can be readily welded by the shielded metal arc, submerged arc, or gas metal arc welding processes with adequate quality.

   e. Process Principles.

   (1) Before welding begins, all oil, grease, paint, rust, dirt, and other contaminants must be removed from the welded areas. This may be accomplished by mechanical means or by the use of vapor or liquid cleaners.

   (2) Striking the arc maybe done by any of the following methods:

      (a) Touching the electrode to the work momentarily and quickly withdrawing it.

      (b) Using an apparatus that will cause a spark to jump from the electrode to the work.

10-9. GAS TUNGSTEN ARC (TIG) WELDING (GTAW) (cont)

(c) Using an apparatus that initiates and maintains a small pilot arc, providing an ionized path for the main arc.

(3) High frequency arc stabilizers are required when alternating current is used. They provide the type of arc starting described in (2)(b) above. High frequency arc initiation occurs when a high frequency, high voltage signal is superimposed on the welding circuit. High voltage (low current) ionizes the shielding gas between the electrode and the workpiece, which makes the gas conductive and initiates the arc. Inert gases are not conductive until ionized. For dc welding, the high frequency voltage is cut off after arc initiation. However, with ac welding, it usually remains on during welding, especially when welding aluminum.

(4) When welding manually, once the arc is started, the torch is held at a travel angle of about 15 degrees. For mechanized welding, the electrode holder is positioned vertically to the surface.

(5) To start manual welding, the arc is moved in a small circle until a pool of molten metal forms. The establishment and maintenance of a suitable weld pool is important and welding must not proceed ahead of the puddle. Once adequate fusion is obtained, a weld is made by gradually moving the electrode along the parts to be welded to melt the adjoining surfaces. Solidification of the molten metal follows progression of the arc along the joint, and completes the welding cycle.

(6) The welding rod and torch must be moved progressively and smoothly so the weld pool, hot welding rod end, and hot solidified weld are not exposed to air that will contaminate the weld metal area or heat-affected zone. A large shielding gas cover will prevent exposure to air. Shielding gas is normally argon.

(7) The welding rod is held at an angle of about 15 degrees to the work surface and slowly fed into the molten pool. During welding, the hot end of the welding rod must not be removed from the inert gas shield. A second method is to press the welding rod against the work, in line with the weld, and melt the rod along with the joint edges. This method is used often in multiple pass welding of V-groove joints. A third method, used frequently in weld surfacing and in making large welds, is to feed filler metal continuously into the molten weld pool by oscillating the welding rod and arc from side to side. The welding rod moves in one direction while the arc moves in the opposite direction, but the welding rod is at all times near the arc and feeding into the molten pool. When filler metal is required in automatic welding, the welding rod (wire) is fed mechanically through a guide into the molten weld pool.

(8) The selection of welding position is determined by the mobility of the weldment, the availability of tooling and fixtures, and the welding cost. The minimum time, and there fore cost, for producing a weld is usually achieved in the flat position. Maximum joint wet-ration and deposition rate are obtained in this position, because a large volume of molten metal can be supported. Also, an acceptably shaped reinforcement is easily obtained in this position.

(9) Good penetration can be achieved in the vertical-up position, but the rate of welding is slower because of the effect of gravity on the molten weld metal. Penetration in vertical-dam welding is poor. The molten weld metal droops

and lack of fusion occurs unless high welding speeds are used to deposit thin layers of weld metal. The welding torch is usually pointed forward at an angle of about 75 degrees from the weld surface in the vertical-up and flat positions. Too great an angle causes aspiration of air into the shielding gas and consequent oxidation of the molten weld metal.

(10) Joints that may be welded by this process include all the standard types, such as square-groove and V-groove joints, T-joints, and lap joints. As a rule, it is not necessary to bevel the edges of base metal that is 1/8 in. (3.2 mm) or less in thickness. Thicker base metal is usually beveled and filler metal is always added.

(11) The gas tungsten arc welding process can be used for continuous welds, intermittent welds, or for spot welds. It can be done manually or automatically by machine.

(12) The major operating variables summarized briefly are:

(a) Welding current, voltage, and power source characteristics.

(b) Electrode composition, current carrying capacity, and shape.

(c) Shielding gas--welding grade argon, helium, or mixtures of both.

(d) Filler metals that are generally similar to the metal being joined and suitable for the intended service.

(13) Welding is stopped by shutting off the current with foot-or-hand-controlled switches that permit the welder to start, adjust. and stop the welding current. They also allow the welder to control the welding current to obtain good fusion and penetration. Welding may also be stopped by withdrawing the electrode from the current quickly, but this can disturb the gas shielding and expose the tungsten and weld pool to oxidation.

f. Filler Metals. The base metal thickness and joint design determine whether or not filler metal needs to be added to the joints. When filler metal is added during manual welding, it is applied by manually feeding the welding rod into the pool of molten metal ahead of the arc, but to one side of the center line. The technique for manual TIG welding is shown in figure 10-34.

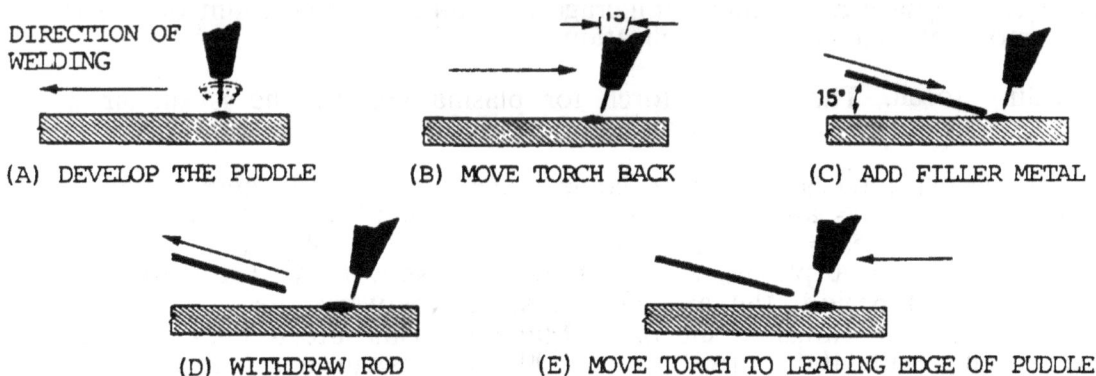

DIRECTION OF
WELDING

(A) DEVELOP THE PUDDLE   (B) MOVE TORCH BACK   (C) ADD FILLER METAL

(D) WITHDRAW ROD   (E) MOVE TORCH TO LEADING EDGE OF PUDDLE

Figure 10-34. Technique for manual gas tungsten arc (TIG) welding.

TC 9-237

10-10. PLASMA ARC WELDING (PAW)

a. <u>General</u>. Plasma arc welding (PAW) is a process in which coalescence, or the joining of metals, is produced by heating with a constricted arc between an electrode and the workpiece (transfer arc) or the electrode and the constricting nozzle (nontransfer arc). Shielding is obtained from the hot ionized gas issuing from the orifice, which may be supplemented by an auxiliary source of shielding gas.Shielding gas may be an inert gas or a mixture of gases.Pressure may or may not be used, and filler metal may or may not be supplied.The PAW process is shown in figure 10-35.

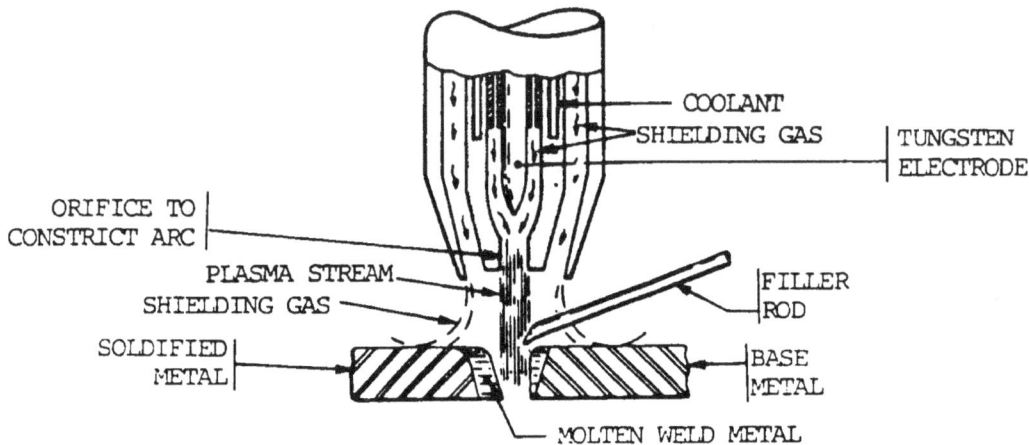

Figure 10-35. Process diagram - keyhole mode - PAW.

b. <u>Equipment</u>.

(1) <u>Power source</u>. A constant current drooping characteristic power source supplying the dc welding current is recommended; however, ac/dc type power source can be used. It should have an open circuit voltage of 80 volts and have a duty cycle of 60 percent. It is desirable for the power source to have a built-in contactor and provisions for remote control current adjustment.For welding very thin metals, it should have a minimum amperage of 2 amps. A maximum of 300 is adequate for most plasma welding applications.

(2) Welding torch. The welding torch for plasma arc welding is similar in appearance to a gas tungsten arc torch, but more complex.

(a) All plasma torches are water cooled, even the lowest—current range torch. This is because the arc is contained inside a chamber in the torch where it generates considerable heat. If water flow is interrupted briefly, the nozzle may melt. A cross section of a plasma arc torch head is shown by figure 10-36. During the nontransferred period, the arc will be struck between the nozzle or tip with the orifice and the tungsten electrode.Manual plasma arc torches are made in various sizes starting with 100 amps through 300 amperes.Automatic torches for machine operation are also available.

10-38

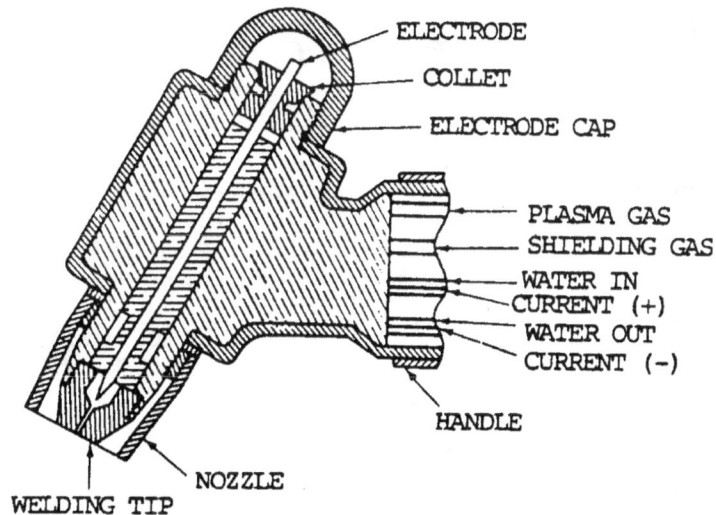

Figure 10-36. Cross section of plasma arc torch head.

(b) The torch utilizes the 2 percent thoriated tungsten electrode similar to that used for gas tungsten welding. Since the tungsten electrode is located inside the torch, it is almost impossible to contaminate it with base metal.

(3) <u>Control console</u>. A control console is required for plasma arc welding. The plasma arc torches are designed to connect to the control console rather than the power source. The console includes a power source for the pilot arc, delay timing systems for transferring from the pilot arc to the transferred arc, and water and gas valves and separate flow meters for the plasma gas and the shielding gas. The console is usually connected to the power source and may operate the contactor. It will also contain a high-frequency arc starting unit, a nontransferred pilot arc power supply, torch protection circuit, and an ammeter. The high frequency generator is used to initiate the pilot arc. Torch protective devices include water and plasma gas pressure switches which interlock with the contactor.

(4) <u>Wire feeder.</u> A wire feeder may be used for machine or automatic welding and must be the constant speed type. The wire feeder must have a speed adjustment covering the range of from 10 in. per minute (254 mm per minute) to 125 in. per minute (3.18 m per minute) feed speed.

c. <u>Advantages and Major Uses</u>.

(1) Advantages of plasma arc welding when compared to gas tungsten arc welding stem from the fact that PAW has a higher energy concentration. Its higher temperature, constricted cross-sectional area, and the velocity of the plasma jet create a higher heat content. The other advantage is based on the stiff columnar type of arc or form of the plasma, which doesn't flare like the gas tungsten arc. These two factors provide the following advantages:

(a) The torch-to-work distance from the plasma arc is less critical than for gas tungsten arc welding. This is important for manual operation, since it gives the welder more freedom to observe and control the weld.

10-10. PLASMA ARC WELDING (PAW) (cont)

(b) High temperature and high heat concentration of the plasma allow for the keyhole effect, which provides complete penetration single pass welding of many joints.   In this operation, the heat affected zone and the form of the weld are more desirable. The heat-affected zone is smaller than with the gas tungsten arc, and the weld tends to have more parallel sides which reduces angular distortion.

(c) The higher heat concentration and the plasma jet allow for higher travel speeds. The plasma arc is more stable and is not as easily deflected to the closest point of base metal. Greater variation in joint alignment is possible with plasma arc welding. This is important when making root pass welds on pipe and other one-side weld joints. Plasma welding has deeper penetration capabilities and produces a narrower weld. This means that the depth-to-width ratio is more advantageous.

(2) Uses.

(a) Some of the major uses of plasma arc are its application for the manufacture of tubing. Higher production rates based on faster travel speeds result from plasma over gas tungsten arc welding. Tubing made of stainless steel, titanium, and other metals is being produced with the plasma process at higher production rates than previously with gas tungsten arc welding.

(b) Most applications of plasma arc welding are in the low—current range. from 100 amperes or less. The plasma can be operated at extremely low currents to allow the welding of foil thickness material.

(c) Plasma arc welding is also used for making small welds on weldments for instrument manufacturing and other small components made of thin metal It is used for making butt joints of wall tubing.

This process is also used to do work similar to electron beam welding, but with a much lower equiment cost.

(3) Plasma arc welding is nornally applied as a manual welding process, but is also used in automatic and machine applications Manual application is the most popular. Semiautomatic methods of application are not useful. The normal methods of applying plasma arc welding are manual (MA), machine (ME), and automatic (AU).

(4) The plasma arc welding process is an all-position welding process. Table 10-2 shows the welding position capabilities.

Table 10-2. **Welding Position Capabilities**

Welding Position	Rating
1. Flat	A
Horizontal fillet	A
2. Horizontal	A
3. Vertical	A
4. Overhead	A
5. Pipe - fixed	A

(5) The plasma arc welding process is able to join practically all commercially available metals. It may not be the best selection or the most economical process for welding some metals. The plasma arc welding process will join all metals that the gas tungsten arc process will weld. This is illustrated in table 10-3.

(6) Regarding thickness ranges welded by the plasma process, the keyhole mode of operation can be used only where the plasma jet can penetrate the joint. In this mode, it can be used for welding material from 1/16 in. (1.6 mm through 1/4 in. (12.0 mm). Thickness ranges vary with different metals. The melt-in mode is used to weld material as thin as 0.002 in. (0.050 mm) up through 1/8 in. (3.2 mm). Using multipass techniques, unlimited thicknesses of metal can be welded. Note that filler rod is used for making welds in thicker material. Refer to table 10-4 for base metal thickness ranges.

10-10. PLASMA ARC WELDING (PAW) (cont)

Table 10-3. Base Metals Weldable by the Plasma Arc Process

Base Metal	Weldability
Aluminums	Weldable
Bronzes	Possible but not popular
Copper	Weldable
Copper nickel	Weldable
Cast, malleable, nodular	Possible but not popular
Wrought iron	Possible but not popular
Lead	Possible but not popular
Magnesium	Possible but not popular
Inconel	Weldable
Nickel	Weldable
Monel	Weldable
Precious metals	Weldable
Low carbon steel	Weldable
Low alloy steel	Weldable
High and medium carbon	Weldable
Alloys steel	Weldable
Stainless steel	Weldable
Tool steels	Weldable
Titanium	Weldable
Tungsten	Weldable

Table 10-4. Base Metal Thickness Range

Thickness / Factor	in.	.005	.015	.062	.125	3/16	1/4	3/8	1/2	3/4	1	2	4	8
	mm	.13	.4	1.6	3.2	4.8	6.4	10	12.7	19	25	51	102	203
Single pass melt in mode		←		→										
Single pass keyhole mode				←			→							
Multipass melt in mode						←				— — —			→	

d. Limitations of the Process. The major limitations of the process have to do more with the equipment and apparatus. The torch is more delicate and complex than a gas tungsten arc torch. Even the lowest rated torches must be water cooled. The tip of the tungsten and the alignment of the orifice in the nozzle is extremely important and must be maintained within very close limits. The current level of the torch cannot be exceeded without damaging the tip. The water-cooling passages in the torch are relatively small and for this reason water filters and deionized water are recommended for the lower current or smaller torches. The control console adds another piece of equipment to the system. This extra equipment makes the system more expensive and may require a higher level of maintenance.

e. Principles of Operation.

(1) The plasma arc welding process is normally compared to the gas tungsten arc process. If an electric arc between a tungsten electrode and the work is constricted in a cross-sectional area, its temperature increases because it carries the same amount of current. This constricted arc is called a plasma, or the fourth state of matter.

(2) Two modes of operation are the non-transferred arc and the transferred arc.

(a) In the non-transferred mode, the current flow is from the electrode inside the torch to the nozzle containing the orifice and back to the power supply. It is used for plasma spraying or generating heat in nonmetals.

(b) In transferred arc mode, the current is transferred from the tungsten electrode inside the welding torch through the orifice to the workpiece and back to the power supply.

(c) The difference between these two modes of operation is shown by figure 10-37. The transferred arc mode is used for welding metals. The gas tungsten arc process is shown for comparison.

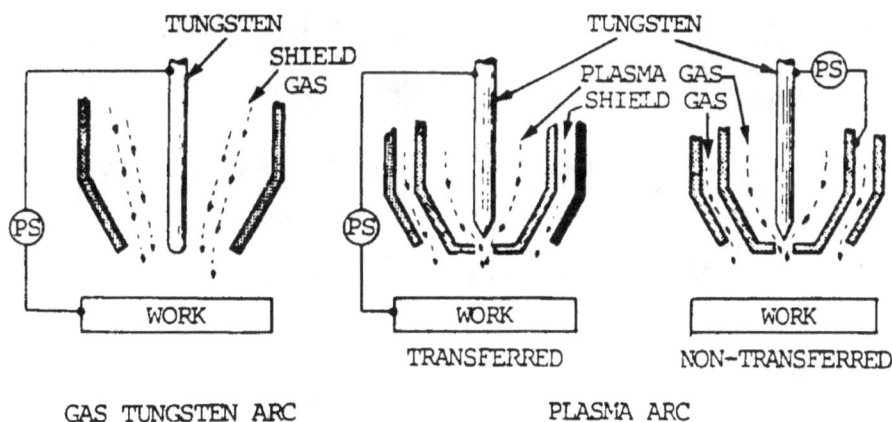

Figure 10-37. Transferred and nontransferred plasma arcs.

10-10. PLASMA ARC WELDING (PAW) (cont)

(3) The plasma is generated by constricting the electric arc passing through the orifice of the nozzle. Hot ionized gases are also forced through this opening. The plasma has a stiff columnar form and is parallel sided so that it does not flare out in the same manner as the gas tungsten arc. This high temperature arc, when directed toward the work, will melt the base metal surface and the filler metal that is added to make the weld. In this way, the plasma acts as an extremely high temperature heat source to form a molten weld puddle. This is similar to the gas tungsten arc. The higher-temperature plasma, however, causes this to happen faster, and is known as the melt-in mode of operation. Figure 10-36, p 10-39, shows a cross-sectional view of the plasma arc torch head.

(4) The high temperature of the plasma or constricted arc and the high velocity plasma jet provide an increased heat transfer rate over gas tungsten arc welding when using the same current. This results in faster welding speeds and deeper weld penetration. This method of operation is used for welding extremely thin material. and for welding multipass groove and welds and fillet welds.

(5) Another method of welding with plasma is the keyhole method of welding. The plasma jet penetrates through the workpiece and forms a hole, or keyhole. Surface tension forces the molten base metal to flow around the keyhole to form the weld. The keyhole method can be used only for joints where the plasma can pass through the joint. It is used for base metals 1/16 to 1/2 in. (1.6 to 12.0 mm) in thickness. It is affected by the base metal composition and the welding gases. The keyhole method provides for full penetration single pass welding which may be applied either manually or automatically in all positions.

(6) Joint design.

(a) Joint design is based on the metal thicknesses and determined by the two methods of operation. For the keyhole method, the joint design is restricted to full-penetration types. The preferred joint design is the square groove, with no minimum root opening. For root pass work, particularly on heavy wall pipe, the U groove design is used. The root face should be 1/8 in. (3.2 mm) to allow for full keyhole penetration.

(b) For the melt-in method of operation for welding thin gauge, 0.020 in. (0.500 mm) to 0.100 in. (2.500 mm) metals, the square groove weld should be utilized. For welding foil thickness, 0.005 in. (0.130 mm) to 0.020 in. (0.0500 mm), the edge flange joint should be used. The flanges are melted to provide filler metal for making the weld.

(c) When using the melt-in mode of operation for thick materials, the same general joint detail as used for shielded metal arc welding and gas tungsten arc welding can be employed. It can be used for fillets, flange welds, all types of groove welds, etc., and for lap joints using arc spot welds and arc seam welds. Figure 10-38 shows various joint designs that can be welded by the plasma arc process.

(7) Welding circuit and current. The welding circuit for plasma arc welding is more complex than for gas tungsten arc welding. An extra component is required as the control circuit to aid in starting and stopping the plasma arc. The same power source is used. There are two gas systems, one to supply the plasma gas and the second for the shielding gas. The welding circuit for plasma arc welding is shown by figure 10-39. Direct current of a constant current (CC) type is used. Alternating current is used for only a few applications.

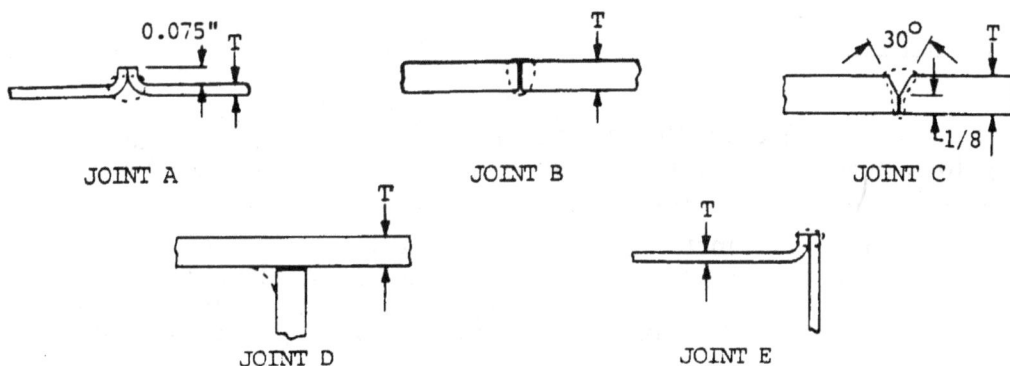

Figure 10-38. Various joints for plasma arc.

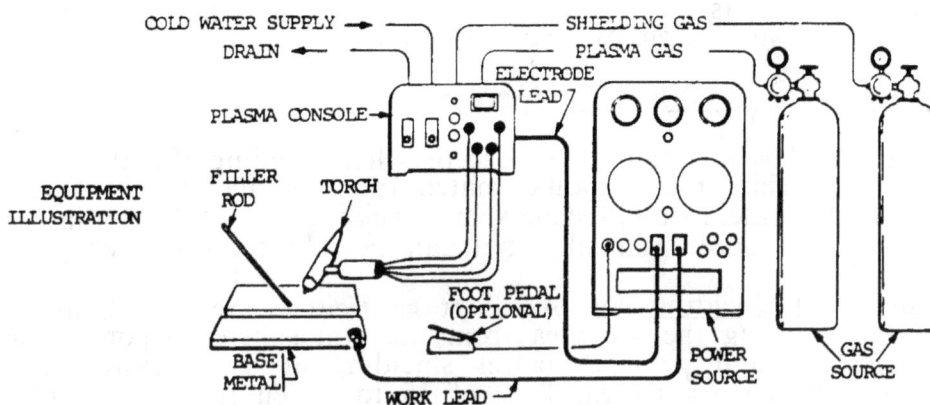

Figure 10-39. Circuit diagram - PAW

Figure 10-40. Quality and common faults.

10-10. PLASMA ARC WELDING (PAW) (cont)

(8) Tips for Using the Process.

(a) The tungsten electrode must be precisely centered and located with respect to the orifice in the nozzle. The pilot arc current must be kept sufficiently low, just high enough to maintain a stable pilot arcWhen welding extremely thin materials in the foil range,the pilot arc may be all that is necessary.

(b) When filler metal is used, it is added in the same manner as gas tungsten arc welding. However, with the torch–to-work distance a little greater there is more freedom for adding filler metal. Equipment must be properly adjusting so that the shielding gas and plasma gas are in the right proportions. Proper gases must also be used.

(c) Heat input is imprtant. Plasma gas flow also has an important effect. These factors are shown by figure 10-40.

e. Filler Metal and Other Equipment.

(1) Filler metal is normally used except when welding the thinnest metals. Composition of the filler metal should match the base metal. The filler metal rod size depends on the base metal thickness and welding current.The filler metal is usually added to the puddle manually, but can be added automatically.

(2) Plasma and shielding gas. An inert gas, either argon, helium, or a mixture, is used for shielding the arc area from the atmosphere. Argon is more common because it is heavier and provides better shielding at lower flow rates. For flat and vertical welding, a shielding gas flow of 15 to 30 cu ft per hour (7 to 14 liters per minute) is sufficient. Overhead position welding requires a slightly higher flow rate. Argon is used for plasma gas at the flew rate of 1 cu ft per hour (0.5 liters per minute) up to 5 cu ft per hour (2.4 liters per minute) for welding, depending on torch size and application.Active gases are not reccommended for plasma gas. In addition, cooling water is required.

f. Quality, Deposition Rates, and Variables.

(1) The quality of the plasma arc welds is extremely high and usually higher than gas tungsten arc welds because there is little or no possibility of tungsten inclusions in the weld. Deposition rates for plasma arc welding are somewhat higher than for gas tungsten arc welding and are shown by the curve in figure 10-41. Weld schedules for the plasma arc process are shown by the data in table 10-5.

(2) The process variables for plasma arc welding are shown by figure 10-41. Most of the variables shown for plasma arc are similar to the other arc welding processes. There are two exceptions: the plasma gas flow and the orifice diameter in the nozzle. The major variables exert considerable control in the process. The minor variables are generally fixed at optimum conditions for the given application. All variables should appear in the welding procedure. Variables such as the angle and setback of the electrode and electrode type are considered fixed for the application. The plasma arc process does respond differently to these variables than does the gas tungsten arc process. The standoff, or torch-to-work distance, is less sensitive with plasma but the torchangle when welding parts of unequal thicknesses is more important than with gatungsten arc.

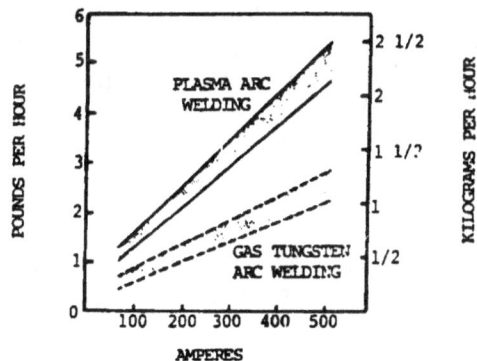

Figure 10-41. Deposition rates.

g. Variations of the Process.

(1) The welding current may be pulsed to gain the same advantages pulsing provides for gas tungsten arc welding. A high current pulse is used for maximum penetration but is not on full time to allow for metal solidification. This gives a more easily controlled puddle for out-of-position work Pulsing can be accomplished by the same apparatus as is used for gas tungsten arc welding.

(2) Programmed welding can also be employed for plasma arc welding in the same manner as it is used for gas tungsten arc welding. The same power source with programming abilities is used and offers advantages for certain types of work. The complexity of the programming depends on the needs of the specific application. In addition to programming the welding current, it is often necessary to program the plasma gas flow. This is particularly important when closing a keyhole which is required to make the root pass of a weld joining two pieces of pipe.

(3) The method of feeding the filler wire with plasma is essentially the same as for gas tungsten arc welding. The "hot wire" concept can be used. This means that low-voltage current is applied to the filler wire to preheat it prior to going into the weld puddle.

Table 10-5. Weld Procedure Schedule—Plasma Arc Welding—Manual Application

Material	Material Thickness in.	Type of Weld	Orifice Dia. in.	Filler Dia. in.	Shield Gas at 20 CFH	Plasma Gas Flow CFH Argon	Weld Current Amps	No. of Passes	Travel Speed ipm
Stainless steel (1)	0.008	Edge butt	0.093	–	A	0.50	12 DCEN	1	7
	0.008	Edge butt	0.093	–	A-5H$_2$	0.50	10 DCEN	1	13
	0.020	Square groove	0.046	–	A-5H$_2$	0.50	12 DCEN	1	21
	0.030	Square groove	0.046	–	A-5H$_2$	0.50	34 DCEN	1	17
	0.062	Square groove	0.081	–	A-5H$_2$	0.70	65 DCEN	1	14
	0.093	Square groove	0.081	–	A	2.00	85 DCEN	1	12
	0.093	Square groove	0.081	–	A-5H$_2$	2.00	85 DCEN	1	16
	0.125	Square groove	0.081	–	A	2.50	100 DCEN	1	10
	0.125	Square groove	0.081	–	A-5H$_2$	2.50	100 DCEN	1	16
	0.187	Square groove	0.081	–	A-5H$_2$	3.50	100 DCEN	1	7
	0.250	V-groove	0.081	–	A-5H$_2$	3.00	100 DCEN	First	5
	0.250	V-groove	0.081	3/32 (2.381)	A-5H$_2$	1.40	100 DCEN	Second	2
Copper (1) Mild steel Aluminum	0.030	Square groove	0.081	–	A	0.50	45 DCEN	1	26
	0.080	Square groove	0.081	–	A	1.00	55 DCEN	1	17
	0.016	Edge butt	0.093	–	He	0.50	18 DCEN	1	24
	0.036	Square groove	0.081	1/16 (1.588)	He	0.05	47 DCEP	1	24
	0.050	Edge joint	0.081	–	He	0.50	48 DCEP	1	22
	0.090	Fillet	0.081	3/32 (2.381)	He	1.40	34 DCEP	1	4

(1) Backing gas 5 to 10 CFH argon.

10-11. CARBON ARC WELDING (CAW)

a. General Carbon arc welding is a process in which the joining of metals is produced by heating with an arc between a carbon electrode and the work. No shielding is used. Pressure and/or filler metal may or may not be used.

b. Equipment.

(1) Electrodes Carbon electrodes range in size from 1/8 to 7/8 in. (3.2 to 22.2 mm) in diameter. Baked carbon electrodes last longer than graphite electrodes. Figure 10-42 shows typical air-cooled carbon electrode holders. Water-cooled holders are available for use with the larger size electrodes, or adapters can be fitted to regular holders to permit accommodation of the larger electrodes.

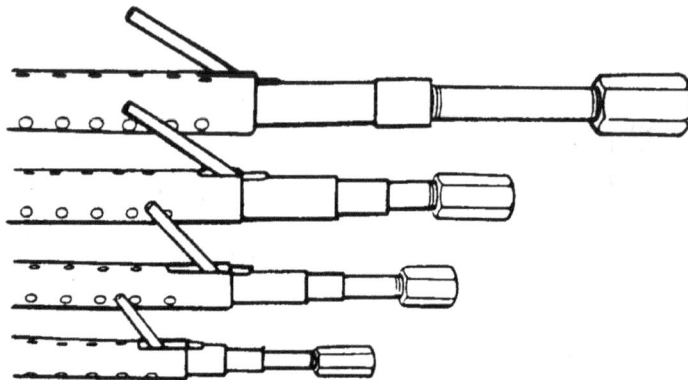

Figure 10-42. Typical air cooled carbon electrode holders.

(2) Machines. Direct current welding machines of either the rotating or rectifier type are power sources for the carbon arc welding process.

(3) Welding circuit and welding current.

(a) The welding circuit for carbon arc welding is the same as for shielded metal arc welding. The difference in the apparatus is a special type of electrode holder used only for holding carbon electrodes.This type of holder is used because the carbon electrodes become extremely hot in use, and the conventional electrode holder will not efficiently hold and transmit current to the carbon electrode. The power source is the conventional or constant current type with drooping volt-amp characteristics. Normally, a 60 percent duty cycle power source is utilized. The power source should have a voltage rating of 50 volts, since this voltage is used when welding copper with the carbon arc.

(b) Single electrode carbon arc welding is always used with direct current electrode negative (DCEN), or straight polarity. In the carbon steel arc, the positive pole (anode) is the pole of maximum heat.If the electrode were positive, the carbon electrode would erode very rapidly because of the higher heat, and would cause black carbon smoke and excess carbon, which could be absorbed by the weld metal. Alternating current is not recommended for single-electrcde carbon arc welding. The electrode should be adjusted often to compensate for the erosion of carbon. From 3.0 to 5.0 in. (76.2 to 127.0 mm) of the carbon electrode should protrude through the holder towards the arc.

c. Advantages and Major Uses.

(1) The single electrode carbon arc welding process is no longer widely used. It is used for welding copper, since it can be used at high currents to develop the high heat usually required. It is also used for making bronze repairs on cast iron parts. When welding thinner materials, the process is used for making autogenous welds, or welds without added filler metal. Carbon arc welding is also used for joining galvanized steel. In this case, the bronze filler rod is added by placing it between the arc and the base metal.

(2) The carbon arc welding process has been used almost entirely by the manual method of applying. It is an all-position welding process. Carbon arc welding is primarily used as a heat source to generate the weld puddle which can be carried in any position. Table 10-6 shows the normal method of applying carbon-arc welding. Table 10-7 shows the welding position capabilities.

Table 10-6. Method of Applying Carbon Arc Processes

Method of Applying	Rating
Manual (MA)	A
Semiautomatic (SA)	No
Machine (ME)	B
Automatic (AU)	No

Table 10-7. Welding Position Capabilities

Welding Position	Rating
1. Flat	A
Horizontal fillet	A
2. Horizontal	A
3. Vertical	A
4. Overhead	A
5. Pipe - fixed	--

d. Weldable Metals. Since the carbon arc is used primarily as a heat source to generate a welding puddle, it can be used on metals that are not affected by carbon pickup or by the carbon monoxide or carbon dioxide arc atmosphere. It can be used for welding steels and nonferrous metals, and for surfacing.

10-11. CARBON ARC WELDING (CAW) (cont)

(1) <u>Steels.</u> The main use of carbon arc welding of steel is making edge welds without the addition of filler metal. This is done mainly in thin gauge sheet metal work, such as tanks, where the edges of the work are fitted closely together and fused using an appropriate flux. Galvanized steel can be braze welded with the carbon arc. A bronze welding rod is used. The arc is directed on the rod so that the galvanizing is not burned off the steel sheet. The arc should be started on the welding rod or a starting block Low current, a short arc length, and. rapid travel speed should be used. The welding rod should melt and wet the galvanized steel.

(2) <u>Cast iron.</u> Iron castings may be welded with the carbon arc and a cast iron welding rod. The casting should be preheated to about 1200 °F (649 °C) and slowly cooled if a machinable weld is desired.

(3) <u>Copper.</u> Straight polarity should always be used for carbon arc welding of copper. Reverse polarity will produce carbon deposits on the work that inhibit fusion. The work should be preheated in the range of 300 to 1200 °F (149 to 649 °C) depending upon the thickness of the parts. If this is impractical, the arc should be used to locally preheat the weld area. The high thermal conductivity of copper causes heat to be conducted away from the point of welding so rapidly that it is difficult to maintain welding heat without preheating. A root opening of 1/8 in. (3.2 mm) is recommended. Best results are obtained at high travel speeds with the arc length directed on the welding rod. A long arc length should be used to permit carbon from the electrode to combine with oxygen to form carbon dioxide, which will provide some shielding of weld metal.

e. <u>Principles of Operation.</u>

(1) Carbon arcwelding, as shown in figure 10-43, uses a single electrode with the arc betweenit and the base metal. It is the oldest arc process, and is not popular today.

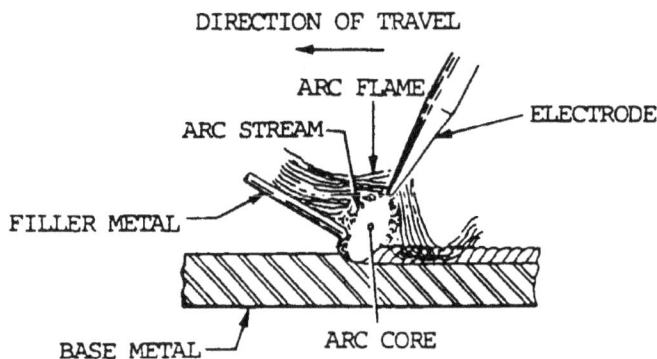

Figure 10-43. Process diagram - CAW.

(2) In carbon arc welding, the arc heat between the carbon electrode and the work melts the base metal and, when required, also melts the filler rod. As the molten metal solidifies, a weld is produced. The nonconsumable graphite electrode erodes rapidly and, in disintegrating, produces a shielding atmosphere of carbon monoxide and carbon dioxide gas. These gases partially displace air from the arc

atmosphere and prohibit the oxygen and nitrogen from coming in contact with molten metal. Filler metal, when used, is of the same composition as the base metal. Bronze filler metal can be used for brazing and braze welding.

(3) The workpieces must be free from grease, oil, scale, paint, and other foreign matter. The two pieces should be clamped tightly together with no root opening. They may be tack welded together.

(4) Carbon electrodes 1/8 to 5/16 in. (3.2 to 7.9 mm) in diameter may be used, depending upon the current required for welding. The end of the electrode should be prepared with a long taper to a point. The diameter of the point should be about half that of the electrode. For steel, the electrode should protrude about 4.0 to 5.0 in. (101.6 to 127.0 mm) from the electrode holder.

(5) A carbon arc may be struck by bringing the tip of the electrode into contact with the work and immediately withdrawing it to the correct length for welding. In general, an arc length between 1/4 and 3/8 in. (6.4 and 9.5 mm) is best. If the arc length is too short, there is likely to be excessive carburization of the molten metal resulting in a brittle weld.

(6) When the arc is broken for any reason, it should not be restarted directly upon the hot weld metal. This could cause a hard spot in the weld at the point of contact. The arc should be started on cold metal to one side of the joint, and then quickly returned to the point where welding is to be resumed.

(7) When the joint requires filler metal, the welding rod is fed into the molten weld pool with one hand while the arc is manipulated with the other. The arc is directed on the surface of the work and gradually moved along the joint, constantly maintaining a molten pool into which the welding rod is added in the same manner as in gas tungsten arc welding. Progress along the weld joint and the addition of a welding rod must be tired to provide the size and shape of weld bead desired. Welding vertically or overhead with the carbon arc is difficult because carbon arc welding is essentially a puddling process. The weld joint should be backed up, especially in the case of thin sheets, to support the molten weld pools and prevent excessive melt-thru.

(8) For outside corner welds in 14 to 18 gauge steel sheet, the carbon arc can be used to weld the two sheets together without a filler metal. Such welds are usually smother and more economical to make than shielded metal arc welds made under similar conditions.

f. _Welding schedules_. The welding schedule for carbon arc welding galvanized iron using silicon bronze filler metal is given in table 10-8. A short arc should be used to avoid damaging the galvanizing. The arc must be directed on the filler wire which will melt and flow on to the joint. For welding copper, use a high arc voltage and follow the schedule given in table 10-9. Table 10-10 shows the welding current to be used for each size of the two types of carbon electrodes.

10-11. CARBON ARC WELDING (CAW) (cont)

Table 10-8. Welding Procedure Schedule--Galvanized Steel--Braze Welding

Material Thickness			Electrode Size		Filler Rod Size		Welding Current	Arc Voltage
Gauge	in.	mm	in.	mm	in.	mm	Amps dc	Electrode Neg.
24	0.024	0.6	3/16	4.8	3/32	2.4	25-30	13-15
22	0.030	0.8	3/16	4.8	3/32	2.4	25-30	13-15
20	0.036	0.9	3/16	4.8	3/32	2.4	30-35	14-16
18	0.048	1.2	1/4	6.4	1/8	3.2	30-35	14-16
16	0.060	1.5	1/4	6.4	1/8	3.2	30-35	14-16
14	0.075	1.9	1/4	6.4	1/8	3.2	30-35	14-16
12	0.105	2.7	1/4	6.4	1/8	3.2	35-40	15-17

Table 10-9. Welding Procedure Schedule for Carbon Arc Welding Copper

THICKNESS OF COPPER			DIAMETER OF ELECTRODE AND FILLER ROD				Welding Current	Voltage
Decimal Inches	Fraction Inches on US Gauge		Electrode Carbon		Filler Rod			Electrode
	in.	mm	in.	mm	in.	mm	dc Amps	Negative
0.05	18				3/32		80	
0.0563	17		3/16	4.8	3/32	2.4	90	35
0.0625	1/16	1.6			1/8		90	
0.07	15		3/16	4.8	1/8	3.2	100	40
0.078	5/64	2.0			5/32	4.0	120	
0.094	3/32	2.4	1/4	6.4	5/32	4.0	135	
0.109	7/64	2.8			5/32	4.0	140	40
0.125	1/8	3.2			3/16		150	
0.141	9/64	3.6			3/16		160	
0.156	5/32	4.0	1/4	6.4	3/16		165	
0.172	11/64	4.4			3/16		170	
0.1875	3/16	4.8			3/16	4.8	185	45
0.203	13/64	5.2			1/4		200	
0.219	7/32	5.6			1/4		200	
0.234	15/64	6.0	5/16	7.9	1/4		205	
0.25	1/4	6.4			1/4		215	
0.266	17/64	6.7			1/4	6.4	225	45
0.281	9/32	7.1			5/16		250	
0.3125	5/16	7.9			5/16		250	
0.344	11/32	8.7	5/16	7.9	5/16		255	
0.375	3/8	9.5			5/16		270	
0.406	13/32	10.3			5/16	7.9	290	50
0.4375	7/16	11.1			3/8		300	
0.4688	15/32	11.9	3/8	9.5	3/8		310	
0.5	1/2	12.7			3/8	9.5	325	50

Table 10-10. Welding Current for Carbon Electrode Types

Electrode Diameter in.	mm	Carbon Electrodes DCEN Amps	Graphite Electrodes DCEN Amps
1/8	3.2	15-30	15-35
3/16	4.8	25-55	25-60
1/4	6.4	50-85	50-90
5/16	7.9	75-115	80-125
3/8	9.5	100-165	110-165
7/16	11.1	125-185	140-210
1/2	12.7	150-225	170-260
5/8	15.9	200-310	230-370
3/4	19.0	250-400	290-490
7/8	22.2	300-500	400-750

g. Variations of the Process.

(1) There are two important variations of carbon arc welding. One is twin carbon arc welding. The other is carbon arc cutting and gouging.

(2) Twin carbon arc welding is an arc welding process in which the joining of metals is produced, using a special electrode holder, by heating with an electric arc maintained between two carbon electrodes.Filler metal may or may not he used. The process can also be used for brazing.

(a) The twin carbon electrode holder is designed so that one electrode is movable and can be touched against the other to initiate the arc.The carbon electrodes are held in the holder by means of set screws and are adjusted so they protrude equally from the clamping jaws. When the two carbon electrodes are brought together, the arc is struck and established between them.The angle of the electrodes provides an arc that forms in front of the apex angle and fans out as a soft source of concentrated heat or arc flame.It is softer than that of the single carbon arc. The temperature of this arc flame is between 8000 and 9000 °F (4427 and 4982 °C).

(b) Alternating current is used for the twin carbon welding arc. With alternating current, the electrodes will burn off or disintegrate at equal rates. Direct current power can be used, but when it is, the electrode connected to the positive terminal should be one size larger than the electrode connected to the negative terminal to ensure even disintegration of the carbon electrodes.The arc gap or spacing between the two electrodes most be adjusted more or less continuously to provide the fan shape arc.

10-11.   CARBON ARC WELDING (CAW) (cont)

(c) The twin carbon arc can be used for many applications in addition to welding, brazing, and soldering. It can be used as a heat source to bend or form metal. The welding current settings or schedules for different size of electrodes is shown in table 10-11.

The twin carbon electrode method is relatively slow and does not have much use as an industrial welding process.

Table 10-11. Welding current for carbon electrode (twin torch).

Carbon Electrode Diameter		Welding Current		Base Metal Thickness	
in.	mm	Amperes ac	Arc Voltage	in.	mm
1/4	6.4	55	35-40	1/16	1.6
5/16	7.9	75	35-40	1/8	3.2
3/8	9.5	95	35-40	1/4	6.4
3/8	9.5	120	35-40	over 1/4	over 6.4

(3) Carbon arc cutting is an arc cutting process in which metals are severed by melting them with the heat of an arc between a carbon electrode and the base metal. The process depends upon the heat input of the carbon arc to melt the metal. Gravity causes the molten metal to fall away to produce the cut. The process is relatively slow, results in a ragged cut, and is used only when other cutting equipment is not available.

10-12.   GAS METAL-ARC WELDING (GMAW OR MIG WELDING)

a.   General.

(1) Gas metal arc welding (GMAW or MIG welding) is an electric arc welding process which joins metals by heating them with an arc established beween a continuous filler metal (consumable) electrode and the work.Shielding of the arc and molten weld pool is obtained entirely from an externally supplied gas or gas mixture, as shown in figure 10-44. The process is sometimes referred to as MIG or CO welding. Recent developments in the process include operation at low current densities and pulsed directcurrent, application to a broader range of materials, and the use of reactive gases, particularly CO or gas mixtures.   This latter development has led to the formal acceptance of the term gas metal arc welding (GMAW) for the process because both inert and reactive gases are used. The term MIG welding is still more commonly used.

(2) MIG welding is operated in semiautomatic, machine, and automatic modes. It is utilized particularly in high production welding operations. All commercially important metals such as carbon steel, stainless steel, aluminum, and copper can be welded with this process in all positions by choosing the appropriate shielding gas, electrode, and welding conditions.

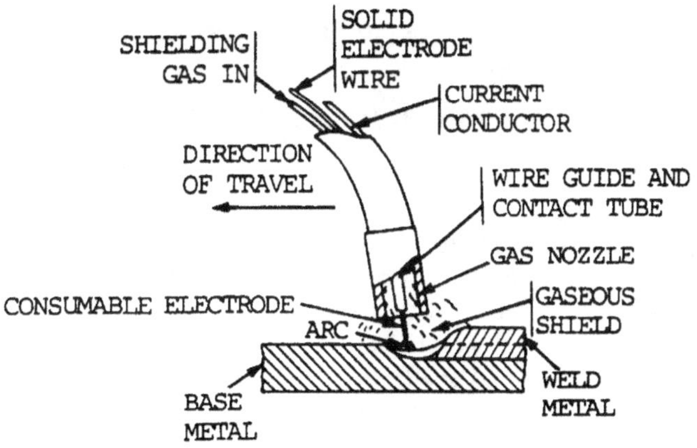

Figure 10-44. Gas metal arc welding process.

b. Equipment.

(1) Gas metal arc welding equipment consists of a welding gun, a power supply, a shielding gas supply, and a wire-drive system which pulls the wire electrode from a spool and pushes it through a welding gun. A source of cooling water may be required for the welding gun. In passing through the gun, the wire becomes energized by contact with a copper contact tube, which transfers current from a power source to the arc. While simple in principle, a system of accurate controls is employed to initiate and terminate the shielding gas and cooling water, operate the welding contactor, and control electrode feed speed as required. The basic features of MIG welding equipment are shown in figure 10-45. The MIG process is used for semiautomatic, machine, and automatic welding. Semiautomatic MIG welding is often referred to as manual welding.

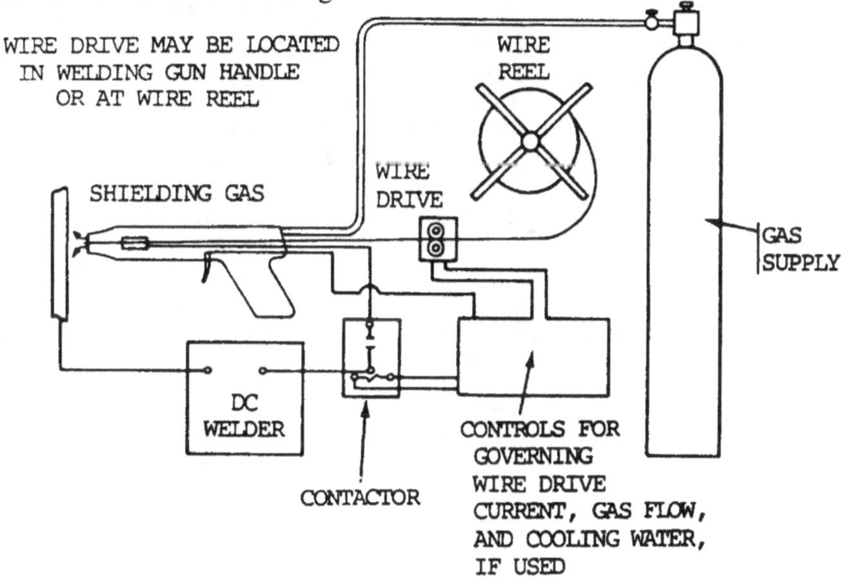

Figure 10-45. MIG welding process.

10-12. GAS METAL-ARC WELDING (GMAW OR MIG WELDING) (cont)

(2) Two types of power sources are used for MIG welding: constant current and constant voltage.

(a) <u>Constant current power supply</u>. With this type, the welding current is established by the appropriate setting on the power supply. Arc length (voltage) is ax-trolled by the automatic adjustment of the electrode feed rate. This type of welding is best suited to large diameter electrodes and machine or automatic welding, where very rapid change of electrode feed rate is not required. Most constant current power sources have a drooping volt-ampere output characteristic. However, true constant current machines are available. Constant current power sources are not normally selected for MIG welding because of the control needed for electrode feed speed. The systems are not self-regulating.

(b) <u>Constant voltage power supply</u>. The arc voltage is established by setting the output voltage on the power supply. The power source will supply the necessary amperage to melt the welding electrode at the rate required to maintain the present voltage or relative arc length. The speed of the electrode drive is used to control the average welding current. This characteristic is generally preferred for the welding of all metals. The use of this type of power supply in conjunction with a constant wire electrode feed results in a self–correcting arc length system.

(3) Motor generator or dc rectifier power sources of either type may be used. With a pulsed direct current power supply, the power source pulses the dc output from a low background value to a high peak value. Because the average power is lower, pulsed welding current can be used to weld thinner sections than those that are practical with steady dc spray transfer.

(4) <u>Welding guns</u>. Welding guns for MIG welding are available for manual manipulation (semiautomatic welding) and for machine or automatic welding. Because the electrode is fed continuously, a welding gun must have a sliding electrical contact to transmit the welding current to the electrode. The gun must also have a gas passage and a nozzle to direct the shielding gas around the arc and the molten weld pool. Cooling is required to remove the heat generated within the gun and radiated from the welding arc and the molten weld metal. Shielding gas, internal circulating water, or both, are used for cooling. An electrical switch is needed to start and stop the welding current, the electrode feed system, and shielding gas flow.

(a) <u>Semiautomatic guns</u>. Semiautomatic, hand-held guns are usually similar to a pistol in shape. Sometimes they are shaped similar to an oxyacetylene torch, with electrode wire fed through the barrel or handle. In some versions of the pistol design, where the most cooling is necessary, water is directed through passages in the gun to cool both the contact tube and the metal shielding gas nozzle. The curved gun uses a curved current-carrying body at the front end, through which the shielding gas is brought to the nozzle. This type of gun is designed for small diameter wires and is flexible and maneuverable. It is suited for welding in tight, hard to reach corners and other confined places. Guns are equipped with metal nozzles of various internal diameters to ensure adequate gas shielding. The orifice usually varies from approximately 3/8 to 7/8 in. (10 to 22 mm), depending upon welding requirements. The nozzles are usually threaded to make replacement easier. The conventional pistol type holder is also used for arc spot welding applications where filler metal is required. The heavy nozzle of the holder is slotted to exhaust the gases away from the spot. The pistol grip handle pennits easy manual loading of the holder against the work. The welding control is designed to reguate the flow of cooling water and the supply of shielding gas. It is also designed to prevent the wire freezing freezing to the weld by timing the weld over a preset interval. A typical semiautomatic gas-cooled gun is shown in figure 10-46.

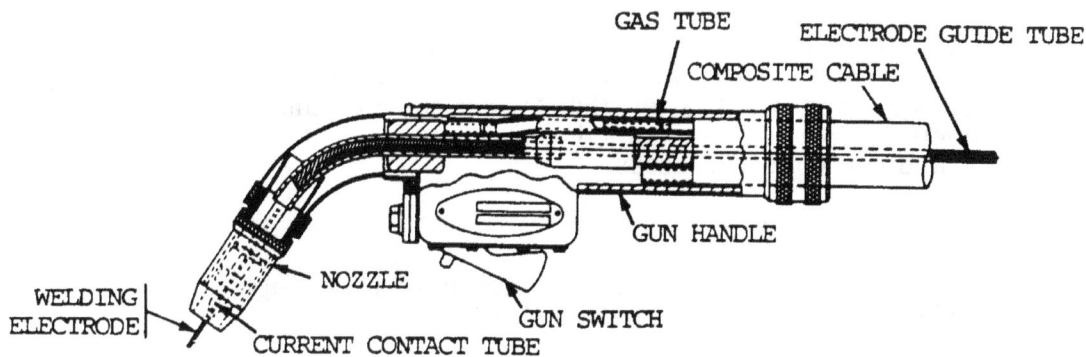

Figure 10-46. Typical semiautomatic gas-cooled, curved-neck gas metal arc welding gun.

(b) <u>Air cooled guns</u>. Air-cooled guns are available for applications where water is not readily obtainable as a cooling medium. These guns are available for service up to 600 amperes, intermittent duty, with carbon dioxide shielding gas. However, they are usually limited to 200 amperes with argon or heliun shielding. The holder is generally pistol-like and its operation is similar to the water-cooled type. Three general types of air-cooled guns are available.

<u>1.</u> A gun that has the electrode wire fed to it through a flexible conduit from a remote wire feeding mechanism. The conduit is generally in the 12 ft (3.7 m) length range due to the wire feeding limitations of a push-type system. Steel wires of 7/20 to 15/16 in. (8.9 to 23.8 mm) diameter and aluminum wires of 3/64 to 1/8 in. (1.19 to 3.18 mm) diameter can be fed with this arrangement.

<u>2.</u> A gun that has a self-contained wire feed mechanism and electrode wire supply. The wire supply is generally in the form of a 4 in. (102 mm) diamter, 1 to 2-1/2 lb (0.45 to 1.1 kg) spool. This type of gun employs a pull-type wire feed system, and it is not limited by a 12 ft (3.7 m) flexible conduit. Wire diameters of 3/10 to 15/32 in. (7.6 to 11.9 mm) are normally used with this type of gun.

<u>3.</u> A pull-type gun that has the electrode wire fed to it through a flexible conduit from a remote spool. This incorporates a self-contained wire feeding mechanism. It can also be used in a push-pull type feeding system. The system permits the use of flexible conduits in lengths up to 50 ft (15 m) or more from the remote wire feeder. Aluminum and steel electrodes with diameters of 3/10 to 5/8 in. (7.6 to 15.9 mm) can be used with these types of feed mechanisms.

10-12. GAS METAL-ARC WELDING (GMAW OR MTG WELDING) (cont)

(c) Water-cooled guns for manual MIG welding similar to gas-cooled types with the addition of water cooling ducts. The ducts circulate water around the contact tube and the gas nozzle. Water cooling permits the gun to operate continuously at rated capacity and at lower temperatures. Water-coded guns are used for applications requiring 200 to 750 amperes. The water in and out lines to the gun add weight and reduce maneuverability of the gun for welding.

(d) The selection of air- or water-cooled guns is based on the type of shielding gas, welding current range, materials, weld joint design, and existing shop practice. Air-cooled guns are heavier than water-cooled guns of the same welding current capacity. However, air-cooled guns are easier to manipulate to weld out-of-position and in confined areas.

c. Advantages.

(1) The major advantage of gas metal-arc welding is that high quality welds can be produced much faster than with SMAW or TIG welding.

(2) Since a flux is not used, there is no chance for the entrapment of slag in the weld metal.

(3) The gas shield protects the arc so that there is very little loss of alloying elements as the metal transfers across the arc. Only minor weld spatter is produced, and it is easily removed.

(4) This process is versatile and can be used with a wide variety of metals and alloys, including aluminum, copper, magnesium, nickel, and many of their alloys, as well as iron and most of its alloys. The process can be operated in several ways, including semi- and fully automatic. MIG welding is widely used by many industries for welding a broad variety of materials, parts, and structures.

d. Disadvantages.

(1) The major disadvantage of this process is that it cannot be used in the vertical or overhead welding positions due to the high heat input and the fluidity of the weld puddle.

(2) The equipment is complex compared to equipment used for the shielded metal–arc welding process.

e. Process Principles.

(1) Arc power and polarity.

(a)   The vast majority of MIG welding applications require the use of direct current reverse polarity (electrode positive). This type of electrical connection yields a stable arc, smooth metal transfer, relatively low spatter loss, and good weld bead characteristics for the entire range of welding currents used. Direct current straight polarity (electrode negative) is seldom used, since the arc can become unstable and erratic even though the electrode melting rate is higher than that achieved with dcrp (electrode positive).When emloyed, dcsp (electrode negative) is used in conjunction with a "buried arc or short circuiting metal transfer. Penetration is lower with straight polarity than with reverse polarity direct current.

(b)   Alternating current has found no commercial acceptance with the MIG welding process for two reasons: the arc is extinguished during each half cycle as the current reduces to zero, and it may not reignite if the cathode cools sufficiently; and rectification of the reverse polarity cycle promotes the erratic arc operation.

(2) Metal transfer.

(a) Filler metal can be transferred from the electrode to the work in two ways: when the electrode contacts the molten weld pool, thereby establishing a short circuit, which is known as short circuiting transfer (short circuiting arc welding); and when discrete drops are moved across the arc gap under the influence of gravity or electromagnetic forces. Drop transfer can be either globular or spray type.

(b) Shape, size, direction of drops(axial or nonaxial), and type of transfer are determined by a number of factors.The factors having the most influence are:

1.   Magnitude and type of welding current.

2.   Current density.

3.   Electrode composition.

4.   Electrode extension.

5.   Shielding gas.

6.   Power supply characteristics.

(c) Axially directed transfer refers to the movement of drops along a line that is a continuation of the longitudinal axis of the electrode. Nonaxially directed transfer refers to movement in any other direction.

(3) Short circuiting transfer.

(a) Short circuiting arc welding uses the lowest range of welding currents and electrode diameters associated with MIG welding.This type of transfer produces a small, fast-freezing weld pool that is generally suited for the joining of thin sections, out-of-position welding, and filling of large root openings. When weld heat input is extremely low, plate distortion is small. Metal is transferred from the electrode to the work only during a period when the electrode is in contact with the weld pool. There is no metal transfer across the arc gap.

10-12. GAS METAL-ARC WELDING (GMAW OR MIG WELDING) (cont)

(b) The electrode contacts the molten weld pool at a steady rate in a range of 20 to over 200 times each second. As the wire touches the weld metal, the current increases. It would continue to increase if an arc did not form. The rate of current increase must be high enough to maintain a molten electrode tip until filler metal is transferred. It should not occur so fast that it causes spatter by disintegration of the transferring drop of filler metal. The rate of current increase is controlled by adjustment of the inductance in the power source. The value of inductance required depends on both the electrical resistance of the welding circuit and the temperature range of electrode melting. The open circuit voltage of the power source must be low enough so that an arc cannot continue under the existing welding conditions. A portion of the energy for arc maintenance is provided by the inductive storage of energy during the period of short circuiting.

(c) As metal transfer only occurs during short circuiting, shielding gas has very little effect on this type of transfer. Spatter can occur. It is usually caused either by gas evolution or electromagnetic forces on the molten tip of the electrode.

(4) Globular transfer.

(a) With a positive electrode (dcrp), globular transfer takes place when the current density is relatively low, regardless of the type of shielding gas. However, carbon dioxide (CO_2) shielding yields this type of transfer at all usable welding currents. Globular transfer is characterized by a drop size of greater diameter than that of the electrode.

(b) Globular, axially directed transfer can be achieved in a substantially inert gas shield without spatter. The arc length must be long enough to assure detachment of the drop before it contacts the molten metal. However, the resulting weld is likely to be unacceptable because of lack of fusion, insufficient penetration, and excessive reinforcement.

(c) Carbon dioxide shielding always yields nonaxially directed globular transfer. This is due to an electromagnetic repulsive force acting upon the bottom of the molten drops. Flow of electric current through the electrode generates several forces that act on the molten tip. The most important of these are pinch force and anode react ion force. The magnitude of the pinch force is a direct function of welding current and wire diameter and is usually responsible for drop detachment. With CO_2 shielding, the wire electrode is melted by the arc heat conducted through the molten drop. The electrode tip is not enveloped by the arc plasma. The molten drop grows until it detaches by short circuiting or gravity.

(5) Spray transfer.

(a) In a gas shield of at least 80 percent argon or helium, filler metal transfer changes from globular to spray type as welding current increases for a given size electrode. For all metals, the change takes place at a current value called the globular-to-spray transition current.

(b) Spray type transfer has a typical fine arc column and pointed wire tip associated with it. Molten filler metal transfers across the arc as fine droplets. The droplet diameter is equal to or less than the electrode diameter. The metal spray is axially directed. The reduction in droplet size is also accompanied by an increase in the rate of droplet detachment, as illustrated in figure 10-47. Metal transfer rate may range from less than 100 to several hundred droplets per second as the electrode feed rate increases from approximately 100 to 800 in./min (42 to 339 mm/s).

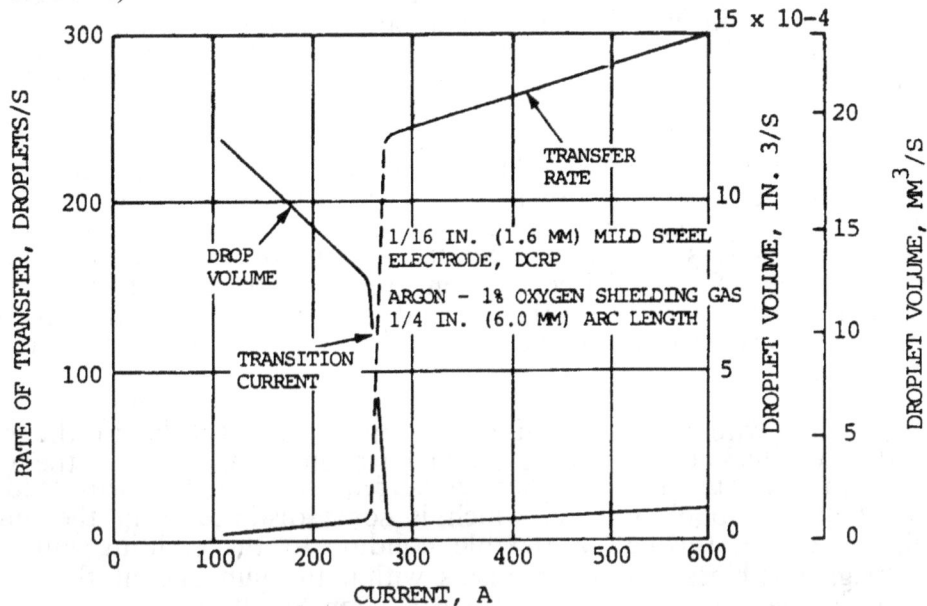

Figure 10-47. Variation in volumes and transfer rate of drops with welding current (steel electrode).

(6) Free flight transfer.

(a) In free-flight transfer, the liquid drops that form at the tip of the consumable electrode are detached and travel freely across the space between the electrode and work piece before plunging into the weld pool. When the transfer is gravitational, the drops are detached by gravity alone and fall slowly through the arc column. In the projected type of transfer, other forces give the drop an initial acceleration and project it independently of gravity toward the weld pool. During repelled transfer, forces act on the liquid drop and give it an initial velocity directly away from the weld pool. The gravitational and projected ties of free-flight metal transfer may occur in the gas metal-arc welding of steel, nickel alloys, or aluminum alloys using a direct current, electrode-positive (reverse polarity) arc and properly selected types of shielding gases.

10-12. GAS METAL-ARC WELDING (GMAW OR MIG WELDING) (cont)

(b) At low currents, wires of these alloys melt slowly. A large spherical drop forms at the tip and is detached when the force due to gravity exceeds that of surface tension. As the current increases, the electromagnetic force becomes significant and the total. separating force increases.The rate at which drops are formed and detached also increases. At a certain current, a change occurs in the character of the arc and metal transfer.The arc column, previously bell-shaped or spherical and having relatively low brightness, becomes narrower and more conical and has a bright central core. The droplets that form at the wire tip become elongated due to magnetic pressure and are detached at a much higher rate. When carbon dioxide is used as the shielding gas,the type of metal transfer is much different. At low and medium reversed-polarity currents, the drop appears to be repelled from the work electrode and is eventually detached while moving away from the workpiece and weld pool. This causes an excessive amount of spatter. At higher currents, the transfer is less irregular because other forces, primarily electrical, overcame the repelling forces. Direct current reversed-polarity is recommended for the MIG welding process. Straight polarity and alternating current can be used, but require precautions such as a special coating on the electrode wire or special shield gas mixtures.

(c) The filler wire passes through a copper contact tube in the gun, where it picks up the welding current. Some manual welding guns contain the wire-driving mechanism within the gun itself. Other guns require that the wire-feeding mechanism be located at the spool of wire, which is some distance from the gun.In this case, the wire is driven through a flexible conduit to the welding gun. Another manual gun design combines feed mechanisms within the gun and at the wire supply itself. Argon is the shielding gas used most often. Small amounts of oxygen (2 to 5 percent) frequently are added to the shielding gas when steel is welded. This stabilizes the arc and promotes a better wetting action, producing a more uniform weld bead and reducing undercut. Carbon dioxide is also used as a shielding gas because it is cheaper than argon and argon-oxygen mixtures. Electrodes designed to be used with carbon dioxide shielding gas require extra deoxidizers in their formulation because in the heat of the arc, the carbon dioxide dissociates to carbon monoxide and oxygen, which can cause oxidation of the weld metal.

(7) Welding parameters. Figures 10-48 through 10-54 show the relationship between the voltage and the current levels and the type of transfer across the arcs.

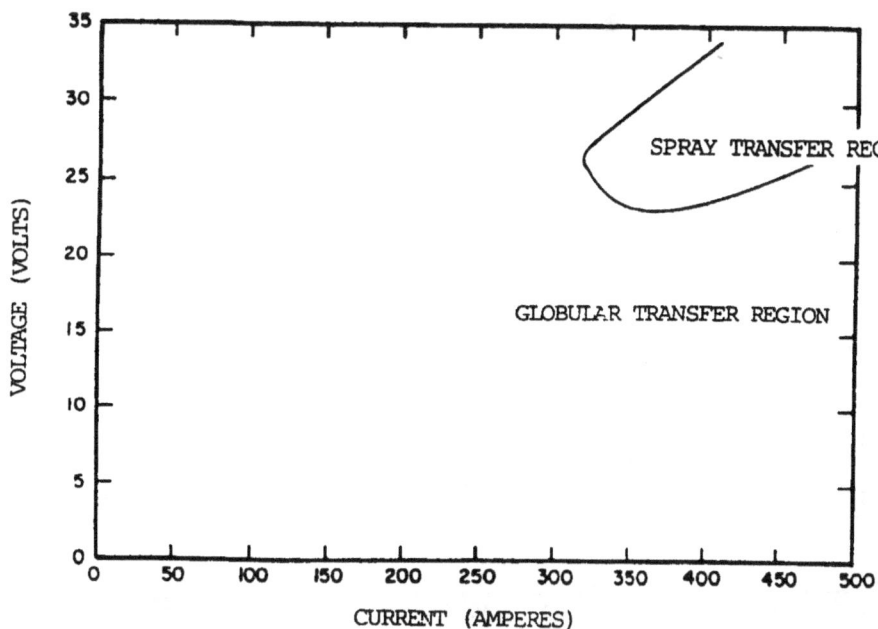

Figure 10-48. Voltage versus current for E70S-2 1/16-inch-diameter electrode and shield gas of argon with 2-percent oxygen addition.

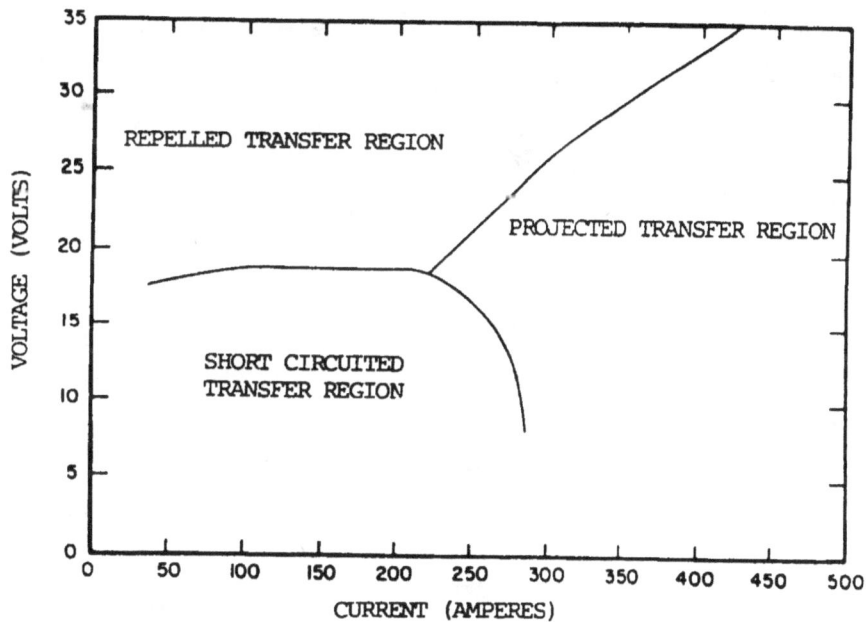

Figure 10-49. Voltage versus current for E70S-2 1/16-inch-diameter electrode and carbon dioxide shield gas.

10-12. GAS METAL-ARC WELDING (GMAW OR MIG WELDING) (cont)

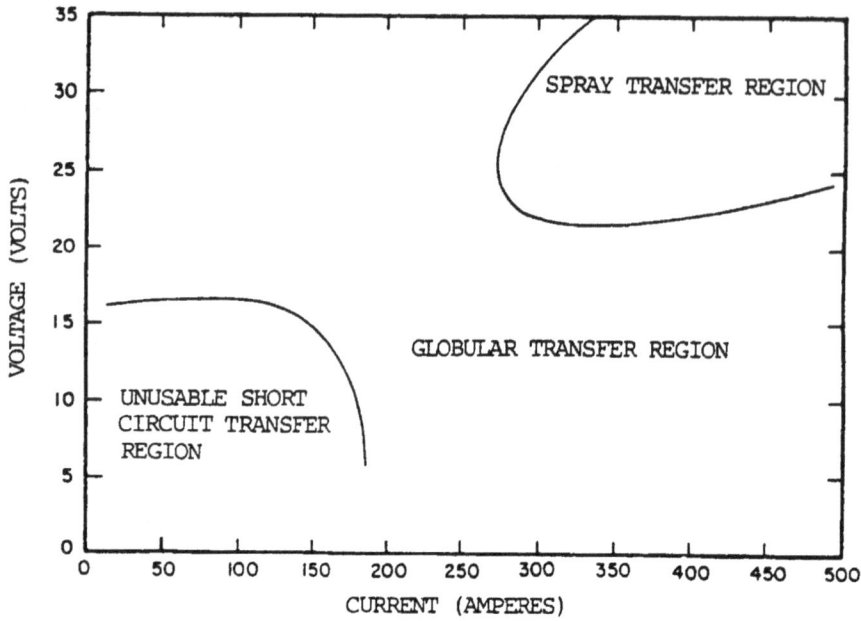

Figure 10-50. Voltage versus current for E70S-3 1/16-inch-diameter electrode and shield gas of argon with 2-percent oxygen addition

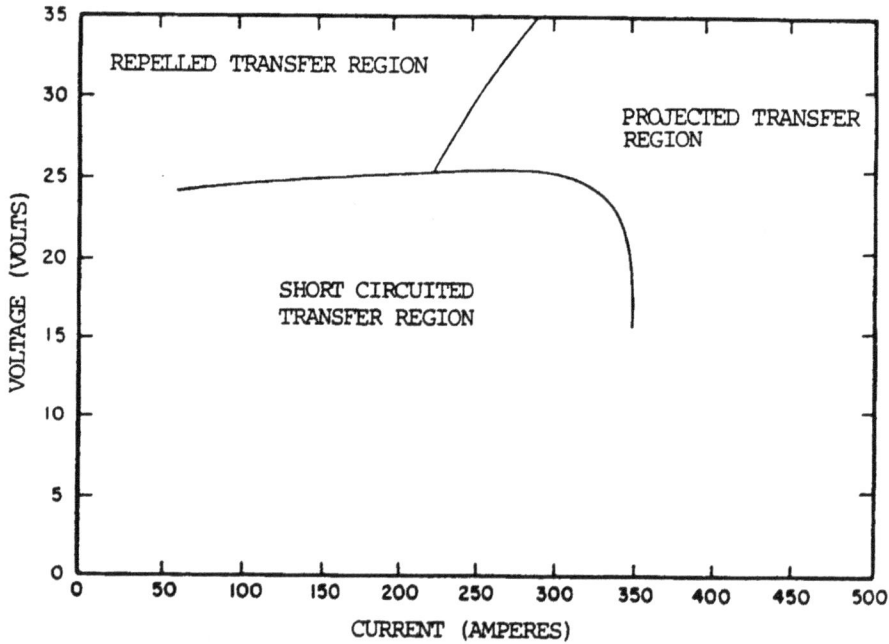

Figure 10-51. Voltage versus current for E70S-3 1/16-inch-diameter electrode and carbon dioxide shield gas.

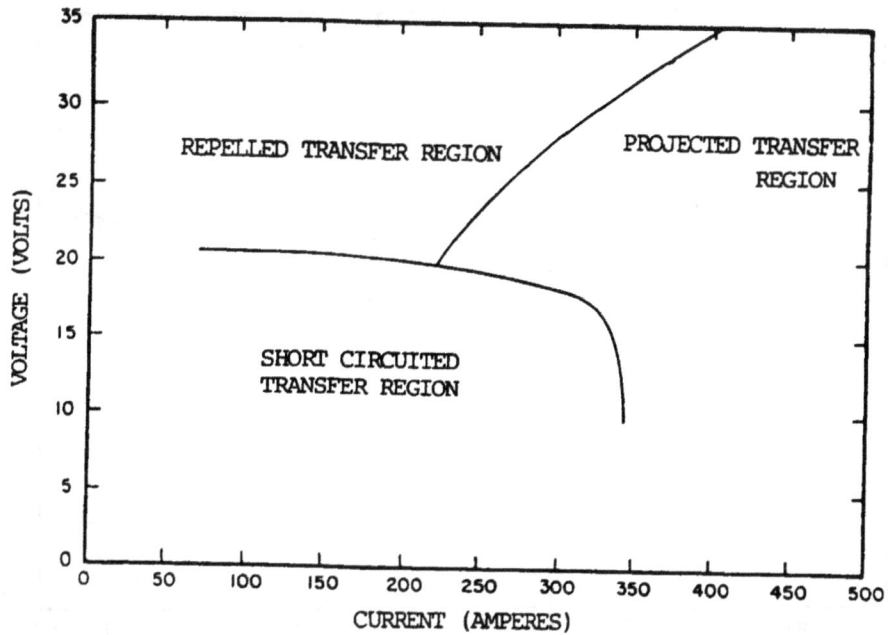

Figure 10-52. Voltage versus current for E70S-4 1/16-inch-diameter electrode and carbon dioxide shield gas.

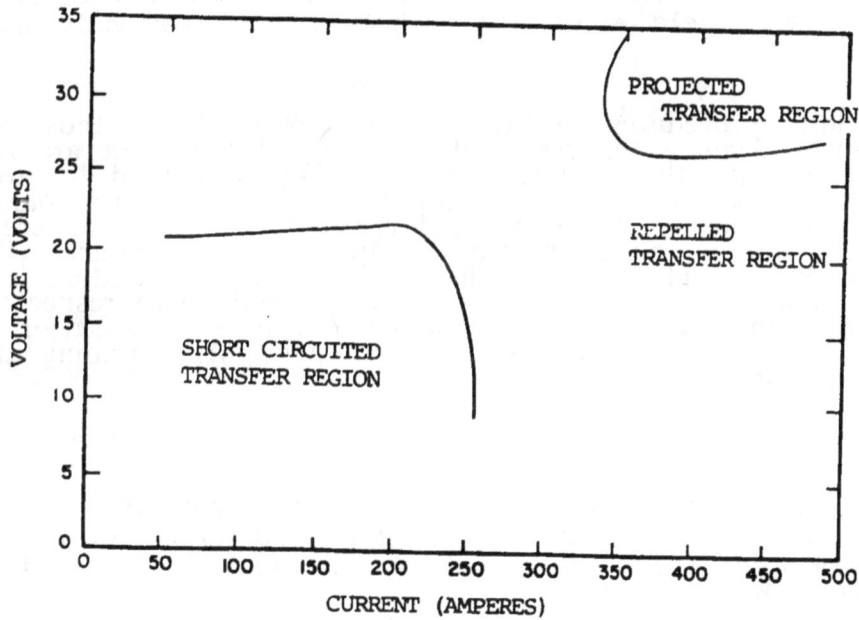

Figure 10-53. Voltage versus current for E70S-6 1/16-inch-diameter electrode and carbon dioxide shield gas.

10-12. GAS METAL-ARC WELDING (GMAW OR MIG WELDING) (cont)

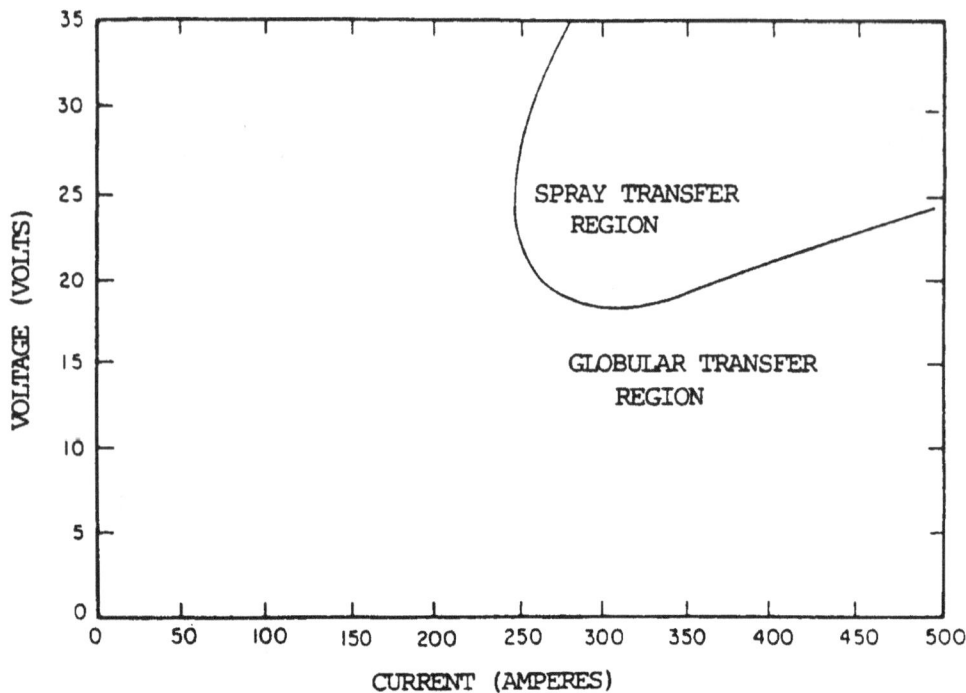

Figure 10-54. Voltage versus current for E110S 1/16-inch-diameter electrode
and shield gas of argon with 2-percent oxygen addition.

f. Welding Procedures.

(1) The welding procedures for MIG welding are similar to those for other arc welding processes. Adequate fixturing and clamping of the work are required with adequate accessibility for the welding gun. Fixturing must hold the work rigid to minimize distortion from welding. It should be designed for easy loading and un-loading. Good connection of the work lead (ground) to the workpiece or fixturing is required. Location of the connectio is important, particularly when welding ferromagnetic materials such as steel. The best direction of welding is away from the work lead connection. The position of the electrode with respect to the weld joint is important in order to obtain the desired joint penetration, fusion, and weld bead geometry. Electrode positions for automatic MIG welding are similar to those used with submerged arc welding.

(2) When complete joint penetration is required, somemethod of weld backing will help to control it. A backing strip, backing weld, or copper backing bar can be used. Backing strips and backing welds usually are left in place. Copper back-ing bars are removable.

(3) The assembly of the welding equipment should be done according to the manufacturer's directions. All gas and water connections should be tight; there should be no leaks. Aspiration of water or air into the shielding gas will result in erractic arc operation and contamination of the weldPorosity may also occur.

(4) The gun nozzle size and the shielding gas flow rate should be set according to the recommended welding procedure for the material and joint design to be welded. Joint designs that require long nozzle-to-work distances will need higher gas flow rates than those used with normal nozzle-to-work distances. The gas nozzle should be of adequate size to provide good gas coverage of the weld area. When welding is done in confined areas or in the root of thick weld joints, small size nozzles are used.

(5) The gun contact tube and electrode feed drive rolls are selected for the particular electrode composition and diameter as specified by the equipment manufacturer. The contact tube will wear with usage, and must be replaced periodically if good electrical contact with electrode is to be maintained and heating of the gun is to be minimized.

(6) Electrode extension is set by the distance between the tip of the contact tube and the gas nozzle opening. The extension used is related to the type of MIG welding, short circuiting or spray type transfer. It is important to keep the electrode extension (nozzle-to-work distance) as uniform as-possible during welding. Therefore, depending on the application, the contact tube may be inside, flush with, or extending beyond the gas nozzle.

(7) The electrode feed rate and welding voltage are set to the recommended values for the electrode size and material. With a constant voltage power source, the welding current will be establish by the electrode feed rate. A trial bead weld should be made to establish proper voltage (arc length) and feed rate values. Other variables, such as slope control, inductance, or both, should be adjusted to give good arc starting and smooth arc operation with minimum spatter. The optimum settings will depend on the equipment design and controls, electrode material and size, shielding gas, weld joint design, base metal composition and thickness, welding position, and welding speed.

10-13. FLUX-CORED ARC WELDING FCAW)

a. <u>General</u>.

(1) Flux-cored, tubular electrode welding has evolved from the MIG welding process to improve arc action, metal transfer, weld metal properties, and weld appearance. It is an arc welding process in which the heat for welding is provided by an arc between a continuously fed tubular electrode wire and the workpiece. Shielding is obtained by a flux contained within the tubular electrode wire or by the flux and an externally supplied shielding gas. A diagram of the process is shown in figure 10-55.

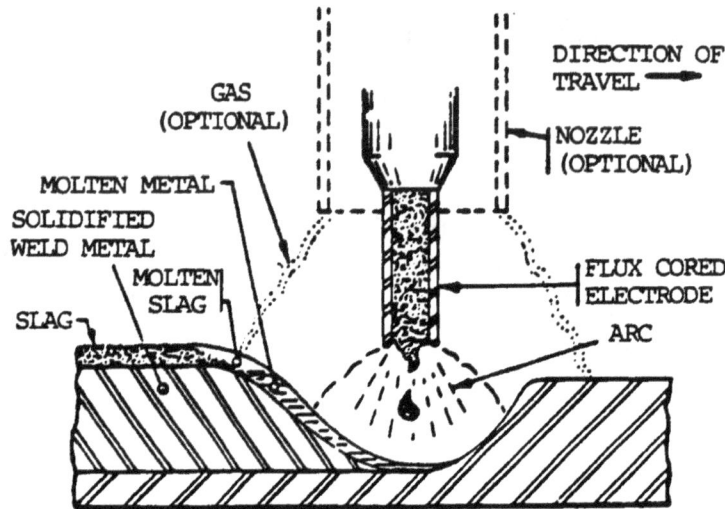

Figure 10-55. Flux-cored arc welding process.

(2) Flux-cored arc welding is similar to gas metal arc welding in many ways. The flux-cored wire used for this process gives it different characteristics. Flux-cored arc welding is widely used for welding ferrous metals and is particularly good for applications in which high deposition rates are needed. At high welding currents, the arc is smooth and more manageable when compared in using large diameter gas metal arc, welding electrodes with carbon dioxide. The arc and weld pool are clearly visible to the welder. A slag coating is left on the surface of the weld bead, which must beremoved. Since the filler metal transfers across the arc, some spatter is createdand some smoke produced.

b. <u>Equipment</u>.

(1) The equipment usedfor flux-cored arc welding is similar to that used for gas metal arc welding. The basic arc welding equipment consists of a power source, controls, wire feeder, welding gun, and welding cables. A major difference between the gas shielded electrodes and the self -shielded electrodes is that the gas shielded wires also require a gas shielding system. This may also have an effect on the type of welding gun used. Fume extractors are often used with this process. For machines and automatic welding, several items, such as seam followers and motion devices, are added to the basic equipment.Figure 10-56 shows a diagram of the equipment used for semiautomatic flux-cored arc welding.

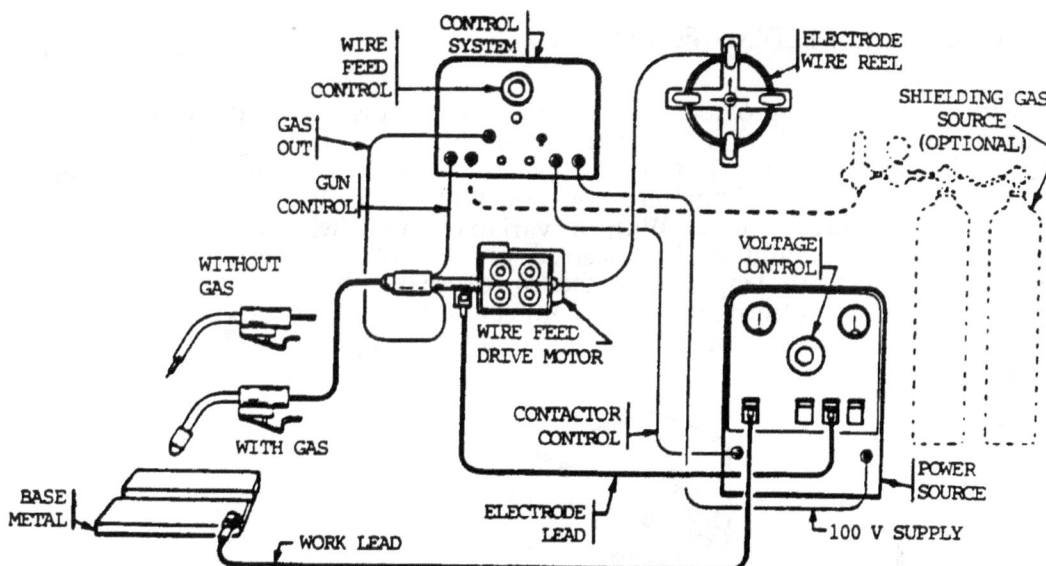

Figure 10-56. Equipment for flux-cored arc welding.

(2) The power source, or welding machine, provides the electric power of the proper voltage and amperage to maintain a welding arc. Most power sources operate on 230 or 460 volt input power, but machines that operate on 200 or 575 volt input are also available. Power sources may operate on either single phase or three-phase input with a frequency of 50 to 60 hertz. Most power sources used for flux-cored arc welding have a duty cycle of 100 percent, which indicates they can be used to weld contiuously. Some machines used for this process have duty cycles of 60 percent, which means that they can be used to weld 6 of every 10 minutes. The power sources generally recommended for flux-cored arc welding are direct current constant voltage type. Both rotating (generator) and static (single or three-phase transformer-rectifiers) are used. The same power sources used with gas metal arc welding are used with flux-cored arc welding. Flux-cored arc welding generally uses higher welding currents than gas metal arc welding, which sometimes requires a larger power source. It is important to use a power source that is capable of producing the maximum current level required for an application.

(3) Flux-cored arc welding uses direct current. Direct current can be either reverse or straight polarity. Flux-cored electrode wires are designed to operate on either DCEP or DCEN. The wires designed for use with an external gas shielding system are generally designed for use with DCEP. Some self-shielding flux-cored ties are used with DCEP while others are developed for use with DCEN. Electrode positive current gives better penetration into the weld joint. Electrode negative current gives lighter penetration and is used for welding thinner metal or metals where there is poor fit-up. The weld created by DCEN is wider and shallower than the weld produced by DCEP.

(4) The generator welding machines used for this process can be powered by an electric rotor for shop use, or by an internal combustion engine for field applications. The gasoline or diesel engine-driven welding machines have either liquid- or air-cooled engines. Motor-driven generators produce a very stable arc, but are noisier, more expensive, consume more power, and require more maintenance than transformer-rectifier machines.

10-13.   FLUX-CORED ARC WELDING (FCAW) (cont)

(5) A wire feed motor provides power for driving the electrode through the cable and gun to the work. There are several different wire feeding systems available. System selection depends upon the application. Most of the wire feed systems used for flux-cored arc welding are the constant speed type, which are used with constant voltage power sources. With a variable speed wire feeder, a voltage sensing circuit is used to maintain the desired arc length by varying the wire feed speed. Variations in the arc length increase or decrease the wire feed speed. A wire feeder consists of an electrical rotor connected to a gear box containing drive rolls. The gear box and wire feed motor shown in figure 10-57 have form feed rolls in the gear box.

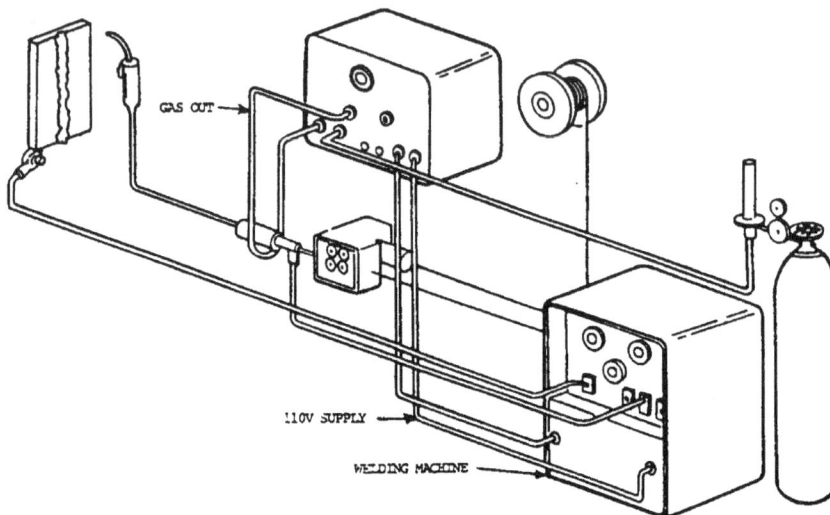

Figure 10-57. Wire feed assembly.

(6) Both air-cooled and water-cooled guns are used for flux-cored arc welding. Air-cooled guns are cooled primarily by the surrounding air, but a shielding gas, when used, provides additional cooling effects. A water-cooled gun has ducts to permit water to circulate around the contact tube and nozzle. Water-cooled guns permit more efficient cooling of the gun. Water-cooled guns are recommended for use with welding currents greater than 600 amperes and are preferred for many applications using 500 amperes. Welding guns are rated at the maximum current capacity for continuous operation. Air-cooled guns are preferred for most applications less than 500 amperes, although water-cooled guns may also be used. Air-cooled guns are lighter and easier to manipulate.

(7) Shielding gas equipment and electrodes.

(a) Shielding gas equipment used for gas shielded flux-cored wires consists of a gas supply hose, a gas regulator, control valves, and supply hose to the welding gun.

(b) The shielding gases are supplied in liquid form when they are in storage tanks with vaporizers, or in a gas form in high pressure cylinders. An exception to this is carbon dioxide. When put in high pressure cylinders, it exists in both liquid and gas forms.

(c) The primary purpose of the shielding gas is to protect the arc and weld puddle from contaminating effects of the atmosphere. The nitrogen and oxygen of the atmosphere, if allowed to come in contact with the molten weld metal, cause porosity and brittleness. In flux-cored arc welding, shielding is accomplished by the decomposition of the electrode core or by a combination of this and surrounding the arc with a shielding gas supplied from an external source. A shielding gas displaces air in the arc area. Welding is accomplished under a blanket of shielding gas. Inert and active gases may both be used for flux-cored arc welding. Active gases such as carbon dioxide, argon-oxygen mixture, and argon-carbon dioxide mixtures are used for almost all applications. Carbon dioxide is the most common. The choice of the proper shielding gas for a specific application is based on the type of metal to be welded, arc characteristics and metal transfer, availability, cost of the gas, mechanical property requirements, and penetration and weld bead shape. The various shielding gases are summarized below.

1. Carbon dioxide. Carbon dioxide is manufactured from fuel gases which are given off by the burning of natural gas, fuel oil, or coke. It is also obtained as a by-product of calcining operation in lime kilns, from the manufacturing of ammonia and from the fermentation of alcohol, which is almost 100 percent pure. Carbon dioxide is made available to the user in either cylinder or bulk containers. The cylinder is more common. With the bulk system, carbon dioxide is usually drawn off as a liquid and heated to the gas state before going to the welding torch. The bulk system is normally only used when supplying a large number of welding stations. In the cylinder, the carbon dioxide is in both a liquid and a vapor form with the liquid carbon dioxide occupying approximately two thirds of the space in the cylinder. By weight, this is approximately 90 percent of the content of the cylinder. Above the liquid, it exists as a vapor gas. As carbon dioxide is drawn from the cylinder, it is replaced with carbon dioxide that vaporizes from the liquid in the cylinder and therefore the overall pressure will be indicated by the pressure gauge. When the pressure in the cylinder has dropped to 200 psi (1379 kPa), the cylinder should be replaced with a new cylinder. A positive pressure should always be left in the cylinder in order to prevent moisture and other contaminants from backing up into the cylinder. The normal discharge rate of the CO_2 cylinder is about 10 to 50 cu ft per hr (4.7 to 24 liters per min). However, a maximum discharge rate of 25 cu ft per hr (12 liters per min is recommended when welding using a single cylinder. As the vapor pressure drops from the cylinder pressure to discharge pressure through the CO_2 regulator, it absorbs a great deal of heat. If flow rates are set too high, this absorption of heat can lead to freezing of the regulator and flowmeter which interrupts the shielding gas flow. When flow rate higher than 25 cu ft per hr (12 liters per rein) is required, normal practice is to manifold two CO_2 cylinders in parallel or to place a heater between the cylinder and gas regulator, pressure regulator, and flowmeter. Excessive flow rates can also result in drawing liquid from the cylinder. Carbon dioxide is the most widely used shielding gas for flux-cored arc welding. Most active gases cannot be used for shielding, but carbon dioxide provides several advantages for use in welding steel. These are deep penetration and low cost. Carbon dioxide promotes a globular transfer. The carbon dioxide shielding gas breaks down into components such as carbon monoxide and oxygen. Because carbon dioxide is an oxidizing gas, deoxidizing elements are added to the core of the electrode wire to remove oxygen. The oxides formed by the deoxidizing elements float to the surface of the weld and become part of the slag covering. Some of the carbon dioxide gas will break down to carbon and oxygen. If the carbon content of the weld pool is below

10-13. FLUX-CORED ARC WELDING (FCAW) (cont)

about 0.05 percent, carbon dioxide shielding will tend to increase the carbon content of the weld metal. Carbon, which can reduce the corrosion resistance of some stainless steels, is a problem for critical corrosion application. Extra carbon can also reduce the toughness and ductility of some low alloy steels. If the carbon content in the weld metal is greater than about 0.10 percent, carbon dioxide shielding will tend to reduce the carbon content. This loss of carbon can be attributed to the formation of carbon monoxide, which can be trapped in the weld as porosity deoxidizing elements in the flux core reducing the effects of carbon monoxide formation.

2. Argon-carbon dioxide mixtures. Argon and carbon dioxide are sometimes mixed for use with flux-cored arc welding. A high percentage of argon gas in the mixture tends to promote a higher deposition efficiency due to the creation of less spatter. The most commonly used gas mixture in flux-cored arc welding is a 75 percent argon-25 percent carbon dioxide mixture. The gas mixture produces a fine globular metal transfer that approaches a spray. It also reduces the amount of oxidation that occurs, compared to pure carbon dioxide. The weld deposited in an argon-carbon dioxide shield generally has higher tensile and yield strengths. Argon-carbon dioxide mixtures are often used for out-of-position welding, achieving better arc characteristics. These mixtures are often used on low alloy steels and stainless steels. Electrodes that are designed for use with CO_2 may cause an excessive buildup of manganese, silicon, and other deoxidizing elements if they are used with shielding gas mixtures containing a high percentage of argon. This will have an effect on the mechanical properties of the weld.

3. Argon-oxygen mixtures. Argon-oxygen mixtures containing 1 or 2 percent oxygen are used for some applications. Argon-oxygen mixtures tend to promote a spray transfer which reduces the amount of spatter produced. A major application of these mixtures is the welding of stainless steel where carbon dioxide can cause corrosion problems.

(d) The electrodes used for flux-cored arc welding provide the filler metal to the weld puddle and shielding for the arc. Shielding is required for sane electrode types. The purpose of the shielding gas is to provide protection from the atmosphere to the arc and molten weld puddle. The chemical composition of the electrode wire and flux core, in combination with the shielding gas, will determine the weld metal composition and mechanical properties of the weld. The electrodes for flux-cored arc welding consist of a metal shield surrounding a core of fluxing and/or alloying compounds as shown in figure 10-58. The cores of carbon steel and low alloy electrodes contain primarily fluxing compounds. Some of the low alloy steel electrode cares contain high amounts of alloying compounds with a low flux content. Most low alloy steel electrodes require gas shielding. The sheath comprises approximately 75 to 90 percent of the weight of the electrode. Self-shielded electrodes contain more fluxing compounds than gas shielded electrodes. The compounds contained in the electrode perform basically the same functions as the coating of a covered electrode used in shielded metal arc welding. These functions are:

1. To form a slag coating that floats on the surface of the weld metal and protects it during solidification.

<u>2.</u> To provide deoxidizers and scavengers which help purify and produce solid weld-metal.

<u>3.</u> To provide arc stabilizers which produce a smooth welding arc and keep spatter to a minimum.

<u>4.</u> To add alloying elements to the weld metal which will increase the strength and improve other properties in the weld metal.

<u>5.</u> To provide shielding gas. Gas shielded wires require an external supply of shielding gas to supplement that produced by the core of the electrode.

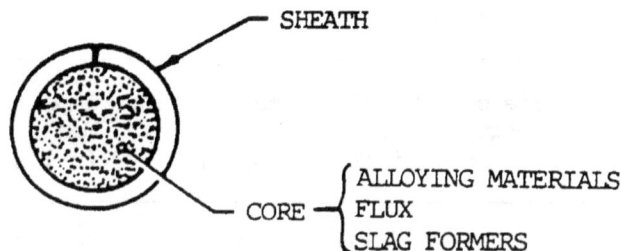

Figure 10-58. Cross-section of a flux-cored wire.

(e) The classification system used for tubular wire electrodes was devised by the American Welding Society. Carbon and low alloy steels are classified on the basis of the following items:

<u>1.</u> Mechanical properties of the weld metal.

<u>2.</u> Welding position.

<u>3.</u> Chemical composition of the weld metal.

<u>4.</u> Type of welding current.

<u>5.</u> Whether or not a CO_2 shielding gas is used.

An example of a carbon steel electrode classification is E70T-4 where:

<u>1.</u> The "E" indicates an electrode.

<u>2.</u> The second digit or "7" indicates the minimum tensile strength in units of 10,000 psi (69 MPa). Table 10-12, p 10-74, shows the mechanical property requirements for the various carbon steel electrodes.

<u>3.</u> The third digit or "0" indicates the welding positions. A "0" indicates flat and horizontal positions and a "1" indicates all positions.

10-13. FLUX-CORED ARC WELDING (FCAW) (cont)

    4. The "T" stands for a tubular of flux cored wire classification.

    5. The suffix "4" gives the performance and usability capabilities as shown in table 10-13. When a "G" classification is used, no specific performance and usability requirements are indicated. This classification is intended for electrodes not covered by another classification. The chemical composition requirements of the deposited weld metal for carbon steel electrodes are shown in table 10-14. Single pass electrodes do not have chemical composition requirements because checking the chemistry of undiluted weld metal does not give the true results of normal single pass weld chemistry.

Table 10-12. Mechanical Property Requirements of Carbon Steel
Flux-Cored Electrodes

AWS Classification	Shielding Gas	Yield Strength ksi (Mpa)	Tensile Strength ksi (Mpa)	%Elongation Min in 1 in. (50 mm)	Impact Strength Min ft-lbs @ $^{\circ}$F (J @$^{\circ}$C)
E6XT-1	CO_2	50 (345)	62 (428)	22	20 @ 0 (27 @ -18)
E6XT-4	None	50 (345)	62 (428)	22	-
E6XT-5	CO_2	50 (345)	62 (428)	22	20 @ -20 (27 @ -29)
E6XT-6	None	50 (345)	62 (428)	22	20 @ -20 (27 @ -29)
E7XT-7	None	50 (345)	62 (428)	22	-
E6XT-8	None	50 (345)	62 (428)	22	20 @ -20 (27 @ -29)
E6XT-11	None	50 (345)	62 (428)	22	-
E6XT-G	*	50 (345)	62 (428)	22	-
E6XT-GS	*	-	62 (428)	-	-
E7XT-1	CO_2	60 (414)	72 (497)	22	20 @ 0 (27 @ -18)
E7XT-2	CO_2	-	72 (497)	-	-
E7XT-3	None	-	72 (497)	-	-
E7XT-4	None	60 (414)	72 (497)	22	-
E7XT-5	CO_2	60 (414)	72 (497)	22	20 @ -20 (27 @ -29)
E7XT-6	None	60 (414)	72 (497)	22	20 @ -20 (27 @ -29)
E7XT-7	None	60 (414)	72 (497)	22	-
E7XT-8	None	60 (414)	72 (497)	22	20 @ -20 (27 @ -29)
E7XT-10	None	-	72 (497)	-	-
E7XT-11	None	60 (414)	72 (497)	22	-
E7XT-G	*	60 (414)	72 (497)	22	-
E7XT-GS	*	-	72 (497)	-	-

* As agreed upon between supplier and user.

Table 10-13. Performance and Usability Characteristics of Carbon Steel
Flux Cored Electrodes

AWS Classification	Welding Current	Shielding	Single or Multiple Pass
EXXT-1	DCEP	CO_2	Multiple
EXXT-2	DCEP	CO_2	Single
EXXT-3	DCEP	None	Single
EXXT-4	DCEP	None	Multiple
EXXT-5	DCEP	CO_2	Multiple
EXXT-6	DCEP	None	Multiple
EXXT-7	DCEN	None	Multiple
EXXT-8	DCEN	None	Multiple
EXXT-10	DCEN	None	Single
EXXT-11	DCEN	None	Multiple
EXXT-G	*	*	Multiple
EXXT-GS	*	*	Single

* As agreed between purchaser and supplier

Table 10-14. Chemical Composition Requirements of Carbon Steel
Flux Cored Electrodes

AWS Classification	Chemical Composition (% max.)[a]									
	C	Mn	Si	P	S	Cr	Ni	Mo	V	Al
EXXT-1										
EXXT-4										
EXXT-5										
EXXT-6										
EXXT-7	b	1.75	0.90	0.04	0.03	0.20	0.50	0.30	0.08	1.8
EXXT-8										
EXXT-11										
EXXT-G										
EXXT-2										
EXXT-3		NO CHEMICAL REQUIREMENTS[c]								
EXXT-10										
EXXT-GS										

a Chemical compositions are based on the analysis of the deposited weld metal.
b No requirement, but the amount of carbon shall be determined.
c Since these are single pass analysis, the analysis of the undiluted weld metal
   is not meaningful.

10-13. FLUX-CORED ARC WELDING (FCAW) (cont)

The classification of low alloy steel electrodes is similar to the classification of carbon steel electrodes. An example of a low alloy steel classification is E81T1-NI2 where:

    <u>1.</u> The "E" indicates electrode.

    <u>2.</u> The second digit or "8" indicates the minimum tensile in strength in units of 10,000 psi (69 MPa). In this case it is 80,000 psi (552 MPa). The mechanical property requirements for low alloy steel electrodes are shown in table 10-15. Impact strength requirements are shown in table 10-16.

    <u>3.</u> The third digit or "1" indicates the welding position capabilities of the electrode. A "1" indicates all positions and an "0" flat and horizontal position only.

    <u>4.</u> The "T" indicates a tubular or flux-cored electrode used in flux cored arc welding.

    <u>5.</u> The fifth digit or "1" describes the usability and performance characteristics of the electrode. These digits are the same as used in carbon steel electrode classification but only EXXT1-X, EXXT4-X, EXXT5-X and EXXT8-X are used with low alloy steel flux-cored electrode classifications.

    <u>6.</u> The suffix or "Ni2" tells the chemical compsition of the deposited weld metal as shown in table 10-17, p 10-78.

Table 10-15. Mechanical Property Requirements of Low Alloy Flux-Cored Electrodes

AWS Classification	Tensile Strength Range psi	MPa	Yield Strength psi	MPa	Percent Elongation in 2 in. (50 mm) min
E6XTX-X	60,000 to 80,000	410 to 550	50,000	340	22
E7XTX-X	70,000 to 90,000	490 to 620	58,000	400	20
E8XTX-X	80,000 to 100,000	550 to 690	68,000	470	19
E9XTX-X	90,000 to 110,000	620 to 760	78,000	540	17
E10XTX-X	100,000 to 120,000	690 to 830	88,000	610	16
E11XTX-X	110,000 to 130,000	760 to 900	98,000	680	15
E12XTX-X	120,000 to 140,000	830 to 970	108,000	750	14
EXXXTX/G	AS AGREED BETWEEN SUPPLIER AND PURCHASER				

Table 10-16. Impact Requirements For Low Alloy Flux-Cored Electrodes

Classification	Condition*	Impact Strength
E80T1-A1	PWHT	Not Required
E81T1-A1	PWHT	Not required
E70T5-A1	PWHT	20 ft-lb @ -20°F (27 J @ -29°C)
E81T1-B1	PWHT	Not required
E81T1-B2	PWHT	Not required
E80T1-B2	PWHT	Not required
E80T5-B2	PWHT	Not required
E80T1-B2H	PWHT	Not required
E80T5-B2L	PWHT	Not required
E90T1-B3	PWHT	Not required
E91T1-B3	PWHT	Not required
E90T5-B3	PWHT	Not required
E100T1-B3	PWHT	Not required
E90T1-B3L	PWHT	Not required
E90T1-B3H	PWHT	Not required
E71T8-Ni1	A.W.	20 ft-lb @ -20°F (27 J @ -29°C)
E80T1-Ni1	A.W.	20 ft-lb @ -20°F (27 J @ -29°C)
E81T1-Ni1	A.W.	20 ft-lb @ -20°F (27 J @ -29°C)
E80T5-Ni1	PWHT	20 ft-lb @ -60°F (27 J @ -51°C)
E71T8-Ni2	A.W.	20 ft-lb @ -20°F (27 J @ -29°C)
E80T1-Ni2	A.W.	20 ft-lb @ -40°F (27 J @ -40°C)
E81T1-Ni2	A.W.	20 ft-lb @ -40°F (27 J @ -40°C)
E80T5-Ni2**	PWHT	20 ft-lb @ -75°F (27 J @ -59°C)
E90T1-Ni2	A.W.	20 ft-lb @ -40°F (27 J @ -40°C)
E91T1-Ni2	A.W.	20 ft-lb @ -40°F (27 J @ -40°C)
E80T5-Ni3**	PWHT	20 ft-lb @ -100°F (27 J @ -73°C)
E90T5-Ni3**	PWHT	20 ft-lb @ -100°F (27 J @ -73°C)
E91T1-D1	A.W.	20 ft-lb @ -40°F (27 J @ -40°C)
E90T5-D2	PWHT	20 ft-lb @ -60°F (27 J @ -51°C)
E100T5-D2	PWHT	20 ft-lb @ -40°F (27 J @ -40°C)
E90T1-D3	A.W.	20 ft-lb @ -20°F (27 J @ -29°C)
E80T5-K1	A.W.	20 ft-lb @ -40°F (27 J @ -40°C)
E70T4-K2	A.W.	20 ft-lb @ 0°F (27 J @ -18°C)
E71T8-K2	A.W.	20 ft-lb @ -20°F (27 J @ -29°C)
E80T1-K2	A.W.	20 ft-lb @ -20°F (27 J @ -29°C)
E90T1-K2	A.W.	20 ft-lb @ 0°F (27 J @ -18°C)
E91T1-K2	A.W.	20 ft-lb @ 0°F (27 J @ -18°C)
E80T5-K2	A.W.	20 ft-lb @ -20°F (27 J @ -29°C)
E90T5-K2	A.W.	20 ft-lb @ -60°F (27 J @ -51°C)
E100T1-K3	A.W.	20 ft-lb @ 0°F (27 J @ -18°C)
E110T1-K3	A.W.	20 ft-lb @ 0°F (27 J @ -18°C)
E100T5-K3	A.W.	20 ft-lb @ -60°F (27 J @ -51°C)
E110T5-K3	A.W.	20 ft-lb @ -60°F (27 J @ -51°C)
E110T5-K4	A.W.	20 ft-lb @ -60°F (27 J @ -51°C)
E111T1-K4	A.W.	20 ft-lb @ -60°F (27 J @ -51°C)
E120T5-K4	A.W.	20 ft-lb @ -60°F (27 J @ -51°C)
E120T1-K5	A.W.	Not required
E61T8-K6	A.W.	20 ft-lb @ -20°F (27 J @ -29°C)
E71T8-K6	A.W.	20 ft-lb @ -20°F (27 J @ -29°C)
E101T1-K7	A.W.	20 ft-lb @ -60°F (27 J @ -51°C)
E80T1-W	A.W.	20 ft-lb @ -20°F (27 J @ -29°C)
EXXXTX-G	Properties as agreed upon between supplier and purchaser	

* A.W. = As welded
PWHT = Postweld heat treated in accordance with AWS A5.29 Specification
** PWHT = Temperatures in excess of 1150 °F (621 °C) will decrease the impact value.

10-13. FLUX-CORED ARC WELDING (FCAW) (cont)

Table 10-17. Chemical Composition Requirements for Low Alloy Flux-Cored Electrodes

Chemical Composition, percent[a]

AWS Classification	C	Mn	P	S	Si	Ni	Cr	Mo	V	Al[b]
Carbon-Molybdenum Steel Electrodes										
E70T5-A1 E80T1-A1 E81T1-A1	0.12	1.25	0.03	0.03	0.80	-	-	0.40/ 0.65	-	-
Chromium-Molybdenum Steel Electrodes										
E81T1-B1	0.12	1.25	0.03	0.03	0.80	-	0.40/ 0.65	0.40/ 0.65	-	-
E80T5-B2L	0.05	1.25	0.03	0.03	0.80	-	1.00/ 1.50	0.40/ 0.65	-	-
E80T1-B2 E81T1-B2 E80T5-B2	0.12	1.25	0.03	0.03	0.80	-	1.00/ 1.50	0.40/ 0.65	-	-
E80T1-B2H	0.10/ 0.15	1.25	0.03	0.03	0.80	-	1.00/ 1.50	0.40/ 0.65	-	-
E90T1-B3L	0.05	1.25	0.03	0.03	0.80	-	2.00/ 2.50	0.90/ 1.20	-	-
E91T1-B3	0.12	1.25	0.03	0.03	0.80	-	2.00/ 2.50	0.90/ 1.20	-	-
E90T5-B3 E100T1-B3 E90T1-B3H	0.10/ 0.15	1.25	0.03	0.03	0.80	-	2.00/ 2.50	0.90/ 1.20	-	-
Nickel-Steel Electrodes										
E71T8-Ni1 E80T1-Ni1	0.12	1.50	0.03	0.03	0.80	0.80/ 1.10	0.15	0.35	0.05	1.8
E81T1-Ni1 E80T5-Ni1 E71T8-Ni2 E80T1-Ni2 E81T1-Ni2 E80T5-Ni2	0.12	1.50	0.03	0.03	0.80	1.75/ 2.75	-	-	-	1.8
E91T1-Ni2 E80T5-Ni3 E90T5-Ni3	0.12	1.50	0.03	0.03	0.80	2.75/ 3.75	-	-	-	-
Manganese Molybdenum Steel Electrodes										
E91T1-D1	0.12	1.25/ 2.00	0.03	0.03	0.80	-	-	0.25/ 0.55	-	-
E90T5-D2 E100T5-D2	0.15	1.65/ 2.25	0.03	0.03	0.80	-	-	0.25/ 0.55	-	-
E100T5-D2 E90T1-D3	0.12	1.00/ 1.75	0.03	0.03	0.80	-	-	0.40/ 0.85	-	-

Table 10-17. Chemical Composition Requirements for Low Alloy Flux-Cored
Electrodes (cont)
Chemical Composition, percent[a]

AWS Classification	C	Mn	P	S	Si	Ni	Cr	Mo	V	Al[b]
		All Other Low Alloy Steel Electrodes								
E80T5-K1	0.15	0.80/1.40	0.03	0.03	0.80	0.80/1.10	0.15	0.20/0.65	0.05	-
E70T4-K2										
E71T8-K2										
E80T1-K2										
E90T1-K2	0.15	0.50/	0.03	0.03	0.80	1.00/2.00	0.15	0.35	0.05	1.8
E91T1-K2										
E80T5-K2										
E90T5-K2										
E100T1-K3										
E110T1-K3	0.15	0.75/2.25	0.03	0.03	0.80	1.25/2.60	0.15	0.25/0.65	0.05	-
E100T5-K3										
E110T5-K3										
E110T5-K4										
E111T1-K4	0.15	1.20/2.25	0.03	0.03	0.80	1.75/2.60	0.20/0.60	0.30/0.65	0.05	-
E120T5-K4										
E120T1-K5	0.10/0.25	0.60/1.60	0.03	0.03	0.80	0.75/2.00	0.20/0.70	0.15/0.55	0.05	-
E61T8-K6										
E71T8-K6	0.15	0.50/1.50	0.03	0.03	0.80	0.40/1.10	0.15	0.15	0.05	1.8
E101T1-K7	0.15	1.00/1.75	0.03	0.03	0.80	2.00/2.75	-	-	-	-
EXXXTX-G[c]	-	1.00 Min.	0.03	0.03	0.80 min.	0.50 min.	0.20 min.	0.10	1.8	
E80T1-W[d]	0.12	0.50/1.30	0.03	0.03	0.35/0.80	0.40/0.80	0.45/	-	-	-

a Single values are maximums unless otherwise noted.
b For self-shielded electrodes only.
c In order to meet the alloy requirements of the G group, the weld deposit need
   have the minimum, as specified in the table of only one of the elements listed.
d The E80T1-W classification also contains 0.30-0.75 percent copper.

The classification system for stainless steel electrodes is based on the chemical composition of the weld metal and the type of shielding to be employed during welding. An example of a stainless steel electrode classification is E308T–1 where:

   1. The "E" indicates the electrode.
   2. The digits between the "E" and the "T" indicates the chemical composition of the weld as shown in table 10-18, p 10-80.
   3. The "T" designates a tubular or flux cored electrode wire.
   4. The suffix of "1" indicates the type of shielding to be used as shown in table 10-19, p 10-81.

10-13. FLUX-CORED ARC WELDING (FCAW) (cont)

Table 10-18. Weld Metal Chemical Composition Requirements for
Stainless Steel Electrodes

AWS Classification	C	Cr	Ni	Mo	CB + Ta	Mn	Si	Ti
E307T-1 or 2	0.13	18.0-20.5	9.0-10.5	0.5-1.5	-	3.3-4.475	1.0	-
E308T-1 or 2	0.08	18.0-21.0	9.0-11.0	0.5	-	0.5-2.5	1.0	-
E308LT-1 or 2	a	18.0-21.0	9.0-11.0	0.5	-	0.5-2.5	1.0	-
E308MoT-1 or 2	0.08	18.0-21.0	9.0-12.0	2.0-3.0	-	0.5-2.5	1.0	-
E308MolT-1 or 2	a	18.0-21.0	9.0-12.0	2.0-3.0	-	0.5-2.5	1.0	-
E309T-1 or 2	0.10	22.0-25.0	12.0	0.5	-	0.5-2.5	1.0	-
E309CblT-1 or 2	a	22.0-25.0	12.0-14.0	0.5	0.70-1.00	0.5-2.5	1.0	-
E309LT-1 or 2	a	22.0-25.0	12.0-14.0	0.5	-	0.5-2.5	1.0	-
E310T-1 or 2	0.20	25.0-28.0	20.0-22.5	0.5	-	1.0-2.5	1.0	-
E312T-1 or 2	0.15	28.0-32.0	8.0-10.5	0.5	-	0.5-2.5	1.0	-
E316T-1 or 2	0.08	17.0-20.0	11.0-14.0	2.0-3.0	-	0.5-2.5	1.0	-
E316LT-1 or 2	a	17.0-20.0	11.0-14.0	2.0-3.0	-	0.5-2.5	1.0	-
E317LT-1 or 2	a	18.0-21.0	12.0-14.0	3.0-4.0	-	0.5-2.5	1.0	-
E347T-1 or 2	0.08	18.0-21.0	9.0-11.0	0.5	8 x C min to 1.0 max	0.5-2.5	1.0	-
E409T-1 or 2	0.10	10.5-13.0	0.60	0.5	-	0.80	1.0	10 x C min to 1.50 max
E410T-1 or 2	0.12	11.0-13.5	0.60	0.5	-	1.2	1.0	-
E41NiMoT-1 or 2	0.06	11.0-12.5	4.0-5.0	0.40-0.70	-	1.0	1.0	-
E410NiTiT-1 or 2	a	11.0-12.0	3.6-4.5	0.05	-	0.70	0.50	10 x C min to 1.50 max
E430T-1 or 2	0.10	15.0-18.0	0.60	0.5	-	1.2	1.0	-
E502T-1 or 2	0.10	4.0-6.0	0.40	0.45-0.65	-	1.2	1.0	-
E505T-1 or 2	0.10	8.0-10.5	0.40	0.85-1.20	-	1.2	1.0	-
E307T-3	0.13	19.5-22.0	9.0-10.5	0.5-1.5	-	3.3-4.75	1.0	-
E308T-3	0.08	19.5-22.0	9.0-11.0	0.5	-	0.5-2.5	1.0	-
E308LT-3	0.03	19.5-22.0	9.0-11.0	0.5	-	0.5-2.5	1.0	-
E308MOT-3	0.08	18.0-21.0	9.0-12.0	2.0-3.0	-	0.5-2.5	1.0	-
E308MolT-3	0.03	18.0-21.0	9.0-12.0	2.0-3.0	-	0.5-2.5	1.0	-
E309T-3	0.10	23.0-25.5	12.0-14.0	0.5	-	0.5-2.5	1.0	-
E309LT-3	0.03	23.0-25.5	12.0-14.0	0.5	-	0.5-2.5	1.0	-
E309CbLT-3	0.03	23.0-25.5	12.0-14.0	0.5	0.70-1.00	0.5-2.5	1.0	-

Table 10-18. Weld Metal Chemical Composition Requirements for
Stainless Steel Electrodes (cont)

AWS Classifi-cation	C	Cr	Ni	Mo	CB + Ta	Mn	Si	Ti
E310T-3	0.20	25.0-28.0	20.0-22.5	0.5	-	1.0-2.5	1.0	-
E312T-3	0.15	28.0-32.0	8.0-10.5	0.5	-	0.5-2.5	1.0	-
E316T-3	0.08	18.0-20.5	11.0-14.0	2.0-3.0	-	0.5-2.5	1.0	-
E316LT-3	0.03	18.0-20.5	11.0-14.0	2.0-3.0	-	0.5-2.5	1.0	-
E317LT-3	0.03	18.5-21.0	13.0-15.0	3.0-4.0	-	0.5-2.5	1.0	-
E347T-3	0.06	19.0-21.5	9.0-11.0	0.5	8 x C min to 1.0 max	0.5-2.5	1.0	-
E409T-3	0.10	10.5-13.0	0.60	0.5	-	0.80	1.0	10 x C min to 1.50 max
E410T-3	0.12	11.0-13.5	0.60	0.5	-	1.0	1.0	-
E410NiMo T-3	0.06	11.0-12.5	4.0-5.0	0.40-0.70	-	1.0	1.0	-
F410NiTi T-3	0.04	11.0-12.0	3.6-4.5	0.5	-	0.70	0.50	10 x C min to 1.50 max
E430T-3	0.10	15.0-18.0	0.60	0.5	-	1.0	1.0	-
EXXXT-G	As agreed upon between supplier and purchaser							

NOTE 1--Single values indicate maximum percentage.
NOTE 2--All electrode classifications contain a maximum of 04% P 03% S and 5% Cu.

a   The carbon content is .04 percent maximum when the suffix is "1".
     The carbon content is .03 percent maximum when the suffix is "2".

Table 10-19. Shielding

Classification	Shielding Gas	Welding Current
EXXXT-1	CO_2	DCEP
EXXXT-2	$Ar-CO_2$	DCEP
EXXXT-3	NONE	DCEP
EXXXT-G	NOT SPECIFIED	NOT SPECIFIED

(8) Welding Cables.

(a) The welding cables and connectors are used to connect the power source to the welding gun and to the work. These cables are normally made of copper. The cable consists of hundreds of wires that are enclosed in an insulated casing of natural or synthetic rubber. The cable that connects the power source to the welding gun is called the electrode lead. In semiautomatic welding, this cable is often part of the cable assembly, which also includes the shielding gas hose and the conduit that the electrode wire is fed through. For machine or automatic welding, the electrode lead is normally separate. The cable that connects the work to the power source is called the work lead. The work leads are usually connected to the work by pinchers, clamps, or a bolt.

10-13.   FLUX-COREDARC WELDING (FCAW) (cont)

(b) The size of the welding cables used depends on the output capacity of the welding machine, the duty cycle of the machine, and the distance between the welding machine and the work. Cable sizes range from the smallest AWG No 8 to AWG No 4/0 with amperage ratings of 75 amperes on up. Table 10-20 shows recommended cable sizes for use with different welding currents and cable lengths. A cable that is too small may become too hot during welding.

Table 10-20.   Recommended Cable Sizes for Different Welding Currents and Cable Lengths

Weld Type	Weld Current	Length of Cable Circuit in Feet - Cable Size A.W.G.					
		60'	100'	150'	200'	300'	400'
Manual	100	4	4	4	2	1	1/0
(Low	150	2	2	2	1	2/0	3/0
Duty	200	2	2	1	1/0	3/0	4/0
Cycle)	250	2	2	1/0	2/0		
	300	1	1	2/0	3/0		
	350	1/0	1/0	3/0	4/0		
	400	1/0	1/0	3/0			
	450	2/0	2/0	4/0			
	500	2/0	2/0	4/0			
Automatic	400	4/0	4/0				
(High	800	4/0 (2)	4/0 (2)				
Duty	1200	4/0 (3)	4/0 (3)				
Cycle)							

c.   Advantages. The major advantages of flux-cored welding are reduced cost and higher deposition rates than either SMAW or solid wire GMAW. The cost is less for flux-cored electrodes because the alloying agents are in the flux, not in the steel filler wire as they are with solid electrodes. Flux-cored welding is ideal where bead appearance is important and no machining of the weld is required. Flux-cored welding without carbon dioxide shielding can be used for most mild steel construction applications. The resulting welds have higher strength but less ductility than those for which carbon dioxide shielding is used. There is less porosity and greater penetration of the weld with carbon dioxide shielding. The flux-cored process has increased tolerances for scale and dirt. There is less weld spatter than with solid-wire MIG welding. It has a high deposition rate, and faster travel speeds are often used. Using small diameter electrode wires, welding can be done in all positions. Some flux-cored wires do not need an external supply of shielding gas, which simplifies the equipment. The electrode wire is fed continuously so there is very little time spent on changing electrodes. A higher percentage of the filler metal is deposited when compared to shield metal arc welding. Finally, better penetration is obtained than from shielded metal arc welding.

d.   Disadvantages. Most low-alloy or mild-steel electrodes of the flux-cored type are more sensitive to changes in welding conditions than are SMAW electrodes. This sensitivity, called voltage tolerance, can be decreased if a shielding gas is used, or if the slag-forming components of the core material are increased. A constant-potential power source and constant-speed electrode feeder are needed to maintain a constant arc voltage.

e. <u>Process Principles</u>. The flux-cored welding wire, or electrode, is a hollow tube filled with a mixture of deoxidizers, fluxing agents, metal powders, and ferro-alloys. The closure seam, which appears as a fine line, is the only visible difference between flux-cored wires and solid cold-drawn wire. Flux-cored electrode welding can be done in two ways: carbon dioxide gas can be used with the flux to provide additional shielding, or the flux core alone can provide all the shielding gas and slagging materials. The carbon dioxide gas shield produces a deeply penetrating arc and usually provides better weld than is possible without an external gas shield. Although flux-cored arc welding may be applied semiautomatically, by machine, or automatically, the process is usually applied semiautomatically. In semiautomatic welding, the wire feeder feeds the electrode wire and the power source maintains the arc length. The welder manipulates the welding gun and adjusts the welding parameters. Flux-cored arc welding is also used in machine welding where, in addition to feeding the wire and maintaining the arc length, the machinery also provides the joint travel. The welding operator continuously monitors the welding and makes adjustments in the welding parameters. Automatic welding is used in high production applications.

10-14. SUBMERGED ARC WELDING (SAW)

a. <u>General</u>. Submerged arc welding is a process in which the joining of metals is produced by heating with an arc or arcs between a bare metal electrode or electrodes and the work. The arc is shielded by a blanket of granular fusible material on the work. Pressure is not used. Filler metal is obtained from the electrode or from a supplementary welding rod.

b. <u>Equipment</u>.

(1) The equipment components required for submerged arc welding are shown by figure 10-59. Equipment consists of a welding machine or power source, the wire feeder and control system, the welding torch for automatic welding or the welding gun and cable assembly for semiautomatic welding, the flux hopper and feeding mechanism, usually a flux recovery system, and a travel mechanism for automatic welding.

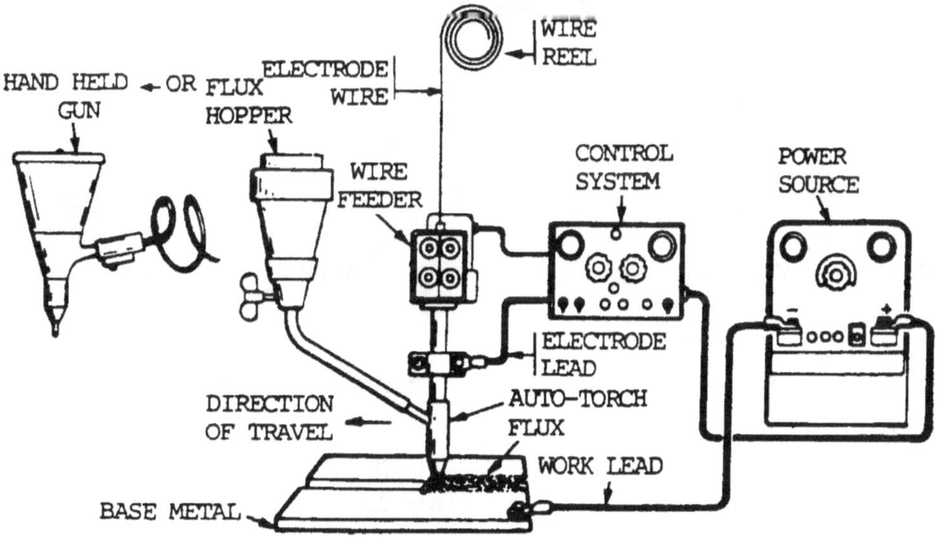

Figure 10-59. Block diagram - SAW.

10-14. SUBMERGED ARC WELDING (SAW) (cont)

(2) The power source for submerged arc welding must be rated for a 100 percent duty cycle, since the submerged arc welding operations are continuous and the length of time for making a weld may exceed 10 minutes. If a 60 percent duty cycle power source is used, it must be derated according to the duty cycle curve for 100 percent operation.

(3) When constant current is used, either ac or dc, the voltage sensing electrode wire feeder system must be used. When constant voltage is used, the simpler fixed speed wire feeder system is used. The CV system is only used with direct current.

(4) Both generator and transformer-rectifier power sources are used, but the rectifier machines are more popular. Welding machines for submerged arc welding range in size from 300 amperes to 1500 amperes. They may be connected in parallel to provide extra power for high current applications. Direct current power is used for semiautomatic applications but alternating current power is used primarily with the machine or the automatic method. Multiple electrode systems require specialized types of circuits, especially when ac is employed.

(5) For semiautomatic application, a welding gun and cable assembly are used to carry the electrode and current and to provide the flux at the arc. A small flux hopper is attached to the end of the cable assembly. The electrode wire is fed through the bottom of this flux hopper through a current pickup tip to the arc. The flux is fed from the hopper to the welding area by mans of gravity. The amount of flux fed depends on how high the gun is held above the work. The hopper gun may include a start switch to initiate the weld or it may utilize a "hot" electrode so that when the electrode is touched to the work, feeding will begin automatically.

(6) For automatic welding, the torch is attached the wire feed motor and includes current pickup tips for transmitting the welding current to the electrode wire. The flux hopper is normally attached to the torch and may have magnetically operated valves which can be opened or closed by the control system.

(7) Other pieces of equipment sometimes used may include a travel carriage, which can be a simple tractor or a complex moving specialized fixture. A flux recovery unit is normally provided to collect the unused submerged arc flux and return it to the supply hopper.

(8) Submerged arc welding system can become quite complex by incorporating additional devices such as seam followers, weavers, and work rovers.

c. Advantages and Major Uses.

(1) The major advantages of the submerged arc welding process are:

(a) high quality of the weld metal.
(b) extremely high deposition rate and speed.
(c) smooth, uniform finished weld with no spatter.
(d) little or no smoke.
(e) no arc flash, thus minimal need for protective clothing.
(f) high utilization of electrode wire.
(g) easy automation for high-operator factor.
(h) normally, no involvement of manipulative skills.

(2) The submerged arc process is widely used in heavy steel plate fabrication work. This includes the welding of structural shapes, the longitudinal seam of larger diameter pipe, the manufacture of machine components for all types of heavy industry, and the manufacture of vessels and tanks for pressure and storage use. It is widely used in the shipbuilding industry for splicing and fabricating subassemblies, and by many other industries where steels are used in medium to heavy thicknesses. It is also used for surfacing and buildup work, maintenance, and repair.

    d. Limitations of the Process.

    (1) A major limitation of submerged arc welding its limitation of welding positions. The other limitation is that it is primarilused only to weld mild and low-alloy high-strength steels.

    (2) The high-heat input, slow-cooling cycle can be a problem when welding quenched and tempered steels. The heat input limitationof the steel in question must be strictly-adhered to when using submerged arc welding.This may require the making of multipass welds where a single pass weld would be acceptable in mild steel. In some cases, the economic advantages may be reduced to the point where flux-cored arc welding or some other process should be considered.

    (3) In semiautomatic submerged arc welding, the inability to see the arc and puddle can be a disadvantage in reaching the root of a groove weld and properly filling or sizing.

    e. Principles of Operation.

    (1) The submerged arc welding process is shown by figure 10-60. It utilizes the heat of an arc between a continuously fed electrode and the work. The heat of the arc melts the surface of the base metal and the end of the electrode. The metal melted off the electrode is transferred through the arc to the workpiece, where it becomes the deposited weld metal. Shielding is obtained from a blanket of granular flux, which is laid directly over the weld area. The flux close to the arc melts and intermixes with the molten weld metal, helping to purify and fortify it. The flux forms a glass-like slag that is lighter in weight than the deposited weld metal and floats on the surface as a protective cover.The weld is submerged under this layer of flux and slag, hence the name submerged arc welding. The flux and slag normally cover the arc so that it is not visible.The unmelted portion of the flux can be reused. The electrode is fed into the arc automatically from a coil. The arc is maintained automatically. Travel can be manual or by machine. The arc is initiated by a fuse type start or by a reversing or retrack system.

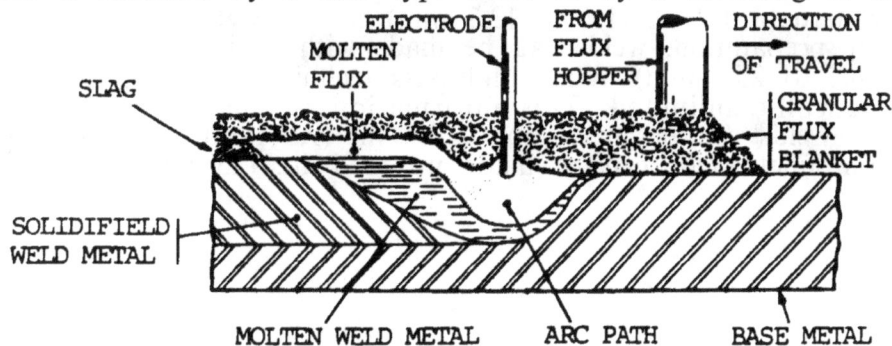

Figure 10-60. Process diagram--submerged arc welding.

10-14. SUBMERGED   ARC   WELDING (SAW) (cont)

(2) Normal method of application and position capabilities. The most popular method of application is the machine method, where the operator monitors the welding operation. Second in popularity is the automatic method, where welding is a pushbutton operation. The process can be applied semiautomatically; however, this method of application is not too popular.The process cannot be applied manually because it is impossible for a welder to control an arc that is not visible. The submerged arc welding process is a limited-position welding process. The welding positions are limited because the large pool of molten metal and the slag are very fluid and will tend to run out of the joint.Welding can be done in the flat position and in the horizontal fillet position with ease. Under special controlled procedures, it is possible to weld in the horizontal position, sometimes called 3 o' clock welding. This requires special devices to hold the flux up so that the molten slag and weld metal cannot run away.The process cannot be used in the vertical or overhead position.

(3) Metals weldable and thickness range. Submerged arc welding is used to weld low- and medium-carbon steels, low-alloy high-strength steels, quenched and tempered steels, and many stainless steels. Experimentally, it has been used to weld certain copper alloys, nickel alloys, and even uranium. This information is summarized in table 10-21.

Table 10-21. Base Metals Weldable by the Submerged Arc Process

Base Metal	Weldability
Wrought iron	Weldable
Low carbon steel	Weldable
Low alloy steel	Weldable
High and medium carbon	Possible but not popular
Alloys steel	Possible but not popular
Stainless steel	Weldable

Metal thicknesses from 1/16 to 1/2 in. (1.6 to 12.7 mm) can be welded with no edge preparation. With edge preparation, welds can be made with a single pass on material from 1/4 to 1 in. (6.4 to 25.4 mm). When multipass technique is used, the maximum thickness is practically unlimited. This information is summarized in table 10-22. Horizontal fillet welds can be made up to 3/8 in. (9.5 mm) in a single pass and in the flat position, fillet welds can be made up to 1 in. (25 mm) size.

Table 10-22. Base Metal Thickness Range

Thickness / Factor	inch	0.005	0.015	0.062	0.125	3/16	1/4	3/8	1/2	3/4	1	2	4	8
	mm	0.13	0.4	01.6	3.2	4.8	6.4	10	12.7	19	25	51	102	203
Single pass no prep.					←				→					
Single pass prep.								←		→				
Multi pass										←				→

(4) Joint design. Although the submerged arc welding process can utilize the same joint design details as the shielded metal arc welding process, different joint details are suggested for maximum utilization and efficiency of submerged arc welding. For groove welds, the square groove design can be used up to 5/8 in. (16 mm) thickness. Beyond this thickness, bevels are required. Open roots are used but backing bars are necessary since the molten metal will run through the joint. When welding thicker metal, if a sufficiently large root face is used, the backing bar may be eliminate. However, to assure full penetration when welding from one side, backing bars are recommended. Where both sides are accessible, a backing weld can be made which will fuse into the original weld to provide full penetration. Recommended submerged arc joint designs are shown by figure 10-61, p 10-88.

(5) Welding circuit and current.

(a) The welding circuit employedfor single electrode submerged arc welding is shown by figure 10-59, p 10-83. Thisrequires a wire feeder system and a power supply.

(b) The submerged arc welding process uses either direct or alternating current for welding power. Direct current is used for most applications which use a single arc. Both direct current electrode positive (DCEP) and electrode negative (DCEN) are used.

(c) The constant voltage type of direct current power is more popular for submerged arc we welding with 1/8 in. (3.2 mm) and smaller diameter electrode negative

(d) The constant current power system is normally used for welding with 5/3 2 in. (4 mm) and larger-diameter electrode wires. The control circuit for CC power is more complex since it attempts to duplicate the actions of the welder to retain a specific arc length. The wire feed system must sense the voltage across the arc and feed the electrode wire into the arc to maintain this voltage. As conditions change, the wire feed must slow down or speed up to maintain the pre-fixed voltage across the arc. This adds complexity to the control system. The system cannot react instantaneously.Arc starting is mare complicated with the constant current system since it requires the use of a reversing system to strike the arc, retract, and then maintain the preset arc voltage.

10-14. SUBMERGED ARC WELDING (SAW) (cont)

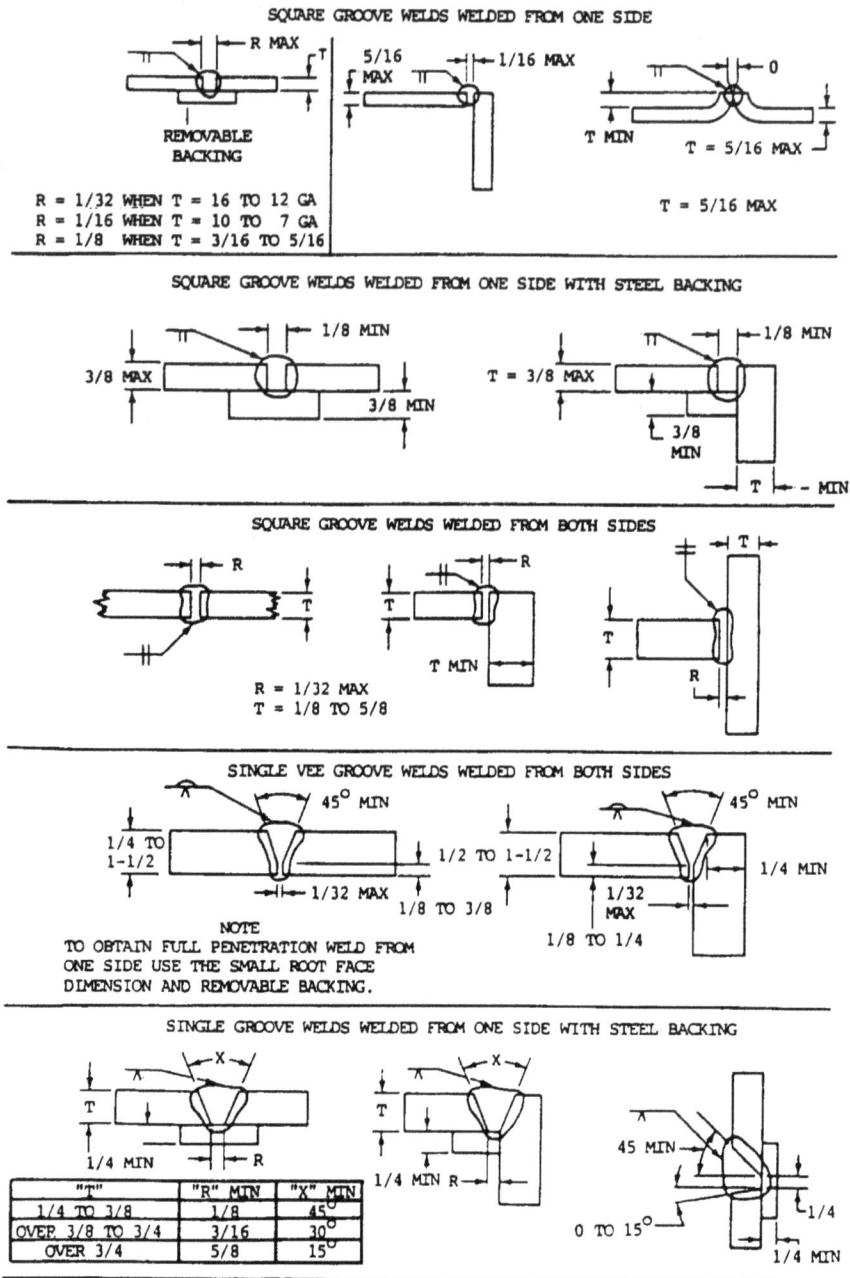

SQUARE GROOVE WELDS WELDED FROM ONE SIDE

R MAX

REMOVABLE
BACKING

R = 1/32 WHEN T = 16 TO 12 GA
R = 1/16 WHEN T = 10 TO 7 GA
R = 1/8 WHEN T = 3/16 TO 5/16

5/16
MAX

1/16 MAX

T = 5/16 MAX

0

T MIN

T = 5/16 MAX

SQUARE GROOVE WELDS WELDED FROM ONE SIDE WITH STEEL BACKING

3/8 MAX

1/8 MIN

3/8 MIN

T = 3/8 MAX

1/8 MIN

3/8
MIN

T — MIN

SQUARE GROOVE WELDS WELDED FROM BOTH SIDES

R

T

R

T

T MIN

T

R

T

R = 1/32 MAX
T = 1/8 TO 5/8

SINGLE VEE GROOVE WELDS WELDED FROM BOTH SIDES

45° MIN

1/4 TO
1-1/2

1/32 MAX

1/2 TO 1-1/2

1/8 TO 3/8

NOTE
TO OBTAIN FULL PENETRATION WELD FROM
ONE SIDE USE THE SMALL ROOT FACE
DIMENSION AND REMOVABLE BACKING.

45° MIN

1/4 MIN

1/32
MAX

1/8 TO 1/4

SINGLE GROOVE WELDS WELDED FROM ONE SIDE WITH STEEL BACKING

X

T

1/4 MIN

R

X

T

1/4 MIN R

45 MIN

0 TO 15°

1/4

1/4 MIN

"T"	"R" MIN	"X" MIN
1/4 TO 3/8	1/8	45°
OVER 3/8 TO 3/4	3/16	30°
OVER 3/4	5/8	15°

Figure 10-61. Weld joint designs for submerged arc welding (sheet 1 of 2).

DOUBLE VEE GROOVE WELDS WELDED FROM BOTH SIDES

DOUBLE BEVEL GROOVE WELDS WELDED FROM BOTH SIDES

SINGLE U GROOVE WELDS WELDED FROM ONE OR BOTH SIDES

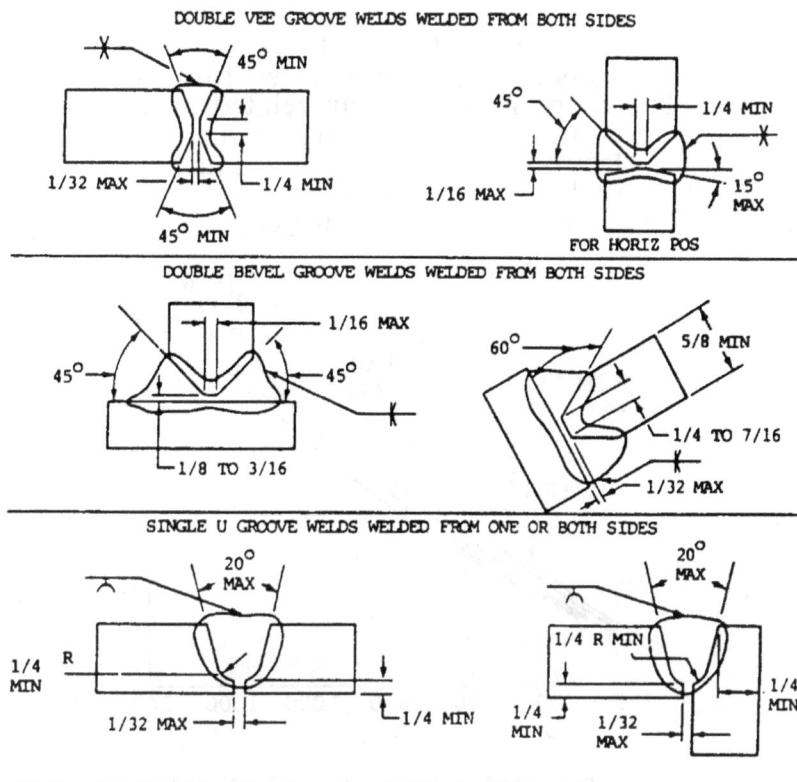

Figure 10-61. Weld joint designs for submerged arc welding (sheet 2 of 2).

(e) For ac welding, the constant current power is always used. when multi ple electrode wire systems are used with both ac and dc arcs, the constant current power system is utilized. The constant voltage system, however, can be applied when two wires are fed into the arc supplied by a single power source. Welding current for submerged arc welding can vary from as low as 50 amperes to as high as 2000 amperes. Most submerged arc welding is done in the range of 200 to 1200 amperes.

10-14. SUBMERGED ARC WELDING (cont)

(6) Deposition rates and weld quality.

(a) The deposition rates of the submerged arc welding process are higher than any other arc welding process. Deposition rates for single electrodes are shown by figure 10-62. There are at least four related factors that control the deposition rate of submerged arc welding: polarity, long stickout, additives in the flux, and additional electrodes. The deposition rate is the highest for direct current electrode negative (DCEN). The deposition rate for alternating current is between DCEP and DCEN. The polarity of maximum heat is the negative pole.

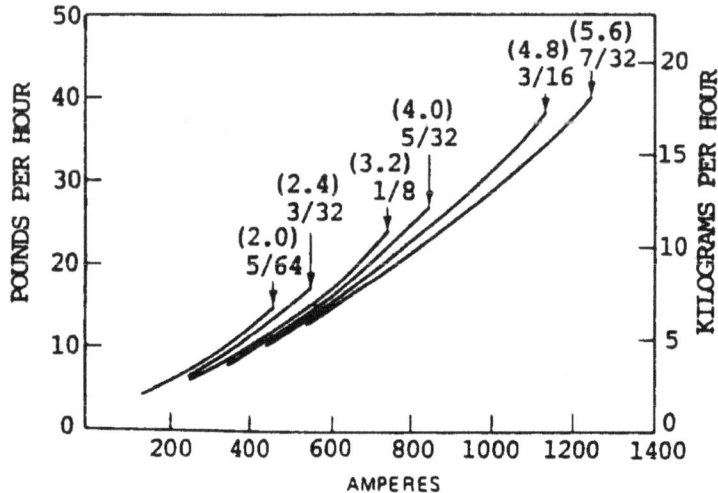

Figure 10-62. Deposition rates for single electrodes.

(b) The deposition rate with any welding current can be increased by extending the "stickout." This is the distance from the point where current is introduced into the electrode to the arc. When using "long stickout" the amount of penetration is reduced. The deposition rates can be increased by metal additives in the submerged arc flux. Additional electrodes can be used to increase the overall deposition rate.

(c) The quality of the weld metal deposited by the submerged arc welding process is high. The weld metal strength and ductility exceeds that of the mild steel or low-alloy base material when the correct combination of electrode wire and submerged arc flux is used. When submerged arc welds are made by machine or automatically, the human factor inherent to the manual welding processes is eliminated. The weld will be more uniform and free from inconsistencies. In general, the weld bead size per pass is much greater with submerged arc welding than with any of the other arc welding processes. The heat input is higher and cooling rates are slower. For this reason, gases are allowed more time to escape. Additionally, since the submerged arc slag is lower in density than the weld metal, it will float out to the top of the weld. Uniformity and consistency are advantages of this process when applied automatically.

(d) Several problems may occur when using the semiautomatic application method. The electrode wire may be curved when it leaves the nozzle of the welding gun. This curvature can cause the arc to be struck in a location not expected by the welder. When welding in fairly deep grooves, the curvature may cause the arc to be against one side of the weld joint rather than at the root. This will cause incomplete root fusion. Flux will be trapped at the root of the weld. Another problem with semiautomatic welding is that of completely filling the weld groove or maintaining exact size, since the weld is hidden and cannot be observed while it is being made. This requires making an extra pass. In some cases, too much weld is deposited. Variations in root opening affect the travel speed. If travel speed is uniform, the weld may be under- or overfilled in different areasHigh operator skill will overcome this problem.

(e) There is another quality problem associated with extremely large single-pass weld deposits. When these large welds solidify, the impurities in the melted base metal and in the weld metal all collect at the last point to freeze, which is the centerline of the weld. If there is sufficient restraint and enough impurities are collected at this point, centerline cracking may occur. This can happen when making large single-pass flat fillet welds if the base metal plates are 45°from flat. A simple solution is to avoid placing the parts at a true 45 angle. It should be varied approximately 10 so that the root of the joint is not in line with the centerline of the fillet weld. Another solution is to make multiple passes rather than attempting to make a large weld in a single pass.

(f) Another quality problem has to do with the hardness of the deposited weld metal . Excessively hard weld deposits contribute to cracking of the weld during fabrication or during service. A maximum hardness level of 225 Brinell is recommended. The reason for the hard weld in carbon and low-alloy steels is too rapid cooling, inadequate postweld treatment, or excessive alloy pickup in the weld metal. Excessive alloy pickup is due to selecting an electrode that has too much alloy, selecting a flux that introduces too much alloy into the weld, or the use of excessively high welding voltages.

(g) In automatic and machine welding, defects may occur at the start or at the end of the weld. The best solution is to use runout tabs so that starts and stops will be on the tabs rather than on the product.

(7) Weld schedules. The submerged arc welding process applied by machine or fully automatically should be done in accordance with welding procedure schedules. Table 10-23, p 10-93, and figure 10-63, p 10-92, show the recommended welding schedules for submerged arc welding using a single electrode on mild and low-alloy steels. The table can be used for welding other ferrous materials, but was developed for mild steel. All of the welds made by this procedure should pass qualification, tests, assuming that the correct electrode and flux have been selected. If the schedules are varied more than 10 percent, qualification tests should be performed to determine the weld quality.

10-14. SUBMERGED ARC WELDING (cont)

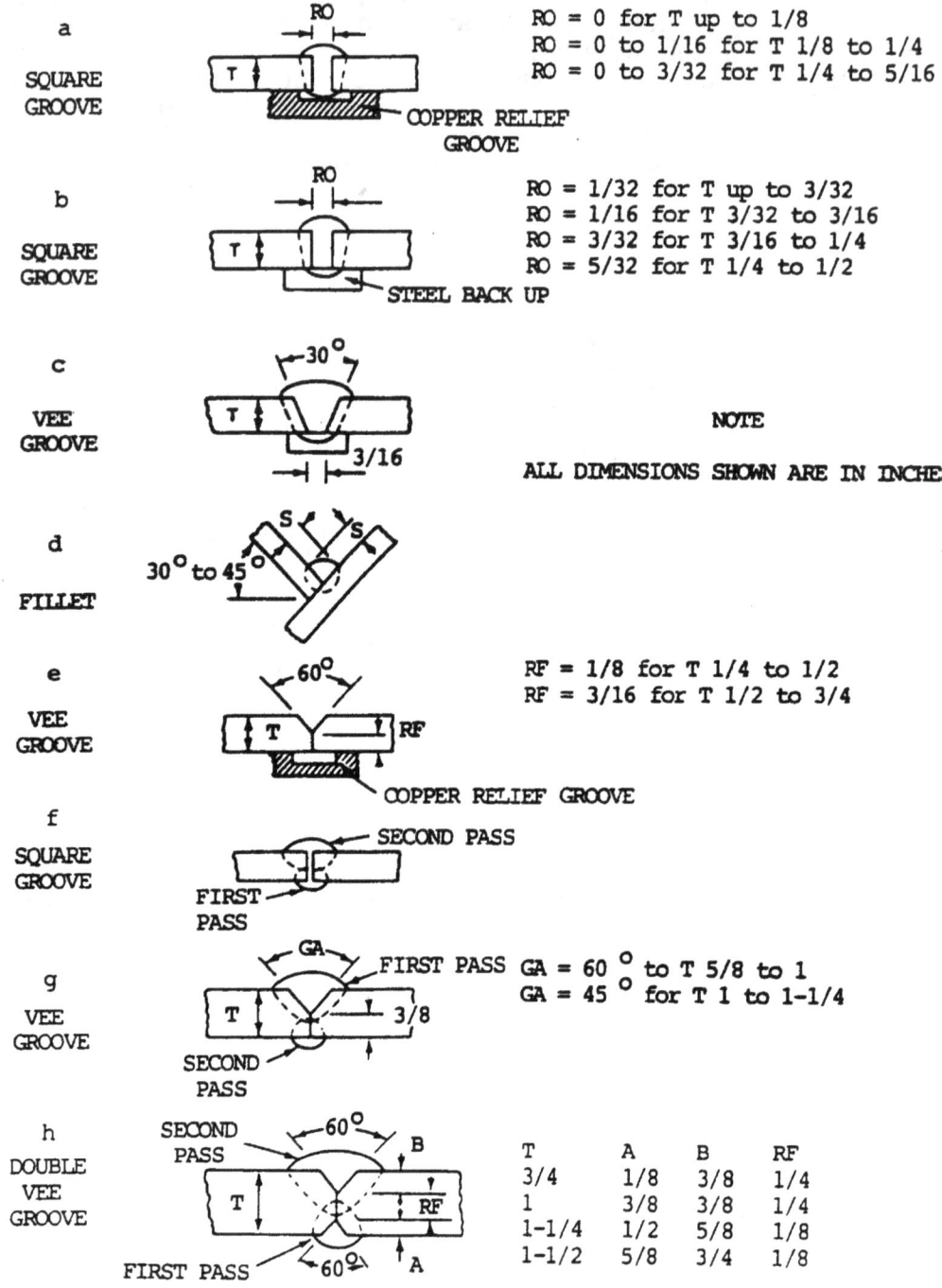

a

SQUARE
GROOVE

RO = 0 for T up to 1/8
RO = 0 to 1/16 for T 1/8 to 1/4
RO = 0 to 3/32 for T 1/4 to 5/16

COPPER RELIEF
GROOVE

b

SQUARE
GROOVE

RO = 1/32 for T up to 3/32
RO = 1/16 for T 3/32 to 3/16
RO = 3/32 for T 3/16 to 1/4
RO = 5/32 for T 1/4 to 1/2

STEEL BACK UP

c

VEE
GROOVE

NOTE

ALL DIMENSIONS SHOWN ARE IN INCHES.

d

FILLET

e

VEE
GROOVE

RF = 1/8 for T 1/4 to 1/2
RF = 3/16 for T 1/2 to 3/4

COPPER RELIEF GROOVE

f

SQUARE
GROOVE

SECOND PASS

FIRST
PASS

g

VEE
GROOVE

FIRST PASS GA = 60° to T 5/8 to 1
GA = 45° for T 1 to 1-1/4

SECOND
PASS

h

DOUBLE
VEE
GROOVE

SECOND
PASS

FIRST PASS

T	A	B	RF
3/4	1/8	3/8	1/4
1	3/8	3/8	1/4
1-1/4	1/2	5/8	1/8
1-1/2	5/8	3/4	1/8

Figure 10-63. Welds corresponding to table 10-23.

Table 10-23. Welding Procedure Schedules for SAW

Material Thickness (Gauges, Inches)	Type of Weld See Figure 10-63	Electrode Dia. (2)	Welding Current Amps-dc	Arc Voltage Elec. Pos.	Wire Feed ipm	Travel Speed ipm
16 0.063	a Square groove	3/32	300	22	68	100-140
	b Square groove	1/8	425	26	53	95-120
14 0.078	a Square groove	3/32	375	23	85	100-140
	b Square groove	1/8	500	27	65	75-85
12 0.109	a Square groove	1/8	400	23	51	70-90
	b Square groove	1/8	550	27	65	50-60
	d Fillet	1/8	400	25	51	40-60
10 0.140	a Square groove	1/8	425	26	53	50-80
	b Square groove	5/32	650	27	55	40-45
3/16 0.188	a Square groove	5/32	600	26	50	40-75
	b Square groove	3/16	875	31	55	35-40
	d Fillet	1/8	525	26	67	35-40
1/4 0.250	a Square groove	3/16	800	28	50	30-35
	b Square groove	3/16	875	31	56	22-25
	d Fillet	5/32	650	28	56	30-35
	e Vee groove	3/16	750	30	47	25-40
3/8 0.375	b Square groove	3/16	950	32	61	20-25
	f Square groove	3/16	1st pass 500	32	27	30
			2nd pass 750	33	47	30
	e Vee groove	3/16	900	33	57	23-25
	d Fillet	3/16	950	31	61	30-35
1/2 0.500	c Vee groove	3/16	975	33	63	12-17
	f Square groove	3/16	1st pass 650	34	40	25
			2nd pass 850	35	54	23-27
	e Vee groove	3/16	950	35	61	18-20
	d Fillet	3/16	950	33	61	14-17

10-14. SUBMERGED ARC WELDING (cont)

Table 10-23. Welding Procedure Schedules for SAW (cont)

Material Thickness (Gauges, Inches)	Type of Weld See Figure 10-63	Electrode Dia. (2)	Welding Current Amps-dc	Arc Voltage Elec. Pos.	Wire Feed ipm	Travel Speed ipm
3/4 0.750	c Vee groove	7/32	1000	35	49	68
	f Square groove	3/16	1st pass 925	37	59	12
			2nd pass 1000	40	65	11
	e Vee groove	7/32	950	36	46	10-12
	d Fillet	7/32	1000	35	49	6-8
	g Vee groove	7/32	1st pass 950	34	46	15
			2nd pass 750	34	25	22
	h Double vee groove	3/16	1st pass 700	35	42	20-22
			2nd pass 1000	36	65	14-16
1 1.000	g Vee groove	7/32	1st pass 1150	36	58	11
			2nd pass 850	36	40	20
	h Double vee groove	7/32.	1st pass 900	36	42	13-15
			2nd pass 1075	36	52	12-14
1-1/4 1.250	h Double vee groove	7/32	1st pass 1000	36	50	13
			2nd pass 1125	37	56	8
1-1/2 1.500	h Double vee groove	7/32	1st pass 1050	36	51	9
			2nd pass 1125	37	56	7

(8) Welding variables.

(a) The welding variables for submerged arc welding are similar to the other arc welding processes, with several exceptions.

(b) In submerged arc welding, the electrode type and the flux type are usually based on the mechanical properties required by the weld. The electrode and flux combination selection is based on table 10-24, p 10-103, to match the metal being welded. The electrode size is related to the weld joint size and the current recommended for the particular joint. This must also be considered in determining the number of passes or beads for a particular joint. Welds for the same joint dimension can be made in many or few passes, depending on the weld metal metallurgy desired. Multiple passes usually deposit higher-quality weld metal. Polarity is established initially and is based on whether maximum penetration or maximum deposition rate is required.

(c) The major variables that affect the weld involve heat input and include the welding current, arc voltage, and travel speed. Welding current is the most important. For single-pass welds, the current should be sufficient for the desired penetration without burn-through. The higher the current, the deeper the penetration. In multi-pass work, the current should be suitable to produce the size of the weld expected in each pass. The welding current should be selected based on the electrode size. The higher the welding current, the greater the melt-off rate (deposition rate).

(d) The arc voltage is varied within narrower limits than welding current. It has an influence on the bead width and shape. Higher voltages will cause the bead to be wider and flatter. Extremely high arc voltage should be avoided, since it can cause cracking. This is because an abnormal amount of flux is melted and excess deoxidizers may be transferred to the weld deposit, lowering its ductility. Higher arc voltage also increases the amount of flux consumed. The low arc voltage produces a stiffer arc that improves penetration, particularly in the bottom of deep grooves. If the voltage is too low, a very narrow bead will result. It will have a high crown and the slag will be difficult to remove.

(e) Travel speed influences both bead width and penetration. Faster travel speeds produce narrower beads that have less penetration. This can be an advantage for sheet metal welding where small beads and minimum penetration are required. If speeds are too fast, however, there is a tendency for undercut and porosity, since the weld freezes quicker. If the travel speed is too slow, the electrode stays in the weld puddle too long. This creates poor bead shape and may cause excessive spatter and flash through the layer of flux.

10-14. SUBMERGED ARC WELDING (cont)

    (f) The secondary variables include the angle of the electrode to the work, the angle of the work itself, the thickness of the flux layer, and the distance between the current pickup tip and the arc.This latter factor, called electrode "stickout," has a considerable effect on the weld. Normally, the distance between the contact tip and the work is 1 to 1-1/2 in. (25 to 38 mm). If the stickout is increased beyond this amount, it will cause preheating of the electrode wire, which will greatly increase the deposition rate. As stickout increases, the penetration into the base metal decreases. This factor must be given serious consideration because in some situations the penetration is required. The relationship between stickout and deposition rate is shown by figure 10-64.

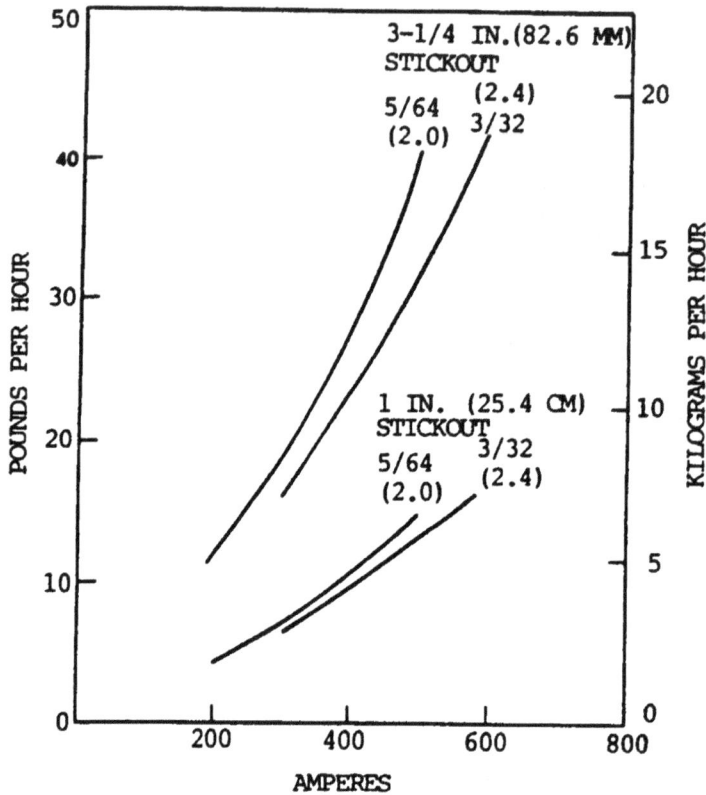

Figure 10-64. Stickout vs. deposition rate.

    (g) The depth of the flux layer must also be considered. If it is too thin, there will be too much arcing through the flux or arc flash. This also may cause porosity. If the flux depth is too heavy, the weld may be narrow and humped. Too many small particles in the flux can cause surface pitting since the gases generated in the weld may not be allowed to escape. These are sometimes called peck marks on the bead surface.

(9) Tips for using the process.

(a) One of the major applications for submerged arc welding is on circular welds where the parts are rotated under a fixed head.These welds can be made on the inside or outside diameter. Submerged arc welding produces a large molten weld puddle and molten slag which tends to run. This dictates that on outside diameters, the electrode should be positioned ahead of the extreme top, or 12 o'clock position, so that the weld metal will begin to solidify before it starts the downside slope. This becomes more of a problem as the diameter of the part being welded gets smaller. Improper electrode position will increase the possibility of slag entrapment or a poor weld surface.The angle of the electrode should also be changed and pointed in the direction of travel of the rotating part. When the welding is done on the inside circumference, the electrode should be angled so that it is ahead of bottom center, or the 6 o'clock positionFigure 10-65 illustrates these two conditions.

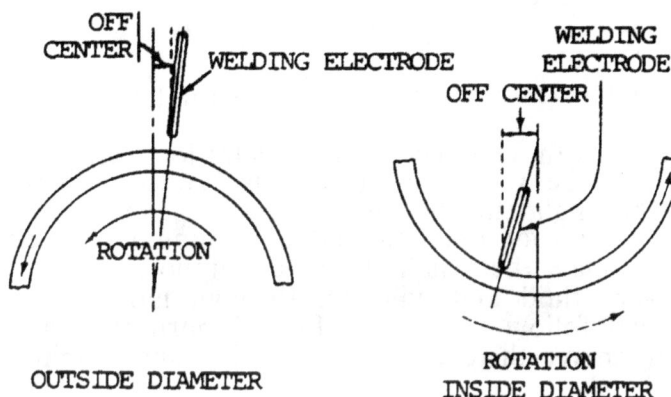

Figure 10-65. Welding on rotating circular parts.

(b) Sometimes the work being welded is sloped downhill or uphill to provide different types of weld bead contours. If the work is sloped downhill, the bead will have less penetration and will be wider. If the weld is sloped uphill, the bead will have deeper penetration and will be narrower. This is based on all other factors remaining the same. This information is shown by figure 10-66.

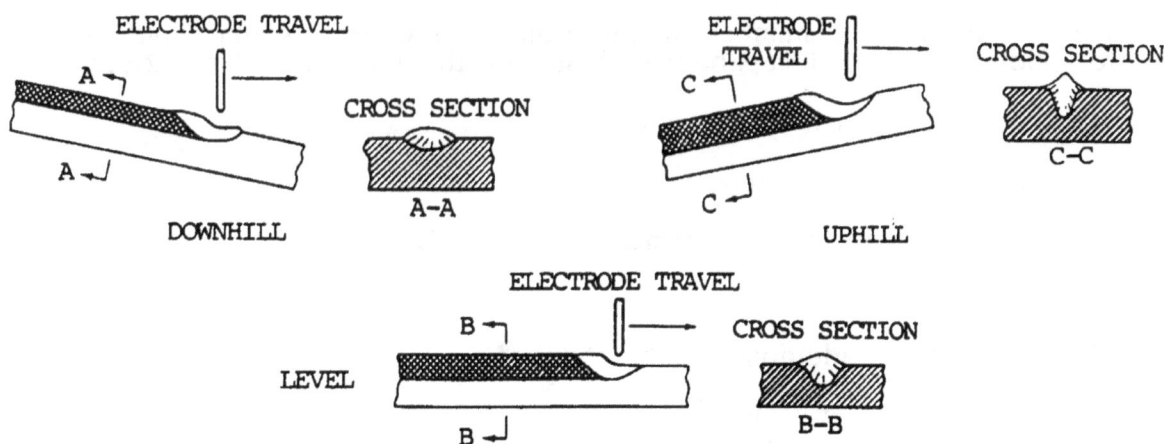

Figure 10-66. Angle of slope of work vs. weld.

10-14. SUBMERGED ARC WELDING (cont)

(c) The weld will be different depending on the angle of the electrode with respect to the work when the work is level. This is the travel angle, which can be a drag or push angle. It has a definite effect on the bead contour and weld metal penetration. Figure 10-67 shows the relationship.

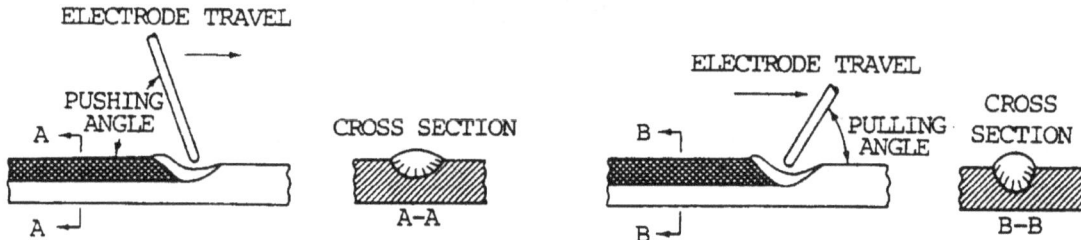

Figure 10-67. Angle of electrode vs weld.

(d) One side welding with complete root penetration can be obtained with submerged arc welding. When the weld joint is designed with a tight root opening and a fairly large root face, high current and electrode positive should be used. If the joint is designed with a root opening and a minimum root face, it is necessary to use a backing bar, since there is nothing to support the molten weld metal. The molten flux is very fluid and will run through narrow openings. If this happens, the weld metal will follow and the weld will burn through the joint. Backing bars are needed whenever there is a root opening and a minimum root face.

(e) Copper backing bars are useful when welding thin steel. Without backing bars, the weld would tend to melt through and the weld metal would fall away from the joint. The backing bar holds the weld metal in place until it solidifies. The copper backing bars may be water cooled to avoid the possibility of melting and copper pickup in the weld metal. For thicker materials, the backing may be submerged arc flux or other specialized type flux.

(10) Variations of the process.

(a) There are a large number of variations to the process that give submerged arc welding additional capabilities. Some of the more popular variations are:

1. Two-wire systems — same power source.

2. Two-wire systems — separate power source.

3. Three-wire systems — separate power source.

4. Strip electrode for surfacing.

5. Iron powder additions to the flux.

6. Long stickout welding.

7. Electrically "cold" filler wire.

(b) The multi-wire systems offer advantages since deposition rates and travel speeds can be improved by using more electrodes. Figure 10-68 shows the two methods of utilizing two electrodes, one with a single-power source and one with two power sources. When a single-power source is used, the same drive rolls are used for feeding both electrodes into the weld. When two power sources are used, individual wire feeders must be used to provide electrical insulation between the two electrodes. With two electrodes and separate power, it is possible to utilize different polarities on the two electrodes or to utilize alternating current on one and direct current on the other. The electrodes can be placed side by side. This is called transverse electrode position. They can also be placed one in front of the other in the tandem electrode position.

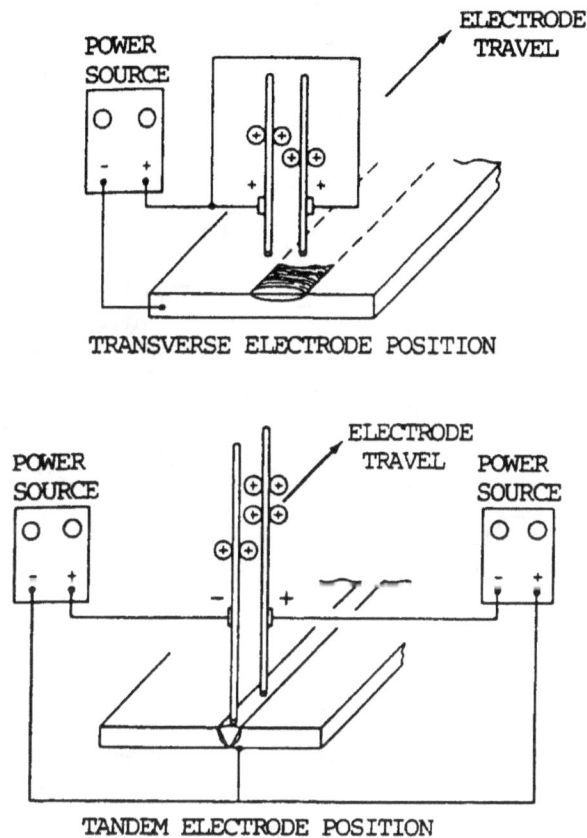

Figure 10-68. Two electrode wire systems.

(c) The two-wire tandem electrode position with individual power sources is used where extreme penetration is required. The leading electrode is positive with the trailing electrode negative. The first electrode creates a digging action and the second electrode fills the weld joint. When two dc arcs are in close proximity, there is a tendency for arc interference between them. In some cases, the second electrode is connected to alternating current to avoid the interaction of the arc.

10-14. SUBMERGED ARC WELDING (SAW) (cont)

(d) The three-wire tandem system normally uses ac power on all three electrodes connected to three-phase power systems. These systems are used for making high-speed longitudinal seams for large-diameter pipe and for fabricated beams. Extremely high currents can be used with correspondingly high travel speeds and deposition rates.

(e) The strip welding system is used to overlay mild and alloy steels usually with stainless steel. A wide bead is produced that has a uniform and minimum penetration. This process variation is shown by figure 10-69. It is used for overlaying the inside of vessels to provide the corrosion resistance of stainless steel while utilizing the strength and economy of the low-alloy steels for the wall thickness. A strip electrode feeder is required and special flux is normally used. When the width of the strip is over 2 in. (51 mm), a magnetic arc oscillating device is used to provide for even burnoff of the strip and uniform penetration.

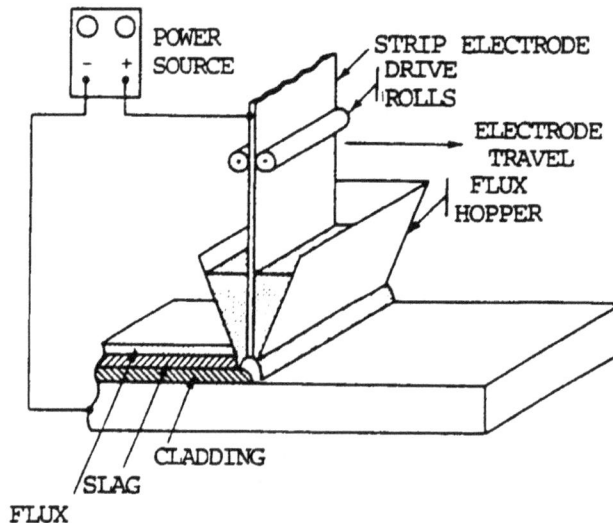

Figure 10-69. Strip electrode on surfacing.

(f) Another way of increasing the deposition rate of submerged arc welding is to add iron base ingredients to the joint under the flux. The iron in this material will melt in the heat of the arc and will become part of the deposited weld metal. This increases deposition rates without decreasing weld metal properties. Metal additives can also be used for special surfacing applications. This variation can be used with single-wire or multi-wire installations. Figure 10-70 shins the increased deposition rates attainable.

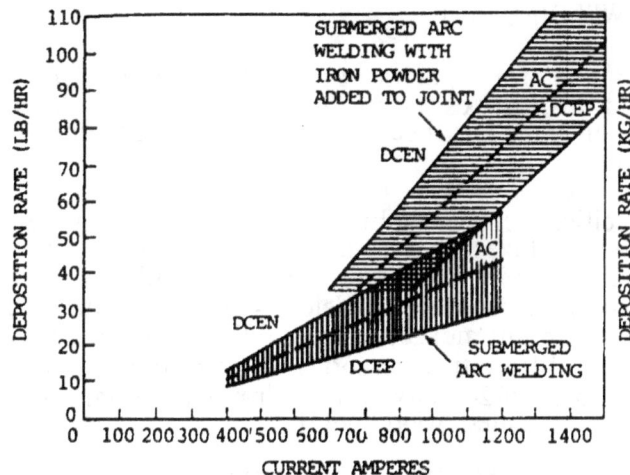

Figure 10-70. Welding with iron powder additives.

(g) Another variation is the use of an electrically "cold" filler wire fed into the arc area. The "cold" filler rod can be solid or flux-cored to add special alloys to the weld metal. By regulating the addition of the proper material, the properties of the deposited weld metal can be improved. It is possible to utilize a flux-cored wire for the electrode, or for one of the multiple electrodes to introduce special alloys into the weld metal deposit. Each of these variations requires special engineering to ensure that the proper material is added to provide the desired deposit properties.

(11) Typical applications. The submerged arc welding process is widely used in the manufacture of most heavy steel products. These include pressure vessels, boilers, tanks, nuclear reactors, chemical vessels, etc. Another use is in the fabrication of trusses and beams. It is used for welding flanges to the web. The heavy equipment industry is a major user of submerged arc welding.

f. Materials Used.

(1) Two materials are used in submerged arc welding: the welding flux and the consumable electrode wire.

(2) Submerged arc welding flux shields the arc and the molten weld metal from the harmful effects of atmospheric oxygen and nitrogen. The flux contains deoxidizers and scavengers which help to remove impurities from the molten weld metal. Flux also provides a means of introducing alloys into the weld metal. As this molten flux cools to a glassy slag, it forms a covering which protects the surface of the weld. The unmelted portion of the flux does not change its form and its properties are not affected, so it can be recovered and reused. The flux that does melt and forms the slag covering must be removed from the weld bead. This is easily done after the weld has cooled. In many cases, the slag will actually peel without requiring special effort for removal. In groove welds, the solidified slag may have to be removed by a chipping hammer.

10-14.   SURMERGED ARC WELDING (cont)

(3) Fluxes are designed for specific applications and for specific types of weld deposits. Submerged arc fluxes come in different particle sizes. Many fluxes are not marked for size of particles because the size is designed and produced for the intended application.

(4) There is no specification for submerged arc fluxes in use in North America. A method of classifying fluxes, however, is by means of the deposited weld metal produced by various combinations of electrodes and proprietary submerged arc fluxes. This is covered by the American Welding Society Standard. Bare carbon steel electrodes and fluxes for submerged arc welding. In this way, fluxes can be designated to be used with different electrodes to provide the deposited weld metal analysis that is desired. Table 10-24 shows the flux wire combination and the mechanical properties of the deposited weld metal.

Section III. RELATED PROCESSES

10-15.   PLASMA ARC CUTTING (PAC)

a.   General.   The plasma arc cutting process cuts metal by melting a section of metal with a constricted arc. A high velocity jet flow of hot ionized gas melts the metal and then removes the molten material to form a kerf. The basic arrangement for a plasma arc cutting torch similar to the plasma arc welding torch, is shown in figure 10-71. Three variations of the process exist: low current plasma cutting, high current plasma cutting, and cutting with water added. Low current arc cutting, which produces high-quality cuts of thin materials, uses a maximum of 100 amperes and a much smaller torch than the high current version. Modifications of processes and equipment have been developed to permit use of oxygen in the orifice gas to allow efficient cutting of steel. All plasma torches constrict the arc by passing it through an orifice as it travels away from the electrode toward the workpiece. As the orifice gas passes through the arc, it is heated rapidly to a high temperature, expands and accelerates as it passes through the constricting orifice. The intensity and velocity of the arc plasma gas are determined by such variables as the type of orifice gas and its entrance pressure, constricting orifice shape and diameter, and the plasma energy density on the work.

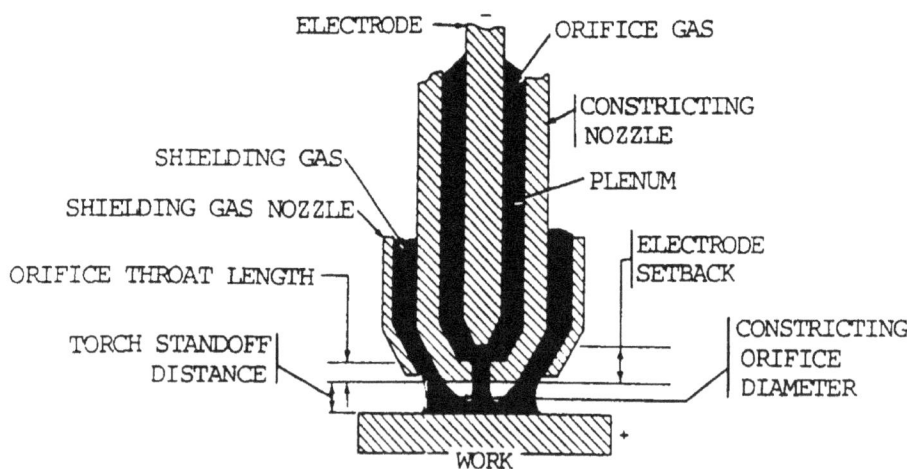

Figure 10-71. Plasma arc torch terminology.

Table 10-24. Typical Analysis and Mechanical Properties of Submerged Arc Flux-Wire Combinations

Wire/Flux Classification	Typical Deposit Chemistry					Typical Mechanical Properties				Charpy V-Notch Impact Value	
	C	Mn	Si	P	S	Tensile Strength psi	Yield Strength psi	Elong. % in. 2"	Reduction of area %	Ft-lb	°F
EL12	0.09	0.50	0.01	0.020	0.025						
F60-EL12	0.06	0.70	0.75	0.025	0.020	70,300	60,100	27.0	48.0	--	
F63-EL12	0.04	1.18	0.48	0.036	0.011	72,000	56,000	35.0	67.0	30	-40
EH14	0.14	1.85	0.04	0.010	0.018						
F62-EH14	0.08	1.05	0.55	0.020	0.016	72,000	58,000	29.0	58.0	30	-20
F72-EH14	0.08	1.80	0.65	0.016	0.018	88,000	73,000	28.0	56.0	24	-20
F64-EH14	0.12	1.17	0.24	0.022	0.021	71,000	57,000	31.0	59.1	24	-60
EM15K	0.15	1.10	0.25	0.022	0.025						
F70-EM15K	0.09	0.93	0.94	0.027	0.022	81,700	67,000	30.0	61.0	--	
F72-EM15K	0.08	1.54	0.79	0.025	0.021	82,000	58,500	30.0	59.0	26	-20
F64-EM15K	0.11	0.78	0.30	0.022	0.025	70,000	55,000	29.5	56.5	21	-60
(same as above)	0.13	1.95	0.04	0.010	Mo.53						
Weld stress	0.07	1.95	0.70	0.020	Mo.35	99,250	84,000	25.0	57.0	23	-20
relieved	0.08	1.17	0.23	0.017	Mo.38	80,000	65,500	27.0	66.2	22	-60
EM13K	0.11	1.20	0.50	0.020	0.019						
F70-EM13K	0.09	1.74	1.17	0.017	0.026	86,000	66,500	26.0	53.2	--	
F64-EM13K	0.10	0.90	0.54	0.016	0.020	70,500	54,000	31.0	62.8	29	-60

10-15. PLASMA ARC CUTTING (PAC) (cont)

b. Equipment. Plasma arc cutting requires a torch, a control unit, a power supply, one or more cutting gases, and a supply of clean cooling water. Equipment is available for both manual and mechanized PAC.

(1) Cutting torch. A cutting torch consists of an electrode holder which centers the electrode tip with respect to the orifice in the constricting nozzle. The electrode and nozzle are water cooled to prolong their lives. Plasma gas is injected into the torch around the electrode and exits through the nozzle orifice. Nozzles with various orifice diameters are available for each type of torch. Orifice diameter depends on the cutting current. Larger diameters are required at higher currents. Nozzle design depends on the type of PAC and the metal being cut. Both single and multiple port nozzles may be used for PAC. Multiple port nozzles have auxiliary gas ports arranged in a circle around the main orifice. All of the arc plasma passes through the main orifice with a high gas flew rate per unit area. These nozzles produce better quality cuts than single port nozzles at equivalent travel speeds. However, cut quality decreases with increasing travel speed. Torch designs for introducing shielding gas or water around the plasma flame are available. PAC torches are similar in appearance to gas tungsten arc welding electrode holders, both manual and machine types. Mechanized PAC torches are mounted on shape cutting machines similar to mechanized oxyfuel gas shape cutting equipment. Cutting may be controlled by photoelectric tracing, numerical control, or computer.

(2) Controls. Control consoles for PAC may contain solenoid valves to turn gases and cooling water on and off. They usually have flowmeters for the various types of cutting gases used and a water flow switch to stop the operation if cooling water flow falls below a safe limit. Controls for high-power automatic PAC may also contain programming features for upslope and downslope of current and orifice gas flow.

(3) Power sources. Power sources for PAC are specially designed units with open-circuit voltages in the range of 120 to 400 V. A power source is selected on the basis of the design of PAC torch to be used, the type and thickness of the work metal, and the cutting speed range. Their volt-ampere output characteristic must be the typical drooping type.

(a) Heavy cutting requires high open-circuit voltage (400 V) for capability of piercing material as thick as 2 in. (51 mm). Low current, manual cutting equipment uses lower open-circuit voltages (120 to 200 V). Some power sources have the connections necessary to change the open-circuit voltage as required for specific applications.

(b) The output current requirements range from about 70 to 1000 A depending on the material, its thickness, and cutting speed. The unit may also contain the pilot arc and high frequency power source circuitry.

(4) Gas selection.

(a) Cutting gas selection depends on the material being cut and the cut surface quality requirements. Most nonferrous metals are cut by using nitrogen, nitrogen-hydrogen mixtures, or argon-hydrogen mixtures. Titanium and zirconium are

(b) Carbon steels are cut by using compressed air (80 percent$_2$, N20 percent 0$_2$) or nitrogen for plasma gas. Nitrogen is used with the water injection method of PAC. Some systems use nitrogen for the plasma forming gas with oxygen injected into the plasma downstream of the electrode.This arrangement prolongs the life of the electrode by not exposing it to oxygen.

(c) For some nonferrous cutting with the dual flow system, nitrogen is used for the plasma gas with carbon dioxide (CO$_2$) for shielding. For better quality cuts, argon-hydrogen plasma gas and nitrogen shielding are used.

c. Principles of Operation.

(1) The basic plasma arc cutting circuitry is shown in figure 10-72. The process operates on direct current, straight polarity (dcsp), electrode negative, with a constricted transferred arc. In the transferred arc mode, an arc is struck between the electrode in the torch and the workpiece.The arc is initiated by a pilot arc between the electrode and the constricting nozzle. The nozzle is connected to ground (positive) through a current limiting resistor and a pilot arc relay contact. The pilot arc is initiated by a high frequency generator connected to the electrode and nozzle. The welding power supply then maintains this low current arc inside the torch. Ionized orifice gas from the pilot arc is blown through the constricting nozzle orifice. This forms a low resistance path to ignite the main arc between the electrode and the workpiece.When the main arc ignites, the pilot arc relay may be opened automatically to avoid unnecessary heating of the constricting nozzle.

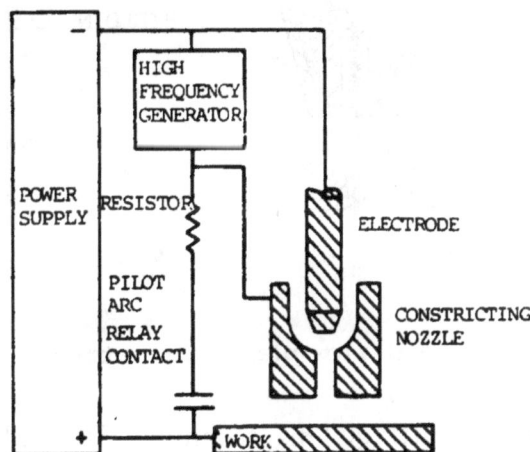

Figure 10-72. Basic plasma arc cutting circuitry.

10-15. PLASMA ARC CUTTING (PAC) (cont)

cut with pure argon because of their susceptibility to embrittlement by reactive gases.

(2) Because the plasma constricting nozzle is exposed to the high plasma flare temperatures (estimated at 18,032 to 25,232 °F (10,000 to 14,000° C)), the nozzle must be made of water-cooled copper. In addition, the torch should be designed to produce a boundary layer of gas between the plasma and the nozzle.

(3) Several process variations are used to improve the PAC quality for particular applications. They are generally applicable to materials in the 1/8 to 1-1/2 in. (3 to 38 mm) thickness range. Auxiliary shielding, in the form of gas or water, is used to improve cutting quality.

(a) Dual flow plasma cutting. Dual flow plasma cutting provides a secondary gas blanket around the arc plasma, as shown in figure 10-73. The usual orifice gas is nitrogen. The shielding gas is selected for the material to be cut. For mild steel, it may be carbon dioxide (CO_2) or air; for stainless steals, CO_2 and an argon-hydrogen mixture for aluminum. For mild steel, sutting speeds are slightly faster than with conventional PAC, but the cut quality is not satisfactory for many applications.

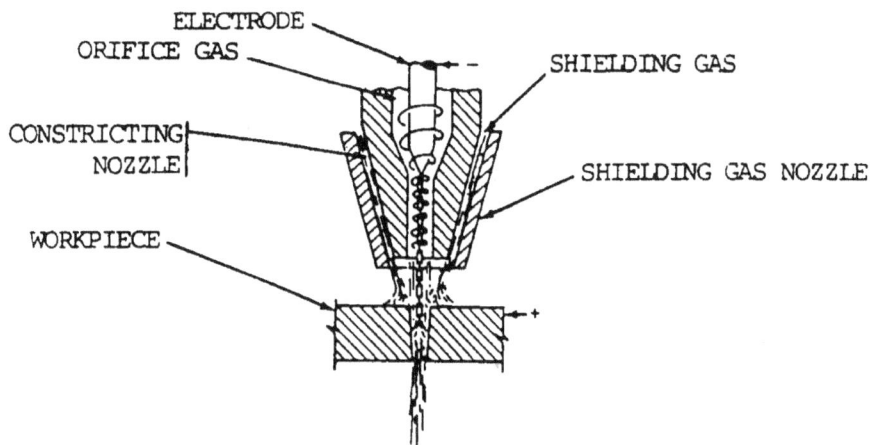

Figure 10-73. Dual flow plasma arc cutting.

(b) Water shield plasma cutting. This technique is similar to dual flow plasma cutting. Water is used in place of the auxiliary shielding gas. Cut appearance and nozzle life are improved by the use of water in place of gas for auxiliary shielding. Cut squareness and cutting speed are not significantly improved over conventional PAC.

(c) Water injection plasma cutting. This modification of the PAC process uses a symmetrical impinging water jet near the constricting nozzle orifice to further constrict the plasma flame. The arrangement is shown in figure 10-74. The water jet also shields the plasma from mixing with the surrounding atmosphere. The end of the nozzle can be made of ceramic,which helps to prevent double arcing. The water constricted plasma produces a narrow, sharply defined cut at speeds above those of conventional PAC. Because most of the water leaves the nozzle as a liquid spray, it cools the kerf edge, producing a sharp corner. The kerf is clean. When the orifice gas and water are injected in tangent, the plasma gas swirls as it emerges from the nozzle and water jet. This can produce a high quality perpendicular face on one side of the kerf. The other side of the kerf is beveled. In shape cutting applications, the direction of travel must be selected to produce a perpendicular cut on the part and the bevel cut on the scrap.

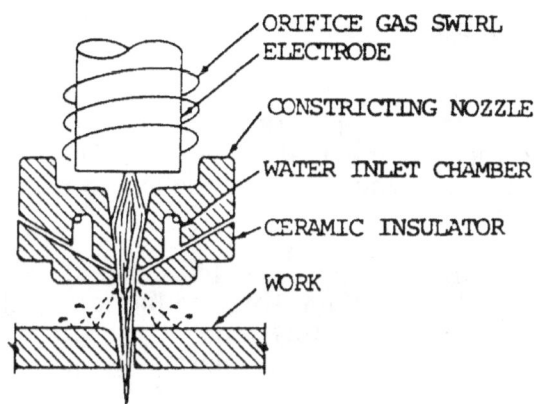

Figure 10-74. Water injection plasma arc arrangement.

(4) For high current cutting, the torch is mounted on a mechanical carriage. Automatic shape cutting can be done with the same equipment used for oxygen cutting, if sufficiently high travel speed is attainable. A water spray is used surrounding the plasma to reduce smoke and noise.Work tables containing water which is in contact with the underside of the metal being cut will also reduce noise and smoke.

(5) The plasma arc cutting torch can be used in all positions. It can also be used for piercing holes and for gouging. The cutting torch is of special design for cutting and is not used for welding.

(6) The metals usually cut with this process are the aluminums and stainless steels. The process can also be used for cutting carbon steels, copper alloys, and nickel alloys.

(7) Special controls are required to adjust both plasma and secondary gas flow. Torch-cooling water is required and is monitored by pressure or flow switches for torch protection. The cooling system should be self-contained, which includes a circulating pump and a heat exchanger.

10-15.   PLASMA ARC CUTTING (PAC) (cont)

(8) Plasma cutting torches will fit torch holders in automatic flame cutting machines.

(9) The amount of gases and tines generated requires the use of local exhaust for proper ventilation. Cutting should be done over a water reservoir so that the particles removed from the cut will fall in the water. This will help reduce the amount of fumes released into the air.

d.   Applications.   Plasma arc cutting can be used to cut any metal. Most applications are for carbon steel, aluminum, and stainless steel. It can be used for stack cutting, plate beveling, shape cutting, and piercing.

WARNING

Ear protection must be worn when working with high-powered equipment.

(1) The noise level generated by the high-powered equipment is uncomfortable. The cutter must wear ear protection. The normal protective clothing to protect the cutter from the arc must also be worn. This involves protective clothing, gloves, and helmet. The helmet should be equipped with a shade no. 9 filter glass lens.

(2) There are many applications for low-current plasma arc cutting, including the cutting of stainless and aluminum for production and maintenance. Plasma cutting can also be used for stack cutting and it is more efficient than stack cutting with the oxyacetylene torch. Low current plasma gouging can also be used for upgrading castings.

10-16.   AIR CARBON ARC CUTTING (AAC)

a.   General.   Air carbon arc cutting is an arc cutting process in which metals to be cut are melted by the heat of a carbon arc. The molten metal is removed by a blast of air. This is a method for cutting or removing metal by melting it with an electric arc and then blowing away the molten metal with a high velocity jet of compressed air. The air jet is external to the consumable carbon-graphite electrode. It strikes the molten metal immediately behind the arc. Air carbon arc cutting and metal removal differ from plasma arc cutting in that they employ an open (unconstricted) arc, which is independent of the gas jet. The air carbon arc process is shown in figure 10-75.

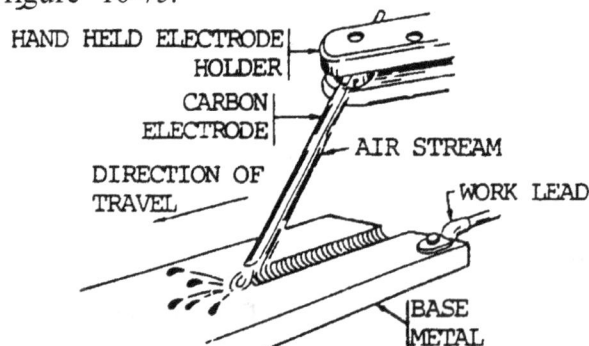

Figure 10-75. Process diagram for air carbon arc cutting.

b. Equipment.

(1) The circuit diagram for air carbon arc cutting or gouging is shown by figure 10-76. Normally, conventional welding machines with constant current are used. Constant voltage can be used with this process. When using a CV power source, precautions must be taken to operate it within its rated output of current and duty cycle. Alternating current power sources having conventional drooping characteristics can also be used for special applications. AC type carbon electrodes must be used.

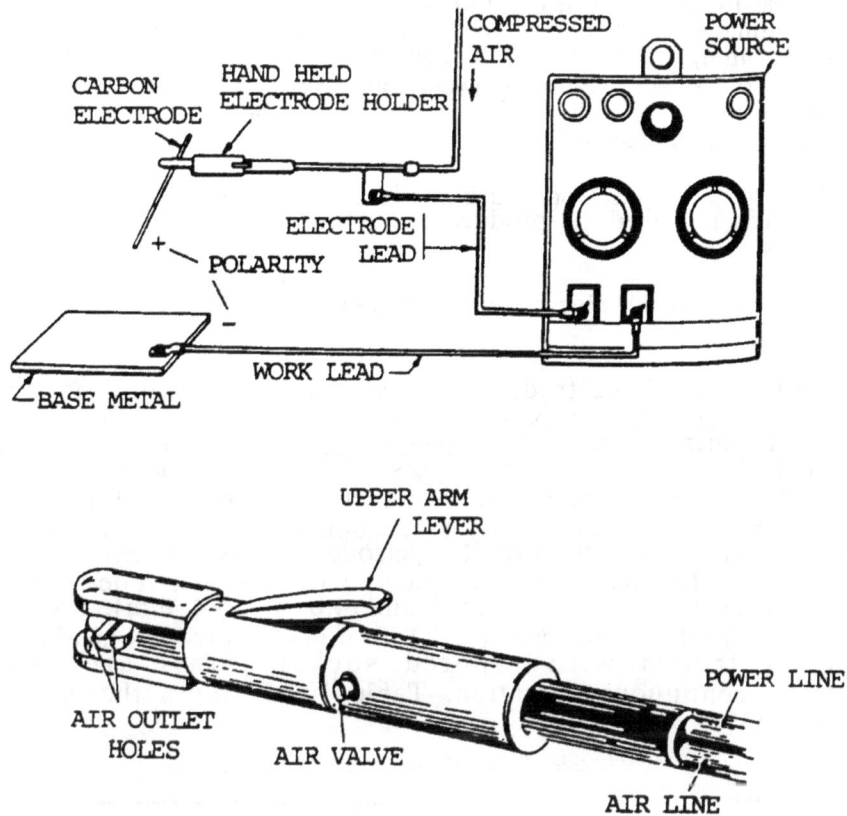

Figure 10-76. Air carbon arc cutting diagram.

(2) Equipment required is shown by the block diagram. Special heavy duty high current machines have been made specifically for the air carbon arc process. This is because of extremely high currents used for the large size carbon electrodes.

(3) The electrode holder is designed for the air carbon arc process. The holder includes a small circular grip head which contains the air jets to direct compressed air along the electrode. It also has a groove for gripping the electrode. This head can be rotated to allow different angles of electrode with respect to the holder. A heavy electrical lead and an air supply hose are connected

10-16. AIR CARBON ARC CUTTING (AAC) (cont)

to the holder through a terminal block. A valve is included in th holder for turning the compressed air on and off. Holders are available in several sizes depending on the duty cycle of the work performed, the welding current, and size of carbon electrode used. For extra heavy duty work, water-cooled holders are used.

(4) The air pressure is not critical but should range from 80 to 100 psi (552 to 690 kPa). The volume of compressed air required ranges from as low as 5 cu ft per min (2.5 liter per min) up to 50 cu ft per min (24 liter per min) for the largest-size carbon electrodes. A one-horsepower compressor will supply sufficient air for smaller-size electrodes. It will require up to a ten-horsepower compressor when using the largest-size electrodes.

(5) The carbon graphite electrodes are made of a mixture of carbon and graphite plus a binder which is baked to produce a homogeneous structure. Electrodes come in several types.

(a) The plain uncoated electrode is less expensive, carries less current, and starts easier.

(b) The copper-coated electrode provides better electrical conductivity between it and the holder. The copper-coated electrode is better for maintaining the original diameter during operation. It lasts longer and carries higher current. Copper-coated electrodes are of two types, the dc type and the ac type. The composition ratio of the carbon and graphite is slightly different for these two types. The dc type is more common. The ac type contains special elements to stabilize the arc. It is used for direct current electrode negative when cutting cast irons. For normal use, the electrode is operated with the electrode positive. Electrodes range in diameter from 5/32 to 1 in. (4.0 to 25.4 mm). Electrodes are normally 12 in. (300 mm) long; however, 6 in. (150 mm) electrodes are available Copper-coated electrodes with tapered socket joints are available operation, and allow continuous operation. Table 10-25 shows the electrode types and the arc current range for different sizes.

Table 10-25. Electrode Type—Size and Current Range

Electrode Type	Electrode Size in.	mm	Current Min	Max.
	5/32	4.0	90	150
	3/16	4.8	150	200
DC	1/4	6.4	200	400
(Plain)	5/16	7.9	250	450
or	3/8	9.5	350	600
AC	1/2	12.7	600	1000
(Copper	5/8	15.9	800	1200
Covered)	3/4	19.1	1200	1600
	1	25.4	1800	2200

Polarity of electrode is positive (reverse polarity).

Note: For DC copper covered electrodes current can be increased percent.

c. Advantages and Major Uses.

(1) The air carbon arc cutting process is used to cut metal, to gouge out defective metal, to remove old or inferior welds, for root gouging of full penetration welds, and to prepare grooves for welding. Air carbon arc cutting is used when slightly ragged edges are not objectionable. The area of the cut is small and, since the metal is melted and removed quickly, the surrounding area does not reach high temperatures. This reduces the tendency towards distortion and cracking.

(2) The air carbon arc cutting and gouging process is normally manually operated. The apparatus can be mounted on a travel carriage. This is considered machine cutting or gouging. Special applications have been made where cylindrical work has been placed on a lathe-like device and rotated under the air carbon arc torch. This is machine or automatic cutting, depending on operator involvement.

(3) The air carbon arc cutting process can be used in all positions. It can also be used for gouging in all positions. Use in the overhead position requires a high degree of skill.

(4) The air carbon arc process can be used for cutting or gouging most of the common metals. Metals include: aluminums, copper, iron, magnesium, and carbon and stainless steels.

(5) The process is not recommended for weld preparation for stainless steel, titanium, zirconium, and other similar metals without subsequent cleaning. This cleaning, usually by grinding, must remove all of the surface carbonized material adjacent to the cut. The process can be used to cut these materials for scrap for remelting.

d. Process Principles.

(1) The procedure schedule for making grooves in steel is shown in table 10-26, p 10-112.

(2) To make a cut or a gouging operation, the cutter strikes an arc and almost immediately starts the air flow. The electrode is pointed in the direction of travel with a push angle approximately 45° with the axis of the groove. The speed of travel, the electrode angle, and the electrode size and current determine the groove depth. Electrode diameter determines the groove width.

(3) The normal safety precautions similar to carbon arc welding and shielded metal arc welding apply to air carbon arc cutting and gouging. However, two other precautions must be observed. First, the air blast will cause the molten metal to travel a very long distance. Metal deflection plates should be placed in front of the gouging operation. All combustible materials should be moved away from the work area. At high-current levels, the mass of molten metal removed is quite large and will become a fire hazard if not properly contained.

(4) The second factor is the high noise level. At high currents with high air pressure a very loud noise occurs. Ear protection, ear muffs or ear plugs should be worn by the arc cutter.

10-16. AIR CARBON ARC CUTTING (AAC) (cont)

(5) The process is widely used for back gouging, preparing joints, and removing defective weld metal.

Table 10-26. Air Carbon Arc Gouging Procedure Schedule

Groove Width in.	mm	Groove Depth in.	mm	Electrode Dia. in.	mm	Amperes Direct Current	Volts Electrode Positive	Electrode Feed ipm	mm/min.	Travel Speed ipm	mm/min.
1/4	6.4	1/16	1.6	3/16	4.8	200	43	6.2	157.4	82.0	2028.8
9/32	7.1	1/8	3.2	3/16	4.8	200	40	6.7	170.2	38.2	970.3
5/16	7.9	3/16	4.8	3/16	4.8	190	42	6.7	170.2	27.2	690.9
5/16	7.9	1/4	6.4	3/16	4.8	(To make 1/4 in. (64 mm) deep groove, make two 1/8 in. (32 mm) deep passes.)					
5/16	7.9	3/32	2.4	1/4	6.4	270	40	4.0	101.6	54.0	1371.6
5/16	7.9	1/8	3.2	1/4	6.4	300	42	4.0	101.6	51.0	1295.4
5/16	7.9	3/16	4.8	1/4	6.4	300	40	6.7	170.2	38.2	970.3
5/16	7.9	1/4	6.4	1/4	6.4	320	42	6.2	157.4	29.5	749.3
5/16	7.9	3/8	9.5	1/4	6.4	320	46	3.6	91.4	15.0	381.0
3/8	9.5	1/8	3.2	5/16	7.9	320	40	3.0	76.2	65.5	1663.7
3/8	9.5	3/16	4.8	5/16	7.9	400	46	4.3	109.2	46.0	1168.4
3/8	9.5	1/4	6.4	5/16	7.9	420	42	3.8	96.5	31.2	792.5
3/8	9.5	1/2	12.7	5/16	7.9	540	42	5.6	142.2	27.2	690.9
7/16	11.1	1/8	3.2	3/8	9.5	560	42	4.2	106.7	82.0	2082.8
7/16	11.1	1/8	3.2	3/8	9.5	560	42	3.3	83.8	65.0	1651.0
7/16	11.1	3/16	4.8	3/8	9.5	560	42	2.6	66.0	41.0	1041.4
7/16	11.1	1/4	6.4	3/8	9.5	560	42	3.0	76.2	29.5	749.3
7/16	11.1	1/2	12.7	3/8	9.5	560	42	3.2	81.3	15.0	381.0
7/16	11.1	11/16	17.5	3/8	9.5	560	42	3.5	88.9	12.2	309.9
9/16	14.3	1/8	3.2	1/2	12.7	1200	45	3.0	76.2	34.0	863.6
9/16	14.3	1/4	6.4	1/2	12.7	1200	45	3.0	76.2	22.0	558.8
9/16	14.3	3/8	9.5	1/2	12.7	1200	45	3.0	76.2	20.7	525.8
9/16	14.3	1/2	12.7	1/2	12.7	1200	45	3.0	76.2	18.5	469.9
9/16	14.3	5/8	15.9	1/2	12.7	1200	45	3.0	76.2	15.0	381.0
9/16	14.3	3/4	19.1	1/2	12.7	1200	45	3.0	76.2	12.5	317.5
13/16	20.6	1/8	3.2	5/8	15.9	1300	42	2.5	63.5	44.5	1130.3
13/16	20.6	1/4	6.4	5/8	15.9	1300	42	2.5	63.5	29.5	749.3
13/16	20.6	3/8	9.5	5/8	15.9	1300	42	2.5	63.5	20.0	508.0
13/16	20.6	1/2	12.7	5/8	15.9	1300	42	2.5	63.5	14.5	368.3
13/16	20.6	5/8	15.9	5/8	15.9	1300	42	2.5	63.5	13.0	330.2
13/16	20.6	3/4	19.1	5/8	15.9	1300	42	2.5	63.5	11.0	279.4
13/16	20.6	1	25.4	5/8	15.9	1300	42	2.5	63.5	10.0	254.0

NOTES: 1 Air pressures 80 to 100 psi (552 to 690 kPa) is recommended for 1/2 and 5/8 in. (13 and 16 mm) electrodes.
2 Combination of settings and multiple passes may be used for grooves deeper than 3/4 in. (19 mm).

10-17.   RESISTANCE WELDING

a.   General. Resistance welding is a group of welding processes in which coalescence is produced by the heat obtained from resistance of the work to electric current in a circuit of which the work is a part and by the application of pressure. There are at least seven important resistance-welding processes. These are flash welding, high frequency resistance welding, percussion welding, projection welding, resistance seam welding, resistance spot welding, and upset welding.

b.   Principles of the Process.

(1) The resistance welding processes differ from all those previously mentioned. Filler metal is rarely used and fluxes are not employed. Three factors involved in making a resistance weld are the amount of current that passes through the work, the pressure that the electrodes transfer to the work, and the time the current flows through the work. Heat is generated by the passage of electrical current through a resistance circuit. The force applied before, during, and after the current flow forces the heated parts together so that coalescence will occur. Pressure is required throughout the entire welding cycle to assure a continuous electrical circuit through the work.

(2) This concept of resistance welding is most easily understood by relating it to resistance spot welding. Resistance spot welding, the most popular, is shown by figure 10-77. High current at a low voltage flows through the circuit and is in accordance with Ohm's law,

$$I = \frac{E}{R} \text{ or } R = \frac{E}{I}$$

(a) I is the current in amperes, E is the voltage in volts, and R is the resistance of the material in ohms. The total energy is expressed by the formula: Energy equals I x E x T in which T is the time in seconds during which current flows in the circuit.

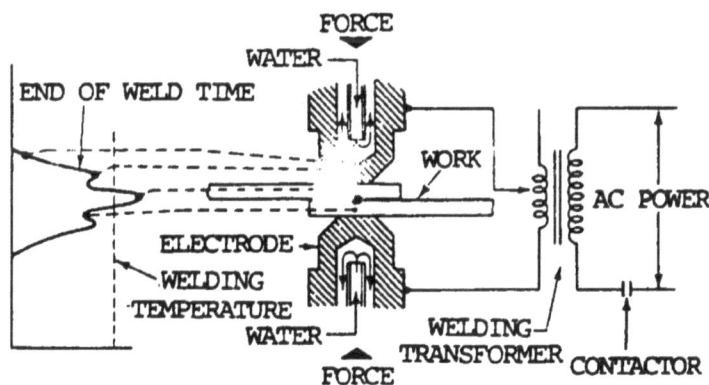

Figure 10-77. Resistance spot welding process.

10-17.   RESISTANCE WELDING (cont)

(b) Combining these two equations gives H (heat energy) $= I^2 \times R \times T$. For practical reasons a factor which relates to heat losses should be included; therefore, the actual resistance welding formula is

$$H(\text{heat energy}) = I^2 \times R \times T \times K$$

(c) In this formula, I = current squared in amperes, R is the resistance of the work in ohms, T is the time of current flow in seconds, and K represents the heat losses through radiation and conduction.

(3) Welding heat is proportional to the square of the welding current. If the current is doubled, the heat generated is quadrupled. Welding heat is proportional to the total time of current flow, thus, if current is doubled, the time can be reduced considerably. The welding heat generated is directly proportional to the resistance and is related to the material being welded and the pressure applied. The heat losses should be held to a minimum. It is an advantage to shorten welding tire. Mechanical pressure which forces the parts together helps refine the grain structure of the weld.

(4) Heat is also generated at the contact between the welding electrodes and the work. This amount of heat generated is lower since the resistance between high conductivity electrode material and the normally employed mild steel is less than that between two pieces of mild steel. In most applications, the electrodes are water cooled to minimize the heat generated between the electrode and the work.

(5) Resistance welds are made very quickly; however, each process has its own time cycle.

(6) Resistance welding operations are automatic. The pressure is applied by mechanical, hydraulic, or pneumatic systems. Motion, when it is involved, is applied mechanically. Current control is completely automatic once the welding operator initiates the weld. Resistance welding equipment utilizes programmers for controlling current, time cycles, pressure, and movement. Welding programs for resistance welding can become quite complex. In view of this, quality welds do not depend on welding operator skill but more on the proper set up and adjustment of the equipment and adherence to weld schedules.

(7) Resistance welding is used primarily in the mass production industries where long production runs and consistent conditions can be maintained. Welding is performed with operators who normally load and unload the welding machine and operate the switch for initiating the weld operation. The automotive industry is the major user of the resistance welding processes, followed by the appliance industry. Resistance welding is used by many industries manufacturing a variety of products made of thinner guage metals. Resistance welding is also used in the steel industry for manufacturing pipe, tubing and smaller structural sections. Resistance welding has the advantage of producing a high volume of work at high speeds and does not require filler materials. Resistance welds are reproducible and high-quality welds are normal.

(8) The position of making resistance welds is not a factor, particularly in the welding of thinner material.

c. Weldable Metals.

Table 10-27. Base Metals Weldable by the Resistance Welding Process

Base Metal	Weldability
Aluminums	Weldable
Magnesium	Weldable
Inconel	Weldable
Nickel	Weldable
Nickel silver	Weldable
Monel	Weldable
Precious metals	Weldable
Low carbon steel	Weldable
Low alloy steel	Weldable
High and medium carbon	Possible but not popular
Alloys steel	Possible but not popular
Stainless steel	Weldable

(1) Metals that are weldable, the thicknesses that can be welded, and joint design are related to specific resistance welding processes. Most of the common metals can be welded by many of the resistance welding processes (see table 10-27). Difficulties may be encountered when welding certain metals in thicker sections. Some metals require heat treatment after welding for satisfactory mechanical properties.

(2) Weld ability is controlled by three factors: resistivity, thermal conductivity, and melting temperature.

(3) Metals with a high resistance to current flow and with a low thermal conductivity and a relatively low melting temperature would be easily weldable. Ferrous metals all fall into this category. Metals that have a lower resistivity but a higher thermal conductivity will be slightly more difficult to weld. This includes the light metals, aluminum and magnesium. The precious metals comprise the third group. These are difficult to weld because of very high thermal conductivity. The fourth group is the refractory metals, which have extremely high melting points and are more difficult to weld.

(4) These three properties can be combined into a formula which will provide an indication of the ease of welding a metal. This formula is:

$$W = \frac{R}{FKt} \times 100$$

In this formula, W equals weldability, R is resistivity, and F is the melting temperature of the metal in degrees C, and Kt is the relative thermal conductivity with copper equal to 1.00. If weldability (W) is below 0.25, it is a poor rating.

10-17. RESISTANCE WELDING (cont)

If W is between 0.25 and 0.75, weldability becomes fair. Between 0.75 and 2.0, weldability is good. Above 2.0 weldability is excellent. In this formula, mild steel would have a weldability rating of over 10. Aluminum has a weldability factor of from 1 to 2 depending on the alloy and these are considered having a good weldability rating. Copper and certain brasses have a low weldability factor and are known to be very difficult to weld.

10-18. FLASH WELDING (FW)

    a. General.

       (1) Flash welding is a resistance welding process which produces coalescence simultaneously over the entire area of abutting surfaces by the heat obtained from resistance to electric current between the two surfaces, and by the application of pressure after heating is substantially completed. Flashing and upsetting are accompanied by expulsion of metal from the joint. This is shown by figure 10-78. During the welding operation, there is an intense flashing arc and heating of the metal on the surfaces abutting each other. After a predetermined time, the two pieces are forced together and joining occurs at the interface. Current flow is possible because of the light contact between the two parts being flash welded. Heat is generated by the flashing and is localized in the area between the two parts. The surfaces are brought to the melting point and expelled through the abutting area. As soon as this material is flashed away, another small arc is formed which continues until the entire abutting surfaces are at the melting temperrature. Pressure is then applied. The arcs are extinguished and upsetting occurs.

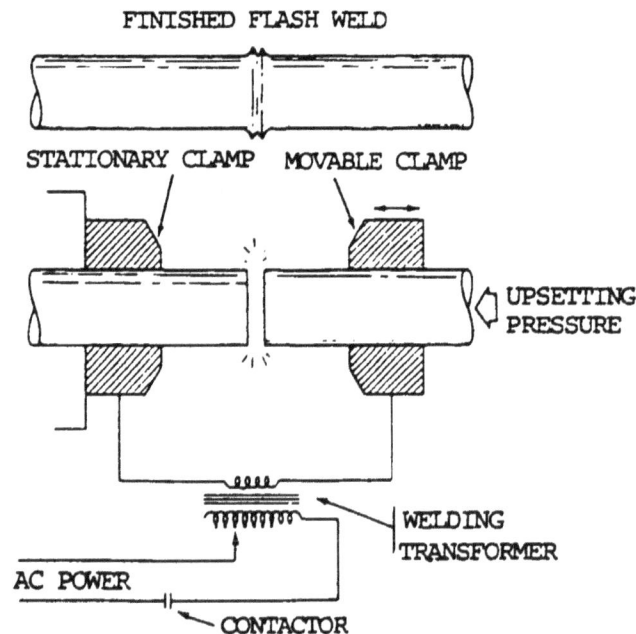

Figure 10-78. Flash welding.

(2) Flash welding can be used on most metals. No special preparation is required except that heavy scale, rust, and grease must be removed. The joints must be cut square to provide an even flash across the entire surface. The material to be welded is clamped in the jaws of the flash welding machine with a high clamping pressure. The upset pressure for steel exceeds 10,000 psi (68, 950 kPa). For high-strength materials, these pressures may be doubled. For tubing or hollow members, the pressures are reduced. As the weld area is more compact, upset pressures are increased. If insufficient upset pressure is used, a porous low strength weld will result. Excess upset pressure will result in expelling too much weld metal and upsetting cold metal. The weld may not be uniform across the entire cross section, and fatigue and impact strength will be reduced. The speed of upset, or the time between the end of flashing period and the end of the upset period, should be extremely short to minimize oxidation of the molten surfaces. In the flash welding operation, a certain amount of material is flashed or burned away. The distance between the jaws after welding compared to the distance before welding is known as the burnoff. It can be from 1/8 in. (3.2 mm) for thin material up to several inches for heavy material. Welding currents are high and are related to the following: 50 kva per square in. cross section at 8 seconds. It is desirable to use the lowest flashing voltage at a desired flashing speed. The lowest voltage is normally 2 to 5 volts per square in. of cross section of the weld.

(3) The upsetting force is usually accomplished by means of mechanical cam action. The design of the cam is related to the size of the parts being welded. Flash welding is completely automatic and is an excellent process for mass-produced parts. It requires a machine of large capacity designed specifically for the parts to be welded. Flash welds produce a fin around the periphery of the weld which is normally removed.

10-19.   FRICTION WELDING (FRW)

   a.   General.

   (1) Friction welding is a solid state welding process which produces coalescence of materials by the heat obtained from mechanically-induced sliding motion between rubbing surfaces. The work parts are held together under pressure. This process usually involves the rotating of one part against another to generate frictional heat at the junction. When a suitable high temperature has keen reached, rotational notion ceases. Additional pressure is applied and coalescence occurs.

   (2) There are two variations of the friction welding process. They are described below.

   (a) In the original process, one part is held stationary and the other part is rotated by a motor which maintains an essentially constant rotational speed. The two parts are brought in contact under pressure for a specified period of time with a specific pressure. Rotating power is disengaged from the rotating piece and the pressure is increased. When the rotating piece stops, the weld is completed. This process can be accurately controlled when speed, pressure, and time are closely regulated.

10-19. FRICTION WELDING (FRW) (cont)

(b) The other variation is inertia welding. A flywheel is revolved by a motor until a preset speed is reached. It, in turn, rotates one of the pieces to be welded. The motor is disengaged from the flywheel and the other part to be welded is brought in contact under pressure with the rotating piece. During the predetermined time during which the rotational speed of the part is reduced, the flywheel is brought to an immediate stop. Additional pressure is provided to complete the weld.

(c) Both methods utilize frictional heat and produce welds of similar quality. Slightly better control is claimed with the original process. The two methods are similar, offer the same welding advantages, and are shown by figure 10-79.

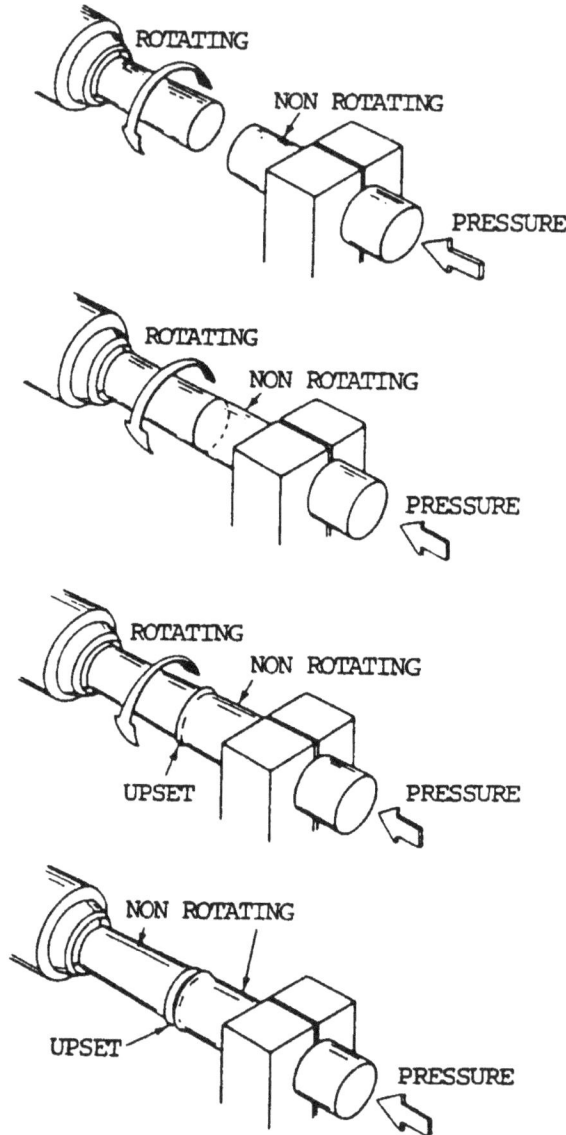

Figure 10-79. Friction welding process.

b. Advantages.

(1) Friction welding can produce high quality welds in a short cycle time.

(2) No filler metal is required and flux is not used.

(3) The process is capable of welding most of the common metals. It can also be used to join many combinations of dissimilar metals. Friction welding requires relatively expensive apparatus similar to a machine tool.

c. Process Principles.

(1) There are three important factors involved in making a friction weld:

(a) The rotational speed which is related to the material to be welded and the diameter of the weld at the interface.

(b) The pressure between the two parts to be welded. Pressure changes during the weld sequence. At the start, pressure is very low, but is increased to create the frictional heat. When the rotation is stopped, pressure is rapidly increased so forging takes place immediately before or after rotation is stopped.

(c) The welding time is related to the shape and the type of metal and the surface area. It is normally a matter of a few seconds. The actual operation of the machine is automatic. It is controlled by a sequence controller, which can be set according to the weld schedule established for the parts to be joined.

(2) Nomally for friction welding, one of the parts to be welded is round in cross section. This is not an absolute necessity. Visual inspection of weld quality can be based on the flash, which occurs around the outside perimeter of the weld. This flash will usually extend beyond the outside diameter of the parts and will curl around back toward the part but will have the joint extending beyond the outside diameter of the part.

(a) If the flash sticks out relatively straight from the joint, it indicates that the welding time was too short, the pressure was too low, or the speed was too high. These joints may crack.

(b) If the flash curls too far back on the outside diameter, it indicates that the the was too long and the pressure was too high.

(c) Between these extremes is the correct flash shape. The flash is normally remved after welding.

10-20. ELECTRON BEAM WELDING

a. General.

(1) Electron beam welding (EBW) is a welding process which produces coalescence of metals with heat from a concentrated beam of high velocity electrons striking the surfaces to be joined. Heat is generated in the workpiece as it is bomarded by a dense stream of high-velocity electrons. Virtually all of the kinetic energy, or the energy of motion, of the electrons is transformed into heat upon impact.

(2) Two basic designs of this process are: the low-voltage electron beam system, which uses accelerating voltages in 30,000-volt (30 kv) to 60,000-volt (60 kv) range, and the high voltage system with accelerating voltages in the 100,000-volt (100 kv) range. The higher voltage system emits more X-rays than the lower voltage system. In an X-ray tube, the beam of electrons is focused on a target of either tungsten or molybdenum which gives off X-rays. The target becomes extremely hot and must be water cooled. In welding, the target is the base metal which absorbs the heat to bring it to the molten stage. In electron beam welding, X-rays may be produced if the electrical potential is sufficiently high. In both systems, the electron gun and the workpiece are housed in a vacuum chamber. Figure 10-80 shows the principles of the electron beam welding process

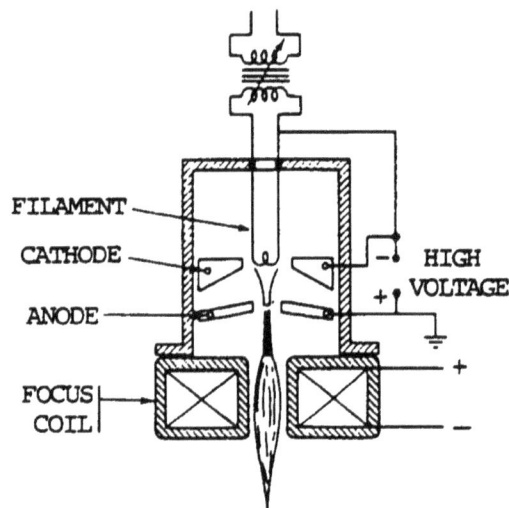

Figure 10-80. Electron beam welding process.

b. Equipment.

(1) There are three basic components in an electron beam welding machine. These are the electron beam gun, the power supply with controls, and a vacuum work chamber with work-handling equipment.

(2) The electron beam gun emits electrons, accelerates the beam of electrons, and focuses it on the workpiece. The electron beam gun is similar to that used in a television picture tube. The electrons are emitted by a heated cathode or filament and accelerated by an anode which is a positively-charged plate with a hole through which the electron beam passes. Magnetic focusing coils located beyond the anode focus and deflect the electron beam.

(3) In the electron beam welding machine, the electron beam is focused on the workpiece at the point of welding. The power supply furnishes both the filament current and the accelerating voltage. Both can be changed to provide different power input to the weld.

(4) The vacuum work chamber must be an absolutely airtight container. It is evacuated by means of mechanical pumps and diffusion pumps to reduce the pressure to a high vacuum. Work-handling equipment is required to move the workpiece under the electron beam and to manipulate it as required to make the weld. The travel mechanisms must be designed for vacuum installations since normal greases, lubricants, and certain insulating varnishes in electric rotors may volatilize in a vacuum. Heretically sealed motors and sealed gearboxes must be used. In some cases, the rotor and gearboxes are located outside the vacuum chamber with shafts operating through sealed bearings.

c. Advantages. One of the major advantages of electron beam welding is its tremendous penetration. This occurs when the highly accelerated electron hits the base metal. It will penetrate slightly below the surface and at that point release the bulk of its kinetic energy which turns to heat energy. The addition of the heat brings about a substantial temperature increase at the point of impact. The succession of electrons striking the same place causes melting and then evaporation of the base metal. This creates metal vapors but the electron beam travels through the vapor much easier than solid metal. This causes the beam to penetrate deeper into the base metal. The width of the penetration pattern is extremely narrow. The depth-to-width can exceed a ratio of 20 to 1. As the power density is increased, penetration is increased. Since the electron beam has tremendous penetrating characteristics, with the lower heat input, the heat affected zone is much smaller than that of any arc welding process. In addition, because of the almost parallel sides of the weld nugget, distortion is very greatly minimized. The cooling rate is much higher and for many metals this is advantageous; however, for high carbon steel this is a disadvantage and cracking may occur.

d. Process Principles.

(1) Recent advances in equipment allow the work chamber to operate at a medium vacuum or pressure. In this system, the vacuum in the work chamber is not as high. It is sometimes called a "soft" vacuum. This vacuum range allowed the same contamination that would be obtained in atmosphere of 99.995 percent argon. Mechanical pumps can produce vacuums to the medium pressure level.

(2) Electron beam welding was initially done in a vacuum because the electron beam is easily deflected by air. The electrons in the beam collide with the molecules of the air and lose velocity and direction so that welding can not be performed.

10-20. ELECTRON BEAM WELDING (cont)

(3) In a high vacuum system, the electron beam can be located as far as 30.0 in. (762.0 mm) away from the workpiece. In the medium vacuum, the working distance is reduced to 12.0 in. (304.8 mm). The thickness that can be welded in a high vacuum is up to 6.0 in. (152.4 mm) thick while in the medium vacuum the thickness that can be welded is reduced to 2.0 in. (50.8 mm). This is based on the same electron gun and power in both cases. With the medium vacuum, pump down time is reduced. The vacuum can be obtained by using mechanical pumps only. In the medium vacuum mode, the electron gun is in its own separate chamber separate from the work chamber by a small orifice through which the electron beam travels. A diffusion vacuum pump is run continuously connected to the chamber containing the electron gun, so that it will operate efficiently.

(4) The most recent development is the nonvacuum electron beam welding system. In this system, the work area is maintained at atmospheric pressure during welding. The electron beam gun is housed in a high vacuum chamber. There are several intermediate chambers between the gun and the atmospheric work area. Each of these intermediate stages is reduced in pressure by means of vacuum pumps. The electron beam passes from one chamber to another through a small orifice large enough for the electron beam but too small for a large volume of air. By means of these differential pressure chambers, a high vacuum is maintained in the electron beam gun chamber. The nonvacuum system can thus be used for the largest weldments, however the workpiece must be positional with 1-1/2 in. (38 mm) of the beam exit nozzle. The maximum thickness that can be welded currently is approximately 2 in. (51 mm). The nonvacuum system utilizes the high-voltage power supply.

(5) The heat input of electron beam welding is controlled by four variables:

(a) Number of electrons per second hitting the workpiece or beam current.

(b) Electron speed at the moment of impact, the accelerating potential.

(c) Diameter of the beam at or within the workpiece, the beam spot size.

(d) Speed of travel or the welding speed.

(6) The first two variables in (5), beam current and accelerating potential, are used in establishing welding parameters. The third factor, the beam spot size, is related to the focus of the beam, and the fourth factor is also part of the procedure. The electron beam current ranges from 250 to 1000 milliamperes, the beam currents can be as low as 25 milliamperes. The accelerating voltage is within the two ranges mentioned previously. Travel speeds can be extremely high and relate to the thickness of the base metal. The other parameter that must be controlled is the gun-to-work distance.

(7) The beam spot size can be varied by the location of the fecal point with respect to the surface of the part. Penetration can be increased by placing the fecal point below the surface of the base metal. As it is increased in depth below the surface, deeper penetration will result. When the beam is focused at the surface, there will be more reinforcement on the surface. When the beam is focused above the surface, there will be excessive reinforcement and the width of the weld will be greater.

(8) Penetration is also dependent on the beam current. As beam current is increased, penetration is increased. The other variable, travel speed, also affects penetration. As travel speed is increased, penetration is reduced.

(9) The heat input produced by electron beam welds is relatively small compared to the arc welding processes. The power in an electron beam weld compared with a gas metal arc weld would be in the same relative amount.The gas metal arc weld would require higher power to produce the same depth of penetration. The energy in joules per inch for the electron beam weld may be only 1/10 as great as the gas metal arc weld. The electron beam weld is equivalent to the SMAW weld with less power because of the penetration obtainable by electron beam welding. The power density is in the range of 100 to 10,000 kw/in

(10) The weld joint details for electron beam welding must be selected with care. In high vacuum chamber welding, special techniques must be used to properly align the electron beam with the joint. Welds are extremely narrow. Preparation for welding must be extremely accurate. The width of a weld in 1/2 in. (12.7 mm) thick stainless steel would only be 0.04 in. (1.00 mm). Small misalignment would cause the electron beam to completely miss the weld joint.Special optical systems are used which allow the operator to align the work with the electron beam. The electron beam is not visible in the vacuum.Welding joint details normally used with gas tungsten arc welding can be used with electron beam welding. The depth to width ratio allows for special lap type joints.Where joint fitup is not precise, ordinary lap joints are used and the weld is an arc seam type of weld. Normally, filler metal is not used in electron beam welding; however, when welding mild steel highly deoxidized filler metal is sometimes used. This helps deoxidize the molten metal and produce dense welds.

(11) In the case of the medium vacuum system, much larger work chambers can be used. Newer systems are available where the chamber is sealed around the part to be welded. In this case, it has to be designed specifically for the job at hand. The latest uses a sliding seal and a movable electron beam gun. In other versions of the medium vacuum system, parts can be brought into and taken out of the vacuum work chamber by means of interlocks so that the process can be made more or less continuous. The automotive industry is using this system for welding gear clusters and other small assemblies of completely machined parts. This can be done since the distortion is minimal.

(12) The non-vacuum system is finding acceptance for other applications. One of the most productive applications is the welding of automotive catalytic converters around the entire periphery of the converter.

(13) The electron beam process is becoming increasingly popular where the cost of equipment can be justified over the production of many parts.It is also used to a very great degree in the automatic energy industry for remote welding and for welding the refractory metals. Electron beam welding is not a cure-all; there are still the possibilities of defects of welds in this process as with any other. The major problem is the welding of plain carbon steel which tends to become porous when welded in a vacuum. The melting of the metal releases gases originally in the metal and results in a porous weld.If deoxidizers cannot be used, the process is not suitable.

10-20. ELECTRON BEAM WELDING (cont)

e. Weldable Metals.

Almost all metals can be welded with the electron beam welding process. The metals that are most often welded are the super alloys, the refractory metals, the reactive metals, and the stainless steels. Many combinations of dissimilar metals can also be welded.

10-21. LASER BEAM WELDING (LBW)

a. General.

(1) Laser beam welding (LBW) is a welding process which produces coalescence of materials with the heat obtained from the application of a concentrate coherent light beam impinging upon the surfaces to be joined.

(2) The focused laser beam has the highest energy concentration of any known source of energy. The laser beam is a source of electromagnetic energy or light that can be pro jetted without diverging and can be concentrated to a precise spot. The beam is coherent and of a single frequency.

(3) Gases can emit coherent radiation when contained in an optical resonant cavity. Gas lasers can be operated continuously but originally only at low levels of power. Later developments allowed the gases in the laser to be cooled so that it could be operated continuously at higher power outputs. The gas lasers are pumped by high radio frequency generators which raise the gas atoms to sufficiently high energy level to cause lasing. Currently, 2000-watt carbon dioxide laser systems are in use. Higher powered systems are also being used for experimental and developmental work. A 6-kw laser is being used for automotive welding applications and a 10-kw laser has been built for research purposes.There are other types of lasers; however, the continuous carbon dioxide laser now available with 100 watts to 10 kw of power seems the most promising for metalworking applications.

(4) The coherent light emitted by the laser can be focused and reflected in the same way as a light beam. The focused spot size is controlled by a choice of lenses and the distance from it to the base metal.The spot can be made as small as 0.003 in. (O. 076 mm) to large areas 10 times as big. A sharply focused spot is used for welding and for cutting. The large spot is used for heat treating.

(S) The laser offers a source of concentrated energy for welding; however, there are only a few lasers in actual production use today. The high-powered laser is extremely expensive. Laser welding technology is still in its infancy so there will be improvements and the cost of equipment will be reduced.Recent use of fiber optic techniques to carry the laser beam to the point of welding may greatly expand the use of lasers in metal-working.

b. Welding with Lasers.

(1) The laser can be compared to solar light beam for welding. It can be used in air. The laser beam can be focused and directed by special optical lenses and mirrors. It can operate at considerable distance from the workpiece.

(2) When using the laser beam for welding, the electromagnetic radiation impinges on the surface of the base metal with such a concentration of energy that the temperature of the surface is melted vapor and melts the metal below. One of the original questions concerning the use of the laser was the possibility of reflectivity of the metal so that the beam would be reflected rather than heat the base metal. It was found, however, that once the metal is raised to its melting temperature, the surface conditions have little or no effect.

(3) The distance from the optical cavity to the base metal has little effect on the laser. The laser beam is coherent and it diverges very little. It can be focused to the proper spot size at the work with the same amount of energy available, whether it is close or far away.

(4) With laser welding, the molten metal takes on a radial configuration similar to convectional arc welding. However, when the power density rises above a certain threshold level, keyholing occurs, as with plasma arc weldingKeyholing provides for extremely deep penetration. This provides for a high depth-to-width ratio. Keyholing also minimizes the problem of beam reflection from the shiny molten metal surface since the keyhole behaves like a black body and absorbs the majority of the energy. In some applications, inert gas is used to shield the molten metal from the atmosphere. The metal vapor that occurs may cause a break-down of the shielding gas and creates a plasma in the region of high-beam intensity just above the metal surface. The plasma absorbs energy from the laser beam and can actually block the beam and reduce melting. Use of an inert gas jet directed along the metal surface eliminates the plasma buildup and shields the surface from the atmosphere.

(5) The welding characteristics of the laser and of the electron beam are similar. The concentration of energy y both beams is similar with the laser having a power density in the order of f0watts per square centimeter. The power density of the electron beam is only slightly greater. This is compared to a current density of only 10watts per square centimeter for arc welding.

(6) Laser beam welding has a tremendous temperature differential between the molten metal and the base metal immediately adjacent to the weldHeating and cooling rates are much higher in laser beam welding than in arc welding, and the heat-affected zones are much smaller. Rapid cooling rates can create problems such as cracking in high carbon steels.

(7) Experimental work with the laser beam welding process indicates that the normal factors control the weld. Maximum penetration occurs when the beam is focused slightly below the surface. Penetration is less when the beam is focused on the surface or deep within the surface. As power is increased the depth of penetration is increased.

c. Weldable Metals. The laser beam has been used to weld carbon steels, high strength low alloy steels, aluminum, stainless steel, and titaniumLaser welds made in these materials are similar in quality to welds made in the same materials by electron beam process. Experimental work using filler metal is being used to weld metals that tend to show porosity when welded with either EB or LB welding. Materials 1/2 in. (12.7 mm) thick are being welded at a speed of 10.0 in. (254.0 mm) per minute.

CHAPTER 11

OXYGEN FUEL GAS WELDING PROCEDURES

Section I. WELDING PROCESSES AND TECHNIQUES

11-1. GENERAL GAS WELDING PROCDURES

a. General.

(1) Oxyfuel gas welding (OEW) is a group of welding processes which join metals by heating with a fuel gas flame or flares with or without the application of pressure and with or without the use of filler metal.OFW includes any welding operation that makes use of a fuel gas combined with oxygen as a heating medium. The process involves the melting of the base metal and a filler metal, if used, by means of the flame produced at the tip of a welding torch. Fuel gas and oxygen are mixed in the proper proportions in mixing chamber which may be part of the welding tip assembly. Molten metal from the plate edges and filler metal, if used, intermix in a common molten pool. Upon cooling, they coalesce to form a continuous piece.

(2) There are three major processes within this group: oxycetylene welding, oxyhydrogen welding, and pressure gas welding. There is one process of minor industrial significance, known as air acetylene welding, in which heat is obtained from the combustion of acetylene with air. Welding with methylacetone-propadiene gas (MAPP gas) is also an oxyfuel procedure.

b. Advantages.

(1) One advantage of this welding process is the control a welder can exercise over the rate of heat input, the temperature of the weld zone, and the oxidizing or reducing potential of the welding atmosphere.

(2) Weld bead size and shape and weld puddle viscosity are also controlled in the welding process because the filler metal is added independently of the welding heat source.

(3) OFW is ideally suited to the welding of thin sheet, tubes, and small diameter pipe. It is also used for repair welding. Thick section welds, except for repair work, are not economical.

c. Equipment.

(1) The equipment used in OFW is low in cost usually portable, and versatile enough to be used for a variety of related operations, such as bending and straightening, preheating, postheating, surface, braze welding, and torch brazing. With relatively simple changes in equipment, manual and mechanized oxygen cutting operations can be performed. Metals normally welded with the oxyfuel process include steels, especially low alloy steels, and most nonferrous metals. The process is generally not used for welding refractory or reactive metals.

11-1. GENERAL GAS WELDING PROCEDURES (cont)

d. Gases.

(1) Commercial fuel gases have one common property: they all require oxygen to support combustion. To be suitable for welding operations, a fuel gas, when burned with oxygen, must have the following:

(a) High flare temperature.

(b) High rate of flame propagation.

(c) Adequate heat content.

(d) Minimum chemical reaction of the flame with base and filler metals.

(2) Among the commercially available fuel gases, acetylene most closely meets all these requirements. Other gases, fuel such as MAPP gas, propylene, propane, natural gas, and proprietary gases based on these, have sufficiently high flame temperatures but exhibit low flame propagation rates. These gas flames are excessively oxidizing at oxygen-to-fuel gas ratios high enough to produce usable heat transfer rates. Flame holding devices, such as counterbores on the tips, are necessary for stable operation and good heat transfer, even the higher ratios. These gases, however, are used for oxygen cutting. They are also used for torch brazing, soldering, and many other operations where the demand upon the flame characteristics and heat transfer rates are not the same as those for welding.

e. Base Metal Preparation.

(1) Dirt, oil, and oxides can cause incomplete fusion, slag inclusions, and porosity in the weld. Contaminants must be removed along the joint and sides of the base metal.

(2) The root opening for a given thickness of metal should permit the gap to be bridged without difficulty, yet it should be large enough to permit full penetration. Specifications for root openings should be followed exactly.

(3) The thickness of the base mteal at the joint determines the type of edge preparation for welding. Thin sheet metal is easily melted completely by the flame. Thus, edges with square faces can be butted-together and welded. This type of joint is limited to material under 3/16 in. (4.8 mm) in thickness For thicknesses of 3/16 to ¼ in. (4.8 to 6.4 mm), a slight root opening or groove is necessary for complete penetration, but filler metal must be added to compensate for the opening.

(4) Joint edges ¼ in. (6.4 mm) and thicker should be beveled. Beveled edges at the joint provide a groove for better penetration and fusion at the sides. The angle of bevel for oxyacetylene welding varies from 35 to 45 degrees, which is equivalent to a variation in the included angle of the joint from 70 to 90 degrees, depending upon the application. A root face 1/16 in. (1.6 mm) wide is normal, but feather edges are sometimes used. Plate thicknesses ¾ in. (19 mm)

and above are double beveled when welding can be done from both sides. The root face can vary from O to 1/8 in. (O to 3.2 mm). Beveling both sides reduces the amount of filler metal required by approximately one-half. Gas consumption per unit length of weld is also reduced.

(5) A square groove edge preparation is the easiest to obtain. This edge can be machined, chipped, ground, or oxygen cut. The thin oxide coating on oxygen-cut surface does not have to be removed, because it is not detrimental to the welding operation or to the quality of the joint. A bevel angle can be oxygen cut.

f. Multiple Layer Welding.

(1) Multiple layer welding is used when maximum ductility of a steel weld in the as-welded or stress-relieved condition is desired, or when several layers are required in welding thick metal. Multiple layer welding is done by depositing filler metal in successive passes along the joint until it is filled. Since the area covered with each pass is small, the weld puddle is reduced in size. This procedure enables the welder to obtain complete joint penetration without excessive penetration and overheating while the first few passes are being deposited. The smaller puddle is more easily controlled. The welder can avoid oxides, slag inclusions, and incomplete fusion with the base metal.

(2) Grain refinement in the underlying passes as they are reheated increases ductility in the deposited steel. The final layer will not have this refinement unless an extra pass is added and removed or the torch is passed over the joint to bring the last deposit up to normalizing temperature.

g. Weld Quality.

(1) The appearance of a weld does not necessarily indicate its quality. Visual examination of the underside of a weld will determine whether there is complete penetration or whether there are excessive globules of metal. Inadequate joint penetration may be due to insufficient beveling of the edges, too wide a root face, too great a welding speed, or poor torch and welding rod manipulation.

(2) Oversized and undersized welds can be observed readily. Weld gauges are available to determine whether a weld has excessive or insufficient reinforcement. Undercut or overlap at the sides of the welds can usually be detected by visual inspection.

(3) Although other discontinuities, such as incomplete fusion, porosity, and cracking may or may not be apparent, excessive grain growth or the presence of hard spots cannot be determined visually. Incomplete fusion may be caused by insufficient heating of the base metal, too rapid travel, or gas or dirt inclusions. Porosity is a result of entrapped gases, usually carbon monoxide, which may be avoided by more careful flame manipulation and adequate fluxing where needed. Hard spots and cracking are a result of metallurgical characteristics of the weldment.

11-1. GENERAL GAS WELDING PROCEDURES (cont)

h. Welding With Other Fuel Gases.

(1) Principles of operation.

(a) Hydrocarbon gases, such as propane, butane, city gas, and natural gas, are not suitable for welding ferrous materials due to their oxidizing characteristics. In some instances, many nonferrous and ferrous metals can be braze welded with care taken in the adjustment of flare and the use of flux. It is important to use tips designed for the fuel gas being employed. These gases are extensively used for brazing and soldering operations, utilizing both mechanized and manual methods .

(b) These fuel gases have relatively low flame propagation rates, with the exception of some manufactured city gases containing considerable amounts of hydrogen. When standard welding tips are used, the maximum flame velocity is so low that it interferes seriously with heat transfer from the flame to the work. The highest flame temperatures of the gases are obtained at high oxygen-to-fuel gas ratios. These ratios produce highly oxidizing flames, which prevent the satisfactory welding of most metals.

(c) Tips should be used having flame-holding devices, such as skirts, counterbores, and holder flames, to permit higher gas velocities before they leave the tip. This makes it possible to use these fuel gases for many heating applications with excellent heat transfer efficiency.

(d) Air contains approximately 80 percent nitrogen by volume. This does not support combustion. Fuel gases burned with air, therefore, produce lower flame temperatures than those burned with oxygen. The total heat content is also lower. The air-fuel gas flame is suitable only for welding light sections of lead and for light brazing and soldering operations.

(2) Equipment.

(a) Standard oxyacetylene equipment, with the exception of torch tips and regulators, can be used to distribute and bum these gases. Special regulators may be obtained, and heating and cutting tips are available. City gas and natural gas are supplied by pipelines; propane and butane are stored in cylinders or delivered in liquid form to storage tanks on the user's property.

(b) The torches for use with air-fuel gas generally are designed to aspirate the proper quantity of air from the atmosphere to provide combustion. The fuel gas flows through the torch at a supply pressure of 2 to 40 psig and serves to aspirate the air. For light work, fuel gas usually is supplied from a small cylinder that is easily transportable.

(c) The plumbing, refrigeration, and electrical trades use propane in small cylinders for many heating and soldering applications. The propane flows through the torch at a supply pressure from 3 to 60 psig and serves to aspirate the air. The torches are used for soldering electrical connections, the joints in copper pipelines, and light brazing jobs.

(3) Applications.

Air-fuel gas is used for welding lead up to approximately ¼ in. (6.4 mm) in thickness. The greatest field of application in the plumbing and electrical industry. The process is used extensively for soldering copper tubing.

11-2. WORKING PRESSURES FOR WELDING OPERATIONS

The required working pressure increases as the tip orifice increases. The relation between the tip number and the diameter of the orifice may vary with different manufacturers. However, the smaller number always indicates the smaller diameter. For the approximate relation between the tip number and the required oxygen and acetylene pressures, see tables 11-1 and 11-2.

Table 11-1. Low Pressure or Injector Type Torch

Tip Size No.	Oxygen psi	Acetylene psi
NOTE		
Tips are provided by a number of manufacturers, and sizes may vary slightly.		
0	9	1
1	9	1
2	10	1
3	10	1
4	11	1
5	12	1
6	14	1
7	16	1
8	19	1
10	21	1
12	25	1
15	30	1

Table 11-2. Balanced Pressure Type Torch

Tip Size No.	Oxygen psi	Acetylene psi
NOTE		
Tips are provided by a number of manufacturers, and sizes may vary slightly.		
1	2	2
3	3	3
4	3	3
5	3.5	3.5
6	3.5	3.5
7	5	5
8	7	7
9	9	9
10	12	12

NOTE
Oxygen pressures are approximately the same as acetylene pressures in the balanced pressure type torch. Pressures for specific types of mixing heads and tips are specified by the manufacturer.

11-3. FLAME ADJUSTMENT AND FLAME TYPES

a. General.

(1) The oxyfuel gas welding torch mixes the combustible and combustion-supporting gases. It provides the means for applying the flame at the desired location. A range of tip sizes is provided for obtaining the required volume or size of welding flame which may vary from a short, small diameter needle flame to a flare 3/16 in. (4.8 mm) or more in diameter and 2 in(51 mm) or more in length.

(2) The inner cone or vivid blue flare of the burning mixture of gases issuing from the tip is called the working flare. The closer the end of the inner cone is to the surface of the metal being heated or welded, the more effective is the heat transfer from flame to metal. The flame can be made soft or harsh by varying the gas flow. Too low a gas flow for a given tip size will result in a soft, ineffective flame sensitive to backfiring. Too high a gas flow will result in a harsh, high velocity flame that is hard to handle and will blow the molten metal from the puddle.

(3) The chemical action of the flame on a molten pool of metal can be altered by changing the ratio of the volume of oxygen to acetylene issuing from the tip. Most oxyacetylene welding is done with a neutral flame having approximately a 1:1 gas ratio. An oxidizing action can be obtained by increasing the oxygen flow, and a reducing action will result from increasing the acetylene flow. Both adjustments are valuable aids in welding.

b. Flare Adjustment.

(1) Torches should be lighted with a friction lighter or a pilot flame. The instructions of the equipment manufacturer should be observed when adjusting operating pressures at the gas regulators and torch valves before the gases issuing from the tip are ignited.

(2) The neutral flame is obtained most easily by adjustment from an excess-acetylene flame, which is recognized by the feather extension of the inner cone. The feather will diminish as the flow of acetylene is decreased or the flow of oxygen is increased. The flame is neutral just at the point of disappearance of the "feather" extension of the inner cone. This flame is actually reducing in nature but is neither carburizing or oxidizing.

(3) A practical method of determining the amount of excess acetylene in a reducing flame is to compare the length of the feather with the length of the inner cone, measuring both from the torch tip. A 2X excess-acetylene flame has an acetylene feather that is twice the length of the inner cone. Starting with a neutral flame adjustment, the welder can produce the desired acetylene feather by increasing the acetylene flow (or by decreasing the oxygen flow). This flame also has a carburizing effect on steel.

(4) The oxidizing flame adjustment is sometimes given as the amount by which the length of a neutral inner cone should be reduced, for example, one tenth. Starting with the neutral flare, the welder can increase the oxygen or decrease the acetylene until the length of the inner cone is decreased the desired amount. See figure 11-1.

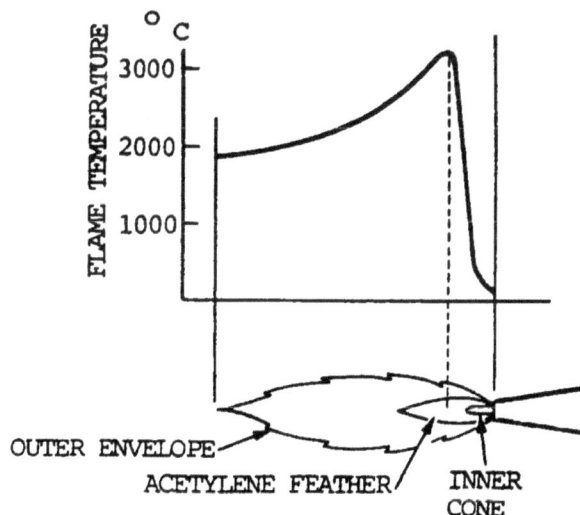

Figure 11-1. The temperature of the flame.

c. Lighting the Torch.

(1) To start the welding torch, hold it so as to direct the flame away from the operator, gas cylinders, hose, or any flammable material. Open the acetylene torch valve ¼-turn and ignite the gas by striking the sparklighter in front of the tip.

(2) Since the oxygen torch valve is closed, the acetylene is burned by the oxygen in the air. There is not sufficient oxygen to provide complete combustion, so the flame is smoky and produces a soot of fine unburned carbon. Continue to open the acetylene valve slowly until the flame burns clean. The acetylene flame is long, bushy, and has a yellowish color. This pure acetylene flame is unsuitable for welding.

(3) Slowly open the oxygen valve. The flame changes to a bluish-white and forms a bright inner cone surrounded by an outer flame. The inner cone develops the high temperature required for welding.

(4) The temperature of the oxyacetylene flame is not uniform throughout its length and the combustion is also different in different parts of the flame. It is so high (up to 6000°F (3316°C)) that products of complete combustion (carbon dioxide and water) are decomposed into their elments. The temperature is the highest just beyond the end of the inner cone and decreases gradually toward the end of the flame. Acetylene burning in the inner cone with oxygen supplied by the torch forms carbon monoxide and hydrogen. As these gases cool from the high temperatures of the inner cone, they burn completely with the oxygen supplied by the surrounding air and form the lower temperature sheath flame. The carbon monoxide burns to form carbon dioxide and hydrogen burns to form water vapor. Since the inner cone contains only carbon monoxide and hydrogen, which are reducing in character (i.e., able to combine with and remove oxygen), oxidation of the metal will not occur within this zone. The chemical reaction for a one-to-one ratio of acetylene and oxygen plus air is as follows:

$$C_2H_2 + O_2 = 2CO + H_2 + Heat$$

11-3.   FLAME ADJUSTMENT AND FLAME TYPES (cont)

This is the primary reaction: however, both carbon monoxide and hydrogen are combustible and will react with oxygen from the air:

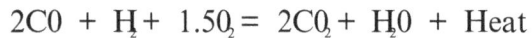

$$2CO + H_2 + 1.5O_2 = 2CO_2 + H_2O + Heat$$

This is the secondary reaction which produces carbon dioxide, heat, and water.

    d.   Types of Flames.

       (1) General. There are three basic flame types: neutral (balanced), excess acetlyene (carburizing), and excess oxygen (oxidizing). They are shown in figure 11-2.

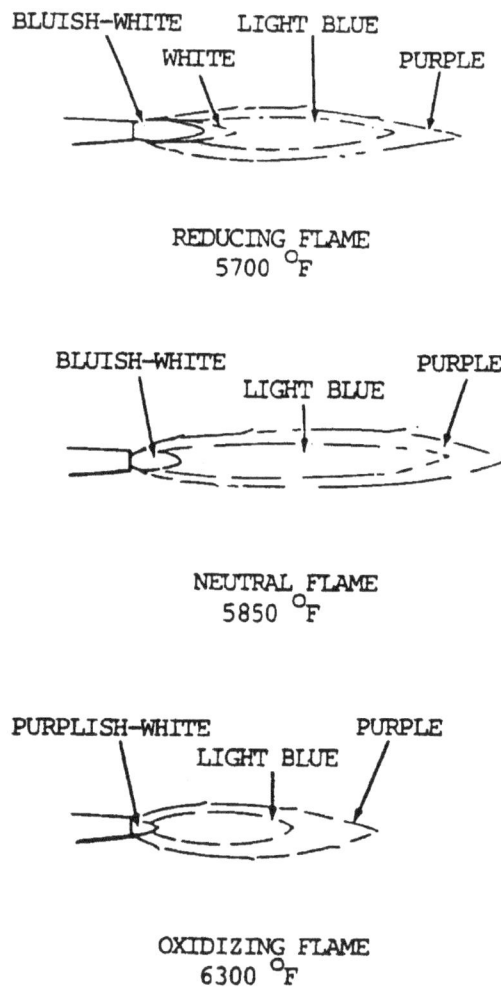

BLUISH-WHITE   LIGHT BLUE
WHITE   PURPLE

REDUCING FLAME
5700 °F

BLUISH-WHITE   PURPLE
LIGHT BLUE

NEUTRAL FLAME
5850 °F

PURPLISH-WHITE   PURPLE
LIGHT BLUE

OXIDIZING FLAME
6300 °F

Figure 11-2. Oxyacetylene flames.

(a) The neutral flame has a one-to-one ratio of acetylene and oxygen. It obtains additional oxygen from the air and provides complete combustion.It is generally preferred for welding. The neutral flame has a clear, well-defined, or luminous cone indicating that combustion is complete.

(b) The carburizing flame has excess acetylene,indicated in the flame when the inner cone has a feathery edge extending beyond it.This white feather is called the acetylene feather. If the acetylene featheris twice as long as the inner cone it is known as a 2X flame, which is a way ofexpressing the amount of excess acetylene. The carburizing flame may add carbonto the weld metal.

(c) The oxidizing flame, which has an excess ofoxygen, has a shorter envelope and a small pointed white cone.The reduction in length of the inner core is a measure of excess oxygen. This flame tends to oxidize the weld metal and is used only for welding specific metals.

(2) Neutral flame.

(a) The welding flame should be adjusted to neutral before either the carburizing or oxidizing flame mixture is set. There are two clearly defined zones in the neutral flame. The inner zone consists of a luminous cone that is bluish-white. Surrounding this is a light blue flame envelope or sheath. This neutral flame is obtained by starting with an excess acetylene flame in which there is a "feather" extension of the inner cone.When the flow of acetylene is decreased or the flow of oxygen increased the feather will tend to disappear. The neutral flame begins when the feather disappears.

(b) The neutral or balanced flame is obtained when the mixed torch gas consists of approximately one volume of oxygen and one volume of acetylene.It is obtained by gradually opening the oxygen valve to shorten the acetylene flame until a clearly defined inner cone is visible. For a strictly neutral flame, no whitish streamers should be present at the end of the cone.In some cases, it is desirable to leave a slight acetylene streamer or"feather" 1/16 to 1/8 in. (1.6 to 3.2 mm) long at the end of the cone to ensure that the flame is not oxidizing. This flame adjustment is used for most welding operations and for preheating during cutting operations. When welding steel with this flare, the molten metal puddle is quiet and clear. The metal flows easily without boiling, foaming, or sparking.

(c) In the neutral flame, the temperature at the inner cone tip is approximately 5850°F (3232°C), while at the end of the outer sheath or envelope the temperature drops to approximately 2300°F (1260°C). This variation within the flame permits some temperature control when making a weld. The position of the flame to the molten puddle can be changed, and the heat controlled in this manner.

(3) Reducing or carburizing flame.

(a) The reducing or carburizing flame is obtained when slightly less than one volume of oxygen is mixed with one volume of acetylene.This flame is obtained by first adjusting to neutral and then slowly opening the acetylene valve until an acetylene streamer or"feather" is at the end of the inner cone.The length of this excess streamer indicates the degree of flame carburization.For most welding operations, this streamer should be no more than half the length of the inner cone.

11-3. FLAME ADJUSTMENT AND FLAME TYPES (cont)

(b) The reducing or carburizing flame can always be recognized by the presence of three distinct flame zones. There is a clearly defined bluish-white inner cone, white intermediate cone indicating the amount of excess acetylene, and a light blue outer flare envelope. This type of flare burns with a coarse rushing sound. It has a temperature of approximately 5700°F (3149°C) at the inner cone tips.

(c) When a strongly carburizing flame is used for welding, the metal boils and is not clear. The steel, which is absorbing carbon from the flame, gives off heat. This causes the metal to boil. When cold, the weld has the properties of high carbon steel, being brittle and subject to cracking.

(d) A slight feather flame of acetylene is sometimes used for back-hand welding. A carburizing flame is advantageous for welding high carbon steel and hard facing such nonferrous alloys as nickel and Monel. When used in silver solder and soft solder operations, only the intermediate and outer flame cones are used. They impart a low temperature soaking heat to the parts being soldered.

(4) Oxidizing flame.

(a) The oxidizing flame is produced when slightly more than one volume of oxygen is mixed with one volume of acetylene. To obtain this type of flame, the torch should first be adjusted to a neutral flame. The flow of oxygen is then increased until the inner cone is shortened to about one-tenth of its original length. When the flame is properly adjusted, the inner cone is pointed and slightly purple. An oxidizing flame can also be recognized by its distinct hissing sound. The temperature of this flame is approximately 6300°F (3482°C) at the inner cone tip.

(b) When applied to steel, an oxidizing flame causes the molten metal to foam and give off sparks. This indicates that the excess oxygen is combining with the steel and burning it. An oxidizing flame should not be used for welding steel because the deposited metal will be porous, oxidized, and brittle. This flame will ruin most metals and should be avoided, except as noted in (c) below.

(c) A slightly oxidizing flame is used in torch brazing of steel and cast iron. A stronger oxidizing flame is used in the welding of brass or bronze.

(d) In most cases, the amount of excess oxygen used in this flame must be determined by observing the action of the flame on the molten metal.

(5) MAPP gas flames.

(a) The heat transfer properties of primary and secondary flames differ for different fuel gases. MAPP gas has a high heat release in the primary flame, and a high heat release in the secondary. Propylene is intermediate between propane and MAPP gas. Heating values of fuel gases are shown in table 11-3.

Table 11-3. Heating Values of Fuel Gases

Fuel	Flame Temp. ($^\circ$F)	Primary Flame (BTU/cu ft)	Secondary Flame (BTU/cu ft)	Total Heat (BTU/cu ft)
MAPP Gas	5301	517	1889	2406
Aceytlene	5589	507	963	1470
Propane	4579	255	2243	2498
Natural Gas	4600	11	989	1000
Propylene	5193	433	1938	2371

(b) The coupling distance between the work and the flame is not nearly as critical with MAPP gas as it is with other fuels.

(c) Adjusting a MAPP gas flame. Flame adjustment is the most important factor for successful welding or brazing with MAPP gas. As with any other fuel gas, there are three basic MAPP gas flames: carburizing, neutral, and oxidizing (fig. 11-3).

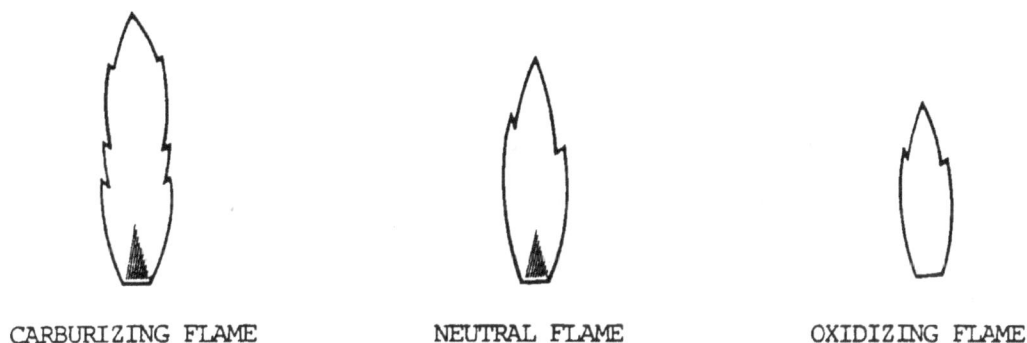

CARBURIZING FLAME   NEUTRAL FLAME   OXIDIZING FLAME

Figure 11-3. What MAPP gas flames should look like.

1. A carburizing flame looks much the same with MAPP gas or acetylene. It has a yellow feather on the end of the primary cone. Carburizing flames are obtained with MAPP gas when oxyfuel ratios are around 2.2:1 or lower. Slightly carburizing or "reducing" flames are used to weld or braze easily oxidized alloys such as aluminum.

2. As oxygen is increased, or the fuel is turned down, the carburizing feather pulls off and disappears. When the feather disappears, the oxyfuel ratio is about 2.3:1. The inner flame is a very deep blue. This is the neutral MAPP gas flame for welding, shown in figure 11-3. The flame remains neutral up to about 2.5:1 oxygen-to-fuel ratio.

TC 9-237

11-3. FLAME ADJUSTMENT AND FLAME TYPES (cont)

3. Increasing the oxygen flame produces a lighter blue flame, a longer inner cone, and a louder burning sound. This is an oxidizing MAPP gas flare. An operator experience with acetylene will immediately adjust the MAPP gas flame to look like the short, intense blue flame typical of the neutral acetylene flame setting. What will be produced, however, is a typical oxidizing MAPP gas flame. With certain exceptions such as welding or brazing copper and copper alloys, an oxidizing flame is the worst possible flame setting, whatever the fuel gas used. The neutral flame is the principle setting for welding or brazing steel. A neutral MAPP gas flame has a primary flame cone abut 1-½ to 2 times as long as the primary acetylene flame cone.

11-4. OXYFUEL WELDING RODS

a. The welding rod, which is melted into the welded joint, plays an important part in the quality of the finished weld. Good welding rods are designed to permit free flowing metal which will unite readily with the base metal to produce sound, clean welds of the correct composition.

b. Welding rods are made for various types of carbon steel, aluminum, bronze, stainless steel, and other metals for hard surfacing.

11-5. OXYFUEL WELDING FLUXES

a. General.

(1) Oxides of all ordinary commercial metal and alloys (except steel) have higher melting points than the metals themselves. They are usually pasty when the metal is quite fluid and at the proper welding temperature. An efficient flux will combine with oxides to form a fusible slag. The slag will have a melting point lower than the metal so it will flow away from the immediate field of action. It combines with base metal oxides and removes them. It also maintains cleanliness of the base metal at the welding area and helps remove oxide film on the surface of the metal. The welding area should be cleaned by any method. The flux also serves as a protection for the molten metal against atmospheric oxidation.

(2) The chemical characteristics and melting points of the oxides of different metals vary greatly. There is no one flux that is satisfactory for all metals, and there is no national standard for gas welding fluxes. They are categorized according to the basic ingredient in the flux or base metal for which they are to be used.

(3) Fluxes are usually in powder form. These fluxes are often applied by sticking the hot filler metal rod in the flux. Sufficient flux will adhere to the rod to provide proper fluxing action as the filler rod is melted in the flame.

(4) Other types of fluxes are of a paste consistency which are usually painted on the filler rod or on the work to be welded.

(5) Welding rods with a covering of flux are also available. Fluxes are available from welding supply companies and should be used in accordance with the directions accompanying them.

11-12

b. The melting point of a flux must be lower than that of either the metal or the oxides formed, so that it will be liquid. The ideal flux has exactly the right fluidity when the welding temperature has been reached. The flux will protect the molten metal from atmospheric oxidation. Such a flux will remain close to the weld area instead of flowing all over the base metal for some distance from the weld.

c. Fluxes differ in their composition according to the metals with which they are to be used. In cast iron welding, a slag forms on the surface of the puddle. The flux serves to break this up. Equal parts of a carbonate of soda and bicarbonate of soda make a good compound for this purpose. Nonferrous metals usually require a flux. Copper also requires a filler rod containing enough phosphorous to produce a metal free from oxides. Borax which has been melted and powdered is often used as a flux with copper alloys. A good flux is required with aluminum, because there is a tendency for the heavy slag formed to mix with the melted aluminum and weaken the weld. For sheet aluminum welding, it is customary to dissolve the flux in water and apply it to the rod. After welding aluminum, all traces of the flux should be removed.

11-6. FOREHAND WELDING

a. In this method, the welding rod precedes the torch. The torch is held at approximately a 45 degree angle from the vertical in the direction of welding, as shown in figure 11-4. The flame is pointed in the direction of welding and directed between the rod and the molten puddle. This position permits uniform preheating of the plate edges immediately ahead of the molten puddle. By moving the torch and the rod in opposite semicircular paths, the heat can be carefully balanced to melt the end of the rod and the side walls of the plate into a uniformly distributed molten puddle. The rod is dipped into the leading edge of the puddle so that enough filler metal is melted to produce an even weld joint. The heat which is reflected backwards from the rod keeps the metal molten. The metal is distributed evenly to both edges being welded by the motion of the tip.

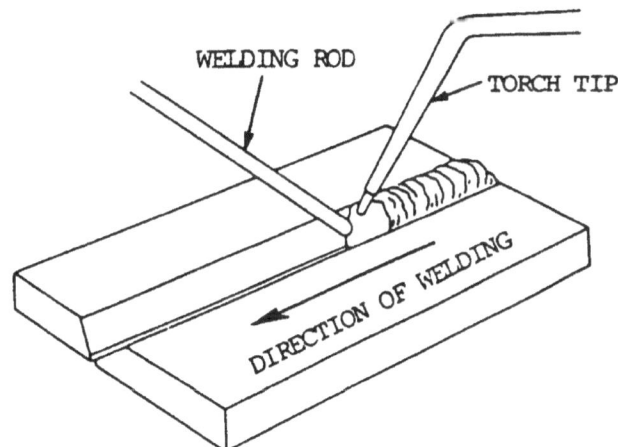

NOTE

TORCH AND ROD ANGLES ARE 45 DEG AS VIEWED BY THE OPERATOR AND PERPENDICULAR (90 DEG) TO THE WORK SURFACE AS VIEWED FROM THE END OF THE WORKPIECE.

Figure 11-4. Forehand welding.

11-6. FOREHAND WELDING (cont)

b. In general, the forehand method is recommended for welding material up to 1/8 in. (3.2 mm) thick, because it provides better control of the small weld puddle, resulting in a smoother weld at both top and bottom. The puddle of molten metal is small and easily controlled. A great deal of pipe welding is done using the forehand technique, even in 3/8 in. (9.5 mm) wall thicknesses. In contrast, some difficulties in welding heavier plates using the forehand metod are:

(1) The edges of the plate must be beveled to provide a wide V with a 90 degree included angle. This edge preparation is necessary to ensure satisfactory melting of the plate edges, good penetration, and fusion of the weld metal to the base metal.

(2) Because of this wide V, a relatively large molten puddle is required. It is difficult to obtain a good joint when the puddle is too large.

11-7. BACKHAND WELDING

a. In this method, the torch precedes the welding rod, as shown in figure 11-5. The torch is held at approximately a 45 degree angle from the vertical away from the direction of welding, with the flame directed at the molten puddle. The welding rod is between the flame and the molten puddle. This position requires less transverse motion than is used in forehand welding.

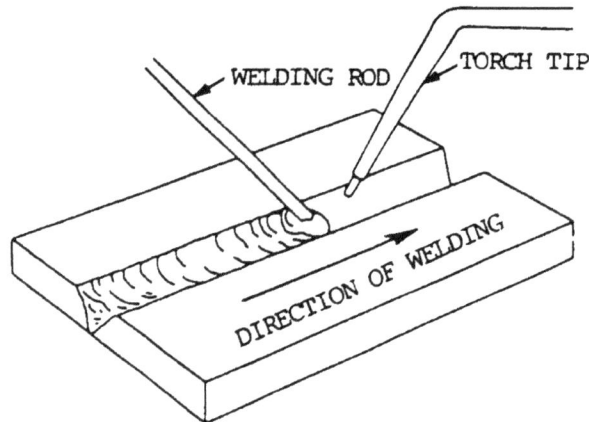

NOTE
TORCH AND ROD ANGLES ARE 45 DEG AS VIEWED BY THE OPERATOR AND PERPENDICULAR (90 DEG) TO THE WORK SURFACE AS VIEWED FROM THE END OF THE WORKPIECE.

Figure 11-5. Backhand welding.

b. Increased speeds and better control of the puddle are possible with backhand technique when metal 1/8 in. (3.2 mm) and thicker is welded, based on the study of speeds normally achieved with this technique and on greater ease of obtaining fusion at the weld root. Backhand welding may be used with a slightly reducing flame (slight acetylene feather) when desirable to melt a minimum amount of steel in making a joint. The increased carbon content obtained from this flame lowers the melting point of a thin layer of steel and increases welding speed. This technique increases speed of making pipe joints where the wall thickness is ¼ to 5/16 in. (6.4 to 7.9 mm) and groove angle is less than normal. Backhand welding is sometimes used in surfacing operations.

11-8.   FILLET WELDING

    a.   General.

        (1) The fillet weld is the most popular of all types of welds because there is normally no preparation required. In some cases, the fillet weld is the least expensive, even though it might require more filler metal than a groove weld since the preparation cost would be less. It can be used for the lap joint, the tee joint, and the corner joint without preparation. Since these are extremely popular, the fillet has wide usage. On corner joints, the double fillet can actually produce a full-penetration weld joint. The use of the fillet for making all five of the basic joints is shown by figure 11-6. Fillet welds are also used in conjunction with groove welds, particularly for corner and tee joints.

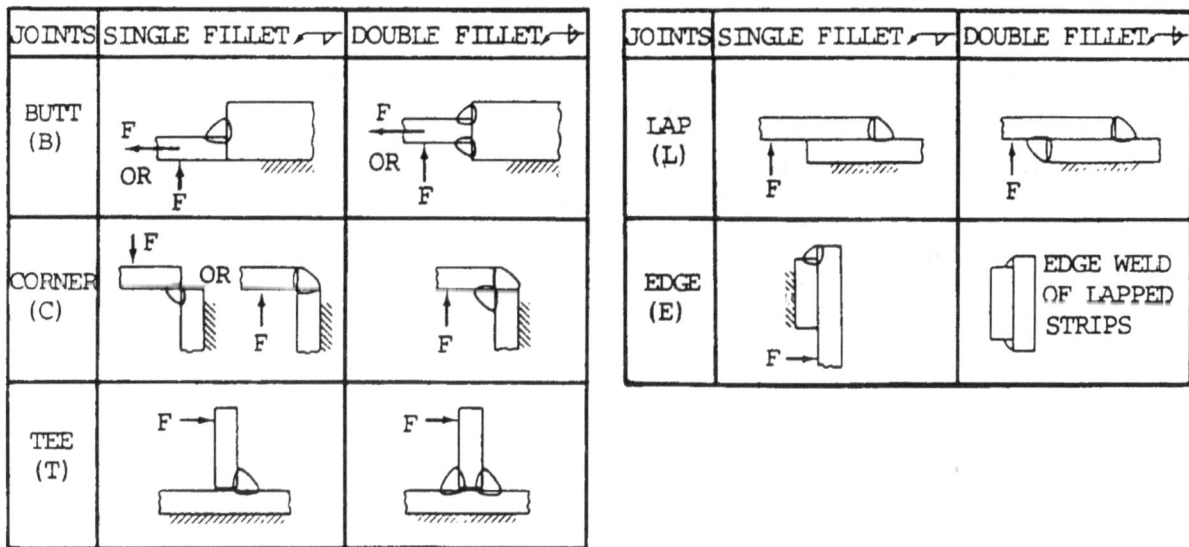

Figure 11-6.   The fillet used to make the five basic joints.

11-8. FILLET WELDING (cont)

(2) The fillet weld is expected to have equal length legs and thus the face of the fillet is on a 45 degree angle. This is not always so, since a fillet may be designed to have a longer base than height, in which case it is specified by the two leg lengths. On the 45 degree or normal type of fillet, the strength of the fillet is based on the shortest or throat dimension which is 0.707 x the leg length. For fillets having unequal legs, the throat length must be calculated and is the shortest distance between the root of the fillet and the theoretical face of the fillet. In calculating the strength of fillet welds, the reinforcement is ignored. The root penetration is also ignored unless a deep penetratingq process is used. If semi- or fully-automatic application is used, the extra penetration can be considered. See figure 11-7 for details about the weld.

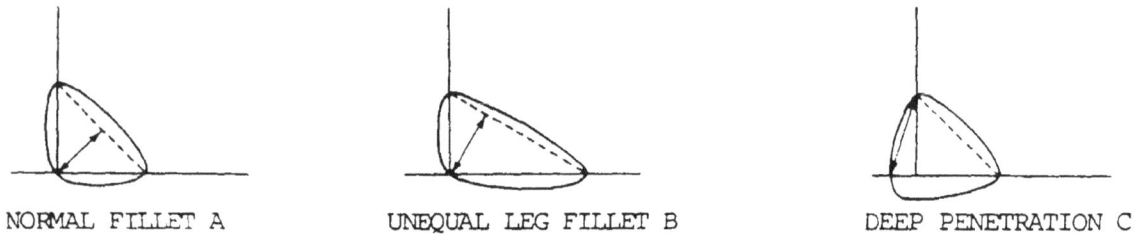

NORMAL FILLET A   UNEQUAL LEG FILLET B   DEEP PENETRATION C

Figure 11-7. Fillet weld throat dimension.

(3) Under these circumstances, the size of the fillet can be reduced, yet equal strength will result. Such reductions can be utilized only when strict welding procedures are enforced. The strength of the fillet weld is determined by its failure area, which relates to the throat dimension. Doubling the size or leg length of a fillet will double its strength, since it doubles the throat dimension and area. However, doubling the fillet size will increase its cross-sectional area and weight four times. This illustrated in figure 11-8, which shows the relationship to throat-versus-cross-sectional area, or weight, of fillet weld. For example, a 3/8 in. (9.5 mm) fillet is twice as strong as a 3/16 in. (4.8 mm) fillet; however, the 3/8 in. (9.5 mm) fillet requires four times as much weld metal.

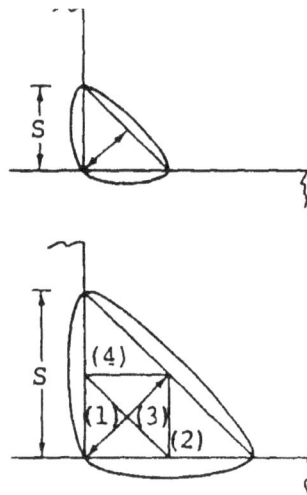

Figure 11-8. Fillet weld size vs strength.

(4) In design work, the fillet size is sometimes governed by the thickness of the metals joined. In some situations, the minimum size of the fillet must be based on practical reasons rather than the theoretical need of the design. Intermittent fillets are sometimes used when the size is minimum, based on code, or for practical reasons, rather than because of strength requirements. Many intermittent welds are based on a pitch and length so that the weld metal is reduced in half. Large intermittent fillets are not recommended because of the volume-throat dimension relationship mentioned previously. For example, a 3/8 in. (9.5 mm) fillet 6 in. (152.4 mm) long on a 12 in. (304.8 mm) pitch (center to center of intermittent welds) could be reduced to a continuous 3/16 in. (4.8 mm) fillet, and the strength would be the same, but the amount of weld metal would be only half as much.

(5) Single fillet welds are extremely vulnerable to cracking if the root of the weld is subjected to tension loading. This applies to tee joints, corner joints, and lap joints. The simple remedy for such joints is to make double fillets, which prohibit the tensile load from being applied to the root of the fillet. This is shown by figure 11-6, page 11-15. Notice the F (force) arrowhead.

b. A different welding technique is required for fillet welding than for butt joints because of the position of the parts to be welded. When welding is done in the horizontal position, there is a tendency for the top plate to melt before the bottom plate because of heat rising. This can be avoided, however, by pointing the flame more at the bottom plate than at the edge of the upper plate. Both plates must reach the welding temperature at the same time.

c. In making the weld, a modified form of backhand technique should be used. The welding rod should be kept in the puddle between the completed portion of the weld and the flame. The flame should be pointed ahead slightly in the direction in which the weld is being made and directed at the lower plate. To start welding, the flame should be concentrated on the lower plate until the metal is quite red. Then the flame should be directed so as to bring both plates to the welding temperature at the same time. It is important that the flame not be pointed directly at the inner corner of the fillet. This will cause excessive amount of heat to build up and make the puddle difficult to control.

d. It is essential in this form of welding that fusion be obtained at the inside corner or root of the joint.

11-9. HORIZONTAL POSITION WELDING

a. Welding cannot always be done in the most desirable position. It must be done in the position in which the part will be used. Often that may be on the ceiling, in the corner, or on the floor. Proper description and definition is necessary since welding procedures must indicate the welding position to be performed, and welding process selection is necessary since some have all-position capabilities whereas others may be used in only one or two positions. The American Welding Society has defined the four basic welding positions as shown in figure 11-9, p 11-8.

11-9. HORZONITAL POSITION WELDING (cont)

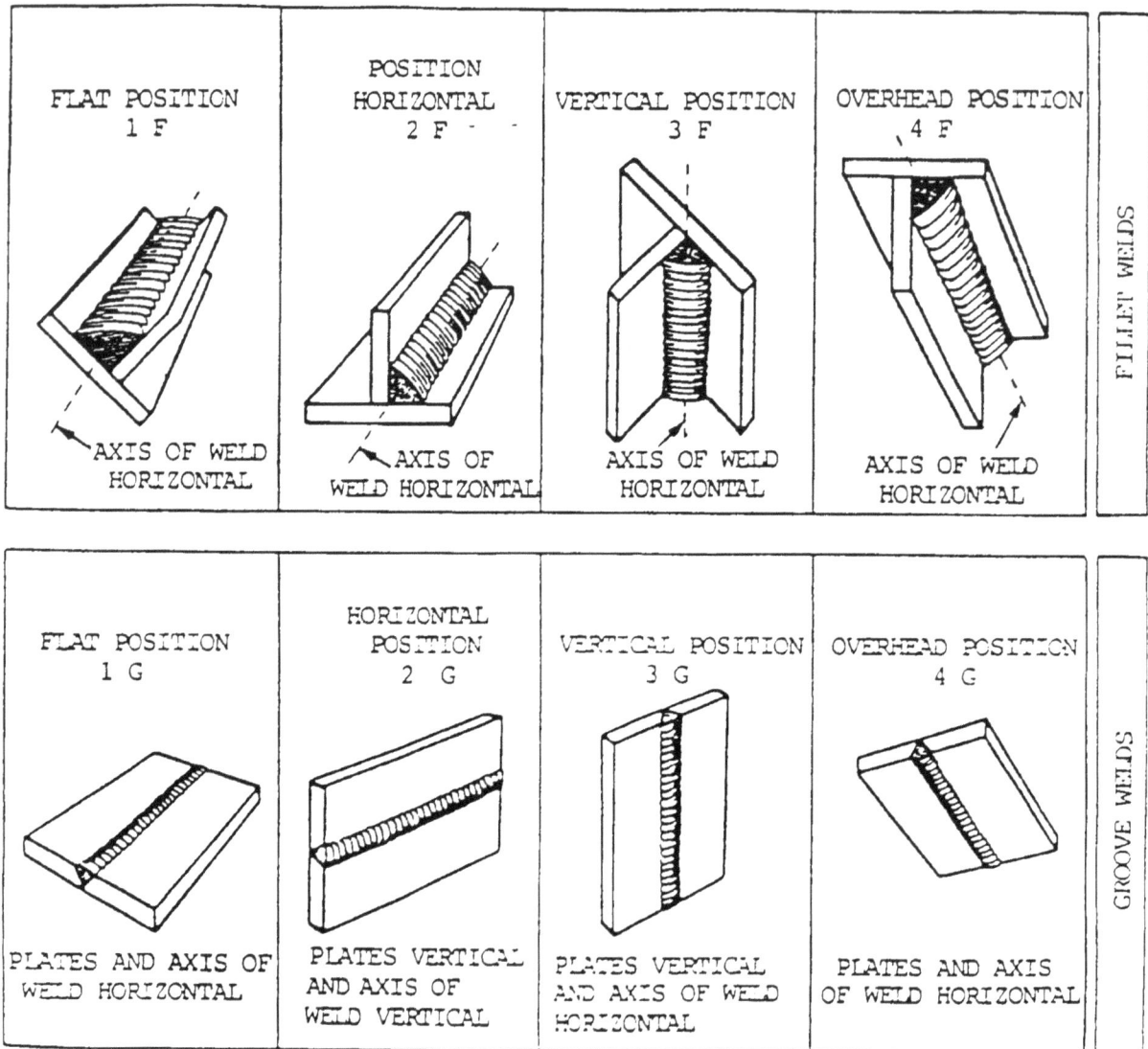

Figure 11-9. Welding position—fillet and groove welds.

b. In horizontal welding, the weld axis is approximately horizontal, but the weld type dictates the complete definition. For a fillet weld, welding is performed on the upper side of an approximately horizontal surface and against an approximately vertical surface. For a groove weld, the face of the weld lies in an approximately vertical plane.

c. Butt welding in the horizontal position is a little more difficult to master than flat position. This is due to the tendency of molten metal to flow to the lower side of the joint. The heat from the torch rises to the upper side of the joint. The combination of these opposing factors makes it difficult to apply a uniform deposit to this joint.

d. Align the plates and tack weld at both ends (fig. 11-10). The torch should move with a slight oscillation up and down to distribute the heat equally to both sides of the joint, thereby holding the molten metal in a plastic state. This prevents excessive flow of the metal to the lower side of the joint, and permits faster solidification of the weld metal. A joint in horizontal position will require considerably more practice than the previous techniques. It is, however, important that the technique be mastered before passing on to other types of weld positions.

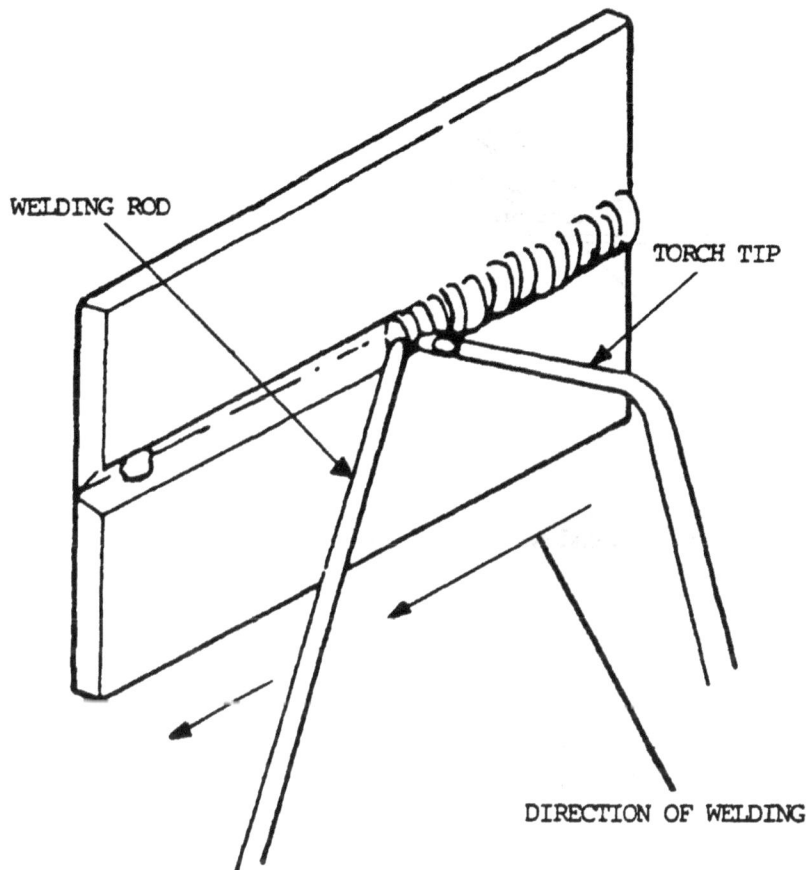

Figure 11-10. Welding a butt joint in the horizontal position.

11-10. FLAT POSITION WELDING

a. General. This type of welding is performed from the upper side of the joint. The face of the weld is approximately horizontal.

b. Bead Welds.

(1) In order to make satisfactory bead welds on a plate surface, the flare motion, tip angle, and position of the welding flame above the molten puddle should be carefully maintained. The welding torch should be adjusted to give the proper type of flame for the particular metal being welded.

11-10.   FLAT POSITION WELDING (cont)

    (2) Narrow bead welds are made by raising and lowering the welding flare with a slight circular motion while progressing forward.   The tip should form an angle of approximately 45 degrees with the plate surface.   The flame will be pointed in the welding direction (figs. 11-11 and 11-12).

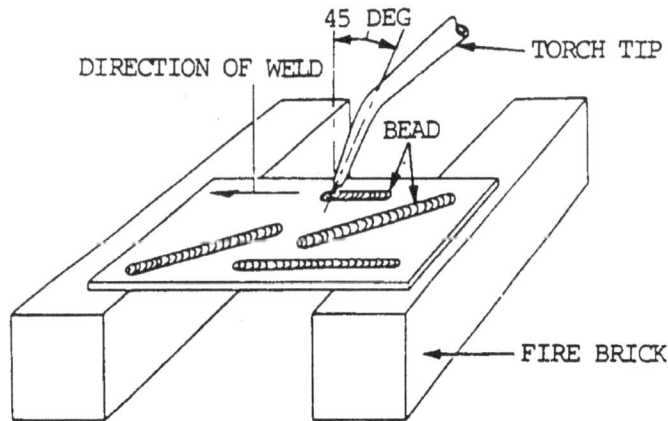

Figure 11-11.   Bead welding without a welding rod.

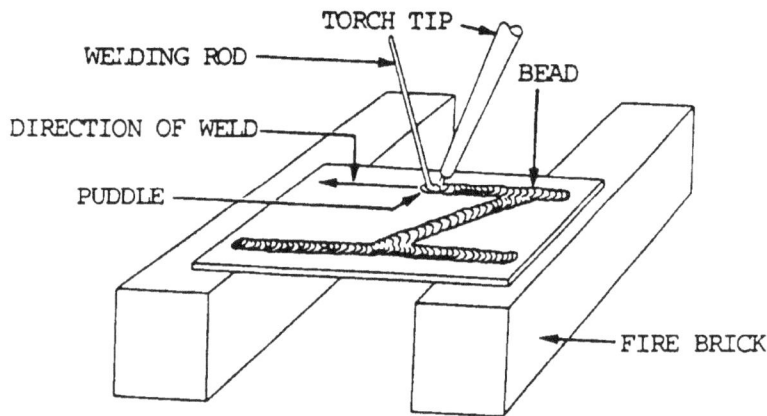

Figure 11-12.   Bead welding with a welding rod.

(3) To increase the depth of fusion, either increase the angle between the tip and the plate surface or decrease the welding speed. The size of the puddle should not be too large because this will cause the flame to burn through the plate. A properly made bead weld, without filler rod, will be slightly below the upper surface of the plate. A bead weld with filler rod shows a buildup on the surface.

(4) A small puddle should be formed on the surface when making a bead weld with a welding rod (fig. 11-12). The welding rod is inserted into the puddle and the base plate and rod are melted together. The torch should be moved slightly from side to side to obtain good fusion. The size of the bead can be controlled by varying the speed of welding and the amount of metal deposited from the welding rod.

c. Butt Welds.

(1) Several types of joints are used to make butt welds in the flat position.

(2) Tack welds should be used to keep the plates aligned. The lighter sheets should be spaced to allow for weld metal contraction and thus prevent warpage.

(3) The following guide should be used for selecting the number of passes (fig. 11-8, p 11-16) in butt welding steel plates:

Plate thickness, in.	Number of passes
1/8 to 1/4	1
1/4 to 5/8	2
5/8 to 7/8	3
7/8 to 1-1/8	4

(4) The position of the welding rod and torch tip in making a flat position butt joint is shown in figure 11-13. The motion of the flame should be controlled so as to melt the side walls of the plates and enough of the welding rod to produce a puddle of the desired size. By oscillating the torch tip, a molten puddle of a given size can be carried along the joint. This will ensure both complete penetration and sufficient filler metal to provide some reinforcement at the weld.

(5) Care should be taken not to overheat the molten puddle. This will result in burning the metal, porosity, and low strength in the completed weld.

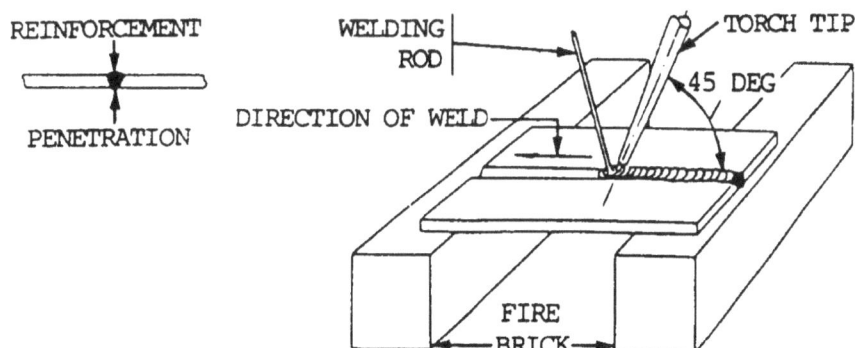

Figure 11-13. Position of rod and torch for a butt weld in a flat position.

11-11. VERTICAL POSITION WELDING

a. General. In vertical position welding, the axis of the weld is approximately vertical.

b. When welding is done on a vertical surface, the molten metal has a tendency to run downward and pile up. A weld that is not carefully made will result in a joint with excessive reinforcement at the lower end and some undercutting on the surface of the plates.

c. The flew of metal can be controlled by pointing the flame upward at a 45 degree angle to the plate, and holding the rod between the flame and the molten puddle (fig. 11-14). The manipulation of the torch and the filler rod keeps the metal from sagging or falling and ensures good penetration and fusion at the joint. Both the torch and the welding rod should be oscillated to deposit a uniform bead. The welding rod should be held slightly above the center line of the joint, and the welding flame should sweep the molten metal across the joint to distribute it evenly.

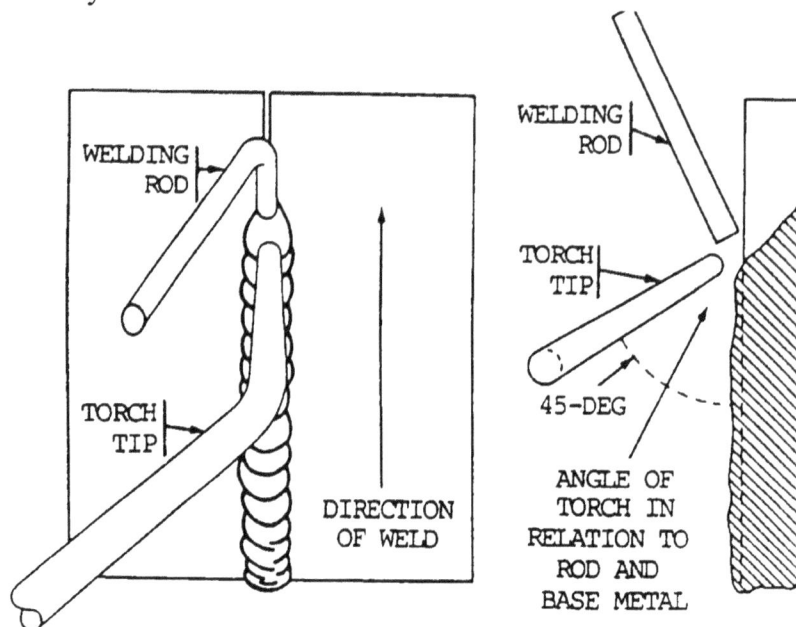

Figure 11-14. Welding a butt joint in the vertical position.

d. Butt joints welded in the vertical position should be prepared for welding in the same manner as that required for welding in the flat position.

11-12. OVERHEAD POSITION WELDING

a. General. Overhead welding is performed from the underside of a joint.

b. Bead welds. In overhead welding, the metal deposited tends to drop or sag on the plate, causing the bead to have a high crown. To overcome this difficulty, the molten puddle should be kept small, and enough filler metal should be added to obtain good fusion with some reinforcement at the bead. If the puddle becomes too large, the flame should be removed for an instant to permit the weld metal to freeze. When welding light sheets, the puddle size can be controlled by applying the heat equally to the base metal and filler rod.

c. <u>Butt Joints</u>. The torch and welding rod position for welding overhead butt joints is shown in figure 11-15. The flame should be directed so as to melt both edges of the joint. Sufficient filler metal should be added to maintain an adequate puddle with enough reinforcement.The welding flame should support the molten metal anddistribute it along the joint. Only a small puddle is required, so a small welding rod should be used. Care should be taken to control the heat to avoid burning through the plates. This is particularly important when welding is done from one side only.

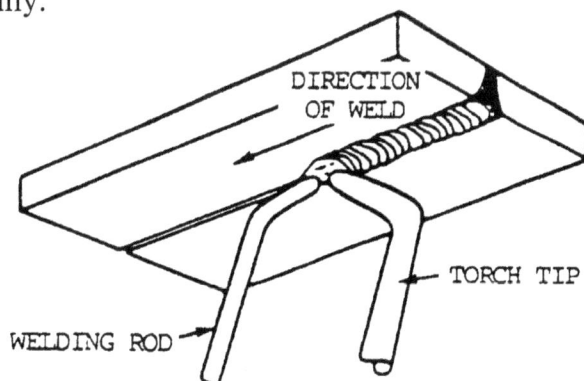

Figure 11-15. Welding a butt joint in the overhead position.

Section II.   WELDING AND BRAZING FERROUS METALS

11-13. GENERAL

a. <u>Welding Sheet Metal</u>.

(1) For welding purposes, the term "sheet metal"is restricted to thicknesses of metals up to and including 1/8 in. (3.2 mm).

(2) Welds in sheet metal up to 1/16 in. (1.6 mm) thick can be made satisfactorily by flanging the edges at the joint. The flanges must be at least equal to the thickness of the metal. The edges should be aligned with the flanges and then tack welded every 5 or 6 in. (127.0 to 152.4 mm). Heavy angles or bars should be clamped on each side of the joint to prevent distortion or buckling. The raised edges are equally melted by the welding flare. This produces a weld nearly flush with the sheet metal surface. By controlling the welding speed and the flame motion, good fusion to the underside of the sheet can he obtained without burning through. A plain square butt joint can also be made on sheet metal up to 1/16 in. (1.6 mm) thick by using a rust-resisting, copper-coated lowcarbon filler rod 1/16 in. (1.6 mm) in diameter. The method of aligning the joint and tacking the edges is the same as that used for welding flanged edge joints.

(3) Where it is necessary to make an inside edge or corner weld, there is danger of burning through the sheet unless special care is taken to control the welding heat. Such welds can be made satisfactorily in sheet metal up to 1/16 in. (1.6 mm) thick by following the procedures below:

(a) Heat the end of a 1/8 in. (3.2 mm) low carbon welding rod until approximately ½ in. (12.7 mm) of the rod is molten.

(b) Hold the rod so that the molten end is above the joint to be welded.

11-13. GENERAL (cont)

(c) By sweeping the flame across the molten end of the rod, the metal can be removed and deposited on the seam. The quantity of molten weld metal is relatively large as compared with the light gauge sheet. Its heat is sufficient to preheat the sheet metal. By passing the flame quickly back and forth, the filler metal is distributed along the joint. The additional heat supplied by the flame will produce complete fusion. This method of welding can be used for making difficult repairs on automobile bodies, metal containers, and similar applications. Consideration should be given to expansion and contraction of sheet metal before welding is stated.

(4) For sheet metal 1/16 to 1/8 in. (1.6 to 3.2 mm) thick, a butt joint, with a space of approximately 1/8 in. (3.2 mm) between the edges, should be prepared. A 1/8 in. (3.2 mm) diameter copper-coated low carbon filler rod should be used. Sheet metal welding with a filler rod on butt joints should be done by the forehand method of welding.

b.

(1) General The term "steel" may be applied to many ferrous metals which differ greatly in both chemical and physical properties. In general, they may be divided into plain carbon and alloy groups. By following the proper procedures, most steels can be successfully welded. However, parts fabricated by welding generally contain less than 0.30 percent carbon. Heat increases the carbon combining power of steel. Care must be taken during all welding processes to avoid carbon pickup.

(2) Welding process. Steel heated with an oxyacetylene flame becomes fluid between 2450 and 2750°F (1343 and 1510°C), depending on its composition. It passes through a soft range between the solid and liquid states. This soft range enables the operator to control the weld. To produce a weld with good fusion, the welding rod should be placed in the molten puddle. The rod and base metal should be melted together so that they will solidify to form a solid joint. Care should be taken to avoid heating a large portion of the joint. This will dissipate the heat and may cause some of the weld metal to adhere to but not fuse with the sides of the welded joint. The flare should be directed against the sides and bottom of the welded joint. This will allow penetration of the lower section of the joint. Weld metal should be added in sufficient quantities to fill the joint without leaving any undercut or overlap. Do not overheat. Overheating will burn the weld metal and weaken the finished joint.

(3) Impurities.

(a) Oxygen, carbon, and nitrogen impurities produce defective weld metal because they tend to increase porosity, blowholes, oxides, and slag inclusions.

(b) When oxygen combines with steel to form iron oxides at high temperatures, care should be taken to ensure that all the oxides formed are removed by proper manipulation of the rod and torch flame. An oxidizing flame causes the steel to foam and give off sparks. The oxides formed are distributed through the metal and cause a brittle, porous weld. Oxides that form on the surface of the finished weld can be removed by wire brushing after cooling.

(c) A carburizing flame adds carbon to the molten steel and causes boiling of the metal. Steel welds made with strongly carburizing flames are hard and brittle.

(d) Nitrogen from the atmosphere will combine with molten steel to form nitrides of iron. These will impair its strength and ductility if included in sufficient quantities.

(e) By controlling the melting rate of the base metal and welding rod, the size of the puddle, the speed of welding, and the flame adjustment, the inclusion of impurities from the above sources may be held to a minimum.

c. Welding Steel Plates.

(1) In plates up to 3/16 in. (4.8 mm) in thickness, joints are prepared with a space between the edges equal to the plate thickness. This allows the flame and welding rod to penetrate to the root of the joint. Proper allowance should be made for expansion and contraction in order to eliminate warping of the plates or cracking of the weld.

(2) The edges of heavy section steel plates (more than 3/16 in. (4.8 mm) thick) should be beveled to obtain full penetration of the weld metal and good fusion at the joint. Use the forehand method of welding.

(3) Plates ½ to ¾ in. (12.7 to 19.1 mm) thick should be prepared for a U type joint in all cases. The root face is provided at the base of the joint to cushion the first bead or layer of weld metal. The backhand method is generally used in welding these plates.

NOTE

Welding of plates ½ to ¾ in. (12.7 to 19.1 mm) thick is not recommended for oxyacetylene welding.

(4) The edges of plates ¾ in. (19.1 mm) or thicker are usually prepared by using the double V or double U type joint when welding can be done from both sides of the plate. A single V or single U joint is used for all plate thicknesses when welding is done from one side of the plate.

d. General Principles in Welding Steel.

(1) A well balanced neutral flame is used for welding most steels. To be sure that the flame is not oxidizing, it is sometimes used with a slight acetylene feather. A very slight excess of acetylene may be used for welding alloys with a high carbon, chromium, or nickel content. However, increased welding speeds are possible by using a slightly reducing flame. Avoid excessive gas pressure because it gives a harsh flame. This often results in cold shuts or laps, and makes molten metal control difficult.

(2) The tip size and volume of flame used should be sufficient to reduce the metal to a fully molten state and to produce complete joint penetration. Care should be taken to avoid the formation of molten metal drip heads from the bottom of the joint. The flame should bring the joint edges to the fusion point ahead of the puddle as the weld progresses.

(3) The pool of the molten metal should progress evenly down the seam as the weld is being made.

11-13. GENERAL (cont)

(4) The inner cone tip of the flame should not be permitted to come in contact with the welding rod, molten puddle, or base metal. The flame should be manipulated so that the molten metal is protected from the atmosphere by the envelope or outer flame.

(5) The end of the welding rod should be melted by placing it in the puddle under the protection of the enveloping flame. The rod should not be melted above the puddle and allowed to drip into it.

11-14. BRAZING

a. General.

(1) Brazing is a group of welding processes which produces coalescence of materials by heating to a suitable temperature and using a filler metal having a liquidus above 840°F (449°C) and below the solidus of the base metals. The filler metal is distributed between the closely fitted surfaces of the joint by capillary attraction. Brazing is distinguished from soldering in that soldering employs a filler metal having a liquidus below 840°F (449°C).

(2) When brazing with silver alloy filler metals (silver soldering), the alloys have liquidus temperatures above 840°F (449°C).

(3) Brazing must meet each of three criteria:

(a) The parts must be joined without melting the base metals.

(b) The filler metal must have a liquidus temperature above 840°F (449°C).

(c) The filler metal must wet the base metal surfaces and be drawn onto or held in the joint by capillary attraction.

(4) Brazing is not the same as braze welding, which uses a brazing filler metal that is melted and deposited in fillets and grooves exactly at the points it is to be used. The brazing filler metal also is distributed by capillary action. Limited base metal fusion may occur in braze welding.

(5) To achieve a good joint using any of the various brazing processes, the parts must be properly cleaned and protected by either flux or the atmosphere during heating to prevent excessive oxidation. The parts must provide a capillary for the filler metal when properly aligned, and a heating process must be selected that will provide proper brazing temperatures and heat distribution.

b. Principles.

(1) Capillary flow is the most important physical principle which ensures good brazements providing both adjoining surfaces molten filler metal. The joint must also be properly spaced to permit efficient capillary action and resulting coalescence. More specifically, capillarity is a result of surface tension between base metal(s), filler metal, flux or atmosphere, and the contact angle between base and filler metals. In actual practice, brazing filler metal flowcharacteristic

are also influenced by considerations involving fluidity, viscosity, vapor pressure, gravity, and by the effects of any metallurgical reactions between the filler and base metals.

(2) The brazed joint, in general, is one of a relatively large area and very small thickness. In the simplest application of the process, the surfaces to be joined are cleaned to remove contaminants and oxide. Next, they are coated with flux or a material capable of dissolving solid metal oxides present and preventing new oxidation. The joint area is then heated until the flux melts and cleans the base metals, which are protected against further oxidation by the liquid flux layer.

(3) Brazing filler metal is then melted at some point on the surface of the joint area. Capillary attraction is much higher between the base and filler metals than that between the base metal and flux. Therefore, the flux is removed by the filler metal. The joint, upon cooling to room temperature, will be filled with solid filler metal. The solid flux will be found on the joint surface.

(4) High fluidity is a desirable characteristic of brazing filler metal because capillary attraction may be insufficient to cause a viscous filler metal to run into tight fitting joints.

(5) Brazing is sometimes done with an active gas, such as hydrogen, or in an inert gas or vacuum. Atmosphere brazing eliminates the necessity for post cleaning and ensures absence of corrosive mineral flux residue. Carbon steels, stainless steels, and super alloy components are widely processed in atmospheres of reacted gases, dry hydrogen, dissociated ammonia, argon, and vacuum. Large vacuum furnaces are used to braze zirconium, titanium, stainless steels, and the refractory metals. With good processing procedures, aluminum alloys can also be vacuum furnace brazed with excellent results.

(6) Brazing is a process preferred for making high strength metallurgical bonds and preserving needed base metal properties because it is economical.

c. Processes.

(1) Generally, brazing processes are specified according to heating methods (sources) of industrial significance. Whatever the process used, the filler metal has a melting point above 840°F (450°C) but below the base metal and distributed in the joint by capillary attraction. The brazing processes are:

(a) Torch brazing.

(b) Furnace brazing.

(c) Induction brazing.

(d) Resistance brazing.

(e) Dip brazing.

(f) Infrared brazing.

11-27

11-14. BRAZING (cont)

(2) Torch brazing.

(a) Torch brazing is performed by heating with a gas torch with a proper tip size, filler metal of required composition, and appropriate flux. This depends on the temperature and heat amount required. The fuel gas (acetylene, propane, city gas, etc.) may be burned with air, compressed air, or oxygen.

(b) Brazing filler metal may be preplaced at the joint in the forms of rings, washers, strips, slugs, or powder, or it may be fed from hand-held filler metal in wire or rod form. In any case, proper cleaning and fluxing are essential.

(c) For manual torch brazing the torch may be equipped with a single tip, either single or multiple flame. Manual torch brazing is particularly useful on assemblies involving sections of unequal mass. Welding machine operations can be set up where the production rate allows, using one or several torches equipped with single or multiple flame tips. The machine may be designed to move either the work or torches, or both. For premixed city gas–air flames, a refractory type burner is used.

(3) Furnance brazing.

(a) Furance brazing is used extensively where the parts to be brazed can be assembled with the brazing filler metal in form of wire, foil, filings, slugs, powder, paste, or tape is preplaced near or in the joint. This process is particularly applicable for high production brazing. Fluxing is employed except when an atmosphere is specifically introduced in the furnace to perform the same function. Most of the high production brazing is done in a reducing gas atomosphere, such as hydrogen and combusted gases that are either exothermic (formed with heat evolution) or endothermic (formed with heat absorption). Pure inert gases, such as argon or helium, are used to obtain special atmospheric properties.

(b) A large volume of furnace brazing is performed in a vacuum, which prevents oxidation and often eliminates the need for flux. Vacuum brazing is widely used in the aerospace and nuclear fields where reactive metals are joined or where entrapped fluxes would be intolerable. If the vacuum is maintained by continuous pumping, it will remove volatile constituents liberated during brazing. There are several base metals and filler metals that should not be brazed in a vacuum because low boiling point or high vapor pressure constituents may be lost. The types of furnaces generally used are either batch or contiguous. These furnaces are usually heated by electrical resistance elements, gas or oil, and should have automatic time and temperature controls. Cooling is sometimes accomplished by cooling chambers, which either are placed over the hot retort or are an integral part of the furnace design. Forced atmosphere injection is another method of cooling. Parts may be placed in the furnace singly, in batches, or on a continuous conveyor.

(c) Vacuum is a relatively economical method of providing an accurately controlled brazing atmosphere. Vacuum provides the surface cleanliness needed for good wetting and flow of filler metals without the use of fluxes. Base metals containing chromium and silicon can be easily vacuum brazed where a very pure, low dew point atmosphere gas would otherwise be required.

(4) Induction brazing.

(a) In this process, the heat necessary to braze metals is obtained from a high frequency electric current consisting of a motor-generator, resonant spark gap, and vacuum tube oscillator. It is induced or produced without magnetic or electric contact in the parts (metals). The parts are placed in or near a water-cooled coil carrying alternating current. They do not form any part of the electrical circuit. The brazing filler metal normally is preplaced.

(b) Careful design of the joint and the coil setup are necessary to assure that the surfaces of all members of the joint reach the brazing temperature at the same time. Flux is employed except when an atmosphere is specifically introduced to perform the same function.

(c) The equipment consists of tongs or clamps with the electrodes attached at the end of each arm. The tongs should preferably be water-cooled to avoid over-heating. The arms are current carrying conductors attached by leads to a transformer. Direct current may be used but is comparatively expensive. Resistance welding machines are also used. The electrodes may be carbon, graphite, refractory metals, or copper alloys according to the required conductivity.

(5) Resistance brazing. The heat necessary for resistance brazing is obtained from the resistance to the flow of an electric current through the electrodes and the joint to be brazed. The parts comprising the joint form a part of the electric circuit. The brazing filler metal, in some convenient form, is preplaced or face fed. Fluxing is done with due attention to the conductivity of the fluxes. (Most fluxes are insulators when dry.) Flux is employed except when an atmosphere is specifically introduced to perform the same function. The parts to be brazed are held between two electrodes, and proper pressure and current are applied. The pressure should be maintained until the joint has solidified. In some cases, both electrodes may be located on the same side of the joint with a suitable backing to maintain the required pressure.

(6) Dip brazing.

(a) There are two methods of dip brazing: chemical bath dip brazing and molten metal bath dip brazing.

(b) In chemical bath dip brazing, the brazing filler metal, in suitable form, is preplaced and the assembly is immersed in a bath of molten salt. The salt bath furnishes the heat necessary for brazing and usually provides the necessary protection from oxidation; if not, a suitable flux should be used. The salt bath is contained in a metal or other suitable pot, also called the furnace, which is heated from the outside through the wall of the pot, by means of electrical resistance units placed in the bath, or by the I^2R loss in the bath itself.

(c) In molten metal bath dip brazing, the parts are immersed in a bath of molten brazing filler metal contained in a suitable pot. The parts must be cleaned and fluxed if necessary. A cover of flux should be maintained over the molten bath to protect it from oxidation. This method is largely confined to brazing small

11-14. BRAZING (cont)

parts, such as wires or narrow strips of metal. The ends of the wires or parts must be held firmly together when they are removed from the bath until the brazing filler metal has fully solidified.

(7) Infrared brazing.

(a) Infrared heat is radiant heat obtained below the red rays in the spectrum. While with every "black" source there is sane visible light, the principal heating is done by the invisible radiation. Heat sources (lamps) capable of delivering up to 5000 watts of radiant energy are commercially available. The lamps do not necessarily need to follow the contour of the part to be heated even though the heat input varies inversely as the square of the distance from the source. Reflectors are used to concentrate the heat.

(b) Assemblies to be brazed are supported in a position that enables the energy to impinge on the part. In some applications, only the assembly itself is enclosed. There are, however, applications where the assembly and the lamps are placed in a bell jar or retort that can be evacuated, or in which an inert gas atmosphere can be maintained. The assembly is then heated to a controlled temperature, as indicated by thermocouples. The part is moved to the cooling platens after brazing.

(8) Special processes.

(a) Blanket brazing is another of the processes used for brazing. A blanket is resistance heated, and most of the heat is transferred to the parts by two methods, conduction and radiation, the latter being responsible for the majority of the heat transfer.

(b) Exothermic brazing is another special process by which the heat required to melt and flow a commercial filler metal is generated by a solid state exothermic chemical reaction. An exothermic chemical reaction is defined as any reaction between two or more reactants in which heat is given off due to the free energy of the system. Nature has provided us with countless numbers of these reactions; however, only the solid state or nearly solid state metal-metal oxide reactions are suitable for use in exothermic brazing units. Exothermic brazing utilizes simplified tooling and equipment. The process employs the reaction heat in bringing adjoining or nearby metal interfaces to a temperature where preplaced brazing filler metal will melt and wet the metal interface surfaces. The brazing filler metal can be a commercially available one having suitable melting and flow temperatures. The only limitations may be the thickness of the metal that must be heated through and the effects of this heat, or any previous heat treatment, on the metal properties.

d. Selection of Base Metal.

(1) In addition to the normal mechanical requirements of the base metal in the brazement, the effect of the brazing cycle on the base metal and the final joint strength must be considered. Cold-work strengthened base metals will be annealed when the brazing process temperature and time are in the annealing range of the base metal being processed. "Hot-cold worked" heat resistant base metals

can also be brazed; however, only the annealed physical properties will be available in the brazement. The brazing cycle will usually anneal the cold worked base metal unless the brazing temperature is very low and the time at heat is very short. It is not practical to cold work the base metal after the brazing operation.

(2) When a brazement must have strength above the annealed properties of the base metal after the brazing operation, a heat treatable base metal should be selected. The base metal can be an oil quench type, an air quench type that can be brazed and hardened in the same or separate operation, or a precipitation hardening type in which the brazing cycle and solution treatment cycle may be combined. Hardened parts may be brazed with a low temperature filler metal using short times at temperature to maintain the mechanical properties.

(3) The strength of the base metal has an effect on the strength of the brazed joint. Some base metals are also easier to braze than others, particularly by specific brazing processes. For example, a nickel base metal containing high titanium or aluminum additions will present special problems in furnace brazing. Nickel plating is sometimes used as a barrier coating to prevent the oxidation of the titanium or aluminum, and it presents a readily wettable surface to the brazing filler metal.

e. Brazing Filler Metals. For satisfactory use in brazing applications, brazing filler metals must possess the following properties:

(1) The ability to form brazed joints possessing suitable mechanical and physical properties for the intended service application.

(2) A melting point or melting range compatible with the base metals being joined and sufficient fluidity at brazing temperature to flow and distribute into properly prepared joints by capillary action.

(3) A composition of sufficient homogeneity and stability to minimize separation of constituents (liquation) under the brazing conditions to be encountered.

(4) The ability to wet the surfaces of the base metals being joined and form a strong, sound bond.

(5) Depending on the requirements, ability to produce or avoid base metal-filler metal interactions.

11-15.   BRAZING GRAY CAST IRON

a.   Gray cast iron can be brazed with very little or no preheating. For this reason, broken castings that would otherwise need to be dismantled and preheated can be brazed in place. A nonferrous filler metal such as naval brass (60 percent copper, 39.25 percent zinc, 0.75 percent tin) is satisfactory for this purpose. This melting point of the nonferrous filler metal is several hundred degrees lower than the cast iron; consequently the work can be accomplished with a lower heat input, the deposition of metal is greater and the brazing can be accomplished faster. Because of the lower heat required for brazing, the thermal stresses developed are less severe and stress relief heat treatment is usually not required.

b. The preparation of large castings for brazing is much like that required for welding with cast iron rods. The joint to be brazed must be clean and the part must be sufficiently warm to prevent chilling of filler metal before sufficient penetration and bonding are obtained. When possible, the joint should be brazed from both sides to ensure uniform strength throughout the weld. In heavy sections, the edges should be beveled to form a 60 to 90 degree V.

11-16. BRAZING MALLEABLE IRON

Malleable iron castings are usually repaired by brazing because the heat required for fusion welding will destroy the properties of malleable iron. Because of the special heat treatment required to develop malleability, it is impossible to completely restore these properties by simply annealing. Where special heat treatment can be performed, welding with a cast iron filler rod and remalleabilizing are feasible.

Section III. RELATED PROCESSES

11-17. SILVER BRAZING (SOLDERING)

a. Silver brazing, frequently called "silver soldering," is a low temperature brazing process with rods having melting points ranging from 1145 to 1650°F (618 to 899°C). This is considerably lower than that of the copper alloy brazing filler metals. The strength of a joint made by this process is dependent on a thin film of silver brazing filler metal. Silver brazing joints are shown in figure 11-16.

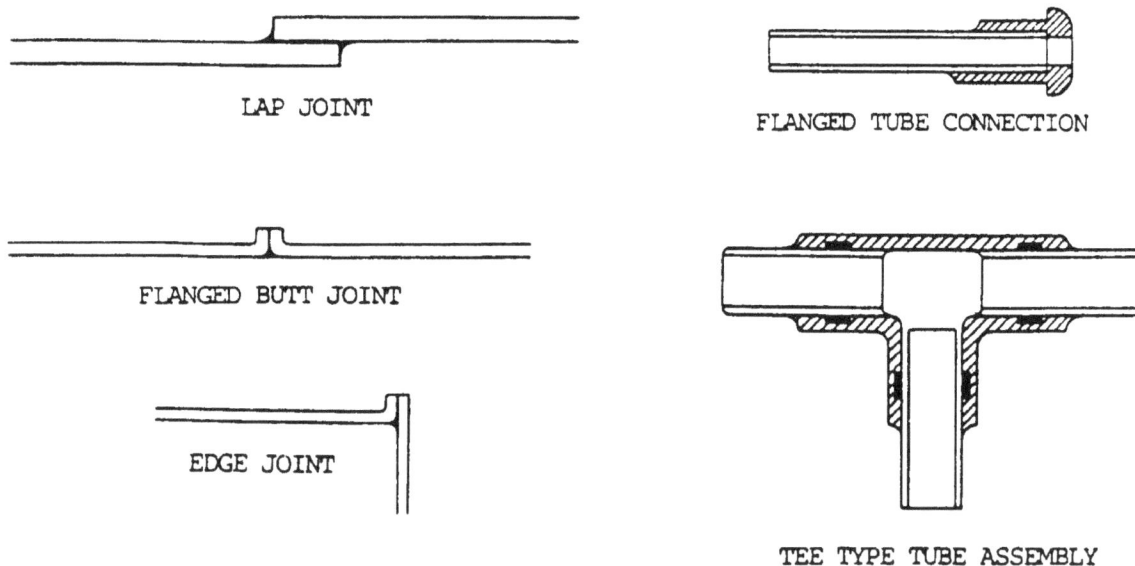

LAP JOINT

FLANGED TUBE CONNECTION

FLANGED BUTT JOINT

EDGE JOINT

TEE TYPE TUBE ASSEMBLY

Figure 11-16. Silver brazing joints.

b. Silver brazing filler metals are composed of silver with varying percentages of copper, nickel, tin, and zinc. They are used for joining all ferrous and non-ferrous metals except aluminum, magnesium, and other metals which have too low a melting point.

WARNING

Cadmium oxide fumes formed by heating and melting of silver brazing alloys are highly toxic. To prevent injury to personnel, personal protective equipment must be worn and adequate ventilation provided.

c. It is essential that the joints be free of oxides, scale, grease, dirt, or other foreign matter. Surfaces other than cadmium plating can be easily cleaned mechanically by wire brushing or an abrasive cloth; chemically by acid pickling or other means. Extreme care must be used to grind all cadmium surfaces to the base metals since cadmium oxide fumes formed by heating and melting of silver brazing alloys are highly toxic.

d. Flux is generally required. The melting point of the flux must be lower than the melting point of the silver brazing filler metal. This will keep the base metal clean and properly flux the molten metal. A satisfactory flux should be applied by means of a brush to the parts to be joined and also to the silver brazing filler metal rod.

e. When silver brazing by the oxyacetylene process, a strongly reducing flame is desirable. The outer envelope of the flame, not the inner cone, should be applied to the work. The cone of the flame is too hot for this purpose. Joint clearances should be between 0.002 and 0.005 in. (0.051 to 0.127 mm) for best filler metal distribution. A thin film of filler metal in a joint is stronger and more effective, and a fillet build up around the joint will increase its strength.

f. The base metal should be heated until the flux starts to melt along the line of the joint. The filler metal is not subjected to the flame, but is applied to the heated area of the base metal just long enough to flow the filler metal completely into the joint. If one of the parts to be joined is heavier than the other, the heavier part should receive the most heat. Also, parts having high heat conductivity should receive more heat.

11-18. OXYFUEL CUTTING

a. General.

(1) If iron or steel is heated to its kindling temperature (not less than 1600°F (871°C)), and is then brought into contact with oxygen, it burns or oxidizes very rapidly. The reaction of oxygen with the iron or steel forms iron oxide (Fe_3O_4) and gives off considerable heat. This heat is sufficient to melt the oxide and some of the base metal; consequently, more of the metal is exposed to the oxygen stream. This reaction of oxygen and iron is used in the oxyacetylene cutting process. A stream of oxygen is firmly fixed onto the metal surface after it has been heated to the kindling temperature. The hot metal reacts with oxygen, generating more heat and melting. The molten metal and oxide are swept away by the rapidly moving stream of oxygen. The oxidation reaction continues and furnishes heat

11-18. OXYFUEL CUTTING (cont)

for melting another layer of metal. The cut progresses in this manner. The principle of the cutting process is shown in figure 11-17.

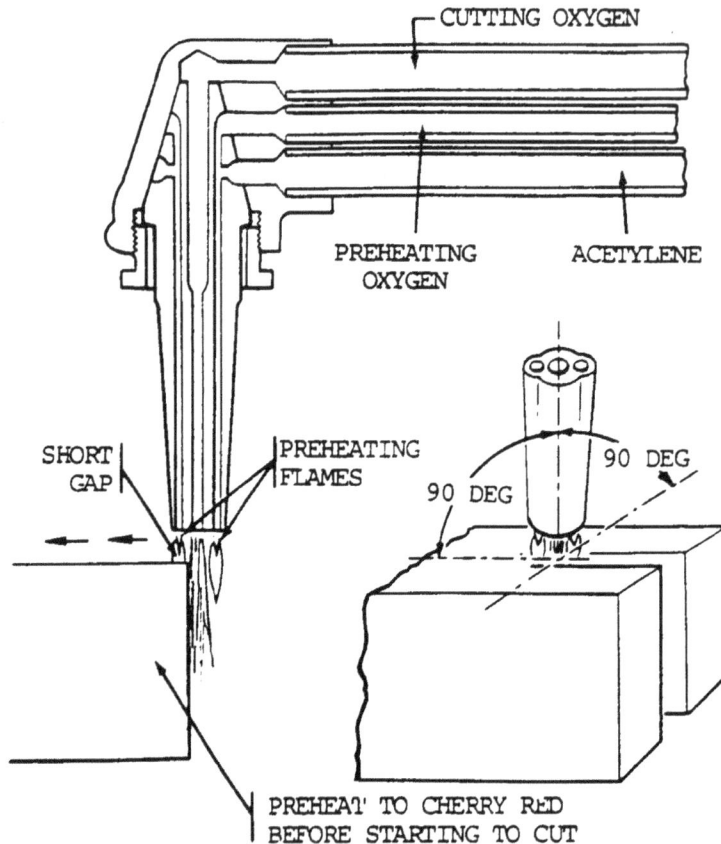

Figure 11-17. Starting a cut and cutting with a cutting torch.

(2)   Theoretically, the heat created by the burning iron would be sufficient to heat adjacent iron red hot, so that once started the cut could be continued indefinitely with oxygen only, as is done with the oxygen lance. In practice, however, excessive heat absorption at the surface caused by dirt, scale, or other substances, make it necessary to keep the preheating flames of the torch burning throughout the operation.

   b. Cutting Steel and Cast Iron.

      (1) General. Plain carbon steels with a carbon content not exceeding 0.25 percent can be cut without special precautions other than those required to obtain cuts of good quality. Certain steel alloys develop high resistance to the action of the cutting oxygen, making it difficult and sometimes impossible to propagate the cut without the use of special techniques. These techniques are described briefly in (2) and (3) which follow:

(2) <u>High carbon steels</u>. The action of the cutting torch on these metals is similar to a flame hardening procedure, in that the metal adjacent to the cutting area is hardened by being heated above its critical temperature by the torch and quenched by the adjacent mass of cold metal. This condition can be minimized or overcome by preheating the part from 500 to 600°F (260 to 316°C) before the cut is made.

(3) <u>Waster plate on alloy steel</u>. The cutting action on an alloy steel that is difficult to cut can be improved by clamping a mild steel "waster plate" tightly to the upper surface and cutting through both thicknesses. This waster plate method will cause a noticeable improvement in the cutting action, because the molten steel dilutes or reduces the alloying content of the base metal.

(4) <u>Chromium and stainless steels</u>. These and other alloy steels that previously could only be cut by a melting action can now be cut by rapid oxidation through the introduction of iron powder or a special nonmetallic powdered flux into the cutting oxygen stream. This iron powder oxidizes quickly and liberates a large quantity of heat. This high heat melts the refractory oxides which normally protect the alloy steel from the action of oxygen. These molten oxides are flushed from the cutting face by the oxygen blast. Cutting oxygen is enabled to continue its reaction with the iron powder and cut its way through the steel plates. The nonmetallic flux, introduced into the cutting oxygen stream, combines chemically with the refractory oxides and produces a slag of a lower melting point, which is washed or eroded out of the cut, exposing the steel to the action of the cutting oxygen.

(5) <u>Cast iron</u>. Cast iron melts at a temperature lower than its oxides. Therefore, in the cutting operation, the iron tends to melt rather than oxidize. For this reason, the oxygen jet is used to wash out and erode the molten metal when cast iron is being cut. To make this action effective, the cast iron must be preheated to a high temperature. Much heat must be liberated deep in the cut. This is done by adjusting the preheating flames so that there is an excess of acetylene. The length of the acetylene streamer and the procedure for advancing the cut are shown in figure 11-18. The use of a mild iron flux to maintain a high temperature in the deeper recesses of the cut, as shown in figure 11-18, is also effective.

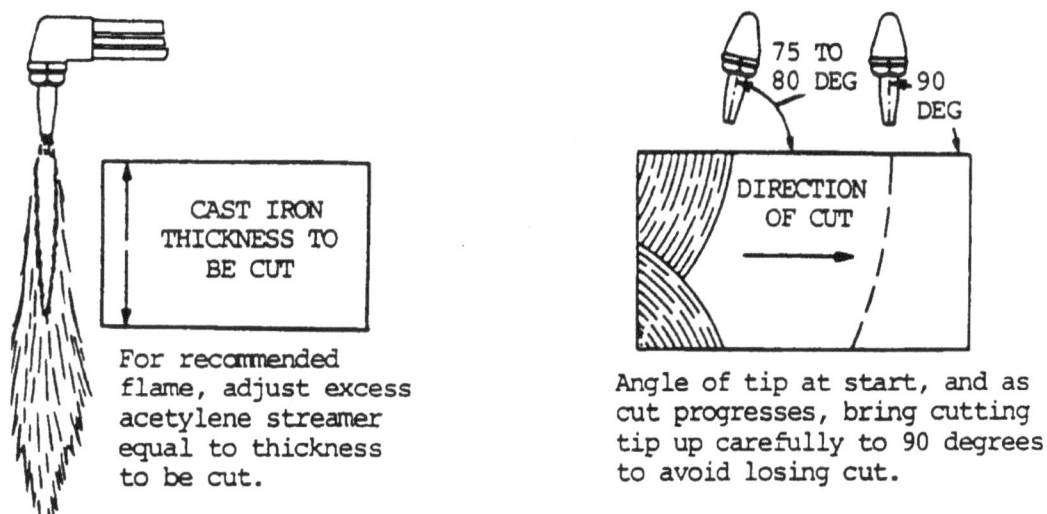

Figure 11-18. Procedure for oxyacetylene cutting of cast iron (sheet 1 of 2).

11-18. OXYFUEL CUTTING (cont)

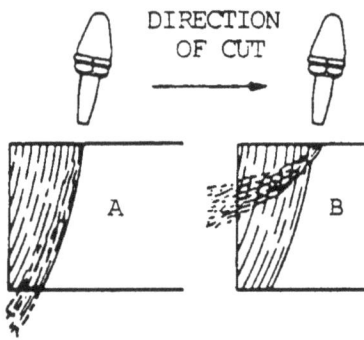

Cutting jet should just sweep edge of cut as shown in A and not advance too deeply as shown in B, otherwise progress of the cut will cease, and black spots will develop under the cutting jet.

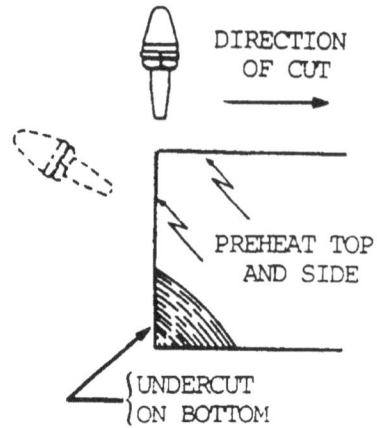

Begin and maintain cut, holding torch tip 1-1/2 to 2 in. (3.81 to 5.08 cm) from cast iron.

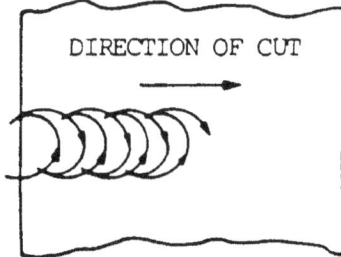

Move torch tip in semi-circular motions 1/2 in. to 3/4 in. (1.27 to 1.91 cm) as required to clear cut in heavy sections. Light sections require reduced oscillations of the torch tip.

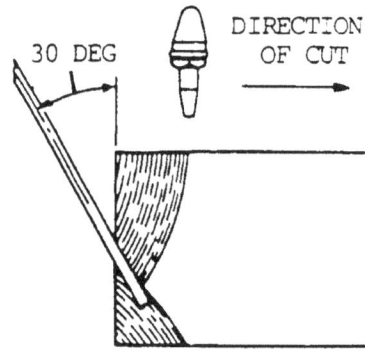

Approximate introduction angle of flux rod or lance to assist cutting operation.

Figure 11-18. Procedure for oxyacetylene cutting of cast iron (sheet 2 of 2).

c. Cutting with MAPP gas.

(1) Quality cuts with MAPP gas require a proper balance between preheat flame adjustment, oxygen pressure, coupling distance, torch angle, travel speed, plate quality, and tip size. Oxyfuel ratios to control flame condition are given in table 11-4.

Table 11-4. Oxyfuel Ratios Control Flame Condition

Flame	Oxy-MAPP Gas Ratio
Very carburizing	2.0 to 1
Slightly carburizing	2.3 to 1
Neutral	2.5 to 1
Oxidizing	3.0 to 1
Very oxidizing	3.5 to 1

(2) MAPP gas is similar to acetylene and other fuel gases in that it can be made to produce carburizing, neutral or oxidizing flames (table 11-4). The neutral flame is the adjust most likely to be used for flame cutting. After lighting the torch, slowly increase the preheat oxygen until the initial yellow flame becomes blue, with some yellow feathers remaining on the end of the preheat cones. This is a slightly carburizing flame. A slight twist of the oxygen valve will cause the feathers to disappear. The preheat cones will be dark blue in color and will be sharply defined. This is a neutral flame adjustment and will remain so, even with a small additional amount of preheat oxygenAnother slight twist of the oxygen valve will cause the flame to suddenly change color from a dark blue to a lighter blue color. An increase in sound also will be noted, and the preheat cones will become longer. This is an oxidizing flame. Oxidizing flames are easier to look at because of their lower radiance.

(3) MAPP gas preheat flame cones are at least one and one-half times longer than acetylene preheat cones when produced by the same basic style of tip.

(4) The situation is reversed for natural gas burners, or for torches with a two-piece tip. MAPP gas flame cones are much shorter than the preheat flame on a natural gas two-piece tip.

(5) Neutral flame adjustments are used most cutting. Carburizing and oxidizing flames also are used in special applications.For exanple, carburizing flame adjustments are used in stack cutting, or where a very square top edge is desired. The "slightly carburizing"flare is used to stack cut light material because slag formation is minimized. If a strongly oxidizing flame is used, enough slag may be produced in the kerf to weld the plates togetherSlag-welded plates often cannot be separated after the cut is completed.

(6) A "moderately oxidizing"flame is used for fast starts when cutting or piercing. It produces a slightly hotter flame temperature, and higher burning velocity than a neutral flame. An oxidizing flame commonly is used with a "high-low" device. The large "high" oxidizing flame is used to obtain a fast start. As soon as the cut has started, the operator drops to the "low" position and continues the cut with a neutral flame.

11-18. OXYFUEL CUTTING (cont)

(7) "Very oxidizing" flames should not be used for fast starting. An overly oxidizing flame will actually increase starting time. The extra oxygen flow does not contribute to combustion, but only cools the flame and oxidizes the steel surface.

(8) The oxygen pressure at the torch not at some remotely located regulator, should be used. Put a low volume, soft flame on the tip. Then turn on the cutting oxygen and vary the pressure to find the best looking stinger (visible oxygen cutting stream).

(a) Low pressures give very short stingers, 20 to 30 in. (50.8 to 76.2 cm) long. Low-pressure stingers will break up at the end. As pressure is increased, the stinger will suddenly become coherent and long.This is the correct cutting oxygen pressure for the given tip. The long stinger will remainover a fairly wide pressure range. But as oxygen pressures are increased, the stinger returns to the short, broken form it had under low pressure.

(b) If too high an oxygen pressure is usedconcavity often will show on the cut surface. Too high an oxygen pressure also can cause notching of the cut surface. The high velocity oxygen stream blows the metal and slag out of the kerf so fast that the cut is continuously being started.If too low a pressure is used, the operation cannot run at an adequate speedExcessive drag and slag formation results, and a wide kerf often is produced at the bottom of the cut.

(9) Cutting oxygen, as well as travel speed, also affects the tendency of slag to stick to the bottom of a cut.This tendency increases as theamount of metallic iron in the slag increases. Two factors cause high iron content in slag: too high a cutting oxygen pressure results in an oxygen velocity through the kerf high enough to blow out molten iron before the metal gets oxidized; and too high a cutting speed results in insufficient time to thoroughly oxidize the molten iron, with the same result as high oxygen pressure.

(10) The coupling distance is the distance between the end of the flame cones and the workpiece. Flame lengths vary with different fuels, and different flame adjusts. Therefore, the distance between the end of the preheat cones and the workpiece is the preferred measure (fig. 11-19). When cutting ordinary plate thicknesses up to 2 to 3 in. (5.08 to 7.62 cm) with MAPP gas, keep the end of the preheat cones abut 1/16 to 1/8 in. (0.16 to 0.32 cm) off the surface of the work. When piercing, or for very fast starts,let the preheat cones impinge on the surface. This will give faster preheating. As plate thicknesses increase above 6 in. (15.24 cm), increase the coupling distance to get more heating from the secondary flame cone. The secondary MAPP gas flame will preheat the thick plate far ahead of the cut. When material 12 in. (30.48 cm) thick or more is cut, use a coupling distance of ¾ to 1¼ in. (1.91 to 3.18 cm) long.

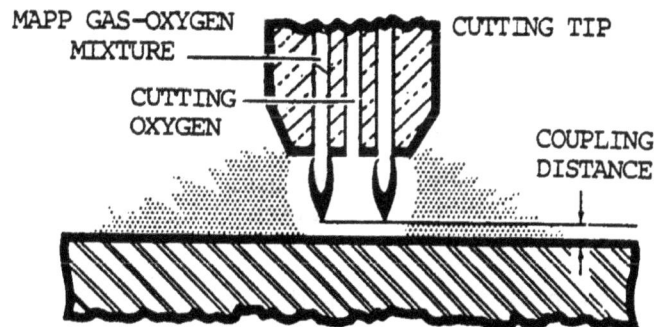

Figure 11-19. Coupling distance.

(11) Torch angle

(a) Torch, or lead angle, is the acute angle between the axis of the torch and the workpiece surface when the torch is pointed in the direction of the cut (fig. 11-20). When cutting light-gauge steel (up to ¼ in. (0.64 cm) thick) a 40 to 50 degree torch angle allows much faster cutting speeds than if the torch were mounted perpendicular to the plate. On plate up to ½ in. (1.27 cm) thick, travel speed can be increased with a torch lead angle, but the angle is larger, about 60 to 70 degrees. Little benefit is obtained from cutting plate over ½ in. (1.27 cm) thick with an acute lead-angle. Plate over this thickness should be cut with the torch perpendicular to the workpiece surface.

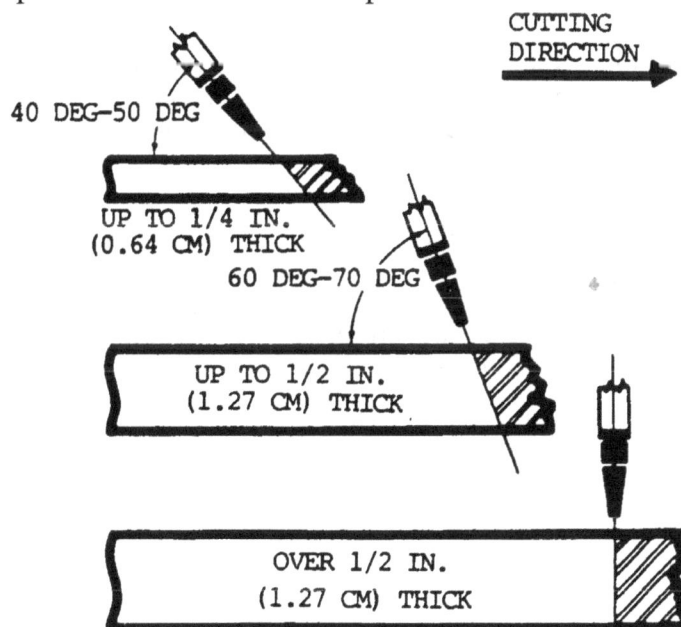

Figure 11-20. Torch angle.

11-18. OXYFUEL CUTTING (cont)

(b) An angled torch cuts faster on thinner-gauge material. The intersection of the kerf and the surface presents a knife edge which is easily ignited. Once the plate is burning, the cut is readily carried through to the other side of the work. When cutting heavy plate, the torch should be perpendicular to the workpiece surface and parallel to the starting edge of the work. This avoids problems of non-drop cuts, incomplete cutting on the opposite side of the thicker plate, gouging cuts in the center of the kerf and similar problems.

(12) There is a best cutting speed for each job. On plate up to about 2 in. (5.08 cm) thick, a high quality cut will be obtained when there is a steady "purring" sound from the torch and the spark stream under the plate has a 15 degree lead angle. This is the angle made by the sparks coming out of the bottom of the cut in the same direction as the torch is traveling. If the sparks go straight down, or even backwards, it means travel speed is too high.

(13) Cut quality.

(a) Variations in cut quality can result from different workpiece surface conditions or plate compositions. For example, rusty or oily plates require more preheat, or slower travel speeds than clean plates. Most variations from the ideal condition of a clean, flat, low-carbon steel plate tend to slow down the cutting action.

(b) One method to use for very rusty plate is to set as big a preheat flame as possible on the torch, then run the flame back and forth over the line to be cut. The extra preheat passes do several things. They spall off much of the scale that would otherwise interfere with the cutting action; and the passes put extra preheat into the plate which usually is beneficial in obtaining improved cut quality and speed.

(c) When working with high strength low alloy plates such as ASTM A-242 steel, or full alloy plates such as ASTM A-514, cut a little bit slower. Also use a low oxygen pressure because these steels are more sensitive to notching than ordinary carbon steels.

(d) Clad carbon-alloy, carbon-stainless, or low-carbon-high-carbon plates require a lower oxygen pressure, and perhaps a lower travel speed than straight low-carbon steel. Ensure the low carbon-steel side is on the same side as the torch. The alloyed or higher carbon cladding will not burn as readily as the carbon steel. By putting the cladding on the bottom, and the carbon steel on the top, a cutting action similar to powder cutting results. The low-carbon steel on top burns readily and forms slag. As the iron-bearing slag passes through the high-carbon or high-alloy cladding, it dilutes the cladding material. The torch, in essence, still burns a lower carbon steel. If the clad or high-carbon steel is on the top surface, the torch is required to cut a material that is not readily oxidizable, and forms refractory slags that can stop the cutting action.

(14) Tip size and style.

(a) Any steel section has a corresponding tip size that gives the most economical operation for a particular fuel. Any fuel will burn in any tip, of course. But the fuel will not burn efficiently, and may even overheat and melt the tip, or cause problems in the cut. For example, MAPP gas will not operate at peak efficiency in most acetylene tips because the preheat orifices are not large enough for MAPP. If MAPP gas is used with a natural-gas tip, there will be a tendency to overheat the tip. The tips also will be susceptible to flash back. A natural-gas tip can be used with MAPP gas, in an emergency, by removing the skirt. Similarly, an acetylene tip can be used if inefficient burning can be tolerated for a short run.

(b) The reasons for engineering different tips for different fuel gases are complex. But the object is to engineer the tip to match the burning velocity, port velocity, and other relationships for each type of gas and orifice size, and to obtain the optimum flame shape and heat transfer properties for each type of fuel. Correct cutting tips cost so little that the cost of conversion is minute compared with the cost savings resulting from efficient fuel use, improved cut quality, and increased travel speed.

Section IV. WELDING, BRAZING, AND SOLDERING NONFERROUS METALS

11-19. ALUMINUM WELDING

a. General.

(1) General. Aluminum is readily joined by welding, brazing, and soldering. In many instances, aluminum is joined with the conventional equipment and techniques used with other metals. However, specialized equipment or techniques may sometimes be required. The alloy, joint configuration, strength required, appearance, and cost are factors dictating the choice of process. Each process has certain advantages and limitations.

(2) Characteristics of aluminum. Aluminum is light in weight and retains good ductility at subzero temperatures. It also has high resistance to corrosion, good electrical and thermal conductivity, and high reflectivity to both heat and light. Pure aluminum melts at 1220°F (660°C), whereas aluminum alloys have an approximate melting range from 900 to 1220°F (482 to 660°C). There is no color change in aluminum when heated to the welding or brazing range.

(3) Aluminum forms. Pure aluminum can be alloyed with many other metals to produce a wide range of physical and mechanical properties. The means by which the alloying elements strengthen alminum is used as a basis to classify alloys into two categories: nonheat treatable and heat treatable. Wrought alloys in the form of sheet and plate, tubing, extruded and rolled shapes, and forgings have similar joining characteristics regardless of the form. Aluminum alloys are also produced as castings in the form of sand, permanent mold, or die castings. Substantially the same welding, brazing, or soldering practices are used on both cast and wrought

11-19. ALUMINUM WELDING (cont)

metal. Die castings have not been widely used where welded construction is required. However, they have been adhesively bonded and to a limited extent soldered. Recent developments in vacuum die casting have improved the quality of the castings to the point where they may be satisfactorily welded for some applications.

(4) Surface preparation. Since aluminum has a great affinity for oxygen, a film of oxide is always present on its surface. This film must be removed prior to any attempt to weld, braze, or solder the material. It also must be prevented from forming during the joining procedure. In preparation of aluminum for welding, brazing, or soldering, scrape this film off with a sharp tool, wire brush, sand paper, or similar means. The use of inert gases or a generous application of flux prevents the forming of oxides during the joining process.

b. Gas Welding.

(1) General. The gas welding processes most commonly used on aluminum and aluminum alloys are oxyacetylene and oxyhydrogen. Hydrogen may be burned with oxygen using the same tips as used with acetylene. However, the temperature is lower and larger tip sizes are necessary (table 11-5). Oxyhydrogen welding permits a wider range of gas pressures than acetylene without losing the desired slightly reducing flame. Aluminum from 1/32 to 1 in. (0.8 to 25.4 mm) thick may be gas welded. Heavier material is seldom gas welded, as heat dissipation is so rapid that it is difficult to apply sufficient heat with a torch. When compared with arc welding, the weld metal freezing rate of gas welding is very slow. The heat input in gas welding is not as concentrated as in other welding processes and unless precautions are taken greater distortion may result. Minimum distortion is obtained with edge or corner welds.

Table 11-5. Approximate Conditions for Gas Welding of Aluminum

| Metal Thickness (in.) | Filler Rod Dia (in.) | Oxyhydrogen * | | | Oxyacetylene | |
		Tip Orifice Dia (in.)	Hydrogen Pressure psi	Tip Orifice Dia (in.)	Oxygen Pressure psi	Acetylene Pressure psi
0.032	3/32	0.025	1	0.021	1	1
0.064	3/32	0.035	1-3	0.031	1+	1+
0.081	1/8	0.040	2-3	0.035	1+	1+
0.125	5/32	0.055	2-4	0.038	1-2	1-2
0.250	3/16	0.070	4-6	0.046	2-4	2-4
0.325	3/16	0.090	6-7	0.065	4-5	4-5
0.375	3/16	0.110	7-9	0.085	5-7	5-7

* Oxygen pressure cannot be given for oxyhydrogen burning conditions. Theoretically, two volumes of hydrogen are used for burning one of oxygen; however, as much as four volumes may be required. Therefore, oxygen pressure must be determined by trial until the best mixture is obtained.

(2) Edge preparation. Sheet or plate edges must be properly prepared to obtain gas welds of maximum strength. They are usually prepared the same as similar thicknesses of steel. However, on material up to 1/16 in. (1.6 mm) thick, the edges can be formed to a 90 degree flange. The flanges prevent excessive warping and buckling. They serve as filler metal during welding. Welding without filler rod is normally limited to the pure aluminum alloys since weld cracking can occur in the higher strength alloys. In gas welding thickness over 3/16 in. (4.8 mm), the edges should be beveled to secure complete penetration. The included angle of bevel may be 60 to 120 degrees. Preheating of the parts is recommended for all castings and plate ¼ in. (6.4 mm) thick or over. This will avoid severe thermal stresses and insure good penetration and satisfactory welding speeds. Common practice is to preheat to a temperature of 700°F (371°C). Thin material should be warmed with the welding torch prior to welding. Even this slight preheat helps to prevent cracks. Heat treated alloys should not be preheated above 800°F (427°C), unless they are to be postweld heat treated. Preheating above 800°F (427°C) will cause a "hot-short" and the metal strength will deteriorate rapidly.

(3) Preheat temperature checking technique. When pyrolytic equipment (temperature gauges) is not available, the following tests can be made to determine the proper preheat temperatures:

(a) Char test. Using a pine stick, rub the end of the stick on the metal being preheated. At the proper temperatures, the stick will char. The darker the char, the higher the temperature.

(b) Carpenter's chalk Mark the metal with ordinary blue carpenter' chalk. The blue line will turn white at the proper preheat temperature

(c) Hammer test. Tap the metal lightly with a hand hammer. The metal loses its ring at the proper preheat temperature.

(d) Carburizing test. Carburize the surface of the metal, sooting the entire surface. As the heat from the torch is applied, the soot disappears. At the point of soot disappearance, the metal surface is slightly above 300°F (149°C). Care should be used not to coat the fluxed area with soot. Soot can be absorbed into the weld, causing porosity.

(4) Welding flame. A neutral or slightly reducing flame is recommended or welding aluminum. Oxidizing flames will cause the formation of aluminum oxide, resulting in poor fusion and a defective weld.

(5) Welding fluxes.

(a) Aluminum welding flux is designed to remove the aluminum oxide film and exclude oxygen from the vicinity of the puddle.

(b) The fluxes used in gas welding are usually in powder form and are mixed with water to form a thin paste.

11-19. ALUMINUM WELDING (cont)

(c) The flux should be applied to the seam by brushing, sprinkling, spraying, or other suitable methods. The welding rod should also be coated. The flux will melt below the welding temperature of the metal and form a protective coating on the surface of the puddle. This coating breaks up the oxides, prevents oxidation, and permits slow cooling of the weld.

WARNING

The acid solutions used to remove aluminum welding and brazing fluxes after welding or brazing are toxic and highly corrosive.Goggles, rubber gloves, and rubber aprons must be worn when handling the acids and solutions. Do not inhale fumes. When spilled on the body or clothing, wash immediately with large quantities of cold water. Seek medical attention. Never pour water into acid when preparing solutions; instead, pour acid into water. Always mix acid and water slowly. These operations should only be performed in well ventilated areas.

(d) The aluminum welding fluxes contain chlorides and flourides. In the presence of moisture, these will attack the base metal. Therefore, all flux remaining on the joints after welding must be completely removed.If the weld is readily accessible, it can be cleaned with boiling water and a fine brush.Parts having joints located so that cleaning with a brush and hot water is not practical may be cleansed by an acid dip and a cold or hot water rinseUse 10 percent sulfuric acid cold water solution for 30 minutes or a 5 percent sulfuric acid hot water (150°F (66°C)) solution for 5 to 10 minutes for this purpose.

(6) Welding technique. After the material to be welded has been properly prepared, fluxed, and preheated,the flame is passed in small circles over the starting point until the flux melts. The filler rod should be scraped over toe surface at three or four second intervals, permitting the filler rod to come clear of the flame each time. The scraping action will reveal when welding can be started without overheating the aluminum.The base metal must be melted before the filler rod is applied. Forehand welding is generally considered best for welding on aluminum, since the flame will preheat the area to be welded.In welding thin aluminum, there is little need for torch movement other than progressing forward. On material 3/16 in. (4.8 mm) thick and over, the torch should be given a uniform lateral motion. This will distribute the weld metal over the entire width of the weld. A slight back and forth motion will assist the flux in the removal of oxide. The filler rod should be dipped into the weld puddle periodically, and withdrawn from the puddle with a forward motion.This method of withdrawal closes the puddle, prevents porosity, and assists the flux in removing the oxide film.

11-20. ALUMINUM BRAZING

a. General. Many aluminum alloys can be brazed. Aluminum brazing alloys are used to provide an all-aluminum structure with excellent corrosion resistance and good strength and appearance.The melting point of the brazing filler metal is relatively close to that of the material being joined.However, the base metal should not be melted; as a result, close temperate control is necessary. The brazing temperature required for aluminum assemblies is determined by the melting points of the base metal and the brazing filler metal.

b. <u>Commercial Filler Metals</u>. Commerical brazing filler metals for aluminum alloys are aluminum base. These filler metals are available as wire or shim stock. A convenient method of preplacing filler metal is by using a brazing sheet (an aluminum alloy base metal coated on one or both sides). Heat treatable or core alloys composed mainly of manganese or magnesium are also usedA third method of applying brazing filler metal is to use a paste mixture of flux and filler metal powder. Common aluminum brazing metals contain silicon as the melting point depressant with or without additions of zinc, copper, and magnesium.

c. <u>Brazing Flux</u>. Flux is required in all aluminum brazing operations. Aluminum brazing fluxes consist of various combinations of fluorides and chlorides and are supplied as a dry powder. For torch and furnace brazing, the flux is mixed with water to make paste. This paste is brushed, sprayed, dipped, or flowed onto the entire area of the joint and brazing filler metal. Torch and furnace brazing fluxes are quite active, may severely attack thin aluminum, and must be used with care. In dip brazing, the bath consists of molten flux. Less active fluxes can be used in this application and thin components can be safely brazed.

d. <u>Brazed Joint Design</u>. Brazed joints should be of lap, flange, lock seam, or tee type. Butt or scarf joints are not generally recommended. Tee joints allow for excellent capillary flow and the formation of reinforcing fillets on both sides of the joint. For maximum efficiency lap joints should have an overlap of at least twice the thickness of the thinnest joint member.An overlap greater than ¼ in. (6.4 mm) may lead to voids or flux inclusions. In this case, the use of straight grooves or knurls in the direction of brazing filler metal flow is beneficial. Closed assemblies should allow easy escape of gases, and in dip brazing easy entry as well as drainage of flux. Good design for long laps requires that brazing filler metal flows in one direction only for maximum joint soundnessThe joint design must also permit complete postbraze flux removal.

e. <u>Brazing Fixtures</u>. Whenever possible, parts should be designed to be self-jigging. When using fixtures, differential expansion can occur between the assembly and the fixture to distort the parts.Stainless steel or Inconel springs are often used with fixtures to accommodate differences in expansionFixture material can be mild steel or stainless steel. However, for repetitive furnace brazing operations and for dip brazing to avoid flux bath contamination, fixtures of nickel, Inconel, or aluminum coated steel are preferred.

f. <u>Precleanig</u>. Precleaning is essential for the production of strong, leaktight, brazed joints. Vapor or solvent cleaningwill usually be adequate for the nonheat treatable alloys. For heat treatable alloys, however, chemical cleaning or manual cleaning with a wire brush or sandpaper is necessary to remove the thicker oxide film.

g. <u>Furnace Brazing</u>. Furnace brazing is performed in gas, oil, or electrically heated furnaces. Temperature regulation within 5°F (2.8°C) is necessary to secure consistent results. Continuous circulation of the furnace atmosphere is desirable, since it reduces brazing time and results in more uniform heating. Products of combustion in the furnace can be detrimental to brazing and ultimate serviceability of brazed assemblies in the heat treatable alloys.

11-20. ALUMINUM BRAZING (cont)

h. Torch Brazing. Torch brazing differs from furnace brazing in that heat is localized. Heat is applied to the part until the flux and brazing filler metal melt and wet the surfaces of the base metal. The process resembles gas welding except that the brazing filler metal is more fluid and flows by capillary action. Torch brazing is often used for the attachment of fittings to previously weld or furnace brazed assemblies, joining of return bends, and similiar applications.

i. Dip Brazing. In dip brazing operations, a large amount of molten flux is held in a ceramic pot at the dip brazing temperature. Dip brazing pots are heated internally by direct resistance heating. Low voltage, high current transformers supply alternating current to pure nickel, nickel alloy, or carbon electrodes immersed in the bath. Such pots are generally lined with high alumina content fire brick and a refractory mortar.

WARNING

The acid solutions used to remove aluminum   welding and brazing fluxes after welding or brazing are toxic and highly corrosive. Goggles, rubber gloves, and rubber aprons must be worn when handling the acids and solutions.   Do not inhale fumes.   When spilled on the body or clothing, wash immediately with large quantities of cold water. Seek medical attention.

Never pour water into acid when preparing solutions: instead, pour acid into water.   Always mix acid and water slowly. These operations should only be performed in well ventilated areas.

j. Postbrazing Cleaning It is always necessary to clean the brazed assemblies, since brazing fluxes accelerate corrosion if left on the parts. The most satisfactory way of removing the major portion of the flux is to immerse the hot parts in boiling water as soon as possible after the brazing alloy has solidified. The steam formed removes a major amount of residual flux. If distortion from quenching is a problem, the part should be allowed to cool in air before being immersed in boiling water. The remaining flux may be removed by a dip in concentrated nitric acid for 5 to 15 minutes. The acid is removed with a water rinse, preferably in boiling water in order to accelerate drying. An alternate cleaning method is to dip the parts for 5 to 10 minutes in a 10 percent nitric plus 0.25 percent hydrofluoric acid solution at room temperature. This treatment is also followed by a hot water rinse. For brazed assemblies consisting of sections thinner than 0.010 in. (0.254 mm), and parts where maximum resistance to corrosion is important. A common treatment is to immerse in hot water followed by a dip in a solution of 10 percent nitric acid and 10 percent sodium dichromate for 5 to 10 minutes. This is followed by a hot water rinse. When the parts emerge from the hot water rinse they are immediately dried by forced hot air to prevent staining.

11-21. SOLDERING

a. General Soldering is a group of processes that join metals by heating them to a suitable temperature. A filler metal that melts at a temperature above 840°F (449°C) and below that of the metals to be joined is used. The filler metal is distributed between the closely fitted surfaces of the joint by capillary attraction.   Soldering uses fusible alloys to join metals. The kind of solder used depends on the metals to be joined. Hard solders are called spelter and hard soldering is called silver solder brazing. This process gives greater strength and will stand more heat than soft solder.

b. <u>Soft Soldering</u>. This process is used for joining most common metals with an alloy that melts at a temperature below that of the base metal. In many respects, this operation is similar to brazing in that the base is not melted, but is merely tinned on the surface by the solder filler metal. For its strength the soldered joint depends on the penetration of the solder into the pores of the base metal surface, along with the consequent formation of a base metal-solder alloy, together with the mechanical bond between the parts. Soft solders are used for airtight or watertight joints which are not exposed to high temperatures.

c. <u>Joint Preparation</u>. The parts to be soldered should be free of all oxide, scale, oil, and dirt to ensure sound joints. Cleaning may be performed by immersing in caustic or acid solutions, filing, scraping, or sandblasting.

d. <u>Flux.</u> All soldering operations require a flux in order to obtain a complete bond and full strength at the joints. Fluxes clean the joint area, prevent oxidations, and increase the wetting power of the solder by decreasing its surface tension. The following types of soft soldering fluxes are in common use: rosin, or rosin and glycerine. These are used on clean joints to prevent the formation of oxides during the soldering operations. Zinc chloride and ammonium chloride may be used on tarnished surfaces to permit good tinning. A solution of zinc cut in hydrochloric (muriacic) acid is commonly used by tin workers as a flux.

e. <u>Application</u>. Soft solder joints may be made by using gas flames, wiping, sweating the joints, or by dipping in solder baths. Dipping is particularly applicable to the repair of radiator cores. Electrical connections and sheet metal are soldered with a soldering iron or gun. Wiping is a method used for joining lead pipe and also the lead jacket of underground and other lead-covered cables. Sweated joints may be made by applying a mixture of solder powder and paste flux to the joints. Then heat the part until this solder mixture liquifies and flows into the joints, or tin mating surfaces of members to be joined, and apply heat to complete the joint.

11-22. **ALUMINUM SOLDERING**

a. <u>General</u>. Aluminum and aluminum base alloys can be soldered by techniques which are similar to those used for other metals. Abrasion and reaction soldering are more commonly used with aluminum than with other metals. However, aluminum requires special fluxes. Rosin fluxes are not satisfactory.

b. <u>Solderability of Aluminum Alloys</u>. The most readily soldered aluminum alloys contain no more than 1 percent magnesium or 5 percent silicon. Alloys containing greater amounts of these constituents have poor flux wetting characteristics. High copper and zinc-containing alloys have poor soldering characteristics because of rapid solder penetration and loss of base metal properties.

c. <u>Joint Design</u>. The joint designs used for soldering aluminum assemblies are similar to those used with other metals. The most commonly used designs are forms of simple lap and T-type joints. Joint clearance varies with the specific soldering method, base alloy composition, solder composition, joint design, and flux

11-22. ALUMINUM SOLDERING (cont)

composition employed. However, as a guide, joint clearance ranging from 0.005 to 0.020 in. (0.13 to 0.51 mm) is required when chemical fluxes are used. A 0.002 to 0.010 in. (0.05 to 0.25 mm) spacing is used when a reaction type flux is used.

d. <u>Preparation for Soldering</u>. Grease, dirt, and other foreign material must be removed from the surface of aluminum before soldering. In most cases, only solvent degreasing is required. However, if the surface is heavily oxidized, wire-brushing or chemical cleaning may be required.

CAUTION
Caustic soda or cleaners with a pH above 10 should not be used on aluminum or aluminum alloys, as they may react chemically.

e. <u>Soldering techniques</u>. The higher melting point solders normally used to join aluminum assemblies plus the excellent thermal conductivity of aluminum dictate that a large capacity heat source must be used to bring the joint area to proper soldering temperature. Uniform, well controlled heating should be provided. Tinning of the aluminum surface can best be accomplished by covering the material with a molten puddle of solder and then scrubbing the surface with a non-heat absorbing item such as a glass fiber brush, serrated wooden stick or fiber block. Wire brush or other metallic substances are not recommended. They tend to leave metallic deposits, absorb heat, and quickly freeze the solder.

f. <u>Solders.</u> The commercial solders for aluminum can be classified into three general groups according to their melting pints:

(1) <u>Lay temperature solders</u>. The melting point of these solders is between 300 and 500°F (149 and 260°C). Solders in this group contain tin, lead, zinc, and/or cadmium and produce joints with the least corrosion resistance.

(2) <u>Intermediate temperature</u> solders. These solders melt between 500 and 700 °F (260 and 371°C). Solders in this group contain tin or cadmium in various combinations with zinc, plus small amounts of aluminum, copper, nickel or silver, and lead.

(3) <u>High temperature solders</u>. These solders melt between 700 and 800°F (371 and 427°C). These zinc base solders contain 3 to 10 percent aluminum and small amounts of other metals such as copper, silver, nickel; and iron to modify their melting and wetting characteristics. The high zinc solders have the highest strength of the aluminum solders, and form the most corrosion-resistant soldered assemblies.

11-23. COPPER WELDING

a. Copper has a high thermal conductivity. The heat required for welding is approximately twice that required for steel of similar thickness. Too offset this heat loss, a tip one or two sizes larger than that required for steel is recommended. When welding large sections of heavy thicknesses, supplementary heating is advisable. This process produces a weld that is less porous.

b. Copper may be welded with a slightly oxidizing flame because the molten metal is protected by the oxide which is formed by the flame. If a flux is used to protect the molten metal, the flame should be neutral.

c. Oxygen-free copper (deoxidized copper red) should be used rather than oxygen-bearing copper for gas welded assemblies. The rod should be of the same composition as the base metal.

d. In welding copper sheets, the heat is conducted away from the welding zone so rapidly that it is difficult to bring the temperature up to the fusion point. It is often necessary to raise the temperature level of the sheet in an area 6.0 to 12.0 in. (152.4 to 304.8 mm) away from the weld. The weld should be started at some point away from the end of the joint and welded back to the end with filler metal being added. After returning to the starting point, the weld should be started and made in the opposite direction to the other end of the seam. During the operation, the torch should be held at approximately a 60 degree angle to the base metal.

e. It is advisable to back up the seam on the underside with carbon blocks or thin sheet metal to prevent uneven penetration. These materials should be channeled or undercut to permit complete fusion to the base of the joint. The metal on each side of the weld should be covered to prevent radiation of heat into the atmosphere. This would allow the molten metal in the weld to solidify and cool slowly.

f. The welding speed should be uniform. The end of the filler rod should be kept in the molten puddle. During the entire welding operation, the molten metal most be protected by the outer flame envelope. If the metal fails to flow freely during the operation, the rod should be raised and the base metal heated to a red heat along the seam. The weld should be started again and continued until the seam weld is completed.

g. When welding thin sheets, the forehand welding method is preferred. The backhand method is preferred for thicknesses of ¼ in (6.4 mm) or more. For sheets up to 1/8 in. (3.2 mm) thick a plain butt joint with squared edges is preferred. For thicknesses greater than 1/8 in. (3.2 mm) the edges should be beveled for an included angle of 60 to 90 degrees. This will ensure penetration with spreading fusion over a wide area.

11-24. COPPER BRAZING

a. Both oxygen-bearing and oxygen-free copper can be brazed to produce a joint with satisfactory properties. The full strength of an annealed copper brazed joint will be developed with a lap joint.

b. The flame used should be slightly carburizing. All of the silver brazing alloys can be used with the proper fluxes. With the copper-phosphorous or copper-phosphorous-silver alloys, a brazed joint can be made without a flux, although the use of flux will result in a joint of better appearance.

c. Butt, lap, and scarf joints are used in brazing operations, whether the joint members are flat, round, tubular, or of irregular cross sections. Clearances to permit the penetration of the filler metal, except in large diameter pipe joints, should not be more than 0.002 to 0.003 in. (0.051 to 0.076 mm). The clearances of large diameter pipe joinings may be 0.008 to 0.100 in. (0.203 to 2.540 mm). The joint may be made with inserts of the filler metal or the filler metal may be fed in from the outside after the joint has been brought up to the proper temperature. The scarf joint is used in joining bandsaws and for joints where the double thickness of the lap is not desired.

11-25.   BRASS AND BRONZE WELDING

a.   General. The welding of brasses and bronzes differs from brazing. This welding process requires the melting of both base metal edges and the welding rod, whereas in brazing only the filler rod is melted.

b.   Low Brasses (Copper 80 to 95 Percent, Zinc 5 to 20 Percent). Brasses of this type can be welded readily in all positions by the oxyacetylene process. Welding rods of the same composition as the base metal are not available. For this reason, 1.5 percent silicon rods are recommended as filler metal. Their weldability differs from copper in that the welding point is progressively reduced as zinc is added. Fluxes are required. Preheating and supplementary heating may also be necessary.

c.   High Brasses (Copper 55 to 80 Percent, Zinc 20 to 45 Percent). These brasses can be readily welded in all positions by the oxyacetylene process. Welding rods of substantially the same composition are available. The welding technique is the same as that required for copper welding, including supplementary heating. Fluxes are required.

d.   Aluminum Bronze. The aluminum bronzes are seldom welded by the oxyacetylene process because of the difficulty in handling the aluminum oxide with the fluxes designed for the brasses. Sane success has been reported by using welding rods of the same content as the base metal and a bronze welding flux, to which has been added a small amount of aluminum welding flux to control the aluminum oxide.

e.   Copper-Beryllium Alloys. The welding of these alloys by the oxyacetylene process is very difficult because of the formation of beryllium oxide.

f.   Copper-Nickel Alloys. From a welding standpoint, these alloys are similar to Monel, and oxyacetylene welding can be used successfully. The flame used should be slightly reducing. The rod must be of the same composition as the base metal. A sufficient deoxidizer (manganese or silicon) is needed to protect the metal during welding. Flux designed specifically for Monel and these alloys must be used to prevent the formation of nickel oxide and to avoid porosity. Limited melting of the base metal is desirable to facilitate rapid solidification of the molten metal. Once started, the weld should be completed without stopping. The rod should be kept within the protective envelope of the flame.

g.   Nickel Silver. Oxyacetylene welding is the preferred method for joining alloys of this type. The filler metal is a high zinc bronze which contains more than 10 percent nickel. A suitable flux must be used to dissolve the nickel oxide and avoid porosity.

h. <u>Phosphor Bronze</u>. Oxyacetylene welding is not commonly used for welding the copper-tin alloys. The heating and slow cooling causes contraction, with consequent cracking and porosity in this hot-short material. However, if the oxyacetylene process must be used the welding rod should be grade E (1.0 to 1.5 percent tin) with a good flux of the type used in braze welding. A neutral flame is preferred unless there is an appreciable amount of lead present. In this case an oxidizing flame will be helpful in producing a sound weld. A narrow heat zone will promote quick solidification and a sound weld.

NOTE
Hot-short is defined as a marked loss in strength at high temperatures below the melting point.

i. <u>Silicon Bronze</u>. Copper-silicon alloys are successfully welded by the oxyacetylene process. The filler metal should be of the same composition as the base metal. A flux with a high boric acid content should be used. A weld pool as small as possible should be maintained to facilitate rapid solidification. This will keep the grain size small and avoid contraction strains during the hot-short temperature range. A slightly oxidizing flare will keep the molten metal clean in oxyacetylene welding of these alloys. This flame is helpful when welding in the vertical or overhead positions.

11-26. MAGNESIUM WELDING

a. <u>General</u>. Gas welding of magnesium is usually used only in emergency repair work. A broken or cracked part can be restored and placed back into use. However, such a repair is only temporary until a replacement part can be obtained. Gas welding has been almost completely phased out by gas-shielded arc welding, which does not require the corrosion-producing flux needed for gas welding.

b. <u>Base Metal Preparation</u>. The base metal preparation is the same as that for arc welding.

c. <u>Welding Fluxes</u>.

(1) The flux protects the molten metal from excessive oxidation and removes any oxidation products from the surfaces. It also promotes proper flow of the weld and proper wetting action between the weld metal and the base metal. Most of the fluxes do not react with magnesium in the fused state, but do react strongly after cooling by taking on moisture. Therefore, all traces of flux and flux residues must be removed immediately after welding.

(2) The fluxes are usually supplied as dry powder in hermetically sealed bottles. They are prepared for use by mixing with water to form a paste. A good paste consistency can be produced with approximately ½ pt (0.24 l) of water to 1 lb (0.45 kg) of powder. Do not prepare any more than a one day supply of flux paste. Keep it in a covered glass container when not in use. The flux paste can be applied to the work and welding rod with a small bristle brush, or when possible, by dipping.

11-26. MAGNESIUM WELDING (cont)

(3) The presence of a large amount of sodium in the welding flux gives an intense glare to the welding flame. Operators must wear proper protective attire. Blue lenses are preferred in the goggles.

d. Welding Rods. The reds should be approximately the same composition as the base metal. When castings or forged fittings are welded to a sheet, it is important that the rod have the same composition as the sheet. If necessary, strips of the base metal may be used instead of regular welding rods. Welding rods may be readily identified by the following characteristics: Dowmetal F is blue; Dowmetal J or J-1 is yellow, green, and aluminum; Dowmetal M is yellow; Mazlo AM 35 is round; Mazlo AM 52S and AM 53S are square; Mazlo AM 57S is triangular; Mazlo AM 88S is oval. Like all magnesium alloys, the welding rods are supplied with a corrosion resistant coating which must be removed before using. After a welding operation, all traces of flux should be removed from the unused portion of the rod.

e. Welding Technique.

(1) A liberal coating of flux should be applied to both sides of the weld seam and onto the welding rod. The torch should be adjusted to a neutral or slightly reducing flame.

(2) Tack welds should be spaced at ½ to 2-½ in. (12.7 to 63.5 mm) intervals along the seam. In making a tack weld, the weld area should be heated gently with the outer flame of the torch to dry and fuse the flux. Do not use a harsh flame which may blow the flux away. When the flux is liquified, the inner cone of the flare is held a distance of 1/16 to 1/8 in. (1.6 to 3.2 mm) from the work and a drop of metal is added from the rod. More flux will be required to finish the weld.

(3) The weld should start in the same manner as the tack welds. The welds should progress in a straight line at a uniform rate of speed with the torch held at a 45 degree angle to the work. The torch should move steadily while the rod is intermittently dipped into the weld puddle. If a decrease of heat is necessary, it is advisable to decrease the angle of the torch from the work. Too hot a flame or too slow a speed increases the activity and viscosity of the flux and causes pitting. If a weld is interrupted, the end of the weld should be refluxed and the flame directed slightly ahead of the weld before restarting the bead. All tack and overlapping welds should be remelted to float away any flux inclusions. To avoid cracking at the start, the weld should be started away from the edge.

(4) Magnesium castings to be welded should be preheated with a torch or in a furnace before welding is started. The entire casting should be brought up to a preheat temperature of abut 650°F (343°C). This temperature can be approximated with blue carpenter's chalk which will turn white at about 600°F (316°C). After welding, the casting should be stress relieved in a furnace for 1 hour at 500°F (260°C). If no furnace is available, a gas flame should be used to heat the entire casting until the stress relieving temperature is reached. The casting should then be allowed to cool slowly, away from all drafts.

WARNING

Precleaning and postcleaning acids used in magnesium welding and brazing are highly toxic and corrosive. Goggles, rubber gloves, and rubber aprons must be worn when handling the acids and solutions. Do not inhale fumes and mists. When spilled on the body or clothing, wash immediately with large quantities of cold water. Seek medical attention.

Never pour water into acid when preparing solutions; instead, pour acid into water. Always mix acid and water slowly. Cleaning operations should be performed only in well ventilated areas.

f. <u>Cleaning After Gas Welding</u>. All traces of flux must be removed from parts immediately after the completion of gas welds. First, scrub with a stiff bristle brush and hot water, and then immerse for 1 to 2 minutes in a chrome pickling solution consisting of 1-½ lb (O.7 kg) sodium dichromate and 1-½ pt (O.7 1) nitric acid with enough water to make a gallon. The temperature of the solution should be 70 to 90°F (21 to 32°C). After chrome pickling, the parts should be washed in cold running water. They should be boiled for 2 hours in a solution of 8 oz (226.8 g) of sodium dichromate in 1 gal. (3.8 1) of water. Parts should then be rinsed and dried.

11-27. MAGNESIUM BRAZING

a. <u>General</u>.

(1) Furnace, torch, and flux dip brazing can be used. Furnace and torch brazing are generally limited to M1A alloys. Flux dip brazing can be used on AX10, AX31B, K1A, M1A, and ZE10A alloys.

(2) Brazed joints are designed to permit the flux to be displaced from the joint by the brazing filler metal as it flows into the joint. The best joints for brazing are butt and lap. Suitable clearances between parts are essential if proper capillary filling action is to take place. The suggested clearance is from 0.004 to 0.010 in. (0.102 to 0.254 mm). In furnace brazing, beryllium is added to the brazing alloy to avoid ignition of the magnesium.

b. <u>Equipment and Materials</u>.

(1) Furnaces and flux pots are equipped with automatic controls to maintain the required temperature within ± 5°F (2.7°C). In torch brazing, the standard type gas welding is used.

(2) Chloride base fluxes similar to those used in gas welding are suitable. A special flux is required for furnace brazing. Fluxes with a water or alcohol base are unsuitable for furnace brazing.

(3) A magnesium base alloy filler metal is used so that the characteristics of the brazed joint are similar to a welded joint and will offer good resistance against corrosion.

11-27. MAGNESIUM BRAZING (cont)

c. Base Metal Preparation.

(1) Parts to be brazed must be thoroughly cleaned and free of such substances as oil, grease, dirt, and surface films such as chromates and oxides.

WARNING
Precleaning and postcleaning acids used in magnesium welding and braz-
ing are highly toxic and corrosive.   Goggles, rubber gloves, and rubber
aprons must be worn when handling the acids and solutions.   Do not
inhale fumes and mists.   When spilled on the body or clothing, wash
immediately with large quantities of cold water. Seek medical atten-
tion.

Never pour water into acid when preparing solutions; instead, pour acid
into water.   Always mix acid and water slowly. Cleaning operations
should be performed only in well ventilated areas.

(2) Mechanical cleaning can be accomplished by sanding with aluminum oxide cloth. Chemical cleaning can be accomplished by vapor decreasing, alkaline cleaning, or acid cleaning. An acid solution consisting of 24 oz (680.4 g) of chromic acid (CrO), 40 oz (1134 g) of sodium nitrate (NaNO and 1/8 oz (3.54 g) of calcium or magnesium fluoride with enough water to make gal. (3.8 1) is suitable for this purpose. Parts are immersed in the solution at 70 to 90°F (21 to 32°C) for 2 minutes and then rinsed thoroughly, first in cold water and then in hot water.

d. Brazing Procedure.

(1) Torch. The equipment and techniques used for gas welding are used in brazing magnesium. A neutral oxyacetylene or a natural gas-air flame may be used. In some operations, natural gas is preferred because of its soft flame and less danger of overheating. The filler metal is placed on the joint and fluxed before melting, or it is added by means of a flux coated filler rod. If a rod is used, the flame is directed at the base metal, and the rod is dipped intermittently into the molten flux puddle.

(2) Furnace. The parts to be brazed are assembled with filler metal placed in or around the joints. A flux, preferably in powder form, is put on the joints. Then the parts are placed in a furnace, which is at brazing temperature. The brazing time is 2 or 3 minutes, depending on the thickness of the parts being brazed. The parts are air cooled after removal from the furnace.

(3) Flux dip. The joints are provided with slots or recessed grooves for the filler metal, to prevent it being washed into the flux bath. The parts are then assembled in a fixture, thoroughly dried, and then immersed for 30 to 45 seconds in a molten bath of flux.

e. Cleaning after Brazing. Removal of all traces of flux is essential. The flux residues are hydroscopic, and will cause a pitting type of corrosion. The parts should be cleaned in the same manner as for gas welded parts.

11-28. MAGNESIUM SOLDERING

a. General. Magnesium and magnesium alloys can be soldered. However, soldering is limited to the filling of small surface defects in castings, small dents in sheet metal, and other minor treatments of surfacesSoldering should not be used in stress areas or to join magnesium to other metals because of low strengths and brittle joints obtained.

b. Soldering Procedure.

(1) Magnesium alloy surfaces must be cleaned to a bright metallic luster before soldering to ensure good fusing between the solder and magnesium. This cleaning can be accomplished by filing, wire brushing, or with aluminum oxide cloth.

(2) The area to be soldered should be heated to just above the melting point of the solder. A small quantity of solder is applied and rubbed vigorously over the area to obtain a uniform tinned surfaceA stiff wire brush or sharp steel tool assists in establishing a bond. After the bond is established, filler metal may be added to the extent desired. Flux is not necessary.

11-29. NICKEL WELDING

a. General. Nickel alloys can, for the most part, be welded with the same processes used for carbon steel. Oxyacetylene welding is preferred to metal arc in some cases. This is true in welding on thin wall pipe or tubing, and tin gauge strip where the arc would penetrate the material.It is also preferred on some high carbon steels because of the lower weld hardening results.

b. Joint Design. Corner and lap joints are satisfactory where high stresses are not to be encountered. Butt joints are used in equipment such as pressure vessels. Beveling is not required for butt joints in material 0.050 to 0.125 in. (1.270 to 3.175 mm) thick. In thicker material, a bevel angle of 37.5 degrees should be made. For sheets 0.43 in. (10.92 mm) and thinner, both butt and corner joints are used. Corner joints are used for thicknesses of 0.037 in. (O.940 mm) and heavier.

c. Fluxes. Flux is not required when welding nickel. However, it is required for Monel and Inconel. The fluxes are used preferably in the form of a thin paste made by mixing the dry flux in water for Monel. A thin solution of shellac and alcohol (approximately 1.0 lb (O.45 kg) of shellac to 1.0 gal (3.8 1) of alcohol) is used for Inconel. For welding K Monel, a flux composed of two parts of Inconel flux and one part of lithium fluoride should be used. The flux is applied with a small brush or swab on both sides of the seam, top and bottom, and on the welding rod.

d. Welding Rods. Welding rods of the same composition as the alloy being welded are available. Rods of the same composition are necessary to insure uniform corrosion resistance without galvanic effects. In some cases, a special silicon Monel rod is used for welding nickel.

e. Welding Technique.

(1) All oil, dirt, and residues must be removed from the area of the weld by machining, sandblasting, grinding, rubbing with abrasive cloth, or chemically by pickling.

11-29.   NICKEL WELDING (cont)

(2) A slightly reducing flame should be used. There is a slight pressure fluctuation in many oxygen and acetylene regulators. The amount of excess acetylene in the flame should be only enough to counteract this fluctuation and prevent the flame from becoming oxidizing in nature.

(3) The tip should be the same size or one size larger than recommended by torch manufacturers for similar thicknesses of steel. The tip should be large enough to permit the use of a soft flame. A high velocity of harsh flame is undesirable.

(4) The parts to be welded should be held firmly in place with jigs or clamps to prevent distortion.

(5) Once started, welding should be continued along the seam without removing the torch from the work. The end of the welding rod should be kept well within the protecting flame envelope to prevent oxidation of the heated rod. The luminous cone tip of the flame should contact the surface of the molten pool in order to obtain concentrated heat. This will also prevent oxidation of the molten metal. The pool should be kept quiet and not puddled or boiled. If surface oxides or slag form on the surface of the molten metal, the rod should be melted into the weld under this surface film.

11-30.   NICKEL SOLDERING

a.   Soft soldering can be used for joining nickel and high nickel alloys only on sheet metal not more than 1/16 in. (1.6 mm) thick and only for those applications where the solder is not readily corroded. Soft solder is inherently of wlo strength. Joint strength must be obtained by rivets, lock seams, or spot welding, with soft solder acting as a sealing medium.

b.   The 50-50 and 60-40 percent tin-lead solders are preferred for joining metals of this type.

The flux used for nickel and Monel is a zinc saturated hydrochloric (muriatic) acid solution. Inconel requires a stronger flux because of its chromium oxide film. All flux and flux residues must be removed from the metal after the soldering operation is completed.

d.   Surfaces of metal parts to be soft soldered must be free from dirt, surface oxide or other discoloration. Where possible, the surfaces to be joined should be tinned with solder to ensure complete bonding during the final soldering operation.

11-31.   LEAD WELDING

a.   General. The welding of lead is similar to welding of other metals except that no flux is required. Processes other than gas welding are not in general use.

b.   Gases Used. Three combinations of gases are commonly used for lead welding. These are oxyacetylene, oxyhydrogen, and oxygen-natural gas. The oxyacetylene and oxyhydrogen processes are satisfactory for all positions. The oxygen-natural gas is not used for overhead welding. A low gas pressure ranging from 1-½ to 5 psi (10.3 to 34.5 kPa) is generally used, depending on the type of weld to be made.

c. Torch. The welding torch is relatively small in size. The oxygen and flammable gas valves are located at the forward end of the handle so that they may be conveniently adjusted by the thumb of the holding hand. Torch tips range in drill size from 78 to 68. The small tips are for 6.0 lb (2.7 kg) lead (i.e., 6.0 lb per sq ft), the larger tips for heavier lead.

d. Welding Rods. The filler rods should be of the same composition as the lead to be welded. They range in size from 1/8 to ¾ in. (3.2 to 19.1 mm) in diameter. The smaller sizes are used for lightweight lead and the larger sizes for heavier lead.

e. Types of Joints. Butt, lap, and edge joints are the types most commonly used in lead welding. Either the butt or lap joint is used on flat position welding. The lap joint is used on vertical and overhead position welding. The edge or flange joint is used only under special conditions.

f. Welding Technique.

(1) The flame must be neutral. A reducing flame will leave soot on the joint. An oxidizing flame will produce oxides on the molten lead and impair fusion. A soft, bushy flame is most desirable for welding in a horizontal position. A more pointed flame is generally used in the vertical and overhead positions.

(2) The flow of molten lead is controlled by the flame, which is usually handled with a semicircular or V-shaped motion. This accounts for the herringbone appearance of the lead weld. The direction of the weld depends on the type of joint and the position of the weld. The welding of vertical position lap joints is started at the bottom of the joint. A welding rod is not generally used. Lap joints are preferred in flat position welding. The torch is moved in a semicircular path toward the lap and then away. Filler metal is used but not on the first pass. Overhead position welding is very difficult. For that position, a lap joint and a sharp flame are used. The molten beads must be small and the welding operation must be completed quickly.

11-32. WHITE METAL WELDING

a. General. White metal is divided into three general classes according to the basic composition, i.e., zinc, aluminium, and magnesium. Most of the castings made are of the zinc alloy type. This alloy has a melting point of 725°F (385°C).

b. Flame Adjustment. The welding flame should be adjusted to carburizing but no soot should be deposited on the joint. The oxyacetylene flame is much hotter than necessary and it is important to select a very small tip.

c. Welding Rod. The welding rod may be of pure zinc or a die-casting alloy of the same type as that to be welded. Metal flux (50 percent zinc chloride and 50 percent ammonium chloride) can be used, but is not mandatory.

11-32. WHITE METAL WELDING (cont)

d. Welding Technique. The castings should be heated until the metal begins to flow. Then turn the flame parallel to the surface allowing the side of the flame to keep the metal soft while heating the welding rod to the same temperature. With both the base metal and the welding rod at the same temperature, the rod should be applied to and thoroughly fused with the walls of the joint. The rod should be manipulated so as to break up surface oxides.

11-33. BRONZE SURFACING

a. General. Bronze surfacing is used for building up surfaces that have been worn down by sliding friction or other types of wear where low heat conditions prevail. This type of repair does not involve the joining of metal parts. It is merely the addition of bronze metal to a part in order that it may be restored to its original size and shape. After bronze surfacing, the piece is machined to the desired finished dimensions. Cast iron, carbon and alloy steels, wrought iron, malleable iron, Monel, and nickel and copper-base alloys are satisfactorily built up by this process. This process is used to repair worn surfaces of rocker-arm rollers, lever bearings, gear teeth, shafts, spindle, yokes, pins, and clevises. Small bushings can be renewed by filling up the hole in the cast iron with bronze and then drilling them out to the required size.

b. Preparation of Surface. The surface to be rebuilt must be machined to remove all scale, dirt, or other foreign matter. If possible, cast iron surfaces should be chipped to clean them. Machining will smear the surface with graphite particles present in cast iron, and make bonding difficult. If the cast iron surface must be machined, an oxidizing flame should be passed over the surface to burn off the surplus graphite and carbon before the bronze coating is applied. Hollow piston heads or castings should be vented by removing the core plugs, or by drilling a hole into the cavities. This will prevent trapped gases from being expanded by the welding heat and cracking the metal.

c. Flame Adjustment. A neutral or slightly oxidizing flame is recommended. An excess acetylene flame will cause porosity and fuming.

d. Fluxes. A suitable brazing flux should be used to obtain good timing and adhesion of the bronze to the base metal.

e. Welding Rods. The bronze rod selected should fulfill the requirements for hardness and/or ductility needed for the particular application.

f. Application. The bronze surfacing metal is usually applied by mechanical means. This is accomplished using two or more flames and with a straight line or an oscillating motion. A layer of bronze 1/16 to ¼ in. (1.6 to 6.4 mm) tick is usually sufficient. It should be slowly cooled to room temperature and then machined to the desired dimensions.

CHAPTER 12

SPECIAL APPLICATIONS

Section I. UNDERWATER CUTTING AND WELDING WITH THE ELECTRIC ARC

12-1. GENERAL

WARNING

Safety precautions must be exercised in underwater cutting and welding. The electrode holder and cable must be insulated, the current must be shut off when changing elecrodes, and the diver should aoid contact between the electrode and grounded work to prevent electrical shock.

a. Underwater Arc Cutting. In many respects, underwater arc cutting is quite similar to underwater gas cutting. An outside jet of oxygen and compressed air is needed to keep the water from the vicinity of the metal being cut. Arc torches for underwater cutting are produced in a variety of types and forms. They are constructed to connect to oxygen-air pressure sources. Electrodes used may be carbon or metal. They are usually hollow in order to introduce a jet of oxygen into the molten crater created by the arc. The current practice is to use direct current for all underwater cutting and welding. In all cases, the electrode is connected to the negative side of the welding generator.

b. Underwater Arc Welding. Underwater arc welding may be accomplished in much the same manner as ordinary arc welding. The only variations of underwater arc welding from ordinary arc welding are that the electrode holder and cable must be well insulated to reduce current leakage and electrolysis, and the coated electrodes must be waterproofed so that the coating will not disintegrate underwater. The waterproofing for the electrode is qenerally a cellulose nitrate in which celluloid has been dissolved. Ordinary airplane dope with 2.0 lb (O.9 kg) of celluloid added per gallon is satisfactory.

12-2. UNDERWATER CUTTING TECHNIQUE

a. Torch. The torch used in underwater cutting is a fully insulated underwater cutting torch that utilizes the electric arc-oxygen cutting process using a tubular steel-covered, insulated, and waterproofed electrode. It utilizes the twist type collet for gripping the electrode and includes an oxygen valve lever and connections for attaching the welding lead and an oxygen hose. It is equipped to handle up to a 5/16-in. (7.9-mm) tubular electrode. In this process, the arc is struck normally and oxygen is fed through the electrode center hole to provide cutting. The same electrical connections mentioned above are employed.

b. The welding techniques involve signaling the surface helper to close the knife switch when the welder begins. The bead technique is employed using the drag travel system. When the electrode is consumed, the welder signals "current off" to the helper who opens the knife switch. "Current on" is signaled when a new electrode is positioned against the work. The current must be connected only when the electrode is against the work.

12-2. UNDERWATER CUTTING TECHNIQUE (cont)

c. Steel electrodes used for underwater cutting should be 14 in. (356 mm) long with a 5/16-in. (7.9-mm) outside diameter and an approximate 0.112-in. (2.845-mm) inside diameter hole. The electrode should have an extruded flux coating and be thoroughly waterproofed for underwater work. A welding current of 275 to 400 amps gives the best result with steel electrodes. When using graphite or carbon electrodes, 600 to 700 amps are required with a voltage setting around 70.

d. When working underwater, the cut is started by placing the tip of the electrode in contact with the work. Depress the oxygen lever slightly and call for current. When the arc is established, the predetermined oxygen pressure (e below) is released and the metal is pierced. The electrode is then kept in continuous contact with the work, cutting at the greatest speed at which complete penetration can be maintained. The electrode should be held at a 90 degree angle to the work. When the electrode is consumed, the current is turned off. A new electrode is then inserted and the same procedure is repeated until the cut is finished.

e. Normal predetermined oxygen pressure required for underwater cutting for a given plate thickness is the normal cutting pressure required in ordinary air cutting plus the depth in feet multiplied by 0.445. As an example, 2-1/4-in. (57.15-mn) plate in normal air cutting requires 20 psi (138 kPa). Therefore, at 10 ft (3 m) underwater, the following result would be reached:

$$20 + (10 \times 0.445) = 24 \text{ psi } (165 \text{ kPa}).$$

NOTE
Allowance for pressure drop in the gas line is 10 to 20 psi (69 to 138 kPa) per 100 ft (30 m) of hose.

12-3. UNDERWATER WELDING TECHNIQUE

a. General. Underwater welding has been restricted to salvage operations and emergency repair work. It is limited to depths helm the surface of not over 30 ft (9 m). Because of the offshore exploration, drilling, and recovery of gas and oil, it is necessary to lay and repair underwater pipelines and the portion of drill rigs and production platforms which are underwater. There are two major categories of underwater welding; welding in a wet environment and welding in a dry environment.

(1) Welding in the wet (wet environment) is used primarily for emergency repairs or salvage operations in shallow water. The pcor quality of welds made in the wet is due to heat transfer, welder visibility, and hydrogen presence in the arc atmosphere during welding. When completely surrounded by water at the arc area, the high temperature reducing weld metal quality is suppressed, and there is no base metal heat buildup at the weld. The arc area is composed of water vapor. The arc atmosphere of hydrogen and the oxygen of the water vapor is absorbed in the molten weld metal. It contributes to porosity and hydrogen cracking. In addition, welders working under water are restricted in manipulating the arc the same as on the surface. They are also restricted by low visibility because of their equipment and the water contaminants, plus those generated in the arc. Under the most ideal conditions, welds produced in the wet with covered electrodes are marginal. They may be used for short periods as needed but should be replaced with quality welds as soon as possible. Underwater in-the-wet welding is shown in figure 12-1. The power source should be a direct current machine rated at 300 or 400 amperes. Motor generator welding machines are most often used for underwater welding in-the-wet.

The welding machine frame must be grounded to the ship. The welding circuit must include a positive type of switch, usually a knife switch operated on the surface and commanded by the welder-diver. The knife switch in the electrode circuit must be capable of breaking the full welding current and is used for safety reasons. The welding power should be connected to the electrode holder only during welding. Direct current with electrode negative (straight polarity) is used. Special welding electrode holders with extra insulation against the water are used. The underwater welding electrode holder utilizes a twist type head for gripping the electrode. It accommodates two sizes of electrodes. The electrode size normally used is 3/16 in. (4.8 mm); however, 5/32-in. (4.0-mm) electrodes can also be used. The electrode types used conform to AWS E6013 classification. The electrodes must be waterproofed prior to underwater welding, which is done by wrapping them with waterproof tape or dipping them in special sodium silicate mixes and allowing them to dry. Commercial electrodes are available. The welding and work leads should be at least 2/0 size, and the insulation must be perfect. If the total length of the leads exceeds 300 ft (91 m), they should be paralleled. With paralleled leads to the electrode holder, the last 3 ft (O.9 m) should be a single cable. All connections must be thoroughly insulated so that the water cannot come in contact with the metal parts. If the insulation does leak, sea water will come in contact with the metal conductor and part of the current will leak away and will not be available at the arc. In addition, there will be rapid deterioration of the copper cable at the point of the leak. The work lead should be connected to the piece being welded within 3 ft (O.9 m) of the point of welding.

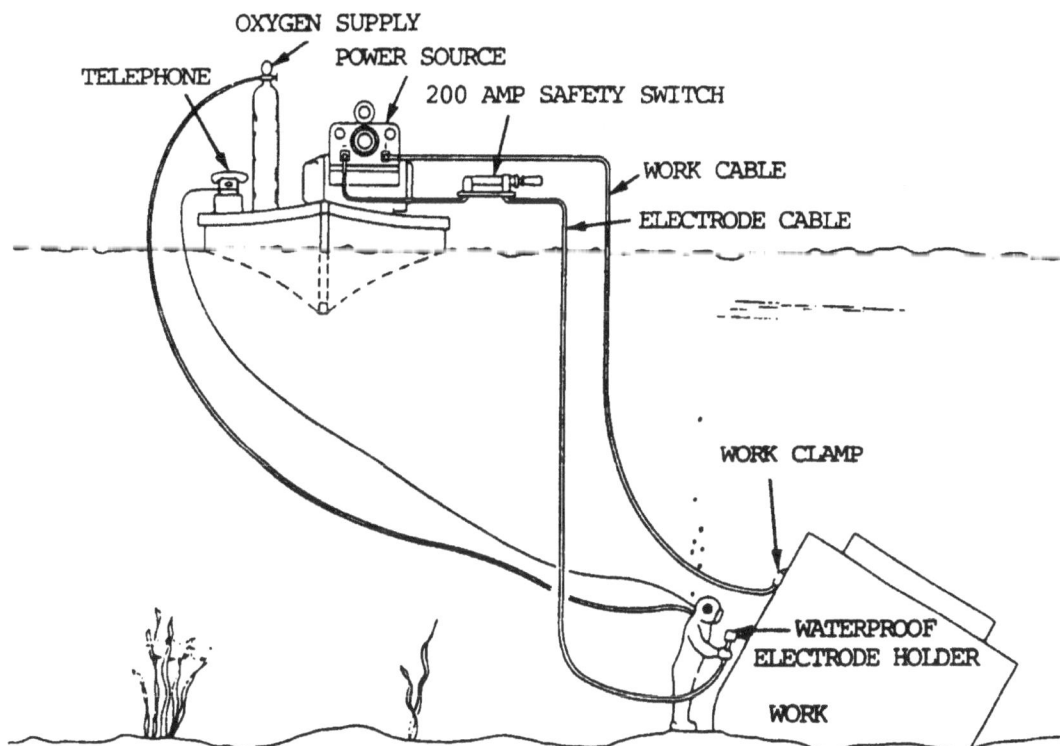

Figure 12-1. Arrangements for underwater welding.

12-3. UNDERWATER CUTTING TECHNIQUE (cont)

(2) Welding in-the-dry (dry environment) produces high-quality weld joints that meet X-ray and code requirements. The gas tungsten arc welding process produces pipe weld joints that meet quality requirements. It is used at depths of up to 200 ft (61 m) for joining pipe. The resulting welds meet X-ray and weld requirements. Gas metal arc welding is the best process for underwater welding in-the-dry. It is an all-position process and can be adopted for welding the metals involved in underwater work. It has been applied successfully in depths as great as 180 ft (55 m). There are two basic types of in-the-dry underwater welding. One involves a large welding chamber or habitat known as hyperbaric welding. It provides the welder-diver with all necessary welding equipment in a dry environment. The habitat is sealed around the welded part. The majority of this work is on pipe, and the habitat is sealed to the pipe. The chamber bottom is exposed to open water and is covered by a grating. The atmosphere pressure inside the chamber is equal to the water pressure at the operating depth.

b. Direct current must be used for underwater welding and a 400 amp welder will generally have ample capacity. To produce satisfactory welds underwater, the voltage must run about 10 volts and the current about 15 amps above the values used for ordinary welding.

c. The procedure recommended for underwater welding is simply a touch technique. The electrode is held in light contact with the work so that the crucible formed by the coating at the end of the electrode acts as an arc spacer. To produce 1/2 in. (12.7 mm) of weld bead per 1.0 in. (25.4 mm) of electrode consumed in tee or lap joint welding, the electrode is held at approximately 45 degrees in the direction of travel and at an angle of about 45 degrees to the surface being welded. To increase or decrease weld size, the lead angle may be decreased or increased. The same procedure applies to welding in any position. No weaving or shipping is employed at any time. In vertical welding, working from the top down is recommended.

d. The touch technique has the following advantages:

(1) It makes travel speed easy to control.

(2) It produces uniform weld surfaces almost automatically.

(3) It provides good arc stability.

(4) It permits the diver to feel his way where visibility is bad or working position is awkward.

(5) It reduces slag inclusions to a minimum.

(6) It assures good penetration.

e. In general, larger electrodes are used in underwater welding than are employed in normal welding. For example, when welding down on a vertical lap weld on 1/8 to 3/16 in. (3.2 to 4.8 mm) material, a 1/8- or 5/32-in. (3.2- or 4.0-mm) electrode would usually be used in the open air. However, a 3/16- or 7/32-in. (4.8- or 5.6-mm) electrode is recommended for underwater work because the cooling action of the water freezes the deposit more quickly. Higher deposition rates are also possible for the same reason. Usually, tee and lap joints are used in salvage operations because they are easier to prepare and they provide a natural groove to guide the electrode. These features are important under the difficult working conditions encountered underwater. Slag is light and has many nonadhering qualities. This means the water turbulence is generally sufficient to remove it.The use of cleaning tools is not necessary. However, where highest quality multipass welds are required, each pass should be thoroughly cleaned before the next is deposited.

f. Amperages given in table 12-1 are for depths up to 50 ft (15.2 m). As depth increases, amperage must be raised 13 to 15 percent for each additional 50 ft (15.2 m). For example, the 3/16-in. (4.8-mm) electrode at 200 ft (61 m) will require approximately 325 amperes to assure proper arc stability.

Table 12-1. Recommended Welding Currents

Electrode diameter (in.)	Amperes	Volts
1/8	130–163	23–26
5/32	180–225	24–28
3/16	225–280	25–30
7/32	260–340	26–30
1/4	330–400	28–32

Section II. UNDERWATER CUTTING WITH OXYFUEL

12-4. GENERAL

Underwater cutting is accomplished by use of the oxyhydrogen torch with a cylindrical tube around the torch tip through which a jet of compressed air is blown. The principles of cutting under water are the same as cutting elsewhere, except that hydrogen is used in preference to acetylene because of the greater pressure required in making cuts at great depths.Oxyacetylene may be used up to 25-ft (7.6-m) depths; however, depths greater than 25.0 ft (7.6 m) require the use of hydrogen gas.

12-5. CUTTING TECHNIQUE

a. Fundamentally, underwater cutting is virtually the same as any hand cutting employed on land. However, the torch used is somewhat different. It requires a tube around the torch tip so air and gas pressure can be used to create a gas pocket. This will induce an extremely high rate of heat at the work area since water dispels heat much faster than air. The preheating flame must be shielded from contact with the water. Therefore, higher pressures are used as the water level deepens (approximately 1.0 lb (0.45 kg) for each 2.0 ft (0.6 m) of depth). Initial pressure adjusments are as follows:

```
Oxygen ............................................. 60-85 psi (413.7-586.1 kPa)
Acetylene .......................................... 12-15 psi  (82.7-103.4 kPa)
Hydrogen ........................................... 35-45 psi (241.3-310.3 kPa)
Compressed air ..................................... 35-50 psi (241.3-344.8 kPa)
```

12-5. CUTTING TECHNIQUE (cont)

b. While the cutting operation itself is similar as on land, a few differences are evident. Same divers light and adjust the flame before descending. There is, however, an electric sparking device which is used for underwater ignition. This device causes somewhat of an explosion, but it is not dangerous to the operator.

c. When starting to preheat the metal to be cut, the torch should be held so the upper rim of the bell touches the metal. When the metal is sufficiently hot to start the cut, the bell should be firmly pressed on the metal since the compressed air will travel with the high pressure oxygen and escape through the kerf. Under these circumstances, the preheated gases will prevent undue "chilling" by the surrounding water. No welder on land would place a hand on the torch tip when cutting. However, this is precisely what the diver does underwater since the tip, bell, or torch will become no more than slightly warm under water. The diver, by placing the left hand around the torch head, can hold the torch steady and manipulate it more easily.

d. Due to the rapid dissipation of heat, it is essential that the cut be started by cutting a hole a distance from the outer edge of the plate. After the hole has been cut, a horizontal or vertical cut can be swiftly continued. A diver who has not previously been engaged in underwater cutting must make test cuts before successfully using an underwater cutting torch.

Section III. METALLIZING

12-6. GENERAL

a. General.

(1) Metallizing is used to spray metal coatings on fabricated workplaces. The coating metal initially is in wire or powder form. It is fed through a special gun and melted by an oxyfuel gas flame, then atomized by a blast of compressed air. The air and combustion gases transport the atomized molten metal onto a prepared surface, where the coating is formed (fig. 12-2).

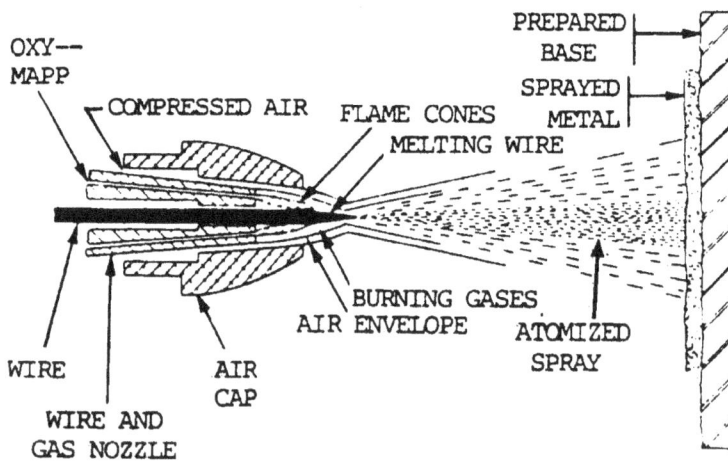

Figure 12-2. The wire metallizing process.

(2) The metallizing process uses a welding spray gun to enable the welder to place precisely as much or as little weld metal as necessary over any desired surface. Metal deposits as thin as 0.003 in. (0.076 mm) to any desired thickness may be made. The process is versatile, time-saving, and, in some cases, more economical than other welding or repair procedures.

(3) Metallized coatings are used to repair worn parts, salvage mismachined components, or to provide special properties to the surface of original equipment. Metallized coatings are used for improving bearing strength, adding corrosion or heat resistance, hard-facing, increasing lubricity, improving thermal and electrical conductivity, and producing decorative coatings.

(4) Corrosion resistant coatings such as aluminum and zinc are applied to ship hulls, bridges, storage tanks, and canal gates, for example. Hard-facing is applied to shafting, gear teeth, and other machine components, as well as to mining equipment, ore chutes, hoppers, tracks, and rails. Coatings with combined bearing and lubricity properties are used to improve the surface life of machine shafting, slides, and ways.

 b. Characteristics of Coated Surfaces.

(1) The chemical properties of sprayed coatings are those of the coating metal. The physical properties often are quite different (table 12-2).

(2) As-sprayed metal coatings are not homogeneous. The first molten droplets from the metallizing gun hit the substrate and flatten out. Subsequent particles overlay the first deposit, building up a porous lamellar coating. Bonding is essentially mechanical, although some metallurgical bonding also may occur.

Table 12-2. Mechanical Properties of Sprayed Coatings

Metal	Rockwell Hardness	Tensile Strength (psi)	Elongation
Aluminum (1100)	H H 72	19,500	0.23
Aluminum bronze (90-9-1)	B 78	29,000	0.46
Babbitt (89% Sn)	H 58	—	—
Bronze (60-40)	B 50	26,500	0.50
Copper	B 33	15,500	
Molybdenum	C 38	7,500	0.30
Monel	B 39	—	—
Nickel	B 49	—	—

12-6. GENERAL (cont)

Table 12-2. Mechanical Properties of Sprayed Coatings (cont)

Metal	Rockwell Hardness	Tensile Strength (psi)	Elongation
Steels:			
1010	B 89	30,000	0.30
1025	B 90	34,700	0.46
1080	C 36	27,500	0.42
Type 202	B 88	30,000	0.50
Type 304	B 78	30,000	0.27
Type 420	C 29	40,000	0.50
Zinc	H 46	13,000	1.43

(3) The small pores between droplets soon became closed as the coating thickness increases. These microscopic pores can hold lubricants and are one of the reasons metallized coatings are used for increased lubricity on wear surfaces.

(4) The tensile strengths of sprayed coatings are high for the relatively low melting point metals used. Ductility is uniformly low. Therefore, parts must be formal first, and then sprayed. Thin coatings of low melting point metals, such as sprayed zinc on steel, are a minor exception to this rule and can withstand limited forming.

c. Workpiece Restrictions.

(1) Metallizing is not limited to any particular size workpiece. The work may vary from a crane boom to an electrical contact. Metallizing may be done on a production line or by hand; in the plant on the field.

(2) Workpiece geometry has an important influence on the process. Cylindrical parts such as shafts, driers, and press rolls that can be rotated in a lathe or fixture are ideal for spraying with a machine-mounted gun. For example, a metallizing gun can be mounted on the carriage of a lathe to spray a workpiece at a predetermined feed rate.

(3) Parts such as cams are usually sprayed by hand. Such parts can be sprayed automatically, but the cost of the elaborate setup for automated spraying may not be justified. The volute part of a small pump casing is difficult to coat because of the backdraft or splash of the metal spray. Small-diameter holes, bores of any depth, or narrow grooves are difficult or impossible to coat because of bridging of the spraying coating.

d. Materials for Metallizing.

(1) A wide range of materials can be flame sprayed. Most of them include metals, but refractory oxides in the form of either powder or reds also can be applied. Wires for flame spraying include the entire range of alloys and metals

from lead, which melts at 618°F (326 °C), to molybdenum with a melting point of 4730 °F (2610 °C). Higher melting point materials also can be sprayed but a plasma-arc spray gun is required.

(2) Between the extremes of lead and molybdenum are common metal coatings such as zinc, aluminum, tin, copper, various brasses, bronzes and carbon steels, stainless steels, and nickel-chromium alloys. Spray coatings may be combined on one workpiece. For example, molybdenum or nickel aluminide often is used as a thin coating on steel parts to increase bond strength. Then another coating metal is applied to build up the deposit.

e. Surface Preparation.

(1) Surfaces for metallizing must be clean. They also require roughening to ensure a good mechanical bond between the workpiece and coating. Grease, oil, and other contaminants are rearmed with any suitable solvent. Cast iron or other porous metals should be preheated at 500 to 800F (260 to 427 °C) to remove entrapped oil or other foreign matter. Sand blasting may be used to remove excessive carbon resulting from preheating cast iron. Chemical cleaning may be necessary prior to preheating.

(2) Undercutting often is necessary on shafts and similar surfaces to permit a uniformly thick buildup on the finished part. The depth of undercutting depends on the diameter of the shaft and on service requirements. If the undercut surface becomes oxidized or contaminated, it should be cleaned before roughening and spraying.

(3) Roughening of the workpiece surface usually is the final step before spraying. Various methods are used, ranging from rough threading or threading and knurling to abrasive blasting and electric bonding.

(4) Thin molybdenum or nickel aluminide spray coatings are often applied to the roughened surface to improve the bond strength of subsequent coatings. Applications that require only a thin coating of sprayed metal often eliminate the roughening step and go directly to a bonding coat. The surface is then built up with some other metal.

f. Coating Thickness.

(1) Cost and service requirements are the basis for determining the practical maximum coating thickness for a particular application, such as building up a worn machine part. Total metallizing cost includes cost of preparation, oxygen, fuel gas and materials application time and finishing operations. If repair costs are too high, it may be more economical to buy a replacement part.

(2) The total thickness for the as-sprayed coating on shafts is determined by the maximum wear allowance, the minimum coating thickness that must be sprayed, and the amount of stock required for the finishing operation. The minimum coating thickness that must be sprayed depends on the diameter of the shaft and is given in table 12-3, p 12-10. For press-fit sections, regardless of diameter, a minimum of 0.005 in. (0.127 mm) of coating is required.

12-6. GENERAL (cont)

Table 12-3. Minimum Thickness of As-Sprayed Coatings on Shafts

Shaft Diameter (in.)	Coating Thickness (in.)
1 or less	0.010
1 to 2	0.015
2 to 3	0.020
3 to 4	0.025
4 to 5	0.030
5 to 6	0.035
6 or more	0.040

(3) Variation in the thickness of deposit depends on the type of surface preparation used. The thickness of a deposit over a threaded surface varies more than that of a deposit over an abrasive blasted surface, or a smooth surface prepared by spray bonding. In general, the total variation in thickness that can be expected for routine production spraying with mounted equipment is 0.002 in. (0.051 mm) for deposits from a metallizing wire.

g. Coating Shrinkage.

(1) The shrinkage of the metal being deposited also must be taken into consideration because it affects the thickness of the final deposit. For example, deposits on inside diameters must be held to a minimum thickness to conform with the shrinkage stresses; coatings of excessive thickness will separate from the workpiece because of excessive stresses and inadequate bond strength.

(2) Table 12-4 gives shrinkage values for the metals commonly used for spray coatings. Thicker coatings can be deposited with metals of lower shrinkage.

(3) All sprayed-metal coatings are stressed in tension to some degree except in those where the substrate material has a high coefficient of expansion, and is preheated to an approximate temperature for spraying. The stresses can cause cracking of thick metal coatings with a high shrinkage value; the austenitic stainless steels are in this category.

Table 12-4. Shrinkage of Commonly Applied Sprayed Coatings

Metal	Shrinkage (in. per in.)
Ferrous metals:	
0.10% C steel	0.0080
0.25% C steel	0.0060
0.80% C steel	0.0014
Type 304 stainless	0.0120
Type 420 stainless	0.0018

Table 12-4. Shrinkage of Commonly Applied Sprayed Coatings (cont)

Metal	Shrinkage (in. per in.)
Nonferrous metals:	
Aluminum (1100)	0.0068
Al-Si (4 to 6% Si)	0.0057
Aluminum bronze	0.0055
Manganese bronze	0.0090
Phosphor bronze	0.0010
Molybdenum	0.0030
Zinc	0.0010

(4) The susceptibility to cracking of thick austenitic stainless steel deposits can be prevented by first spraying a martensitic stainless steel deposit on the substrate, then depositing austenitic stainless steel to obtain the required coating thickness. The martensitic stainless produces a strong bond with the substrate, has good strength in the as-sprayed form, and provides an excellent surface for the austenitic stainless steel.

h. Types of Metallizing.

(1) Electric arc spraying (EASP).

(a) Electric arc spraying is a thermal spraying process that uses an electric arc between two consumable electrodes of the surfacing materials as the heat source. A compressed gas atomizes and propels the molten material to the workpiece. The principle of this process is shown by figure 12-3. The two consumable electrode wires are fed by a wire feeder to bring them together at an angle of approximately 30 degrees and to maintain an arc between them. A compressed air jet is located behind and directly in line with the intersecting wire. The wires melt in the arc and the jet of air atomizes the melted metal and propels the fine molten particles to the workpiece. The power source for producing the arc is a direct-current constant-voltage welding machine. The wire feeder is similar to that used for gas metal arc welding except that it feeds two wires. The gun can be hand held or mounted in a holding and movement mechanism. The part or the gun is moved with respect to the other to provide a coating surface on the part.

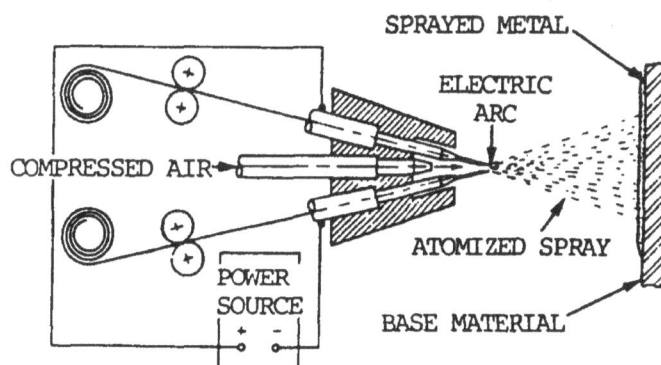

Figure 12-3. Electric arc spraying process.

12-6. GENERAL (cont)

(b) The welding current ranges from 300 to 500 amperes direct current with the voltage ranging from 25 to 35 volts. This system will deposit from 15 to 100 lb/hr of metal. The amount of metal deposited depends on the current level and the type of metal being sprayed. Wires for spraying are sized according to the Brown and Sharp wire gauge system. Normally either 14 gauge (0.064 in. or 1.626 mm) or 11 gauge (0.091 in. or 2.311 mm) is used. Larger diameter wires can be used.

(c) The high temperature of the arc melts the electrode wire faster and deposits particles having higher heat content and greater fluidity than the flame spraying process. The deposition rates are from 3 to 5 times greater and the bond strength is greater. There is coalescence in addition to the mechanical bond. The deposit is more dense and coating strength is greater than when using flame spraying.

(d) Dry compressed air is used for atomizing and propelling the molten metal. A pressure of 80 psi (552 kPa) and from 30 to 80 cu ft/min (0.85 to 2.27 cu m/min) is used. Almost any metal that can be drawn into a wire can be sprayed. Following are metals that are arc sprayed:aluminum, babbitt, brass, bronze, copper, molybdenum, Monel, nickel,stainless steel, carbon steel, tin, and zinc.

(2) Flame spraying (FLSP).

(a) Flame spraying is a thermal spraying process that uses an oxyfuel gas flame as a source of heat for melting the coating material. Compressed air is usually used for atomizing and propelling the material to the workpieceThere are two variations: one uses metal in wire form and the other uses materials in powder form. The method of flame spraying which uses powder is sometimes known as powder flame spraying. The method of flame spraying using wire is known as metallizing or wire flame spraying.

(b) In both versions, the material is fed through a gun and nozzle and melted in the oxygen fuel gas flame. Atomizing, if required, is done by an air jet which propels the atomized particles to the workpiece.When wire is used for surfacing material, it is fed into the nozzle by an air-driven wire feeder and is melted in the gas flame. When powdered materials are used, they may be fed by gravity from a hopper which is a part of the gun. In another system, the powders are picked up by the oxygen fuel gas mixture, carried through the gun where they are melted, and propelled to the surface of the workpiece by the flare.

(c) Figure 12-4 shows the flame spray process using wire. The version that uses wires can spray metals that can be prepared in a wire formThe variation that uses powder has the ability to feed various materials. These include normal metal alloys, oxidation-resistant metals and alloys, and ceramicsIt provides sprayed surfaces of many different characteristics.

(3) Plasma spraying (PSP).

(a) Plasma spraying is a thermal spraying process which uses a nontransferred arc as a source of heat for melting and propelling the surfacing material to the workpiece. The process is shown in figure 12–5.

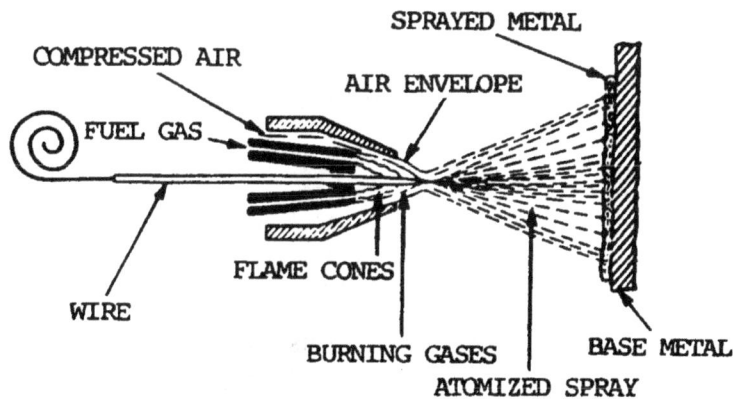

Figure 12-4. Flame spray process.

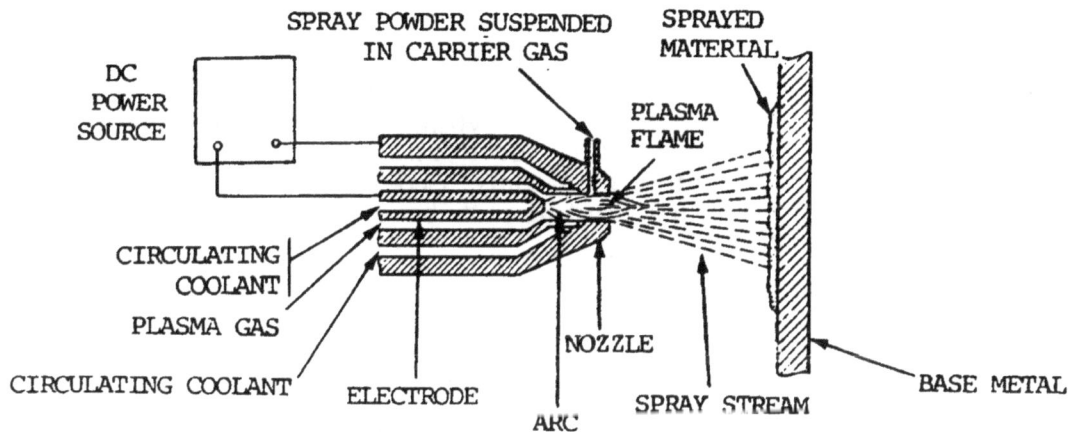

Figure 12-5. Plasma spray process.

(b) The process is sometimes called plasma flame spraying or plasma metallizing. It uses the plasma arc, which is entirely within the plasma spray gun. The temperature is so much higher than either arc spraying or flame spraying that additional materials can be used as the coating. Most inorganic materials, which melt without decomposition, can be used.The material to be sprayed must be in a powder form. It is carried into the plasma spray gun suspended in a gas. The high-temperature plasma immediately melts the powdered material and propels it to the surface of the workpiece. Since inert gas and extra high temperatures are used, the mechanical and metallurgical properties of the coatings are generally superior to either flame spraying or electric arc spraying.This includes reduced porosity and improved bond and tensile strengthsCoating density can reach 95 percent. The hardest metals known, some with extremely high melting temperatures, can be sprayed with the plasma spraying process.

12-6.   GENERAL (cont)

i .   The Spraying Operation.   Spraying should be done immediately after the part
is cleaned.   If the part is not sprayed in-mediately, it should be protected from
the atmosphere by wrapping with paper.If parts are extremely large, it may be
necessary to preheat the part 200 to 400°F (93 to 204 $^\circ$C). Care must be exercised
so that heat does not build up in the workpiece.This increases the possibility of
cracking the sprayed surface. The part to be coated should be preheated to the
approximate temperature that it normally would attain during the spraying operation.   The distance between the spraying gun and the part is dependent on the process and material being sprayed.   Recommendations of the equipment manufacturer
should be followed and modified by experience. Speed and feed of spraying should
be uniform.   The first pass should be applied as quickly as possible. Additional
coats may be applied slcwly.   It is important to maintain uniformity of temperature
throughout the part. When there are areas of the part being sprayed where coating
is not wanted, the area can be protected by masking it with tape.

12-7.   TOOLS AND EQUIPMENT

The major items of equipment used in the process, with the exception of the
eutectic torch and a few fittings, are the same as in a normal oxyacetylene welding
or cutting operation. Oxygen and acetylene cylinders, cylinder-to-regulator fittings, pressure regulators, hoses, striker, torch and regulator wrench, tip cleaners, and goggles are the same as those commonly used by welders.The metallizing
and welding torch, its accessory tips, and the Y hose fittings are the distinct
pieces of equipment used in metallizing.

12-8. METALLIZING AND WELDING TORCH

a.   This torch is a manually operated, powder dispensing, oxyacetylene torch.
There are three sections: the torch body, the mixing chamber and valve assembly,
and the tip assembly. These assemblies are chrome plated to prolong service life
and to prevent corrosion and contamination.

b.   The torch body is also the handle. Like the body of a regular welding
torch, it also has needle valves which control the flew of oxygen and acetylene.

c.   The mixing chamber and valve assembly is the heart of the torch.In this
section, the flow of powder into the oxygen stream is controlled and mixing takes
place.   A lever, like the cutting lever on a cutting torch, controls the flow of
powdered metal. When the lever is held down, powder flows; when released, the powder flow is shut off. The valve and plunger are made of plastic. Should a block-
age occur, no sharp or rough objects should be used to clean it. Occasionally,
material will build up inside the bore. This cuts down the operating efficiency.
If any malfunction occurs or is suspected, the bore is the first item to check.
Just forward of the feed lever is the connection for attaching the powder bellows
modules.   It must be in the UP position while operating.

d.   The tip assembly is made of a low heat-conducting alloy.It can be rotated
and locked at any position or angle from 0 to 360 degrees. The accessory tips are
screwed onto the end of the tip assembly.

12-9. ACCESSORY TIPS

These tips come in three sizes, numbered 45, 48, and 53, according to the size of the drill number used to drill the orifice. The larger the number, the smaller the hole. A number 45 tip would be used for heavy buildup while a number 53 tip would be used for fine, delicate work.

12-10. Y FITTINGS

Two Y fittings are provided with a set: one fitting with left-hand threads for acetylene connections and the other with right-hand threads for oxygen hose connections. These fittings allow the regular welding torch to be used on the same tanks at the same time as the metallizing torch.

12-11. MATERIALS

   a. General. The materials used for making welds and overlays are a little different in form, but not new in purpose. Fluxes are used for hard-to-weld metals. Filler metal in the form of a fine powder is used for the weld or coating material.

   b. Fluxes. Two fluxes in paste form are used in combination with different powdered alloys. One flux is for copper only; the other is for all types of metals.

   c. Welding powders. The metal powders are of high quality, specially formulated alloys developed for a wide variety of jobs. These jobs range from joining copper to copper to putting a hard surface on a gear tooth. Selection of the proper alloy depends on the base metal and surface required. The powder alloys come in plastic containers called bellow modules. They are ready to insert in the connection on the valve assembly after their stoppers have been removed.

12-12. SETUP

The equipment and torch are hooked up to the oxygen and acetylene tanks in the same manner as a regular welding torch. The tip is rotated to the correct angle for the welding position being used, and locked in place.

12-13. OPERATION

   a. To operate the torch, the correct pressure setting is needed. This is determined by the size of the tip being used. Tip number 45 and 48 use 25 to 30 psi (172.4 to 206.9 kPa) of oxygen and 4 to 5 psi (27.6 to 34.5 kPa) of acetylene. Tip number 53 uses 15 to 18 psi (103.4 to 124.1 kPa) of oxygen and 2 to 3 psi (13.8 to 20.7 kPa) of acetylene.

   b. After the pressures have been set, the torch is lit and adjusted to obtain a neutral flame while the alloy feed lever is depressed. This is done before joining the module to the torch.

   c. After proper flame adjustment, the module is attached to the torch. This is done by turning the torch upside down and inserting the end of the module into the mating part located on the valve assembly. A twist to the right locks the module in place. The only time the torch is held upside down is during loading and unloading.

12-13. OPERATION (cont)

d. Before applying the powder to the surface or joint, the area must be preheated. Steel is heated to a straw color while brass and copper are heated to approximately 800 °F (427 °C).

e. After preheating, the powder feed lever is depressed and a thin cover of powder is placed over the desired surface. The lever is then released and the area heated until the powder wets the surface or tinning action takes place. Once tinning is observed, the feed lever is depressed until a layer no more than 1/8 in. (3.2 mm) thick has been deposited. If more metal is desired, the area should be reheated. A light cover of powder should be applied and heated again until tinning takes place to ensure proper bonding. In this manner, any desired thickness may be obtained while depositing the metal. The torch must be kept in a constant circular motion to avoid overheating the metal.

f. To shut down the torch, the same procedures are followed as in regular welding. Once the flame is out, the bellows module is removed by turning the torch upside down and twisting the module to the left. The plug must be replaced to prevent contamination of the alloy.

12-14. MALFUNCTIONS AND CORRECTIVE ACTIONS

As with any piece of equipment, malfunctions can occur. The orifice should be checked first if no observable deposit is made when the feed lever is depressed.

Section IV. FLAME CUTTING STEEL AND CAST IRON

12-15. GENERAL

a. General. Plain carbon steels with a carbon content not exceeding 0.25 percent can be cut without special precautions. Certain steel alloys develop high resistance to the action of the cutting oxygen. This makes it difficult, and sometimes impossible, to propagate the cut without the use of special techniques.

b. Oxygen Cutting. Oxygen cutting (OC) is a group of thermal cutting processes used to sever or remove metals by means of the chemical reaction of oxygen with the base metal at elevated temperatures. In the case of oxidation-resistant metals, the reaction is facilitated by the use of a chemical flux or metal powder. Five basic processes are involved: oxyfuel gas cutting, metal powder cutting, chemical flux cutting, oxygen lance cutting, and oxygen arc cutting. Each of these processes is different and will be described.

c. Oxyfuel Gas Cutting (OFC).

(1) Oxyfuel gas cutting severs metals with the chemical reaction of oxygen with the base metal at elevated temperatures. The necessary temperature is maintained by gas flames from the combustion of a fuel gas and oxygen.

(2) When an oxyfuel gas cutting operation is described, the fuel gas must be specified. There are a number of fuel gases used. The most popular is acetylene. Natural gas is widely used, as is propane, methylacetylene-propadiene stabilized (MAPP gas), and various trade name fuel gases. Hydrogen is rarely used. Each fuel gas has particular characteristics and may require slightly different apparatus.

These characteristics relate to the flame temperatures, heat content, oxygen fuel gas ratios, etc.

(3) The general concept of oxyfuel gas cutting is similar no matter what fuel gas is used. It is the oxygen jet that makes the cut in steel, and cutting speed depends on how efficiently the oxygen reacts with the steel.

(4) Heat is used to bring the base metal steel up to kindling temperature where it will ignite and burn in an atmosphere of pure oxygen. The chemical formulas for three of the oxidation reactions is as follows:

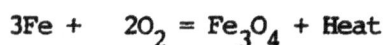

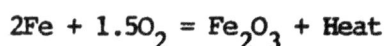

$$Fe + 0.5O_2 = Fe \quad + Heat$$

$$3Fe + \quad 2O_2 = Fe_3O_4 + Heat$$

$$2Fe + 1.5O_2 = Fe_2O_3 + Heat$$

(5) At elevated temperatures, all of the iron oxides are produced in the cutting zone.

(6) The oxyacetylene cutting torch is used to heat steel by increasing the temperature to its kindling point and then introducing a stream of pure oxygen to create the burning or rapid oxidation of the steel. The stream of oxygen also assists in removing the material from the cut. This is shown by figure 12-6.

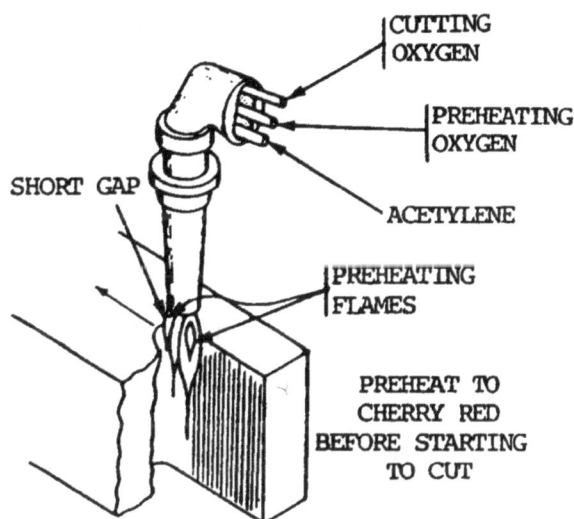

Figure 12-6. Process diagram of oxygen cutting.

(7) Steel and a number of other metals are flame cut with the oxyfuel gas cutting process. The following conditions must apply:

(a) The melting point of the material must be above its kindling temperature in oxygen.

(b) The oxides of the metal should melt at a lower temperature than the metal itself and below the temperature that is developed by cutting.

12-15. GENERAL (cont)

(c) The heat produced by the combustion of the metal with oxygen must be sufficient to maintain the oxygen cutting operation.

(d) The thermal conductivity must be low enough so that the material can be brought to its kindling temperature.

(e) The oxides formed in cutting should be fluid when molten so the cutting operation is not interrupted.

(8) Iron and low-carbon steel fit all of these requirements and are readily oxygen flame cut. Cast iron is not readily flame cut, because the kindling temperature is above the melting point. It also has a refractory silicate oxide which produces a slag covering. Chrome-nickel stainless steels cannot be flame cut with the normal technique because of the refractory chromium oxide formed on the surface. Nonferrous metals such as copper and aluminum have refractory oxide coverings which prohibit normal oxygen flame cutting. They have high thermal conductivity.

(9) When flame cutting, the preheating flame should be neutral or oxidizing. A reducing or carbonizing flame should not be used. The schedule for flame cutting clean mild steel is shown by the table 12-5.

Table 12-5. Welding Procedure Schedule for Oxyfuel Gas Cutting

Material Thickness		Cutting Orifice Dia. (center hole)			Approx. Press. of Gas		Travel Speed Manual	Travel Speed Automatic
in.	mm	Drill Size	in.	mm	Acetylene psi	Oxygen psi	in./min.	in./min.
1/8	3.2	60	0.040	1.0	3	10	20-22	22
1/4	6.4	60	0.040	1.0	3	15	16-18	20
3/8	9.5	55	0.052	1.3	3	20	14-16	19
1/2	12.7	55	0.052	1.3	3	25	12-14	17
3/4	19.0	55	0.052	1.3	4	30	10-12	15
1	25.4	53	0.060	1.5	4	35	8-11	14
1-1/2	38.1	53	0.060	1.5	4	40	6-7 1/2	12
2	50.8	49	0.073	1.9	4	45	5 1/2-7	10
3	76.2	49	0.073	1.9	5	50	5-6 1/2	8
4	101.6	49	0.073	1.9	5	55	4-5	7
5	127.0	45	0.082	2.1	5	603	1/2-4 1/2	6
6	152.4	45	0.082	2.1	6	70	3-4	5
8	203.2	45	0.082	2.1	6	75	3	4

(10) Torches are available for either welding or cutting. By placing the cutting torch attachment on the torch body it is used for manual flare cutting. Figure 12-7 shows a manual oxyacetylene flare-cutting torch. Various sizes of tips can be used for manual flame cutting. The numbering system for tips is not standardized. Most manufacturers use their own tip number system. Each system is,

however, based on the size of the oxygen cutting orifice of the tip. These are related to drill sizes. Different tip sizes are required for cutting different thicknesses of carbon steel.

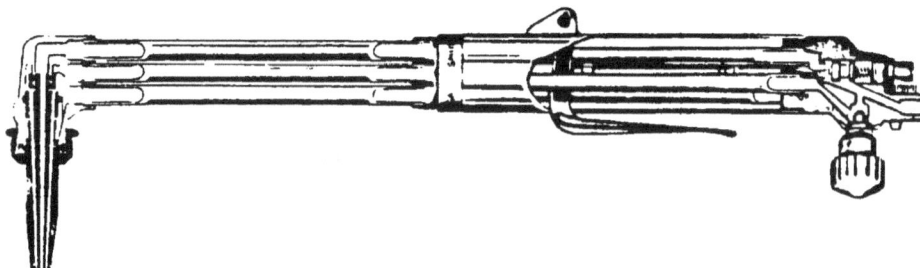

Figure 12-7. Manual oxygen cutting torch.

(11) For automatic cutting with mechanized travel, the same types of tips can be used. High-speed type tips with a specially shaped oxygen orifice provide for higher-speed cutting and are normally used. The schedule shown in table 12-5 provides cutting speeds with normal tips; the speeds can be increased 25 to 50 percent when using high speed tips.

(12) Automatic shape-cutting machines are widely used by the metalworking industry. These machines can carry several torches and cut a number of pieces simultaneously. Multitorch cutting machines are directed by numerically-controlled equipment. Regardless of the tracing control system is used, the cutting operation is essentially the same.

(13) One of the newer advances in automatic flame cutting is the generation of bevel cuts on contour-shaped parts. This breakthrough has made the use of numerically controlled oxygen cutting equipment even more productive.

(14) Many specialized automatic oxygen cutting machines are available for specific purposes. Special machines are available for cutting sprockets and other precise items. Oxygen-cutting machines are available for cutting pipe to fit other pipe at different angles and of different diameters. These are quite complex and have built-in contour templates to accommodate different cuts and bevels on the pipe. Other types of machines are designed for cutting holes in drum heads, test specimens, etc. Two or three torches can be used to prepare groove bevels for straight line cuts as shown by figure 12-8. Extremely smooth oxygen-cut surfaces can be produced when schedules are followed and all equipment is not in proper operating condition.

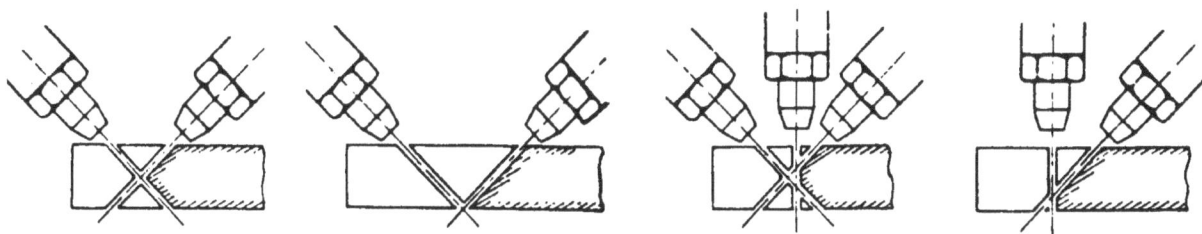

Figure 12-8. Methods of preparing joints.

12-15.   GENERAL (cont)

d.   Metal Powder Cutting (POC).

(1) Metal powder cutting severs metals through the use of powder to facilitate cutting. This process is used for cutting cast iron, chrome nickel stainless steels, and some high-alloy steels.

(2) The process uses finely divided material usually iron powder, added to the cutting-oxygen stream. The powder is heated as it passes through the oxy-acetylene preheat flames and almost immediately oxidizes in the stream of the cutting oxygen. A special apparatus to carry the powder to the cutting tip must be added to the torch. A powder dispenser is also required. Compressed air is used to carry the powder to the torch.

(3) The oxidation, or burning of the iron powder, provides a much higher temperate in the oxygen stream. The chemical reaction in the flame allows the cutting-oxygen stream to oxidize the metal being cut continuously in the same manner as when cutting carbon steels.

(4) With the use of iron powder in the oxygen stream, it is possible to start cuts without preheating the base material.

(5) Powder cutting has found its broadest use in the cutting of cast iron and stainless steel. It is used for removing gates and risers from iron and stainless steel castings.

(6) Cutting speeds and cutting oxygen-pressures are similar to those used when cutting carbon steels. For heavier material over 1 in. (25 mm) thick, a nozzle one size larger should be used. Powder flow requirements vary from 1/4 to 1/2 lb (0.11 to 0.23 kg) of iron powder per minute of cutting. Powder tends to leave a scale on the cut surface which can easily be removed as the surface cools. This is a rather special application process and is used only where required.

(7) Stack cutting is the oxygen cutting of stacked metal sheets or plates arranged so that all the plates are severed by a single cut. In this way, the total thickmess of the stack is considered the same as the equivalent thickness of a solid piece of metal. When stack cutting, particularly thicker material, the cut is often lost because the adjoining plates may not be in intimate contact with each other. The preheat may not be sufficient on the lower plate to bring it to the kindling temperature and therefore the oxygen stream will not cut through the remaining portion of the stack. One way to overcome this problem is to use the metal powder cutting process. By means of the metal powder and its reaction in the oxygen, the cut is completed across separations between adjacent plates.

e.   Chemical Flux Cutting (FOC). Chemical flux cutting is an oxygen-cutting process in which metals are severed using a chemical flux to facilitate cutting. Powdered chemicals are utilized in the same way as iron powder is used in the metal powder cutting process. This process is sometimes called flux injection cutting. Flux is introduced into the cut to combine with the refractory oxides and make them a soluble compound. The chemical fluxes may be salts of sodium such as sodium carbonate. Chemical flux cutting is of minor industrial significance.

f. Oxygen Lance Cutting (LOC).

(1) Oxygen lance cutting severs metals with oxygen supplied through a consumable tube. The preheat is obtained by other means. This is sometimes called oxygen lancing. The oxygen lance is a length of pipe or tubing used to carry oxygen to the point of cutting. The oxygen lance is a small (1/8 or 1/4 in. (3.2 or 6.4 mm) nominal) black iron pipe connected to a suitable handle which contains a shut-off valve. This handle is connected to the oxygen supply hose. The main difference between the oxygen lance and an ordinary flame cutting torch is that there is no preheat flame to maintain the material at the kindling temperature. The lance is consumed as it makes a cut. The principle use of the oxygen lance is the cutting of hot metal in steel mills. The steel is sufficiently heated so that the oxygen will cause rapid oxidation and cutting to occurFor other heavy or deep cuts, a standard torch is used to bring the surface of the metal to kindling temperature. The oxygen lance becomes hot and supplies iron to the reaction to maintain the high temperature.

(2) There are several proprietary specialized oxygen lance type cutting bars or pipes. In these systems, the pipe is filled with wires which may be aluminum and steel or magnesium and steel. The aluminum and magnesium readily oxidize and increase the temperature of the reaction. The steel of the pipe and the steel wires will tend to slow down the reaction whereas the aluminum or magnesium wires tend to speed up the reaction. This type of apparatus will burn in air, under water, or in noncombustible materials. The tremendous heat produced is sufficient to melt concrete, bricks, and other nonmetals. These devices can be used to sever concrete or masonry walls and will cut almost anything.

g. Oxygen Arc Cutting (AOC) .

(1) Oxygen arc cutting severs metals by means of the chemical reaction of oxygen with the base metal at elevated temperatures.The necessary temperature is maintained by means of an arc between a consumable tubular electrode and the base metal.

(2) This process requires a specialized combination electrode holder and oxygen torch. A conventional constant current welding machine and special tubular covered electrodes are used.

(3) This process will cut high chrome nickel stainless steels, high-alloy steels, and nonferrous metals.

(4) The high temperature heat source is an arc between the special covered tubular electrode and the metal to be cut. As soon as the arc is established, a valve on the electrode holder is depressed. Oxygen is introduced through the tubular electrode to the arc. The oxygen causes the material to burn and the stream helps remove the material from the cut.Steel from the electrode plus the flux from the covering assist in making the cut.They combine with the oxides and create so much heat that thermal conductivity cannot remove the heat quickly enough to extinguish the oxidation reaction.

12-15. GENERAL (cont)

(5) This process will routinely cut aluminum, copper, brasses, bronzes, Monel, Inconel, nickel, cast iron,stainless steel, and high-alloy steels. The quality of the cut is not as gocd as the quality of an oxygen cut on mild steel, but sufficient for many applications.Material from 1/4 to 3 in. (6.4 to 76 mm) can be cut with the process. The electric current ranges from 150 to 250 amperes and oxygen pressure of 3 to 60 psi (20.7 to 413.7 kPa) may be used.Electrodes are normally 3/16 in. (4.8 mm) in diameter and 18 in. (457 mm) long. They are suitable for ac or dc use. This process is used for salvage work, as well as for manufacturing and maintenance operations.

12-16. HIGH CARBON STEELS

The action of the cutting torch on high carbon steels is similar to flame hardening processes. The metal adjacent to the cutting area is hardened by being heated above its critical temperature and quenched by the adjacent mass of cold metal. This condition can be minimized by preheating the part from 500 to 60°B (260 to 316 °C) before the cut is made.

12-17. WASTER PLATE ALLOY STEEL

The cutting action on an alloy steel that is difficult to cut can be improved by clamping a mild steel "waster plate"tightly to the upper surface and cutting through both thicknesses. This waster plate method will cause a noticeable improvement in the cutting action. The molten steel dilutes or reduces the alloying content of the base metal.

12-18. CHROMIUM AND STAINLESS STEEL

These and other alloy steels that previously could be cut only by a melting action can now be cut by rapid oxidation.This is done by introducing iron powder or a special nonmetallic powdered flux into the cutting oxygen streamThe iron powder oxidizes quickly and liberates a large quantity of heat. This high heat melts the refractory oxides which normally protect the alloy steel from the action of oxygen . These molten oxides are flushed from the cutting face by the oxygen blast. The cutting oxygen is able to continue its reaction with the iron powder and cut its way through the steel plates. The nonmetallic flux, when introduced into the cutting oxygen stream,combines chemically with the refractory oxides. This produces a slag of a lower melting point,which is washed or eroded out of the cut, exposing the steel to the action of the cutting oxygen.

12-19. CAST IRON

Cast iron melts at a temperature lower than its oxides. Therefore, in the cutting operation, the iron tends to melt rather than oxidizeFor this reason, the oxygen jet is used to wash out and erode the molten metal when cast iron is being cut. To make this action effective, the cast iron must be preheated to a high temperature and much heat must be liberated deep in the cut.This is effected by adjusting the preheating flames so there is an excess of acetylene.The length of the acetylene streamer and the procedure for advancing the cut are shown in figure 12-9The use of a mild iron flux to maintain a high temperaturein the deeper recesses of the cut is also effective (fig. 12–9).

For recommended
flame, adjust excess
acetylene streamer
equal to thickness
to be cut.

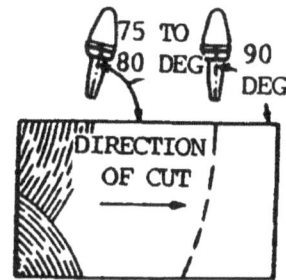

Angle of tip at start and as
cut progresses. Bring cutting
tip up carefully to 90 degrees
to avoid losing cut.

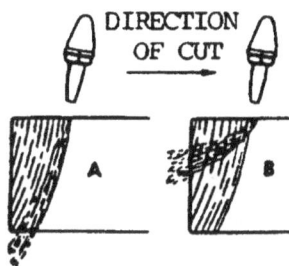

Cutting jet should just sweep
edge of cut as shown in A and
not advance too deeply as
shown in B. Otherwise, progress
of the cut will cease and black
spots will develop under the
cutting jet.

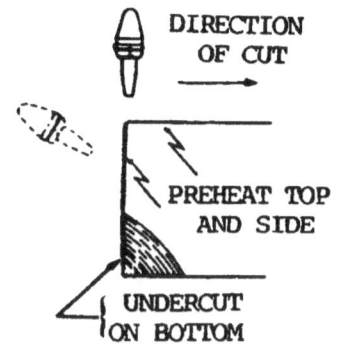

Begin and maintain cut holding torch tip 1-1/2 to 2 in.
from cast iron.

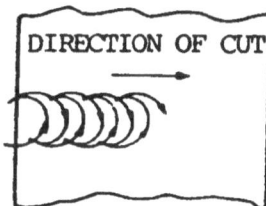

Move torch tip in semi-circular
motions 1/2 in. to 3/4 in., as
required, to clear cut in heavy
sections. Light sections require
reduced oscillations of the
torch tip.

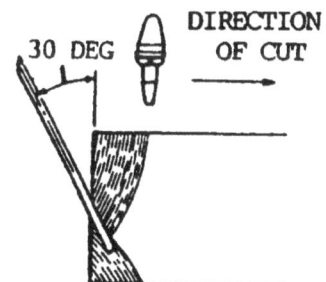

Approximate introduction
angle of flux rod or lance
to assist cutting operation.

Figure 12-9. Procedure for oxyacetylene cutting of cast iron.

Section V. FLAME TREATING METAL

12-20. FLAME HARDENING

a. The oxyacetylene flame can be used to harden the surface of hardenable steel, including stainless steels, to provide better wearing qualities. The carbon content of the steel should be 0.35 percent or higher for appreciable hardening. The best range for the hardening process is 0.40 to 0.50 percent. In this process, the steel is heated to its critical temperature and then quenched, usually with water. Steels containing 0.70 percent carbon or higher can be treated in the same manner, except that compressed air or water sprayed by compressed air, is used to quench the parts less rapidly to prevent surface checking. Oil is used for quenching some steel compositions.

b. The oxyacetylene flame is used merely as a heat source and involves no change in the compsition of the steels as in case hardening where carbon or nitrogen is introduced into the surface. In case hardening, the thickness of the hardened area ranges from 0 to 0.020 in. (0 to 0.508 mm).

c. Ordinary welding torches are used for small work, but for most flame hardening work, water-cooled torches are necessary. Tips or burners are of the multi-flame type. They are water cooled since they must operate for extended periods without backfiring. Where limited areas are to be hardened, the torch is moved back and forth over the part until the area is heated above the critical temperature. Then the area is quenched. The hardening of extended areas is accomplished by steel hardening devices. These consist of a row of flames followed by a row of quenching jets. A means of moving these elements over the surface of the work, or moving the work at the required speed under the flames and jets, is also required.

12-21. FLAME SOFTENING

a. Certain steels, called six-hardening steels, will become hard and brittle when cooled rapidly in the air from a red hot conditionThis hardening action frequently occurs when the steels are flame cut or arc welded. When subsequent machining is required, the hardness must be decreased to permit easier removal of the metal.

b. Oxyacetylene flames adjusted to neutral can be used either to prevent hardening or to soften an already hardened surface.The action of the flame is used to rapidly heat the metal to its critical temperatureHowever, in flare softening, the quench is omitted and the part is cooled slowly, either by still air or by shielding with an insulating material.

c. Standard type torches,tips, and heating heads,like those used for welding equipment, are not applicable. The equipment used in flame hardening is necessary.

12-22. FLAME STRAIGHTENING

a. It is often desirable or necessary to straighten steel that has been expanded or distorted from its original shape by uneven heating. This is especially true if the steel is prevented from expanding by adjacent cold metal.The contraction on cooling tends to shorten the surface dimension on the heated side of the plate. Since some of the metal has been upset permanently, the plate cannot return to its original dimensions and becomes dished or otherwise distorted.

b. Localized heat causes such metal distortion. This principle can be used to remedy warpage, buckling, and other irregularities in plates, shafts, structural members, and other parts. The distorted areas are heated locally and then quenched on cooling. The raised sections of the metal will be drawn down. By repeating this process and carefully applying heat in the proper areas and surfaces, irregularities can be remedied.

12-23. FLAME STRENGTHENING

Flame strengthening differs from flame hardening. The intent is to locally strengthen parts that will have to withstand severe service conditions. This process is used particularly for parts that are subjected to frequently varying stresses that lead to fatigue failure. The section that is to be strengthened is heated to the hardening temperature with the oxyacetylene flame, then quenched either with water, a water-air mixture, or air, depending on the composition of the steel being treated.

12-24. FLAME DESCALING

Flare descaling, sometimes called flame cleaning, is widely used for removing loosely adhering mill scale and rust. It is also used to clean rusted structures prior to painting. The scale and rust crack and flake off because of the rapid expansion under the oxyacetylene flame. The flare also turns any moisture present into steam, which accelerates the scale removal and, at the same time, dries the surface. The loose rust is then removed by wire brushing to prepare the surface for painting (fig. 12-10). This process is also used for burning off old paint. Standard torches equipped with long extensions and multiflame tips of varying widths and shapes are used.

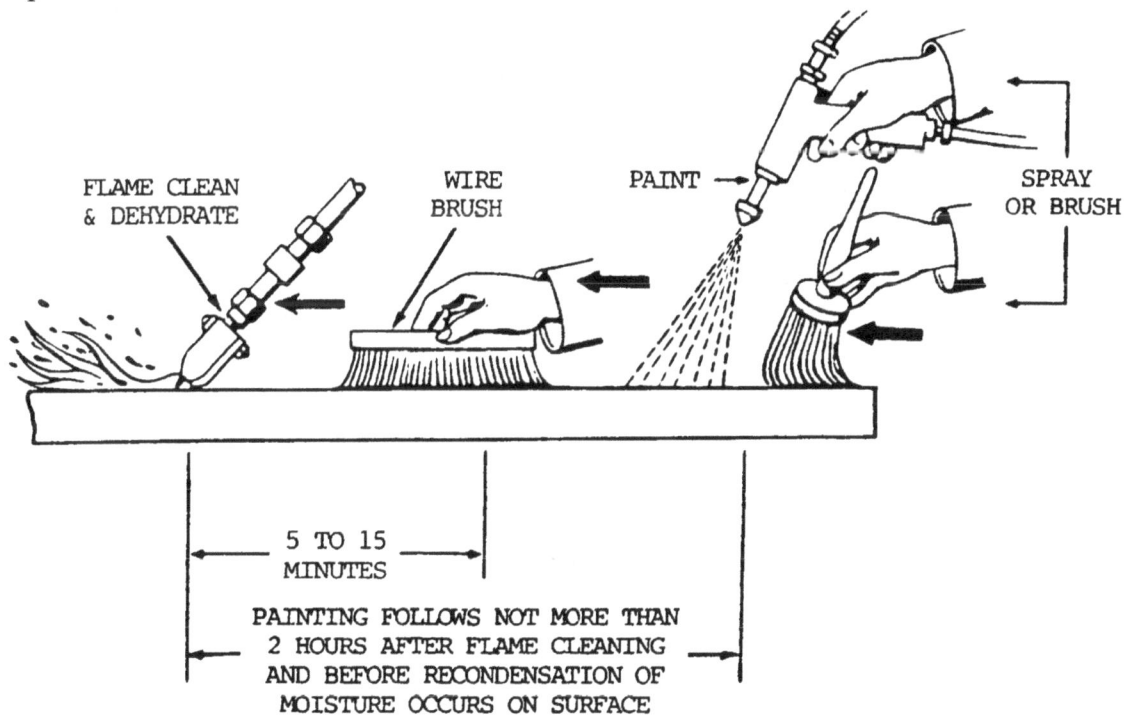

Figure 12-10. Operations and time intervals in flame descaling prior to painting.

12-25.   FLAME MACHINING (OXYGEN MACHINING)

a.   <u>General</u>. Flame machining. or oxygen machining, includes those processes where oxygen and an oxyacetylene flare are used in removing the surfaces of metals. Several of these processes are described below.

b.   <u>Scarfing or Deseaming</u>. This process is used for the removal of cracks, scale, and other defects frcm the surface of blooms, billets, and other unfinished shapes in steel mills. In this process, an area on the surface of the metal is heated to the ignition temperature. Then, a jet or jets of oxygen are applied to the preheated area and advanced as the surface is cut away. The scarfed surface is comparable to that of steel cleaned by chipping.

c.   <u>Gouging</u>. This process is used for the removal of welds, It is also used in the elimination of defects such as cracks, sand inclusions, and porosity from steel castings.

d.   <u>Hogging</u>. This is a flame machining process used for the removal of excess metals, such as risers and sprues, from castings. It is a combination of scarfing and gouging techniques.

e.   <u>Oxygen Turning</u>. This flame machining process is identical in principle to mechanical turning with the substitution of a cutting torch in place of the usual cutting tool.

f.   <u>Surface Planing</u>. Surface planing is a type of flame machining similar to mechanical planing. The metal is removed from flat or round surfaces by a series of parallel and overlapping grooves. Cutting tips with special cutting orifices are used in this operation. The operator controls the width and depth of the cut by controlling the oxygen pressure, the tip angle with relation to the metal surface, and the speed with which the cutting progresses.

12-26.   OXYACETYLENE RIVET CUTTING

a.   <u>Removal of Countersunk Rivets</u>.

(1)   When countersunk rivets are being removed the cutting torch is held so that the cutting nozzle is perpendicular to the plate surface (fig. 12-11). The preheating flames are directed at a point slightly below the center of the rivet head. The tips of the inner cones of the flames should be approximately 1/16 in. (1.6 mm) away from the rivet head.

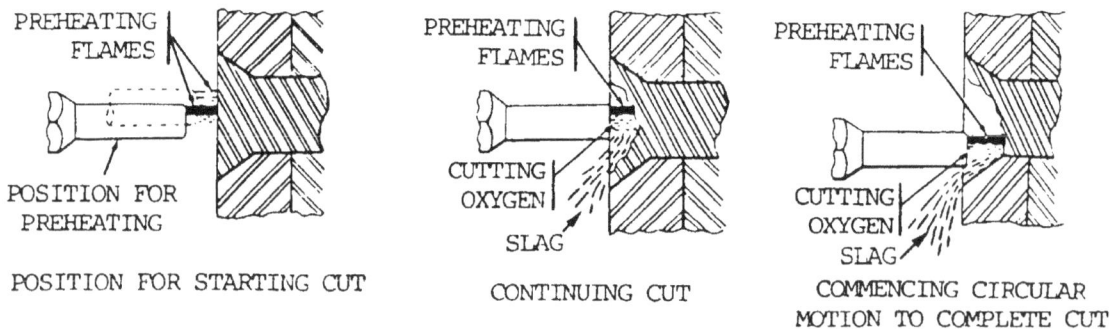

Figure 12-11.   Removal of countersunk rivets.

(2) When the area of the rivet head under the flames becomes bright red, the tip of the torch is raised slightly to direct the cutting oxygen stream to the heated area. The cutting oxygen valve is opened. The torch shield is held steady until the coned head has been burned through and the body or shank of the rivet is reached. The remainder of the head should then be removed in one circular, wiping motion. The torch should be held with the cutting oxygen stream pointed at the base of the countersink, and then moved once around the circumference. After the head has been removed, the shank can be driven out.

b. Removal of Buttonhead Rivets.

(1) Buttonhead rivets can be removed by using the tip size recommended for cutting steel 1.0 in. (25.4 mm) thick. Adjust the oxygen and acetylene pressures accordingly (fig. 12-12). Hold the tip parallel with the plate and cut a slot in the rivet head from the tip of the button to the underside of the head, similar to the screwdriver slot in a roundhead screw. As the cut nears the plate, draw the tip back at least 1-1/2 in. (38.1 mm) from the rivet and swing the tip in a small arc. This slices off half of the rivet head. Immediately swing the tip in the opposite direction and cut off the other half of the rivet head. After the head has been removed, the shank can be driven out.

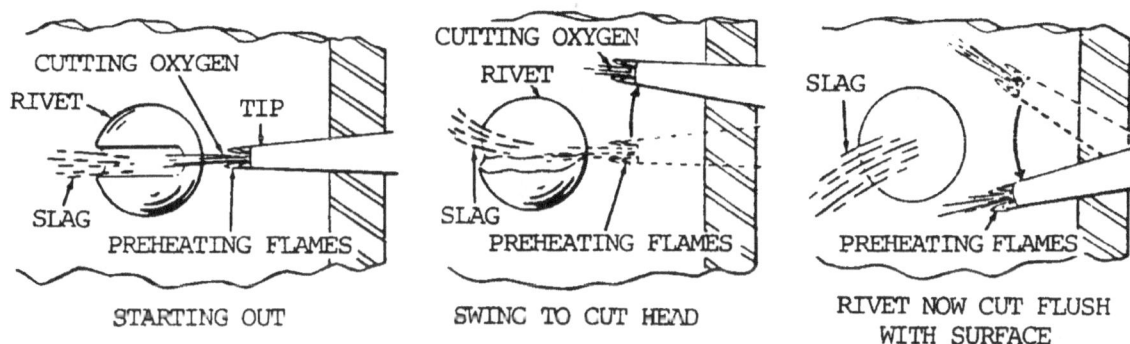

Figure 12-12.   Removal of buttonhead rivets.

(2) By the time the slot is cut, the entire head will be preheated to cutting temperature. While the bottom of the slot is being reached, and just before cutting starts at the surface of the plate, the tip must be drawn back from the rivet a distance of about 1-1/2 in. (38.1 mm). This will permit the oxygen stream to spread out slightly before it strikes the rivet and prevent the jet from breaking through the layer of scale that is always present between the rivet head and the plate. If the tip is not drawn away, the force of the oxygen jet may pierce the film of scale and damage the plate surface.

SECTION VI. CUTTING AND HARD SURFACING WITH THE ELECTRIC ARC

12-27. GENERAL

a. Cutting. Electric arc cutting is a melting process whereby the heat of the electric arc is used to melt the metal along the desired line of the cut. The quality of the cuts produced by arc cutting does not equal that of cuts produced by applications where smooth cuts are essential. Arc cutting is generally confined to the cutting of nonferrous metals and cast iron.

12-27. GENERAL (cont)

b. <u>Hard Surfacing</u>. Hard surfacing is the process of applying extremely hard alloys to the surface of a softer metal to increase its resistance to wear by abrasion, corrosion, or impact. The wearing surfaces of drills, bits, cutters, or other parts, when treated with these special alloys, will outwear ordinary steel parts from 2 to 25 times. This will depend on the hard surfacing alloy and the service to which the part is subjected.

12-28. METAL CUTTING WITH ELECTRIC ARC

a. <u>General</u>. Electric arc cutting can be performed by three methods: carbon-arc, metal-arc, and arc-oxygen.

b. <u>Carbon-Arc Cutting</u>. In carbon-arc cutting, a carbon electrode is utilized to melt the metal progressively by maintaining a steady arc length and a uniform cutting speed. Direct current straight polarity is preferred, because it develops a higher heat at the base metal, which is the positive pole. Direct current also permits a higher cutting rate than alternating current, with easier control of the arc. Air cooled electrode holders are used for currents up to 300 amperes. Water cooled electrode holders are desirable for currents in excess of 300 amperes.

c. <u>Metal-Arc Cutting</u>.

(1) <u>Metal-arc</u>. Metal-arc cutting is a progressive operation with a low carbon steel, covered electrode. The covering on the electrode is a non-conducting refractory material. It permits the electrode to be inserted into the gap of the cut without being short circuited. This insulating coating also stabilizes and intensifies the action of the arc. Direct current straight polarity is preferred, but alternating current can be used. Standard electrode holders are applicable for metal-arc cutting in air.

(2) <u>Air-arc</u>. By slightly converting the standard electrode holder, as described in TB 9-3429-203/1, a stream of air can be directed to the surface of the work, increasing the speed of the cut and holding it to a minimum width.

(3) <u>Underwater cutting</u>. Specially constructed, fully insulated holders must be used for underwater metal-arc cutting.

d. <u>Arc-Oxygen Cutting</u>. Arc-oxygen cutting is a progressive operation in which a tubular electrode is employed. The steel or conducting-type ceramic electrode is used to maintain the arc and series as a conduit through which oxygen is fed into the cut. In this process, the arc provides the heat and the oxygen reacts with the metal in the same manner as in oxyacetylene cutting. Both direct and alternating currents are applicable in this process.

12-29. HARD SURFACING

Hard surfacing is used to apply a layer of metal of a special composition onto the surface of metal of a special composition onto the surface or to a specific section or part of a base metal of another composition. A wide variety of characteristics or performance characteristics can be secured by the selection of proper surfacing metals. The applied layer may be as thin as 1/32 in. (0.79 mm) or as thick as required.

12-30.   METALS THAT CAN BE HARD SURFACED

a.   All plain carbon steels with carbon content up to 0.50 percent can be hard surfaced by either the oxyacetylene or electric arc process.

b.   High carbon steels containing more than O.5O percent carbon can be hard surfaced by any of the arc welding processes. However, preheating to between 300 and 600 °F (149 and 316 °C) is usually advisable. This preheating will prevent cracking due to sudden heating of hardened parts.It will also prevent excessive hardening and cracking of the heat affected zone during cooling.

c.   Low alloy steels can be hard surfaced in the same manner as plain carbon steels of the same hardenability if the steel is not in its hardened state. If it is in the hardened state, it should be annealed before welding. In some cases, heat treatment is required after welding.

d.   The hard surfacing of high speed steels is not generally recommended.This is due to the fact that, regardless of heat treatment, brittleness and shrinkage cracks will develop in the base metal after hard surfacing.Usually there is no need for hard surfacing these steels because surfaced parts of low alloy steels should provide equal service characteristics.

e.   Manganese (Hadfield) steels should be hard surfaced by the shielded metal-arc process only, using the work hardening type of alloys or with alloys that will bond easily with this metal.

f.   Stainless steels, including the high chrome and the 18-8 chrome-nickel steels, can be hard surfaced with most of-the alloys that have suitable melting points. A knowledge of the composition of the stainless steel at hand is needed for the selection of the proper alloy. Otherwise, brittleness or impairment of corrosion resistance may result. The high coefficient of expansion of the 18-8 steels must also be considered.

g.   Gray and alloy cast irons can be hard surfaced with the lower melting point alloys and the austenitic alloys. However, precautions need to be taken to prevent cracking of the cast iron during and after welding.Cobalt base alloys are also applicable to cast iron,although a flux may need to be applied to the cast iron.

h.   White cast iron cannot be successfully surfaced because the welding heat materially alters the properties of the underlying metal.

i.   Malleable iron can be surfaced in the same manner as cast iron.

j.   Copper, brass, and bronze are difficult to surface with ferrous or high alloy nonferrous metals because of the low melting points.However, brass, bronze, and some nickel surfacing alloys can be applied very readily. Fluxes are usually needed in these applications to secure sound welds.

12-31.   ALIOYS USED FOR HARD SURFACING

a.   General. No single hard surfacing material is suitable for all applications. Many types of hard surfacing alloys have been developed to meet the various requirements for hardness,toughness, shock and wear resistance, and other special qualities. These alloys are classified into six groups and are described below.

12-31. ALLOYS USED FOR HARD SURFACING (cont)

b. <u>Group A</u>. These include the low alloy types of surfacing alloys that are air hardened. Most of these electrodes are covered with coatings that supply alloying, deoxidizing, and arc stabilizing elements. Preheating of the base metal may be necessary to prevent cracking when harder types of electrodes are used, but in many applications the presence of small cracks is not important.

c. <u>Group B</u>. These electrodes include the medium alloy and medium-high alloy types. They have a light coating for arc stabilization only. The alloying agents are in the metal of the rod or wire. The electodes in this group have a lower melting point than those in group A. They produce a flatter surface and must be used in the flat position only. Multilayer deposits, with proper preheat, should be free from cracks.

d. <u>Group C</u>. These electrodes include the high speed steel and austenitic steel alloys (other than austenitic manganese steels). The electrodes are either bare or have a light arc stabilizing coating. The bare electrodes should be only used for surfacing manganese steels because their arc characteristics are poor. To avoid embrittlement of the weld metal, the base metal must not be heated over 700 (371 °C). Peening is used in the application of these alloys to reduce stresses and to induce some hardness in the underlying layers.

e. <u>Group D</u>. These electrodes include the cobalt base alloys. They have a moderately heavy coating and are intended for manual welding only. To avoid impairment of metal properties, low welding heat is recommended. Deposits are subject to cracking but this can be prevented by preheating and slow cooling of the workpiece.

f. <u>Group E</u>. These alloys are supplied as tube rods containing granular tungsten carbide inside the tube. Their arc characteristics are poor but porosity and cracks are of little importance in the application for which they are intended. The tungsten carbide granules must not be melted or dissolved in the steel. For this reason, a minimum heat is recommended for welding. The deposits should show a considerable amount of undissolved cubicle particles.

g. <u>Group F</u>. These are nonferrous alloys of copper and nickel base types. They are heavily coated and are intended for direct current reverse polarity welding in the flat position only.

12-32. HARD SURFACING PROCEDURE

a. <u>Preparation of Surface</u>. The surface of the metal to be hard surfaced must be cleaned of all scale, rust, dirt, or other foreign substances by grinding, machining, or chipping. If these methods are not practicable, the surface may be prepared by filing, wire brushing, or sandblasting. The latter methods sometimes leave scale or other foreign matter which must be floated out during the surfacing operation. All edges of grooves, corners or recesses must be well rounded to prevent overheating of the base metal.

b. <u>Hard Surfacing with the Metal Arc</u>. Surfacing by arc welding is done in the same manner and is similar in principle to joining by arc welding, except that the added metal has a composition that is not the same as that of the base metal. The characteristics of the added metal would be changed or impaired if it were excessively diluted by or blended with the base metal. For this reason, penetration into the base metal should be restricted by applying the surfacing metal with the minimum welding heat. In general, the current, voltage, polarity, and other conditions recommended by the manufacturer of the electrodes are based on this factor. An arc as long as possible will give the best results.

c. <u>Hard Surfacing with the Carbon Arc</u>.This process is used principally for the application of group F alloys. The welding machine is set for straight polarity and the heat of the arc is used to weld the particles of carbide to the surface of the base metal.

Section VII. ARMOR PLATE WELDING AND CUTTING

12-33. GENERAL

a. Armor plate is used for the protection of personnel and equipment in combat tanks, self propelled guns, and other combat vehicles against the destructive forces of enemy projectiles. It is fabricated in the forms of castings and rolled plates. These are selectively heat treated, in turn, to develop the desired structural and protective properties. Industrial manufacture of gun turrets and combat tank hulls includes designs using one-piece castings and welded assemblies of cast sections and rolled plates. In certain cases, cast sections of armor are bolted in place to expedite the requirements of maintenance through unit replacement. Welding has replaced riveting as a formative process of structural armor fabrication. Riveting, however, is still used on some vehicles protected by face hardened armor.

b. The development of a suitable technique for welding armor plate is contingent upon a clear understanding of the factors affecting the weldability of armor plates, the structural soundness of the weld, and its ultimate ability to withstand the forces of impact and penetration in service. From the standpoint of field repair by welding, these considerations can be resolved into the factors outlined helm:

(1) Knowledge of the exact type of armor being welded through suitable identification tests.

(2) Knowledge of alternate repair methods which are satisfactory for the particular type of armor and type of defect in question.

(3) Design function of the damaged structure.

(4) Selection of welding materials and repair procedures from the facilities available to produce optimum protective properties and structural strength.

(5) Determination of the need for emergency repair to meet the existing situation.

(6) Careful analysis of the particular defect in the armor to ensure proper disposition of the variables listed below:

(a) Joint preparation and design.

(b) Welding electrodes.

(c) Welding current, voltage, and polarity.

12-33. GENERAL (cont)

      (d) Sequence of welding passes.

      (e) Welding stresses and warpage.

      (f) Proper protection or removal of flammable materials and equipment in the vicinity of the welding operation.

    c. The advantages of welding as an expedient for field repair to damaged armor plate lie principally in the speed and ease with which the operation can be performed. The welding procedures for making repairs in the field are basically the same as those used for industrial fabrication.They must be modified at times because of the varying types of damage due to impact, such as the following:

      (1) Complete shell penetration.

      (2) Bulges or displaced sections.

      (3) Surface gouges.

      (4) Linear cracks of various widths terminating in the armor or extending to its outside edges.

      (5) Linear or transverse cracks in or adjacent to welded seams.

    d. Many repairs made by welding require the selective use of patches obtained by cutting sections from completely disabled armored vehicles having similar armor plate. Also, most of the welding, whether around patches or along linear seams, is performed under conditions that frequently will permit no motion of the base metal sections to yield under contraction stresses produced by the cooling weld metal. The stress problem is further complicated by stresses produced by projectiles physically drifting the edges of the armor at the point of impact or penetration.It is with all these variables in mind that the subsequent plate welding procedures are determined.

12-34. PROPERTIES OR ARMOR PLATE

Armor plate is an air hardening alloy steelwhich means that it will harden by normalizing or heating to its upper critical point and cooling in still air. The base metal quenching effect produced adjacent to a weld in heavy armor plate under normal welding conditions is about halfway between the effects of air cooling and oil quenching. The extremely steep thermal gradients occuring in the region of a weld range from temperatures of 3000°F (1649 °C) or more in the weld metal to the original temperature of the base metal.Therefore, a narrow zone on each side of the deposited weld metal is heated above its critical temperature by the welding heat and quenched by the relatively cold base metal to form a hard brittle zone. It is in this hard, nonductile formation, known as martensite, that cracks are more likely to occur as a result of the sudden application of load. For this reason, special precautions must be taken in all welding operations to minimize the formation of these hard zones and to limit their effect on the structural properties of the welded armor. Care must be taken to prevent rapid cooling of the armor after welding in order to avoid the formation of cracks in these hard zones.

12-35. TYPES OF ARMOR PLATE

a. <u>General</u>. Two types of armor are used on combat vehicles: homogeneous (cast or rolled) and face hardened (rolled). It is essential that the armor be specifically identified before any welding or cutting operations are performed. This is important because the welding procedures for each type of armor are distinctly different and noninterchangeable.

b. <u>Homogeneous Armor</u>. Homogeneous armor is heat treated through its entire thickness to develop good shock or impact resisting properties. As its name indicates, it is uniform in hardness, composition, and structure throughout and can be welded on either side. Aluminum armor plate is in the homogeneous class and welding procedures are the same as gas metal-arc welding (para 10-12, p 10-54).

c. <u>Face Hardened Armor Plate</u>. Face hardened armor plate has an extremely hard surface layer, obtained by carburizing, which extends to a depth of 1/5 to 1/4 of the outward facing thickness of the armor on the tank or armored vehicle. The primary purpose of face hardened armor is to provide good resistance to penetration. The inner side is comparatively soft and has properties similar to those of homogeneous armor. The inside and outside of face hardened armor plate are two different kinds of steel. Face hardened steel UP to 0.5 in. (12.7 mm) in thickness should be welded from the soft side only.

12-36. IDENTIFICATION OF ARMOR PLATE

a. <u>File Test</u>. This test is a simple but accurate method of identifying armor plate. A file will bite into homogeneous armor plate on both sides, but will only bite into the soft side of face hardened armor plate. When applied to the face side, the file will slip, acting in much the same manner as on case hardened steel.

b. <u>Appearance of Fracture</u>. The metal edges of holes or cracks in homogeneous armor plate are ragged and bent, with the metal drifted in the direction of the forces which damaged the armor. Cracks in homogeneous armor are usually caused by stresses and are present at severe bulges or bends in the plate or section. The metal edges of holes and cracks in face hardened armor are relatively clean cut and sharp. The plates do not bulge to any great extent before cracking. By examining the edges of freshly broken face hardened armor, it can be noted that the metal at the face side is brighter and finer in structure than the metal at the soft side. The brighter metal extends to a depth of approximately 1/5 to 1/4 of the thickness from the surface of the side.

12-37. CUTTING ARMOR PLATE

a. <u>Cutting Homogeneous Armor Plate</u>. Either the oxygen cutting torch, which is preferable, or the electric arc can be used to cut homogeneous armor plate. The carbon arc can be used to cut out welds and to cut castings and plates, but the shielded metal-arc is preferred when oxygen and acetylene are not available.

b. <u>Cutting Face Hardened Armor Plate</u>.

(1) <u>General</u>. The procedure for cutting this type of armor is essentially the same as that required for homogeneous armor except that every precaution should be taken to keep as much heat as possible away from the hard face side of the plate. This is accomplishd by performing all cutting operations from the soft side of the armor, thus limiting the extent of heating and consequent softening of the hardened surfaces.

12-37. CUTTING ARMOR PLATE (cont)

(2) <u>Cutting with the oxygen torch</u>.

(a) The general practice used for oxygen torch cutting can be applied for cutting armor plate, but the tip size, cutting oxygen, and preheating gas temperatures should be kept at the minimum consistent with good quality cuts to prevent overheating.

(b) When the cutting of the stainless steel type of weld, such as is used on tanks and other armored vehicles, is performed with an oxygen cutting torch, the cutting process must be modified to suit. This is necessary because stainless steel is a nonoxidizing metal. Cutting is therefore accomplished by using an oxidizable steel rod in conjunction with the oxygen cutting torch. The oxygen combines with the steel rod and the resultant evolution of high temperature creates high temperature molten steel at the end of the rod. Drops of this molten steel are formed at the end of the rod and wash off onto the weld to help melt it. This washing action is accomplished by an oscillating motion of the torch tip which tends to cause the molten weld metal to wash away in thin layers. When thick welds are cut, the steel rod should be held against the side of the weld and fed downward as required to supply sufficient heat. The oscillating motion should also be used to aid in the removal of the metal. The cutting process in which the steel rod is used is illustrate in figure 12-13.

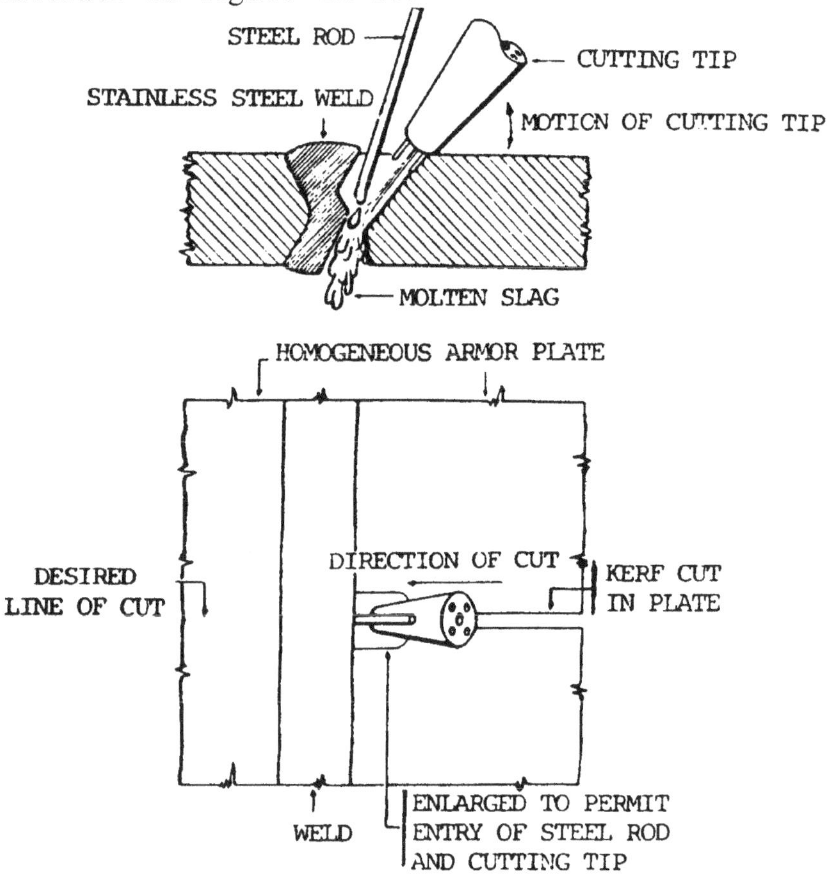

Figure 12-13. Method of cutting stainless steel welds.

(c) Cracks or other defects on the face of stainless steel welds can be removed by holding the cutting tip at a slight angle from the face of the weld as shown in figure 12-14. The reaction between the cutting oxygen and the steel rod develops sufficient heat to melt the weld metal which is washed away. The surface of the joint then can be rewelded.

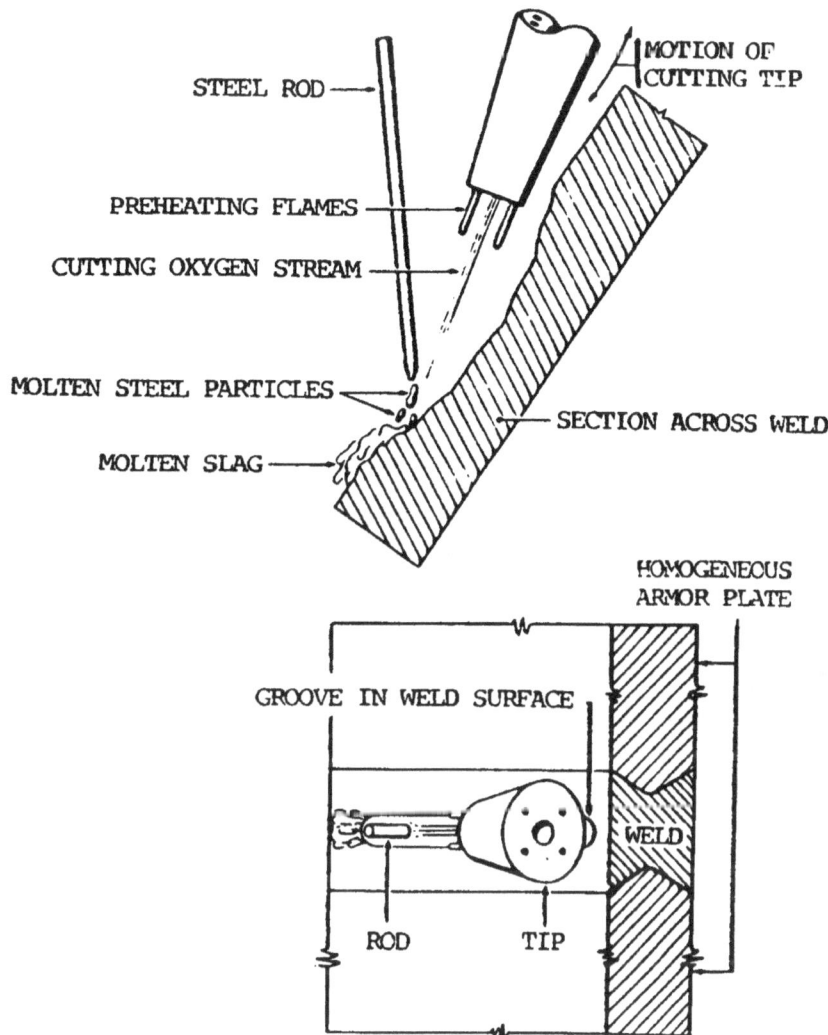

Figure 12-14. Method of removing surface defects from stainless steel welds.

(3) Cutting with the electric arc.

(a) Electric arc cutting is a group of processes whereby metal is cut using the heat of an arc maintained between the electrode and the base metal. Three procedures, described below, are used in cutting with the electric arc.

12-37. CUTTING ARMOR PLATE (cont)

(b) Carbon-arc cutting is a process wherein the cutting of metal is affected by progressive melting with the heat of an electric arc between a metal electrode and the base metal. Direct current straight polarity (electrode negative) is preferred. Under some conditions, the carbon arc is used in conjunction with a jet of compressed air for the removal of defective austenitic weld metal. The carbon arc is utilized for cutting both ferrous and nonferrous metal, but does not produce a cut of particularly good appearance. The electrodes are either carbon or graphite, preferably with a pointed end to reduce arc wandering and produce less erratic cuts.

(c) Metal-arc cutting is a process whereby the cut is produced by progressive melting. Direct current straight polarity is preferred. Coated electrodes ranging in diameter from 1/8 to 1/4 in. (3.2 to 6.4 mm) are used; larger diameters are not satisfactory because of excessive spatter. The thickness of the metal that can be cut by the metal-arc process is limited only by useful length of the electrodes, which are obtainable in 14.0 and 18.0 in. (355.6 and 457.2 mm) lengths. The principal purpose of the electrode coating is to serve as an insulator between the core of the electrode and the side wall of the cut and, consequently, the cut is made with less short-circuiting against the kerf. The cut provided by metal-arc cutting is less ragged than that produced with the carbon-arc. Nevertheless, it is not satisfactory for welding without further preparation by grinding or chiseling. It is used for cutting both ferrous and nonferrous metals.

(d) Oxy-arc cutting is accomplished by directing a stream of oxygen into the molten pool of metal. The pool is made and kept molten by the arc struck between the base metal and the coated tubular cutting rod, which is consumed during the cutting operation. The tubular rod also provides an oxidizing flux and a means of converging oxygen onto the surface being cut. The tubular cutting electrode is made of mild steel. The possibility of contamination is eliminated by the combination of extremely high heat and oxygen under pressure, which act together to oxidize the rod and coating at the point of the arc before the rod metal can fuse with the base metal.

(4) After completing the cut by an arc cutting process, the rough edges and adhering slag should be removed by hammering, chipping, or grinding prior to welding.

12-38. WELDING HOMOGENEOUS ARMOR PLATE

a. General. Welding of damaged armor on vehicles in the field requires, as a preliminay step, that the type of armor be identified by method such as described in paragraph 12-36, p 12-33. Homogeneous armor plate can be satisfactorily welded using the electric arc welding process and 18-8 stainless steel heavy coated electrodes with reverse polarity. Armored vehicles that have been exposed to conditions of extreme cold shall not be welded until the base metal has been sufficiently preheated to bring the temperature of the base metal in the zone of welding up to at least 100 °F (38 °C) At this temperature, the metal will be noticeably warm to the touch. If this preheat is not applied, cracking will occur in the deposited weld metal.

b. Procedure.

(1) When simple cracks (A, fig. 12-15) are welded, the edges of the crack should be beveled by means of flame cutting to produce a double V joint (B, fig. 12-15). Care should be taken to round off the corners at the toe and root of the joint. This is necessary to eliminate excessive dilution of the weld metal by base metal when welding at these points. The included angle of bevel should be approximately 45 degrees to provide electrode clearance for making the root welding beads. The root opening should he from 3/16 to 5/16 in. (4.8 to 7.9 mm) depending on the plate thickness (fig. 12-15).

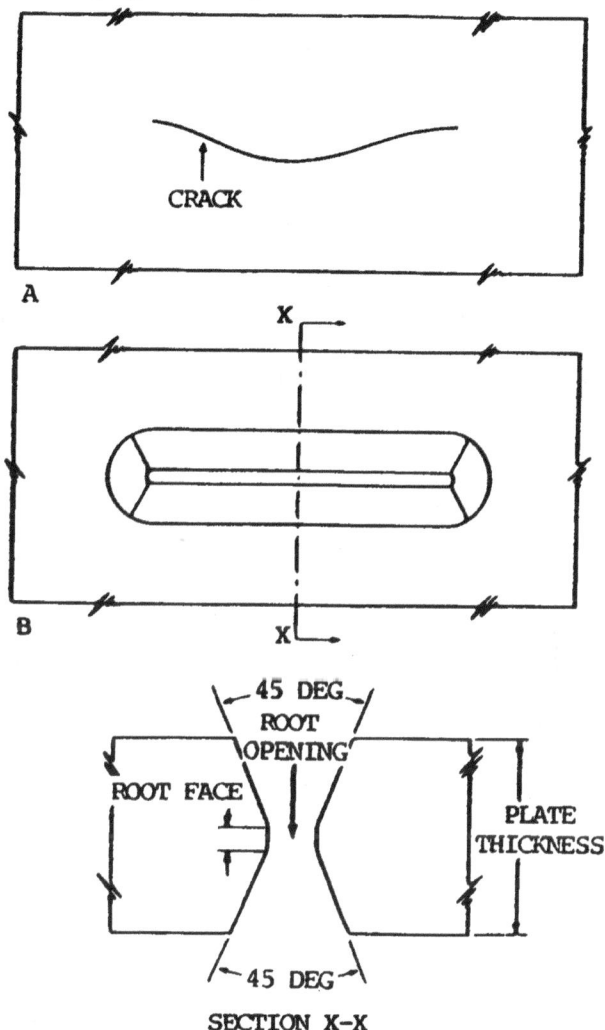

PLATE THICKNESS	ROOT OPENING	ROOT FACE
3/8 TO 7/8	3/16	⌄/0
1 TO 1-1/2	1/4	⌄/0
GREATER THAN 1-1/2	5/16	2⌄/1/16

1 TOLERANCE, PLUS 3/16, MINUS 0
2 TOLERANCE, PLUS 1/16 MINUS 0

NOTE
ALL DIMENSIONS SHOWN
ARE IN INCHES.

Figure 12-15. Preparation for welding cracks in homogeneous armor plate.

12-38. WELDING HOMOGENEOUS ARMOR PLATE (cont)

(2) The weld beads deposited at the root of the weld must be of good quality. It is essential that care be taken to prevent cracks, oxide and slag inclusions, incomplete penetration, or excessive weld metal dilution in this area. Some of the methods recommended as preparatory steps for root head welding are shown in figure 12-16. For narrow root openings, a 3/16-in. (4.8-mm) stainless steel electrode without coating can be tack welded in place (A, fig. 12-16). Welding bead numbers 1, 2, 3, and 4 are then deposited in that order. All slag and oxides should be removed from the joint before beads number 3 and 4 are deposited to insure a sound weld in this zone. If a mild steel rod or strip is used instead of a stainless steel rod (B, fig. 12-16), the back side of the backing rod or strip should be chipped out after beads 1 and 2 are deposited to minimize dilution in beads 3 and 4. The use of a stainless steel strip as a backing for root beads in a wide root opening is shown at C, figure 12-16, together with the sequence of root beads. The alternate method, with a mild steel strip, is shown at D, figure 12-16. When the alternate method is used, the backing rod or strip should be chipped out before depositing beads 3 and 4. Another procedure uses a copper backing bar (E, fig. 12-16). The copper bar is removed after beads 1 and 2 are deposited; the beads will not weld to the bar. Beads 3 and 4 are then deposited. In certain cases where plates of homogeneous armor are cracked along their entire length, thus permitting easy access to the entire cross section of the plate, another method of joint preparation can be used (F, fig. 12-16). The beads deposited at the root of the bevel act as a backing for beads subsequently deposited.

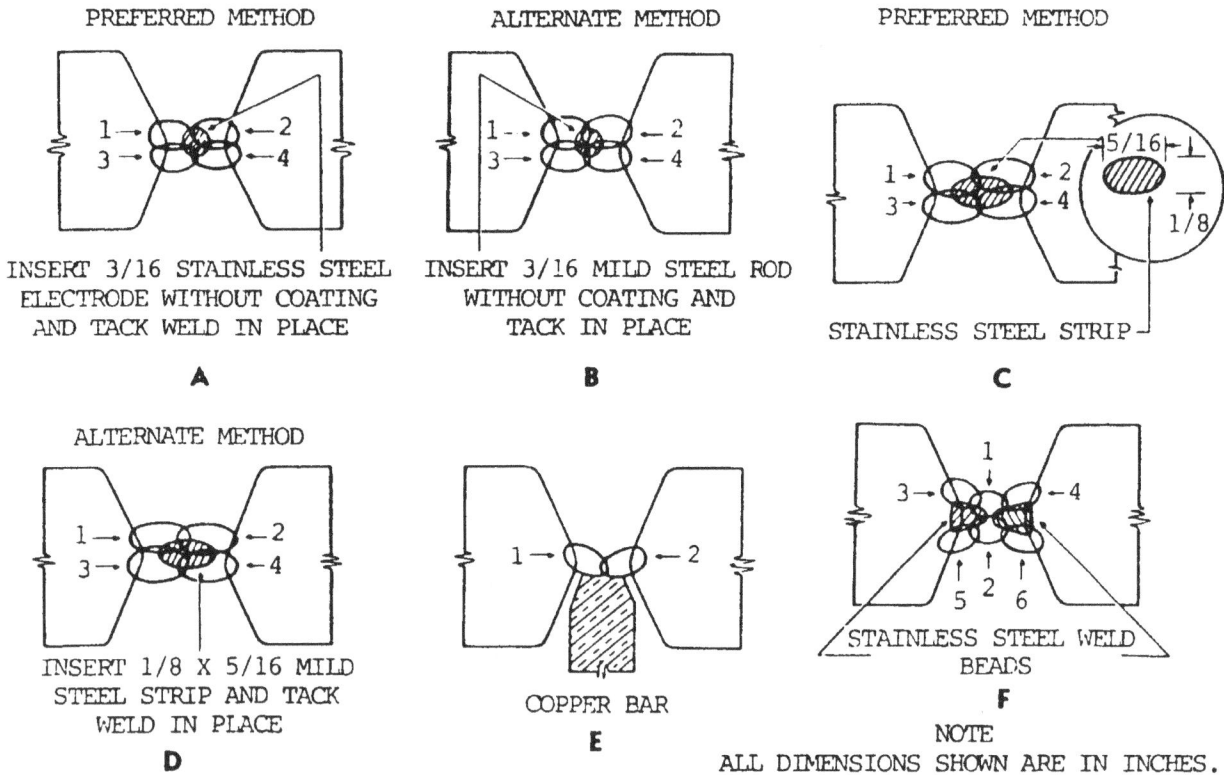

PREFERRED METHOD

ALTERNATE METHOD

PREFERRED METHOD

1 → ← 2
3 → ← 4

INSERT 3/16 STAINLESS STEEL
ELECTRODE WITHOUT COATING
AND TACK WELD IN PLACE

A

1 → ← 2
3 → ← 4

INSERT 3/16 MILD STEEL ROD
WITHOUT COATING AND
TACK IN PLACE

B

1 → ← 2
3 → ← 4
5/16
1/8

STAINLESS STEEL STRIP

C

ALTERNATE METHOD

1 → ← 2
3 → ← 4

INSERT 1/8 X 5/16 MILD
STEEL STRIP AND TACK
WELD IN PLACE

D

1 → ← 2

COPPER BAR

E

1
3 → ← 4
5   2   6

STAINLESS STEEL WELD
BEADS

F

NOTE
ALL DIMENSIONS SHOWN ARE IN INCHES.

Figure 12-16. Backing methods for depositing weld beads at the root
of a double V joint.

(3) A major factor to consider when welding cracks in armor that terminate within the plates is weld crater and fusion zone cracking, especially in the foot beads. An intermittent backstep and overlap procedure (C, fig. 12-17) is recommended to overcome or avoid this hazard. It should be noted that all of the welding steps necessary to complete bead number 1 are completed before bead number 2 is started. By backstepping the passes, the craters at the end of each pass are located on previously deposited metal and are therefore less subject to cracking. All craters on subsequent passes that do not terminte on previously deposited metal should be filled by the hesitation and drawback technique to avoid the formation of star cracks which are caused by the solidification of shallow deposits of molten metal.

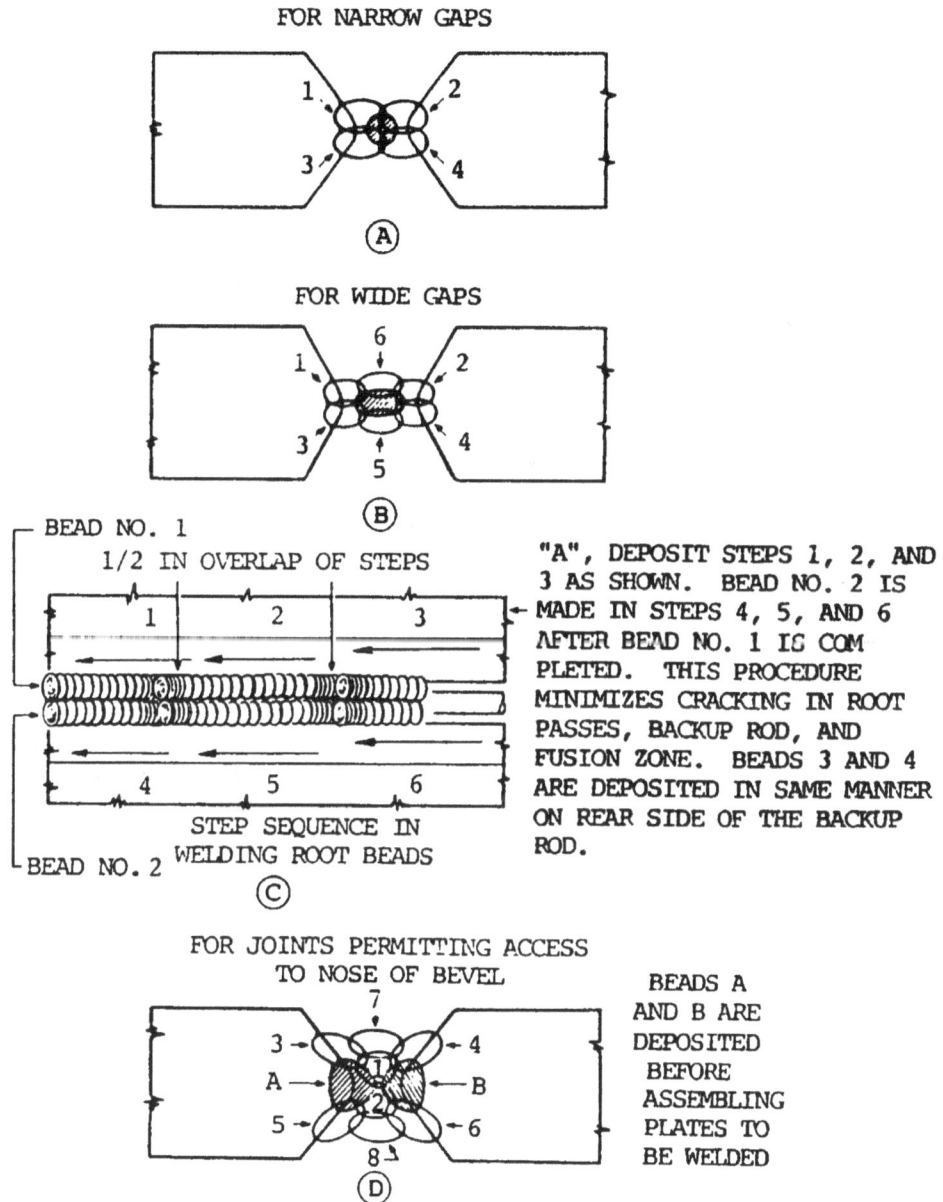

FOR NARROW GAPS

Ⓐ

FOR WIDE GAPS

Ⓑ

BEAD NO. 1

1/2 IN OVERLAP OF STEPS

STEP SEQUENCE IN WELDING ROOT BEADS

BEAD NO. 2

Ⓒ

"A", DEPOSIT STEPS 1, 2, AND 3 AS SHOWN. BEAD NO. 2 IS MADE IN STEPS 4, 5, AND 6 AFTER BEAD NO. 1 IS COMPLETED. THIS PROCEDURE MINIMIZES CRACKING IN ROOT PASSES, BACKUP ROD, AND FUSION ZONE. BEADS 3 AND 4 ARE DEPOSITED IN SAME MANNER ON REAR SIDE OF THE BACKUP ROD.

FOR JOINTS PERMITTING ACCESS TO NOSE OF BEVEL

Ⓓ

BEADS A AND B ARE DEPOSITED BEFORE ASSEMBLING PLATES TO BE WELDED

Figure 12-17.   Sequence of passes when depositing weld beads on homogeneous armor plate.

12-38.   WELDING HOMOGENEOUS ARMOR PLATE (cont)

(4) Each pass in beads 1, 2, 3, and 4 (A and B, fig. 12–17, p 12-39) is limited to 1 to 2 in. (25.4 to 50.8 mm) in length and should be peened while the weld
metal is still hot to help overcome the cooling stresses. No electrode weaving
motion should be used when the root beads are deposited, and the welding should be
performed preferably with a 5/32-in. (4.0-mm) electrode. Peening also tends to
eliminate or minimize warpage in the section being welded. A blow should be controlled by properly adjusting the welding. Some of the more common defects encountered when welding root beads on homogeneous armor plate and the proper remedial
procedures are shown in figure 12-18.

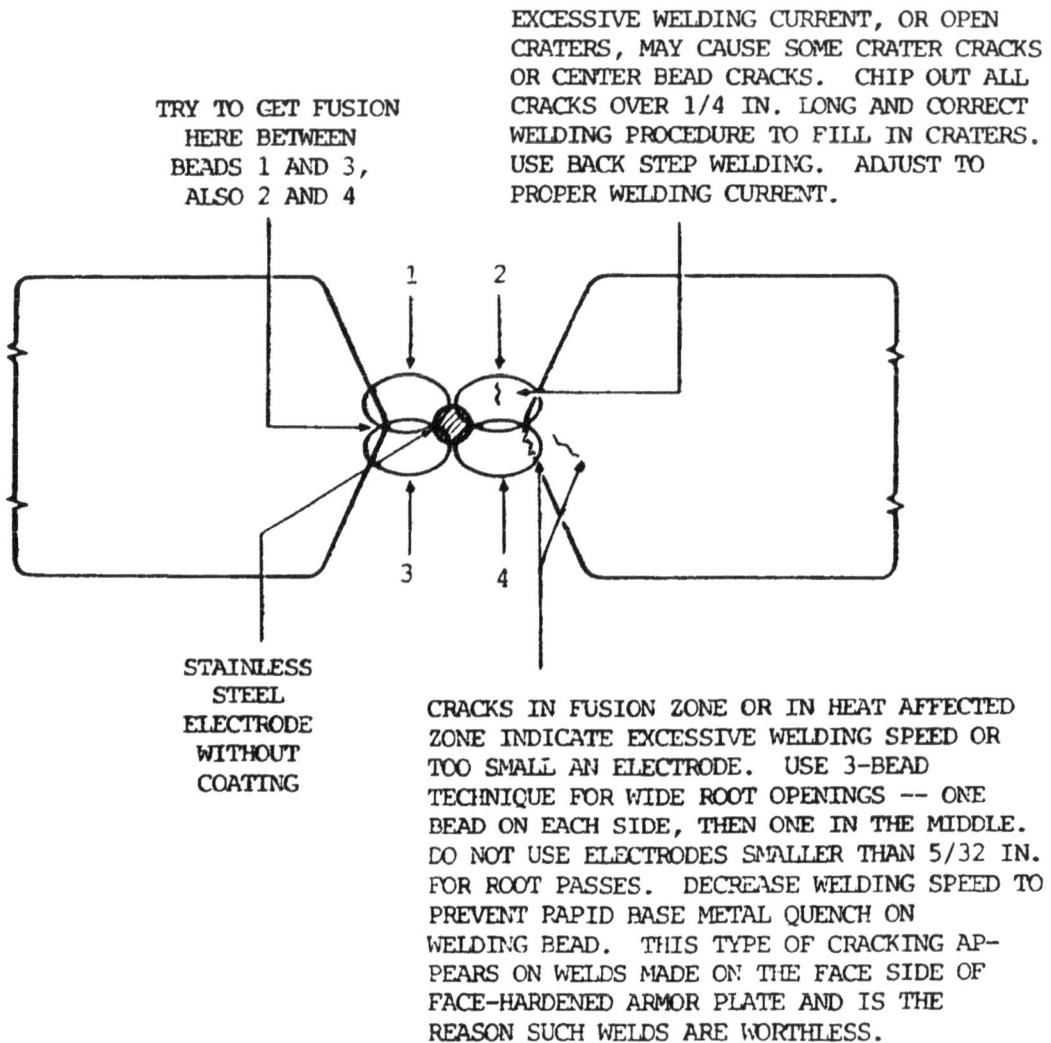

TRY TO GET FUSION
HERE BETWEEN
BEADS 1 AND 3,
ALSO 2 AND 4

EXCESSIVE WELDING CURRENT, OR OPEN
CRATERS, MAY CAUSE SOME CRATER CRACKS
OR CENTER BEAD CRACKS. CHIP OUT ALL
CRACKS OVER 1/4 IN. LONG AND CORRECT
WELDING PROCEDURE TO FILL IN CRATERS.
USE BACK STEP WELDING. ADJUST TO
PROPER WELDING CURRENT.

STAINLESS
STEEL
ELECTRODE
WITHOUT
COATING

CRACKS IN FUSION ZONE OR IN HEAT AFFECTED
ZONE INDICATE EXCESSIVE WELDING SPEED OR
TOO SMALL AN ELECTRODE. USE 3-BEAD
TECHNIQUE FOR WIDE ROOT OPENINGS -- ONE
BEAD ON EACH SIDE, THEN ONE IN THE MIDDLE.
DO NOT USE ELECTRODES SMALLER THAN 5/32 IN.
FOR ROOT PASSES. DECREASE WELDING SPEED TO
PREVENT RAPID BASE METAL QUENCH ON
WELDING BEAD. THIS TYPE OF CRACKING AP-
PEARS ON WELDS MADE ON THE FACE SIDE OF
FACE-HARDENED ARMOR PLATE AND IS THE
REASON SUCH WELDS ARE WORTHLESS.

Figure 12-18. Common defects when welding root beads on homogenous
armor plate and the remedial procedures.

(5) The sequence of welding beads and the procedure recommended to completely weld the single V joint are shown in figure 12-19. This welding should be performed with 5/32- or 3/16-in. (4.0- to 4.8-mm) electrodes. The electrode is directed against the side wall of the joint to form an angle of approximately 20 to 30 degrees with the vertical. The electrode should also be inclined 5 to 15 degrees in the direction of the welding. By this procedure, the side wall penetration can be effectively controlled. The electrode weaving motion should not exceed 2-1/2 electrode core wire diameters. This is important because stainless steel has a coefficient of expansion approximately 1-1/2 times that of mild steel. Consequently, if a weaving motion greater than that recommended is used, longitudinal shrinkage cracks in the weld or fusion zone may develop. The thickness of the layer of metal deposited can be varied by controlling the speed of welding.

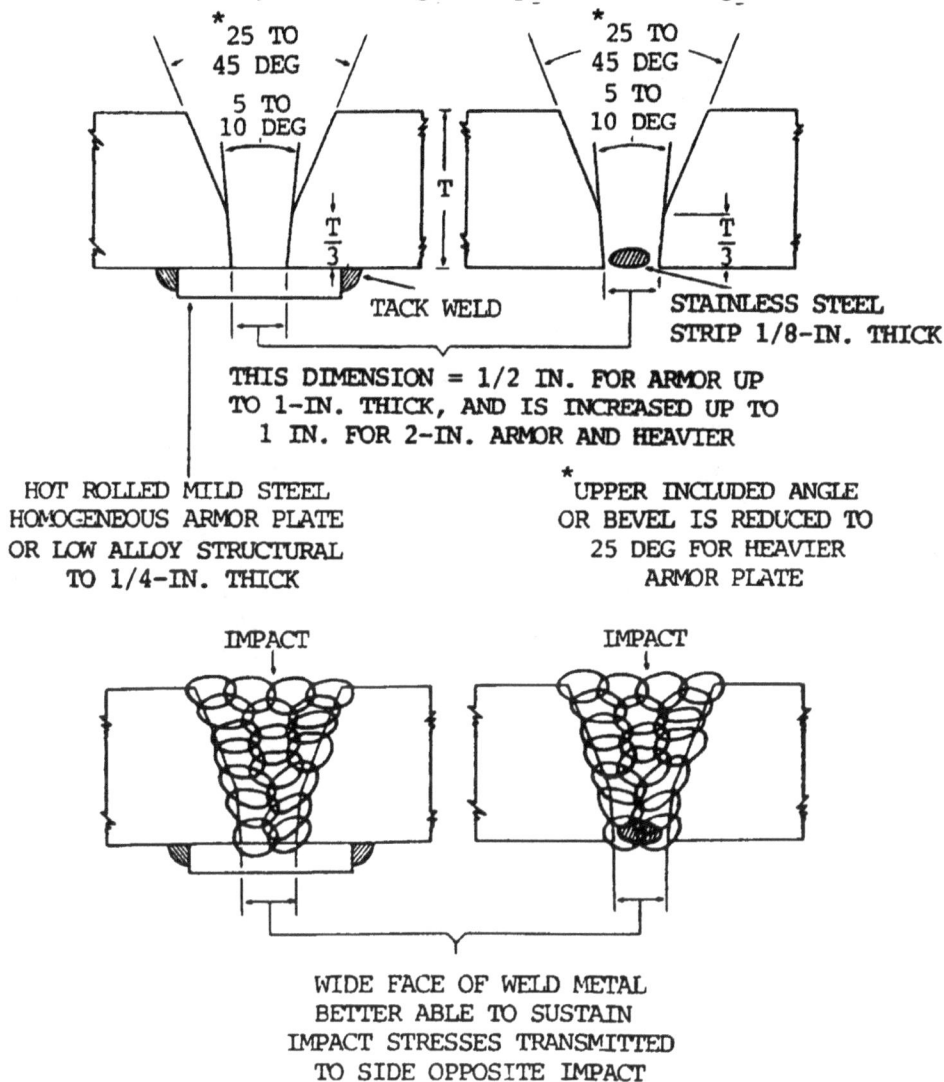

Figure 12-19.   Procedure for welding single V joint on
homogeneous armor plate.

12-38. WELDING HOMOGENEOUS ARMOR PLATE (cont)

(6) The sequence of passes used for completely filling a double V joint (fig. 12-20) was determined after consideration of all the foregoing factors. The depth of penetration of weld metal into base metal should be controlled in order to obtain good fusion without excessive dilution of the weld. Excessive dilution will cause the weld to be non-stainless, brittle, and subject to cracking. Proper penetration will produce long, scalloped heat affected zone on each side of the weld (A and B, fig. 12-20). Insufficient penetration (surface fusion) will produce a fairly straight edged heat affected zone on each side of the weld. This condition is undesirable from the standpoint of good ballistic properties.

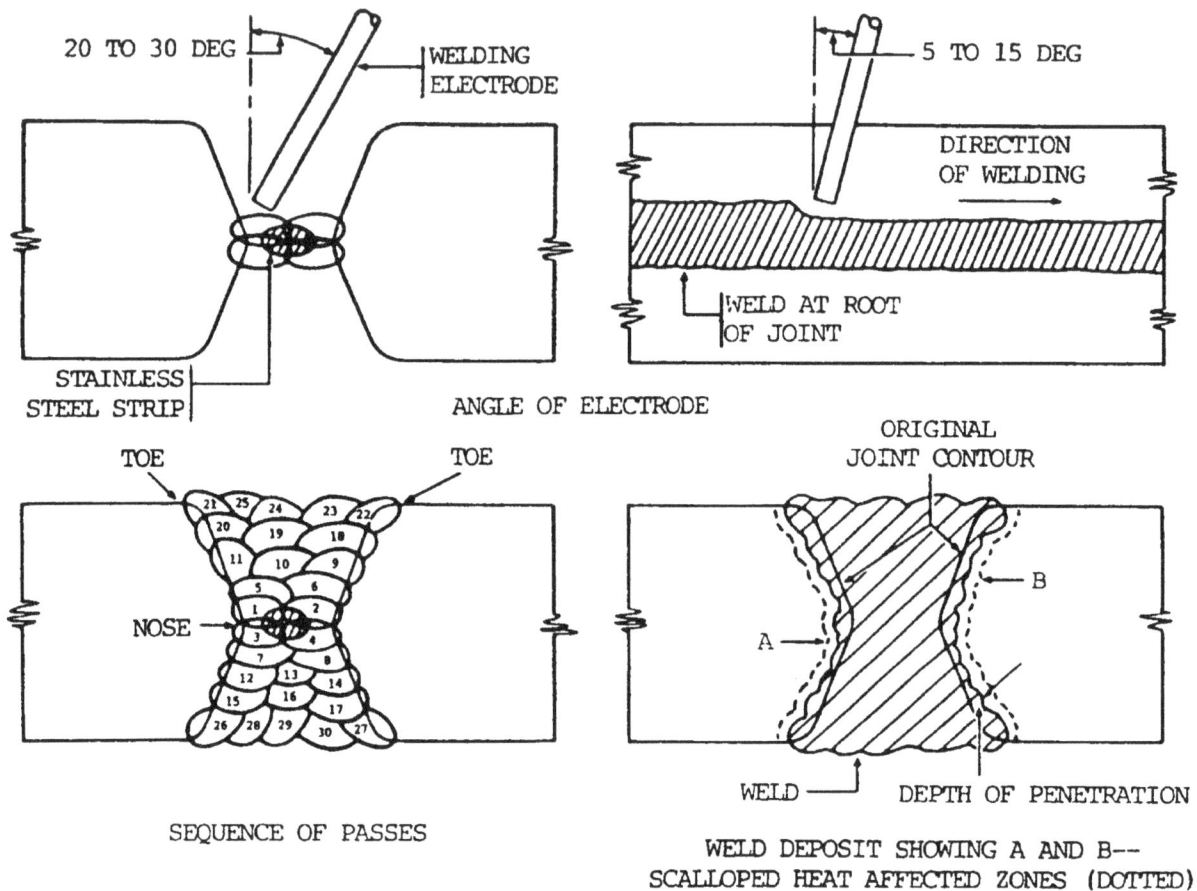

Figure 12-20. Double V weld on homogeneous armor plate.

(7) By alternating the deposition of metal, first on one side of the joint and then the other, a closer control of heat input at the joint is obtained and the shape of the welded structure can be maintained. Each layer of metal deposited serves to stress relieve the weld metal immediately beneath it, and will also partially temper the heat affected zone produced in the base metal by the previous welding bead. The passes at the toe of each weld layer also serve as annealing passes. They are deposited before intermediate passes are added to completely fill

the intervening space (see passes 9 and 11, 12 and 14, 15 and 16, 18 and 20, etc., fig. 12-20). These annealing passes are important factors in the elimination of fusion zone cracks which might start at the surface of the weld. Through careful control of the depth of penetration, a heat affected zone with a scalloped effect is produced.

c. <u>Emergency Repairs</u>. Emergency repairs on cracked armor plate can be made by using butt straps on the back of the cracked armor (fig. 12-21). The primary purpose of these butt straps is to strengthen the section weakened by the crack.

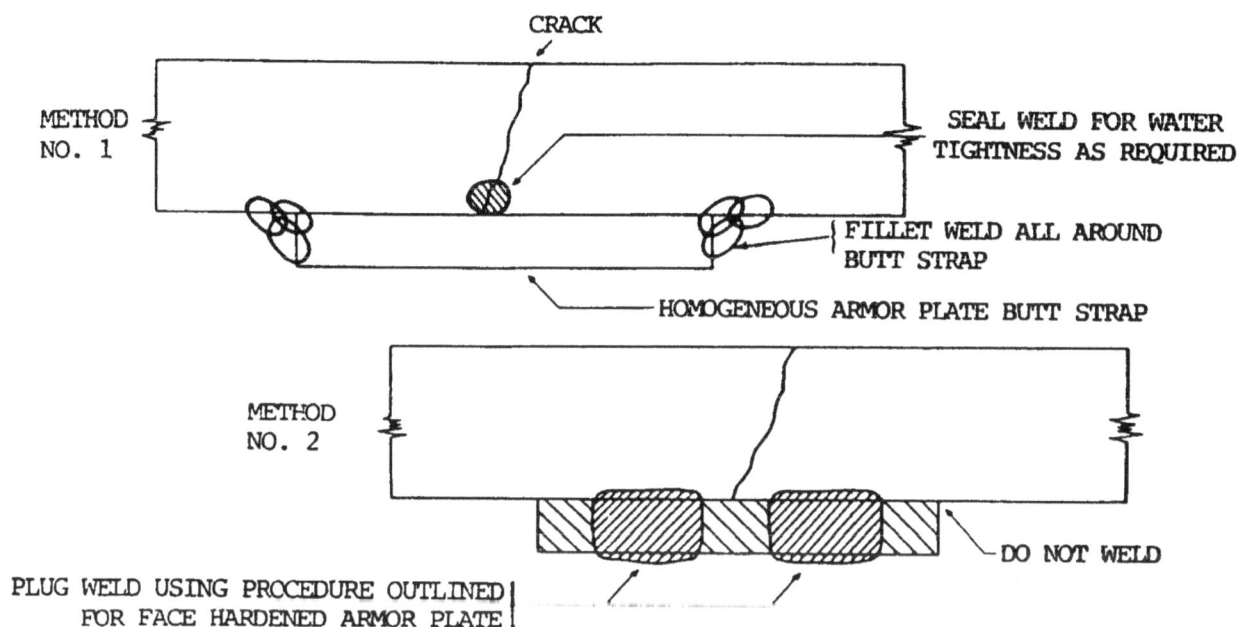

Figure 12-21. Butt strap welds on cracked armor plate.

d. <u>Repairing Penetrations</u>. Complete penetrations in homogeneous armor plate are repaired by using the procedures shown in figures 12-22 through 12-24, p 12-44 and 12-45. Considerable structural damage is done to the metal immediately adjacent to the shell penetration (fig. 12-23, p 12-44). A sufficient amount of metal should be removed to ensure complete freedom from protrusions and subsurface cracks, and good contact between the patch and the base armor plate as shown in figure 12-22, p 12-44. Where the projectile penetration openings are large, relative to the thickness of the plate, a plug patch of homogeneous armor having the same thickness as the base metal should be used.The plug patch should be shaped and welded in place as shown in figure 12-24, p 12-45. Small diameter penetrations in armor can be repaired by plug welding without the use of patches.

12-38.   WELDING HOMOGENEOUS ARMOR PLATE (cont)

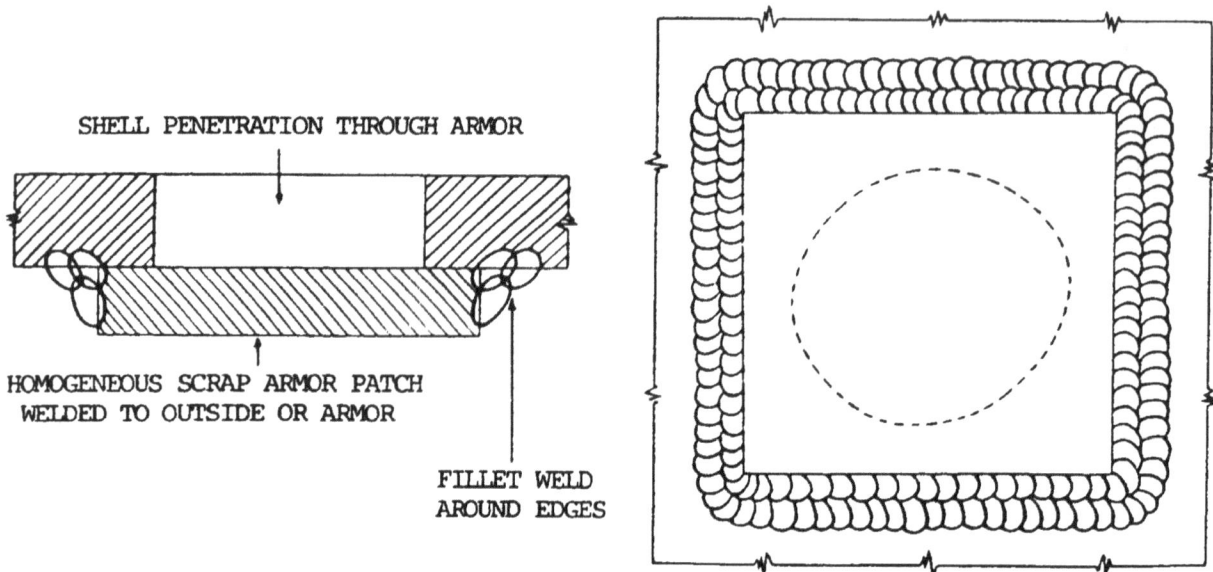

Figure 12-22.   Emergency repair of shell penetration through armor.

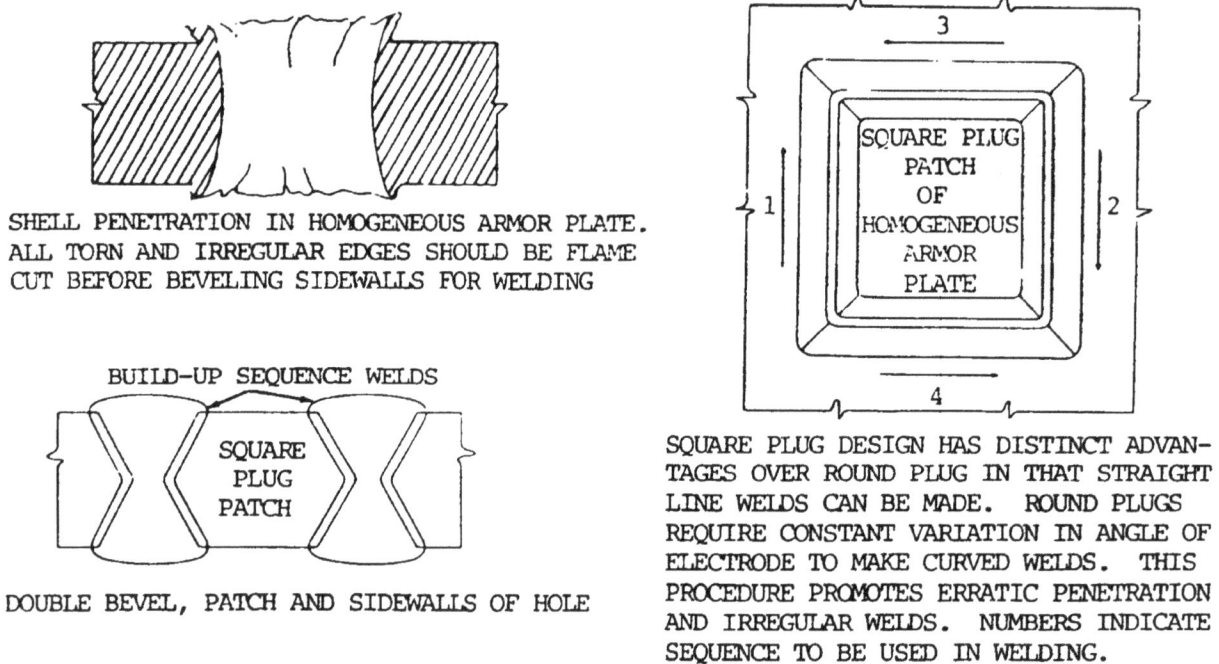

SHELL PENETRATION IN HOMOGENEOUS ARMOR PLATE. ALL TORN AND IRREGULAR EDGES SHOULD BE FLAME CUT BEFORE BEVELING SIDEWALLS FOR WELDING

BUILD-UP SEQUENCE WELDS

SQUARE PLUG PATCH

DOUBLE BEVEL, PATCH AND SIDEWALLS OF HOLE

SQUARE PLUG DESIGN HAS DISTINCT ADVANTAGES OVER ROUND PLUG IN THAT STRAIGHT LINE WELDS CAN BE MADE. ROUND PLUGS REQUIRE CONSTANT VARIATION IN ANGLE OF ELECTRODE TO MAKE CURVED WELDS. THIS PROCEDURE PROMOTES ERRATIC PENETRATION AND IRREGULAR WELDS. NUMBERS INDICATE SEQUENCE TO BE USED IN WELDING.

Figure 12-23.   Double V plug welding procedure for repairing shell penetration in homogeneous armor plate.

PATCH

IMPACT

A

LONG SCALLOPED FUSION ZONE LINES HAVE BETTER
SHOCK ABSORBING PROPERTIES. WIDE FACE OF WELD
METAL "A" IS BETTER ABLE TO SUSTAIN IMPACT
STRESSES TRANSMITTED TO SIDE OPPOSITE IMPACT.

DOUBLE V JOINT--CORRECT

PATCH

IMPACT

FAIRLY STRAIGHT FUSION ZONE LINE
HAS POOR BALLISTIC STRENGTH.

DOUBLE BEVEL JOINT--INCORRECT

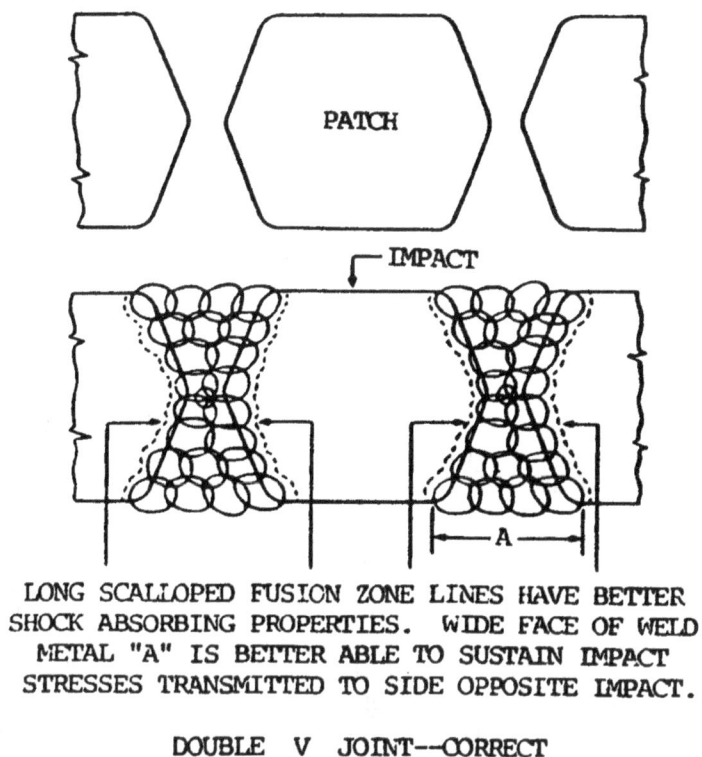

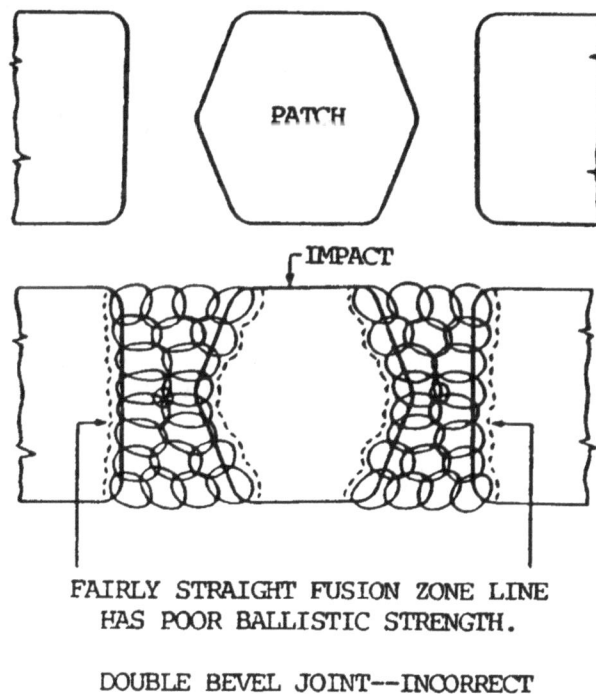

Figure 12-24. Correct and incorrect plug weld preparation for
repairing shell penetration in homogeneous armor plate.

12-38.   WELDING HOMOGENEOUS ARMOR PLATE (cont)

e. <u>Repairing Bulges</u>. Bulges in armor that are also cracked but do not interfere with the operation of internal mechanisms in the vehicle can be repaired by
welding the cracked section, using the procedure previously described in this section. For best repairs, however, the bulge should be cut out and a patch inserted. Where bulges interfere with the operation of internal mechanisms, grinding or
chipping of the bulged surface can be applied to remove the interference.In all
cases, the welds should be made to the full thickness of the plate and all cracks
over 1/4 in. (6.4 mm) in width should be chipped out before rewelding.

f. <u>Repair Made from One Side</u>. Where it is not feasible to make the welding
repair from both sides of the armor, the joint must necessarily be made from one
side (fig. 12-19, p 12-41). Either a butt strap or stainless steel strip can be
used as a backup for the root beads of the weld.

g. <u>Repairs with Nonwelded Butt Strap</u>.For applications where a butt strap
would interfere with the operation of internal mechanisms, a technique is used that
permits removal of the butt strap after welding (fig. 12-25). This welding technique was developed to permit welding a single V joint in homogeneous armor plate
without welding the butt strap to the deposited weld metal.It involves changing
the angle at which the electrode is held during the side to side weaving motion,
which is used in making the root pass. By increasing the electrode angle to approximately 60 degrees from the vertical at the middle of the weave and increasing the
weaving speed simultaneously (A, fig. 12-25), all the deposited metal is welded
only to the previously deposited metal. At each end of the weave, the weaving
speed of the electrode is decreased while simultaneously decreasing the electrode
angle to approximately 15 degrees from the vertical, and the electrode is held
adjacent to the side wall momentarily to ensure good side wall penetration (B, fig. 12-25).
After depositing the root pass, the butt strap can be removed by breaking
the tack welds securing it to the bottom face of the armor. If desired, a finish
pass can be applied to the root of the weld after removing the butt strap.

h.   <u>Repairing Gouges</u>. When armor is struck by a projectile impacting at an
angle and is thus gouged at the surface, the gouge should be prepared in a double V
joint design to allow welding from both sides (fig. 12-26). Merely filling the
gouge with weld metal is an unsatisfactory procedure as this does not remove any
subsurface cracks that may have been caused by the shell impact.Also, the heat
affected zone produced at the base of the filled-in gouge has poor ballistic
strength.

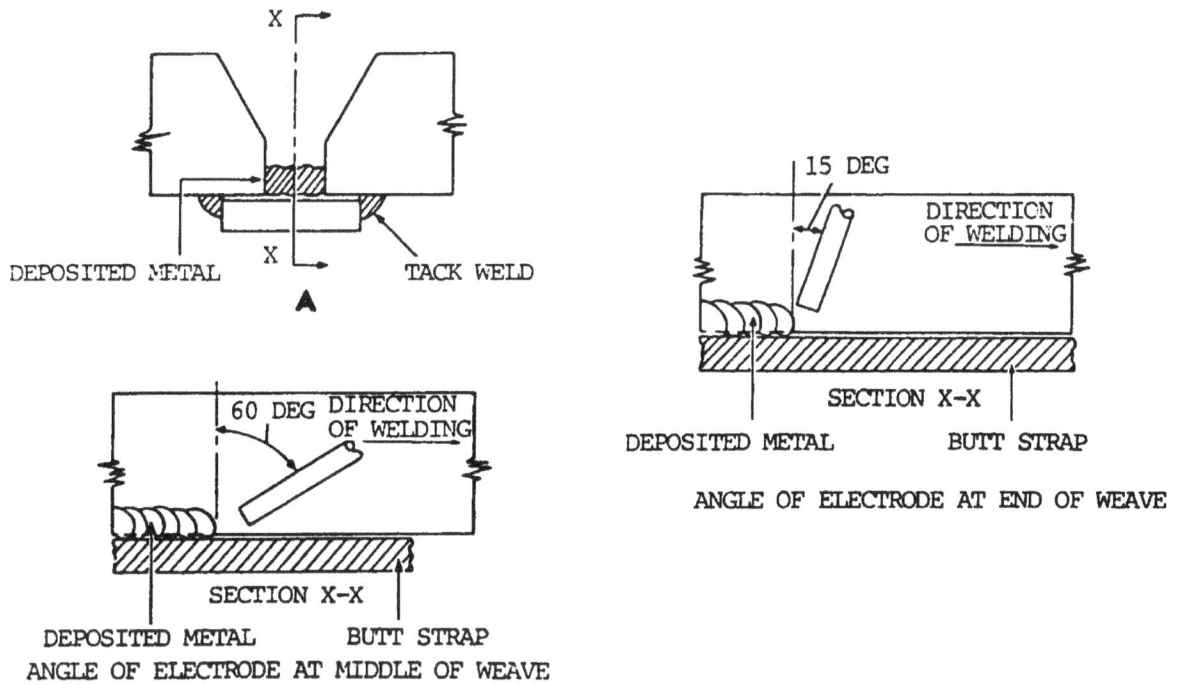

DEPOSITED METAL   TACK WELD

A

60 DEG DIRECTION OF WELDING

SECTION X-X

DEPOSITED METAL   BUTT STRAP
ANGLE OF ELECTRODE AT MIDDLE OF WEAVE

15 DEG

DIRECTION OF WELDING

SECTION X-X

DEPOSITED METAL   BUTT STRAP

ANGLE OF ELECTRODE AT END OF WEAVE

B

Figure 12-25. Welding homogeneous armor without welding butt strap.

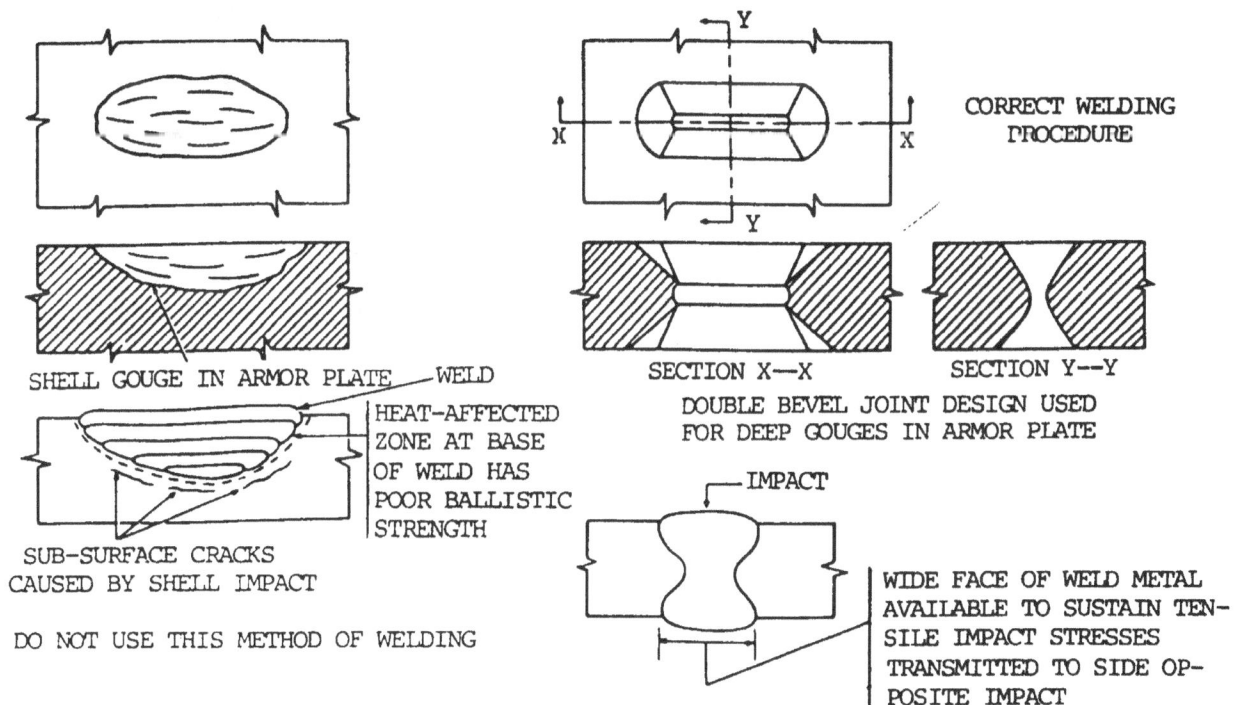

SHELL GOUGE IN ARMOR PLATE

WELD

HEAT-AFFECTED ZONE AT BASE OF WELD HAS POOR BALLISTIC STRENGTH

SUB-SURFACE CRACKS CAUSED BY SHELL IMPACT

DO NOT USE THIS METHOD OF WELDING

CORRECT WELDING PROCEDURE

SECTION X—X   SECTION Y--Y
DOUBLE BEVEL JOINT DESIGN USED FOR DEEP GOUGES IN ARMOR PLATE

IMPACT

WIDE FACE OF WELD METAL AVAILABLE TO SUSTAIN TEN-SILE IMPACT STRESSES TRANSMITTED TO SIDE OP-POSITE IMPACT

Figure 12-26. Welding repair of gouges in surface of homogeneous armor plate.

12-47

12-39.   WELDING FACE HARDENED ARMOR PLATE

a.   General.

(1) Face hardened armor plate can be welded satisfactorily using the arc welding process and 18-8 stainless steel, heavy coated electrodes with reverse polarity. The face side of face hardened armor is extremely hard and brittle. Special precautions must be taken to avoid excessive heating and distortion of the plate to prevent cracking of the face due to the resulting stresses. A satisfactory method for welding this type of armor makes use of the butt strap and plug weld technique. The welding procedure for face hardened armor varying from 1/4 to 1.0 in. (6.4 to 25.4 mm) in thickness is illustrated in figures 12-27 and 12-28, p 12-50. The welding is done from the soft side of the armor plate and the strength of the joint depends on the soundness of the plug welds. The butt strap should be cut to conform to dimensions given for the particular thickness of face hardened armor being welded. The butt strap is tack welded to the soft side of the armor through elongated slots cut into the strap.The plugs should then be welded to completely fill the slots without excessive weld reinforcement or undercutting at the surface of the plug. These precautions are necessary to eliminate surface discontinuities which act as stress raisers and are a source of crack formations under impact loads. To effectively seal the crack in face hardened armor against lead spatter, and where watertightness is required, a seal head weld should be made on the soft side and ground flush before applying the butt strap.All welding should be performed on clean, scale-free surfaces.Previously deposited weld metal should be thoroughly cleaned by chipping and wire brushing to remove slag and oxides and insure sound welds.

(2) Crater cracks can be eliminated by the backstep and overlap procedures, or by using the electrode hesitation and drawback technique.Crater cracks formed in the initial weld passes should be chipped out before additional weld metal is applied. They can be welded out successfully on all subsequent passes of the weld. As a further precaution, string beads should be used for the initial passes. For subsequent passes, do not weave the electrode more than 2-1/2 electrode core wire diameters. The efficiency of the joint welded by this method depends on good fusion to the base metal and side walls of the slots in the butt strap.

(3) If straightening is necessary, do not hammer on the face of the armor; all hammering should be done on the soft side, on the butt strap, or on the plug welds. Force should not be applied to straighten face hardened armor if the applied force will produce tension on the face side.

(4) Where two or more butt straps are used to repair irregular cracks or to make a patch weld, the butt straps are welded together for additional strength (fig. 12-29, p 12-51).

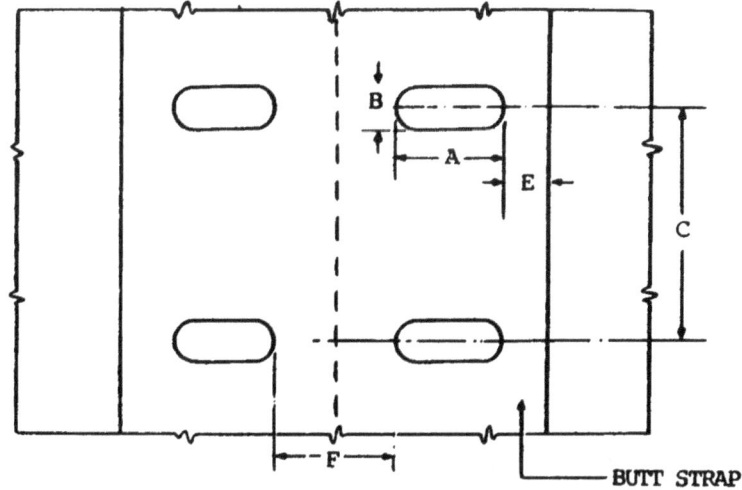

WHERE REQUIRED
DEPOSIT SEAL
BEAD TO INSURE
WATER TIGHT
JOINT. NO
WELDING TO BE
PERFORMED
ON HARD FACE.

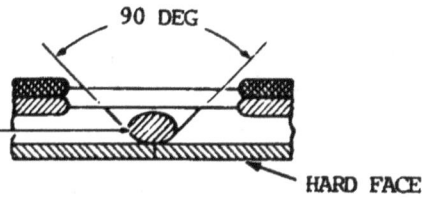

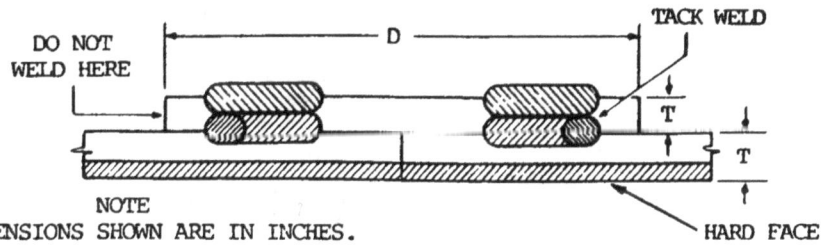

90 DEG

HARD FACE

TACK WELD

DO NOT
WELD HERE

NOTE
ALL DIMENSIONS SHOWN ARE IN INCHES.

HARD FACE

T	t	A	B	C	D (min.)	E (min.)	F
1/4	3/16	3/4	7/16	3	3	1/4	1
3/8	1/4	1-1/8	1/2	3	4	3/8	1
1/2	3/8	1-1/4	5/8	3	4-1/4	3/8	1
5/8	3/8	1-1/4	5/8	3	4-1/4	3/8	1
3/4	1/2	1-1/4	5/8	2	4-1/2	7/16	1
1	5/8	1-3/8	3/4	2	4-3/4	1/2	1

Figure 12-27. Welding joint data for butt welds on face hardened armor.

12-39. WELDING FACE HARDENED ARMOR PLATE (cont)

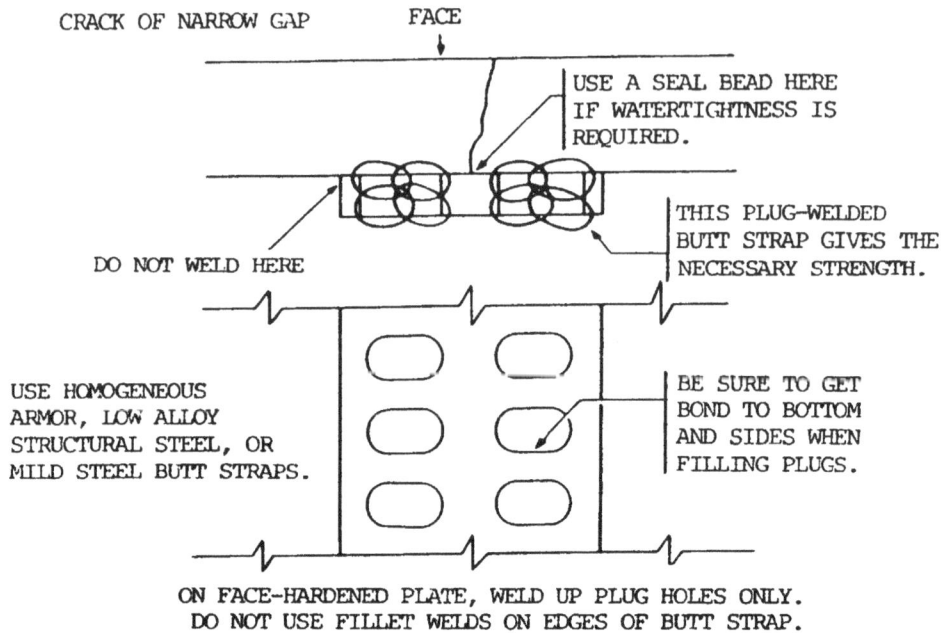

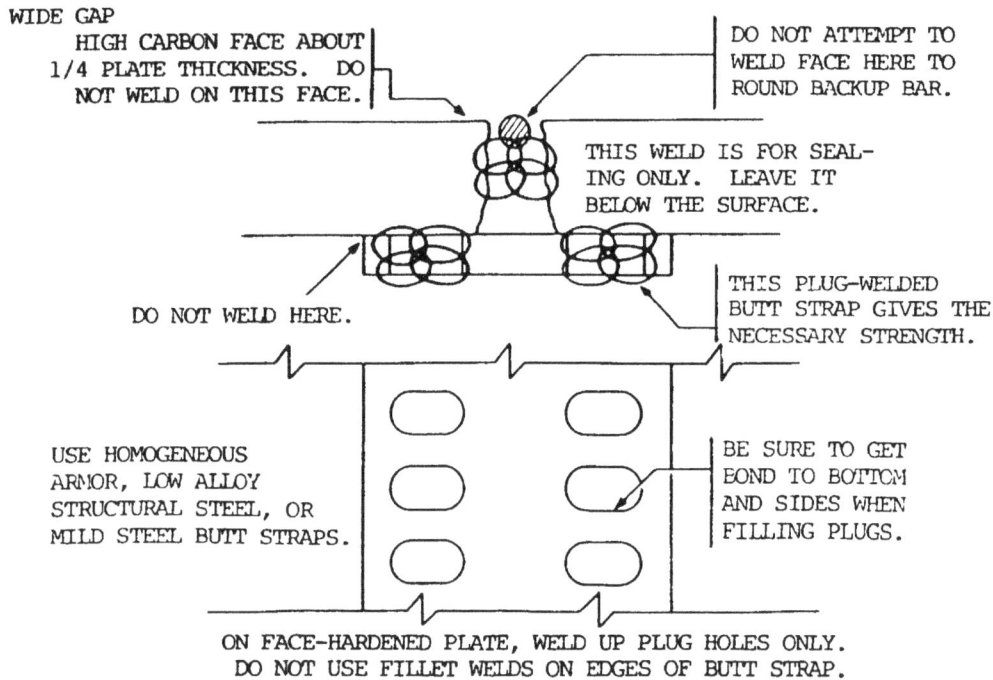

CRACK OF NARROW GAP   FACE

USE A SEAL BEAD HERE
IF WATERTIGHTNESS IS
REQUIRED.

DO NOT WELD HERE

THIS PLUG-WELDED
BUTT STRAP GIVES THE
NECESSARY STRENGTH.

USE HOMOGENEOUS
ARMOR, LOW ALLOY
STRUCTURAL STEEL, OR
MILD STEEL BUTT STRAPS.

BE SURE TO GET
BOND TO BOTTOM
AND SIDES WHEN
FILLING PLUGS.

ON FACE-HARDENED PLATE, WELD UP PLUG HOLES ONLY.
DO NOT USE FILLET WELDS ON EDGES OF BUTT STRAP.

WIDE GAP
HIGH CARBON FACE ABOUT
1/4 PLATE THICKNESS. DO
NOT WELD ON THIS FACE.

DO NOT ATTEMPT TO
WELD FACE HERE TO
ROUND BACKUP BAR.

THIS WELD IS FOR SEAL-
ING ONLY. LEAVE IT
BELOW THE SURFACE.

DO NOT WELD HERE.

THIS PLUG-WELDED
BUTT STRAP GIVES THE
NECESSARY STRENGTH.

USE HOMOGENEOUS
ARMOR, LOW ALLOY
STRUCTURAL STEEL, OR
MILD STEEL BUTT STRAPS.

BE SURE TO GET
BOND TO BOTTOM
AND SIDES WHEN
FILLING PLUGS.

ON FACE-HARDENED PLATE, WELD UP PLUG HOLES ONLY.
DO NOT USE FILLET WELDS ON EDGES OF BUTT STRAP.

Figure 12-28.   Use of butt strap on face hardened armor to repair cracks or gaps.

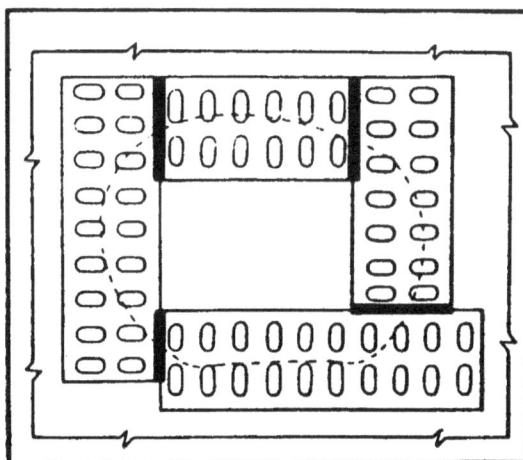

Figure 12-29. Butt strap weld on face hardened armor.

b. Armor Plate Repair Methods.

(1) Corner joints can be repaired by using angle iron for butt straps (fig. 12-30). The same procedures are followed in making plug welds as used for repairing cracked armor.

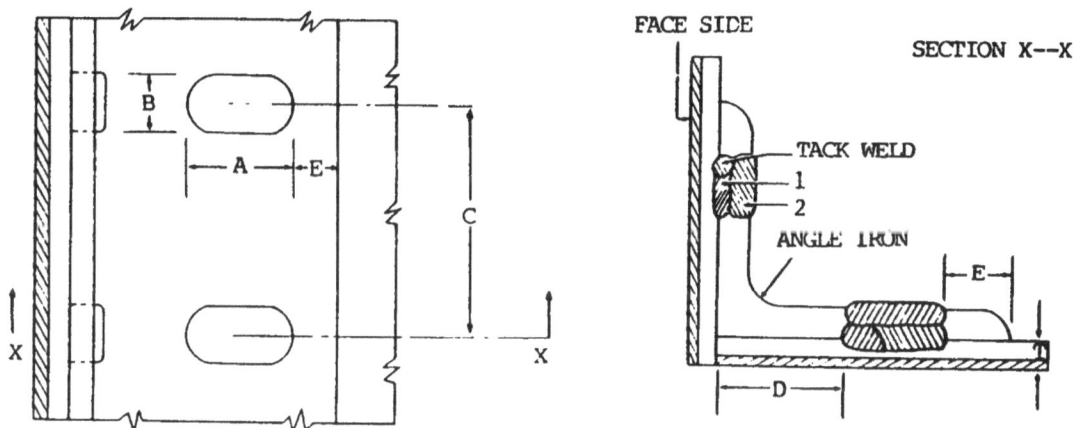

T	A	B	C	D	E (MIN.)	ANGLE IRON
1/4	3/4	7/16	3	5/16	1/4	1-5/16 X 1-5/16 X 3/16
3/8	1-1/8	1/2	3	3/8	3/8	1-7/8 X 1-7/8 X 1/4
1/2	1-1/4	5/8	3	1/2	3/8	2-1/8 X 2-1/8 X 3/8

WHERE REQUIRED-DEPOSIT SEAL BEAD TO ENSURE WATER TIGHT JOINT. NO WELDING TO BE PERFORMED ON HARD FACE.

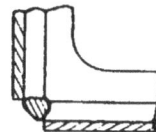

NOTE
ALL DIMENSIONS SHOWN ARE IN INCHES.

Figure 12-30. Weld joint data for corner welds on face hardened armor plate.

12-39. WELDING FACE HARDENED ARMOR PLATE (cont)

(2) Although the butt strap method is satisfactory for repairing damaged face hardened armor up to 1 in.(25.4 mm) thick and heavier, it is usually only used on thicknesses up to and including 1/2 in. (12.7 mm) plate. Another accepted procedure for welding face hardened armor more than 1/2 in thick is a double V joints method requiring that the soft side be completely welded before any welding is attempted on the face side of the plate (fig. 12-31). By using string bead welding and the backstep and overlap procedure for the root passes, the danger of cracking is held to a minimum. Additional passes can be run straight out; however, no weaving should be used on this type of joint in order to keep the structure free from warpage. The depressed joint method is modified procedure for welding face hardened armor up to and including 1/2 in. (12.7 mm) in thickness (fig. 12-32). This joint is made by using a stainless steel bar 1/8 x 1/4 in. (3.2 x 6.4 mm) in cross section. The principal advantages of this joint are its simplicity and good structural and ballistic properties. Care should be taken that no welding is done on the hard face side.

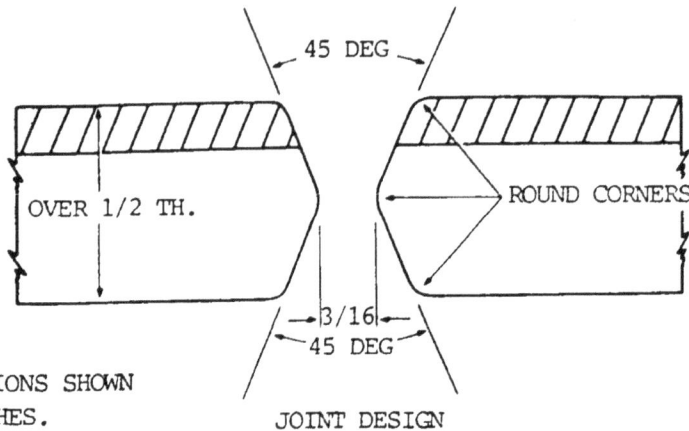

NOTE: ALL DIMENSIONS SHOWN ARE IN INCHES.

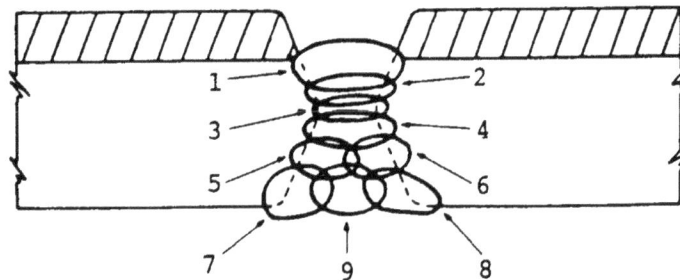

JOINT DESIGN

GENERAL SEQUENCE OF WELDING BEADS

COMPLETELY WELD SOFT SIDE WITH BEADS NO. 1-9.

FOR BEAD NO. 1, USE 1/8 INCH ELECTRODES. FOR ALL OTHER BEADS, USE 5/32 AND 3/16 INCH ELECTRODES.

USE STRING BEAD TECHNIQUE THROUGHOUT.

Figure 12-31. Procedure for welding face hardened armor over 1/2 in. thick, using the double V joint method.

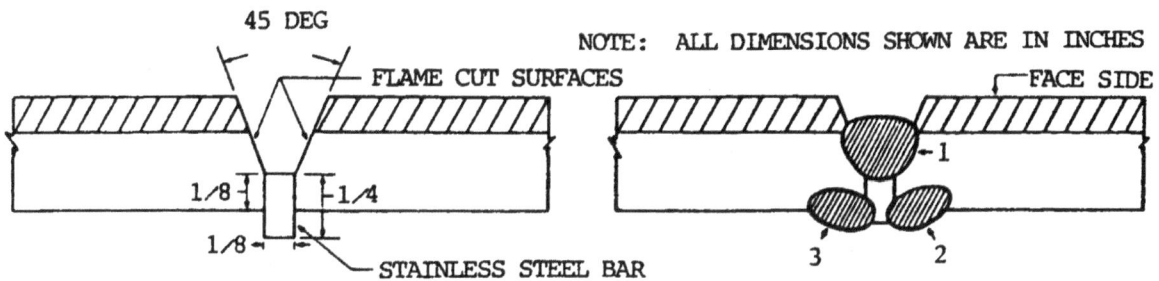

Figure 12-32. Procedure for welding face hardened armor up to
1/2-in., using the depressed joint method.

c. Armor Plate Welding Electrodes.

(1) The most satisfactory method for the repair of homogeneous and face hardened armor plate is the arc welding process with stainless steel electrodes.

(2) The oxyacetylene welding process requires heating of a large section of the base metal on either side of the prepared joint to maintain a welding puddle of sufficient size at the joint to weld satisfactorily. This heating destroys the heat treatment imparted to armor plate, causing large areas to become weak structurally and ballistically. In addition, the procedure is slow and produces considerable warpage in the welded sections.

(3) Initial developments in armor plate welding have specified stainless steel electrodes containing 25 percent chromium and 20 percent nickel. In an effort to conserve chromium and nickel, electrodes containing 18 percent chromium and 8 percent nickel in the core wire and small percentages of either manganese or molybdenum, or both added in the coating produce excellent results. These electrodes are recommended for welding all types of armor plate by the electric arc process without preheating or postheating the structure welded and should be the all position type. By convention, these electrodes are known as manganese modified 18-8 stainless steel and molybdenum modified 18-8 stainless steel electrodes.

d. Current and Polarity. The recommended welding current settings listed are for direct current reserves polarity, all position, heavy coated, modified 18-8 stainless steel electrodes. The exact current requirements will be governed to some extent by the joint type, electrode design, and position of welding.

Electrode diamter (in.)	Current range (amps)
1/8	90 to 100
5/32	110 to 130

e. Electrode Requirements. Field repair units will require the various type electrodes in approximately the following proportions:

Electrode diameter (in.)	Percentage of electrode
1/8	20
5/32	60
3/16	20

12-40. STRENGTHENING RIVETED JOINTS IN ARMOR PLATE

In order to strengthen riveted joints in armor plate which have been made with buttonhead rivets, a seal bead weld is recommended (fig. 12-33). The arc is struck at the top of the rivet with a stainless steel electrode and held there for a sufficient length of time to melt approximately 1/2 in. (12.7 mm) of the electrode. A bead is then deposited along the curved surface of the rivet to the armor plate and continued around the edge of the rivet until the rivet is completely welded to the armor plate. The seal bead weld prevents the rivet head from being sheared off and the shank of the rivet from being punched through the plate. Countersunk rivets are sealed in the same manner. The rivets in joints made in face hardened armor should be seal welded only on the soft side of the plate.

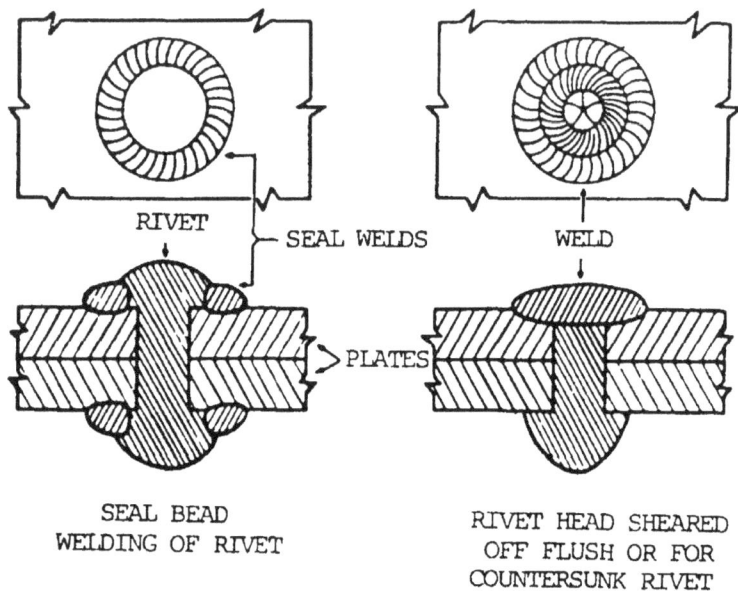

Figure 12-33. Seal bead weld.

Section VIII. PIPE WELDING

12-41. GENERAL

Pipe operating conditions in the handling of oil, gases, water, and other substances range from high vacuum to pressure of several thousand pounds per square inch. Mechanical joints are not satisfactory for many of these services. Electric arc or oxyacetylene welding provide effective joints in these services and also reduce weight, increase the strength, and lower the cost of pipe installations.

12-42. PREPARATION FOR WELDING

a. Pipe Beveled by Manufacturer. Pipe to be welded is usually supplied with a single V bevel of 32-1/2 degrees with a 1/16-in. (1.6-mm) root face for pipe thicknesses up to 3/4 in. (19.1 mm). A single U groove is used for heavier pipe. If the pipe has not been properly beveled or has been cut in the field, it must be beveled prior to welding.

b. <u>Cutting of Pipe</u>. This operation is necessary when pipe must be cut to suit a specific length requirement. To ensure a leak proof welded joint, the pipe must be cut in a true circle in a plane perpendicular to the center line of the pipe. This may be accomplished by using a strip of heavy paper, cardboard, leather belting, or sheet gasket material with a straight edge longer than the circumference of the pipe to be welded. The material is wrapped around the pipe and overlapped and the pipe marked along the edge of the material with a soapstone pencil. Pipe with a wall thickness exceeding 1/8 in. (3. 2 mm) should be cut first with a straight cut, then beveled with a hand torch to a 30 to 35 degree angle, leaving a shoulder of approximately 1/8 in. (3.2 mm).

c. <u>Cleaning of Pipe</u>. After beveling, remove all rust, dirt, scale, or other foreign matter from the outside of the pipe in the vicinity of the weld with a file, wire brush, grinding disk, or other type of abrasive. If the bevels are made by oxyacetylene cutting, the oxide formed must be entirely removed. The inside of the pipe in the vicinity of the weld may be cleaned by a boiler tube and flue cleaner, by sandblasting, by tapping with a hammer with a airblast followup, or by any other suitable method, depending on the inside diameter of the pipe. Care must be taken to clean the scarf faces thoroughly.

d. <u>Aligning the Joint</u>.

(1) A pipe lineup clamp should be used to align and securely hold the pipe ends before tack welding. A spacing tool to separate the pipe ends can be made from an old automobile spring leaf. The spacing for oxyacetylene welding should be approximately 1/8 in. (3. 2 mm); for arc welding, the spacing depends on the size of the electrode used for the root pass.

(2) If a pipe lineup clamp is not available the pipe section must be set in a jig so that their center lines coincide and the spacing of the pipe ends is uniform prior to tack welding. An angle iron (fig 12-34) will serve as a jig for small diameter pipe, while a section of channel or I-beam is satisfactory for larger pipe.

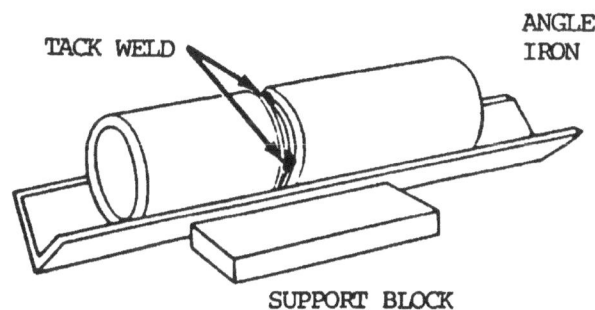

Figure 12-34. Angle iron serving as jig for small diameter pipe.

(3) When a backing ring is used and it is desired to weld to the backing ring, the spacing should not be less than the diameter of the electrode used for the root pass. When welding to the backing ring is not desired, the spacing should not exceed one half the electrode diameter, and varies from this diameter to zero, depending on whether a small or large angle of bevel is used.

12-42. PREPARATION FOR WELDING (cont)

e. Backing Rings and Tack Welding.

(1) The purpose of a backing ring is to make possible the complete penetration of the weld metal to the inside of the pipe without excessive burning through, to prevent spattered metal and slag from entering the pipe at the joint, and to prevent the formation of projections and other irregular shaped formations of metal on the inside of the joint. Backing rings also aid materially in securing proper alignment of the pipe ends and, when used, are inserted during assembly of the joint. Backing rings are not used when the pipe service requires a completely smooth inner pipe surface of uniform internal diameter.

(2) There are several types of backing rings: the plain flat strip rolled to fit the inside of the joint; the forged or pressed type (with or without projections); the circumferential rib which spaces the pipe ends the proper distance apart; and the machined ring. All shapes may be of the continuous or split ring types. Several backing rings are shown in figure 12-35.

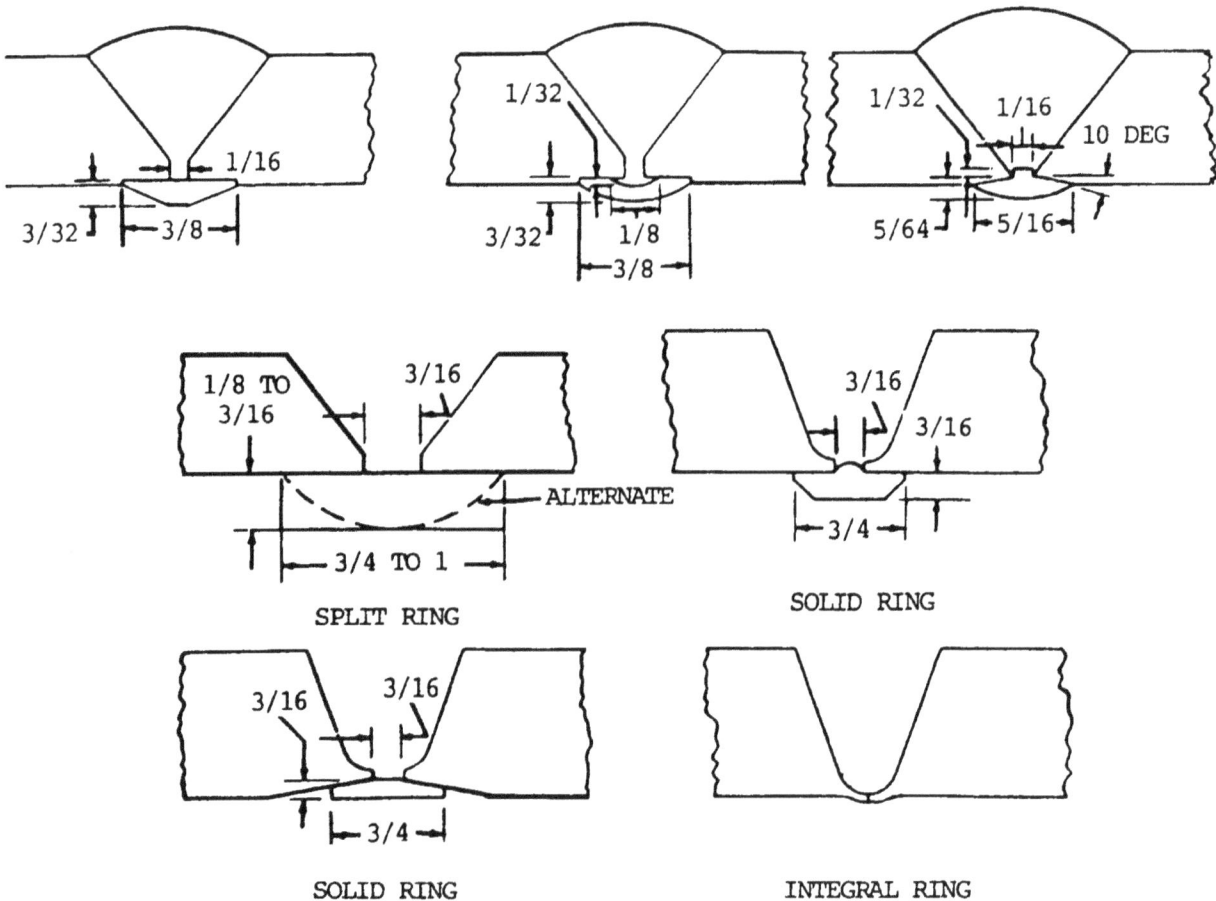

NOTE
ALL DIMENSIONS SHOWN ARE IN INCHES.

Figure 12-35. Types of backing rings.

(3) Backing rings should be made from metal that is readily weldable. Those used when welding steel pipe are usually of low carbon steel.

(4) When the pipe ends have been properly aligned, four tack welds should be made. They should be one-half the thickness of the pipe and equally spaced around the pipe.

12-43. MAKING TEMPLATE PATTERNS

a. General. A template pattern is useful when cutting pipe for a 90 degree bend or other types of joints, such as a tee joint.

b. Material The material necessary for making a template pattern consists of a ruler, a straight-edge, a compass, an angle, a piece of heavy paper, and a pencil.

c. Preparation of a Template.

(1) The information contained in table 12-6 is helpful in preparing a template.

Table 12-6. Template Pattern Data

Size of Pipe (in.) (1)	Outside Dia. of Standard Pipe (in.) (2)	No. of Divisions of Circle (3)	Circumference or Dimension CC (in.) (4)
1-1/4	1.66	12	5.22
1-1/2	1.90	12	5.97
2	2.38	12	7.48
2-1/2	2.88	12	9.05
3	3.50	12	11.00
4	4.50	12	14.14
5	5.56	12	17.47
6	6.63	12	20.83
7	7.63	12	23.97
8	8.63	16	27.11
10	10.75	16	33.77
12	12.75	16	40.05

(2) Lay out the joint full or actual size, with the outside diameter of the pipe (table 12-6, column (2)) represented by the parallel lines (fig. 12-36, p 12-58. Then inscribe a circle of the same diameter, divide it into the correct number of equal parts (column (3)), and number each part beginning with zero.

12-43.   MAKING TEMPLATE PATTERNS (cont)

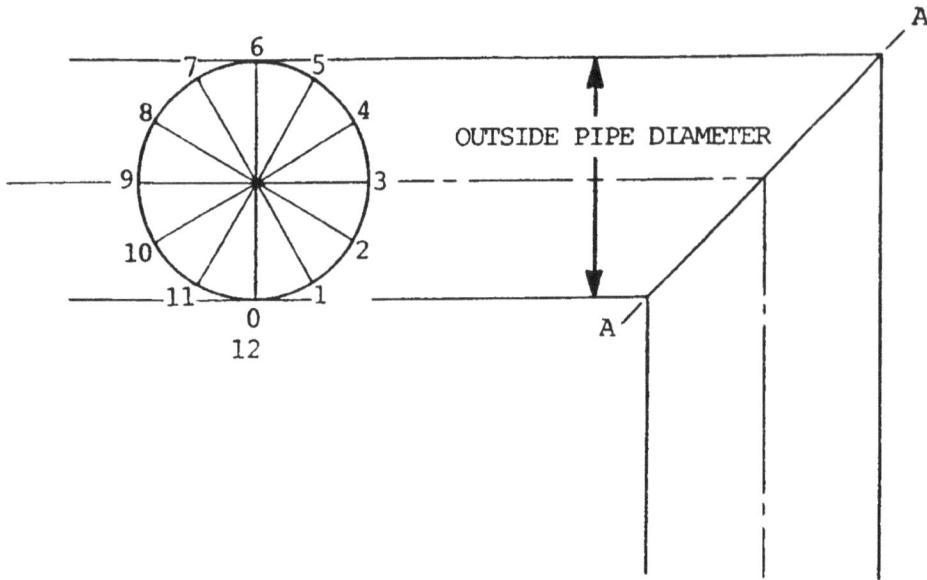

Figure 12-36.   Template pattern, ell joint, first step.

(3)   Extend each point on the circumference of the circle to the line AA, numbering each intersection to correspond with the points on the circle (fig. 12-37). Now draw line BB, as shown, 3.0 in. (76.2 mm) from the corner of the pipe joint.

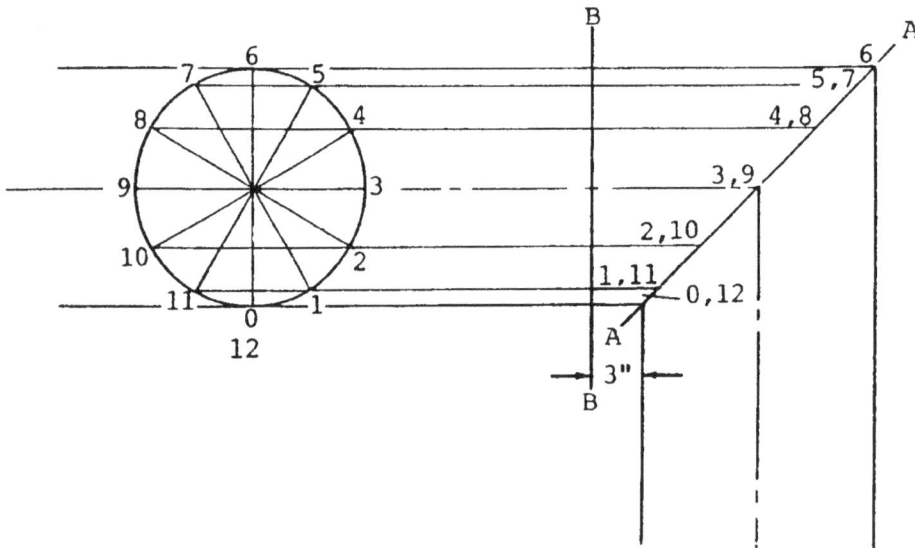

Figure 12-37.   Template pattern, ell joint, second step.

(4) Next, lay off a line, CC, representing the circumference of the circle as determined from table 12-6, column (4). Divide the line into the same number of equal parts as the circle. At each division, draw a line perpendicular to line CC. Beginning at the left, number each division starting with zero, as shown in fig. 12-38.

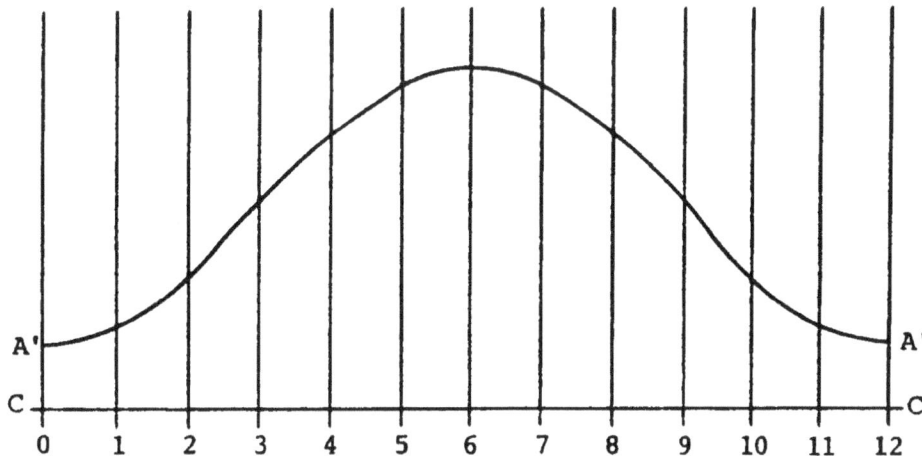

Figure 12-38. Template pattern, ell joint, third step.

(5) Starting at 0, layoff on the vertical line a length equal to B-O. On line 1, lay off a length equal to B-1; on line 2, B-2; and so forth. Join the extremities of these lines. The result will be a curve A'A', corresponding to the line AA in figure 12-36.

(6) Cut out the pattern along edges CA', A'A', A'C and CC. Wrap the pattern around the pipe. Mark the pipe and soapstone or red lead along the line A'A'. This is the cutting edge. Cutting the two pieces of pipe on the line A'A' and fitting them together will result in a 90 degree bend which requires no further trimming. After cleaning and beveling the pipe, it may be welded.

(7) A tee joint, figure 12-39, can be made by applying the above procedure, as shown in figure 12-40, p 12-60. The resulting template pattern, figure 12-41, p 12-60, is cut out and wrapped around the outlet of the tee. A circle will be inscribed upon the outlet, and the outlet is cut. Next, the pattern is placed on the run and the outline marked and cut. After cleaning and beveling, the pipe may be welded.

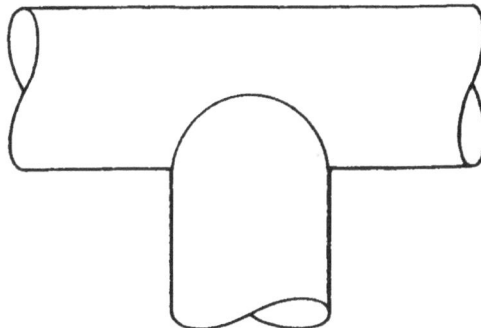

Figure 12-39. Tee joint.

12-43.   MAKING TEMPLATE PATTERNS (cont)

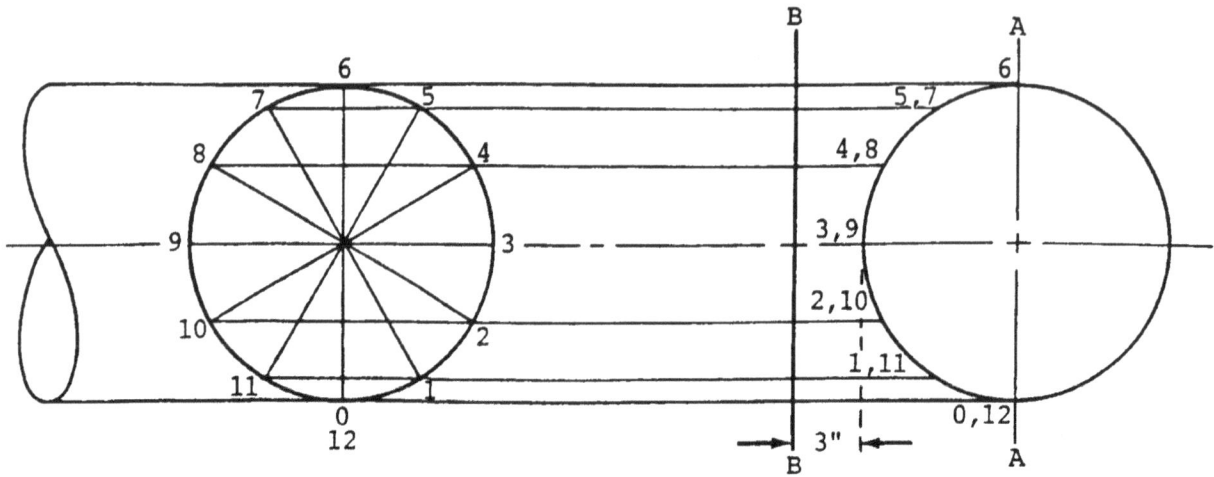

Figure 12-40.   Template pattern, tee joint, first step.

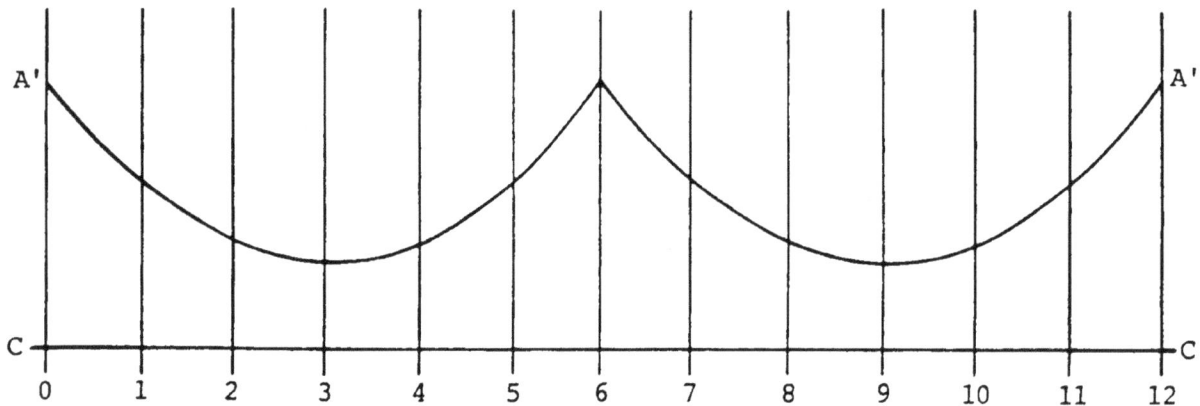

Figure 12-41.   Template pattern, tee joint, second step.

12-44.   PIPE WELDING PROCESSES

a.   General. The most commonly used processes for joining pipe are the manual oxyacetylene process and manual shielded metal-arc process. Automatic and semiautomatic Submerged arc, inert gas metal-arc, and atomic hydrogen welding are also used, particularly in shop operations. The manual shielded metal-arc process may be used for welding all metals used in piping systems, whereas manual oxyacetylene welding is generally limited to small size piping or to welding operations where clearances around tile joints are small.   The equipment required for the oxyacetylene process is also much less expensive and more portable than that required for shielded metal-arc welding.

b. Shielded Metal-Arc Process.

(1) The shielded metal-arc process can be used for welding pipe materials such as aluminum, magnesium, and high chromium-nickel alloys that are difficult to weld by other processes. In shielded metal-arc welding, the number of passes required for welding ferrous metal piping varies with the pipe thickness, the welding position, the size of the electrode, and the welding current used.

(2) The number of passes required for welding low alloy and low carbon steel pipe depends on the thickness of the pipe, the welding position, the size of the electrode, and the current used but, in general, approximately one pass for each 1/8 in. (3.2 mm) of pipe thickness. When welding in the horizontal or rolled position, the number of layers is usually increased 25 to 30 percent. Smaller electrodes are used to lessen the heat concentration and to ensure complete grain refinement of the weld metal.

(3) The electrodes used vary from 1/8 to 5/32 in. (3.2 to 4.0 mm) diameter for the first pass, 5/32 in. (4.0 mm) diameter for the intermediate passes, and up to 3/16 in. (4.8 mm) for the top passes and reinforcement.

c. Manual Oxyacetylene Welding. The number of passes required for pipe welding with the oxyacetylene flame depends on the thickness of the pipe, the position of the pipe, and the size of the welding rod used. The thickness of the deposited layer is somewhat more than that deposited by the shielded metal-arc process.

d. Direction of Welding.

(1) In manual shielded metal-arc welding, as much welding as possible is done in the flat or downhand position using suitable power driven equipment for rotating the pipe at a speed consistent with the speed of welding. When the pipe is in a fixed horizontal position, the weld is usually made from the bottom upward. With thin or medium thickness pipe, the welding is done downward. More metal is deposited when welding upward. Complete grain refinement is easier to achieve, and welding downward requires a much higher degree of manual skill.

(2) When the pipe is in a fixed vertical position, it is customary to deposit the filler metal in a series of overlapping string beads, using 1/8 in. (3.2 mm) maximum electrodes, and allowing 25 to 30 beads per square inch of weld area.

(3) When welding by the oxyacetylene process, the directions of welding as described above will, in general, apply. Backhand welding is used when welding downward, and forehand welding is used when welding upward.

12-45. PIPE WELDING PROCEDURES

a. Horizontal Pipe Rolled Weld.

(1) Align the joint and tack weld or hold in position with steel bridge clamps with the pipe mounted on suitable rollers (fig. 12-42, p 12-62). Start welding at point C, figure 12-42, p 12-62, progressing upward to point B. When B is reached, rotate the pipe clockwise until the stopping point of the weld is at point C and again weld upward to point B. When the pipe is being rotated, the torch should be held between B and C and the pipe rotated past it.

12-45.   PIPE WELDING PROCEDURES (cont)

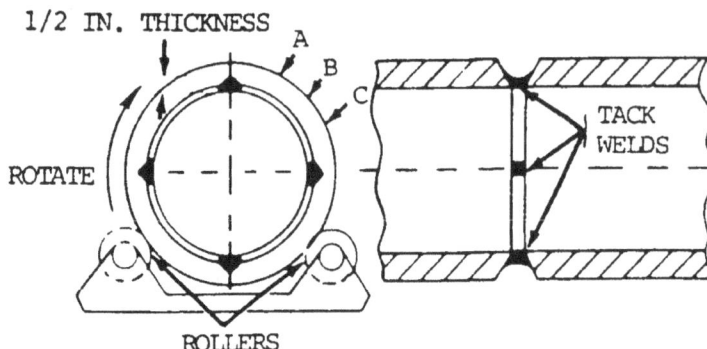

Figure 12-42. Diagram of tack welded pipe on rollers.

(2) The position of the torch at A (fig. 12-42) is similar to that for a vertical weld. As B is approached, the weld assumes a nearly flat position and the angles of application of the torch and rod are altered slightly to compensate for this change.

(3) The weld should be stopped just before the root of the starting point, so that a small opening remains. The starting point is then reheated, so that the area surrounding the junction point is at a uniform temperature. This will ensure a complete fusion of the advancing weld with the starting point.

(4) If the side wall of the pipe is more than 1/4 in. (6.4 mm) in thickness, a multipass weld should be made.

   b. Horizontal Pipe Fixed Position Weld.

(1) After tack welding, the pipe is set up so that the tack welds are oriented approximately as shown in figure 12-43. After welding has been started, the pipe must not be moved in any direction.

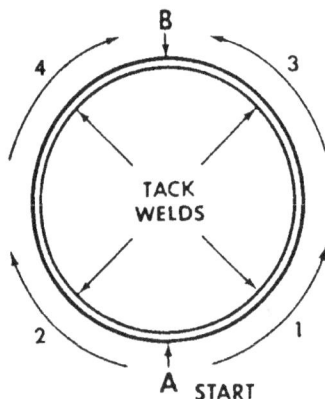

Figure 12-43. Diagram of horizontal pipe weld with uphand method.

(2) When welding in the horizontal fixed position, the pipe is welded in four steps.

Step 1. Starting at the bottom of 6 o'clock position, weld upward to the 3 o'clock position.

Step 2. Starting back at the bottom, weld upward to the 9 o'clock position.

Step 3. Starting back at the 3 o'clock position, weld to the top

Step 4. Starting back at the 9 o'clock position, weld upward to the top. overlapping the bead.

(3) When welding downward, the weld is made in two stages. Start at the top (fig. 12-44) and work down one side (1,fig. 12-44) to the bottom, then returnot the top and work down the other side (2fig. 12-44) to join with the previous weld at the bottom. The welding downward method is particularly effective with arc welding, since the higher temperature of the electric arc makes the use of greater welding speeds possible. With arc welding, the speed is approximately three times that of the upward welding method.

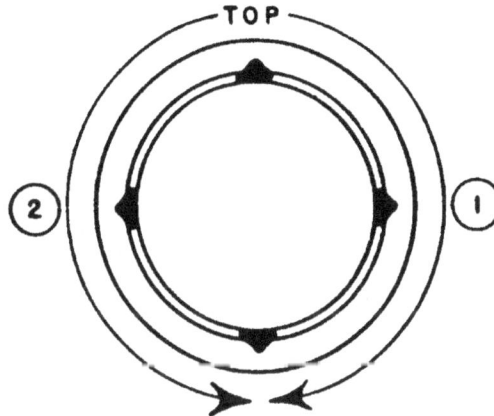

Figure 12-44. Diagram of horizontal pipe weld with downhand method.

(4) Welding by the backhand method is used for joints in low carbon or low alloy steel piping that can be rolled or are in horizontal positionOne pass is used for wall thicknesses not exceeding 3/8 in. (9.5 mm), two passes for wall thicknesses 3/8 to 5/8 in. (9.5 to 15.9 mm), three passes for wall thicknesses 5/8 to 7/8 in. (15.9 to 22.2 mm), and four for wall thicknesses 7/8 to 1-1/8 in. (22.2 to 28.6 mm).

c. Vertical Pipe Fixed Position Weld. Pipe in this position, where the joint is horizontal, is most frequently welded by the backhand method (fig. 12-45, p 12-64). The weld is started at the tack and carried continuously around the pipe.

12-45. PIPE WELDING PROCEDURES (cont)

Figure 12-45. Vertical pipe fixed position weld with backhand method.

d. Multipass Arc Welding

(1) Root beads. If a lineup clamp is used, the root bead (A, fig. 12-46) is started at the bottom of the groove while the clamp is in position When no back– ing ring is used, take care to build up a slight bead on the inside of the pipe. If a backing ring is used, the root bead must be carefully fused to it. As much root bead as the bars of the lineup clamp will permit should be applied before the clamp is removed. Complete the bead after the clamp is removed.

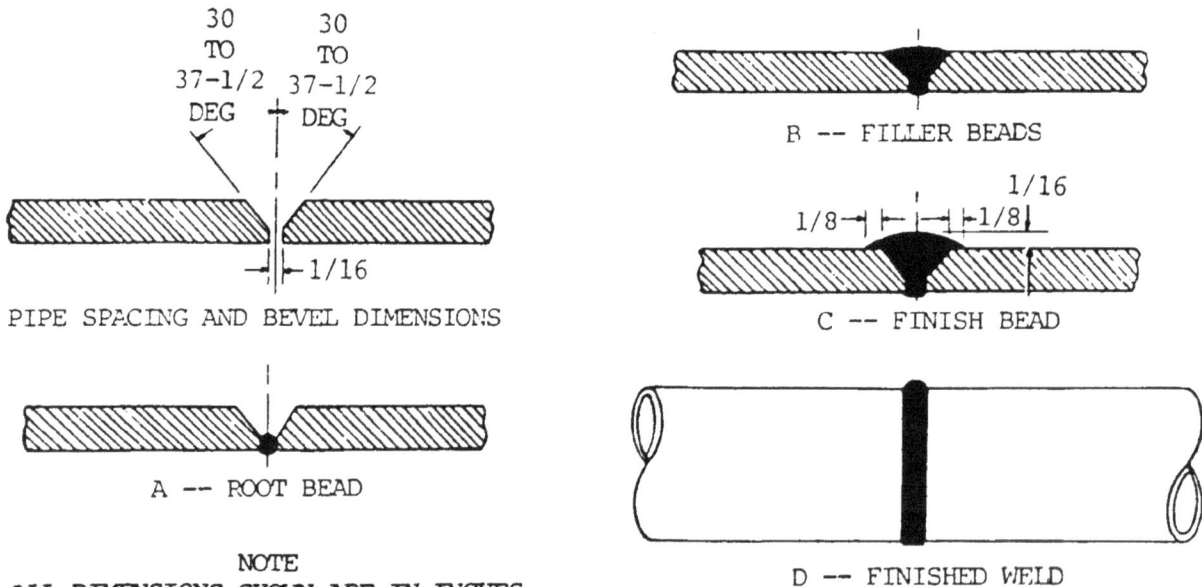

Figure 12-46. Deposition of root, filler, and finish weld beads.

(2) Filler beads. Ensure the filler beads (B, fig. 12-46) are fused into the root bead in order to remove any undercut caused by the deposition of the root bead. One or more filler beads around the pipe will usually be required.

(3) Finish beads. The finish beads (C, fig. 12-46) are applied over the filler beads to complete the joint. Usually, this is a weave bead about 5/8 in. (15.9 mm) wide and approximately 1/16 in. (1.6 mm) above the outside surface of the pipe when complete. The finish weld is shown at D.

Section IX. WELDING CAST IRON, CAST STEEL, CARBON STEEL, AND FORGINGS

12-46. CAST IRON, CAST STEEL, CARBON STEEL, AND FORGINGS

a. In general, parts composed of these metals can be repaired by the same procedure as that used for their assembly. They can also be repaired by brazing or soldering if the joining equipment originally used is not available or suitable for the purpose. For instance, cast iron and cast steel may be repaired by gas welding, arc welding, or by brazing. Parts or sections made of carbon steel originally assembled by spot, projection, or flash welding may be repaired by gas or arc welding. The same is true of forgings.

b. Gray cast iron has a low ductility and therefore will not expand or stretch to any considerable extent before breaking or cracking. Because of this characteristic, preheating is necessary when cast iron is welded by the oxyacetylene process. It can, however, be welded with the metal-arc process without preheating if the welding heat is carefully controlled. Large castings with complicated sections, such as motor blocks, can be welded without dismantling or preheating. Special electrodes designed for this purpose are usually desirable.

c. Generally, the weldability of cast steel is comparable to that of wrought steels. Cast steels are usually welded in order to join one cast item to another or to a wrought steel item, and to repair defects in damaged castings. The weldability of steels is primarily a function of composition and heat treatment. Therefore, the procedures and precautions required for welding wrought steel also apply to cast steels of similar composition, heat treatment, and strength. Welding of cast steels can sometimes be simplified by first considering the load in the area being welded and the actual strength needed in the weld. Castings are often complex; a specific analysis may be required for only part of the entire structure. When welding a section of steel casting that does not require the full strength of the casting, lower-strength weld rods or wires can sometimes be used, or the part being welded to the casting can be of lower strength and leaner analysis than the cast steel part. Under such conditions, the deposited weld metal usually has to match only the strength of the lower-strength member. With heat-treatable electrodes, the welding sometimes can be done before final heat-treating. After being subjected to an austenitizing treatment (heating above the upper critical temperature), weld deposits with carbon contents less than 0.12 percent usually have lower mechanical properties than they have in the as welded or stress-relieved condition.

d. Carbon steels are divided into three groups: low, medium, and high.

(1) Low carbon steels include those with a carbon content up to 0.30 percent. These low carbon steels do not harden appreciably when welded and therefore do not require preheating or postheating except in special cases, such as when heavy sections are to be welded.

12-46. CAST IRON, CAST STEEL, CARBON STEEL, AND FORGINGS (cont)

(2) Medium carbon steels include those that contain from 0.30 to 0.55 percent carbon. These steels are usually preheated to between 300 and 500°F (149 and 260 °C) before welding. Electrodes of the low carbon, heavy coated, straight or re- verse polarity type, similar to those used for metal arc welding of low carbon steels, are satisfactory for steels in this group. The preheating temperature will vary depending on the thickness of the material and its carbon content. After welding, the entire joint should be heated to between 1000 and 1200°F (538 and 649 °C) and slow cooled to relieve stresses in the base metal adjacent to the weld.

(3) High carbon steels include those that have a carbon content exceeding 0.55 percent. Because of the high carbon content and the heat treatment usually given to these steels, their basic properties are impaired by arc welding. Preheat- ing 500 to 800 °F (260 to 427 °C) before welding and stress relieving by heating from 1200 to 1450 °F (649 to 788 °C) with slow cooling should be used to avoid hardness and brittleness in the fusion zone. Either mild steel or stainless steel electrodes can be used with these steels.

e. Parts that were originally forge welded may be repaired by gas or arc weld- ing.

f. High hardness alloy steels are a variety of alloy steels that have been developed to obtain high strength, high hardness, corrosion resistance, and other special properties. Most of these steels depend on a special heat treatment pro- cess in order to develop the desired characteristic in the finished state. Many of these steels can be welded with a heavy coated electrode of the shielded arc type whose composition is similar to that of the base metal. Low carbon electrodes can also be used with some steels and stainless steel electrodes where preheating is not practicable or is undesirable. Heat treated steels should be preheated, if possible, in order to minimize the formation of hard zones or layers in the base metal adjacent to the weld. The molten metal should not be overheated, and for this reason, the welding heat should be controlled by depositing the weld metal in narrow string beads. In many cases, the procedure outlined for medium carbon steels and high carbon steels, including the principles of surface fusion, can be used in the welding of alloy steels.

g. High yield strength, low alloy structural steels are special steels that are tempered to obtain extreme toughness and durability. The special alloys and gener- al makeup of these steels require special treatment to obtain satisfactory weldments.

12-47. PROCEDURES

a. Gray Cast Iron.

(1) Edge preparation. The edges of the joint should be chipped out or ground to form a 60 degree angle or bevel. The V should extend to approximately 1/8 in. (3.2 mm) from the bottom of the crack. A small hole should be drilled at each end of the crack to prevent it from spreading. All grease, dirt, and other foreign substances should be removed by washing with a suitable cleaning material.

(2) Welding technique.

(a) Cast iron can be welded with a coated steel electrode, but this method should be used only as an emergency measure. When using a steel electrode, the contraction of the steel weld metal, the carbon picked up from the cast iron by the weld metal, and the hardness of the weld metal caused by rapid cooling must be considered. Steel shrinks more than cast iron when cooled. When a steel electrode is used, this uneven shrinkage will cause strains at the joint after welding. When a large quantity of filler metal is applied to the joint, the cast iron may crack just back of the line of fusion unless preventive steps are taken. To overcome these difficulties, the prepared joint should be welded by depositing the weld metal in short string beads, 3/4 to 1 in. (19.1 to 25.4 mm) long. These should be made intermittently, and in some cases, by the backstep and skip procedure. To avoid hard spots, the arc should be struck in the V and not on the surface of the base metal. Each short length of weld metal applied to the joint should be lightly peened while hot with a small ball peen hammer and allowed to cool before additional weld metal is applied. The peening action forges the metal and relieves the cooling strains.

(b) The electrodes used should be 1/8 in. (3.2 mm) in diameter to prevent excessive welding heat. The welding should be done with reverse polarity. Weaving of the electrode should be held to a minimum. Each weld metal deposit should be thoroughly cleaned before additional metal is added.

(c) Cast iron electrodes are used where subsequent machining of the welded joint is required. Stainless steel electrodes are used when machining of the weld is not required. The procedure for making welds with these electrodes is the same as that outlined for welding with mild steel electrodes. Stainless steel electrodes provide excellent fusion between the filler and base metals. Great care must be taken to avoid cracking in the weld, because stainless steel expands and contracts approximately 50 percent more than mild steel in equal changes of temperature.

(3) Studding. Cracks in large castings are sometimes repaired by "studding" (fig. 12-47). In this process, the fracture is removed by grinding a V groove. Then holes are drilled and tapped at an angle on each side of the groove. Studs are screwed into these holes for a distance equal to the diameter of the studs, with the upper ends projecting approximately 1/4 in (6.4 mm) above the cast iron surface. The studs should be seal welded in place by one or two beads around each stud and then tied together by weld metal beads. Welds should be made in short lengths and each length peened while hot to prevent high stresses or cracking upon cooling. Each bead should be allowed to cool and be thoroughly cleaned before additional metal is deposited. If the studding method cannot be applied, the edges of the joint should be chipped out or machined. This is done using a round-nosed tool to form a U groove into which the weld metal should be deposited.

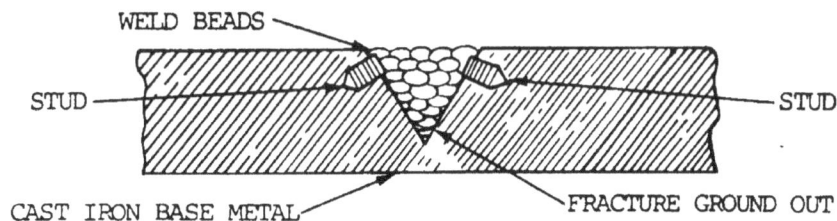

Figure 12-47. Studding method for cast iron repair.

12-47. PROCEDURES (cont)

(4) Metal-arc brazing of cast iron. Cast iron can be brazed with heavy coated, reverse polarity bronze electrodes. The joints made by this method should be prepared in a manner similar to that used for oxyacetylene brazing of cast iron. The strength of the joint depends on the quality of the bond between the filler metal and the cast iron base metal.

(5) Carbon-arc welding of cast iron. Iron castings may be welded with a carbon arc, a cast iron rod, and a cast iron welding flux. The joint should be preheated by moving the carbon electrodes along the surface, thereby preventing too rapid cooling after welding. The molten puddle of metal can be worked with the carbon electrode to remove any slag or oxides that are formed to the surface. Welds made with the carbon arc cool more slowly and are not as hard as those made with the metal arc and a cast iron electrode. The welds are machinable.

b. Cast Steels.

(1) Joint designs for cast steel weldments are similar to those used for wrought steel.

(2) The choice of electrode filler metal is based on the type of cast steel being used, the strength needs of the joint, and the post-weld heat treatment. When welding carbon or low-alloy cast steels, the electrodes recommended for comparable wrought steel plate should be used. When cast austenitic stainless steels are jointed to either cast or wrought ferritic materials, the proper filler metal depends on the service conditions.

c. Carbon Steels.

(1) Low carbon steels.

(a) Metal-arc welding. In metal-arc welding, the bare, thin coated, or heavy coated shielded arc types of electrodes may be used. These electrodes are of low carbon type (O.10 to 0.14 percent). Low carbon sheet or plate materials that have been exposed to low temperatures should be preheated slightly to room temperature before welding. In welding sheet metal up to 1/8 in. (3.2 mm) in thickness, the plain square butt joint type of edge preparation may be used. When long seams are to be welded on this material, the edges should be spaced to allow for shrinkage because the deposited metal tends to pull the plates together This shrinkage is less severe in arc welding than in gas welding. Spacing of approximately 1/8 in. (3.2 mm) per foot of seam will suffice. The backstep or skip welding technique should be used for short seams that are fixed to prevent warpage or distortion and minimize residual stresses. Heavy plates should be beveled to provide an included angle up to 60 degrees, depending on the thickness. The parts should be tack welded in place at short intervals along the seam. The first or root bead should be made with an electrode small enough in diameter to obtain good penetration and fusion at the base of the joint. A 1/8 or 5/32 in. (3.2 to 4.0 mm) electrode is suitable for this purpose. This first bead should be thoroughly cleaned by chipping and wire brushing before additional layers of weld metal are deposited. The additional passes of filler metal should be made with a 5/32 or 3/16 in. (4.0 to 4.8 mm) electrode. For overhead welding, best results are obtained by using string beads throughout the weld. When welding heavy sections that have been beveled from both sides, the weave beads should be deposited alternately on one side and then the other. This will reduce the amount of distortion in the welded structure. Each bead should be cleaned thoroughly to remove all scale, oxides, and slag before additional metal is deposited. The motion of the electrode should be controlled to make the bead uniform in thickness and to prevent undercutting and overlap at the edges of the weld.

(b) Carbon-arc welding. Low carbon sheet and plate up to 3/4 in. (19.1 mm) in thickness can be satisfactorily welded by the carbon-arc welding process.The arc is struck against the plate edges, which are prepared in a manner similar to that required for metal-arc welding. A flux should be used on the joint and filler metal added as in oxyacetylene welding. A gaseous shield should be provided around the molten base and filler metal, by means of a flux coated welding rod. The welding should be done without overheating the molten metal. If these precautions are not taken, the weld metal will absorb an excessive amount of carbon from the electrode and oxygen and nitrogen from the air. This will cause brittleness in the welded joint.

(2) Medium carbon steels. The plates should be prepared for welding in a manner similar to that used for low carbon steels.When welding with low carbon steel electrodes, the welding heat should be carefully controlled to avoid overheating of the weld metal and excessive penetration into the side walls of the joint. This control is accomplished by directing the electrode more toward the previously deposited filler metal adjacent to the side walls than toward the side walls directly. By using this procedure, the weld metal is caused to wash up against the side of the joint and fuse with it without deep or excessive penetration.High welding heats will cause large areas of the base metal in the fusion zone adjacent to the welds to become hard and brittle. The area of these hard zones in the base metal can be kept to a minimum by making the weld with a series of small string or weave beads, which will limit the heat input. Each bead or layer of weld metal will refine the grain in the weld immediately beneath it.This will anneal and lessen the hardness produced in the base metal by the previous bead. When possible, the finished joint should be heat treated after welding.Stress relieving is normally used when joining mild steel. High carbon alloys should be annealed. When welding medium carbon steels with stainless steel electrodes, the metal should be deposited in string beads. This will prevent cracking of the weld metal in the fusion zone. When depositing weld metal in the upper layers of welds made on heavy sections, the weaving motion of the electrode should under no circumstances exceed three electie diameters Each successive bead of weld should be chipped, brushed, and cleaned prior to the laying of another bead.

(3) High carbon steels. The welding heat should be adjusted to provide good fusion at the side walls and root of the joint without excessive penetration. Control of the welding heat can be accomplished by depositing the weld metal in small string beads. Excessive puddling of the metal should be avoided because this will cause carbon to be picked up from the base which, in turn, will make the weld metal hard and brittle. Fusion between the filler metal and the side walls should be confined to a narrow zone. Use the surface fusion procedure prescribed for medium carbon steels. The same procedure for edge preparation, cleaning of the welds, and sequence of welding beads as prescribed for low and medium carbon steels applies to high carbon steels. Small high carbon steel parts are sometimes repaired by building up worn surfaces.When this is done, the piece should be annealed or softened by heating to a red heat and cooling slowly. Then the piece should be welded or built up with medium carbon or high strength electrodes and heat treated after welding to restore its original properties.

12-47. PROCEDURES (cont)

d. Forgings should be welded with the gas or arc processes in a manner similar to parts originally assembled by spot, projection, or flash welding.

e. High hardness alloy steels can be welded with heavy coated electrodes of the shielded arc type whose composition is similar to that of the base metal. Low carbon electrodes can also be used with some steels. Stainless steel electrodes are effective where preheating is not practical or is undesirable. Heat treated steels should be preheated, if possible, in order to minimize formation of hard zones or layers in the base metal adjacent to the weld. The molten metal should not be overheated. For this reason, the welding heat should be controlled by depositing the weld metal in narrow string beads. In many cases, the procedure outlined for medium carbon steels and high carbon steels, including the principles of surface fusion, can be used in the welding of alloy steels.

f. Reliable welding of high yield strength, low alloy structural steels can be performed by using the following guidelines:

(1) Hydrogen is the number one enemy of sound welds in alloy steels. Therefore, use only low hydrogen (MIL-E-18038 or MIL-E-22200/1) electrodes to prevent underbead cracking. Underbead cracking is caused by hydrogen picked up in the electrode coating, released into the arc and absorbed by the molten metal.

(2) If the electrodes are in an airtight container, immediately upon opening the container place the electrodes in a ventilated holding oven set at 250 to 300 °F (121 to 149 °C). In the event that the electrodes are not in an airtight container, put them in a ventilated baking oven and bake for 1 to 1-1/4 hours at 800 °F (427 °C). Baked electrodes should, while still warm, be placed in the holding oven until used. Electrodes must be kept dry to eliminate absorption of hydrogen. Testing for moisture should be in accordance with MIL-E-22200.

NOTE
Moisture stabilizer NSN 3439-00-400-0090 is an ideal holding oven for field use (MIL-M-45558).

(3) Electrodes are identified by classification numbers which are always marked on the electrode containers. For low hydrogen coatings the last two numbers of the classification should be 15, 16, or 18. Electrodes of 5/32 and 1/8 in. (4.0 and 3.2 mm) in diameter are the most commonly used since they are more adaptable to all types of welding of this type steel.

(4) Wire electrodes for submerged arc and gas-shielded arc welding are not classified according to strength. Welding wire and wire-flux combinations used for steels to be stress relieved should contain no more than 0.05 percent vanadium. Weld metal with more than 0.05 percent vanadium may become brittle if stress relieved. when using either the submerged arc or gas metal-arc welding processes to weld high yield strength, low alloy structural steels to lower strength steels, the wire-flux and wire-gas combination should be the same as that recommended for the lower strength steels.

(5) For welding plates under 1 in. (25.4 mm) thick, preheating above 50°F (10 °C) is not required except to remove surface moisture from the base metal.

(6) It is important to avoid excessive heat concentration when welding in order to allow the weld area to cool rather quickly. Either the heat input monograph or the heat input calculator can be used to determine the heat input into the weld.

(7) For satisfactory welds use good welding practices, as defined in section 1, along with the following procedures:

(a) Use a straight stringer bead whenever possible.

(b) Restrict weave to partial weave pattern. Best results are obtained by a slight circular motion of the electrode with the weave area never exceeding two electrode diameters.

(c) Never use a full weave pattern.

(d) Skip weld as practical.

(e) Peening of the weld is sometimes recommended to relieve stresses while cooling larger pieces.

(f) Fillet welds should be smooth and correctly contoured. Avoid toe cracks and undercutting. Electrodes used for fillet welds should be lower strength than those used for butt welding. Air hammer peening of fillet welds can help to prevent cracks, especially if the welds are to be stress relieved. A soft steel wire pedestal can help to absorb shrinkage forces. Butter welding in the toe area before actual fillet welding strengthens the area where a toe crack may start. A bead is laid in the toe area, then ground off prior to the actual fillet welding. This butted weld bead must be located so that the toe passes of the fillet will be laid directly over it during actual fillet welding. Because of the additional material involved in fillet welding, the cooling rate is increased and heat inputs may be extended about 25 percent.

Section X. FORGE WELDING

12-48. GENERAL

a. General. This is a group of welding processes in which a weld is made by heating in a forge or other furnace and by applying pressure or blows.

b. Roll Welding. This is a process in which heat is obtained from a furnace and rolls are used to apply pressure.

c. Die Welding. This is a process in which heat is obtained from a furnace and dies are used to apply pressure.

d. Hammer Welding. This is a process in which heat is obtained from a forge or furnace and hammer blows are used to apply pressure.

e. Forge welding, as performed by the blacksmith is by far the oldest process for joining metal pieces of parts, but hand forge welding is no longer used extensively because of the development of oxyacetylene and electric arc welding. It is, however, an effective process under some field conditions and therefore, equipment and procedures required in hand forge welding are described briefly in this manual.

12-49. APPLICATION

a. In forge welding, metal parts are heated in a forge furnace with fuel such as coal, coke, or charcoal. The parts to be joined are heated until the surface of the metal becomes plastic. When this condition is reached, the parts are quickly superimposed and the weld is made by pressure or hammering The hammering may be done by either hand or machine. The force of the hammering or pressure depends on the size and mass of the parts being joined. In this process, the surfaces to be joined must be free from foreign matter. In some cases, a flux is used (usually sand or borax sprinkled on the surfaces to be joined) just before the metal reaches the welding temperature in order to remove the oxide and dirt. The flux spreads over the metal, prevents further oxidation by keeping out the air, lowers the melting point of the scale, and makes it fluid so that it can be squeezed out of the weld when the metal is hammered. Various types of forge welds are shown in figure 12-48.

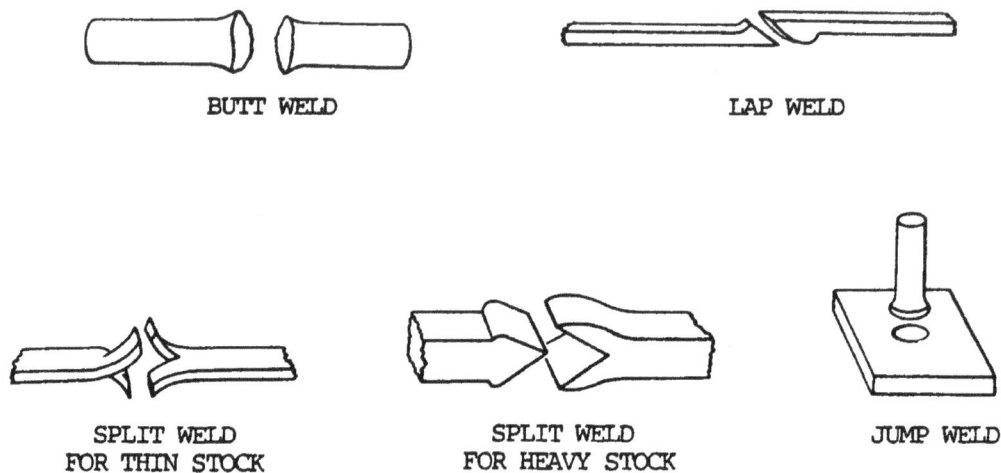

Figure 12-48. Forge welds.

b. Because of the development of machine forge welding, the speed of welding and the size of the parts to be welded have increased greatly. Long seams in lap or butt welded pipe can be made. The quality of the weld is such that its location is almost impossible to detect. This process requires the use of a gas flame or other suitable heating method to bring the edges of the metal up to the welding temperature. Pressure is applied by rolls which press the plastic edges together until another set of rolls roves to the parts being welded along the line of welding .

Section XI. HEAT TREATMENT OF STEEL

12-50. GENERAL

a. Heat treatment of steel may be defined as an operation or combination of operations which involve the heating and cooling of the metal in its solid state in order to obtain certain desirable characteristics or properties. Metal and alloys are primarily heat treated to increase their hardness and strength, to improve ductility, and to soften them for later forming or cutting operations.

b. Alloy steels and plain carbon steels with a carbon content of 0.35 percent or higher can be hardened to the limits attainable for the particular carbon content, or softened as required by controlling the rate of heating, the rate of cooling, and the method of cooling.

c. One of the most important factors in heat treating steels is that the metal should never be heated to a temperature close to its melting point. When this occurs, certain elements in the metal are oxidized (burned out), and the steel becomes coarse and brittle. Steel in this condition usually cannot be restored by any subsequent heat treatment. In general, the lower the carbon content, the higher the temperature to which steels can be heated without being oxidized.

e. The must common problem related to heat treatment are warping, dimensional changes, cracking, failure to harden,soft spots, and excessive brittleness. The following table lists some problems, possible causes, and remedies.

Table 12-7. Common Heat Treating Problems

Problem	Possible Causes	Remedy
A. Warping	1. Non-uniform quenching practice	Employ spray or agitated quench
	2. Improper support during heating	Support with brick, cast iron chips, or spent coke
	3. Release of machining stresses	Machine equal amounts from surface of part or anneal prior to heat treatment
	4. Unbalanced design	Clamp in fixture designed to balance mass
	5. Failure to strain relieve prior to heat treatment	Strain relieve
B. Dimensional changes	1. Release of stresses from previous cold working	Strain relieve prior to to hardening
	2. Unpredicted thermal stresses	Balance mass with quench fixture
	3. Severe quenching practice	Change to less severe quenching media or warm quench bath
	4. Failure to temper or stabilize properly	Employ stabilizing or sub-zero treatment
	5. Dimensional changes for some are predictable and normal	Use table supplied with steel to predict size change
	6. Transformation of retained austenite	Employ multiple tempers or sub-zero treatment
	7. Overheating or underheating	Check furnace control and recommended temperatures

12-50. GENERAL (cont)

Table 12-7. Common Heat Treating Problems (cont)

Problem	Possible Causes	Remedy
C. Cracking	1. Failure to temper immediately after quenching	Temper before it reaches room temperature, approximately 150 $^{\circ}$F
	2. Improper quenching medium	Use less severe quench
	3. Excessive hardening temperature	Check furnace temperature and recommended temperature
	4. Large grain size	Normalize prior to hardening
	5. Poor design, e.g., sharp corners, or unbalanced mass	Discuss with designer
	6. Failure to preheat properly	Preheat is recommended
D. Failure to harden	1. Quench not drastic enough	Employ more drastic quench
	2. Hardening temperature too low or non-uniformly heating	Check recommended temperature
	3. Mislabeled steel	Make test run on sample or get it analyzed
	4. Severe decarburization	Use controlled atmosphere or bath for heating
	5. Tempering temperature too high	Use recommended temperature
E. Soft spots	1. Decarburized case	Use controlled atmosphere or liquid heating bath
	2. Excessive heat treat scale	Same as above
	3. Quench bath too hot	Check temperature
	4. Improper agitation	Review recommended procedures
	5. Contaminated quenching bath	Clean, filter, or change
F. Excessive brittleness	1. Improper quenching medium	Use recommended quench
	2. Failure to temper	Temper immediately after hardening
	3. Excessive hardening temperature	Follow recommended temperature
	4. Coarse grain size	Follow recommended temperature
	5. Mechanical stress raisers, e.g., sharp corners	Discuss with designer or use air hardening steel

12-51. ANNEALNG

a. Annealing is a process involving the heating of a metal above the critical temperature and subsequent slow cooling.The purpose of such heating may be to remove stresses; induce softness; alter ductility, toughness, electrical, magnetic, or other physical properties; refine crystalline structure; remove gases; or produce a definite microstructure.

b. Specific heat treatments which fall under the term annealing are:

(1) Full annealing. This is the heating of iron base alloys above the critical temperature range, holding them above that range for a proper period of time, followed by cooling in a medium which will prolong the time of cooling.

(2) Process annealing. This is the heating of iron base alloys to a temperature below or close to the lower limit of the critical temperature range, followed by cooling as desired.

(3) Normalizing This is the heating of iron base alloys to approximately 100 °F (38 °C) above the critical temperature range, followed by cooling to below that range in still air at ordinary temperature.

(4) Patenting This is the heating of iron base alloys above the critical temperature range, followed by cooling below that range in air, molten lead, a molten mixture of nitrates, or nitrates maintained at a temperature usually between 800 to 1050 °F (427 to 566 °C), depending on the carbon content of the steel and the properties required of the finished product.

(5) Spheroidizing. This is any process of heating and cooling steel that produces a rounded or globular form of carbideMethods of spheroidizing generally used are:

(a) Prolonged heating at a temperature just below the lower critical temperature, usually followed by relatively slow cooling.

(b) In the case of small objects of high carbon steels, the spheroidizing result is achieved more rapidly by prolonged heating to temperatures alternately within and slightly below the critical temperature range.

(6) Tempering (also called drawing). This is reheating hardened steel to some temperature below the lower critical temperature, followed by any desired rate of cooling.

(7) Malleablizing. This is an annealing operation performed on white cast iron to partially or wholly transform the combined carbon to temper carbon, and in some cases, to wholly remove the carbon from the iron by decarburization.

(8) Graphitizing. This is a type of annealing of gray cast iron in which sane or all of the combined carbon is transferred to free graphite carbon.

12-52. HARDENING

a. Plain carbon steel is hardened by heating it above the critical temperature and cooling it rapidly by plunging it into water, iced brine, or other liquid. When heating through the critical temperature range, iron undergoes a transformation and changes from a form with low carbon solubility to one with high carbon solubility. Upon cooling, a reverse transformation occurs. Since these changes are progressive and require time for completion, they may be stopped if the cooling period is shortened.

b. If the cooling is very rapid, as in water quenching, the transformation takes place much below the critical temperature range. The carbon is fixed tied in a highly stressed, finely divided state, and the steel becomes hard, brittle, and much stronger than steel that is slowly cooled.

c. The presence of alloying elements alters the rate of transformation on cooling. Each alloy element shows individuality in its effect; therefore, alloy steels are manufactured and heat treated to meet specific performance requirements.

12-53. TEMPERING

After a steel is hardened, it is too brittle for ordinary purposes. Some of the badness should be removed and toughness induced. This process of reheating quench hardened steel to a temperature below the transformation range and then, cooling it at any rate desired is called tempering. The metal must be heated uniformly to a predetermined temperature, depending on the toughness desired. As the tempering temperature increases, toughness increases and hardness decreases. The tempering range is usually between 370 and 750°F (188 and 399°C), but sometimes is as high as 1100 °F (593 °C).

12-54. SURFACE HARDENING

a. General. A low carbon steel cannot be hardened to any great extent because of its low carbon content, yet the surface can be hardened by means of case hardening. The hardening is accomplished by increasing the carbon content of the surface only.

b. Case Hardening. This process produces a hard surface resistant to wear but leaves a soft, tough core. It is accomplished as follows:

(1) Pack carburizing. The work is placed in a metal container and surrounded by a mixture of carburizing materials. The container is sealed and heated from 1 to 16 hours at 1700 to 1800°F (927 to 982 °C). The approximate penetration is 7/1000 in. per hour. Carburizing is usually followed by quenching to produce a hardened case.

(2) Gas carburizing. The work is placed in a gas tight retort and heated to 1700 (927 °C). Natural or manufactured gas is passed through the retort until proper depth of hardening is obtained.

(3) Nitriding. The work is placed in an atmosphere of ammonia gas at 950 (510 °C) for a period of 10 to 90 hours. The maximum depth of 3/100 in. will be reached at 90 hours. The work is then removed and cooled slowly. Little warpage will result because of the low temperature. The case must then be ground so that it will be corrosion resistant.

WARNING

Cyanide and cyanide fumes are dangerous piosons; therefore, this process requires expert supervision and adequate ventilation .

(4) <u>Cyaniding.</u> The work is preheated and immersed in a cyanide bath at 1550 °F (843 °C). Time of immersion varies from a few minutes to 2 hours with a resulting penetration of 1/100 in. per hour. Parts should be tempered if toughness is desired.

(5) <u>Forge case hardening</u>. This process, usually used in the field, is accomplished by preheating work in a forge or with a torch up to 1650 (899 °C), then dipping the work in potassium cyanide or Kasenite and applying the flame. until the compound melts. Repeat until required depth is attained and then quench.

c. <u>Induction Hardening</u>. This process is accomplished by the use of high frequency current with low voltage and a water spray to quench the work. It is used only on high carbon steels.

d. <u>Flame Hardening</u>. This process is accomplished by heating the surface to be hardened with an oxyacetylene torch and quenching it in water. The steel must be high in carbon.

12-55. USE OF CARBONIZING COMPOUND PASTE, NSN 6580-00-695-9268, AND ISOLATING PASTE, NSN 6850-00-664-0355, FOR SURFACE HARDENING

a. <u>General</u>. The surface hardness of common steel is directly proportional to its carbon content. Low carbon steels may be given a hard exterior shell by increasing the amount of carbon in their surfaces. If the workpiece to be hardened is packed in material of high carbon content and then brought to a relatively high temperature, the carbon of the packing material transfers to the surface of the workpiece and hardens it when it is quenched. The hard case created by quenching is very brittle and may crack and chip easily. Toughness may be imparted to this brittle case by air cooling, reheating the workpiece to a somewhat lower temperature than that used in the initial hardening, and then quickly quenching it. is often desired to have only certain parts of a given workpiece hardened while retaining the basic toughness of the steel in the body of the workpiece; for example, the cutting edges of hand tools. This may be accomplished by packing the surfaces to be hardened with the carbonizing material and the balance of the workpiece with an isolating material. Since the material used to pack the workpiece, either for hardening or for isolating, would dry and peel away in the furnace heat, it must be secured to the workpiece by some method of wrapping or shielding.

b. <u>Workpiece of Preparation</u>.

(1) <u>Packing.</u>

(a) Remove all rust, scale, and dirt from the workpiece to be hardened.

(b) Firmly press the nontoxic carbonizing paste, NSN 6850-00-695-9268, on the surfaces or edges to be hardened. The paste should be approximately 1/2 in. (12.7 mm) thick.

12-55. USE OF CARBONIZING COMPOUND PASTE, NSN 6580-00-695-9268, AND ISOLATING PASTE, NSN 6850-00-664-0355, FOR SURFACE HARDENING (cont)

(c) Firmly press the nontoxic isolating paste, NSN 6850-00-664-0355, over the balance of the workpiece. The paste should be approximately 1/2 in. (12.7 mm) thick.

(d) If the workpiece temperature is to be recognized by its color, leave a small opening where bare metal can be seen. A similar opening must be provided in the workpiece shielding or wrapping ((2) below).

(e) Whenever possible, a suitable pyrometer or temperature measuring instrument must be used on the furnace, as estimating metal temperature by its color is not accurate.

(2) Shielding and wrapping. Wrap the packed workpiece loosely in a piece of thin sheet iron; or insert it into a piece of tubing of suitable dimension. If available, a metal container about 1 in. (25.4 mm) larger than the workpiece can be used. Fill any space between the workpiece and the container with carbonizing paste and/or isolating paste.

NOTE

If a metal container is to be reused, it should be made of a heat resistant material such as 18 percent chromium-8 percent nickel steel. Sheet iron and plain carbon steel will not stand high temperatures for long periods.

c. Heating

(1) Place the container with the workpiece of the shielded workpiece in a furnace.

(a) If a heat treating furnace is not available, heating may be done in a forge or with an acetylene torch. When using the forge, keep the work entirely covered with coal, rotating it periodically to ensure even temperatures at all times in all areas.

(b) If an acetylene torch is used, place the work in a simple muffle jacket of bricks similar to figure 12-49. Care should be taken to keep the temperature as even as possible on all areas. Keep the flame out of contact with the workpiece, or with any one particular portion of it.

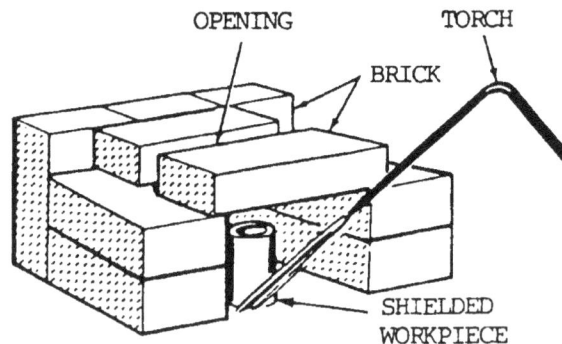

Figure 12-49. Muffle jacket.

(2) The workpiece must be heated to 1700°F (927 °C) (bright orange). The time needed to reach this temperature depends on the size of the workpiece and the furnace.

(3) Note the time when the workpiece reaches 1700°F (927 °C). Hold this heat for the time required to give the desired depth of case hardening. See table 12-8.

(4) Remove the packed workpiece from the furnace after the required heating time.

(5) The heating times listed in table 12-8 are general. They are intended for use with any low carbon steel.

Table 12-8. Time Required in Case Hardening

Depth of Hardened Case (in.)	Heating Time (Minutes)
0.010	5-7
0.015	10-15
0.030	25-30

12-56. QUENCHING AFTER CARBURIZING

After the workpiece has been removed from the furnace or forge, remove the shield and packing and allow the workpiece to cool in the air until it reaches 1405 (763 °C). Then quench by plunging it in water or oil if required by the type of steel alloy. This procedure will not produce as good a grain structure as the procedure outlined in paragraph 12-57 below.

12-57. DRAWING AND QUENCHING AFTER CARBURIZING

a. Normal Drawing and Quenching. For better structure in the finished workpiece, heat the workpiece as outlined in paragraph 12-55 (1) through (3), and remove it to cool to a black heat without removing the paste or the shield. Reheat the still-packed workpiece in a furnace or forge to approximately 1450F (788 °C) (orange in furnace) for a few minutes. Then remove it from the heat, remove the shielding and packing, and quench by plunging in water or oil if required by the type of steel alloy.

b. Drawing and Quenching SAE Steel. For the best grain structure in SAE steel workplaces, follow the procedure outlined in a. above, except reheat the SAE steel to the temperature shown in table 12-9, p 12-80, before quenching. The workpiece is tempered after quenching. This method is generally used when the case hardening penetration is from 1/25 to 3/50 in. (1.0 to 1.5 mm).

12-57. DRAWING ANDQUENCHING AFTER CARBURIZING (cont)

Table 12-9. Approximate Reheating Temperatures after Carburizing of SAE Steel

| SAE No. | Temperature (approximate) | |
	°F	°C
1015	1585	863
1020	1550	843
1117	1520	827
1320	1500	816
3115	1500	816
3310	1435	779
4119	1500	816
4320	1475	802
4615	1485	807
4815	1440	782
8620	1540	838
8720	1540	838

12-58. MUFFLE JACKET

To construct a temporary muffle jacket (fig. 12-49, p 12-78), use enough fire or refractory bricks to build a boxlike structure with a floor, three sides, and a top. The temporary muffle jacket should be located on a level earth base. The interior cavity should be just large enough to comfortably accommodate the workpiece when wrapped in a shield and the flame of the heat source. The top of the jacket must provide an opening to act as a chimney. Pack the sides, back, and bottom of structure with mist earth to help contain the heat. If fire or refractory bricks are not available, use common building bricks. Make every effort to keep the size of the workpiece such that center supports for the top are not required. If this is not possible, use brick for such center supports.

12-59. HEAT SOURCE

When using an oxyacetylene torch for heat, position the torch so that its flare will be completely within the muffle jacket. Do not allow the flame to be in direct contract with any particular portion of the workpiece. Have sufficient fuel available for 2 to 3 hours of operation at full flame. After the workpiece has been packed, shielded, and placed in the muffle jacket, ignite the torch, adjust the flame for maximum heat, and use additional brick to close about one half of the front opening of the jacket.

Section XII. OTHER WELDING PROCESSES

12-60. RESISTANCE WELDING

a. General. Resistance welding is a type of welding process in which the workplaces are heated by the passage of an electric current through the area of contact. Such processes include spot, seam, projection, upset, and flash welding.

b. <u>Resistance Welding Process</u>.

(1) <u>Spot Welding</u>. This is a resistance welding process wherein coalescence is produced by the heat obtained from resistance to the flow of electric current through the workpieces, which are held together under pressure by electrodes. The size and shape of the individually formal welds are limited primarily by the size and contour of the electrodes. Spot welding is particularly adaptable to thin sheet metal construction and has many applications in this type of work. The spot welding principle is illustrated in figure 12-50.

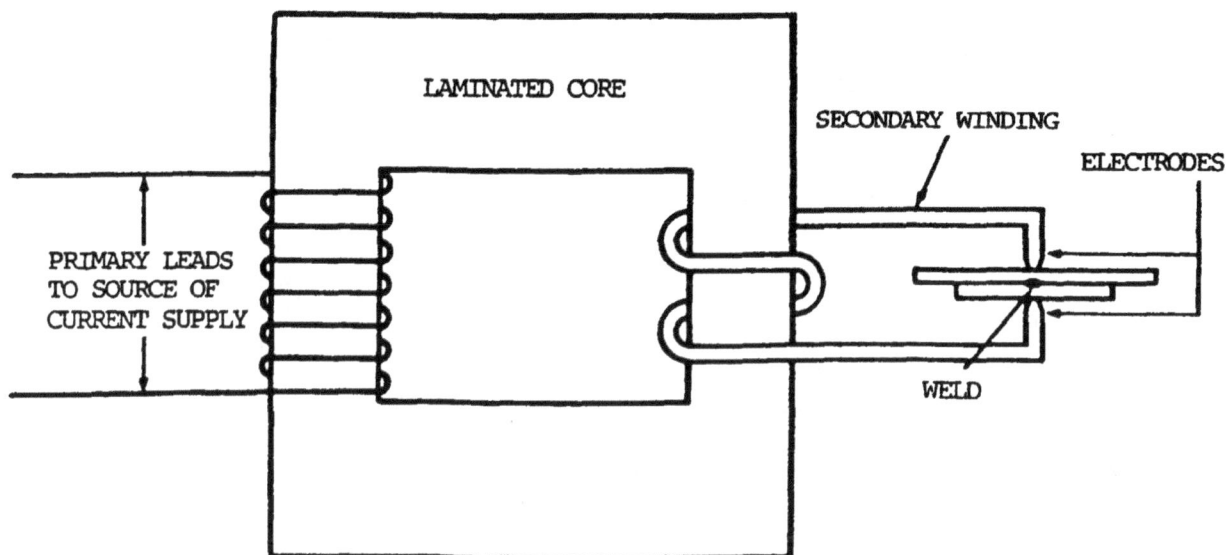

Figure 12-50. Schematic diagram of resistance spot welder.

(2) <u>Roll spot welding</u>. This is a resistance welding process wherein separate spot welds are made without retracting the electrodes. This is accomplished by means of circular electrodes which are in continuous contact with the work.

(3) <u>Seam welding</u>. This is a resistance welding process wherein coalescence is produced by the heat obtained from resistance to the flow of electric current through the workplaces, which are held together under pressure by rotating circular e electrodes. The resulting weld is a series of overlapping spot welds made progressively along a joint. Lapped and flanged joints in cans, buckets, tanks, mufflers, etc., are commonly welded by this process.

(4) <u>Projection welding</u>. This is a process wherein coalescence is produced by the heat obtained from resistance to the flow of electric current through the workpieces, which are held together under pressure by electrodes. The resulting welds are localized at predetermined points by the design of the parts to be welded. This localization is usually accomplished by projections, embossments, or intersections. This process is commonly used in the assembly of punched, formed, and stamped parts.

12-60. RESISTANCE WELDING (cont)

(5) <u>Upset welding</u>. This is a resistance welding process wherein coalescence is produced simultaneously over the entire area of abutting surfaces or progressively along a joint by the heat obtained from resistance to the flow of electric current through the area of contact of these surfaces. Pressure is applied before heating is started and is maintained throughout the heating period. Upsetting is accompanied by expulsion of metal from the joint (A, fig. 12-51).

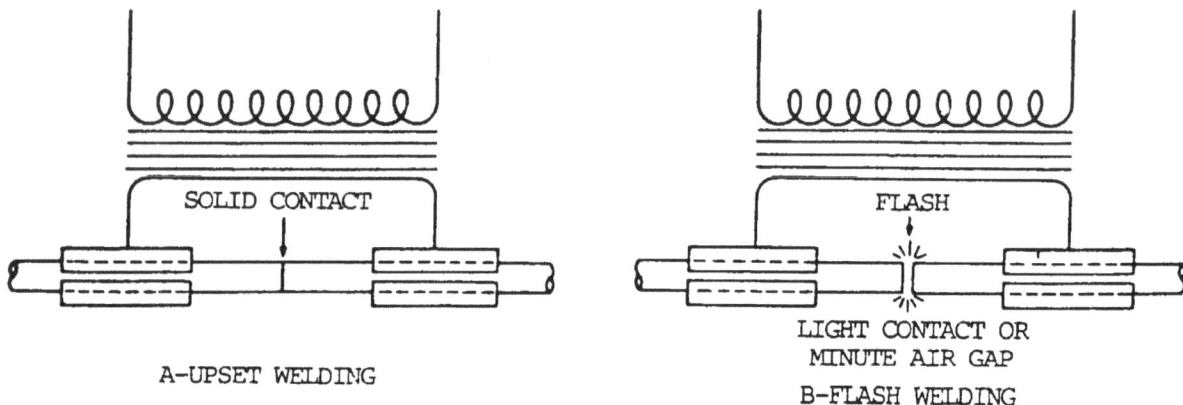

Figure 12-51. Schematic diagram of upset and flash welder.

(6) <u>Flash welding</u>. Flash welding is a resistance welding process wherein coalescence is produced simultaneously over the entire area of abutting surfaces by the heat obtained from resistance to the flow of electric current between the two surfaces, and by the application of pressure after the heating caused by flashing is substantially completed. The final application of pressure is accompanied by expulsion of metal from the joint (B, fig. 12-51).

(7) <u>Percussion welding</u>. This weld is made simultaneously over the entire area of abutting surfaces by the heat obtained from an arc. The arc is produced by a rapid discharge of electrical energy. It is extinguished by pressure applied percussively during the discharge.

c. <u>Welding Procedures</u>.

(1) The operation of spot, seam, and projection welding involves the use of electric current of proper magnitude for the correct length of time. The current and time factors must be coordinated so that the base metal within a confined area will be raised to its melting point and then resolidified under pressure. The temperature obtained must be sufficient to ensure fusion of the base metal elements, but not so high that metal will be forced from the weld zone when the pressure is applied.

(2) In upset welding (A, fig. 12-51), the surfaces to be welded are brought into close contact under pressure. The welding heat is obtained from resistance to the flow of current through the area of contact of the abutting surfaces. When a sufficiently high temperature is obtained, welding of the surfaces is achieved by upsetting with the application of high pressure.

(3) In flash welding (B, fig. 12-51), the fusing of the parts is accomplished in three steps. The surfaces to be joined are brought together under light pressure, then separated slightly to allow arcing to occur. This small arc brings the metals to their melting points at the separated ends and, as a final operation, the molten surfaces are forced together under heavy pressure. As they meet, the molten metal and slag are thrown out and a clean fusion is obtained.

12-61. SPOT WELDING MAGNESIUM

a. General. Magnesium can be joined by spot, seam, or flash welding, but spot welding is the most widely used. Spot welding is used mostly on assemblies subject to low stresses and on those not subjected to vibration. The welding of dissimilar alloys by the spot welding process should be avoided, especially if they are alloys with markedly different properties.

b. Welding Current.

(1) General. Either alternating current or direct current can be used for spot welding magnesium. High currents and short weld duration are required, and both alternating current and direct current spot welders have sufficient capacity and provide the control of current that is necessary in the application of this process.

(2) Alternating current machines. The alternating current spot welding machines, equipped with electronic synchronous timers, heat control, and phase shifting devices to control weld timing and current are suitable for the welding of magnesium. Three types of machines are used; single-phase, three-phase, and dry-disk rectifier type.

(3) Direct current machines. The electrostatic condenser discharge type is the most widely used direct current machine for magnesium welding. The line demand for this type of equipment may be as high as 500 kva when welding sheets approximately 1/8 in. (3.2 mm) thick. Electromagnetic machines are also used. They require lower pressure applied by the electrodes during welding than the electrostatic equipment.

c. Electrodes. Electrodes for spot welding magnesium should be made of high-conductivity copper alloys conforming to Resistance Welder Manufacturer's Association specifications. Hard-rolled copper can be used where special offset electrodes are desired. Electrodes should be water cooled but never to the point where condensation will take place. Intermittent water flow, supplied only when the weld is made, assists in the maintenance of a constant tip temperature. The most common tips are dome-ended with tip radii of curvature ranging from 2.0 to 8.0 in. (50.8 to 203.2 mm) depending on sheet thickness. Four degree flat tips are frequently used. Flat tips with diameters from 3/8 to 1-1/4 in. (9.5 to 31.8 mm) are used on the side of the work where the surface is to be essentially free of marks. Contact surfaces of the electrodes must be kept clean and smooth.

12-61.   SPOT WELDING MAGNESIUM (cont)

d. Cleaning. Magnesium sheets for spot welding should be purchased with an oil coating rather than a chrome pickle finish.Pickled surfaces are hard to clean for spot welding because of surface etch. Satisfactory cleaning can be accomplished by either chemical or mechanical rnethods.Mechanical cleaning is used where the number of parts to be cleaned does not justify a chemical cleaning set-up. Stainless steel wool, stainless steel wire brushes, or aluminum oxide cloth are used for this purpose. Ordinary steel wool and wire brushes leave metallic particles and should not be used, because the magnetic field created in the tip will attract these particles. Chemical cleaning is recommended for high produe tion. It is economical and provides consistently low surface resistance, resulting in more uniform welds and approximately double the number of spot welds between tip cleanings. The allowable time between cleaning and welding is also much longer. Chemically cleaned parts can be welded up to 100 hours after cleaning, while mechanically cleaned parts should be welded at once.

e.   Machine Settings. Spot welding is a machine operation requiring accurate current, timing, and welding force and therefore, the adjustment of the welding machine to the proper setting is the most important step in the production of strong consistent welds. The welding machine manufacturer's operating instructions should be followed closely.   Recommended spacings and edge distances are given in table 12-10.

Table 12-10. Magnesium Spot Weld Data

B & S Gauge No.	Spot Spacing (in.)	Minimum Edge Distance (in.)
24	0.50	0.125
18	0.70	0.187
14	1.00	0.250
12	1.25	0.375
8	1.50	0.625

f.   Pressure. Welding pressures are usually established first, using the liner current or capacitance and voltage values recommended. High pressure provides greater latitude in the currents that can be used for the production of sound welds, but may be limited by excessive sheet separation or the size of the electrodes. After approximating the pressure, the proper weld time, voltage, and weld current or capacitance should be determined to obtain welds of the desired size and strength. If the maximum weld size is too small or cracking is encountered, it may be necessary to increase the pressure and current, or possibly the weld time. After all the settings are fixed, the hold time may need adjustment to make certain that pressure is maintained on the weld until solidification is complete. Insufficient hold time will result in porous welds and is normally indicated by a cracking sound during the contraction of the weld.   Trial welds should be made in material of the same gauge, alloy, hardness,and surface preparation as the metal to be welded.   Test welds between strips crossed at right angles are useful for determining proper welding conditions,because they can be easily twisted apart.

12-62. SPOT AND SEAM WELDING TITANIUM

a. Spot and seam welding procedures for titanium and titanium alloys are very similar to those used on other metals. Welds can be made over a wide range of conditions. Special shielding is not required. The short welding times and proximity of the surfaces being joined prevent embrittlement of the welds by contamination from the air.

b. The spot and seam welding conditions which have the greatest effect on weld quality are welding current and time. With variations in these conditions, the diameter, strength, penetration, and indentation of the spot welds change appreciably. Electrode tip radium and electrode force also have some effect on these properties. For all applications, welding conditions should be established depending on the thicknesses being welded and the properties desired.

c. Most experience in spot welding is available from tests on commercially pure titanium. In these tests, the welding conditions have varied considerably, and it is difficult to determine if there are optimum spot welding conditions for various sheet gauges. One of the major problems encountered is excessive weld penetration. However, penetration can be controlled by selecting suitable welding current and time.

d. Experience with some of the high strength alpha-beta alloys has shown that postweld heat treatments are beneficial to spot and seam weld ductility, but procedures have not been developed to heat treat these welds in the machines. When necessary, furnace heat treatments or an oxyacetylene torch may be used to heat treat spot welds.

e. Specifications have been established for spot and seam welds in commercially pure titanium. The quality control measures of these specifications for stainless steel (MIL-W-6858) are used. Suitable minimum edge distances and spot spacing are listed in table 12-11. These are the same spot spacings and edge distances specified for spot welds in steel.

Table 12-11. Commercially Pure Titanium Spot Weld Data*

B & S Gauge No.	Spot Spacking (in.)	Minimum Edge Distance (in.)
0.008	0.187	0.125
0.012	0.250	0.125
0.016	0.312	0.187
0.020	0.375	0.187
0.025	0.437	0.250
0.030	0.500	0.250
0.035	0.562	0.250
0.042	0.625	0.312
0.050	0.750	0.312
0.062	0.875	0.312
0.078	1.000	0.312
0.093	1.125	0.375
0.125	1.135	0.500

*Values used when not specified in drawings.

12-63. FLASH WELDING TITANIUM

a. Flash welding procedures for titanium are similar to those used for other metals. As was the case for spot and seam welding, special shielding is not necessary to produce satisfactory flash welds. However, inert gas shielding has been used to decrease the possibility of weld contamination and to increase ductility. For many of the high strength alloys, postweld heat treatments are required to prevent cracking and improve weld ductility. These welds are transferred to a furnace for heat treatment.

b. Flash welding conditions have varied considerably.However, short flashing cycles and fast upset speeds similar to those used for aluminum generally are employed. The upset cycle is probably the most important variable, because of its effect on the expulsion of contaminated metal from the jointIn some of the high strength alpha-beta alloys,superior results were obtained by using intermediate pressures (8000 to 10,000 psi (55,160 to 68,950 kPa)).

NOTE
High upset pressure results in high residual stresses that may cause the occurence of microfissures in the hard weld zones in these alloys.

c. An adopted specification requires the tensile strength of the weld area of flash welded joints to be 95 percent minimum of parent metal, and elongation through the weld area to be 50 percent minimum of parent material. With proper welding procedures and postweld treatmentflash welds in titanium and most of the titanium alloys can be held to these criteria.

CHAPTER 13

DESTRUCTIVE AND NONDESTRUCTIVE TESTING

Section I. PERFORMANCE TESTING

13-1. GENERAL

To ensure the satisfactory performance of a welded structure, the quality of the welds must be determined by adequate testing procedures. Therefore, they are proof tested under conditions that are the same or more severe than those encountered by the welded structures in the field. These tests reveal weak or defective sections that can be corrected before the materiel is released for use in the field. The tests also determine the proper welding design for ordnance equipment and forestall injury and inconvenience to personnel and untimely failure of materiel.

13-2. TESTING OF MILITARY MATERIEL

a. Weapons can be proof tested by firing from cover with an extra heavy charge to determine the safety of the welded piece.

b. Automotive materiel can be tested at high speeds over rough ground to determine its road safety.

c. Welded armor plate and other heavy structural members can be tested by gunfire with projectiles of various calibers to determine their strength under shock.

d. Other similar tests are used to check the performance of complex structures; however, because the piece of materiel may consist of several types of metals welded with various filler metals, the successful operation of the entire structure requires that each weld must be able to withstand the particular load for which it is designed. For this reason, a number of physical tests have been devised to determine the strength and other characteristics of the welds used in the structure.

13-3. FIELD INSPECTION OF WELDS AND EQUIPMENT REPAIRED BY WELDING

a. General. A definite procedure for the testing of welds is not set up as a part of the normal routine of ordnance units operating under field conditions. If facilities are available, sane of the physical testing methods may be instituted. In general, however, the item welded is subjected to a thorough visual examination by a qualified inspector, and if found to be satisfactory, it is then returned to the using arm or service.

b. Inspection Procedure. The finished weld should be inspected for undercut, overlap, surface checks, cracks, or other defects. Also, the degree of penetration and side wall fusion, extent of reinforcement, and size and position of the welds are important factors in the determination as to whether a welding job should be accepted or rejected, because they all reflect the qualify of the weld.

13-3.   FIELD INSPECTION OF WELDS AND EQUIPMENT REPAIRED BY WELDING (cont)

   c.   <u>Destructive Tests of Experimental Welds</u>. If special circumstances require the use of a new or novel welding procedure new welding material, or unfamiliar apparatus, and when welding operators lack experience in their use, it is advisable to make experimental welds with scrap or unsalvageable material. These welds or welded materials must be subjected to destructive tests. The required development of procedure and familiarity with equipment can be attained in this manner.

   d.   <u>Performance Tests</u>. When materiel has been repaired by standard welding procedures, visual inspection should be sufficient to determine the efficiency of the weld. However, after the repaired item has been returned to the using arm or service, the item should be subjected to such practical tests as are necessary to prove its ability to withstand the strains and stresses of normal service. This will involve the towing or driving of mobile equipment over terrain that it is normally expected to traverse and the firing of artillery pieces to ensure that the repair will not break down under the forces of recoil. In most cases, the item can be placed in service with instructions to the using personnel to make one or more thorough inspections after the item has been in service a short time and to report signs of possible failure or unsatisfactory performance. Defective repaired parts can, in this way, be detected before serious trouble results.

Section II.   VISUAL INSPECTION AND CORRECTIONS

13-4.   INCOMPLETE PENETRATION

This term is used to describe the failure of the filler and base metal to fuse together at the root of the joint. Bridging occurs in groove welds when the deposited metal and base metal are not fused at the root of the joint. The frequent cause of incomplete penetration is a joint design which is not suitable for the welding process or the conditions of construction. When the groove is welded from one side only, incomplete penetration is likely to result under the following conditions.

   a.   The root face dimension is too big even though the root opening is adequate.

   b.   The root opening is too small.

   c.   The included angle of a V-groove is too small.

   d.   The electrode is too large.

   e.   The rate of travel is too high.

   f.   The welding current is too low.

13-5.   LACK OF FUSION

Lack of fusion is the failure of a welding process to fuse together layers of weld metal or weld metal and base metal. The weld metal just rolls over the plate surfaces. This is generally referred to as overlap. Lack of fusion is caused by the following conditions:

a. Failure to raise to the melting point the temperature of the base metal or the previously deposited weld metal.

b. Improper fluxing, which fails to dissolve the oxide and other foreign material from the surfaces to which the deposited metal must fuse.

c. Dirty plate surfaces.

d. Improper electrode size or type.

e. Wrong current adjustment.

13-6. UNDERCUTTING

Undercutting is the burnin away of the base metal at the toe of the weld. Undercutting may be caused by the following conditions:

a. Current adjustment that is too high.

b. Arc gap that is too long.

c. Failure to fill up the crater completely with weld metal.

13-7. SLAG INCLUSIONS

Slag inclusions are elongated or globular pockets of metallic oxides and other solids compounds. They produce porosity in the weld metal. In arc welding, slag inclusions are generally made up of electrode coating materials or fluxes. In multilayer welding operations, failure to remove the slag between the layers causes slag inclusions. Most slag inclusion can be prevented by:

a. Preparing the groove and weld properly before each bead is deposited.

b. Removing all slag.

c. Making sure that the slag rises to the surface of the weld pool.

d. Taking care to avoid leaving any contours which will be difficult to penetrate fully with the arc.

13-8. POROSITY

a. Porosity is the presence of pockets which do not contain any solid material. They differ from slag inclusions in that the pockets contain gas rather than solid. The gases forming the voids are derived from:

(1) Gas released by the cooling weld because of its reduced volubility as the temperature drops.

(2) Gases formed by chemical reactions in the weld.

13-8. POROSITY (cont)

b. Porosity is best prevented by avoiding:

(1) Overheating and undercutting of the weld metal.

(2) Too high a current setting.

(3) Too long an arc.

13-9. GAS WELDING

a. The weld should be of consistent width throughout.The two edges should form straight parallel lines.

b. The face of the weld should be slightly convex with reinforcement of not more than 1/16 in. (1.6 mm) above the plate surface.The convexity should be even along the entire length of the weld. It should not be high in one place and low in another.

c. The face of the weld should have fine, evenly spaced ripples. It should be free of excessive spatter, scale, and pitting.

d. The edges of the weld should be free of undercut or overlap.

e. Starts and stops should blend together so that it is difficult determine where they have taken place.

f. The crater at the end of the weld should be filled and show porosity, holes, or cracks.

(1) If the joint is a butt joint, check the back side for complete penetration through the root of the joint.A slight bead should form on the back side.

(2) The root penetration and fusion of lap and T-joints can be checked by putting pressure on the upper plate until it is bent double. If the weld has not penetrated through the root, the plate will crack open at the joint as it is being bent. If it breaks, observe the extent of the penetration and fusion at the root. It will probably be lacking in fusion and penetration.

13-10. GAS METAL-ARC WELDING (GMAW)WITH SOLID-CORE WIRE

a. Lack of Penetration. Lack of penetration is the result of too little heat input in the weld area. This can be corrected by:

(1) Increasing the wire-feed speed and reducing the stickout distance.

(2) Reducing the speed of travel.

(3) Using proper welding techniques.

b. Excessive Penetration. Excessive penetration usually causes burn-through. It is the result of too much heat in the weld area. This can be corrected by:

(1) Reducing the wire-feed speed and increasing the speed of travel.

(2) Making sure that the root opening and root face are correct.

(3) Increasing the stickout distance during welding and weaving the gun.

c. Whiskers. Whiskers are short lengths of electrode wire sticking through the weld on the root side of the joint. They are caused by pushing the electrode wire past the leading edge of the weld pool. Whiskers can be prevented by:

(1) Reducing the wire-feed speed and the speed of travel.

(2) Increasing the stickout distance and weaving the gun.

d. Voids. Voids are sometimes referred to as wagon tracks because of their resemblance to ruts in a dirt road. They may be continued along both sides of the weld deposit. They are found in multipass welding. Voids can be prevented by:

(1) Avoiding a large contoured crown and undercut.

(2) Making sure that all edges are filled in.

(3) On succeeding passes, using slightly higher arc voltage and increasing travel speed.

e. Lack of Fusion. Lack of fusion, also referred to as cold lap, is largely the result of improper torch handling, low heat, and higher speed travel. It is important that the arc be directed at the leading edge of the puddle prevent this defect, give careful consideration to the following:

(1) Direct the arc so that it covers all areas of the joint. The arc, not the puddle, should do the fusing.

(2) Keep the electrode at the leading edge of the puddle.

(3) Reduce the size of the puddle as necessary by reducing either the travel speed or wire-feed speed.

(4) Check current values carefully.

f. Porosity. The most common defect in welds produced by any welding process is porosity. Porosity that exists on the face of the weld is readily detected, but porosity in the weld metal below the surface must be determined by x-ray or other testing methods. The causes of most porosity are:

(1) Contamination by the atmosphere and other materials such as oil, dirt, rust, and paint.

(2) Changes in the physical qualities of the filler wire due to excessive current.

(3) Entrapment of the gas evolved during weld metal solidification.

(4) Loss of shielding gas because of too fast travel.

13-10. GAS METAL-ARC WELDING WITH SOLID-CORE WIRE (cont)

(5) Shielding gas flow rate too low, not providing full protection.

(6) Shielding gas flow rate too high, drawing air into the arc area.

(7) Wrong type of shielding gas being used.

(8) Gas shield blown away by wind or drafts.

(9) Defects in the gas system.

(10) Improper welding technique, excessive stickout, improper torch angle, and too fast removal of the gun and the shielding gas at the end of the weld.

g. Spatter. Spatter is made up of very fine particles of metal on the plate surface adjoining the weld area. It is usually caused by high current, a long arc, an irregular and unstable arc, improper shielding gas, or a clogged nozzle.

h. Irregular Weld Shape. Irregular welds include those that are too wide or too narrow, those that have an excessively convex or concave surface, and those that have coarse, irregular ripples. Such characteristics may be caused by poor torch manipulation, a speed of travel that is too slow, current that is too high or low, improper arc voltage, improper stickout, or improper shielding gas.

i. Undercutting. Undercutting is a cutting away of the base material along the edge of the weld. It may be present in the cover pass weld bead or in multipass welding. This condition is usually the result of high current, high voltage, excessive travel speed, low wire-feed speed, poor torch technique, improper gas shielding or the wrong filler wire. To correct undercutting, move the gun from side to side in the joint. Hesitate at each side before returning to the opposite side.

13-11. GAS METAL-ARC WELDING (GMAW) WITH FLUX-CORED WIRE

a. Burn-Through. Burn-through may be caused by the following:

(1) Current too high.

(2) Excessive gap between plates.

(3) Travel speed too slow.

(4) Bevel angle too large.

(5) Nose too small.

(6) Wire size too small.

(7) Insufficient metal hold-down or clamping.

b. Crown Too High or Too Low. The crown of the weld may be incorrect due to the following:

(1) Current too high or low.

(2) Voltage too high or low.

(3) Travel speed too high.

(4) Improper weld backing.

(5) Improper spacing in welds with backing.

(6) Workpiece not level.

c. <u>Penetration Too Deep or Too Shallow</u>.Incorrect penetration may be caused by any of the following:

(1) Current too high or low.

(2) Voltage too high or low.

(3) Improper gap between plates.

(4) Improper wire size.

(5) Travel speed too slow or fast.

d. <u>Porosity and Gas Pocket</u>s. These defects may be the results of any of the following:

(1) Flux too shallow.

(2) Improper cleaning.

(3) Contaminated weld backing.

(4) Improper fitup in welds with manual backing.

(5) Insufficient penetration in double welds.

e. <u>Reinforcement Narrow and Steep-Sloped (Pointe</u>d)Narrow and pointed rein-forcements may be caused by the following:

(1) Insufficient width of flux.

(2) Voltage too low.

f. <u>Mountain Range Reinforcemen</u>t.If the reinforcement is ragged, the flux was too deep.

g. <u>Undercuttin</u>g. Undercutting may be caused by any of the following:

(1) Travel speed too high.

(2) Improper wire position (fillet welding).

(3) Improper weld backing.

13-11. GAS METAL-ARC WELDING WITH FLUX-CORED WIRE (cont)

h. Voids and Cracks. These weld deficiencies may be caused by any of the following:

(1) Improper cooling.

(2) Failure to preheat.

(3) Improper fitup.

(4) Concave reinforcement (fillet weld).

Section III. PHYSICAL TESTING

13-12. GENERAL

a. The tests described in this section have been developed to check the skill of the welding operator as well as the quality of the weld metal and the strength of the welded joint for each type of metal used in ordnance materiel.

b. Some of these tests, such as tensile and bending tests, are destructive, in that the test specimens are loaded until they fail, so the desired information can be gained. Other testing methods, such as the X-ray and hydrostatic tests, are not destructive.

13-13. ACID ETCH TEST

a. This test is used to determine the soundness of weld. The acid attacks or reacts with the edges of cracks in the base or weld metal and discloses weld defects, if present. It also accentuates the boundary between the base and weld metal and, in this manner, shows the size of the weld which may otherwise be indistinct. This test is usually performed on a cross section of the joint.

b. Solutions of hydrochloric acid, nitric acid, ammonium per sulfate, or iodine and potassium iodide are commonly used for etching carbon and low alloy steels.

13-14. GUIDED BEND TEST

The quality of the weld metal at the face and root of the welded joint, as well as the degree of penetration and fusion to the base metal, are determined by means of guided bend tests. These tests are made in a jig (fig. 13-1). These test specimens are machined from welded plates, the thickmess of which must be within the capacity of the bending jig. The test specimen is placed across the supports of the die which is the lower portion of the jig. The plunger, operated from above by a hydraulic jack or other device, causes the specimen to be forced into and to assure the shape of the die. To fulfill the requirements of this test, the specimens must bend 180 degrees and, to be accepted as passable, no cracks greater than 1/8 in. (3.2 mm) in any dimension should appear on the surface. The face bend tests are made in the jig with the face of the weld in tension (i.e., on the outside of the bend) (A, fig. 13–2). The root bend tests are made with the root of the weld in tension (i.e., on the outside of the bend) (B, fig. 13-2). Guided bend test specimens are also shown in figure 13-3, p 13-10.

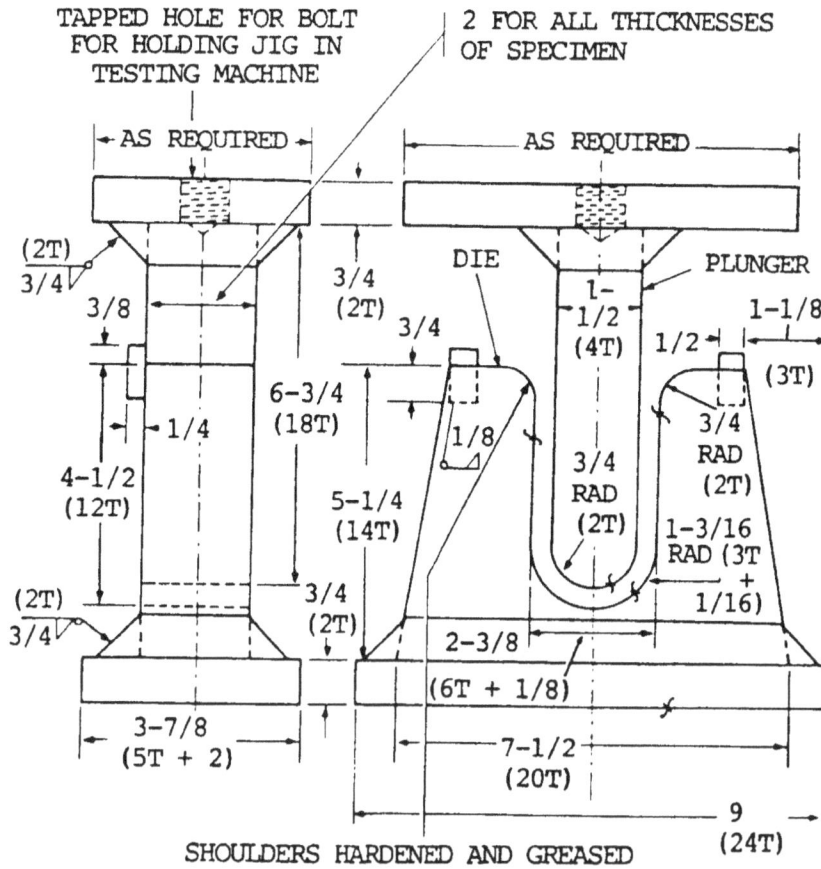

TAPPED HOLE FOR BOLT
FOR HOLDING JIG IN
TESTING MACHINE

2 FOR ALL THICKNESSES
OF SPECIMEN

← AS REQUIRED →

AS REQUIRED

(2T)
3/4

3/8

3/4
(2T)

3/4

DIE

PLUNGER

1-
1/2
(4T)

1-1/8

1/2

(3T)

6-3/4
(18T)

1/8

3/4
RAD
(2T)

3/4
RAD
(2T)

1/4

4-1/2
(12T)

5-1/4
(14T)

3/4
RAD
(2T)

1-3/16
RAD (3T
+
1/16)

(2T)
3/4

3/4
(2T)

2-3/8

(6T + 1/8)

3-7/8
(5T + 2)

7-1/2
(20T)

9
(24T)

SHOULDERS HARDENED AND GREASED

NOTES
1— T=TEST PLATE THICKNESS.
2—HARDENED ROLLS MAY BE USED ON SHOULDERS IF DESIRED.
3—SPECIFIC DIMENSIONS FOR 3/8 PLATE.
4 –ALL DIMENSIONS SHOWN ARE IN INCHES.

Figure 13-1. Guided bend test jig.

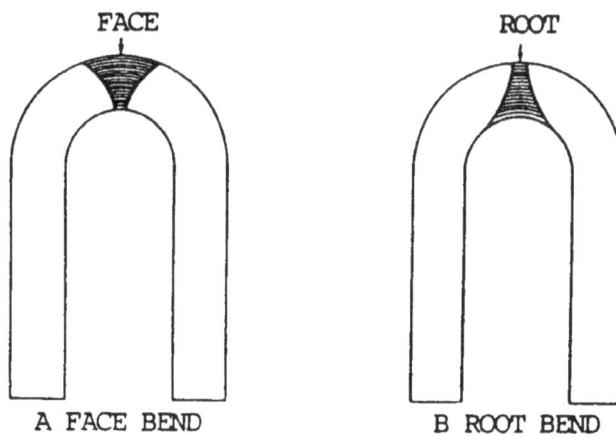

FACE

ROOT

A FACE BEND

B ROOT BEND

Figure 13-2. Guided bend test specimens.

13-14. GUIDED BEND TEST (cont)

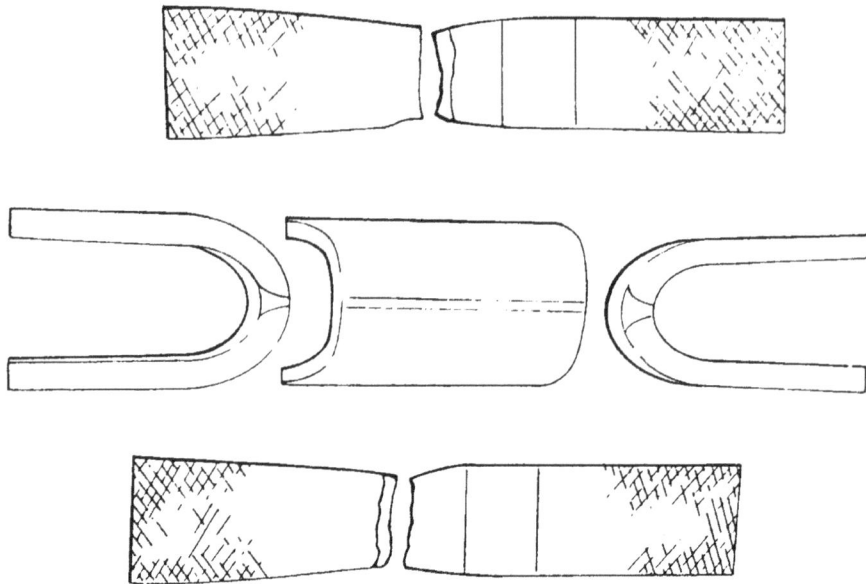

Figure 13-3. Guided bend and tensile strength test specimens.

13-15. FREE BEND TEST

a. The free bend test has been devised to measure the ductility of the weld metal deposited in a weld joint. A test specimen is machined from the welded plate with the weld located as shown at A, figure 13-4. Each corner lengthwise of the specimen shall be rounded in a radius not exceeding one-tenth of the thickness of the specimen. Tool marks, if any, shall be lengthwise of the specimen. Two scribed lines are placed on the face 1/16 in. (1.6 mm) in from the edge of the weld. The distance between these lines is measured in inches and recorded as the initial distance X (B, fig. 13-4). The ends of the test specimen are then bent through angles of about 30 degrees, these bends being approximately one-third of the length in from each end. The weld is thus located centrally to ensure that all of the bending occurs in the weld. The specimen bent initially is then placed in a machine capable of exerting a large compressive force (C, fig. 13-4) and bent until a crack greater than 1/16 in. (1.6 mm) in any dimension appears on the face of the weld. If no cracks appear, bending is continued until the specimens 1/4 in. (6.4 mm) thick or under can be tested in vise. Heavier plate is usually tested in a press or bending jig. Whether a vise or other type of compression device is used when making the free bend test, it is advisable to machine the upper and lower contact plates of the bending equipment to present surfaces parallel to the ends of the specimen (E, fig. 13-4). This will prevent the specimen from slipping and snapping out of the testing machine as it is bent.

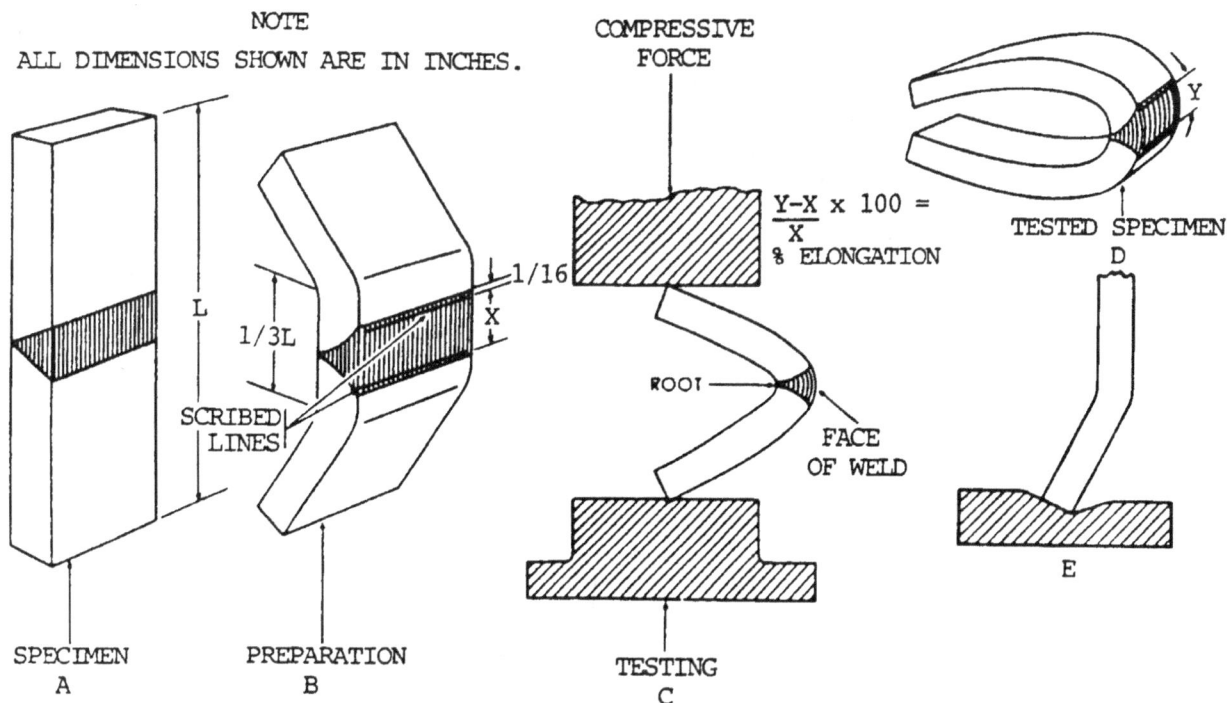

Figure 13-4. Free bend test of welded metal.

b. After bending the specimen to the point where the test bend is concluded, the distance between the scribed lines on the specimen is again measured and recorded as the distance Y. To find the percentage of elongation, subtract the initial from the final distance, divide by the initial distance, and multiply by 100 (fig. 13-4). The usual requirements for passing this test are that the minimum elongation be 15 percent and that no cracks greater than 1/16 in. (1.6 mm) in any dimension exist on the face of the weld.

c. The free bend test is being largely replaced by the guided bend test where the required testing equipment is available.

13-16. BACK BEND TEST

The back bend test is used to determine the quality of the weld metal and the degree of penetration into the root of the Y of the welded butt joint. The specimens used are similar to those required for the free bend test (para 13-15) except they are bent with the root of the weld on the tension side or outside. The specimens tested are required to bend 90 degrees without breaking apart. This test is being largely replaced by the guided bend test (para 13-14, 13-18).

13-17. NICK BREAK TEST

a. The nick break test has been devised to determine if the weld metal of a welded butt joint has any internal defects such as slag inclusions, gas pockets, poor fusion, and/or oxidized or burnt metal. The specimen is obtained from a welded butt joint either by machining or by cutting with an oxyacetylene torch. Each edge of the weld at the joint is slotted by means of a saw cut through the center (fig. 13-5). The piece thus prepared is bridged across two steel blocks (fig. 13-5) and stuck with a heavy hammer until the section of the weld between the slots fractures. The metal thus exposed should be completely fused and free from slag inclusions. The size of any gas pocket must not be greater than 1/16 in. (1.6 mm) across the greater dimension and the number of gas pockets or pores per square inch (64.5 sq mm) should not exceed 6.

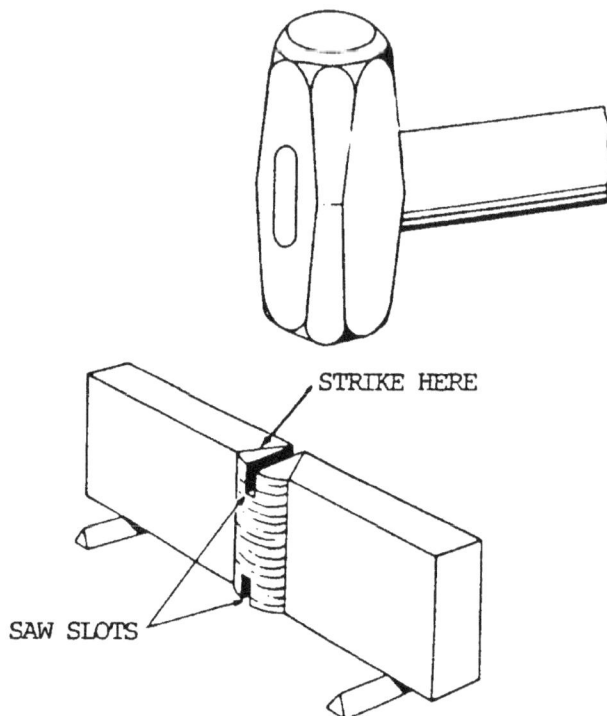

Figure 13-5. Nick break test.

b. Another break test method is used to determine the soundness of fillet welds. This is the fillet weld break test. A force, by means of a press, a testing machine, or blows of a hammer is applied to the apex of the V shaped specimen until the fillet weld ruptures. The surfaces of the fracture will then be examined for soundess.

13-18. TENSILE STRENGTH TEST

a. This test is used to measure the strength of a welded joint. A portion of a welded plate is machined to locate the weld midway between the jaws of the testing machine (fig. 13–6). The width and thickness of the test specimen are measured before testing, and the area in square inches is calculated by multiplying these

before testing, and the area in square inches is calculated by multiplying these two figures (see formula, fig. 13-6). The tensile test specimen is then mounted in a machine that will exert enough pull on the piece to break the specimen. The testing machining may he either a stationary or a portable type. A machine of the portable type, operating on the hydraulic principle and capable of pulling as well as bending test specimens, is shown in figure 13-7. As the specimen is being tested in this machine, the load in pounds is registered on the gauge. In the stationary types, the load applied may be registered on a balancing beam. In either case, the load at the point of breaking is recorded. Test specimens broken by the tensile strength test are shown in figure 13-3, p 13-10.

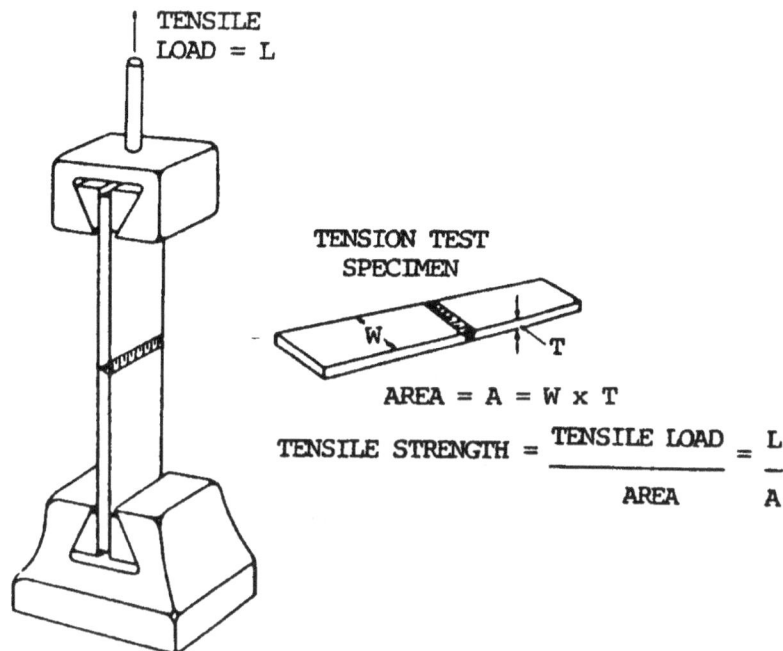

Figure 13-6. Tensile strength test specimen and test method.

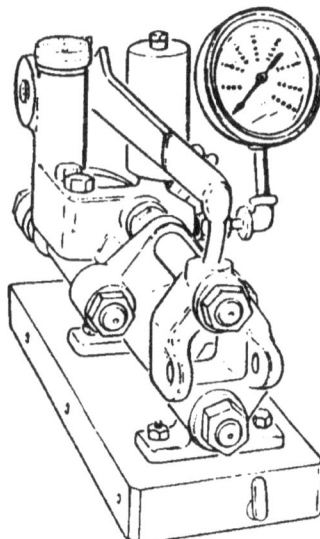

Figure 13-7. Portable tensile strength and bend testing machine.

13-18. TENSILE STRENGTH TEST (cont)

b. The tensile strength, which is defined as stress in pounds per square inch, is calculated by dividing the breaking load of the test piece by the original cross section area of the specimen. The usual requirements for the tensile strength of welds is that the specimen shall pull not less than 90 percent of the base metal tensile strength.

c. The shearing strength of transverse and longitudinal fillet welds is determined by tensile stress on the test specimens. The width of the specimen is measured in inches. The specimen is ruptured under tensile load, and the maximum load in pounds is determined. The shearing strength of the weld in pounds per linear inch is determined by dividing the maximum load by the length of fillet weld that ruptured. The shearing strength in pounds per square inch is obtained by dividing the shearing strength in pounds per linear inch by the average throat dimension of the weld in inches. The test specimens are made wider than required and machined down to size.

13-19. HYDROSTATIC TEST

This is a nondestructive test used to check the quality of welds on closed containers such as pressure vessels and tanks. The test usually consists of filling the vessel with water and applying a pressure greater than the working pressure of the vessel. Sometimes, large tanks are filled with water which is not under pressure to detect possible leakage through defective welds. Another method is to test with oil and then steam out the vessel. Back seepage of oil from behind the liner shows up visibly.

13-20. MAGNETIC PARTICLE TEST

This is a test or inspection method used on welds and parts made of magnetic alloy steels. It is applicable only to feffomagnetic materials in which the deposited weld is also ferromagnetic. A strong magnetic field is set up in the piece being inspected by means of high amperage electric currents. A leakage field will be set up by any discontinuity that intercepts this field in the part Local poles are produced by the leakage field. These poles attract and hold magnetic particles that are placed on the surface for this purpose. The particle pattern produced on the surface indicates the presence of a discontinuity or defect on or close to the surface of the part.

13-21. X-RAY TEST

This is a radiographic test method used to reveal the presence and nature of internal defects in a weld, such as cracks, slag, blowholes, and zones where proper fusion is lacking. In practice, an X-ray tube is placed on one side of the welded plate and an X-ray film, with a special sensitive emulsion, on the other side. When developed, the defects in the metal show up as dark spots and bands, which can be interpreted by an operator experienced in this inspection methcd. Porosity and defective root penetration as disclosed by X-ray inspection are shown in figure 13-8.

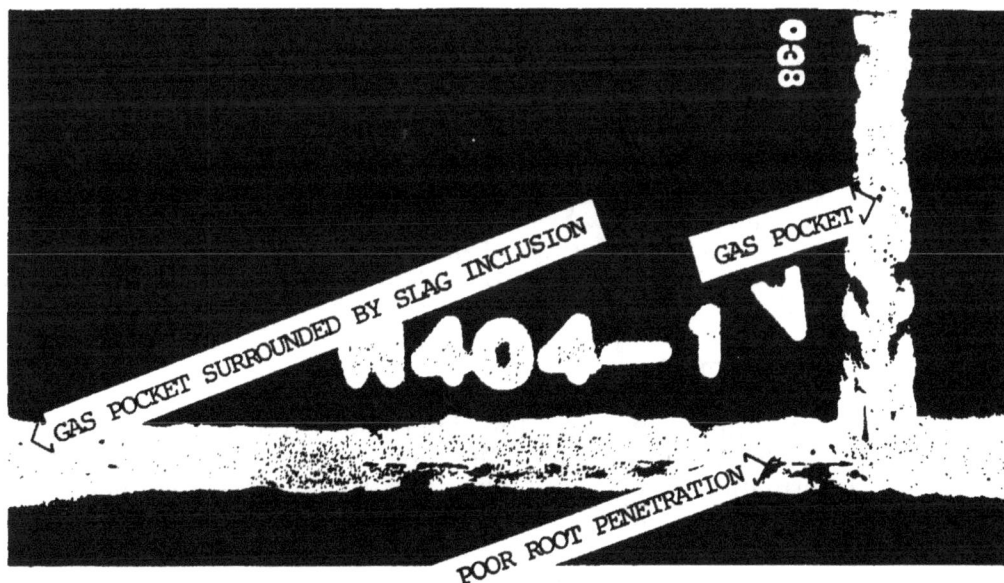

Figure 13-8. Internal weld defects disclosed by X-ray inspection.

NOTE
Instructions for handling X-ray apparatus to aviod harm to operating personnel are found in the "American Standard Code for the Industrial Use of X-rays".

13-22. GAMMA RAY TEST

This test is a radiographic inspection method similar to the X-ray method described in paragraph 13-13, p 13-8, except that the gamma rays emanate from a capsule of radium sulfate instead of an X-ray tube. Because of the short wave lengths of gamma rays, the penetration of sections of considerable thickness is possible, but the time required for exposure for any thickness of metal is much longer than that required for X-rays because of the slower rate at which the gamma rays are produced. X-ray testing is used for most radiographic inspections, but gamma ray equipment has the advantage of being extremely portable.

13-23. FLUORESCENT PENETRANT TEST

Fluorescent penetrant inspection is a nondestructive test method by means of which cracks, pores, leaks, and other discontinuities can be located in solid materials. It is particularly useful for locating surface defects in nonmagnetic materials such as aluminum, magnesium, and austenitic steel welds and for locating leaks in all types of welds. This method makes use of a water washable, highly fluorescent material that has exceptional penetration qualitiesThis material is applied to the clean dry surface of the metal to be inspected by brushing, spraying, or dipping. The excess material is removed by rinsing, wiping with clean water-soaked cloths, or by sandblasting. A wet or dry type developer is then applied. Discontinuities in surfaces which have been properly cleaned, treated with the penetrant, rinsed, and treated with developer show brilliant fluorescent indications under black light.

13-24. HARDNESS TESTS

a. General. Hardness may be defined as the ability of a substance to resist indentation of localized displacement. The hardness test usually applied is a nondestructive test, used primarily in the laboratory and not to any great extent in the field. Hardness tests are used as a means of controlling the properties of materials used for specific purposes after the desired hardness has been established for the particular application. A hardness test is used to determine the hardness of weld metal. By careful testing of a welded joint, the hard areas can be isolated and the extent of the effect of the welding heat on the properties of the base metal determined.

b. Hardness Testing Equipment.

(1) File test. The simplest method for determining comparative hardness is the file test. It is performed by running a file under manual pressure over the piece being tested. Information may be obtained as to whether the metal tested is harder or softer than the file or other materials that have been given the same treatment.

(2) Hardness testing machines.

(a) General There are several types of hardness testing machines. Each of them is singular in that its functional design best lends itself to the particular field or application for which the machine is intended. However, more than one type of machine can be used on a given metal, and the hardness values obtained can be satisfactorily correlated. Two types of machines are used most commonly in laboratory tests for metal hardness: the Brinell hardness tester and the Rockwell hardness tester.

(b) Brinell hardness tester. In the Brinell tests, the specimen is mounted on the anvil of the machine and a load of 6620 lb (3003 kg) is applied against a hardened steel ball which is in contact with the surface of the specimen being tested. The steel ball is 0.4 in. (10.2 mm) in diameter. The load is allowed to remain 1/2 minute and is then released, and the depth of the depression made by the ball on the specimen is measured. The resultant Brinell hardness number is obtained by the following formula:

$$Bhn = \frac{P}{\frac{\pi D}{2}(D - \sqrt{D^2 - d^2})}$$

Bhn: Brinell hardness number
P: applied load in kilograms
D: diameter of steel ball in millimeters
d: diameter of impression in millimeters

It should be noted that, in order to facilitate the determination of Brinell hardness, the diameter of the depression rather than the depth is actually measured. Charts of Brinell hardness numbers have been prepared for a range of impression diameters. These charts are commonly used to determine Brinell numbers.

(c) Rockwell hardness tester. The principle of the Rockwell tester is essentially the same as the Brinell tester. It differs from the Brinell tester in that a lesser load is impressed on a smaller ball or cone shaped diamond. The depth of the indentation is measured and indicated on a dial attached to the machine. The hardness is expressed in arbitrary figures called "Rockwell numbers." These are prefixed with a letter notation such as "B" or "C" to indicate the size of the ball used, the impressed load and the scale used in the test.

13-25. MAGNAFLUX TEST

a. <u>General</u>. This is a rapid, non-destructive method of locating defects at or near the surface of steel and its magnetic alloys by means of correct magnetization and the application of ferromagnetic particles.

b. <u>Basic Principles</u>. For all practical purposes, magnaflux inspection may be likened to the use of a magnifying glass. Instead of using a glass, however, a magnetic field and ferromagnetic powders are employed.The method of magnetic particle inspection is based upon two principles:one, that a magnetic field is produced in a piece of metal when an electric current is flowed through or around it; two, that minute poles are set up on the surface of the metal wherever this magnetic field is broken or distorted.

c. When ferromagnetic particles are brought into the vicinity of a magnetized part, they are strongly attracted by these poles and are held more firmly to them than to the rest of the surface of the part, thereby forming a visible indication.

13-26. EDDY CURRENT (ELECTROMAGNETIC) TESTING

a. General. Eddy current (electromagnetic) testing is a nondestructive test methcd based on the principle that an electric current will flow in any conductor subjected to a changing magnetic field. It is used to check welds in magnetic and nonmagnetic materials and is particularly useful in testing bars, fillets, welded pipe, and tubes. The frequency may vary from 50 Hz to 1 MHz, depending on the type and thickness of material current methods.The former pertains to tests where the magnetic permeability of a material is the factor affecting the test results and the latter to tests where electrical conductivity is the factor involved.

b. Nondestructive testing by eddy current methods involves inducing electric currents (eddy or foucault currents) in a test piece and measuring the changes produced in those currents by discontinuities or other physical differences in the test piece. Such tests can be used not only to detect discontinuties, but also to measure variations in test piece dimensions and resistivity.Since resistivity is dependent upon such properties as chemical composition (purity and alloying), crys-tal orientation, heat treatment, and hardness, these properties can also be deter-mined indirectly. Electromagnetic methods are classified as magnetoinductive and eddy current methods. The former pertains to tests where the magnetic permea-ability of a material is the factor affecting the test results and the latter to tests where electrical conductivity is the factor involved.

c. One method of producing eddy currents in a test specimen is to make the specimen the core of an alternating current (at) induction coilThere are two ways of measuring changes that occur in the magnitude and distribution of these currents. The first is to measure the resistive component of impedance of the exciting coil (or of a secondary test coil) and the second is to measure the induc-tive component of impedance of the exciting (or of a secondary) coil. Electronic equipment has been developed for measuring either the resistive or inductive imped-ance components singly or both simultaneously.

13-26. EDDY CURRENT (ELECTROMAGNETIC) TESTING (cont)

d. Eddy currents are induced into the conducting test specimen by alternating electromagnetic induction or transformer actionEddy currents are electrical in nature and have all the properties associated with electric currentsIn generating eddy currents, the test piece, which must be a conductor, is brought into the field of a coil carrying alternating current.The coil may encircle the part, may be in the form of a probe or in the case of tubular shapes, may be wound to fit inside a tube or pipe. An eddy current in the metal specimen also sets up its own magnetic field which opposes the original magnetic field. The impedance of the exciting coil, or of a second coil coupled to the first, in close proximity to the specimen, is affected by the presence of the induced eddy currents.This second coil is often used as a convenience and is called a sensing or pick up coil.The path of the eddy current is distorted by the presence of a discontinuity. A crack both diverts and crowds eddy currents.In this manner, the apparent impedance of the coil is changed by the presence of the defect. This change can be measured and is used to give an indication of defects or differences in physical, chemical, and metallurgical structure. Subsurface discontinuities may also be detected, but the current falls off with depth.

13-27. ACOUSTIC EMISSION TESTING

a. Acoustic emission testing (AET') methods are currently considered supplementary to other nondestructive testing methods.They have been applied, however, during proof testing, recurrent inspections, service, and fabrication.

b. Acoustic emission testing consists of the detection of acoustic signals produced by plastic deformation or crack formation during loading. These signals are present in a wide frequency spectrum along with ambient noise from many other sources. Transducers, strategically placed on a structure, are activated by arriving signals. By suitable filtering methods,ambient noise in the composite signal is notably reduced. Any source of significant signals is plotted by triangulation based on the arrival times of these signals at the different transducers.

13-28. FERRITE TESTING

a. Effects of Ferrite Content. Fully austenitic stainless steel weld deposits have a tendency to develop small fissures even under conditions of minimal restraint These small fissures tend to be located transverse to the weld fusion line in weld passes and base metal that were reheated to near the melting point of the material by subsequent weld passes.Cracks are clearly injurious defects and cannot be tolerated. On the other hand, the effect of fissures on weldment performance is less clear, since these micro-fissures are quickly blurted by the very tough austenitic matrix. Fissured weld deposits have performed satisfactorily under very severe conditions. However, a tendency to form fissures generally goes hand-in-hand with a tendency for larger cracking, so it is often desirable to avoid fissure-sensitive weld metals.

b. The presence of a small fraction of the magnetic delta ferrite phase in an otherwise austenitic (nonmagnetic) weld deposit has an influence in the prevention of both centerline cracking and fissuring.The amount of delta ferrite in as-welded material is largely controlled by a balance in the weld metal composition be—

tween the ferrite-promoting elements (chromium, silicon, molybdenum, and columbium are the most common) and the austenite-promoting elements (nickel, manganese, carbon, and nitrogen are the most common).Excessive delta ferrite, however, can have adverse effects on weld metal properties. The greater the amount of delta ferrite, the lower will be the weld metal ductility and toughness. Delta ferrite is also preferentially attacked in a few corrosive environments, such as urea. In extended exposure to temperatures in the range of 900 to 1700°F (482 to 927 °C), ferrite tends to transform in part to a brittle intermetallic compound that severely embrittles the weldment.

c. Portable ferrite indicators are designed for on-site use. Ferrite content of the weld deposit maybe indicated in percent ferrite and may be bracketed between two values. This provides sufficient control in most applications where minimum ferrite content or a ferrite range is specified.

APPENDIX A

REFERENCES

A-1. PUBLICATION INDEXES

The following indexes should be consulted for latest changes or revisions of references given in this appendix and for new publications relating to information contained in this manual:

DA Pam 108-1 . . . . Index of Motion Pictures, Film Strips, Slides, and Phono-Recordings

DA Pam 310-1 . . . . Index of Administrative Publications

DA Pam 310-2 . . . . Index of Blank .Forms

DA Pam 310-3 . . . . Index of Training Publications

DA Pam 310-4 . . . . Index of Technical Manuals, Supply Manuals, Supply Bulletins, Lubrication Orders, and Modification Work Orders

A-2. SUPPLY MANUALS

The Department of the Army supply manuals pertaining to the materials contained in this manual are as follows:

SC 3433-95-CL-A03 . . . . Torch Outfit, Cutting and Welding

SC 3433-95-CL-A04 . . . . Tool Kit, Welding

SC 3439-IL . . . . FSC Group 34, Class 3439: Metal Working Machinery, Miscellaneous Welding, Soldering and Brazing Supplies and Accessories

SC 3470-95-CL-A07 . . . . Shop Set, Welding and Blacksmith

SC 3470-95-CL-A10 . . . . Shop Equipment, Welding

A-3. TECHNICAL MANUALS AND TECHNICAL BULLETINS

The following DA publications contain information pertinent to this manual:

TB ENG 53 . . . . Welding and Metal Cutting at NIKE Sites

TB MED 256 . . . . Toxicology of Ozone

TB TC 11 . . . .Arc Welding on Water-Borne Vessels

TB 34-91-167 . . . . Welding Terms and Definitions Glossary

TB 9-2300-247-40 . . . . Transport Wheeled Vehicles: Repair of Frames

A-3. TECHNICAL MANUALS AND TECHNICAL BULLETINS (cont)

TB 9-3439-203/1 . . . . Conversion of Welding Electrode Holder for Supplemental Air-Arc Metal cutting

TM 10-270 . . . . General Repair of Quartermaster Items of General Equipment

TM 38-750 . . . . The Army Equipment Record System and Procedure

TM 5-805-7 . . . . Welding Design, Procedures, and Inspection

TM 5-3431-209-5 . . . . Operator, Organizational, Direct Support and General Support Maintenance Manual: Welding Machine, Arc, Generator, Power Take-off Driven, 200 Amp, DC, Single Operator, Base Mounted (Valentine Model 26381)

TM 5-3431-211-15 . . . . Operator, Organizational, Direct Support, General Support, and Depot Maintenance Manual (Including Repair Parts and Special Tools Lists): Welding Set, Arc, Inert Gas Shielded Consumable Metal Electrode for 3/4 Inch Wire, DC 115 V (Air Reduction Model 2351-0685)

TM 5-3431-213-15 . . . . Organizational, Direct Support, General Support, and Depot Maintenance Manual with Repair Parts and Special Tools Lists: Welding Machine, Arc, General and Inert Gas Shielded, Transformer-Rectifier Type AC and DC; 300 Ampere Rating at 60% Duty Cycle (Harnischfeger Model DAR-300HFSG)

TM 5-3431-221-15 . . . . Operator, Organizational, Direct Support and General Support Maintenance Manual: Welding Machine, Arc, Generator, Gasoline Driven, 300 Amp at 20 V Min, 375 Amp at 40 V Max, 115 V, DC, 3 KW, Skid Mounted, Winterized (Libby Model LEW-300)

TM 9 -213 . . . . Painting Instructions, Field Use

TM 9-2920 . . . . Shop Mathematics

TM 9-3433 -206-10 . . . . Spray Gun, Metallizing (Metaillizing Co. of America "Turbo-Jet")

A-4.   OTHER FORMS AND PUBLICATIONS

A-4.   OTHER FORMS AND PUBLICATIONS (cont)

a. The following explanatory publications contain information pertinent to this material and associated equipment:

AWS A2.0-58 . . . . Welding Symbols

DA FORM 2028 . . . . Recommended Changes to publications and Blank Forms

MIL-E-17777C . . . . Electrodes Cutting and Welding Carbon-Graphite Uncoated and Copper Coated

MIL-E-18038 . . . . Electrodes, Welding, Mineral Covered, Low Hydrogen, Medium and High Tensile Steel as Welded or Stress and Relieved Weld Application and Use

MIL-E-22200/1 . . . . Electrodes, Welding, Covered
t h r u
MIL-E-22200/7

MIL-M-45558 . . . . Moisture Stabilizer, Welding Electrode

MIL-STD-21 . . . . Weld Joint Designs, Armored Tank Type

MIL-STD-22 . . . . Weld Joint Designs

MIL-STD-101 . . . . Color Code for Pipe Lines and Compressed Gas Cylinders

MIL-W-12332 . . . . Welding, Resistance, Spot and Projection, for Fabricating Assemblies of Low Carbon Steel

MIL-W-18326 . . . . Welding of Magnesium Alloys, Gas and Electric, Manual and Machine, Process for

MIL-W-21157 . . . . Weldments, Steel, Carbon and Low Alloy; Yield Strength 30,000-60,000 PSI

MIL-W-22248 . . . . Weldments, Aluminum and Aluminum Alloys

MIL-W-27664 . . . . Welding, Spot, Inert Gas Shielded Arc

MIL-W-41 . . . . Welding of Armor, Metal-Arc, Manual, with Austentic Electrodes for Aircraft

MIL-W-6858 . . . . Welding, Resistance, Aluminum, Magnesium, Non-Hardening Steels or Alloys, and Titanium Alloys, Spot and Seam

MIL-W-6873 . . . . Welding, Flash, Carbon and Alloy Steel

MIL-W-8604 . . . . Welding of Aluminum Alloys, Process for

MIL-W-8611 . . . . Welding, Metal-Arc and Gas, Steels and Corrosion and Heat Resisting Alloys, Process for

MIL-W-45205 . . . . Welding, Inert Gas, Metal-Arc, Aluminum Alloys Readily Weldable for Structures, Excluding Armor

MIL-W-45206 . . . . Welding, Aluminum Alloy Armor

A-4. OTHER PUBLICATIONS (cont)

MIL-W-45210 . . . . Welding, Resistance, Spot, Weldable Aluminum Alloys

MIL-W-45223 . . . . Welding, Spot, Hardenable Steel

b. The following health and safety standards are pertinent to this material and associated equipment:

ANSI (American National Standards Institute) Z49.1-1973, Safety in Welding and Cutting

ANSI Z87.1-1968, American National Standard Practice for Occupational and Educational Eye and Face Protection

ANSI 788.12, Practices for Respiratory Protection

AWS (American Welding Society), Bare Mild Steel Electrodes and Fluxes for Submerged Arc Welding

AWS, Carbon Steel Electrodes for Flux Cored Arc Welding

AWS, Flux Cored Corrosion Resisting Chromium and Chromium-Nickel Steel Electrodes

41 Code of Federal Regulations 50-204.7

29 Code of Federal Regulations 1910

National Bureau of Standards, Washington DC, National Safety Code for the Protection of Hands and Eyes of Industrial Workers

NFPA (National Fire Protection Association) 51-1969, Welding and Cutting Oxygen Fuel Gas systems

NFPA 51B-1962, Standard for Fire Prevention in Use of Cutting and Welding Processes

NFPA 566-1965, Standard for Bulk Oxygen Systems at Consumer Sites

Public Law 91-596, Occupational Safety and Health Act of 1970; especially Subpart 1, Personal Protective Equipment, paragraph 1910.132; and Subpart Q, Welding, Cutting, and Brazing, paragraph 1910.252

c. The following commercial publications are available in technical libraries:

Welding Data Book. . . . . Welding Design & Fabrication (Industrial Publishing Co.) Cleveland, OH 44115

The Welding
Encyclopedia . . . . Welding Engineers Publications Inc. Morton Grove, IL 60053

d. The following commercial and military publications are provided as a bibliography:

Modern Welding Technology, Prentice-Hall, 1979, Englewood Cliffs, NJ

ST 9-187, Properties and Identification of Metal and Heat Treatment of Steel, 1972

Symbols for Welding and Nondestructive Testing Including Brazing, American Welding Society, 9179, Miami, FL

TM 5-805-7, Welding Design, Procedures and Inspection, 1976

TM 9-237, Welding Theory and Application, 1976

Welding Encyclopedia, Monticello Books, 1976, Lake Zurich, IL

Welding Handbook, Seventh Edition, Volume 1:Fundmentals of Welding, 1981, American Welding Society, Miami, FL

Welding Handbook, Seventh Edition, Volume 2: Welding Processes - Arc and Gas Welding, Cutting, and Brazing, 1981, American Welding Society, Miami, FL

Welding Handbook, Seventh Edition, Volume 3: Welding Processes - Resistance and Solid-State Welding and Other Joining Processes, 1981, American Welding Society, Miami, FL

Welding Handbook, Seventh Edition, Volume 4:Metals and their Weldability, 1981, American Welding Society, Miami, FL

Welding Handbook, Sixth Edition, Volume 5:Applications of Welding, 1973, American Welding Society, Miami, FL

Welding Inspection, 1980, American Welding Society, Miami, FL

Welding Terms and Definitions,1976, American Welding Society, Miami, FL

APPENDIX B
PROCEDURE GUIDES FOR WELDING

Table B-1. Guide for Welding Automotive Equipment

(See explanation of symbols at end of table)

Automotive Part	Usual Metal Composition											Recommended Welding Method						
	Gray cast iron	Malleable iron	Cast steel	Steel forgings	To 0.40 carbon steel	Over 0.40 carbon steel	Alloy steels	Aluminum	Brass, copper or bronze	Miscellaneous	Babbitt	Brazing	Welding with rod of similar composition	No. 1 HT	Soldering	Heating	Haynes stellite	Welding not recommended
DIVISION I – CYLINDERS																		
Group 1 – Cylinder Parts																		
Cylinder block	x											x						
Cylinder head	x											x						
Water jacket covers	x											x						
Valve spring cover	u									p		p						u
Valve stem guide					x													
Group 2 – Crank Case Parts																		
Crank case (various types) .	1							2				1	2					
Oil pan			x									x						
Breather					x							x						
Crankshaft bearings					x						2	1						
Crankshaft bearing cap ..					x				1			x			2			
Crankshaft bushing supports ...					x							x						
Handhole cover					x							x						
Timing gear cover					x							x						
Flywheel housing	1							2				1	2					
Generator bracket	x			u								x						u
Group 3 – Crankshaft Parts																		
Crankshaft				x														
Flywheel	x		x									x						

Table B-1. Guide for Welding Automotive Equipment (cont)

(See explanation of symbols at end of table)

Automotive Part	Usual Metal Composition											Recommended Welding Method						
	Gray cast iron	Malleable iron	Cast steel	Steel forgings	To 0.40 carbon steel	Over 0.40 carbon steel	Alloy steels	Aluminum	Brass, copper or bronze	Miscellaneous	Babbitt	Brazing	Welding with rod of similar composition	No. 1 HT	Soldering	Heating	Haynes stellite	Welding not recommended
DIVISION I – CYLINDERS (cont)																		
Crankshaft timing gear					x	u						x						u
Flywheel starter gear					x	x	x					x						
Crankshaft starter sprocket					x		x					x						
Crankshaft starting jaw (or pin)					x									x				
Group 4 – Starting Crank Parts																		
Starting crank jaw					x							x						
Starting crankshaft					x	u	u					x						u
Starting crankshaft spring																		
Starting crank handle					x							x						
Group 5 – Connecting Rods																		
Connecting rod				u			u	u										u
Connecting rod cap				u				u				1						u
Connecting rod bushing					1				1		2	1	2					
Connecting rod dipper								2				1				2		u
Piston pin bushing									u						u			u
Group 6 – Pistons and Parts																		
DIVISION II – VALVES																		
Group 1 – Camshaft Parts																		
Camshaft				u	u	u	u					u						u
Eccentric shaft				u														u
Camshaft timing gear					x		x					x						
Camshaft idler gear					x		x					x						
Camshaft oil pump gear					x		x					x						
Camshaft ignition distributor gear					x		x					x						
Camshaft time drive gear					x		x					x						
Oil pump eccentric (or cam)				u	u							u						u

Table B-1. Guide for Welding Automotive Equipment (cont)

(See explanation of symbols at end of table)

Automotive Part	Gray cast iron	Malleable iron	Cast steel	Steel forgings	To 0.40 carbon steel	Over 0.40 carbon steel	Alloy steels	Aluminum	Brass, copper or bronze	Miscellaneous	Babbitt	Brazing	Welding with rod of similar composition	No. 1 HT	Soldering	Heating	Haynes stellite	Welding not recommended
DIVISION II – VALVES (cont)																		
Group 2 – Valves																		
Poppet valve	n						n											n
Inlet valve	n						n											n
Exhaust valve	n						n											n
Valve spring							n											n
Valve spring retainer					x		x											
Valve lifter					x								x					
Valve lifter guide					n	n							x					
Valve rocker	n				x		x					x					x	n
Valve push rod					x	x						x						
DIVISION III – COOLING SYSTEM																		
Group 1 – Fan Parts																		
Fan bracket	x		x		x							x						
Fan spindle					x							x						
Fan hub												x						
Fan hub bushing (or bearing)	x											x						
Fan blades					x				x			x						
Fan pulley					x							x						
Fan driving pulley	x		x		x							x						
Group 2 – Radiator Parts																		
Radiator core					x				x						x			
Radiator core header sheets															x			
Radiator upper tank					x													
Radiator filler neck					x				x						x			
Radiator filler neck sleeve					x				x			x			x			
Radiator filler cap					1		2		3	d,n		1,2,						
Radiator tie rod fitting					x							x						n

Table B-1. Guide for Welding Automotive Equipment (cont)

(See explanation of symbols at end of table)

Automotive Part	Usual Metal Composition											Recommended Welding Method						
	Gray cast iron	Malleable iron	Cast steel	Steel forgings	To 0.40 carbon steel	Over 0.40 carbon steel	Alloy steels	Aluminum	Brass, copper or bronze	Miscellaneous	Babbitt	Brazing	Welding with rod of similar composition	No. 1 HT	Soldering	Heating	Haynes stellite	Welding not recommended
DIVISION III - COOLING SYSTEM - (cont)																		
Group 2 - Radiator Parts (cont)																		
Radiator baffle					x				x						x			
Radiator inlet fitting	x								x			x						
Radiator lower tank					x				x						x			
Radiator outlet fitting	x											x						
Radiator drain flange					x				x			x						
Radiator anchor plate					x							x						
Radiator overflow tube					x				x						x			
Radiator side bolting member					x							x						
Radiator shell anchorage clips					x [1]		[2]		[3]			x [1]	[2,3]					
Radiator shell					x							x						
Radiator supports					x							x						
Radiator support reinforcement					x							x						
Radiator hinge rod fitting					x							x						
Radiator brace rod fitting					x							x						
Radiator hood ledge liner strip					x							x						
Radiator starting crank hole cover					[1]		[2]					[1]	[2]					
Group 3 - Water Pump Parts																		
Water pump impeller	x [1]								x	[d,n]		x [1]	[1]					[n]
Water pump body					x							[1]						
Water pump cover					x		x					x						
Water pump shaft					x		x					x					x	

Table B-1. Guide for Welding Automotive Equipment (cont)

(See explanation of symbols at end of table)

Automotive Part	Usual Metal Composition											Recommended Welding Method						
	Gray cast iron	Malleable iron	Cast steel	Steel forgings	To 0.40 carbon steel	Over 0.40 carbon steel	Alloy steels	Aluminum	Brass, copper or bronze	Miscellaneous	Babbitt	Brazing	Welding with rod of similar composition	No. 1 HT	Soldering	Heating	Haynes stellite	Welding not recommended
DIVISION III – COOLING SYSTEM – (cont)																		
Water pump gland					x							x						
Water pump shaft gear					x		x					x						
Water pump shaft bushing ...	x							x				x						
Group 4 – Pipes																		
Engine water outlet	x											x						
Engine water inlet	x											x						
Radiator water fitting	x				x							x						
Water pump outlet pipe	x											x						
DIVISION IV – FUEL SYSTEM																		
Group 1 – Carburetor and Inlet Pipe																		
Carburetor	1		2						3	d,n		1,2 3						n
Inlet manifold	x											x						
Inlet pipe									x			x						
Group 2 – Carburetor Control Parts																		
Accelerator pedal					1			2				1	2					
Accelerator pedal bracket ..	x		x									x						
Accelerator pedal rod					x							x						
Carburetor mixture hand regulator																		
Carburetor choke					x							x						
Group 3 – Carburetor Air Heater Parts																		
Carburetor air heater	x				x							x						
Carburetor hot air pipe	x				x							x						

Table B-1. Guide for Welding Automotive Equipment (cont)

(See explanation of symbols at end of table)

Automotive Part	Gray cast iron	Malleable iron	Cast steel	Steel forgings	To 0.40 carbon steel	Over 0.40 carbon steel	Alloy steels	Aluminum	Brass, copper or bronze	Miscellaneous	Babbitt	Brazing	Welding with rod of similar composition	No. 1 HT	Soldering	Heating	Haynes stellite	Welding not recommended
DIVISION IV – FUEL SYSTEM – (cont)																		
Group 4 – Fuel Tank																		
Fuel tank					x							x						
Fuel tank outlets					x							x						
Group 5 – Fuel Pipes and Feed Systems																		
Fuel pipes					x				x			x						
Fuel pressure pump					x							x						
Fuel hand pump					x							x						
Fuel pressure pipes												x						
DIVISION V – EXHAUST SYSTEM																		
Group 1 – Exhaust Manifold																		
Exhaust manifold	x												x					
Group 2 – Exhaust Pipe and Muffler																		
Muffler					x							x						
Exhaust pipe					x							x						
Muffler outlet pipe					x							x						
DIVISION VI – LUBRICATION SYSTEM																		
Group 1 – Oil Pan or Reservoir																		
Oil pan					x							x						
Oil filler strainer									x						x			
Oil filler cap					x							x						
Group 2 – Oil Pump Parts																		
Oil pump body	x											x						
Oil pump plunger	x				x				x			x						
Oil pump plunger spring						u						u						u
Oil pump valves					x	u						x						
Oil pump shaft					x	x	x					x						

Table B-1. Guide for Welding Automotive Equipment (cont)

(See explanation of symbols at end of table)

Automotive Part	Usual Metal Composition											Recommended Welding Method						
	Gray cast iron	Malleable iron	Cast steel	Steel forgings	To 0.40 carbon steel	Over 0.40 carbon steel	Alloy steels	Aluminum	Brass, copper or bronze	Miscellaneous	Babbitt	Brazing	Welding with rod of similar composition	No. 1 HT	Soldering	Heating	Haynes stellite	Welding not recommended
DIVISION VI – LUBRICATION (cont)																		
Oil pump shaft gears					n		n					n						n
Oil pump following gear					n							n						n
Oil pump cover					x		x					x						
Group 3 – Oil Pipes, Strainers, Gauges																		
Oil pipes									x			x						
Circulating oil strainer									x					x				
Oil strainer cap					x							x						
Oil level gauge					x				x			x						
DIVISION VII – IGNITION SYSTEM																		
Group 1 – Spark Plug Cables																		
Spark plug cables									x						x			
Coil high-tension cable									x							x		
Low-tension cables									x						x			
Group 2 – Battery Ignition Equipment Parts																		
Timer-distributor shaft					1		n					1,n						
Timer-distributor shaft gear					1		n					1,n						
Ignition drive shaft					x		x					x						
Ignition drive shaft gear					x		x					x						
Manual advance arm					x							x						n
Automatic advance element					x							x						n
Ignition unit, magneto-base mounting	x		x									x						

Table B-1. Guide for Welding Automotive Equipment (cont)

(See explanation of symbols at end of table)

Automotive Part	Usual Metal Composition											Recommended Welding Method						
	Gray cast iron	Malleable iron	Cast steel	Steel forgings	To 0.40 carbon steel	Over 0.40 carbon steel	Alloy steels	Aluminum	Brass, copper or bronze	Miscellaneous	Babbitt	Brazing	Welding with rod of similar composition	No. 1 HT	Soldering	Heating	Haynes stellite	Welding not recommended
DIVISION VIII – STARTING AND GENERATOR EQUIPMENT																		
Group 1 – Generator Parts																		
Generator driving gear or sprocket					x				x			x						
Generator shaft					x							x						
Generator coupling												x						
Group 2 – Starting Motor Parts																		
Starting motor pinions						n	n					u						n
Starting motor gear					n	n	n					u						n
Starting motor gear shaft							n					u						n
Group 3 – Starter Generator (See VIII – 1,2)																		
Group 4 – Ignition Generator (See VII – 2, VIII – 1)																		
Group 5 – Ignition Starter Generator (See VII – 2, VIII – 1,2)																		
Group 6 – Storage Battery Parts																		
Terminal post										1, n			x						n
Plates										1								
Post straps					x					1			x						
Battery holddown					x								x						
Handles										1			x						
Terminals													x						
Through bolt					x							x	x						

Table B-1. Guide for Welding Automotive Equipment (cont)

(See explanation of symbols at end of table)

Automotive Part	Usual Metal Composition											Recommended Welding Method						
	Gray cast iron	Malleable iron	Cast steel	Steel forgings	To 0.40 carbon steel	Over 0.40 carbon steel	Alloy steels	Aluminum	Brass, copper or bronze	Miscellaneous	Babbitt	Brazing	Welding with rod of similar composition	No. 1 HT	Soldering	Heating	Haynes stellite	Welding not recommended
DIVISION IX – MISCELLANEOUS ELECTRICAL EQUIPMENT																		
Group 1 – Lamps and Wiring																		
Head lamp housing					1		2		3			1,3	2					
Head lamp housing flange					x				x			x						
Head lamp door					1		2		3			1,3	2					
Head lamp reflector									n									n
(Auxiliary light parts are similar to head lamp parts.)																		
Head lamp support tie rod					x							x						
Taillight support					x							x						
Group 2 – Switches and Instruments																		
Starting switch lever					x				x			x						
Switches and instruments															x			
Group 3 – Horn																		
Horn projector					x							x						n
DIVISION X – CLUTCH																		
Group 1 – Clutching Parts																		
Clutch case (rotating member)	x		2		1			3			x	1,2	3					
Clutch housing	1				1			2				1	2					
Clutch cover					n							x						
Clutch housing cover					n													
Clutch driving disk			x		x							x						
Clutch pressure plates	x											x						
Clutch driver spider (or drum)					x							x						
Clutch facing spring							n	n										n
Clutch spring							n	n										n
Clutch shaft					x						x	x						
Clutch pilot bearing							n											n

Table B-1. Guide for Welding Automotive Equipment (cont)

(See explanation of symbols at end of table)

Automotive Part	Gray cast iron	Malleable iron	Cast steel	Steel forgings	To 0.40 carbon steel	Over 0.40 carbon steel	Alloy steels	Aluminum	Brass, copper or bronze	Miscellaneous	Babbitt	Brazing	Welding with rod of similar composition	No. 1 HT	Soldering	Heating	Haynes stellite	Welding not recommended
DIVISION X – CLUTCH (cont)																		
Clutch driven plate	X		X									X					X	n
Clutch driving plate		X	X	X	n													
Clutch pressure levers					X							X						
Group 2 – Releasing Parts																		
Clutch release sleeve	X											X						
Clutch release bearing housing												X						
Clutch release bearing					X							X					X	n
Clutch release yoke		X	X		1		n					1,n					X	n
Clutch release yoke shaft					1		n					1,n						n
Clutch pedal shaft																		
Clutch pedal adjusting link					X				X			X						
Clutch release yoke lever				X								X						
Clutch pedal		X		X								X						
Clutch brake					X							X						
DIVISION XI – TRANSMISSION																		
Group 1 – Transmission Parts																		
Transmission case and cover	1					n		2				1	2					
Transmission gears						n	n					n						n
Transmission bearings and bearing parts							n											n
Transmission shafts and counter shafts						n	n											n
Transmission shaft pilot bushings					n		n		1			1						n
Group 2 – Shifting Mechanism Parts																		
Control housing	X											X						

Table B-1. Guide for Welding Automotive Equipment (cont)

(See explanation of symbols at end of table)

Automotive Part	Gray cast iron	Malleable iron	Cast steel	Steel forgings	To 0.40 carbon steel	Over 0.40 carbon steel	Alloy steels	Aluminum	Brass, copper or bronze	Miscellaneous	Babbitt	Brazing	Welding with rod of similar composition	No. 1 HT	Soldering	Heating	Haynes stellite	Welding not recommended
DIVISION XI – TRANSMISSION																		
Control shift frame		x		x								x					x	
Transmission shift forks		x		x	M		x					x						n
Transmission shift rails					1		n					1,n						n
Transmission interlock rail				1	2		n					1,2						n
Group 3 – Control Parts																		
Control lever					x							x						
Control lever fulcrum ball					x	x	x					x					x	n
Group 4 – Propeller Shaft Parts																		
Propeller shaft					1									1				
Propeller shaft universal joints				x	x		n					n						n
Propeller shaft bearings and bearing parts			n			n	x											n
Transmission shaft					x		n					x						
Universal joint flange		n		x	x							x						n
Universal joint yoke				n								n						n
Universal joint center cross, ring or block																		n
Universal joint bearing bushing									x			x						
Universal joint pin						n	n					x						n
Universal joint casings			x		x		n					x						n
Universal joint trunnion					n		n					n					n	n
Universal joint trunnion block					n		n					n						n

Table B-1. Guide for Welding Automotive Equipment (cont)

(See explanation of symbols at end of table)

Automotive Part	Gray cast iron	Malleable iron	Cast steel	Steel forgings	To 0.40 carbon steel	Over 0.40 carbon steel	Alloy steels	Aluminum	Brass, copper or bronze	Miscellaneous	Babbitt	Brazing	Welding with rod of similar composition	No. 1 HT	Soldering	Heating	Haynes stellite	Welding not recommended
DIVISION XII – REAR AXLE																		
Group 1 – Housing Parts																		
Rear axle housing	x				x							x						
Bevel or worm gear housing	x		x		x							x						
Rear axle tubes					x		x					x						
Differential carrier	x		x	x	x							x						
Rear axle spring seat			x	x								x						
Axle brake shaft bracket		x	x	x	x							x						
Brake support			x	x	x							x						
Brake shield			x	x	x							x						
Group 2 – Torque Arm and Radius Rod Parts																		
Radius rods					x									x				
Group 3 – Drive Pinion Parts																		
Axle drive bevel pinion					u	u	u					u						u
Axle drive pinion shaft					u		u					u						u
Axle drive pinion bearings and bearing parts						u	u											u
Axle drive pinion adjusting sleeves							u											u
Axle drive pinion (or worm) carrier	x	x	x		x							x						
Group 4 – Differential Parts																		
Bevel drive pinion				u	u	u	u					u						u
Bevel drive gear					u		u					u						u
Differential case flange half			u															u
Differential case plain half			u				u					u						u
Differential bearing							u					u						u
Differential sleeve						u	u											u
Differential side gear					u		u					u						u

Table B-1. Guide for Welding Automotive Equipment (cont)

(See explanation of symbols at end of table)

Automotive Part	Usual Metal Composition											Recommended Welding Method						
	Gray cast iron	Malleable iron	Cast steel	Steel forgings	To 0.40 carbon steel	Over 0.40 carbon steel	Alloy steels	Aluminum	Brass, copper or bronze	Miscellaneous	Babbitt	Brazing	Welding with rod of similar composition	No. 1 HT	Soldering	Heating	Haynes stellite	Welding not recommended
DIVISION XII – REAR AXLE (cont)																		
Differential spider pinion					n	n	n					n						n
Differential spider					n	n	n					n						n
Differential cross pin pinion					n	n	n					n						n
Differential cross pin					n	n	n					n						n
Differential side gear					x							x						
spacer					n		n		n			n						n
Worm or worm gear					n		n					x						n
Group 5 – Axle Shafts																		
Axle shaft					x		n					x						n
Axle shaft wheel flange		x	x									x						
DIVISION XIII – BRAKES																		
Group 1 – Outer Brake Parts																		
Outer brake band					n		n					x						
Outer brake band lever		x		x	x							1,2						
Outer brake lever shaft				1	2							x						
Outer brake shaft end levers				x	x							x						
Group 2 – Inner Brake Parts																		
Inner brake shoe	x											x						
Inner brake toggle		x		x	x							x						
Inner brake toggle lever		x		x	x							x						
Inner brake toggle shaft				x	x							x						n
Inner brake cam					1	2						1						
Inner brake camshaft						2						2,n						
Inner brake camshaft lever				x	1		n					1,n						n
Group 3 – Pedal (or Outer) Brake Control Parts																		
Pedal brake rod					x							x						
Pedal brake rod yoke					x							x						

Table B-1. Guide for Welding Automotive Equipment (cont)

(See explanation of symbols at end of table)

Automotive Part	Gray cast iron	Malleable iron	Cast steel	Steel forgings	To 0.40 carbon steel	Over 0.40 carbon steel	Alloy steels	Aluminum	Brass, copper or bronze	Miscellaneous	Babbitt	Brazing	Welding with rod of similar composition	No. 1 HT	Soldering	Heating	Haynes stellite	Welding not recommended
DIVISION XIII – BRAKES (cont)																		
Pedal brake intermediate shafts				x								x						
Pedal brake equalizer levers				x	x							x						
Pedal brake equalizer				x	x							x						
Brake pedal				x	x							x						
Brake pedal rod					x							x						
Brake pedal rod yokes				x	x							x						
Brake pedal shaft					1		n					1,n						n
Group 4 – Handbrake (or Inner Brake) Control Parts																		
Handbrake rod					x							x						
Handbrake rod yoke				x	x							x						
Handbrake intermediate shafts				x	x							x						
Handbrake equalizer levers				x	x							x						
Handbrake equalizer				x	x							x						
Brake hand lever rod					x							x						
Brake hand lever rod yoke				x	x							x						
Brake hand lever				x	x							x						
DIVISION XIV – FRONT AXLE AND STEERING																		
Group 1 – Axle Center Parts																		
Front axle center				n								n						n
Front spring seats						n	n					1						n
Front axle bushing						n	n		1			1						n
Wheel spindles			n									n						n
Group 2 – Steering Knuckles																		
Steering knuckles		n	n	n		n	n					n						n

Table B-1. Guide for Welding Automotive Equipment (cont)

(See explanation of symbols at end of table)

Automotive Part	Gray cast iron	Malleable iron	Cast steel	Steel forgings	To 0.40 carbon steel	Over 0.40 carbon steel	Alloy steels	Aluminum	Brass, copper or bronze	Miscellaneous	Babbitt	Brazing	Welding with rod of similar composition	No. 1 HT	Soldering	Heating	Haynes stellite	Welding not recommended
DIVISION XIV – FRONT AXLE AND STEERING (cont)																		
Steering knuckle bushing	x								x			x						n
Steering knuckle pivot					n													n
Steering knuckle thrust bearing																		n
Steering knuckle arms							n											n
Steering knuckle gear rod arm						n	n											n
Group 3 – Steering Rods																		
Steering knuckle tie rod				n	n		n											n
Steering gear connecting rod				n	n		n											n
Group 4 – Steering Gear Parts																		
Steering gear case		x	x		x							x						n
Steering gear bracket		x			x							x						n
Steering gear arm		n		n														
Steering gear shaft					n													
Steering wheel spider			x		x							x						
Steering wheel tube (or shaft)	x		x		x							x						
Spark and throttle sector					M													
Spark and throttle sector tube					M M							x x						
Spark hand lever					M M							x x						
Spark hand lever tube (or rod)					x x							x x						
Throttle hand lever					x x							x x						
Throttle hand lever tube (or rod)					x							x						

Table B-1. Guide for Welding Automotive Equipment (cont)

(See explanation of symbols at end of table)

Automotive Part	Usual Metal Composition											Recommended Welding Method						
	Gray cast iron	Malleable iron	Cast steel	Steel forgings	To 0.40 carbon steel	Over 0.40 carbon steel	Alloy steels	Aluminum	Brass, copper or bronze	Miscellaneous	Babbitt	Brazing	Welding with rod of similar composition	No. 1 HT	Soldering	Heating	Haynes stellite	Welding not recommended
DIVISION XIV - FRONT AXLE AND STEERING (cont)																		
Steering column bracket			1		2					d,n		1,2						n
Steering worm					n		n		n									n
Steering worm sector (or gear)					n		n		n									n
Steering worm shaft																		
DIVISION XV - WHEELS																		
Group 1 - Wheels																		
Wheel rims					x							n		x				
Wheel hub			n		x							x						n
Wheel hub flanges					x							x						
Wheel bearings and bearing parts						n	n											n
Wheel brake drums	x		x		x							x						
DIVISION XVI - FRAME AND SPRINGS																		
Group 1 - Frame parts																		
Frame members					x							x		x		x		
Gussets					x							x		x		x		
Group 2 - Frame Brackets and Sockets																		
Spring brackets					x							x						
Running board brackets					x													
Engine support brackets					x													
Torque arm bracket					x							x		x				
Radius rod bracket					x									x				
Group 3 - Front Springs																		
Front springs						n	n											
Front spring shackle					x								x					
Front spring seat					x								x	x			n	
Front spring clip plate					x								x	x				

Table B-1. Guide for Welding Automotive Equipment (cont)

(See explanation of symbols at end of table)

Automotive Part	Usual Metal Composition											Recommended Welding Method						
	Gray cast iron	Malleable iron	Cast steel	Steel forgings	To 0.40 carbon steel	Over 0.40 carbon steel	Alloy steels	Aluminum	Brass, copper or bronze	Miscellaneous	Babbitt	Brazing	Welding with rod of similar composition	No. 1 HF	Soldering	Heating	Haynes stellite	Welding not recommended
DIVISION XVI – FRAME AND SPRINGS (cont)																		
Group 4 – Rear Springs																		
Rear springs						u	u											u
Rear spring pivot seat					x							x						
Rear spring double shackle					x									x				
(Other parts as for front spring)																		
DIVISION XVII – HOOD, FENDERS, AND SHIELDS																		
Group 1 – Hood Parts																		
Hood					x			x					x					
Hood sill					x							x						
Hood handle					x							x						
Hood fastener					x							x						
Hood fastener bracket			x									x						
Group 2 – Engine Shield Parts																		
Engine shield					x							x						
Engine shield bracket					x							x						
Group 3 – Fenders and Running Boards																		
Running boards					x								x					
Running board shields					x								x					
Fenders					x								x					
Fender support socket					x							x						
Fender supports					x							x						
DIVISION XVIII – BODY																		
Group 1 – Floorboards and Dash																		
Floorboards (metal parts)					x								x					
Dash					x								x					
Instrument board													x					
Group 2 – Body Parts (Metal)					x								x					

Table B-1. Guide for Welding Automotive Equipment (cont)

(See explanation of symbols at end of table)

Automotive Part	Usual Metal Composition										Recommended Welding Method							
	Gray cast iron	Malleable iron	Cast steel	Steel forgings	To 0.40 carbon steel	Over 0.40 carbon steel	Alloy steels	Aluminum	Brass, copper or bronze	Miscellaneous	Babbitt	Brazing	Welding with rod of similar composition	No. 1 HT	Soldering	Heating	Haynes stellite	Welding not recommended
DIVISION XVIII – BODY (cont)																		
All metal panels					x								x			x		
Body posts and braces					x								x			x		
Window frames					x								x			x		
Group 3 – Seat Frames					x								x			x		
DIVISION XIX – ACCESSORIES																		
Group 1 – Speedometer (and Parts)										n								n
Group 2 – Tire Pump Parts																		
Tire pump driving gear					x							x						
Tire pump shaft gear					x		x					x						
Tire pump idler gear					x		x					x						
Group 3 – Body Furnishings																		
Door and window handles					1		2		3	d,n			1,2,3					n
Bumpers						x	x					x						
Bumper brackets					x							x						

x – Indicates the metal composition and the recommended welding method.

1,2,3 – Indicates corresponding compositions and methods.

n – Welding not recommended. Minor areas may be built up if an "n" is placed in one of the welding method columns. Otherwise do not weld and do not build up.

1 – Lead.

d – Die cast metal.

p – Indicates corresponding method for composition other than "Gray cast iron" or "to 0.40 carbon steel."

Table B-2. Guide for Oxyacetylene Welding
(See footnotes at the end of the table)

Base Metal or Alloy	Welding Process[1]	Flame Adjustment	Welding Rod	Flux Required	Preheating Required
IRON					
1. Wrought iron	FW	Neutral	Low carbon or high strength steel	No	No
	B	Sl oxidizing	Bronze	Brazing	No
2. Low carbon iron	FW	Neutral	Low carbon steel	No	No
	B	Sl oxidizing	Bronze	Brazing	No
CARBON STEELS					
1. Low carbon (up to 0.30 percent C)	FW	Neutral	Low carbon steel	No	No
	B	Sl oxidizing	Bronze	Brazing	No
2. Medium carbon (0.30 to 0.55 percent C)	FW	Sl carburizing	Low carbon or high strength steel	No	300 to 500 °F (149 to 260 °C)
	B	Sl oxidizing	Bronze	Brazing	200 to 400 °F (93 to 204 °C)
3. High carbon (exceeding 0.55 percent C)	FW	Carburizing	Medium or high carbon steel	No	500 to 800 °F (260 to 427 °C)
	B	Sl oxidizing	Bronze	Brazing	300 to 500 °F (149 to 260 °C)
4. Tool steel (exceeding 0.83 percent C)	FW	Carburizing	Drill rod	Some cast iron flux	Up to 1000 °F (538 °C)
	B	Sl oxidizing	Bronze	Brazing	500 to 600 °F (260 to 316 °C)
CAST STEELS					
1. Plain carbon (Up to 0.25 percent C)	FW	Neutral	Low carbon	No	200 °F (93 °C)
	B	Sl oxidizing	Bronze	Brazing	200 °F (93 °C)
2. High manganese (12 percent Mn)	FW	Sl carburizing	Nickel-manganese steel	Wrap rod with Al wire	No
	B	Sl oxidizing	Bronze	Brazing	No
3. Other alloys	FW	Neutral to Sl carburizing	Same as base metal	No	In some cases
	B	Sl oxidizing	Bronze	Brazing	In some cases
CAST IRONS					
1. Gray cast iron	FW	Neutral	Cast iron	Cast iron flux	750 to 900 °F (399 to 482 °C)
	B	Sl oxidizing	Bronze	Brazing	Locally to 500 °F (260 °C)
2. Malleable iron	FW[2]	Neutral	White cast iron	Cast iron flux	750 to 900 °F (399 to 482 °C)
	B[3]	Sl oxidizing	Bronze	Brazing	Locally to 500 °F (260 °C)

Table B-2. Guide for Oxyacetylene Welding (cont)
(See footnotes at the end of the table)

Base Metal or Alloy	Welding Process[1]	Flame Adjustment	Welding Rod	Flux Required	Preheating Required
3. Alloy cast irons	FW	Neutral	Same as base metal, or cast iron	Cast iron flux	500 to 1000 °F (260 to 538 °C)
	B	Sl oxidizing	Bronze	Brazing	Locally to 500 °F (260 °C)
LOW ALLOY HIGH TENSILE STEELS (General)	FW	Neutral to sl carburizing	Same as base metal, or high strength steel	No	Yes
1. Nickel alloy steel (3 to 3-1/2 percent Ni) (Up to 0.25 percent C)	FW	Neutral to sl carburizing	Same as base metal, or high strength steel	No	No preheating, slow cool
(More than 0.25 percent C)	FW	Neutral to sl carburizing	Same as base metal, or high strength steel	No	300 to 600 °F (149 to 316 °C), slow cool
2. Nickel-copper alloy steels	FW	Neutral to sl carburizing	Same as base metal, or high strength steel	No	250 to 300 °F (121 to 149 °C)
3. Manganese-molybdenum alloy steels	FW	Neutral to sl carburizing	Carbon-molybdenum or high strength rod	No	250 to 300 °F (121 to 149 °C)
4. Carbon-molybdenum alloy steels (0.10 to 0.20 percent C)	FW	Neutral to sl carburizing	Carbon-molybdenum or high strength rod	No	300 to 400 °F (149 to 204 °C)
(0.20 to 0.30 percent C)	FW	Neutral to sl carburizing	Carbon-molybdenum or high strength rod	No	400 to 500 °F (204 to 260 °C), slow cool
5. Nickel-chromium alloy steels (up to 0.20 percent C)	FW	Neutral to sl carburizing	Same as base metal, or high strength rod	No	200 to 300 °F (93 to 149 °C), slow cool
6. Chrome-molybdenum alloy steels	FW	Neutral to sl carburizing	Same as base metal, or high strength rod	No	300 to 800 °F (149 to 427 °C), slow cool
7. Chromium alloy steels	FW	Neutral to sl carburizing	Same as base metal, or high strength rod	No	300 to 800 °F (149 to 427 °C)
8. Chromium-vanadium alloy steels	FW	Neutral to sl carburizing	Same as base metal, or high strength rod	No	200 to 800 °F (93 to 427 °C)

Table B-2. Guide for Oxyacetylene Welding (cont)
(See footnotes at the end of the table)

Base Metal or Alloy	Welding Process[1]	Flame Adjustment	Welding Rod	Flux Required	Preheating Required
9. Manganese alloy steels (1.6 percent-1.9 percent Mn)	FW	Neutral to sl carburizing	Same as base metal, or high strength rod	No	300 to 800 °F (149 to 427 °C)
STAINLESS STEELS					
1. Chromium alloys (12 percent to 28 percent Cr) (Stainless irons)	FW	Neutral	Same as base metal, or 18-8 stainless steel	Stainless	No
2. Chromium nickel alloys	FW	Neutral to sl carburizing	(18-8) stainless steel	Stainless	No
COPPER AND COPPER ALLOYS					
1. Deoxidized copper	FW	Sl oxidizing	Deoxidized copper	No	500 to 800 °F (260 to 427 °C)
	B	Sl oxidizing	Silver, copper-phosphorous, or copper-phosphorous-silver alloys	Brazing	400 to 600 °F (204 to 316 °C)
2. Commercial bronze and low brass	FW	Oxidizing	Same as base metal	Brazing	200 to 300 °F (93 to 149 °C)
	B	Sl oxidizing	Bronze	Brazing	200 to 300 °F (93 to 149 °C)
3. Spring, admiralty, and yellow brass	FW	Oxidizing	Same as base metal, or bronze	Brazing	200 to 300 °F (93 to 149 °C)
4. Muntz metal, Tobin bronze, naval brass, manganese bronze	FW	Oxidizing	Bronze	Brazing	200 to 300 °F (93 to 149 °C)
5. Nickel silver	FW	Neutral	Nickel silver	Brazing	200 to 300 °F (93 to 149 °C)
6. Phosphor bronze	FW	Neutral	Bronze	Brazing	300 to 500 °F (149 to 260 °C)
	B	Neutral or sl oxidizing	Bronze	Brazing	200 to 300 °F (93 to 149 °C)
7. Aluminum bronze	FW	Sl carburizing	Aluminum bronze	Brazing	200 to 300 °F (93 to 149 °C)
8. Beryllium copper	Oxyacetylene welding or brazing not recommended; use silver solder and flux.				
ALUMINUM AND ALUMINUM ALLOYS					
1. Pure aluminum (1100)	FW	Neutral	Pure aluminum	Aluminum	500 to 800 °F (260 to 427 °C)

Table B-2. Guide for Oxyacetylene Welding (cont)
(See footnotes at the end of the table)

Base Metal or Alloy	Welding Process[1]	Flame Adjustment	Welding Rod	Flux Required	Preheating Required
2. Aluminum alloys (General)	FW	Neutral	Same as base metal, or 95 percent aluminum-5 percent silicon	Aluminum	500 to 800 °F (260 to 427 °C)
3. Aluminum-manganese alloy (3003)	FW	Neutral	95 percent aluminum-5 percent silicon	Aluminum	500 to 800 °F (260 to 427 °C)
4. Aluminum-magnesium-chromium alloy (5052)	FW	Neutral	95 percent aluminum-5 percent silicon	Aluminum	500 to 800 °F (260 to 427 °C)
5. Aluminum-manganese-magnesium alloy (3004)	FW	Neutral	95 percent aluminum-5 percent silicon	Aluminum	500 to 800 °F (260 to 427 °C)
6. Aluminum-magnesium-silicon alloy (6151) (6053)	FW[4]	Neutral	95 percent aluminum-5 percent silicon	Aluminum	Up to 400 °F (204 °C)
7. Aluminum-copper-magnesium-manganese alloy (duraluminum) (2017) (2024)		Welding not recommended.			
8. Aluminum clad		Welding not recommended.			
NICKEL AND NICKEL ALLOYS					
1. Nickel	FW	Sl carburizing	Nickel	No	200 to 300 °F (93 to 149 °C)
	B	Sl oxidizing	Bronze	Brazing	200 to 300 °F (93 to 149 °C)
2. Monel (67 percent Ni-29 percent Cu)	FW	Sl carburizing	Monel	Brazing	200 to 300 °F (93 to 149 °C)
	B	Sl oxidizing	Bronze	Brazing	200 to 300 °F (93 to 149 °C)
3. Inconel (79 percent Ni-13 percent Cr-6 percent Fe)	FW	Sl carburizing	Inconel	Brazing	200 to 300 °F (93 to 149 °C)
	B	Sl oxidizing	Bronze	Brazing	200 to 300 °F (93 to 149 °C)
LEAD	FW	Neutral	Same as base metal	No	No
	FW	Neutral to Sl carburizing	Same as base metal	Special flux	500 to 650 °F (260 to 343 °C)
MAGNESIUM ALLOYS[5]					
WHITE METAL	FW	Carburizing	Same as base metal	No	No

In general, in welding low alloy, high tensile steels, it is recommended that the filler metal used should be of the same composition as the base metal to obtain good corrosion resistance at the welded joint.

In welding low alloy, high tensile steels in the heat treated condition, it is recommended that the filler metal used should be of the austenitic type, such as the 18 percent chromium-8 percent nickel stainless steel welding rod.

In all cases where the low alloy, high tensile steels are to be heat treated after welding, the filler metals used should be of the same composition as the base metal or other suitable high strength welding rod.

[1] In the welding process column, FW indicates fusion welding and B indicates brazing. In the flame adjustment column, Sl indicates "slightly."

[2] Welded as white cast iron only and should be followed by heat treatment to induce malleability. Fusion welding is not recommended for malleable iron.

[4] Brazing, rather than fusion welding, is the preferred method for repairing malleable iron.

[4] Heat treat (6151) and (6053) after welding. Properties of (2017) and (2024) alloys cannot be restored by heat treatment after welding.

[5] Welding is not recommended on some magnesium alloys because of their porous nature, and such welds are made only as emergency repairs until a replacement can be obtained.

Table B-3. Guide for Electric Arc Welding
(See footnotes at the end of table)

Base Metal or Alloy	Welding Process[1]	Polarity	Welding Electrode or Filler Metal		Preheating Required
			Material	Type	
IRON					
1. Wrought iron	MAW	Reverse	Mild steel	Shielded arc[2]	No
	CAW	Straight	Mild steel	Use a flux	No
	MAB	Reverse	Bronze	Shielded arc	No
2. Low carbon iron	MAW	Reverse	Mild steel	Shielded arc	No
CARBON STEELS					
1. Low carbon (Up to 0.30 percent C)	MAW	Straight	Mild steel	Bare or light coated	Up to 300 °F (149 °C)
	MAW	Reverse	Mild steel	Shielded arc	Up to 300 °F (149 °C)
	CAW	Straight	Mild steel	Use a flux	Up to 300 °F (149 °C)
	MAB	Reverse	Bronze	Shielded arc	Up to 300 °F (149 °C)
2. Medium carbon (0.30 to 0.55 percent C)	MAW	Reverse	25-20 or modified 18-8 stainless steel	Shielded arc	No
	MAW	Reverse	Mild steel or high strength steel	Shielded arc	300 to 500 °F (149 to 260 °C)
3. High carbon (0.55 to 0.83 percent C)	MAW[3]	Reverse	25-20 modified 18-8 stainless steel	Shielded arc	No
	MAW	Reverse	Mild or high strength steel	Shielded arc	500 to 800 °F (260 to 427 °C)
4. Tool steel (0.83 to 1.55 percent C)	MAW	Reverse	25-20 or modified 18-8 stainless steel	Shielded arc	Up to 800 °F (427 °C)
	MAW	Reverse	Mild or high strength steel	Shielded arc	Up to 1000 °F (538 °C)

Table B-3. Guide for Electric Arc Welding (cont)
(See footnotes at the end of table)

Base Metal or Alloy	Welding Process[1]	Polarity	Welding Electrode or Filler Metal		Preheating Required
			Material	Type	
CAST STEELS					
1. Plain carbon (Up to 0.25 percent C)	MAW	Reverse	Mild steel	Shielded arc	200 °F (93 °C)
2. High manganese (12 percent Mn)	MAB	Reverse	Bronze	Shielded arc	200 °F (93 °C)
	MAW	Reverse	Weld with 25-20 stainless steel and surface with nickel-manganese steel	Shielded arc	No
	To build up sections	Reverse	Nickel-manganese steel	Shielded arc	No preheating; quench and peen weld
3. Other alloys	MAW	Reverse	Mild steel	Shielded arc	In some cases
CAST IRONS					
1. Gray cast iron (Machinable welds)	MAW	Reverse	Cast iron or monel	Shielded arc	700 to 800 °F (371 to 427 °C), or no preheating but peen weld
(Nonmachinable welds)	MAW	Reverse	18-8 stainless steel or mild steel	Shielded arc	700 to 800 °F (371 to 427 °C)
	MAB	Reverse	Bronze	Shielded arc	Up to 500 °F (260 °C)
	CAB	Straight	Bronze	Shielded arc	Up to 500 °F (260 °C)
2. Malleable iron (Machinable welds)	MAW	Reverse	Cast iron or monel	Shielded arc	700 to 800 °F (371 to 427 °C), anneal weld
(Nonmachinable welds)	MAW	Reverse	18-8 stainless steel or mild steel	Shielded arc	700 to 800 °F (371 to 427 °C), anneal weld
	MAB	Reverse	Bronze	Shielded arc	Up to 500 °F (260 °C)
	CAB	Straight	Bronze	Shielded arc	Up to 500 °F (260 °C)
3. Alloy cast irons			(Same as gray cast iron)		
LOW ALLOY HIGH TENSILE STEELS (General)	MAW[4]	Reverse	Same as base metal; or high strength or mild steel, or 25-20 stainless steel	Shielded arc	Yes

Table B-3. Guide for Electric Arc Welding (cont)
(See footnotes at the end of table)

Base Metal or Alloy	Welding Process[1]	Polarity	Welding Electrode or Filler Metal		Preheating Required
			Material	Type	
1. Nickel alloy steel (3 to 3-1/2 percent Ni) (Up to 0.25 percent C)	MAW	Reverse	Nickel alloy or 25-20 stainless steel	Shielded arc	No preheating, slow cool
(More than 0.25 percent C)	MAW	Reverse	Nickel alloy or 25-20 stainless steel	Shielded arc	300 to 600 °F (149 to 316 °C)
2. Nickel-copper alloy steels	MAW	Reverse	Nickel alloy or 25-20 stainless steel	Shielded arc	250 to 300 °F (121 to 149 °C)
3. Manganese-molybdenum alloy steels	MAW	Reverse	Carbon-molybdenum or special electrode	Shielded arc	250 to 300 °F (121 to 149 °C)
4. Carbon-molybdenum alloy steels (0.10 to 0.20 percent C)	MAW	Straight or reverse	Carbon-molybdenum steel	Shielded arc	300 to 400 °F (149 to 204 °C)
(0.20 to 0.30 percent C)	MAW	Straight or reverse	Carbon-molybdenum steel	Shielded arc	400 to 500 °F (204 to 260 °C), slow cool
5. Nickel-chromium alloy steels (1 to 3-1/2 percent Ni) (Up to 0.20 percent C)	MAW	Reverse	Same as base metal, or 25-20 stainless steel	Shielded arc	200 to 300 °F (93 to 149 °C), slow cool
(0.20 to 0.55 percent C)	MAW	Reverse	Same as base metal, or 25-20 stainless steel	Shielded arc	600 to 800 °F (316 to 427 °C), slow cool
(High alloy content)	MAW	Reverse	Same as base metal, or 25-20 stainless steel	Shielded arc	900 to 1000 °F (482 to 538 °C), slow cool
6. Chrome-molybdenum alloy steels	MAW	Straight or reverse	Chrome-molybdenum or carbon-molybdenum steel	Shielded arc	300 to 800 °F (149 to 427 °C), slow cool
	CAW	Straight	Same as base metal	Use a flux	300 to 800 °F (149 to 427 °C), slow cool
7. Chromium alloy steels	MAW	Reverse	Same as base metal, or 25-20 or 18-8 stainless steel	Shielded arc	300 to 800 °F (149 to 427 °C), slow cool
8. Chromium-vanadium alloy steels	MAW	Reverse	Chrome-molybdenum or carbon-molybdenum steel	Shielded arc	200 to 800 °F (93 to 427 °C)

Table B-3. Guide for Electric Arc Welding (cont)
(See footnotes at the end of table)

Base Metal or Alloy	Welding Process[1]	Polarity	Welding Electrode or Filler Metal		Preheating Required
			Material	Type	
9. Manganese alloy steels (1.6 to 1.9 percent Mn)	MAW	Reverse	Carbon-molybdenum or mild steel	Shielded arc	300 to 800 °F (149 to 427 °C)
STAINLESS STEELS					
1. Chromium alloys (12 to 28 percent Cr) (Stainless irons)	MAW	Reverse	25-20 or columbium-bearing 18-8 stainless steel	Shielded arc	No
2. Chromium-nickel alloys	MAW	Reverse	25-20 or columbium-bearing 18-8 stainless steel	Shielded arc	No
COPPER AND COPPER ALLOYS					
1. Deoxidized copper	MAW	Reverse	Deoxidized copper, phosphor bronze, or silicon copper	Shielded arc	500 to 800 °F (260 to 427 °C)
	CAW	Straight	Deoxidized copper, phosphor bronze, or silicon copper	Use of flux optional	500 to 800 °F (260 to 427 °C)
2. Commercial bronze and low brass	MAW	Reverse	Phosphor bronze or silicon bronze	Shielded arc	200 to 300 °F (93 to 149 °C)
	CAW	Straight	Phosphor bronze or silicon bronze	Use a flux	200 to 300 °F (93 to 149 °C)
3. Spring, admiralty, and yellow brass	CAW	Straight	Phosphor bronze	Use a flux	200 to 300 °F (93 to 149 °C)
4. Muntz metal, Tobin bronze, naval bronze, manganese bronze	CAW	Straight	Phosphor bronze	Use a flux	200 to 300 °F (93 to 149 °C)
5. Nickel silver	MAW	Reverse	High nickel alloy, phosphor bronze, or silicon copper	Use a flux	300 to 500 °F (149 to 260 °C)
	CAW	Straight	High nickel alloy, phosphor bronze, or silicon copper	Use a flux	300 to 500 °F (149 to 260 °C)
6. Phosphor bronze	MAW	Reverse	Phosphor bronze	Shielded arc	200 to 300 °F (93 to 149 °C)
	CAW	Straight	Phosphor bronze	Use a flux	200 to 300 °F (93 to 149 °C)
7. Aluminum bronze	MAW	Reverse	Aluminum bronze or phosphor bronze	Shielded arc	200 to 300 °F (93 to 149 °C)
	CAW	Straight	Aluminum bronze or phosphor bronze	Use of flux optional	200 to 300 °F (93 to 149 °C)
8. Beryllium copper	CAW	Straight	Beryllium copper	Use of flux optional	500 to 700 °F (260 to 371 °C)

Table B-3. Guide for Electric Arc Welding (cont)
(See footnotes at the end of table)

Base Metal or Alloy	Welding Process[1]	Polarity	Welding Electrode or Filler Metal		Preheating Required
			Material	Type	
ALUMINUM AND ALUMINUM ALLOYS					
1. Pure aluminum (1100)	MAW	Reverse	Pure aluminum or 95 percent aluminum-5 percent silicon	Shielded arc	500 to 800 °F (260 to 427 °C)
	CAW	Straight	Pure aluminum or 95 percent aluminum-5 percent silicon	Flux-coated welding rod	500 to 800 °F (260 to 427 °C)
2. Aluminum alloys (General)	MAW	Reverse	95 percent aluminum-5 percent silicon	Shielded arc	500 to 800 °F (260 to 427 °C)
	CAW	Straight	95 percent aluminum-5 percent silicon	Flux-coated welding rod	500 to 800 °F (260 to 427 °C)
3. Aluminum-manganese alloy (3003)	MAW	Reverse	95 percent aluminum-5 percent silicon	Shielded arc	500 to 800 °F (260 to 427 °C)
	CAW	Straight	95 percent aluminum-5 percent silicon	Flux-coated welding rod	500 to 800 °F (260 to 427 °C)
4. Aluminum-magnesium-chromium alloy (5052)	MAW	Reverse	95 percent aluminum-5 percent silicon	Shielded arc	500 to 800 °F (260 to 427 °C)
	CAW	Straight	95 percent aluminum-5 percent silicon	Flux-coated welding rod	500 to 800 °F (260 to 427 °C)
5. Aluminum-magnesium-manganese alloy (3004)	MAW	Reverse	95 percent aluminum-5 percent silicon	Shielded arc	500 to 800 °F (260 to 427 °C)
	CAW	Straight	95 percent aluminum-5 percent silicon	Flux-coated welding rod	500 to 800 °F (260 to 427 °C)
6. Aluminum-silicon-magnesium alloys (6151) (6053)	MAW	Reverse	95 percent aluminum-5 percent silicon	Shielded arc	Up to 400 °F (204 °C)
7. Aluminum-copper-magnesium-manganese alloys — Duraluminum (2017) (2024)	CAW	Straight	95 percent aluminum-5 percent silicon	Flux-coated welding rod	Up to 400 °F (204 °C)
8. Aluminum clad	Arc welding not recommended				
NICKEL AND NICKEL ALLOYS					
1. Nickel	MAW	Reverse	Nickel	Shielded arc	200 to 300 °F (93 to 149 °C)
	CAW	Straight	Nickel	Lightly flux-coated welding rod	200 to 300 °F (93 to 149 °C)
2. Monel (67 percent Ni-29 percent Cu)	MAW	Reverse	Monel	Shielded arc	200 to 300 °F (93 to 149 °C)
	CAW	Straight	Monel	Lightly flux-coated welding rod	200 to 300 °F (93 to 149 °C)

Table B-3. Guide for Electric Arc Welding (cont)
(See footnotes at the end of table)

Base Metal or Alloy	Welding Process[1]	Polarity	Welding Electrode or Filler Metal		Preheating Required
			Material	Type	
3. Inconel (79 percent Ni-13 percent Cr-6 percent Fe)	MAW	Reverse	Same as base metal	Shielded arc	200 to 300 °F (93 to 149 °C)
LEAD			Lead cannot be arc welded		
MAGNESIUM ALLOYS	MAW	Reverse	Tungsten	Shielded arc	No
	MAW	Reverse	Magnesium	Shielded arc	No

[1]In the welding process column, MAW indicates metal-arc welding, CAW indicates carbon-arc welding, MAB indicates metal-arc brazing, and CAB indicates carbon-arc brazing.

[2]Shielded arc electrodes are heavy-coated and usually require reverse polarity; however, manufacturer's recommendations specify the preferred polarity for special electrodes, which may differ from the polarity recommended above in some cases.

[3]Stress relieve by heating to between 1200 and 1450 °F (649 and 788 °C), for 1 hour per inch of thickness and cooling slowly.

[4]A large number and variety of low alloy high tensile steels are used in ordnance construction. In arc welding these steels, certain special precautions are required, such as preheating before welding, use of special electrodes, and a postheating treatment. In general, where good corrosion resistance is required or when the welded joint is to be heat treated after welding, electrodes having the same composition or properties as the base metal are used. Where these steels are in the heat treated condition, it is recommended that the filler metal used should be of the austenitic type, such as 25 percent chromium-12 percent nickel, 25 percent chromium-20 percent nickel or 18 percent chromium-8 percent nickel stainless steel, in order to obtain good weld metal properties. Some of these stainless steel electrodes have columbium or other alloying elements added to retain their properties after welding. An example of this is the so-called modified 18-8 stainless steel electrode, which contains small percentages of either manganese or molybdenum. This electrode may be used in place of the 25-20 type of electrode in any of the welding processes for which 25-20 electrodes are specified. Usually no preheating is required in welding with these electrodes.

APPENDIX C

TROUBLESHOOTING PROCEDURES

MALFUNCTION
    TEST OR INSPECTION
        CORRECTIVE ACTION

OXYACETYLENE WELDING

1. **DISTORTION** (fig. C-1, p C-18)

Step 1. Check to see whether shrinkage of deposited metal has pulled welded parts together.

    a. Properly clamp or tack weld parts to resist shrinkage.

    b. Separate or preform parts sufficiently to allow for shrinkage of welds.

    c. Peen the deposited metal while still hot.

Step 2. Check for uniform heating of parts during welding.

    a. Support parts of structure to be welded to prevent buckling in heated sections due to weight of parts themselves.

    b. Preheating is desirable in some heavy structures.

    c. Removal of rolling or forming strain before welding is sometimes helpful.

Step 3. Check for proper welding sequence.

    a. Study the structure and develop a definite sequence of welding.

    b. Distribute welding to prevent excessive local heating.

2. **WELDING STRESSES**

Step 1. Check the joint design for excessive rigidity.

    a. Slight movement of parts during welding will reduce welding stresses.

    b. Develop a welding procedure that permits all parts to be free to move as long as possible.

MALFUNCTION
  TEST OR INSPECTION
    CORRECTIVE ACTION

OXYACETYLENE WELDING (cont)

2. WELDING STRESSES (cont)

   Step 2. Check for proprer welding procedure.

      a. Make weld in as few passes as practical.

      b. Use special intermittent or alternating welding sequence and backstep or skip welding procedure.

      c. Properly clamp parts adjacent to the joint. Use backup fixtures to cool parts rapidly.

   Step 3. If no improper conditions exist, stresses could merely be those inherent in any weld, especially in heavy parts.

      Peen each deposit of weld metal. Stress relieve finished product at 1100 to 1250°F (593 to 677°C) 1 hour per 1.0 in. (25.4 cm) of thickness.

3. WARPING OF THIN PLATES (fig. C-2, p C-18)

   Step 1. Check for shrinkage of deposited weld metal.

      Distribute heat input more evenly over full length of seam.

   Step 2. Check for excessive local heating at the joint.

      Weld rapidly with a minimum heat input to prevent excessive local heating of the plates adjacent to the weld.

   Step 3. Check for proper preparation of the joint.

      a. Do not have excessive space between the part to be welded. Prepare thin plate edges with flanged joints, making offset approximately equal to the thickness of the plates. No filler rod is necessary for this type of joint.

      b. Fabricate a U-shaped corrugation in the plates parallel to and approximately 1/2 in. (12.7 mm) away from the seam. This will serve as an expansion joint to take up movement during and after the welding operation.

MALFUNCTION		
	TEST OR INSPECTION	
		CORRECTIVE ACTION

Step 4. Check for proper welding procedure.

    a. Use special welding sequence and backstep or skip Procedure.

    b. Preheat material to relieve stress.

Step 5. Check for properclamping of parts.

    Properly clamp parts adjacento the joint. Use backup fixtures to cool parts rapidly.

4. POOR WELD APPEARANCE (fig. C-3, p C-19)

Step 1. Check the welding technique, flame adjustment, and welding rod manipulation.

    a. Ensure the use of the proper welding technique for the welding rod used.

    b. Donot use excessive heat.

    c. Use a uniform weave and welding speed at all times.

Step 2. Check the welding rod used, as the poor appearance may be due to the inherent characteristics of the particular rod.

    Use a welding rod designed for the type of weld being made.

Step 3. Check for proper joint preparation.

    Prepare all joints properly.

5. CRACKED WELDS (fig. C-4, p C-19)

Step 1. Check the joint design for excessive rigidity.

    Redesign the structure or modify the welding procedure in order to eliminate rigid joints.

Step 2. Check to see if the welds are too small for the size of the parts joined.

    Do not use too small a weldbetween heavy plates. Increase the size of welds by addingmore filler metal.

TROUBLESHOOTING PROCEDURES (cont)

MALFUNCTION
      TEST OR INSPECTION
            CORRECITVE ACTION

OXYACETYLENE WELDING (cont)

5. CRACKED WELDS (cont)

    Step 3. Check for proper welding procedure.

        a. Do not make welds in string beads. Deposit weld metal full size in short sections 8.0 to 10.0 in. (203.2 to 254.0 mm) long. (This is called block sequence.)

        b. Welding sequence should be such as to leave ends free to move as long as possible.

        c. Preheating parts to be welded sometimes helps to reduce high contraction stresses caused by localized high temperatures.

    Step 4. Check for poor welds.

        Make sure welds are sound and the fusion is good.

    Step 5. Check for proper preparation of joints.

        Prepare joints with a uniform and proper free space. In some cases a free space is essential. In other cases a shrink or press fit may be required.

6. UNDERCUT

    Step 1. Check for excessive weaving of the bead, improper tip size, and insufficient welding rod added to molten puddle.

        a. Modify welding procedure to balance weave of bead and rate of welding rod deposition, using proper tip size.

        b. Do not use too small a welding rod.

    Step 2. Check for proper manipulation of the welding.

        a. Avoid excessive and nonuniform weaving.

        b. A uniform weave with unvarying heat input will aid greatly in preventing undercut in butt welds.

Step 3. Check for proper welding technique- improper welding rod deposition with nonuniform heating.

Do not hold welding rod too near the lower edge of the vertical plate when making a horizontal fillet weld, as undercut on the vertical plate will result.

7. INCOMPLETE PENETRATION (fig. C-5, p C-19)

Step 1. Check for proper preparation of joint.

a. Be sure to allow the proper free space at the bottom of the weld.

b. Deposit a layer of weld metal on the back side of the joint, where accessible, to ensure complete fusion at the root of the joint.

Step 2. Check the size of the welding rod used.

a. Select proper sized welding rod to obtain a balance in the heat requirements for melting welding rod, breaking down side walls, and maintaining the puddle of molten metal at the desired size.

b. Use small diameter welding rods in a narrow welding groove.

Step 3. Check to see if welding tip is too small, resulting in insufficient heat input.

Use sufficient heat input to obtain proper penetration for the plate thickness being welded.

Step 4. Check for an excessive welding speed.

Welding speed should be slow enough to allow welding heat to penetrate to the bottom of the joint.

8. POROUS WELDS (fig. C-6, p C-20)

Step 1. Check the inherent properties of the particular type of welding rod.

Use welding rod of proper chemical analysis.

TROUBLESHOOTING PROCEDURES (cont)

MALFUNCTION		
	TEST OR INSPECTION	
		CORRECTIVE ACTION

OXYACETYLENE WELDING (cont)

8. POROUS WELDS (cont)

Step 2. Check the welding procedure and flame adjustment.

a. Avoid overheating molten puddle of weld metal.

b. Use the proper flame adjustment and flux, if necessary, to ensure sound welds.

Step 3. Check to see if puddling time is sufficient to allow entrapped gas, oxides, and slag inclusions to escape to the surface.

a. Use the multilayer welding technique to avoid carrying too large a molten puddle of weld metal.

b. Puddling keeps the weld metal longer and often ensures sounder welds.

Step 4. Check for poor base metal.

Modify the normal welding procedure to weld poor base metals of a given type.

9. BRITTLE WELDS

Step 1. Check for unsatisfactory welding rod, producing air-hardening weld metal.

Avoid welding rods producing air-hardening weld metal where ductility is desired. High tensile strength, low alloy steel rods are air-hardened and require proper base metal preheating, postheating, or both to avoid cracking due to brittleness.

Step 2. Check for excessive heat input from oversized welding tip, causing coarse-grained and burnt metal.

Do not use excessive heat input, as this may cause coarse grain structure and oxide inclusions in weld metal deposits.

MALFUNCTION
    TEST OR INSPECTION
        CORRECTIVE ACTION

Step 3. Check for high carbon or alloy base metal which has not been taken into consideration.

Welds may absorb alloy elements from the patent metal and become hard. Do not weld a steel unless the conposition and characteristics are known.

Step 4. Check for proper flame adjustment and welding procedure.

a. Adjust the flare so that the molten metal does not boil, foam, or spark.

b. A single pass weld maybe more brittle than multilayer weld, because it has not been refined by successive layers of weld metal.

10. POOR FUSION (fig. C-7, p C-20)

Step 1. Check the welding rod size.

When welding in narrow grooves, use a welding rod small enough to reach the bottom.

Step 2. Check the tip size and heat input.

Use sufficient heat to melt welding rod and to break down sidewalls of plate edges.

Step 3. Check the welding technique.

Be sure the weave is wide enough to melt the sides of the joint thoroughly.

Step 4. Check for proper preparation of the joint.

The deposited metal should completely fuse with the side walls of the plate to form a consolidated joint of base and weld metal.

11. CORROSION

Step 1. Check the type of welding rod used.

Select welding rods with the proper corrosion resistance properties which are not changed by the welding process.

TROUBLESHOOTING PROCUDURE(cont)

MALFUNCTION
    TEST OR INSPECTION
        CORRECTIVE ACTION

OXYACETYLENE WELDING (cont)

11. CORROSION (cont)

Step 2. Check whether the weld deposit is proper for the corrosive fluid or atmosphere.

 a. Use the proper flux on both parent metal and welding rod to produce welds with the desired corrosion resistance.

 b. DO not expect more from the weld than from the parent metal. On stainless steels, use welding rods that are equal to or better than the base metal in corrosion resistance.

 c. For best corrosion resistance, use a filler rod whose composition is the same as the base metal.

Step 3. Check the metallurgical effect of welding.

When welding 18-8 austenitic stainless steel, be sure the analysis of the steel and the welding procedure are correct, so that the welding process does not cause carbide precipitation. This condition can be corrected by annealing at 1900 to 2100 °F (1038 to 1149 °C).

Step 4. Check for proper cleaning of weld.

Certain materials such as aluminum require special procedures for thorough cleaning of all slag to prevent corrosion.

12. BRITTLE JOINTS

Step 1. Check base metal for air hardening characteristics.

In welding on medium carbon steel or certain alloy steels, the fusion zone may be hard as the result of rapid cooling. Preheating at 300 to 500 °F (149 to 260 °C) should be resorted to before welding.

Step 2. Check welding procedure.

Multilayer welds will tend to anneal hard zones. Stress-relieving at 1000 to 1250 °F (538 to 677 °C) after welding will generally reduce hard areas formed during welding.

MALFUNCTION		
	TEST OR INSPECTION	
		CORRECTIVE ACTION

Step 3. Check type of welding rod used.

The use of austenitic welding rods will often work on special steels, but the fusion zone will generally contain an alloy which is hard.

ARC WELDING

13. DISTORTION (fig. C-1, p C-18)

Step 1. Check for shrinkage of deposited metal.

    a. Properly tack weld or clamp parts to resist shrinkage.

    b. Separate or preform parts so as to allow for shrinkage of welds.

    c. Peen the deposited metal while still hot.

Step 2. Check for uniform heating of parts.

    a. Preheating is desirable in some heavy structures.

    b. Removal of rolling or forming strain by stress relieving before welding is sometimes helpful.

Step 3. Check the welding sequence.

    a. Study structure and develop a definite sequence of welding.

    b. Distribute welding to prevent excessive local heating.

14. WELDING STRESSES

Step 1. Check for excessive rigidity of joints.

    a. Slight movement of parts during welding will reduce welding stresses.

    b. Develop a welding procedure that permits all parts to be free to move as long as possible.

MALFUNCTION
    TEST OR INSPECTION
        CORRECTIVE ACTION

ARC WELDING (cont)

14. WELDING STRESSES (cont)

Step 2. Check the welding procedure.

a. Make weld in as few passes as practical.

b. Use special intermittent or alternating welding sequence and backstep or skip procedures.

c. Properly clamp parts adjacent to the joint. Use backup fixtures to cool parts rapidly.

Step 3. If no improper conditions exist, stresses could merely be those inherent in any weld, especially in heavy parts.

a. Peen each deposit of weld metal.

b. Stress relieve finished product at 1100 to 1250°F (593 to 677 °C) 1 hour per 1.0 in. (25.4 cm) of thickness.

15. WARPING OF THIN PLATES (fig. C-2, p C-18)

Step 1. Check for shrinkage of deposited weld metal.

Select electrode with high welding speed and moderate penetrating properties.

Step 2. Check for excessive local heating at the joint.

Weld rapidly to prevent excessive local heating of the plates adjacent to the weld.

Step 3. Check for proper preparation of joint.

a. Do not have excessive root opening in the joint between the parts to be welded.

b. Hammer joint edges thinner than the rest of the plates before welding. This elongates the edges and the weld shrinkage causes them to pull back to the original shape.

MALFUNCTION
TEST OR INSPECTION
CORRECTIVE ACTION

Step 4. Check the welding procedure.

a. Use special intermittent or alternating welding sequence and backstep or skip procedure.

b. Preheat material to achieve stress.

Step 5. Check the clamping of parts.

Properly clamp parts adjacento the joint. Use backup fixtures to cool parts rapidly.

16. POOR WELD APPEARANCE(fig. C-3, p C-19)

Step 1. Check welding technique for propecurrent and electrode manipulation.

a. Ensure the use of the proper welding technique for the electrode used.

b. Do not use excessive welding current.

c. Use a uniform weave or rate of travel at all times.

Step 2. Check characteristics of type of electrode used.

Use an electrode designed for the type of weld and base metal and the position in which the weld is to be made.

Step 3. Check welding position for which electrode is designed.

Do not make fillet welds with downhandflat position) electrodes unless the parts are positionedroperly.

Step 4. Check for proper joint preparation.

Prepare all joints properly.

17. CRACKED WELDS (fig. C-4, p C-19)

Step 1. Check for excessive rigidity of joint.

Redesign the structure and modifythe welding procedure in order toeliminate rigid joints.

MALFUNCTION
    TEST OR INSPECTION
        CORRECTIVE ACTION

ARC WELDING (cont)

17. CRACKED WELDS (cont)

Step 2. Check to see if the welds are too small for the size of the parts joined.

> DO not use too small a weld between heavy plates. Increase the size of welds by adding more filler metal.

Step 3. Check the welding procedure.

a. Do not make welds in string beads. Deposit weld metal full size in short sections 8.0 to 10.0 in. (203.2 to 254.0 mm) long. (This is called block sequence.)

b. Welding sequence should be such as to leave ends free to move as long as possible.

c. Preheating parts to be welded sometimes helps to reduce high contraction stresses caused by localized high temperature.

d. Fill all craters at the end of the weld pass by moving the electrode back over the finished weld for a short distance equal to the length of the crater.

Step 4. Check for poor welds.

> Make sure welds are sound and the fusion is good. Be sure arc length and polarity are correct.

Step 5. Check for proper preparation of joints.

> Prepare joints with a uniform and proper root opening. In some cases, a root opening is essential. In other cases, a shrink or press fit may be required.

18. UNDERCUT

Step 1. Check the welding current setting.

> Use a moderate welding sent and do not try to weld at too high a speed.

MALFUNCTION
    TEST OR INSPECTION
        CORRECTIVE ACTION

Step 2. Check for proper manipulation of the electrode.

    a. Do not use too large an electrode. If the puddle of molten metal becomes too large, undercut may result.

    b. Excessive width of weave will cause undercut and should not be used. A uniform weave, not over three times the electrode diameter, will aid greatly in preventing undercut in butt welds.

    c. If an electrode is held to near the vertical plate in making a horizontal fillet weld, undercut on the vertical plate will result.

19. POOR PENETRATION (fig. C-5, p C-19)

Step 1. Check to see if the electrode is designed for the welding position being used.

    a. Electrodes should be used for welding in the position for which they were designed.

    b. Be sure to allow the proper root openings at the bottom of a weld.

    c. Use a backup bar if possible.

    d. Chip or cut out the back of the joint and deposit a bead of weld metal at this point.

Step 2. Check size of electrode used.

    a. Do not expect excessive penetration from an electrode.

    b. Use small diameter electrodes in a narrow welding groove.

Step 3. Check the welding current setting.

Use sufficient welding current to obtain proper penetration. Do not weld too rapidly.

Step 4. Check the welding speed.

Control the welding speed to penetrate to the bottom of the welded joint.

TROUBLESHOOT PROCEDURES (cont)

MALFUNCTION
    TEST OR INSPECTION
        CORRECTIVE ACTION

ARC WELDING (cont)

20. POROUS WELDS (fig. C-6, p C-20)

    Step 1. Check the properties of the electrode used.

        Some electrodes inherently produce sounder welds than others. Be sure that proper electrodes are used.

    Step 2. Check welding procedure and current setting.

        A weld made of a series of string beads may contain small pinholes. Weaving will often eliminate this trouble.

    Step 3. Check puddling time to see whether it is sufficient to allow entrapped gas to escape.

        Puddling keeps the weld metal molten longer and often insures sounder welds.

    Step 4. Check for dirty base metal.

        In some cases, the base metal may be at fault. Check this for segregations and impurities.

21. BRITTLE WELDS

    Step 1. Check the type of electrode used.

        Bare electrodes produce brittle welds. Shielded arc electrodes must be used if ductile welds are required.

    Step 2. Check the welding current setting.

        Do not use excessive welding current, as this may cause coarse-grained structure and oxidized deposits.

Step 3. Check for high carbon or alloy base metal which has not been taken into consideration.

   a. A single pass weld may be more brittle than a multilayer weld because its microstructure has not been refined by successive layers of weld metal.

   b. Welds may absorb alloy elements from the parent metal and become hard.

   c. Do not weld a metalunless the composition and characteristics are known.

22. POOR FUSION (fig. C-7, p C-20)

Step 1. Check diameter of electrode.

   When welding in narrow groove joints use an electrode small enough to properly reach the bottom of the joint.

Step 2. Check the welding current setting.

   a. Use sufficient welding current to deposit the metal and penetrate into the plates.

   b. Heavier plates require higher current for a given electrode than light plates.

Step 3. Check the welding technique.

   Be sure the weave is wide enough to melt the sidewalls of the joint thoroughly.

Step 4. Check the preparation of the joint.

   The deposited metal should fuse with the base metal and not curl away from it or merely adhere to it.

MALFUNCTION
     TEST OR INSPECTION
          CORRECTIVE ACTION

ARC WELDING (cont)

23. CORROSION

    Step 1. Check the type of electrode used.

          a. Bare electrodes produce welds that are less resistant to corrosion than the parent metal.

          b. Shield arc electrodes produce welds that are more resistant to corrosion than the parent metal.

          c. For the best corrosion resistance, use a filler rod whose composition is similar to that of the base metal.

    Step 2. Check to see if the weld metal deposited is proper for the corrosive fluid or atmosphere to be encountered.

          Do not expect more from the weld than you do from the parent metal. On stainless steels, use electrodes that are equal to or better than the parent metal in corrosion Resistance.

    Step 3. Check on the metallurgical effect of the welding.

          When welding 18-8 austenitic stainless steel, be sure the analysis of the steel and welding procedure is correct, so that the welding does not cause carbide precipitations. Carbide precipitation is the rising of carbon to the surface of the weld zone. This condition can be corrected by annealing at 1900 to 2100 °F (1038 to 1149 °C) after welding. By doing this corrosion in the form of iron oxide, or rust, can be eliminated.

    Step 4. Check for proper cleaning of the weld.

          Certain materials, such as aluminum require careful cleaning of all slag after welding to prevent corrosion in service.

24. BRITTLE JOINTS

    Step 1. Check for air hardening of the base metal.

          In medium carbon steel or certain alloy steals, the heat affected zone may be hard as a result of rapid cooling. Preheating at 300 to 500 °F (149 to 260 °C) should be resorted to before welding.

MALFUNCTION
  TEST OR INSPECTION
    CORRECTIVE ACTION

Step 2. Check the welding procedure.

    a. Multilayer welds will tend to anneal hard heat affected zones.

    b. Stress relieving at 1100 to 1250 °F (593 to 677 °C) after welding will generally reduce hard areas formed during welding.

Step 3. Check the type of electrode used.

The use of austenitic electrodes will often be successful on special steels, but the heat-affected zone will generally contain an alloy which is hard.

25. MAGNETIC BLOW

Step 1. Check for deflection of the arc from its normal path, particularly at the ends of joints and in corners.

    a. Make sure the ground is properly located on the work. Placing the ground in the direction of the arc deflection is often helpful.

    b. Separating the ground into two or more parts is helpful.

    c. Weld toward the direction in which the arc blows.

    d. Hold a short arc.

    e. Changing the angle of the electrode relative to the work may help to stabilize the arc.

    f. Magnetic blow is held to a minimum in alternating current welding.

26. SPATTER

Step 1. Check the properties of the electrode used.

Select the proper type of electrode.

Step 2. Check to see if the welding current is excessive for the type and diameter of electrode used.

Use a short arc but do not use excessive welding current.

TROUBLESHOOTING PROCEDURES (cont)

MALFUNCTION
      TEST OR INSPECTION
            CORRECTIVE ACTION

ARC WELDING (cont)

26.   SPATTER (cont)

     Step 3. Check for spalls.

             a. Paint parts adjacent to welds with whitewash or other protective coating. This prevents spalls from welding to parts, and they can be easily removed.

             b. Coated electrodes produce larger spalls than bare electrodes.

WHY:
1. Overheating at joint
2. Welding too slow
3. Rod too small
4. Improper sequence

CORRECTION:
1. Allow each bead to cool
2. Weld at constant speed--use speed tip
3. Use larger sized or triangular shaped rod
4. Offset pieces before welding

Figure C-1. Distortion.

WHY:
1. Shrinkage of material
2. Overheating
3. Faulty preparation
4. Faulty clamping of parts

CORRECTION:
1. Preheat material to relieve stress
2. Weld rapidly--use back-up weld
3. Too much root gap
4. Clamp parts properly--back-up to cool
5. For multilayer welds--allow time for each bead to cool

Figure C-2. Warping.

WHY:
1. Uneven pressure
2. Excessive stretching
3. Uneven heating

CORRECTION:
1. Practice starting, stopping, and finger manipulation on rod
2. Hold rod at proper angle
3. Use slow uniform fanning motion, heat both rod and material

(For speedwelding: use only moderate pressure, constant speed, keep shoe free of residue)

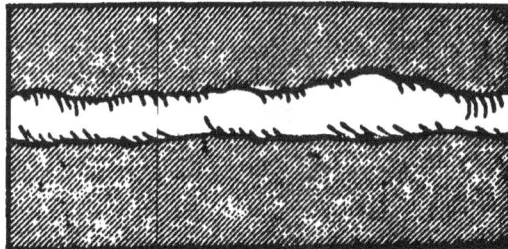

Figure C-3. Poor appearance.

WHY:
1. Improper welding temperature
2. Undue stress on weld
3. Chemical attack
4. Rod and base material not same composition
5. Oxidation or degradation of weld

CORRECTION:
1. Use recommended welding temperature
2. Allow for expansion and contraction
3. Stay within known chemical resistance and working temperatures of material
4. Use similar materials and inert gas for welding.
5. Refer to recommended application

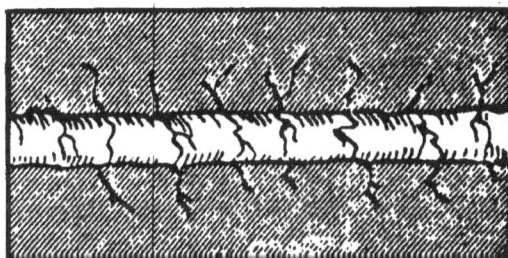

Figure C-4. Stress cracking.

WHY:
1. Faulty preparation
2. Rod too large
3. Welding too fast
4. Not enough root gap

CORRECTION:
1. Use 60 degree bevel
2. Use small rod at root
3. Check for flow lines while welding
4. Use tacking tip or leave 1/32-in. root gap and clamp pieces

Figure C-5. Poor penetration.

WHY:
1. Porous weld rod
2. Balance of heat on rod
3. Welding too fast
4. Rod too large
5. Improper starts or stops
6. Improper crossing of beads
7. Stretching rod

CORRECTION:
1. Inspect rod
2. Use proper fanning motion
3. Check welding temperature
4. Weld beads in proper sequence
5. Cut rod at angle, but cool before releasing
6. Stagger starts and overlap splices 1/2 in.

Figure C-6. Porous weld.

WHY:
1. Faulty preparation
2. Improper welding techniques
3. Wrong speed
4. Improper choice of rod size
5. Wrong temperature

CORRECTION:
1. Clean materials before welding
2. Keep pressure and fanning motion con-constant
3. Take more time by welding at lower temperatures
4. Use small rod at root and large rods at top—practice proper sequence
5. Preheat materials when necessary
6. Clamp parts securely

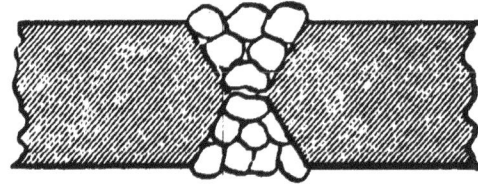

Figure C-7. Poor fusion.

APPENDIX D

MATERIALS USED FOR BRAZING, WELDING, SOLDERING
CUTTING, AND METALLIZING

D-1. GENERAL

This appendix contains listings of common welding equipment and materials used in connection with the equipment to perform welding operations. These lists are published to inform using personnel of those materials available for brazing, welding, soldering, cutting, and metallizing. These materials are used to repair, rebuild, and/or produce item requiring welding procedures.

D-2. SCOPE

The data provided in this appendix is for information and guidance. The listings contained herein include descriptions, identifying references, and specific use of common welding materials available in the Army supply system.

Table D-1. Common Welding Equipment
By Commercial and Government Entity Code (CAGEC)

CAGEC	Equipment
3436	ALIGNMENT TOOL, WELDING, PIPE
6830	ACETYLENE, TECHNICAL
8415	APRON, WELDER'S
6830	ARGON, TECHNICAL
3439	BAG, WELDING ROD
3439	BLOCK, CARBON (CARBON BLOCK)
3431	BONDING MACHINE, METALLIZING
3439	BRAZING ALLOY
3433	BRAZING & SOLDERING SET
7920	BRUSH, WIRE, SCRATCH
6151	CABLE ASSY, POWER, ELECTRICAL
3439	CARBON BLOCK
3439	CARBON PASTE
3439	CARBON ROD
3431	CHEST, WELDING
3436	CLAMP, PIPE WELDING (ALIGNMENT TOOL, WELDING, PIPE)
3439	CLEANER SET, WELDING & CUTTING TIPS
4940	CRANKSHAFT RECONDITIONING OUTFIT
3433	CUTTING ATTACHMENT, WELDING TORCH
3433	CUTTING MACHINE, OXYGEN
3439	DESOLDERING & RESOLDERING SET
3439	ELECTRODE, CHAMFERING
3439	ELECTRODE, CUTTING
3439	ELECTRODE, HEATING
3439	ELECTRODE, OVERLAY
3439	ELECTRODE, WELDING
5120	FLINT TIP, FRICTION IGNITER

Table D-1. Common Welding Equipment
By Commercial and Government Entity Code (CAGEC) (cont)

CAGEC	Equipment
5120	FRICTION IGNITER (IGNITER, FRICTION)
3439	FLUX (for brazing, soldering, welding)
8415	GLOVES (cloth or leather)
4240	GOGGLES, INDUSTRIAL
5120	HAMMER, WELDER'S
6830	HELIUM, TECHNICAL
4240	HELMET, WELDER'S
3439	HOLDER, ELECTRODE, WELDING
5120	IGNITER, FRICTION
6150	LEAD, ELECTRICAL
4240	LENS, GOGGLES, INDUSTRIAL
4240	LENS, HELMET, WELDER'S
3432	MANIFOLD, GAS CYLINDER
3433	METALLIZING GUN (SPRAY GUN, METALLIZING)
3431	METALLIZING MACHINE (BONDING MACHINE)
3433	METALLIZING OUTFIT
8415	MITTENS (cloth or leather)
3439	MOISTURE STABILIZER, WELDING ELECTRODE
6830	NITROGEN, TECHNICAL
3439	OVEN, WELDING ELECTRODES (MOISTURE STABILIZER)
6830	OXYGEN, TECHNICAL
3439	PASTE, CARBON (CARBON PASTE)
3439	REEL, WIRE, METALLIZING
....	REGULATOR, COMPRESSED GAS
....	REGULATOR, FLUID PRESSURE
....	REGULATOR, ARGON-HELIUM-NITROGEN-ETC (See VALVE, REGULATING, FLUID PRESSURE)
3433	REGULATOR & FILTER UNIT, AIR CONTROL (For use with metallizing gun)
3439	ROD, WELDING
3431	SCREEN, WELDING
4240	SHIELD, ARC VIEWING, HAND HELD
8415	SLEEVE, WELDER'S
3439	SOLDER
3433	SPRAY GUN, METALLIZING
3439	TAPE, WELD BACKUP
3433	TORCH, ARC-OXYGEN CUTTING
3431	TORCH, ARC WELDING, GAS SHIELDED (TIG TORCH set)
3433	TORCH, CUTTING
3433	TORCH, WELDING
3433	TORCH OUTFIT, AIR-ARC CUTTING
3433	TORCH OUTFIT, CUTTING-WELDING
3433	TORCH SET, CUTTING-WELDING

Table D-1. Common Welding Equipment
By Commercial and Government Entity Code (CAGEC) (cont)

CAGEC	Equipment
3433	TORCH OUTFIT, SOLDERING & HEATING
....	VALVE, REGULATING, FLUID PRESSURE (REGULATORS and VALVES are under the following Federal stock classes: 3431, 3432, 3433, 6685, and 6920)
3431	WELDING MACHINE, ARC
3432	WELDING MACHINE, RESISTANCE
3431	WELDING SET, METAL-ARC, GAS SHIELDED (MIG Gun set)
3439	WIRE, SPRAY GUN, METALLIZING
5120	WRENCH, TORCH & REGULATOR

Table D-2. Metallizing Wire

Wire material	Dia (inch)	Coil weight (pounds)	(CAGEC 3439) NIIN	Identifying Reference	Use
18% Cr, 8% Ni	1/8	25	00-223-3695	MIL-W-6712, type 1 (18-8)	Metallizing Spray Gun
High carbon steel	"	50	00-265-7096	" " (0.80C)	"
Medium carbon steel	"	50	00-223-3703	" " (0.25C)	"
Mild steel	"	50	00-223-3707	" " (0.10C)	"
99% Molybdenum	0.0907	20	00-903-7703	" type 2 (Molybdenum)	"
99% Copper	1/8	50	00-223-3735	" " (Copper)	"
60% Cu, 40% Zn	"	25	00-223-3731	" " (Naval brass)	"
99% Aluminum	"	25	00-223-3728	" " (Aluminum)	"

Table D-3. Welding Electrodes

Positions: Flat	Horizontal	Vertical	Overhead	Current: AC	DC Straight	DC Reverse	DC S & R	Shielded Yes	Shielded No	Material	Dia	Length	(CAGEC 3439) NIIN	Identifying Reference	Use
X	X	X	X			X		X		Steel	3/32	12	00-262-2669	MIL-E-15599, type 6010, C11	Welding of zinc-coated, low & medium carbon steels; also medium carbon, high tensile steel plate up to 5/8-in. thickness
X	X	X	X			X		X		"	1/8	14	00-262-2670	" " " C11	
X	X	X	X			X		X		"	5/32	14	00-262-2671	" " "	
X	X	X	X			X		X		"	3/16	14	00-262-2672	" " "	
	X					X		X		"	1/4	18	00-262-2674	" " " C12	
X	X	X	X	X			X	X		Steel	3/32	12	00-262-2652	MIL-E-15599, type 6011, C11	
X	X	X	X	X			X	X		"	1/8	14	00-262-2653	" " " C11	
X	X	X	X	X			X	X		"	5/32	14	00-262-2654	" " "	
X	X	X	X	X			X	X		"	3/16	14	00-262-2655	" " " C12	Welding of uncoated mild & medium carbon steels; electrodes suitable for poorly fitted joints
X	X	X	X	X			X	X		"	1/4	18	00-262-2657	" " " C11	
X	X			X	X			X		Steel	3/16	14	00-273-3719	MIL-E-15599, type 6012, C11	
X	X			X	X			X		"	1/4	18	00-262-3876	" " " C12	

Table D-3. Welding Electrodes (cont)

Flat	Horizontal	Vertical	Overhead	AC	DC Straight	DC Reverse	DC S & R	Shielded Yes	Shielded No	Material	Dia	Length	NIIN (CAGEC 3439)	Identifying Reference	Use
X	X	X	X	X			X	X		Steel	1/8	14	00-262-2648	MIL-E-15599, type 6012, C11	Aircraft welding of mild and low alloy sheet steels; shallow penetration
X	X	X	X	X			X	X		Steel	5/32	14	00-262-2649	MIL-E-15599, type 6027, C12	Uncoated, medium carbon, high tensile steel; deep penetration; fast weld speed
X	X	X	X	X			X	X		"	3/16	14	00-262-2650	"	
X	X			X			X	X		Steel	5/32	14	00-061-2896	MIL-E-15599, type 6027, C12	
X	X			X			X	X		"	3/16	18	00-061-2897	"	
X	X			X			X	X		"	1/4	18	00-061-2898	"	
X	X	X	X	X		X		X		Steel	3/16	14	00-853-2719	MIL-E-22200/1, type 7018, C11	Low hydrogen electrode; medium & high tensile steels of up to 5/8-in. thickness
X	X			X		X		X		"	7/32	18	00-542-0964	" C12	
X	X	X	X	X		X		X		Steel	1/8	14	00-853-2716	MIL-E-22200/1, type 9018, C11	Low hydrogen electrode; low alloy, medium & high tensile steels (HY-80); fillet welds in high yield strength, low alloy structural steels (T-1 and RQ-100A)
X	X	X	X	X		X		X		Steel	5/32	14	00-853-2718	"	
X	X	X	X	X		X		X		Steel	1/8	14	00-587-2412	MIL-E-22200/1, type 11018, C11	Low hydrogen electrode; groove butt joints in high yield strength, low alloy structural steels (HY-80, T-1, & RQ-100A)
X	X	X	X	X		X		X		"	5/32	14	00-587-2413	"	
X	X	X	X	X		X		X		"	3/16	14	00-878-2158	"	
X	X	X	X	X		X		X		Steel	5/32	14	00-287-7089	MIL-E-18038, type 10016, C11	Low hydrogen electrode; welding of low alloy, high tensile steels (HY-80)
X	X	X	X	X		X		X		"	3/16	14	00-287-7090	"	
X	X	X	X	X		X		X		"	1/8	14	00-287-7088	"	
X	X	X	X	X		X		X		Steel	5/32	14	00-984-4786	MIL-E-22200/1, type 10018, C11	Low hydrogen electrode; welding of low alloy, high tensile steels (HY-80)
X				X					X	Steel	1/8	Coil	00-200-1583	MIL-E-18193, type A-1	Welding & surfacing carbon and low alloy steels

Table D-3. Welding Electrodes (cont)

Flat	Horizontal	Vertical	Overhead	AC	Straight	Reverse	S & R	Yes	No	Material	Dia	Length	NIIN (CAGEC 3439)	Identifying Reference	Use
X						X		X		Steel	3/16	14	00-262-2639	MIL-E-19141, type A2C, C13	using submerged arc machines (use type "A" fluxes per MIL-F-18251)
X	X	X	X	X		X		X		"	3/16	14	00-752-7818	" type FeMn-A, C13	Corrosion & abrasion resistant surfacing (severe impact)
X	X	X	X			X		X		Steel	1/8	14	00-262-2746	MIL-E-16589, type 52-15, C11	
X	X	X	X	X		X		X		"	1/8	14	00-204-3140	" type 202-16, C11	Low hydrogen; 1-1/4% Cr, 1/2% Mo; high temperature service (950 °F)
X	X	X	X	X		X		X		"	5/16	14	00-204-4512	" "	Low hydrogen; 5% Cr, 1/2% Mo; high temperature service (1200 °F)
X	X	X	X			X		X		Steel	3/16	14	00-204-3277	" type 202-16, C12	
X	X	X	X			X		X		"	1/8	14	00-984-4778	" type 94-15, C11	Low hydrogen; 2-1/4% Cr, 1% Mo; high temperature service (1050 °F)
X	X	X	X			X		X		"	5/32	14	00-984-4779	" "	
X	X	X	X			X		X		"	3/16	14	00-984-4780	" "	
X	X	X	X			X		X		Steel	1/8	14	00-246-9544	MIL-E-13080, type 307L-15, C11	High hardenability steels & armor
X	X	X	X			X		X		"	5/32	14	00-246-9545	00- "	
X	X	X	X			X		X		"	3/16	14	00-266-9752	" C12	
X	X	X	X			X		X		Steel	1/16	9	00-245-6630	MIL-E-22200/2, type 308-15, C11	Weld 18% Cr, 8% Ni corrosion resistant steel (18-8)
X	X	X	X			X		X		"	1/8	14	00-277-7550	" "	
X	X	X	X	X		X		X		"	5/32	14	00-528-9064	" type 308-16, C11	
X	X	X	X			X		X		"	5/32	14	00-262-2696	" type 347-15, C12	
X	X	X	X	X		X		X		"	1/8	14	00-262-2695	" "	
X	X	X				X		X		"	3/16	14	00-262-2697	type 347-15	
X	X	X	X			X		X		Monel	5/32	10-14	00-204-3247	MIL-E-22200/3, type 3N10, C12	Wrought nickel-copper alloys
X	X	X	X			X		X		"	3/16	14	00-262-2644	" "	
X	X	X	X			X		X		Nickel	1/8	14	00-901-7637	type 4N11, C11	Wrought commercially pure Ni to steel
X	X	X	X			X		X		Steel	5/32	14	00-984-4768	MIL-E-22200/6, type 7015, C11	Weld mild & medium strength steels
X	X	X	X			X		X		"	3/16	14	00-984-4770	" "	

Table D-3. Welding Electrodes (cont)

Flat	Horizontal	Vertical	Overhead	AC	DC Straight	DC Reverse	DC S&R	Shielded Yes	Shielded No	Material	Dia	Length	NIIN (CAGEC 3439)	Identifying Reference	Use
X	X	X	X			X		X		Steel	5/32	14	00-465-1923	MIL-E-22200/6, type 8015-C3, C11	Weld low alloy, medium strength steels
X	X	X	X			X		X		"	3/16	14	00-984-4776	"	
X	X	X	X			X		X		"	5/32	14	00-262-2678	MIL-E-22200/7, type 7010-A1, C11	Molybdenum alloy steel pipe, forging & casting
X	X	X	X			X		X		"	3/16	14	00-262-2679	"	
X	X	X	X	X			X	X		"	1/4	18	00-262-2681	" C12	
X	X	X	X	X				X		High Speed steel	1/8	14	00-255-8922	ASTM A339-56T, type EFe5-A	Cutting tool repair & buildup; for 1100°F use
X	X			X				X		Mild Steel	1/8	14	00-293-4716	ASTM A398-65T, type EST	Cast iron nonmachinable weld
X	X	X	X	X		X		X		Nickel 97% Ni, 1% Co	1/8	14	00-640-2351	ASTM A398-65T, type ENiFe-Cl	Grooving and roughing prior to metallizing
									X		1/8	18	00-449-6558	METCO Co. "FUSE BOND"	
X	X					X		X		Bronze	1/8	14	00-200-1376	MIL-E-13191, type CuSn-A	Welding of phosphor bronze, brass, copper & cast iron
X	X					X		X		Bronze	3/16	14	00-255-8910	"	
X	X					X		X		"	1/8	14	00-262-2738	" type CuSn-C	
X	X					X		X		Copper	5/32	14	00-262-2739	"	
X	X					X		X		"	3/16	14	00-262-2740	"	
X				X				X		"	5/32	14	00-618-5797	MIL-E-278, type CuAl-B	Aluminum bronzes, high strength Cu-Zn alloys
X				X				X		Copper	5/32	14	00-247-5157	MIL-E-278, type CuAl-D	Wear-resistant bearing surfaces
X	X					X		X		Aluminum	3/32	14	00-262-2597	MIL-E-15597, type 4043, C12	Welding of aluminum & aluminum alloys
X	X					X		X		"	1/8	14	00-262-2598	"	
X	X					X		X		"	5/32	14	00-262-2599	"	
X	X					X		X		"	3/16	14	00-262-2600	"	
X	X	X	X			X		X		Aluminum	3/16	14	00-974-7079	EUTECTIC #2101E	Heavy aluminum castings, long joints & filler

Table D-4. Overlay, Welding and Cutting, Chamfering, and Heating Electrodes

Material	Dia	Length	NIIN (CAGEC 3439)	Identifying Reference	Use
Chrome	1/4		00-902-4215	EUTECTIC Co. "EUTECTRODE 10"	Overlay on ferrous metals
"	3/16		00-902-4216	"	
"	3/16		00-902-4208	EUTECTIC Co. "EUTECTRODE #680"	
"	3/32		00-902-4209	"	Overlay on tool & die steels
"	3/16	14	00-262-2639	MIL-E-19141, type A2C, C13	Corrosion and abrasion resistant surfacing (tough, forgeable)
"	3/16	14	00-752-7818	type FeMn-A, C13	(severe impact)
Tubular Steel	3		00-255-7711	MIL-E-17764	Underwater arc-oxygen cutting
Carbon [4]	3/16	12	00-262-4227	MIL-E-17777, type C	Carbon-arc welding process
	3/8	12	00-262-4228	"	
Carbon [4]	1/2	12	00-262-4229	MIL-E-17777, type C	
"	3/4	12	00-262-4230	"	
"	1/4	12	00-262-4294	"	
99% Tungsten	1/16	7	00-814-6030	ASTM B297-55T, class EWP	Welding aluminum, magnesium, copper, or titanium
"	3/32	7	00-814-6031	"	
"	1/8	7	00-814-6029	"	
"	5/32	7	00-814-6028	"	
5/32			00-766-7749	EUTECTIC Co. "CUTTRODE #1"	Cutting without air
	3/16		00-902-4213	EUTECTIC Co. "CHAMFERTRODE"	Chamfering & grooving

Table D-4. Overlay, Welding and Cutting, Chamfering, and Heating Electrodes (cont)

Positions				Current					Shielded		Material	Dia	Length	NIIN (CAGEC 3439)	Identifying Reference	Use
Flat	Horizontal	Vertical	Overhead	AC	DC Straight	Reverse	S & R	Yes	No						
X	X	X	X	X				X		Carbon	6		00-296-9891	IDEAL INDUSTRIES #L-3321	(Pliers No. 12-067)
X	X	X	X	X				X		"	7		00-296-9892	#L-2925	Spares for IDEAL soldering (Pliers #12-067)
X	X	X	X	X				X		"	1/2	3	00-242-2599	#L-2926	(Fork #12-068)
X	X	X	X	X				X		"	1/2	3-1/2	00-242-2600	#L-3322	(Pencil #12-069)
X	X	X	X	X				X		"	1/8	3	00-765-5395	#L-4848	(Pencil #12-166)
X	X	X	X	X				X		Metal	1/16	3-1/2	00-818-5859	#L-5241	(Pencil #12-167)

1 Flux-coated.
2 Covered for underwater use.
3 5/16 od, 0.112 id, 14 in. long.
4 See TB 9-3439-201/1 for application using standard electrode holders.
5 Copper-coated carbon electrode. Exothermic coating effects arc blow without air source.
6 Flat shape, 1/2 in. wide by 1-1/2 in. long.
7 Curved surface, 1/2 in. wide by 1-1/2 in. long.

Table D-5. Welding Rods

Positions				Process			Coated		Material	Dia	Length	NIIN (CAGEC 3439)	Identifying Reference	Use
Flat	Horizontal	Vertical	Overhead	Oxyacetylene	Carbon & Tungsten-Arc	Metal-Arc	Yes	No						
X	X	X	X	X		X		X	Steel	1/16	36	00-246-0564	MIL-R-908, C11	Welding of low & medium carbon steels (not air-craft)
X	X	X	X	X			Cu		"	3/32	36	00-246-0565	"	
X	X	X	X	X				X	"	1/8	36	00-246-0566	"	
X	X	X	X	X			Cu		"	5/32	36	00-246-0567	"	

Table D-5. Welding Rods (cont)

Positions				Process			Coated		Material	Dia	Length	NIIN (CAGEC 3439)	Identifying Reference	Use
Flat	Horizontal	Vertical	Overhead	Oxyacetylene	Carbon & Tungsten-Arc	Metal-Arc	Yes	No						
X	X	X	X	X				X	"	3/16	36	00-246-0568	"	Aircraft & welding of low and medium carbon steels
X	X	X	X	X				X	"	1/4	36	00-246-0569	"	
X	X	X	X	X	X		Cu	X	Cast Iron	1/8	24	00-247-2981	C12	
X	X	X	X	X	X	X		X	Medium Carbon Steel	1/16	36	00-294-6910	MIL-R-908, C12	Aircraft welding of low alloy steels (heat treat after weld)
X	X	X	X	X	X	X		X	"	1/8	36	00-163-4362	MIL-R-5632, C11	
X	X	X	X	X	X	X	Cu	X	"	1/8	36	00-204-3592	MIL-R-5632, C12	
X	X	X	X	X	X	X		X	AISI #309	1/4	36	00-262-4279	"	Welding of stainless steel 309
X	X	X	X	X	X	X		X	"	1/8	36	00-288-1469	MIL-R-5031, C13	
X	X	X	X	X	X	X	Cu	X	AISI #316	1/16	36	00-246-0575	C15	" "
X	X	X	X	X	X	X		X	AISI #316	3/32	36	00-246-0576	"	" 316
X	X	X	X	X	X	X		X	AISI #316	1/8	36	00-246-0577	"	"
X	X	X	X	X	X	X		X	AISI #316-L	1/16	36	00-163-4360	MIL-R-5031, C16	Welding of stainless steel 316-L
X	X	X	X	X				X	75Ni, 18Cr	1/4	18	00-542-0411	MIL-R-17131, C1 NiCrC	Corrosion & abrasion resistant overlays
X	X	X	X	X				X	75Ni, 18Cr	5/16	18	00-542-0412		
X	X	X	X	X				X	70Ni, 15Cr, 1-0Fe	1/16	36	00-273-8824	QQ-R-571, type 2, C1 NiCrFe-4	Nickel-chrome-iron alloy (use flux)
X	X	X	X	X				X	30Cu	1/8	36	00-246-0560	"C1 NiCu-5	Nickel-copper alloy (use flux)
X	X	X	X	X				X	30Cu	3/16	36	00-246-0562	"	
X	X	X	X	X	X	X		X	30Cu	1/4	36	00-254-5039	QQ-R-571, type 1, C1 CuSn-A	
X	X	X	X	X	X	X		X	93Cu, 6Sn	3/16	36	00-255-8943	"	Phosphor bronze (use flux for carbon-arc)
X	X	X	X	X				X	93Cu, 6Sn	1/16	36	00-255-8944	C1 CuZn-A	
X	X	X	X	X				X	60Cu, 40Zn	3/32	36	00-268-9668	"	Steel, cast iron, malleable iron (use flux)
X	X	X	X	X				X	40Zn	1/8	36	00-262-7565	"	
X	X	X	X	X				X	40Zn	1/16	36	00-247-2978	"	
X	X	X	X	X				X	60Cu, 30Zn	3/16	36	00-255-7757	C1 CuZn-B	Steel & cast iron; build-up surfaces (use flux)
X	X	X	X	X				X	30Zn	1/8	36	00-254-5033	"	
X	X	X	X	X				X	60Cu, 40Zn	3/16	36	00-244-4540	C1 CuZn-C	Welding of steel & cast iron (use flux)
X	X	X	X	X				X	40Zn	1/4	36	00-244-4541	"	
X	X	X	X	X				X	40Zn	1/8	36	00-244-4542	"	
X	X	X	X	X	X	X		X	99% Al	1/8	36	00-268-9652	QQ-R-566, C1 1100	Pure aluminum & manganese aluminum

Table D-5. Welding Rods (cont)

Positions				Process			Coated		Material	Dia	Length	NIIN (CAGEC 3439)	Identifying Reference	Use
Flat	Horizontal	Vertical	Overhead	Oxyacetylene	Carbon & Tungsten-Arc	Metal-Arc	Yes	No						
X	X	X	X	X	X	X		X	93% Al, 6 Si	1/16	36	00-178-8590	QQ-R-566, Cl 4043	Pure aluminum & silicon & manganese aluminums
X	X	X	X	X	X	X		X	"	3/32	36	00-268-9654	"	
X	X	X	X	X	X	X		X	"	1/8	36	00-247-2982	"	
X	X	X	X	X	X	X		X	"	3/16	36	00-247-2983	"	
X	X	X	X	X	X	X		X	"	1/4	36	00-255-8942	"	
X	X	X	X	X	X	X		X	87Mg, 9Al, 2Zn	3/32	36	00-204-3280	MIL-R-6944, Cl AZ92A	Magnesium-aluminum-zinc alloys
X	X	X	X	X	X	X		X	"	1/8	36	00-204-3203	"	
X	X	X	X	X	X	X		X	"	3/16	36	00-262-4285	"	
X	X	X	X	X	X	X		X	"	1/4	36	00-204-3279	"	

Table D-6. Brazing Alloys

Temperatures (degrees F)		Chemistry						Dimensions				Coil Spool Pkg Lb Oz	(CAGEC 3439) NIIN	Identifying Reference	Use
Melting	Brazing	Copper	Silver	Zinc	Cadmium	Phosphorus	Nickel	Dia	Length	Width	Thickness				
1250	1370	30	45	25					A/A	3/4	0.003	1 oz pkg	00-238-3077	MIL-B-15395, Gr 1	Small delicate parts—OK for dissimilar metals
1280	1325	20	65	15						3/4	0.003		00-247-6926	MIL-B-15395, Gr 2	
1185	1500	80	15			5			20	1/8	0.050		00-188-6982	QQ-B-650, Cl CuP-5	Copper and copper alloys only—not for ferrous metals—joint clearance of 0.003 to 0.005
1185	1500	80	15	35		5			36	1/8	1/8		00-204-2555	QQ-B-650, Cl CuP-5	
1430	1500	45	20	25						3/4	0.003	1 oz pkg	00-247-6927	QQ-S-561, C10	Brazing joint has high physical properties
1250	1370	30	45					1/16				1-1/2 oz coil	00-224-3573	QQ-S-561, C11	General silver soldering
1160	1175	15	50	17	18			1/32				1 oz coil	00-184-8952	QQ-S-561, C14	Brazing on dissimilar metals
1160	1175	15	50	17	18			1/16				1 lb spool	00-184-8948	QQ-S-561, C14	
1170	1270	15.5	50	15.5	16		3	3/32				1 lb coil	00-224-3561	QQ-S-561, C15	Fillet joints & brazing carbide tool tips to tool

Table D-7. Soldering Materials

Flow Temp (deg F)	Tin	Lead	Zinc	Antimony	Bismuth	Silver	Solid	Cored Rosin	Cored Acid	Diameter	Length	Width	Thickness	Spool or bar	NIIN (CAGEC 3439)	Identifying reference	Use
490	30	67		2							14	5/8	3/8		00-247-6970	QQ-S-571, SN-30, BS	Automotive dents & seams
460	40	60									14	5/8	3/8		00-247-6968	QQ-S-571, SN-40, RS	Dip & wiping solder
460	40	60									(Any Shape)			1 lb bar	00-247-6921	QQ-S-571, SN-40, BS	"
420	50	50									(Any Shape)			1-1/4 lb bar	00-163-4347	QQ-S-571, SN-50, BS	**Sweated joint;** copper, cast iron, steel
375	60	40									14	5/8	3/8		00-254-8437	QQ-S-571, SN-60, BS	Electrical connections & coating
475	35	60		2							5-1/2	2	1-1/2		00-247-6969	QQ-S-571, SN-35, IS	Plumber's wiping solder
260	25	38			37						14	1/4	1/4		00-239-8506	(L.B. ALLEN Co., "BIS-MUTH-LFAD")	Low temperature melting application
570	10	87				2		P		1/16				1 lb spool	00-265-7102	QQ-S-571, SN-10, WRP2	Electrical connections, high temperature
460	40	60						P		0.090				1 lb spool	00-188-6988	QQ-S-571, SN-40, WRAP3	Dip & wiping solder
460	40	60							P	3/32				5 lb spool	00-188-6986	QQ-S-571, SN-40, WACP6	"

Table D-7. Soldering Materials (cont)

Temp Flow (deg F)	Chemistry Tin	Lead	Zinc	Antimony	Bismuth	Silver	Solid	Form Cored (plastic or dry) Acid	Rosin	Diameter	Length	Width	Thickness	Spool or bar	NIIN (CAGEC 3439)	Identifying reference	Use
460	40	60							P	3/32				1 lb spool	00-224-3575	QQ-S-571, SN-40, WACP3	"
460	40	60						P		1/8				1 lb spool	00-184-8960	QQ-S-571, SN-40, WACP6	"
460	40	60							P	1/8				1 lb spool	00-243-1882	QQ-S-571, SN-40, WRP3	"
460	40	60					X			1/8				5 lb spool	00-247-6967	QQ-S-571, SN-40, WS	"
420	50	50					X			1/16				5 lb spool	00-141-8244	QQ-S-571, SN-50, WS	Sweated joint; copper, cast iron, steel
420	50	50							P	1/16				1 lb spool	00-640-2404	QQ-S-571, SN-50, WRAP3	"
420	50	50						P		0.090				1 lb spool	00-184-8953	QQ-S-571, SN-50, WRAP3	"
460	40	60						P		1/16				1 lb spool	00-243-1888	QQ-S-571, SN-40, WRP2	"
420	50	50						P		3/32				1 lb spool	00-727-0489	QQ-S-571, SN-50, WRP2	"
420	50	50					X			1/8				1 lb spool	00-239-8505	QQ-S-571, SN-50, WS	Sweated joint; copper, cast iron, steel

Table D-7. Soldering Materials (cont)

Temp Flow (deg F)	Tin	Lead	Zinc	Antimony	Bismuth	Silver	Solid	Acid	Rosin	Diameter	Length	Width	Thickness	Spool or bar	(CAGEC 3439) NIIN	Identifying reference	Use
375	60	40							P	1/32				1 lb spool	00-555-4629	QQ-S-571, SN-60, WRAP3	Electrical connections & coating
375	60	40							D	1/16				5 lb spool	00-163-4351	QQ-S-571, SN-60, WRD3	"
375	60	40								3/32				5 lb spool	00-224-3567	QQ-S-571, SN-60, WRAP3	"
375	60	40							P	1/8				1 lb spool	00-273-2536	QQ-S-571, SN-60, WRP2	"
375	60	40		2				P		0.162				1 lb spool	00-254-8439	QQ-S-571, SN-60, WACP3	Electrical connections & coating
475	35	60	4					P				3/16	1/8	5 lb spool	00-224-3562	QQ-S-571, SN-35, RACP6	Plumber's wiping solder
500	35	60					X			1/8				1 lb spool	00-528-9616	MIL-S-12204, type 1, comp A	Aluminum & aluminum alloys
N/A	Plastic Aluminum													(Paste solder—5-1/2 oz. tube)	00-726-9822	(WOODHILL CHEMICAL Co. "P-AL")	Aluminum and aluminum alloys

Table D-8. Fluxes, Welding, Brazing, and Soldering

Process	Liquid	Paste	Stick	Powder	Granular	Mesh size	Unit of issue	(CAGEC 3439) NIIN	Identifying reference	Use	Type of solder used with (solder flux only)
Gas Welding				X		80	1 lb	00-255-4580	MIL-F-16136, type C	Cast iron	
Gas Welding		X					5 lb	00-255-9940	MIL-F-16136, type C	Cast iron & corrosion resistant steels	
Gas & Arc Welding				X	X	80	1 lb	00-255-4577	MIL-F-16136, type A, C11	Copper	
Arc Welding				X		60	100 lb	00-068-5058	MIL-F-18251, A760	Steel	
Arc Welding				X		60	100 lb	00-200-1581	MIL-F-18251, 840	Steel	
Brazing		X					1 lb	00-255-4572	(Alcoa #33)	Aluminum	
Brazing	X						1 lb	00-640-3713	O-F-499, type B	All except aluminum bronze	
Soldering	X						4 oz	00-250-2629	O-F-506, type 2, form B	Heat resisting steel	Sn-Pb
Soldering	X						4 oz	00-250-2635	O-F-506, type 1, form B	All except aluminum & heat resisting alloys	Sn-Pb
Soldering		X					1/4 lb	00-255-4566	O-F-506, type 1, form A	All except aluminum & heat resisting alloys	Sn-Pb
Soldering		X					2 oz	00-260-1264	O-F-506, type 1, form A	All except aluminum & heat resisting alloys	Sn-Pb
Soldering			X				8 oz	00-288-0868	O-F-499, type B	All except aluminum bronze	Ag-Cu
"		X					2 oz	00-529-0621	O-F-506, type 1, form A	All except aluminum & heat resisting alloys	Sn-Pb
"		X					1/4 lb	00-270-6050	MIL-F-12784, Comp IC-3	Lead joints of telephone cable splices (see TM 11-372)	Sn-Pb

Table D-9. Carbon Blocks, Rods, and Paste

Form	Measurements	Identifying Reference	(CAGEC 3439) NIIN	Use
Carbon Block	1/2 T, 6W, 12L	MIL-C-1143	00-262-4159	Block, plug, or dike to restrict flow of molten metal
Carbon Block	1 T, 6W, 12L	"	00-262-4163	"
Carbon Rod	1/4 diam, 12 long	MIL-C-1143	00-262-4160	"
Carbon Rod	1/2 "	"	00-262-4161	"
Carbon Rod	5/8 "	"	00-262-4164	"
Carbon Rod	3/4 "	"	99-262-4162	"
Carbon Rod	7/8 "	"	00-262-4165	"
Carbon Rod	1-1/4 "	"	00-262-4166	"
Carbon Rod	1-1/2 "	"	00-262-4167	"
Carbon Paste	5 lb pail	MIL-C-1143	00-255-9943	"

APPENDIX E
MISCELLANEOUS DATA

Table E-1. Temperature Ranges for Processing Metals

Process	Temperature Range °F	°C
Joining temperature		
Brazing (copper and copper alloys)	1300 to 2150	704 to 1177
Brazing (silver alloys)	1100 to 1650	593 to 899
Forging	1700 to 2150	927 to 1177
Soft soldering	300 to 700	149 to 371
Welding (ferrous metals)	1800 to 2800	982 to 1538
Welding (nonferrous metals)	600 to 3300	316 to 1816
Hardening		
Carbon steel	1350 to 1550	732 to 843
Alloy steel	1400 to 1850	760 to 1010
High speed steel	2150 to 2400	1177 to 1316
Tempering		
Carbon steel	300 to 1050	149 to 566
Alloy steel	300 to 1300	149 to 704
High speed steel	350 to 1100	177 to 593

Table E-2. Combustion Constants of Fuel Gases

Name of Gas	Heat Value Btu per ft^3	Flame Temperature with Oxygen °F	°C
Acetylene	1433 net	6300	3482
Butane	2999 net	5300	2927
City gas	300 to 800 net	4600	2538
Coke oven gas	500 to 550 net	4600	2538
Ethane	1631 net	5100	2816
Ethylene	1530 net	5100	2816
Hydrogen	275.1 net	5400	2982
Methane	913.8 net	5000	2760
Natural gas	800 to 1200 net	4600	2538
MAPP gas	2406 net	5300	2927

TC 9-237

Table E-3. Melting Points of Metals and Alloys

Metal or Alloy	Melting point °F	°C
Aluminum, cast (8 percent copper)	1175	635
Aluminum, pure	1220	660
Aluminum (5 percent silicon)	1118	603
Brass, naval	1625	885
Brass, yellow	1660	904
Bronze, aluminum	1905	1041
Bronze, manganese	1600	871
Bronze, phosphor	1830 to 1922	999 to 1050
Bronze, tobin	1625	885
Chromium	2740	1504
Copper	1981	1083
Iron, cast	2300	1260
Iron, malleable	2300	1260
Iron, pure	2786	1530
Iron, wrought	2750	1510
Lead	620	327
Manganese	2246	1230
Magnesium	1200	649
Molybdenum	4532	2500
Monel metal	2480	1360
Nickel	2646	1452
Nickel silver (18 percent nickel)	2030	1110
Silver, pure	1762	961
Silver solders (50 percent silver)	1160 to 1275	627 to 691
Solder (50-50)	420	216
Stainless steel (18-8)	2550	1399
Stainless steel, low carbon (18-8)	2640	1449
Steel, high carbon (0.55-0.83) percent carbon)	2500 to 2550	1371 to 1399
Steel, low carbon (maximum 0.30 percent carbon)	2600 to 2750	1427 to 1510
Steel, medium carbon (0.30-0.55 percent carbon)	2550 to 2600	1399 to 1427
Steel, manganese	2450	1343
Steel, cast	2600 to 2750	1427 to 1510
Steel, nickel (3.5 percent nickel)	2600	1427
Tantalum	5160	2849
Tin	420	216
Titanium	3270	1799
Tungsten	6152	3400
Vanadium	3182	1750
White metal	725	385
Zinc	786	419

E-2

Table E-4. Temper Colors and Temperatures

Temper Color	Temperatures °F	°C	Uses
Faint straw	400	204	
Straw	440	227	Scrapers, hammer faces, lathe, shaper, and planer tools
Dark straw	460	238	Milling cutters, taps, and dies
Very deep straw	480	249	Punches, dies, knifes, and reamers
Brown yellow	500	260	Stone-cutting tools and twist drills
Bronze or brown purple	520	271	Drift pins
Peacock or full purple	540	282	Augers, cold chisels for steel
Bluish purple	550	288	Axes, cold chisels for iron, screwdrivers, and springs
Blue	570	299	Saws for wood
Full blue	590	310	
Very dark blue	600	316	
Light blue	640	338	

Table E-5. Heat Colors with Approximate Temperature

Color	Temperature °F	°C
White	2200	1204
Light yellow	1975	1079
Lemon	1825	996
Orange	1725	941
Salmon	1650	899
Bright red	1550	843
Bright cherry or dull red	1450	788
Cherry or full red	1375	746
Medium cherry	1250	677
Dark cherry	1175	635
Blood red	1050	566
Faint red	900	482
Faint red (visible in dark)	750	399

Table E-6. Stub Steel Wire Gauges

Gauge No.	Dia	Gauge No.	Dia	Gauge No.	Dia	Gauge No.	Dia
7/0	....	16	0.175	38	0.101	61	0.038
6/0	....	17	0.172	39	0.099	62	0.037
5/0	....	18	0.168	40	0.097	63	0.036
4/0	....	19	0.164	41	0.095	64	0.035
3/0	....	20	0.161	42	0.092	65	0.033
2/0	....	21	0.157	43	0.088	66	0.032
0	....	22	0.155	44	0.085	67	0.031
1	0.227	23	0.153	45	0.081	68	0.030
2	0.219	24	0.151	47	0.077	69	0.029
3	0.212	25	0.148	48	0.075	70	0.027
4	0.207	26	0.146	49	0.072	71	0.026
5	0.204	27	0.143	50	0.069	72	0.024
6	0.201	28	0.139	51	0.066	73	0.023
7	0.199	29	0.134	52	0.063	--	--
8	0.197	30	0.127	53	0.058	74	0.022
9	0.194	31	0.120	54	0.055	75	0.020
10	0.191	32	0.115	55	0.050	76	0.018
11	0.188	33	0.112	56	0.045	77	0.016
12	0.185	34	0.110	57	0.042	78	0.015
13	0.182	35	0.108	58	0.041	79	0.014
14	0.180	36	0.106	59	0.040	80	0.013
15	0.178	37	0.103	60	0.039	--	--

Table E-7. Standard Gauge Abbreviations

Standard gauge	Abbreviation
American wire gauge	AWG
Brown & Sharpe gauge	B&S
. . . .	
American steel wire gauge	Stl WG
National wire gauge	NATL
Roebling wire gauge	ROEBL
Washburn & Moen gauge	W&M
. . . .	
Standard wire gauge	SWG
English standard gauge	SWG
English legal standard gauge	SWG
Imperial wire gauge	IWG
British Imperial wire gauge	IWG
British standard wire gauge	SWG
New British standard gauge	NBS
Olde English gauge	OEG
London wire gauge	Lon WG
. . . .	
1914 Birmingham gauge	BWG
. . . .	
Birmingham wire gauge	BWG
Stub iron wire gauge (Peters Stubbs)	STUB IRON GA
Stub steel wire gauge	STUB STL
U.S. standard gauge	US STD
. . . .	

NOTE
Gauges grouped within broken lines (...) are identical.

Table E-8. Metal Gauge Comparisons

Gauge No.	U.S. Standard Gauge (obsolete)	Manufacturer's Standard	Music Wire	American Wire (AWG)	American Steel Wire (StlWG)	Standard Wire (SWG)	Olde English (OEG)	1914 Birmingham (BG)	Birmingham Wire (BWG)
7/0	0.5000	......	0.003	......	0.4900	0.5000	......	0.6666	......
6/0	0.4687	......	0.004	0.5800	0.4600	0.4640	......	0.6250	......
5/0	0.4370	......	0.005	0.5165	0.4300	0.4320	0.4540	0.5883	......
4/0	0.4063	......	0.006	0.4600	0.3938	0.4000	0.4250	0.5416	0.454
3/0	0.3750	......	0.007	0.4096	0.3625	0.3720	0.3800	0.5000	0.425
2/0	0.3437	......	0.008	0.3648	0.3310	0.3480	0.3400	0.4452	0.380
0	0.3125	......	0.009	0.3249	0.3065	0.3240	0.3000	0.3964	0.340
1	0.2813	......	0.010	0.2893	0.2830	0.3000	0.2840	0.3532	0.300
2	0.2656	0.2391	0.011	0.2576	0.2625	0.2760	0.2590	0.3147	0.284
3	0.2500	0.2242	0.012	0.2294	0.2437	0.2520	0.2380	0.2804	0.259
4	0.2344	0.2092	0.013	0.2043	0.2253	0.2320	0.2200	0.2500	0.238
5	0.2188	0.1943	0.014	0.1819	0.2070	0.2120	0.2030	0.2225	0.220
6	0.2031	0.1793	0.016	0.1620	0.1920	0.1920	0.1800	0.1981	0.203
7	0.1875	0.1644	0.018	0.1443	0.1770	0.1760	0.1650	0.1764	0.180
8	0.1719	0.1495	0.020	0.1285	0.1620	0.1600	0.1480	0.1570	0.165
9	0.1563	0.1345	0.022	0.1144	0.1483	0.1440	0.1340	0.1398	0.148
10	0.1406	0.1196	0.024	0.1019	0.1350	0.1280	0.1200	0.1250	0.134
11	0.1250	0.1046	0.026	0.0907	0.1205	0.1160	0.1090	0.1113	0.120
12	0.1094	0.0897	0.029	0.0808	0.1055	0.1040	0.0950	0.0991	0.109
13	0.0937	0.0747	0.031	0.0720	0.0915	0.0920	0.0830	0.0882	0.095
14	0.0781	0.0673	0.033	0.0641	0.0800	0.0800	0.0720	0.0785	0.083
15	0.0703	0.0598	0.035	0.0571	0.0720	0.0720	0.0650	0.0699	0.072
16	0.0625	0.0538	0.037	0.0508	0.0625	0.0640	0.0580	0.0625	0.065
17	0.0563	0.0478	0.039	0.0453	0.0540	0.0560	0.0490	0.0556	0.058
18	0.0500	0.0418	0.041	0.0403	0.0475	0.0480	0.0400	0.0495	0.049
19	0.0438	0.0359	0.043	0.0359	0.0410	0.0400	0.0360	0.0440	0.042
20	0.0373	0.0329	0.045	0.0320	0.0348	0.0360	0.0350	0.0392	0.035
21	0.0344	0.0299	0.047	0.0284	0.0318	0.0320	0.0315	0.0349	0.032
22	0.0313	0.0269	0.049	0.0254	0.0286	0.0280	0.0295	0.0312	0.028
23	0.0281	0.0239	0.051	0.0226	0.0258	0.0240	0.0270	0.0278	0.025
24	0.0250		0.055	0.0201	0.0230	0.0220	0.0250	0.0248	0.022

Table E-8. Metal Gauge Comparisons (cont)

Gauge No.	U.S. Standard Gauge (obsolete)	Manufacturer's Standard	Music Wire	American Wire (AWG)	American Steel Wire (St1WG)	Standard Wire (SWG)	Olde English (OEG)	1914 Birmingham (BG)	Birmingham Wire (BWG)
25	0.0219	0.0209	0.059	0.0179	0.0204	0.0200	0.0230	0.0220	0.020
26	0.0188	0.0179	0.063	0.0159	0.0181	0.0180	0.0205	0.0196	0.018
27	0.0172	0.0164	0.067	0.0142	0.0173	0.0164	0.0187	0.0174	0.016
28	0.0156	0.0149	0.071	0.0126	0.0162	0.0148	0.0165	0.0156	0.014
29	0.0141	0.0135	0.075	0.0113	0.0150	0.0136	0.0155	0.0139	0.013
30	0.0125	0.0120	0.080	0.0100	0.0140	0.0124	0.0137	0.0123	0.012
31	0.0109	0.0105	0.085	0.0089	0.0132	0.0116	0.0122	0.0110	0.010
32	0.0102	0.0097	0.090	0.0080	0.0128	0.0108	0.0112	0.0098	0.009
33	0.0094	0.0090	0.095	0.0071	0.0118	0.0100	0.0102	0.0087	0.008
34	0.0086	0.0082	0.100	0.0063	0.0104	0.0092	0.0095	0.0077	0.007
35	0.0078	0.0075	0.106	0.0056	0.0095	0.0084	0.0090	0.0069	0.005
36	0.0070	0.0067	0.112	0.0050	0.0090	0.0076	0.0075	0.0061	0.004
37	0.0066	0.0064	0.118	0.0045	0.0085	0.0068	0.0065	0.0054	......
38	0.0063	0.0060	0.124	0.0040	0.0080	0.0060	0.0057	0.0048	......
39	......	......	0.130	0.0035	0.0075	0.0052	0.0050	0.0043	......
40	......	......	0.138	0.0031	0.0070	0.0048	0.0045	0.0039	......
41	......	......	......	0.0028	......	0.0044	......	0.0034	......
42	......	......	......	0.0025	......	0.0040	......	0.0031	......
43	......	......	......	0.0022	......	0.0036	......	0.0027	......
44	......	......	......	0.0020	......	0.0032	......	0.0024	......
45	......	......	......	0.0018	......	0.0028	......	0.0021	......
46	......	......	......	0.0016	......	0.0024	......	0.0019	......
47	......	......	......	0.0014	......	0.0020	......	0.0017	......
48	......	......	......	0.0012	......	0.0016	......	0.0015	......
49	......	......	......	0.0010	......	0.0012	......	0.0013	......
50	......	......	......	0.00098	......	0.0010	......	0.0012	......

Table E-9. Sheet Metal Gauge

Sheet Copper		Sheet Zinc		Tin Plate		Stainless Steel		
Oz per sq foot	Thickness	Gauge No.	Thickness	Gauge No.	Thickness	Gauge No.	Average Sheet Thickness 4 x 8 foot	6 x 12 foot
96 oz	0.1296	28	1.000	6X	0.028	6	*	*
88 oz	0.1188	27	0.500	4X	0.022	7	0.187	*
80 oz	0.1080	26	0.375	3X	0.019	8	0.165	*
72 oz	0.0972	25	0.250	2X	0.017	9	*	0.141
64 oz	0.0864	24	0.125	1X	0.016	10	0.135	0.125
56 oz	0.0756	23	0.100	1C	0.013	11	0.120	0.109
52 oz	0.0702	22	0.090			12	0.105	0.094
48 oz	0.0648	21	0.080			13	0.090	0.078
44 oz	0.0594	20	0.070			14	0.075	*
40 oz	0.0540	19	0.060			15	*	0.063
36 oz	0.0486	18	0.055			16	0.060	*
32 oz	0.0432	17	0.050			17	*	0.050
28 oz	0.0378	16	0.045			18	0.048	0.044
26 oz	0.0351	15	0.040			19	0.042	0.038
24 oz	0.0324	14	0.036			20	0.036	*
20 oz	0.0270	13	0.032			21	*	0.031
18 oz	0.0243	12	0.028			22	0.030	*
16 oz	0.0216	11	0.024			23	*	0.025
14 oz	0.0189	10	0.020			24	0.024	*
13 oz	0.0175	9	0.018			25	*	0.019
12 oz	0.0162	8	0.016			26	0.018	*
10 oz	0.0135	7	0.014			27	*	0.016
9 oz	0.0121	6	0.012			28	0.015	
8 oz	0.0108	5	0.010					
7 oz	0.0094	4	0.008					
6 oz	0.0081	3	0.006					
5 oz	0.0067							
4 oz	0.0054							

*Not normally manufactured in these gauges.

Table E-10. Elements and Related Chemical Symbols

Chemical Symbol	Element	Chemical Symbol	Element
Ar	Argon	Mo	Molybdenum
Ac	Actinium	Md	Mendelevium
Ag	Silver	N	Nitrogen
Al	Aluminum	Na	Sodium
Am	Americium	Nb	Niobium
As	Arsenic	Nd	Neodymium
At	Astatine	Ne	Neon
Au	Gold	Ni	Nickel
B	Boron	No	Nobelium
Ba	Barium	Np	Neptunium
Be	Beryllium	O	Oxygen
Bi	Bismuth	Os	Osmium
Bk	Berkelium	P	Phosphorus
Br	Bromine	Pa	Protactinium
C	Carbon	Pb	Lead
Ca	Calcium	Pd	Palladium
Cd	Cadmium	Pm	Promethium
Ce	Cerium	Po	Polonium
Cf	Californium	Pr	Praseodymium
Cl	Chlorine	Pt	Platinum
Cm	Curium	Pu	Plutonium
Co	Cobalt	Ra	Radium
Cr	Chromium	Rb	Rubidium
Cs	Cesium	Re	Rhenium
Cu	Copper	Rh	Rhodium
Dy	Dysprosium	Rn	Radon
Es	Einsteinium	Ru	Ruthenium
Er	Erbium	S	Sulfur
Eu	Europium	Sb	Antimony
F	Fluorine	Sc	Scandium
Fe	Iron	Se	Selenium
Fm	Fermium	Si	Silicon
Fr	Francium	Sm	Samarium
Ga	Gallium	Sn	Tin
Gd	Gadolinium	Sr	Strontium
Ge	Germanium	Ta	Tantalum
H	Hydrogen	Tb	Terbium
He	Helium	Tc	Technetium
Hf	Hafnium	Te	Tellurium
Hg	Mercury	Th	Thorium
Ho	Holmium	Ti	Titanium
I	Iodine	Tl	Thallium
In	Indium	Tm	Thulium
Ir	Iridium	U	Uranium
K	Potassium	V	Vanadium
Kr	Krypton	W	Tungsten
La	Lanthanum	Xe	Xenon
Li	Lithium	Y	Yttrium
Lu	Lutetium	Yb	Ytterbium
Lr	Lawrencium	Zn	Zinc
Mg	Magnesium	Zr	Zirconium
Mn	Manganese		

Table E-11. Decimal Equivalents of Fractions of an Inch

Inch Fraction	Decimal Equivalent	Inch Fraction	Decimal Equivalent
1/64	0.015625	33/64	0.515625
1/32	0.031250	17/32	0.531250
3/64	0.046875	35/64	0.546875
1/16	0.062500	9/16	0.562500
5/64	0.078125	37/64	0.578125
3/32	0.093750	19/32	0.593750
7/64	0.109375	39/64	0.609375
1/8	0.125000	5/8	0.625000
9/64	0.140625	41/64	0.640625
5/32	0.156250	21/32	0.656250
11/64	0.171875	43/64	0.671875
3/16	0.187500	11/16	0.687500
13/64	0.203125	45/64	0.703125
7/32	0.218750	23/32	0.718750
15/64	0.234375	47/64	0.734375
1/4	0.250000	3/4	0.750000
17/64	0.265625	49/64	0.765625
9/32	0.281250	25/32	0.781250
19/64	0.296875	51/64	0.796875
5/16	0.312500	13/16	0.812500
21/64	0.328125	53/64	0.828125
11/32	0.343750	27/32	0.843750
23/64	0.359375	55/64	0.859375
3/8	0.375000	7/8	0.875000
25/64	0.390625	57/64	0.890625
13/32	0.406250	29/32	0.906250
27/64	0.421875	59/64	0.921875
7/16	0.437500	15/16	0.937500
29/64	0.453125	61/64	0.953125
15/32	0.468750	31/32	0.968750
31/64	0.484375	63/64	0.984375
1/2	0.500000	1	1.000000

Table E-12.   Inches and Equivalents in Millimeter
(1/64 Inch to 100 Inches)

Inches	MM	Inches	MM	Inches	MM
1/64	0.397	7/8	22.225	48	1219.200
1/32	0.794	57/64	22.622	49	1244.600
3/64	1.191	29/32	23.019	50	1270.000
1/16	1.588	59/64	23.416	51	1295.400
5/64	1.984	15/16	23.813	52	1320.800
3/32	2.381	61/64	24.209	53	1346.200
7/64	2.778	31/32	24.606	54	1371.600
1/8	3.175	63/64	25.003	55	1397.000
9/64	3.572	1	25.400	56	1422.400
5/32	3.969	2	50.800	57	1447.800
11/64	4.366	3	76.200	58	1473.200
3/16	4.763	4	101.600	59	1498.600
13/64	5.159	5	127.000	60	1524.000
7/32	5.556	6	152.400	61	1549.400
15/64	5.953	7	177.800	62	1574.800
1/4	6.350	8	203.200	63	1600.200
17/64	6.747	9	228.600	64	1625.600
9/32	7.144	10	254.000	65	1651.000
19/64	7.541	11	279.400	66	1676.400
5/16	7.938	12	304.800	67	1701.800
21/64	8.334	13	330.200	68	1727.200
11/32	8.731	14	355.600	69	1752.600
23/64	9.128	15	381.000	70	1778.000
3/8	9.525	16	406.400	71	1803.400
25/64	9.922	17	431.800	72	1828.800
13/32	10.319	18	457.200	73	1854.200
27/64	10.716	19	482.600	74	1879.600
7/16	11.113	20	508.000	75	1905.000
29/64	11.509	21	533.400	76	1930.400
15/32	11.906	22	558.800	77	1955.800
31/64	12.303	23	584.200	78	1981.200
1/2	12.700	24	609.600	79	2006.600
33/64	13.097	25	635.000	80	2032.000
17/32	13.494	26	660.400	81	2057.400
35/64	13.891	27	685.800	82	2082.800
9/16	14.288	28	711.200	83	2108.200
37/64	14.684	29	736.600	84	2133.600
19/32	15.081	30	762.000	85	2159.000
39/64	15.478	31	787.400	86	2184.400
5/8	15.875	32	812.800	87	2209.800
41/64	16.272	33	838.200	88	2235.200
21/32	16.669	34	863.600	89	2260.600
43/64	17.066	35	889.000	90	2286.000
11/16	17.463	36	914.400	91	2311.400
45/64	17.859	37	939.800	92	2336.800
23/32	18.256	38	965.200	93	2362.200
47/64	18.653	39	990.600	94	2387.600
3/4	19.050	40	1016.000	95	2413.000
49/64	19.447	41	1041.400	96	2438.400
25/32	19.844	42	1066.800	97	2463.800
51/64	20.241	43	1092.200	98	2489.200
13/16	20.638	44	1117.600	99	2514.600
53/64	21.034	45	1143.000	100	2540.000
27/32	21.431	46	1168.400		
55/64	21.828	47	1193.800		

GLOSSARY

Section I. GENERAL

G-1. GENERAL

This glossary of welding terms has been prepared to acquaint welding personnel with nomenclatures and definitions of common terms related to welding and allied processes, methods, techniques, and applications.

G-2. SCOPE

The welding terms listed in section II of this chapter are those terms used to describe and define the standard nomenclatures and language used in this manual. This glossary is a very important part of the manual and should be carefully studied and regularly referred to for better understanding of common welding terms and definitions. Terms and nomenclatures listed herein are grouped in alphabetical order.

Section II. WELDING TERMS

G-3. WELDING TERMS

A

ACETONE: A flammable, volatile liquid used in acetylene cylinders to dissolve and stabilize acetylene under high pressure.

ACETYLENE: A highly combustible gas composed of carbon and hydrogen. Used as a fuel gas in the oxyacetylene welding process.

ACTUAL THROAT: See THROAT OF FILLET WELD.

AIR-ACETYLENE: A low temperature flame produced by burning acetylene with air instead of oxygen.

AIR-ARC CUTTING: An arc cutting process in which metals to be cut are melted by the heat of the carbon arc.

ALLOY: A mixture with metallic properties composed of two or more elements, of which at least one is a metal.

ALTERNATING CURRENT: An electric current that reverses its direction at regularly recurring intervals.

AMMETER: An instrument for measuring electrical current in amperes by an indicator activated by the movement of a coil in a magnetic field or by the longitudinal expansion of a wire carrying the current.

ANNEALING: A comprehensive term used to describe the heating and cooling cycle of steel in the solid state. The term annealing usually implies relatively slow cooling. In annealing, the temperature of the operation, the rate of heating and cooling, and the time the metal is held at heat depend upon the composition, shape, and size of the steel product being treated, and the purpose of the treatment. The more important purposes for which steel is annealed are as follows: to remove stresses; to induce softness; to alter ductility, toughness, electric, magnetic, or other physical and mechanical properties; to change the crystalline structure; to remove gases; and to produce a definite microstructure.

ARC BLOW: The deflection of an electric arc from its normal path because of magnetic forces.

ARC BRAZING: A brazing process wherein the heat is obtained from an electric arc

G-3. WELDING TERMS (cont)

A (cont)

formed between the base metal and an electrode, or between two electrodes.

ARC CUTTING: A group of cutting processes in which the cutting of metals is accomplished by melting with the heat of an arc between the electrode and the base metal. See CARBON-ARC CUTTING, METAL-ARC CUTTING, ARC-OXYGEN CUTTING, AND AIR-ARC CUTTING.

ARC LENGTH: The distance between the tip of the electrode and the weld puddle.

ARC-OXYGEN CUTTING: An oxygen-cutting process used to sever metals by a chemical reaction of oxygen with a base metal at elevated temperatures.

ARC VOLTAGE: The voltage across the welding arc.

ARC WELDING: A group of welding processes in which fusion is obtained by heating with an electric arc or arcs, with or without the use of filler metal.

AS WELDED: The condition of weld metal, welded joints, and weldments after welding and prior to any subsequent thermal, mechanical, or chemical treatments.

ATOMIC HYDROGEN WELDING: An arc welding process in which fusion is obtained by heating with an arc maintained between two metal electrodes in an atmosphere of hydrogen. Pressure and/or filler metal may or may not be used.

AUSTENITE: The non-magnetic form of iron characterized by a face-centered cubic lattice crystal structure. It is produced by heating steel above the upper critical temperature and has a high solid solubility for carbon and alloying elements.

AXIS OF A WELD: A line through the length of a weld, perpendicular to a cross section at its center of gravity.

B

BACK FIRE: The momentary burning back of a flame into the tip, followed by a snap or pop, then immediate reappearance or burning out of the flame.

BACK PASS: A pass made to deposit a back weld.

BACK UP: In flash and upset welding, a locator used to transmit all or a portion of the upsetting force to the workpieces.

BACK WELD: A weld deposited at the back of a single groove weld.

BACKHAND WELDING: A welding technique in which the flame is directed towards the completed weld.

BACKING STRIP: A piece of material used to retain molten metal at the root of the weld and/or increase the thermal capacity of the joint so as to prevent excessive warping of the base metal.

BACKING WELD: A weld bead applied to the root of a single groove joint to assure complete root penetration.

BACKSTEP: A sequence in which weld bead increments are deposited in a direction opposite to the direction of progress.

BARE ELECTRODE: An arc welding electrode that has no coating other than that incidental to the drawing of the wire.

BARE METAL-ARC WELDING: An arc welding process in which fusion is obtained by heating with an unshielded arc between a bare or lightly coated electrode and the work. Pressure is not used and filler metal is obtained from the electrode.

BASE METAL: The metal to be welded or cut. In alloys, it is the metal present in the largest proportion.

BEAD WELD: A type of weld composed of one or more string or weave beads deposited on an unbroken surface.

BEADING: See STRING BEAD WELDING and WEAVE BEAD.

BEVEL ANGLE: The angle formed between the prepared edge of a member and a plane perpendicular to the surface of the member.

BLACKSMITH WELDING: See FORGE WELDING.

BLOCK BRAZING: A brazing process in which bonding is produced by the heat obtained from heated blocks applied to the parts to be joined and by a nonferrous filler metal having a melting point above 800 °F (427 °C), but below that of the base metal. The filler metal is distributed in the joint by capillary attraction.

BLOCK SEQUENCE: A building up sequence of continuous multipass welds in which separated lengths of the weld are completely or partially built up before intervening lengths are deposited. See BUILDUP SEQUENCE.

BLOW HOLE: See GAS POCKET.

BOND: The junction of the welding metal and the base metal.

BOXING: The operation of continuing a fillet weld around a corner of a member as an extension of the principal weld.

BRAZING: A group of welding processes in which a groove, fillet, lap, or flange joint is bonded by using a nonferrous filler metal having a melting point above 800 °F (427 °C), but below that of the base metals. Filler metal is distributed in the joint by capillary attraction.

BRAZE WELDING: A method of welding by using a filler metal that liquifies above 450 °C (842 °F) and below the solid state of the base metals. Unlike brazing, in braze welding, the filler metal is not distributed in the joint by capillary action.

BRIDGING: A welding defect caused by poor penetration. A void at the root of the weld is spanned by weld metal.

BUCKLING: Distortion caused by the heat of a welding process.

BUILDUP SEQUENCE: The order in which the weld beads of a multipass weld are deposited with respect to the cross section of a joint. See BLOCK SEQUENCE.

BUTT JOINT: A joint between two workpieces in such a manner that the weld joining the parts is between the surface planes of both of the pieces joined.

BUTT WELD: A weld in a butt joint.

BUTTER WELD: A weld composed of one or more string or weave beads laid down on an unbroken surface to obtain desired properties or dimensions.

C

CAPILLARY ATTRACTION: The phenomenon by which adhesion between the molten filler metal and the base metals, together with surface tension of the molten filler metal, causes distribution of the filler metal between the properly fitted surfaces of the joint to be brazed.

CARBIDE PRECIPITATION: A condition occurring in austenitic stainless steel which contains carbon in a supersaturated solid solution. This condition is unstable. Agitation of the steel during welding causes the excess carbon in solution to precipitate. This effect is also called weld decay.

CARBON-ARC CUTTING: A process of cutting metals with the heat of an arc between a carbon electrode and the work.

CARBON-ARC WELDING: A welding process in which fusion is produced by an arc between a carbon electrode and the work. Pressure and/or filler metal and/or shielding may or may not be used.

CARBURIZING FLAME: An oxyacetylene flame in which there is an excess of acetylene. Also called excess acetylene or reducing flame.

CASCADE SEQUENCE: Subsequent beads are stopped short of a previous bead, giving a cascade effect.

CASE HARDENING: A process of surface hardening involving a change in the composition of the outer layer of an iron base alloy by inward diffusion from a gas or liquid, followed by appropriate thermal treatment. Typical hardening processes are carburizing, cyaniding, carbonitriding, and nitriding.

CHAIN INTERMITTENT FILLET WELDS: Two lines of intermittent fillet welds in a T or

G-3. WELDING TERMS (cont)

C (cont)

lap joint in which the welds in one line are approximately opposite those in the other line.

CHAMFERING: The preparation of a welding contour, other than for a square groove weld, on the edge of a joint member.

COALESCENCE: The uniting or fusing of metals upon heating.

COATED ELECTRODE: An electrode having a flux applied externally by dipping, spraying, painting, or other similar methods. Upon burning, the coat produces a gas which envelopes the arc.

COMMUTATORY CONTROLLED WELDING: The making of a number of spot or projection welds in which several electrodes, in simultaneous contact with the work, progressively function under the control of an electrical commutating device.

COMPOSITE ELECTRODE: A filler metal electrode used in arc welding, consisting of more than one metal component combined mechanically. It may or may not include materials that improve the properties of the weld, or stabilize the arc.

COMPOSITE JOINT: A joint in which both a thermal and mechanical process are used to unite the base metal parts.

CONCAVITY: The maximum perpendicular distance from the face of a concave fillet weld to a line joining the toes.

CONCURRENT HEATING: Supplemental heat applied to a structure during the course of welding.

CONE: The conical part of a gas flame next to the orifice of the tip.

CONSUMABLE INSERT: Preplaced filler metal which is completely fused into the root of the joint and becomes part of the weld.

CONVEXITY: The maximum perpendicular distance from the face of a convex fillet weld to a line joining the toes.

CORNER JOINT: A joint between two members located approximately at right angles to each other in the form of an L.

COVER GLASS: A clear glass used in goggles, hand shields, and helmets to protect the filter glass from spattering material.

COVERED ELECTRODE: A metal electrode with a covering material which stabilizes the arc and improves the properties of the welding metal. The material may be an external wrapping of paper, asbestos, and other materials or a flux covering.

CRACK: A fracture type discontinuity characterized by a sharp tip and high ratio of length and width to opening displacement.

CRATER: A depression at the termination of an arc weld.

CRITICAL TEMPERATURE: The transition temperature of a substance from one crystalline form to another.

CURRENT DENSITY: Amperes per square inch of the electrode cross sectional area.

CUTTING TIP: A gas torch tip especially adapted for cutting.

CUTTING TORCH: A device used in gas cutting for controlling the gases used for preheating and the oxygen used for cutting the metal.

CYLINDER: A portable cylindrical container used for transportation and storage of a compressed gas.

D

DEFECT: A discontinuity or discontinuities which, by nature or accumulated effect (for example, total crack length), render a part or product unable to meet minimum applicable acceptance standards or specifications. This term designates rejectability.

DEPOSITED METAL: Filler metal that has been added during a welding operation.

DEPOSITION EFFICIENCY: The ratio of the weight of deposited metal to the net weight of electrodes consumed, exclusive of stubs.

DEPTH OF FUSION: The distance from the original surface of the base metal to that point at which fusion ceases in a welding operation.

DIE:

a. Resistance Welding. A member, usually shaped to the work contour, used to clamp the parts being welded and conduct the welding current.

b. Forge Welding. A device used in forge welding primarily to form the work while hot and apply the necessary pressure.

DIE WELDING: A forge welding process in which fusion is produced by heating in a furnace and by applying pressure by means of dies.

DIP BRAZING: A brazing process in which bonding is produced by heating in a molten chemical or metal bath and by using a nonferrous filler metal having a melting point above 800 ^{O}F (427 ^{O}C), but below that of the base metals. The filler metal is distributed in the joint by capillary attraction. When a metal bath is used, the bath provides the filler metal.

DIRECT CURRENT ELECTRODE NEGATIVE (DCEN): The arrangement of direct current arc welding leads in which the work is the positive pole and the electrode is the negative pole of the welding arc.

DIRECT CURRENT ELECTRODE POSITIVE (DCEP): The arrangement of direct current arc welding leads in which the work is the negative pole and the electrode is the positive pole of the welding arc.

DISCONTINUITY: An interruption of the typical structure of a weldment, such as lack of homogeneity in the mechanical, metallurgical, or physical characteristics of the material or weldment. A discontinuity is not necessarily a defect.

DRAG: The horizontal distance between the point of entrance and the point of exit of a cutting oxygen stream.

DUCTILITY: The property of a metal which allows it to be permanently deformed, in tension, before final rupture. Ductility is commonly evaluated by tensile testing in which the amount of elongation and the reduction of area of the broken specimen, as compared to the original test specimen, are measured and calculated.

DUTY CYCLE: The percentage of time during an arbitrary test period, usually 10 minutes, during which a power supply can be operated at its rated output without overloading.

E

EDGE JOINT: A joint between the edges of two or more parallel or nearly parallel members.

EDGE PREPARATION: The contour prepared on the edge of a joint member for welding.

EFFECTIVE LENGTH OF WELD: The length of weld throughout which the correctly proportioned cross section exits.

ELECTRODE:

a. Metal-Arc. Filler metal in the form of a wire or rod, whether bare or covered, through which current is conducted between the electrode holder and the arc.

b. Carbon-Arc. A carbon or graphite rod through which current is conducted between the electrode holder and the arc.

c. Atomic Hydrogen. One of the two tungsten rods between the points of which the arc is maintained.

d. Electrolytic Oxygen-Hydrogen Generation. The conductors by which current enters and leaves the water, which is decomposed by the passage of the current.

e. Resistance Welding. The part or parts of a resistance welding machine through which the welding current and the pressure are applied directly to the work.

G-3. WELDING TERMS (cont)

E (cont)

ELECTRODE FORCE:
   a. _Dynamic_. In spot, seam, and projection welding, the force (pounds) between the electrodes during the actual welding cycle.
   b. _Theoretical_. In spot, seam, and projection welding, the force, neglecting friction and inertia, available at the electrodes of a resistance welding machine by virtue of the initial force application and the theoretical mechanical advantage of the system.
   c. _Static_. In spot, seam, and projection welding, the force between the electrodes under welding conditions, but with no current flowing and no movement in the welding machine.
ELECTRODE HOLDER: A device used for mechanically holding the electrode and conducting current to it.
ELECTRODE SKID: The sliding of an electrode along the surface of the work during spot, seam, or projection welding.
EMBOSSMENT: A rise or protrusion from the surface of a metal.
ETCHING: A process of preparing metallic specimens and welds for macrographic or micrographic examination.

F

FACE REINFORCEMENT: Reinforcement of weld at the side of the joint from which welding was done.
FACE OF WELD: The exposed surface of a weld, made by an arc or gas welding process, on the side from which welding was done.
FAYING SURFACE: That surface of a member that is in contact with another member to which it is joined.
FERRITE: The virtually pure form of iron existing below the lower critical temperature and characterized by a body-centered cubic lattice crystal structure. It is magnetic and has very slight solid solubility for carbon.
FILLER METAL: Metal to be added in making a weld.
FILLET WELD: A weld of approximately triangular cross section, as used in a lap joint, joining two surfaces at approximately right angles to each other.
FILTER GLASS: A colored glass used in goggles, helmets, and shields to exclude harmful light rays.
FLAME CUTTING: See OXYGEN CUTTING.
FLAME GOUGING: See OXYGEN GOUGING.
FLAME HARDENING: A method for hardening a steel surface by heating with a gas flame followed by a rapid quench.
FLAME SOFTENING: A method for softening steel by heating with a gas flame followed by slow cooling.
FLASH: Metal and oxide expelled from a joint made by a resistance welding process.
FLASH WELDING: A resistance welding process in which fusion is produced, simultaneously over the entire area of abutting surfaces, by the heat obtained from resistance to the flow of current between two surfaces and by the application of pressure after heating is substantially completed. Flashing is accompanied by expulsion of metal from the joint.
FLASHBACK: The burning of gases within the torch or beyond the torch in the hose, usually with a shrill, hissing sound.
FLAT POSITION: The position in which welding is performed from the upper side of the joint and the face of the weld is approximately horizontal.
FLOW BRAZING: A process in which bonding is produced by heating with a molten

nonferrous filler metal poured over the joint until the brazing temperature is attained. The filler metal is distributed in the joint by capillary attraction. See BRAZING.

FLOW WELDING: A process in which fusion is produced by heating with molten filler metal poured over the surfaces to be welded until the welding temperature is attained and the required filler metal has been added. The filler metal is not distributed in the joint by capillary attraction.

FLUX: A cleaning agent used to dissolve oxides, release trapped gases and slag, and to cleanse metals for welding, soldering, and brazing.

FOREHAND WELDING: A gas welding technique in which the flame is directed against the base metal ahead of the completed weld.

FORGE WELDING: A group of welding processes in which fusion is produced by heating in a forge or furnace and applying pressure or blows.

FREE BEND TEST: A method of testing weld specimens without the use of a guide.

FULL FILLET WELD: A fillet weld whose size is equal to the thickness of the thinner member joined.

FURNACE BRAZING: A process in which bonding is produced by the furnace heat and a nonferrous filler metal having a melting point above 800 ^{O}F (427 ^{O}C), but below that of the base metals. The filler metal is distributed in the joint by capillary attraction.

FUSION: A thorough and complete mixing between the two edges of the base metal to be joined or between the base metal and the filler metal added during welding.

FUSION ZONE (FILLER PENETRATION): The area of base metal melted as determined on the cross section of a weld.

G

GAS CARBON-ARC WELDING: An arc welding process in which fusion is produced by heating with an electric arc between a carbon electrode and the work. Shielding is obtained from an inert gas such as helium or argon. Pressure and/or filler metal may or may not be used.

GAS METAL-ARC (MIG) WELDING (GMAW): An arc welding process in which fusion is produced by heating with an electric arc between a metal electrode and the work. Shielding is obtained from an inert gas such as helium or argon. Pressure and/or filler metal may or may not be used.

GAS POCKET: A weld cavity caused by the trapping of gases released by the metal when cooling.

GAS TUNGSTEN-ARC (TIG) WELDING (GTAW): An arc welding process in which fusion is produced by heating with an electric arc between a tungsten electrode and the work while an inert gas flows around the weld area to prevent oxidation. No flux is used.

GAS WELDING: A process in which the welding heat is obtained from a gas flame.

GLOBULAR TRANSFER (ARC WELDING): A type of metal transfer in which molten filler metal is transferred across the arc in large droplets.

GOGGLES: A device with colored lenses which protect the eyes from harmful radiation during welding and cutting operations.

GROOVE: The opening provided between two members to be joined by a groove weld.

GROOVE ANGLE: The total included angle of the groove between parts to be joined by a groove weld.

GROOVE FACE: That surface of a member included in the groove.

GROOVE RADIUS: The radius of a J or U groove.

GROOVE WELD: A weld made by depositing filler metal in a groove between two members to be joined.

GROUND CONNECTION: The connection of the work lead to the work.

G-3. WELDING TERMS (cont)

G (cont)

GROUND LEAD: See WORK LEAD.

GUIDED BEND TEST: A bending test in which the test specimen is bent to a definite shape by means of a jig.

H

HAMMER WELDING: A forge welding process.

HAND SHIELD: A device used in arc welding to protect the face and neck. It is equipped with a filter glass lens and is designed to be held by hand.

HARD FACING: A particular form of surfacing in which a coating or cladding is applied to a surface for the main purpose of reducing wear or loss of material by abrasion, impact, erosion, galling, and cavitation.

HARD SURFACING: The application of a hard, wear-resistant alloy to the surface of a softer metal.

HARDENING:

a. The heating and quenching of certain iron-base alloys from a temperature above the critical temperature range for the purpose of producing a hardness superior to that obtained when the alloy is not quenched. This term is usually restricted to the formation of martensite.

b. Any process of increasing the hardness of metal by suitable treatment, usually involving heating and cooling.

HEAT AFFECTED ZONE: That portion of the base metal whose structure or properties have been changed by the heat of welding or cutting.

HEAT TIME: The duration of each current impulse in pulse welding.

HEAT TREATMENT: An operation or combination of operations involving the heating and cooling of a metal or an alloy in the solid state for the purpose of obtaining certain desirable conditions or properties. Heating and cooling for the sole purpose of mechanical working are excluded from the meaning of the definition.

HEATING GATE: The opening in a thermit mold through which the parts to be welded are preheated.

HELMET: A device used in arc welding to protect the face and neck. It is equipped with a filter glass and is designed to be worn on the head.

HOLD TIME: The time that pressure is maintained at the electrodes after the welding current has stopped.

HORIZONTAL WELD: A bead or butt welding process with its linear direction horizontal or inclined at an angle less than 45 degrees to the horizontal, and the parts welded being vertically or approximately vertically disposed.

HORN: The electrode holding arm of a resistance spot welding machine.

HORN SPACING: In a resistance welding machine, the unobstructed work clearance between horns or platens at right angles to the throat depth. This distance is measured with the horns parallel and horizontal at the end of the downstroke.

HOT SHORT: A condition which occurs when a metal is heated to that point, prior to melting, where all strength is lost but the shape is still maintained.

HYDROGEN BRAZING: A method of furnace brazing in a hydrogen atmosphere.

HYDROMATIC WELDING: See PRESSURE CONTROLLED WELDING.

HYGROSCOPIC: Readily absorbing and retaining moisture.

I

IMPACT TEST: A test in which one or more blows are suddenly applied to a specimen. The results are usually expressed in terms of energy absorbed or number of blows of a given intensity required to break the specimen.

IMPREGNATED-TAPE METAL-ARC WELDING: An arc welding process in which fusion is produced by heating with an electric arc between a metal electrode and the work. Shielding is obtained from decomposition of an impregnated tape wrapped around the electrode as it is fed to the arc. Pressure is not used, and filler metal is obtained from the electrode.

INDUCTION BRAZING: A process in which bonding is produced by the heat obtained from the resistance of the work to the flow of induced electric current and by using a nonferrous filler metal having a melting point above 800 $^{\circ}$F (427 $^{\circ}$C), but below that of the base metals. The filler metal is distributed in the joint by capillary attraction.

INDUCTION WELDING: A process in which fusion is produced by heat obtained from resistance of the work to the flow of induced electric current, with or without the application of pressure.

INERT GAS: A gas which does not normally combine chemically with the base metal or filler metal.

INTERPASS TEMPERATURE: In a multipass weld, the lowest temperature of the deposited weld metal before the next pass is started.

J

JOINT: The portion of a structure in which separate base metal parts are joined.

JOINT PENETRATION: The maximum depth a groove weld extends from its face into a joint, exclusive of reinforcement.

K

KERF: The space from which metal has been removed by a cutting process.

L

LAP JOINT: A joint between two overlapping members.

LAYER: A stratum of weld metal, consisting of one or more weld beads.

LEG OF A FILLET WELD: The distance from the root of the joint to the toe of the fillet weld.

LIQUIDUS: The lowest temperature at which a metal or an alloy is completely liquid.

LOCAL PREHEATING: Preheating a specific portion of a structure.

LOCAL STRESS RELIEVING: Stress relieving heat treatment of a specific portion of a structure.

M

MANIFOLD: A multiple header for connecting several cylinders to one or more torch supply lines.

MARTENSITE: Martensite is a microconstituent or structure in quenched steel characterized by an acicular or needle-like pattern on the surface of polish. It has the maximum hardness of any of the structures resulting from the decomposition products of austenite.

MASH SEAM WELDING: A seam weld made in a lap joint in which the thickness at the lap is reduced to approximately the thickness of one of the lapped joints by applying pressure while the metal is in a plastic state.

MELTING POINT: The temperature at which a metal begins to liquefy.

MELTING RANGE: The temperature range between solidus and liquidus.

G-3. WELDING TERMS (cont)

M (cont)

MELTING RATE: The weight or length of electrode melted in a unit of time.

METAL-ARC CUTTING: The process of cutting metals by melting with the heat of the metal arc.

METAL-ARC WELDING: An arc welding process in which a metal electrode is held so that the heat of the arc fuses both the electrode and the work to form a weld.

METALLIZING: A method of overlay or metal bonding to repair worn parts.

MIXING CHAMBER: That part of a welding or cutting torch in which the gases are mixed for combustion.

MULTI-IMPULSE WELDING: The making of spot, projection, and upset welds by more than one impulse of current. When alternating current is used each impulse may consist of a fraction of a cycle or a number of cycles.

N

NEUTRAL FLAME: A gas flame in which the oxygen and acetylene volumes are balanced and both gases are completely burned.

NICK BREAK TEST: A method for testing the soundness of welds by nicking each end of the weld, then giving the test specimen a sharp hammer blow to break the weld from nick to nick. Visual inspection will show any weld defects.

NONFERROUS: Metals which contain no iron. Aluminum, brass, bronze, copper, lead, nickel, and titanium are nonferrous.

NORMALIZING: Heating iron-base alloys to approximately 100 ^{O}F (38 ^{O}C) above the critical temperature range followed by cooling to below that range in still air at ordinary temperature.

NUGGET: The fused metal zone of a resistance weld.

O

OPEN CIRCUIT VOLTAGE: The voltage between the terminals of the welding source when no current is flowing in the welding circuit.

OVERHEAD POSITION: The position in which welding is performed from the underside of a joint and the face of the weld is approximately horizontal.

OVERLAP: The protrusion of weld metal beyond the bond at the toe of the weld.

OXIDIZING FLAME: An oxyacetylene flame in which there is an excess of oxygen. The unburned excess tends to oxidize the weld metal.

OXYACETYLENE CUTTING: An oxygen cutting process in which the necessary cutting temperature is maintained by flames obtained from the combustion of acetylene with oxygen.

OXYACETYLENE WELDING: A welding process in which the required temperature is attained by flames obtained from the combustion of acetylene with oxygen.

OXY-ARC CUTTING: An oxygen cutting process in which the necessary cutting temperature is maintained by means of an arc between an electrode and the base metal.

OXY-CITY GAS CUTTING: An oxygen cutting process in which the necessary cutting temperature is maintained by flames obtained from the combustion of city gas with oxygen.

OXYGEN CUTTING: A process of cutting ferrous metals by means of the chemical action of oxygen on elements in the base metal at elevated temperatures.

OXYGEN GOUGING: An application of oxygen cutting in which a chamfer or groove is formed.

OXY-HYDROGEN CUTTING: An oxygen cutting process in which the necessary cutting temperature is maintained by flames obtained by the combustion of hydrogen with oxygen.

OXY-HYDROGEN WELDING: A gas welding process in which the required welding temperature is attained by flames obtained from the combustion of hydrogen with oxygen.

OXY-NATURAL GAS CUTTING: An oxygen cutting process in which the necessary cutting temperature is maintained by flames obtained from the combustion of natural gas with oxygen.

OXY-PROPANE CUTTING: An oxygen cutting process in which the necessary cutting temperature is maintained by flames obtained from the combustion of propane with oxygen.

P

PASS: The weld metal deposited in one general progression along the axis of the weld.

PEENING: The mechanical working of metals by means of hammer blows. Peening tends to stretch the surface of the cold metal, thereby relieving contraction stresses.

PENETRANT INSPECTION:

a. Fluorescent. A water washable penetrant with high fluorescence and low surface tension. It is drawn into small surface openings by capillary action. When exposed to black light, the dye will fluoresce.

b. Dye. A process which involves the use of three noncorrosive liquids. First, the surface cleaner solution is used. Then the penetrant is applied and allowed to stand at least 5 minutes. After standing, the penetrant is removed with the leaner solution and the developer is applied. The dye penetrant, which has remained in the surface discontinuity, will be drawn to the surface by the developer resulting in bright red indications.

PERCUSSIVE WELDING: A resistance welding process in which a discharge of electrical energy and the application of high pressure occurs simultaneously, or with the electrical discharge occurring slightly before the application of pressure.

PERLITE: Perlite is the lamellar aggregate of ferrite and iron carbide resulting from the direct transformation of austenite at the lower critical point.

PITCH: Center to center spacing of welds.

PLUG WELD: A weld is made in a hole in one member of a lap joint, joining that member to that portion of the surface of the other member which is exposed through the hole. The walls of the hole may or may not be parallel, and the hole may be partially or completely filled with the weld metal.

POKE WELDING: A spot welding process in which pressure is applied manually to one electrode. The other electrode is clamped to any part of the metal much in the same manner that arc welding is grounded.

POROSITY: The presence of gas pockets or inclusions in welding.

POSITIONS OF WELDING: All welding is accomplished in one of four positions: flat, horizontal, overhead, and vertical. The limiting angles of the various positions depend somewhat as to whether the weld is a fillet or groove weld.

POSTHEATING: The application of heat to an assembly after a welding, brazing, soldering, thermal spraying, or cutting operation.

POSTWELD INTERVAL: In resistance welding, the heat time between the end of weld time, or weld interval, and the start of hold time. During this interval, the weld is subjected to mechanical and heat treatment.

PREHEATING: The application of heat to a base metal prior to a welding or cutting operation.

PRESSURE CONTROLLED WELDING: The making of a number of spot or projection welds in which several electrodes function progressively under the control of a pressure sequencing device.

PRESSURE WELDING: Any welding process or method in which pressure is used to complete the weld.

G-3. WELDING TERMS (cont)

P (cont)

PREWELD INTERVAL: In spot, projection, and upset welding, the time between the end of squeeze time and the start of weld time or weld interval during which the material is preheated. In flash welding, it is the time during which the material is preheated.

PROCEDURE QUALIFICATION: The demonstration that welds made by a specific procedure can meet prescribed standards.

PROJECTION WELDING: A resistance welding process between two or more surfaces or between the ends of one member and the surface of another. The welds are localized at predetermined points or projections.

PULSATION WELDING: A spot, projection, or seam welding process in which the welding current is interrupted one or more times without the release of pressure or change of location of electrodes.

PUSH WELDING: The making of a spot or projection weld in which the force is ap-ing current is interrupted one or more times without the release of pressure or change of location of electrodes.

PUSH WELDING: The making of a spot or projection weld in which the force is applied manually to one electrode and the work or a backing bar takes the place of the other electrode.

Q

QUENCHING: The sudden cooling of heated metal with oil, water, or compressed air.

R

REACTION STRESS: The residual stress which could not otherwise exist if the members or parts being welded were isolated as free bodies without connection to other parts of the structure.

REDUCING FLAME: See CARBURIZING FLAME.

REGULATOR: A device used to reduce cylinder pressure to a suitable torch working pressure.

REINFORCED WELD: The weld metal built up above the surface of the two abutting sheets or plates in excess of that required for the size of the weld specified.

RESIDUAL STRESS: Stress remaining in a structure or member as a result of thermal and/or mechanical treatment.

RESISTANCE BRAZING: A brazing process in which bonding is produced by the heat obtained from resistance to the flow of electric current in a circuit of which the workpiece is a part, and by using a nonferrous filler metal having a melting point above 800 ^{O}F (427 ^{O}C), but below that of the base metals. The filler metal is distributed in the joint by capillary attraction.

RESISTANCE BUTT WELDING: A group of resistance welding processes in which the weld occurs simultaneously over the entire contact area of the parts being joined.

RESISTANCE WELDING: A group of welding processes in which fusion is produced by heat obtained from resistance to the flow of electric current in a circuit of which the workpiece is a part and by the application of pressure.

REVERSE POLARITY: The arrangement of direct current arc welding leads in which the work is the negative pole and the electrode is the positive pole of the welding arc.

ROCKWELL HARDNESS TEST: In this test a machine measures hardness by determining the depth of penetration of a penetrator into the specimen under certain arbitrary fixed conditions of test. The penetrator may be either a steel ball or a diamond spherocone.

ROOT: See ROOT OF JOINT and ROOT OF WELD.

ROOT CRACK: A crack in the weld or base metal which occurs at the root of a weld.

ROOT EDGE: The edge of a part to be welded which is adjacent to the root.

ROOT FACE: The portion of the prepared edge of a member to be joined by a groove weld which is not beveled or grooved.

ROOT OF JOINT: That portion of a joint to be welded where the members approach closest to each other. In cross section, the root of a joint may be a point, a line, or an area.

ROOT OF WELD: The points, as shown in cross section, at which the bottom of the weld intersects the base metal surfaces.

ROOT OPENING: The separation between the members to be joined at the root of the joint.

ROOT PENETRATION: The depth a groove weld extends into the root of a joint measured on the centerline of the root cross section.

S

SCARF: The chamfered surface of a joint.

SCARFING: A process for removing defects and checks which develop in the rolling of steel billets by the use of a low velocity oxygen deseaming torch.

SEAL WELD: A weld used primarily to obtain tightness and to prevent leakage.

SEAM WELDING: Welding a lengthwise seam in sheet metal either by abutting or over-lapping joints.

SELECTIVE BLOCK SEQUENCE: A block sequence in which successive blocks are completed in a certain order selected to create a predetermined stress pattern.

SERIES WELDING: A resistance welding process in which two or more welds are made simultaneously by a single welding transformer with the total current passing through each weld.

SHEET SEPARATION: In spot, seam, and projection welding, the gap surrounding the weld between faying surfaces, after the joint has been welded.

SHIELDED WELDING: An arc welding process in which protection from the atmosphere is obtained through use of a flux, decomposition of the electrode covering, or an inert gas.

SHOULDER: See ROOT FACE.

SHRINKAGE STRESS: See RESIDUAL STRESS.

SINGLE IMPULSE WELDING: The making of spot, projection, and upset welds by a single impulse of current. When alternating current is used, an impulse may consist of a fraction of a cycle or a number of cycles.

SIZE OF WELD:

   a. Groove weld. The joint penetration (depth of chamfering plus the root penetration when specified).

   b. Equal leg fillet welds. The leg length of the largest isosceles right triangle which can be inscribed within the fillet weld cross section.

   c. Unequal leg fillet welds. The leg length of the largest right triangle which can be inscribed within the fillet weld cross section.

   d. Flange weld. The weld metal thickness measured at the root of the weld.

SKIP SEQUENCE: See WANDERING SEQUENCE.

SLAG INCLUSION: Non-metallic solid material entrapped in the weld metal or between the weld metal and the base metal.

SLOT WELD: A weld made in an elongated hole in one member of a lap or tee joint joining that member to that portion of the surface of the other member which is exposed through the hole. The hole may be open at one end and may be partially

G-3. WELDING TERMS (cont)

S (cont)

or completely filled with weld metal. (A fillet welded slot should not be construed as conforming to this definition.)

SLUGGING: Adding a separate piece or pieces of material in a joint before or during welding with a resultant welded joint that does not comply with design drawing or specification requirements.

SOLDERING: A group of welding processes which produce coalescence of materials by heating them to suitable temperature and by using a filler metal having a liquidus not exceeding 450 $^{\circ}$C (842 $^{\circ}$F) and below the solidus of the base materials. The filler metal is distributed between the closely fitted surfaces of the joint by capillary action.

SOLIDUS: The highest temperature at which a metal or alloy is completely solid.

SPACER STRIP: A metal strip or bar inserted in the root of a joint prepared for a groove weld to serve as a backing and to maintain the root opening during welding.

SPALL: Small chips or fragments which are sometimes given off by electrodes during the welding operation. This problem is especially common with heavy coated electrodes.

SPATTER: The metal particles expelled during arc and gas welding which do not form a part of the weld.

SPOT WELDING: A resistance welding process in which fusion is produced by the heat obtained from the resistance to the flow of electric current through the workpieces held together under pressure by electrodes. The size and shape of the individually formed welds are limited by the size and contour of the electrodes.

SPRAY TRANSFER: A type of metal transfer in which molten filler metal is propelled axially across the arc in small droplets.

STAGGERED INTERMITTENT FILLET WELD: Two lines of intermittent welding on a joint, such as a tee joint, wherein the fillet increments in one line are staggered with respect to those in the other line.

STORED ENERGY WELDING: The making of a weld with electrical energy accumulated electrostatically, electromagnetically, or electrochemically at a relatively low rate and made available at the required welding rate.

STRAIGHT POLARITY: The arrangement of direct current arc welding leads in which the work is the positive pole and the electrode is the negative pole of the welding arc.

STRESS RELIEVING: A process of reducing internal residual stresses in a metal object by heating to a suitable temperature and holding for a proper time at that temperature. This treatment may be applied to relieve stresses induced by casting, quenching, normalizing, machining, cold working, or welding.

STRING BEAD WELDING: A method of metal arc welding on pieces 3/4 in. (19 mm) thick or heavier in which the weld metal is deposited in layers composed of strings of beads applied directly to the face of the bevel.

STUD WELDING: An arc welding process in which fusion is produced by heating with an electric arc drawn between a metal stud, or similar part, and the other workpiece, until the surfaces to be joined are properly heated. They are brought together under pressure.

SUBMERGED ARC WELDING: An arc welding process in which fusion is produced by heating with an electric arc or arcs between a bare metal electrode or electrodes and the work. The welding is shielded by a blanket of granular, fusible material on the work. Pressure is not used. Filler metal is obtained from the electrode, and sometimes from a supplementary welding rod.

SURFACING: The deposition of filler metal on a metal surface to obtain desired properties or dimensions.

T

TACK WELD: A weld made to hold parts of a weldment in proper alignment until the final welds are made.

TEE JOINT: A joint between two members located approximately at right angles to each other in the form of a T.

TEMPER COLORS: The colors which appear on the surface of steel heated at low temperature in an oxidizing atmosphere.

TEMPER TIME: In resistance welding, that part of the postweld interval during which a current suitable for tempering or heat treatment flows. The current can be single or multiple impulse, with varying heat and cool intervals.

TEMPERING: Reheating hardened steel to some temperature below the lower critical temperature, followed by a desired rate of cooling. The object of tempering a steel that has been hardened by quenching is to release stresses set up, to restore some of its ductility, and to develop toughness through the regulation or readjustment of the embrittled structural constituents of the metal. The time-temperature conditions for tempering may be selected for a given composition of steel to obtain almost any desired combination of properties.

TENSILE STRENGTH: The maximum load per unit of original cross-sectional area sustained by a material during the tension test.

TENSION TEST: A test in which a specimen is broken by applying an increasing load to the two ends. During the test, the elastic properties and the ultimate tensile strength of the material are determined. After rupture, the broken specimen may be measured for elongation and reduction of area.

THERMIT CRUCIBLE: The vessel in which the thermit reaction takes place.

THERMIT MIXTURE: A mixture of metal oxide and finely divided aluminum with the addition of alloying metals as required.

THERMIT MOLD: A mold formed around the parts to be welded to receive the molten metal.

THERMIT REACTION: The chemical reaction between metal oxide and aluminum which produces superheated molten metal and aluminum oxide slag.

THERMIT WELDING: A group of welding processes in which fusion is produced by heating with superheated liquid metal and slag resulting from a chemical reaction between a metal oxide and aluminum, with or without the application of pressure. Filler metal, when used, is obtained from the liquid metal.

THROAT DEPTH: In a resistance welding machine, the distance from the centerline of the electrodes or platens to the nearest point of interference for flatwork or sheets. In a seam welding machine with a universal head, the throat depth is measured with the machine arranged for transverse welding.

THROAT OF FILLET WELD:
a. Theoretical. The distance from the beginning of the root of the joint perpendicular to the hypotenuse of the largest right triangle that can be inscribed within the fillet-weld cross section.
b. Actual. The distance from the root of the fillet weld to the center of its face.

TOE CRACK: A crack in the base metal occurring at the toe of the weld.

TOE OF THE WELD: The junction between the face of the weld and the base metal.

TORCH: See CUTTING TORCH or WELDING TORCH.

TORCH BRAZING: A brazing process in which bonding is produced by heating with a gas flame and by using a nonferrous filler metal having a melting point above 800 °F (427 °C), but below that of the base metal. The filler metal is distributed in the joint of capillary attraction.

TRANSVERSE SEAM WELDING: The making of a seam weld in a direction essentially at right angles to the throat depth of a seam welding machine.

G-3. WELDING TERMS (cont)

T (cont)

TUNGSTEN ELECTRODE: A non-filler metal electrode used in arc welding or cutting, made principally of tungsten.

U

UNDERBEAD CRACK: A crack in the heat affected zone not extending to the surface of the base metal.

UNDERCUT: A groove melted into the base metal adjacent to the toe or root of a weld and left unfilled by weld metal.

UNDERCUTTING: An undesirable crater at the edge of the weld caused by poor weaving technique or excessive welding speed.

UPSET: A localized increase in volume in the region of a weld, resulting from the application of pressure.

UPSET WELDING: A resistance welding process in which fusion is produced simultaneously over the entire area of abutting surfaces, or progressively along a joint, by the heat obtained from resistance to the flow of electric current through the area of contact of those surfaces. Pressure is applied before heating is started and is maintained throughout the heating period.

UPSETTING FORCE: The force exerted at the welding surfaces in flash or upset welding.

V

VERTICAL POSITION: The position of welding in which the axis of the weld is approximately vertical. In pipe welding, the pipe is in a vertical position and the welding is done in a horizontal position.

W

WANDERING BLOCK SEQUENCE: A block welding sequence in which successive weld blocks are completed at random after several starting blocks have been completed.

WANDERING SEQUENCE: A longitudinal sequence in which the weld bead increments are deposited at random.

WAX PATTERN: Wax molded around the parts to be welded by a thermit welding process to the form desired for the completed weld.

WEAVE BEAD: A type of weld bead made with transverse oscillation.

WEAVING: A technique of depositing weld metal in which the electrode is oscillated. It is usually accomplished by a semicircular motion of the arc to the right and left of the direction of welding. Weaving serves to increase the width of the deposit, decreases overlap, and assists in slag formation.

WELD: A localized fusion of metals produced by heating to suitable temperatures. Pressure and/or filler metal may or may not be used. The filler metal has a melting point approximately the same or below that of the base metals, but always above 800 $^{\circ}$F (427 $^{\circ}$C).

WELD BEAD: A weld deposit resulting from a pass.

WELD GAUGE: A device designed for checking the shape and size of welds.

WELD METAL: That portion of a weld that has been melted during welding.

WELD SYMBOL: A picture used to indicate the desired type of weld.

WELDABILITY: The capacity of a material to form a strong bond of adherence under pressure or when solidifying from a liquid.

WELDER CERTIFICATION: Certification in writing that a welder has produced welds

meeting prescribed standards.
WELDER PERFORMANCE QUALIFICATION: The demonstration of a welder's ability to produce welds meeting prescribed standards.
WELDING LEADS:
   a. Electrode lead. The electrical conductor between the source of the arc welding current and the electrode holder.
   b. Work lead. The electrical conductor between the source of the arc welding
current and the workpiece.
WELDING PRESSURE: The pressure exerted during the welding operation on the parts
   being welded.
WELDING PROCEDURE: The detailed methods and practices including all joint welding
   procedures involved in the production of a weldment.
WELDING ROD: Filler metal in wire or rod form, used in gas welding and brazing
   processes and in those arc welding processes in which the electrode does not provide the filler metal.
WELDING SYMBOL: The assembled symbol consists of the following eight elements, or
   such of these as are necessary: reference line, arrow, basic weld symbols, dimension and other data, supplementary symbols, finish symbols, tail, specification,
   process, or other references.
WELDING TECHNIQUE: The details of a manual, machine, or semiautomatic welding
   operation which, within the limitations of the prescribed joint welding procedure, are controlled by the welder or welding operator.
WELDING TIP: The tip of a gas torch especially adapted to welding.
WELDING TORCH: A device used in gas welding and torch brazing for mixing and controlling the flow of gases.
WELDING TRANSFORMER: A device for providing current of the desired voltage.
WELDMENT: An assembly whose component parts are formed by welding.
WIRE FEED SPEED: The rate of speed in mm/sec or in./min at which a filler metal is
   consumed in arc welding or thermal spraying.
WORK LEAD: The electric conductor (cable) between the source of arc welding cur-
   rent and the workpiece.

X

X-RAY: A radiographic test method used to detect internal defects in a weld.

Y

YIELD POINT: The yield point is the load per unit area value at which a marked
   increase in deformation of the specimen occurs with little or no increase of
   load; in other words, the yield point is the stress at which a marked increase in
   strain occurs with little or no increase in stress.

ALPHABETICAL INDEX

A

D

E

F

H

N

O

O (cont)

P

W (cont)

X

Y

METRIC CHART
THE METRIC SYSTEM AND EQUIVALENTS

LINEAR MEASURE

1 Centimeter = 10 Millimeters = 0.01 Meters = 0.3937 Inches
1 Meter = 100 Centimeters = 1000 Millimeters = 39.37 Inches
1 Kilometer = 1000 Meters = 0.621 Miles

WEIGHTS

1 Gram = 0.001 Kilograms = 1000 Milligrams = 0.035 Ounces
1 Kilogram = 1000 Grams = 2.2 Lb
1 Metric Ton = 1000 Kilograms = 1 Megagram = 1.1 Short Tons

LIQUID MEASURE

1 Milliliter = 0.001 Liters = 0.0338 Fluid Ounces
1 Liter = 1000 Milliliters = 33.82 Fluid Ounces

SQUARE MEASURE

1 Sq Centimeter = 100 Sq Millimeters = 0.155 Sq Inches
1 Sq Meter = 10,000 Sq Centimeters = 10.76 Sq Feet
1 Sq Kilometer = 1,000,000 Sq Meters = 0.386 Sq Miles

CUBIC MEASURE

1 Cu Centimeter = 1000 Cu Millimeters = 0.06 Cu Inches
1 Cu Meter = 1,000,000 Cu Centimeters = 35.31 Cu Feet

TEMPERATURE

5/9 (°F -32) = °C
212° Fahrenheit is equivalent to 100° Celsius
90° Fahrenheit is equivalent to 32.2° Celsius
32° Fahrenheit is equivalent to 0° Celsius
9/5 C° + 32 = F°

APPROXIMATE CONVERSION FACTORS

TO CHANGE	TO	MULTIPLY BY
Inches	Centimeters	2.540
Feet	Meters	0.305
Yards	Meters	0.914
Miles	Kilometers	1.609
Square Inches	Square Centimeters	6.451
Square Feet	Square Meters	0.093
Square Yards	Square Meters	0.836
Square Miles	Square Kilometers	2.590
Acres	Square Hectometers	0.405
Cubic Feet	Cubic Meters	0.028
Cubic Yards	Cubic Meters	0.765
Fluid Ounces	Milliliters	29.573
Pints	Liters	0.473
Quarts	Liters	0.946
Gallons	Liters	3.785
Ounces	Grams	28.349
Pounds	Kilograms	0.454
Short Tons	Metric Tons	0.907
Pound-Feet	Newton-Meters	1.356
Pounds per Square Inch	Kilopascals	6.895
Miles per Gallon	Kilometers per Liter	0.425
Miles per Hour	Kilometers per Hour	1.609

TO CHANGE	TO	MULTIPLY BY
Centimeters	Inches	0.394
Meters	Feet	3.280
Meters	Yards	1.094
Kilometers	Miles	0.621
Square Centimeters	Square Inches	0.155
Square Meters	Square Feet	10.764
Square Meters	Square Yards	1.196
Square Kilometers	Square Miles	0.386
Square Hectometers	Acres	2.471
Cubic Meters	Cubic Feet	35.315
Cubic Meters	Cubic Yards	1.308
Milliliters	Fluid Ounces	0.034
Liters	Pints	2.113
Liters	Quarts	1.057
Liters	Gallons	0.264
Grams	Ounces	0.035
Kilograms	Pounds	2.205
Metric Tons	Short Tons	1.102
Newton-Meters	Pound-Feet	0.738
Kilopascals	Pounds per Square Inch	0.145
Kilometers per Liter	Miles per Gallon	2.354
Kilometers per Hour	Miles per Hour	0.621

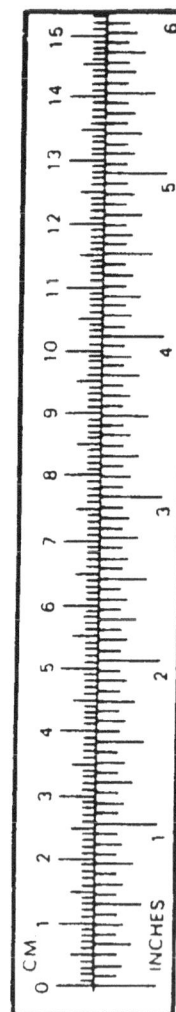

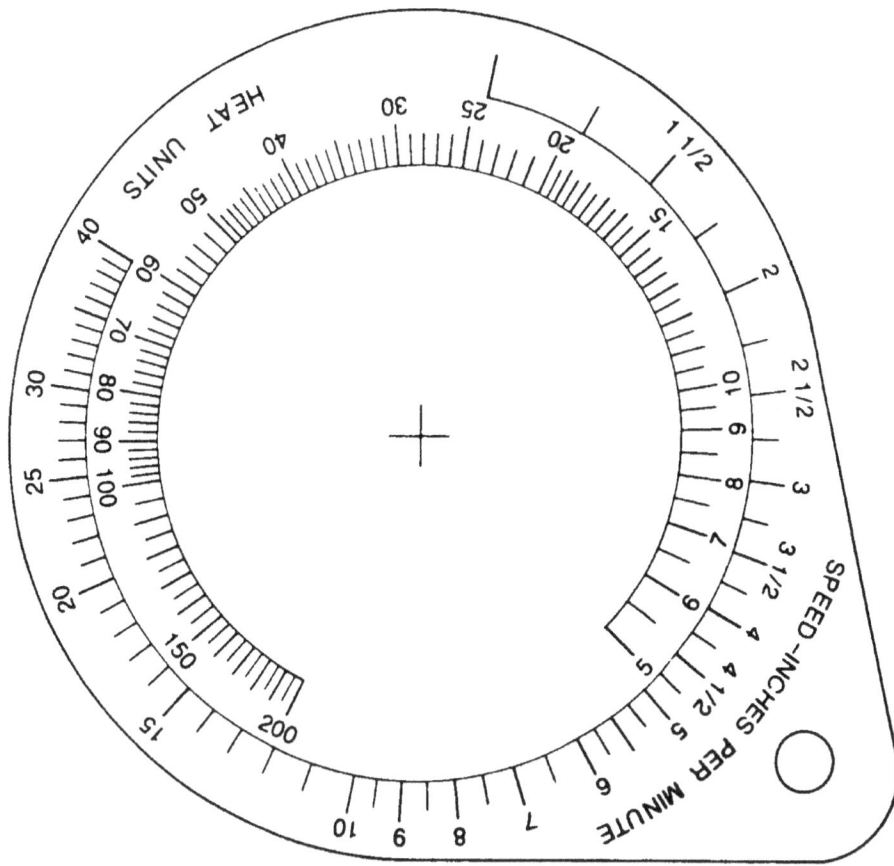

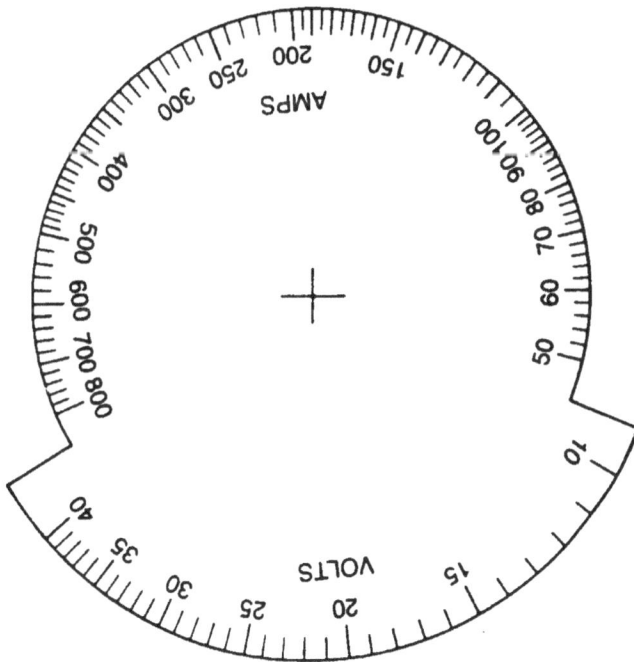

By Order of the Secretary of the Army:

GORDON R. SULLIVAN
General, United States Army
Chief of Staff

Official:

MILTON H. HAMILTON
Administrative Assistant to the
Secretary of the Army
03870

DISTRIBUTION:

Active Army, USAR, and ARNG: To be distributed in accordance with
DA Form 12-11E, requirements for TC 9-237, Operator's Circular for
Welding Theory and Application (Qty rqr block no. 4623).

☆U.S. GOVERNMENT PRINTING OFFICE: 1995 - 388-421/02577

PIN: 071226-000

www.ingramcontent.com/pod-product-compliance
Lightning Source LLC
Chambersburg PA
CBHW080413030426
42335CB00020B/2437